Second Edition

Data Networks

DIMITRI BERTSEKAS

Massachusetts Institute of Technology

ROBERT GALLAGER

Massachusetts Institute of Technology

PRENTICE HALL, Upper Saddle River, NJ 07458

Library of Congress Cataloging-in-Publication Data

Bertsekas, Dimitri P.
Data networks / Dimitri Bertsekas, Robert Gallager. -- 2nd ed.
p. cm.
Includes bibliographical references and index.
ISBN 0-13-200916-1
1. Data transmission systems. I. Gallager, Robert G. II. Title.
TK5105.B478 1992
004.6--dc20 91-35561
CIP

Acquisitions editor: Pete Janzow
Production editor: Bayani Mendoza de Leon
Copy editor: Zeiders & Associates
Cover designer: Butler/Udell
Prepress buyer: Linda Behrens
Manufacturing buyer: Dave Dickey
Editorial assistant: Phyllis Morgan

A Simon & Schuster Company
Upper Saddle River, New Jersey 07458

Printed in the United States of America

10

ISBN 0-13-200916-1

Prentice-Hall International (UK) Limited, *London*
Prentice-Hall of Australia Pty. Limited, *Sydney*
Prentice-Hall Canada Inc., *Toronto*
Prentice-Hall Hispanoamericana, S. A., *Mexico*
Prentice-Hall of India Private Limited, *New Delhi*
Prentice-Hall of Japan, Inc., *Tokyo*
Simon & Schuster Asia Pte. Ltd., *Singapore*
Editora Prentice-Hall do Brasil, Ltda., *Rio de Janeiro*

To Joanna and Marie

Contents

PREFACE xv

1 INTRODUCTION AND LAYERED NETWORK ARCHITECTURE 1

1.1 Historical Overview **1**

1.1.1 Technological and Economic Background, 5
1.1.2 Communication Technology, 6
1.1.3 Applications of Data Networks, 7

1.2 Messages and Switching **9**

1.2.1 Messages and Packets, 9
1.2.2 Sessions, 11
1.2.3 Circuit Switching and Store-and-Forward Switching, 14

1.3 Layering **17**

1.3.1 The Physical Layer, 20
1.3.2 The Data Link Control Layer, 23

The MAC sublayer, 24

1.3.3 The Network Layer, 25

The Internet sublayer, 28

1.3.4 The Transport Layer, 29
1.3.5 The Session Layer, 30

1.3.6 The Presentation Layer, 31
1.3.7 The Application Layer, 31

1.4 A Simple Distributed Algorithm Problem **32**

Notes and Suggested Reading **35**

Problems **35**

2 **POINT-TO-POINT PROTOCOLS AND LINKS** 37

2.1 Introduction **37**

2.2 The Physical Layer: Channels and Modems **40**

2.2.1 Filtering, 41
2.2.2 Frequency Response, 43
2.2.3 The Sampling Theorem, 46
2.2.4 Bandpass Channels, 47
2.2.5 Modulation, 48
2.2.6 Frequency- and Time-Division Multiplexing, 52
2.2.7 Other Channel Impairments, 53
2.2.8 Digital Channels, 53

ISDN, 54

2.2.9 Propagation Media for Physical Channels, 56

2.3 Error Detection **57**

2.3.1 Single Parity Checks, 58
2.3.2 Horizontal and Vertical Parity Checks, 58
2.3.3 Parity Check Codes, 59
2.3.4 Cyclic Redundancy Checks, 61

2.4 ARQ: Retransmission Strategies **64**

2.4.1 Stop-and-Wait ARQ, 66

Correctness of stop and wait, 69

2.4.2 Go Back n ARQ, 72

Rules followed by transmitter and receiver in go back n, *74*
Correctness of go back n, *76*
Go back n *with modulus* $m > n$, *78*
Efficiency of go back n *implementations, 80*

2.4.3 Selective Repeat ARQ, 81
2.4.4 ARPANET ARQ, 84

2.5 Framing **86**

2.5.1 Character-Based Framing, 86
2.5.2 Bit-Oriented Framing: Flags, 88
2.5.3 Length Fields, 90
2.5.4 Framing with Errors, 92
2.5.5 Maximum Frame Size, 93

Variable frame length, 93
Fixed frame length, 97

2.6 Standard DLCs **97**

2.7 Initialization and Disconnect for ARQ Protocols **103**

2.7.1 Initialization in the Presence of Link Failures, 103
2.7.2 Master–Slave Protocol for Link Initialization, 104
2.7.3 A Balanced Protocol for Link Initialization, 107
2.7.4 Link Initialization in the Presence of Node Failures, 109

2.8 Point-to-Point Protocols at the Network Layer **110**

2.8.1 Session Identification and Addressing, 111

Session identification in TYMNET, 112
Session identification in the Codex networks, 113

2.8.2 Packet Numbering, Window Flow Control, and Error Recovery, 114

Error recovery, 115
Flow control, 116
Error recovery at the transport layer versus the network layer, 117

2.8.3 The X.25 Network Layer Standard, 118
2.8.4 The Internet Protocol, 120

2.9 The Transport Layer **123**

2.9.1 Transport Layer Standards, 123
2.9.2 Addressing and Multiplexing in TCP, 124
2.9.3 Error Recovery in TCP, 125
2.9.4 Flow Control in TCP/IP, 127
2.9.5 TP Class 4, 128

2.10 Broadband ISDN and the Asynchronous Transfer Mode **128**

2.10.1 Asynchronous Transfer Mode (ATM), 132
2.10.2 The Adaptation Layer, 135

Class 3 (connection-oriented) traffic, 136
Class 4 (connectionless) traffic, 137
Class 1 and 2 traffic, 137

2.10.3 Congestion, 138

Summary **139**

Notes, Sources, and Suggested Reading **140**

Problems **141**

3 **DELAY MODELS IN DATA NETWORKS** **149**

3.1 Introduction **149**

3.1.1 Multiplexing of Traffic on a Communication Link, 150

3.2 Queueing Models: Little's Theorem **152**

3.2.1 Little's Theorem, 152
3.2.2 Probabilistic Form of Little's Theorem, 154
3.2.3 Applications of Little's Theorem, 157

3.3 The $M/M/1$ Queueing System **162**

3.3.1 Main Results, 164

Arrival statistics—the Poisson process, 164
Service statistics, 165
Markov chain formulation, 166
Derivation of the stationary distribution, 167

3.3.2 Occupancy Distribution upon Arrival, 171
3.3.3 Occupancy Distribution upon Departure, 173

3.4 The $M/M/m$, $M/M/\infty$, $M/M/m/m$, and Other Markov Systems **173**

3.4.1 $M/M/m$: The m-Server Case, 174
3.4.2 $M/M/\infty$: The Infinite-Server Case, 177
3.4.3 $M/M/m/m$: The m-Server Loss System, 178
3.4.4 Multidimensional Markov Chains: Applications in Circuit Switching, 180

Truncation of independent single-class systems, 182
Blocking probabilities for circuit switching systems, 185

3.5 The $M/G/1$ System **186**

3.5.1 $M/G/1$ Queues with Vacations, 192
3.5.2 Reservations and Polling, 195

Single-user system, 196
Multi-user system, 198
Limited service systems, 201

3.5.3 Priority Queueing, 203

Nonpreemptive priority, 203

Preemptive resume priority, 205

3.5.4 An Upper Bound for the $G/G/1$ System, 206

3.6 Networks of Transmission Lines **209**

3.6.1 The Kleinrock Independence Approximation, 211

3.7 Time Reversibility—Burke's Theorem **214**

3.8 Networks of Queues—Jackson's Theorem **221**

Heuristic explanation of Jackson's Theorem, 227

3.8.1 Extensions of Jackson's Theorem, 229

State-dependent service rates, 229
Multiple classes of customers, 230

3.8.2 Closed Queueing Networks, 233
3.8.3 Computational Aspects—Mean Value Analysis, 238

Summary **240**

Notes, Sources, and Suggested Reading **241**

Problems **242**

Appendix A: Review of Markov Chain Theory **259**

3A.1 Discrete-Time Markov Chains, 259
3A.2 Detailed Balance Equations, 261
3A.3 Partial Balance Equations, 262
3A.4 Continuous-Time Markov Chains, 262
3A.5 Drift and Stability, 264

Appendix B: Summary of Results **265**

4 MULTIACCESS COMMUNICATION 271

4.1 Introduction **271**

4.1.1 Satellite Channels, 273
4.1.2 Multidrop Telephone Lines, 274
4.1.3 Multitapped Bus, 274
4.1.4 Packet Radio Networks, 275

4.2 Slotted Multiaccess and the Aloha System **275**

4.2.1 Idealized Slotted Multiaccess Model, 275

Discussion of assumptions, 276

4.2.2 Slotted Aloha, 277
4.2.3 Stabilized Slotted Aloha, 282

Stability and maximum throughput, 282
Pseudo-Bayesian algorithm, 283
Approximate delay analysis, 284
Binary exponential backoff, 286

4.2.4 Unslotted Aloha, 287

4.3 Splitting Algorithms **289**

4.3.1 Tree Algorithms, 290

Improvements to the tree algorithm, 292
Variants of the tree algorithm, 293

4.3.2 First-Come First-Serve Splitting Algorithms, 293

Analysis of FCFS splitting algorithm, 297
Improvements in the FCFS splitting algorithm, 301
Practical details, 302
Last-come first-serve (LCFS) splitting algorithm, 302
Delayed feedback, 303
Round-robin splitting, 304

4.4 Carrier Sensing **304**

4.4.1 CSMA Slotted Aloha, 305
4.4.2 Pseudo-Bayesian Stabilization for CSMA Aloha, 307
4.4.3 CSMA Unslotted Aloha, 309
4.4.4 FCFS Splitting Algorithm for CSMA, 310

4.5 Multiaccess Reservations **312**

4.5.1 Satellite Reservation Systems, 313
4.5.2 Local Area Networks: CSMA/CD and Ethernet, 317

Slotted CSMA/CD, 317
Unslotted CSMA/CD, 318
The IEEE 802 standards, 320

4.5.3 Local Area Networks: Token Rings, 320

IEEE 802.5 token ring standard, 323
Expected delay for token rings, 324
FDDI, 326
Slotted rings and register insertion rings, 330

4.5.4 Local Area Networks: Token Buses and Polling, 331

IEEE 802.4 token bus standard, 332
Implicit tokens: CSMA/CA, 333

4.5.5 High-Speed Local Area Networks, 333

Distributed queue dual bus (IEEE 802.6), 335

Expressnet, 339
Homenets, 341

4.5.6 Generalized Polling and Splitting Algorithms, 342

4.6 Packet Radio Networks **344**

4.6.1 TDM for Packet Radio Nets, 346
4.6.2 Collision Resolution for Packet Radio Nets, 347
4.6.3 Transmission Radii for Packet Radio, 349
4.6.4 Carrier Sensing and Busy Tones, 350

Summary **351**

Notes, Sources, and Suggested Reading **352**

Problems **353**

5 ROUTING IN DATA NETWORKS 363

5.1 Introduction **363**

5.1.1 Main Issues in Routing, 365
5.1.2 Wide-Area Network Routing: An Overview, 367

Flooding and broadcasting, 368
Shortest path routing, 370
Optimal routing, 372
Hot potato (deflection) routing schemes, 372
Cut-through routing, 373
ARPANET: An example of datagram routing, 374
TYMNET: An example of virtual circuit routing, 376
Routing in SNA, 378
Routing in circuit switching networks, 379

5.1.3 Interconnected Network Routing: An Overview, 379

Bridged local area networks, 382
Spanning tree routing in bridged local area networks, 383
Source routing in bridged local area networks, 385

5.2 Network Algorithms and Shortest Path Routing **387**

5.2.1 Undirected Graphs, 387
5.2.2 Minimum Weight Spanning Trees, 390
5.2.3 Shortest Path Algorithms, 393

The Bellman-Ford algorithm, 396
Bellman's equation and shortest path construction, 399
Dijkstra's algorithm, 401
The Floyd-Warshall algorithm, 403

5.2.4 Distributed Asynchronous Bellman–Ford Algorithm, 404

5.2.5 Stability of Adaptive Shortest Path Routing Algorithms, 410

Stability issues in datagram networks, 410
Stability issues in virtual circuit networks, 414

5.3 Broadcasting Routing Information: Coping with Link Failures **418**

5.3.1 Flooding: The ARPANET Algorithm, 420
5.3.2 Flooding without Periodic Updates, 422
5.3.3 Broadcast without Sequence Numbers, 425

5.4 Flow Models, Optimal Routing, and Topological Design **433**

5.4.1 Overview of Topological Design Problems, 437
5.4.2 Subnet Design Problem, 439

Capacity assignment problem, 439
Heuristic methods for capacity assignment, 442
Network reliability issues, 445
Spanning tree topology design, 447

5.4.3 Local Access Network Design Problem, 448

5.5 Characterization of Optimal Routing **451**

5.6 Feasible Direction Methods for Optimal Routing **455**

5.6.1 The Frank–Wolfe (Flow Deviation) Method, 458

5.7 Projection Methods for Optimal Routing **464**

5.7.1 Unconstrained Nonlinear Optimization, 465
5.7.2 Nonlinear Optimization over the Positive Orthant, 467
5.7.3 Application to Optimal Routing, 468

5.8 Routing in the Codex Network **476**

Summary **477**

Notes, Sources, and Suggested Reading **478**

Problems **479**

6 FLOW CONTROL 493

6.1 Introduction **493**

6.1.1 Means of Flow Control, 494
6.1.2 Main Objectives of Flow Control, 496

Limiting delay and buffer overflow, 496
Fairness, 498

6.2 Window Flow Control **500**

6.2.1 End-to-End Windows, 501

Limitations of end-to-end windows, 502

6.2.2 Node-by-Node Windows for Virtual Circuits, 506
6.2.3 The Isarithmic Method, 508
6.2.4 Window Flow Control at Higher Layers, 508
6.2.5 Dynamic Window Size Adjustment, 510

6.3 Rate Control Schemes **510**

Queueing analysis of the leaky bucket scheme, 513

6.4 Overview of Flow Control in Practice **515**

Flow control in the ARPANET, 515
Flow control in the TYMNET, 517
Flow control in SNA, 517
Flow control in a Codex network, 518
Flow control in the PARIS network, 518
Flow control in X.25, 519

6.5 Rate Adjustment Algorithms **519**

6.5.1 Combined Optimal Routing and Flow Control, 519
6.5.2 Max-Min Flow Control, 524

Summary **530**

Notes, Sources, and Suggested Reading **530**

Problems **531**

REFERENCES **537**

INDEX **552**

Preface

In the five years since the first edition of *Data Networks* appeared, the networking field has changed in a number of important ways. Perhaps the most fundamental has been the rapid development of optical fiber technology. This has created almost limitless opportunities for new digital networks of greatly enhanced capabilities. In the near term, the link capacities available for data networks, both wide area and local, are increasing by many orders of magnitude. In the longer term, public broadband integrated service networks that provide integrated data, voice, and video on a universal scale are now technologically feasible. These networks of the future appear at first to have almost nothing in common with the data networks of the last 20 years, but in fact, many of the underlying principles are the same. This edition is designed both to provide a fundamental understanding of these common principles and to provide insight into some of the new principles that are evolving for future networks.

Our approach to helping the reader understand the basic principles of networking is to provide a balance between the description of existing networks and the development of analytical tools. The descriptive material is used to illustrate the underlying concepts, and the analytical material is used to generate a deeper and more precise understanding of the concepts. Although the analytical material can be used to analyze the performance of various networks, we believe that its more important use is in sharpening one's conceptual and intuitive understanding of the field; that is, analysis should precede design rather than follow it.

The book is designed to be used at a number of levels, varying from a senior undergraduate elective, to a first year graduate course, to a more advanced graduate course, to a reference work for designers and researchers in the field. The material has been tested in a number of graduate courses at M.I.T. and in a number of short courses at varying levels. The book assumes some background in elementary probability and some background in either electrical engineering or computer science, but aside from this, the material is self-contained.

Throughout the book, major concepts and principles are first explained in a simple non-mathematical way. This is followed by careful descriptions of modelling issues and then by mathematical analysis. Finally, the insights to be gained from the analysis are explained and examples are given to clarify the more subtle issues. Figures are liberally used throughout to illustrate the ideas. For lower-level courses, the analysis can be glossed over; this allows the beginning and intermediate-level student to grasp the basic ideas, while enabling the more advanced student to acquire deeper understanding and the ability to do research in the field.

Chapter 1 provides a broad introduction to the subject and also develops the layering concept. This layering allows the various issues of data networks to be developed in a largely independent fashion, thus making it possible to read the subsequent chapters in any desired depth (including omission) without seriously hindering the ability to understand other chapters.

Chapter 2 covers the two lowest layers of the above layering and also discusses a number of closely related aspects of the higher layers. The treatment of the lowest, or physical, layer provides a brief overview of how binary digits are transmitted over physical communication media. The effort here is to provide just enough material so that the student can relate the abstraction of digital transmission to physical phenomena. The next layer, data link control, allows packets to be transmitted reliably over communication links. This provides an introduction into the distributed algorithms, or protocols, that must be used at the ends of the link to provide the desired reliability. These protocols are less important in modern high speed networks than in older networks, but the concepts are used repeatedly at many layers in all kinds of networks. The remainder of the chapter focuses on other point to point protocols that allow the end points of a link, or the end points of a network session, to cooperate in providing some required service.

Chapter 3 develops the queueing theory used for performance analysis of multiaccess schemes (Chapter 4) and, to a lesser extent, routing algorithms (Chapter 5). Less analytical courses will probably omit most of this chapter, simply adopting the results on faith. Little's theorem and the Poisson process should be covered however, since they are simple and greatly enhance understanding of the subsequent chapters. This chapter is rich in results, often developed in a far simpler way than found in the queueing literature. This simplicity is achieved by considering only steady-state behavior and by sometimes sacrificing rigor for clarity and insight. Mathematically sophisticated readers will be able to supply the extra details for rigor by themselves, while for most readers the extra details would obscure the line of argument.

Chapter 4 develops the topic of multiaccess communication, including local area networks, metropolitan area networks, satellite networks, and radio networks. Less theoretical courses will probably skip the last half of section 4.2, all of section 4.3, and most of section 4.4, getting quickly to local area networks in section 4.5. Conceptually, one gains a great deal of insight into the nature of distributed algorithms in this chapter.

Chapter 5 develops the subject of routing. The material is graduated in order of increasing difficulty and depth, so readers can go as far as they are comfortable. Along with routing itself, which is treated in greater depth than elsewhere in the literature, further insights are gained into distributed algorithms. There is also a treatment of topological design and a section on recovery from link failures.

Chapter 6 deals with flow control (or congestion control as it is sometimes called). The first three sections are primarily descriptive, describing first the objectives and the problems in achieving these objectives, and then two general approaches, window flow control, and rate control. The fourth section describes the ways that flow control is handled in several existing networks. The last section is more advanced and analytical, describing various algorithms to select session rates in rate control schemes.

A topic that is not treated in any depth in the book is that of higher-layer protocols, namely the various processes required in the computers and devices using the network to communicate meaningfully with each other given the capability of reliable transport of packets through the network provided by the lower layers. This topic is different in nature than the other topics covered and would have doubled the size of the book if treated in depth.

We apologize in advance for the amount of jargon and acronyms in the book. We felt it was necessary to include at least the most commonly used acronyms in the field, both to allow readers to converse with other workers in the field and also for the reference value of being able to find out what these acronyms mean.

An extensive set of problems are given at the end of each chapter except the first. They range from simple exercises to gain familiarity with the basic concepts and techniques to advanced problems extending the results in the text. Solutions of the problems are given in a manual available to instructors from Prentice-Hall.

Each chapter also contains a brief section of sources and suggestions for further reading. Again, we apologize in advance to the many authors whose contributions have not been mentioned. The literature in the data network field is vast, and we limited ourselves to references that we found most useful, or that contain material supplementing the text.

The stimulating teaching and research environment at M.I.T. has been an ideal setting for the development of this book. In particular we are indebted to the many students who have used this material in courses. Their comments have helped greatly in clarifying the topics. We are equally indebted to the many colleagues and advanced graduate students who have provided detailed critiques of the various chapters. Special thanks go to our colleague Pierre Humblet whose advice, knowledge, and deep insight have been invaluable. In addition, Erdal Arikan, David Castanon, Robert Cooper, Tony Ephremides, Eli Gafni, Marianne Gardner, Inder Gopal, Paul Green, Ellen Hahne, Bruce Hajek, Michael Hluchyi, Robert Kennedy, John Spinelli, and John Tsitsiklis have all been very helpful. We are also grateful to Nancy Young for typing the many revisions. Our editors at Prentice-Hall have also been very helpful and cooperative in producing the final text under a very tight schedule. Finally we wish to acknowledge the research support of DARPA under grant ONR-N00014-84-K-0357, NSF under grants ECS-8310698, ECS-8217668, 8802991-NCR, and DDM-8903385, and ARO under grants DAAG 29-84-K-000 and DAAL03-86-K-0171.

Dimitri Bertsekas

Robert Gallager

ABOUT THE AUTHORS

Dimitri Bertsekas

DIMITRI P. BERTSEKAS received a B.S. in Mechanical and Electrical Engineering from the National Technical University of Athens, Greece in 1965, and a Ph.D. in system science from the Massachusetts Institute of Technology in 1971. He has held faculty positions with the Engineering-Economic Systems Department, Stanford University, and the Electrical Engineering Department of the University of Illinois, Urbana. In 1979 he joined the faculty of the Massachusetts Institute of Technology where he is currently Professor of Electrical Engineering and Computer Science. He consults regularly with private industry, and has held editorial positions in several journals. He has been elected Fellow of the IEEE.

Professor Bertsekas has done research in control of stochastic systems, and in linear, nonlinear, and dynamic programming. He has written numerous papers in each of these areas. His current interests are primarily in data networks, distributed computation, and large-scale optimization. He is the author of *Dynamic Programming and Stochastic Control,* Academic Press, 1976, *Constrained Optimization and Lagrange Multiplier Methods,* Academic Press, 1982, *Dynamic Programming: Deterministic and Stochastic Models,* Prentice Hall, 1987, *Linear Network Optimization: Algorithms and Codes,* MIT Press, 1991, and the co-author of *Stochastic Optimal Control: The Discrete-Time Case,* Academic Press, 1978, and *Parallel and Distributed Computation: Numerical Methods,* Prentice Hall, 1989.

Robert Gallager

ROBERT G. GALLAGER received the S.B. degree in electrical engineering from the University of Pennsylvania, Philadelphia, PA, in 1953, and the S.M. and Sc.D. degrees in electrical engineering from the Massachusetts Institute of Technology, Cambridge, MA, in 1957 and 1960, respectively.

Following several years as a member of the Technical Staff at Bell Telephone Laboratories and service in the U.S. Army Signal Corps, he has been with the Massachusetts Institute of Technology since 1956. He is currently the Fujitsu Professor of Electrical Engineering and Computer Science, Co-Director of the Laboratory for Information and Decision Systems, and Co-Chairman of the Department Area I (Control, Communication, and Operations Research). He is a consultant to Codex Corporation. He is the author of the textbook, *Information Theory and Reliable Communication* (New York: Wiley, 1968). His major research interests are data communication networks, information theory, and communication theory.

Dr. Gallager was awarded the IEEE Baker Prize Paper Award in 1966 for the paper "A simple derivation of the coding theorem and some applications." He has been a member of the Board of Governors of the IEEE Information Theory Society from 1965 to 1970 and from 1979 to 1986, and was President of the Society in 1971. He is a Fellow of the IEEE and a member of the National Academy of Engineering. He was awarded the 1990 IEEE Medal of Honor for his contributions to communication coding techniques.

1

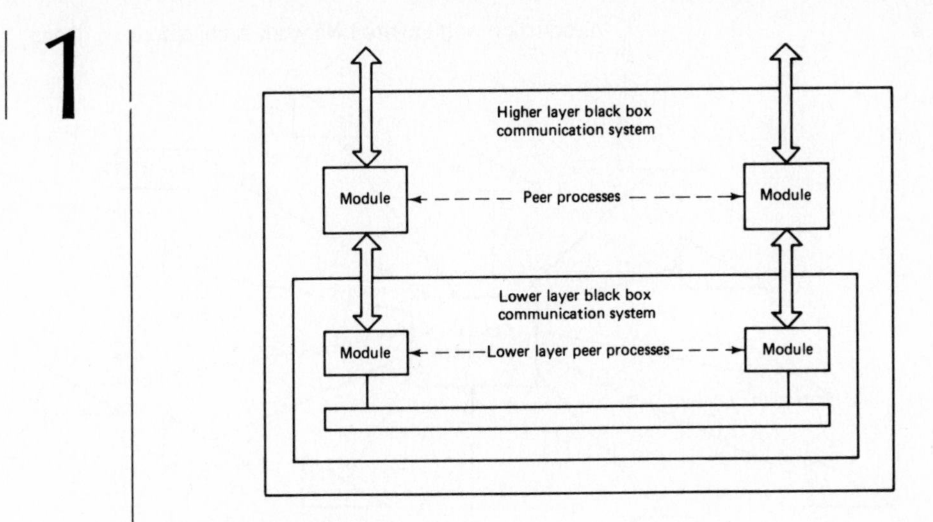

Introduction and Layered Network Architecture

1.1 HISTORICAL OVERVIEW

Primitive forms of data networks have a long history, including the smoke signals used by primitive societies, and certainly including nineteenth-century telegraphy. The messages in these systems were first manually encoded into strings of essentially binary symbols, and then manually transmitted and received. Where necessary, the messages were manually relayed at intermediate points.

A major development, in the early 1950s, was the use of communication links to connect central computers to remote terminals and other peripheral devices, such as printers and remote job entry points (RJEs) (see Fig. 1.1). The number of such peripheral devices expanded rapidly in the 1960s with the development of time-shared computer systems and with the increasing power of central computers. With the proliferation of remote peripheral devices, it became uneconomical to provide a separate long-distance communication link to each peripheral. Remote multiplexers or concentrators were developed to collect all the traffic from a set of peripherals in the same area and to send it on a single link to the central processor. Finally, to free the central processor from handling all this communication, special processors called *front ends* were developed to

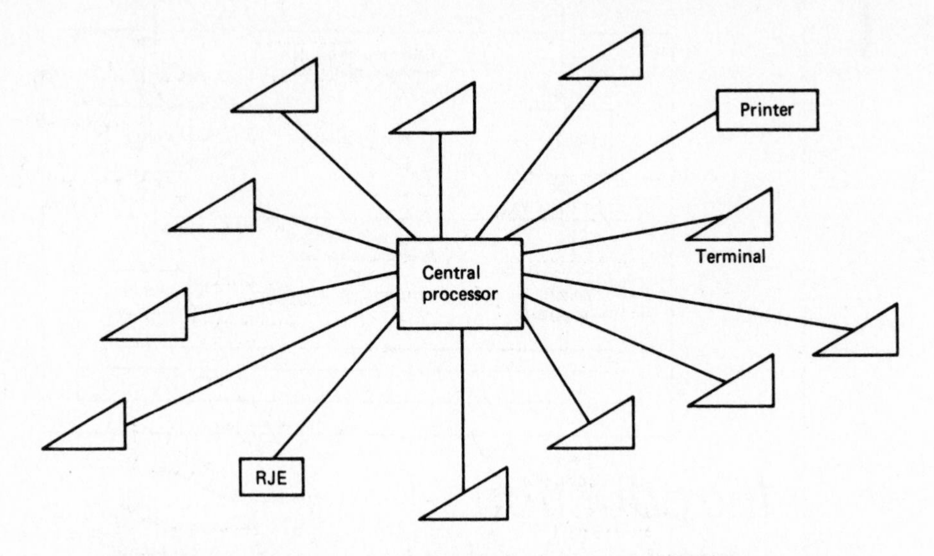

Figure 1.1 Network with one central processor and a separate communication link to each device.

control the communication to and from all the peripherals. This led to the more complex structure shown in Fig. 1.2. The communication is automated in such systems, in contrast to telegraphy, for example, but the control of the communication is centrally exercised at the computer. While it is perfectly appropriate and widely accepted to refer to such a system as a data network or computer communication network, it is simpler to view it as a computer with remote peripherals. Many of the interesting problems associated with data networks, such as the distributed control of the system, the relaying of messages over multiple communication links, and the sharing of communication links between many users and processes, do not arise in these centralized systems.

The ARPANET and TYMNET, introduced around 1970, were the first large-scale, general-purpose data networks connecting geographically distributed computer systems, users, and peripherals. Figure 1.3 shows such networks. Inside the "subnet" are a set of nodes, various pairs of which are connected by communication links. Outside the subnet are the various computers, data bases, terminals, and so on, that are connected via the subnet. Messages originate at these external devices, pass into the subnet, pass from node to node on the communication links, and finally pass out to the external recipient. The nodes of the subnet, usually computers in their own right, serve primarily to route the messages through the subnet. These nodes are sometimes called IMPs (interface message processors) and sometimes called switches. In some networks (*e.g.*, DECNET), nodes in the subnet might be physically implemented within the external computers using the network. It is helpful, however, to view the subnet nodes as being logically distinct from the external computers.

It is important to observe that in Figs. 1.1 and 1.2 the computer system is the center of the network, whereas in Fig. 1.3 the subnet (*i.e.*, the communication part of the network) is central. Keeping this picture of external devices around a communication

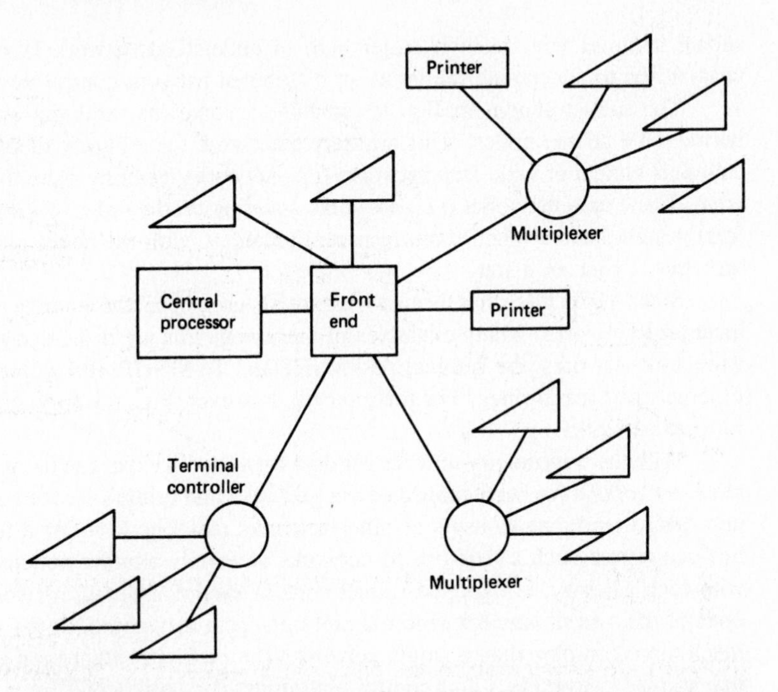

Figure 1.2 Network with one central processor but with shared communication links to devices.

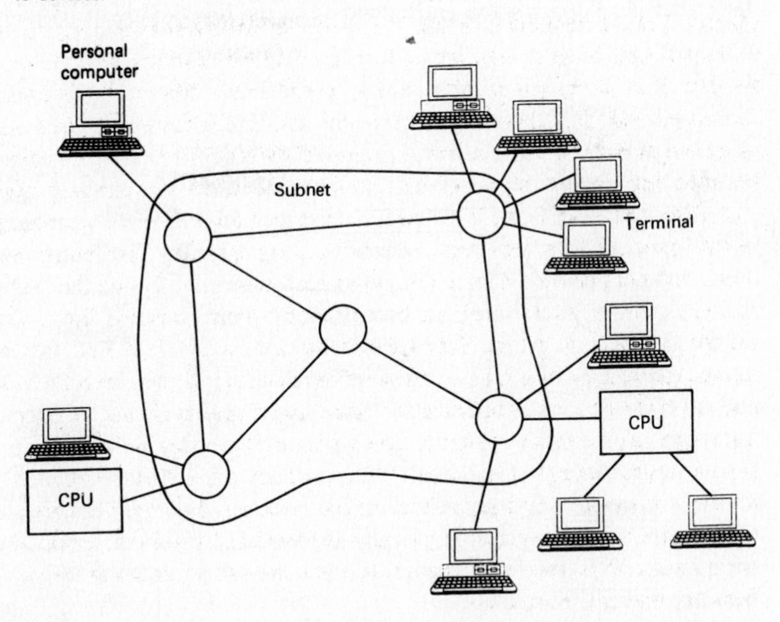

Figure 1.3 General network with a subnet of communication links and nodes. External devices are connected to the subnet via links to the subnet nodes.

subnet in mind will make it easier both to understand network layering later in this chapter and to understand the issues of distributed network control throughout the book.

The subnet shown in Fig. 1.3 contains a somewhat arbitrary placement of links between the subnet nodes. This arbitrary placement (or arbitrary topology as it is often called) is typical of wide area networks (*i.e.*, networks covering more than a metropolitan area). Local area networks (*i.e.*, networks covering on the order of a square kilometer or less) usually have a much more restricted topology, with the nodes typically distributed on a bus, a ring, or a star.

Since 1970 there has been an explosive growth in the number of wide area and local area networks. Many examples of these networks are discussed later, including as wide area networks, the seminal ARPANET and TYMNET, and as local area networks, Ethernets and token rings. For the moment, however, Fig. 1.3 provides a generic model for data networks.

With the multiplicity of different data networks in existence in the 1980s, more and more networks have been connected via gateways and bridges so as to allow users of one network to send data to users of other networks (see Fig. 1.4). At a fundamental level, one can regard such a network of networks as simply another network, as in Fig. 1.3, with each gateway, bridge, and subnet node of each constituent network being a subnet node of the overall network. From a more practical viewpoint, a network of networks is much more complex than a single network. The problem is that each constituent subnet has its own conventions and control algorithms (*i.e.*, protocols) for handling data, and the gateways and bridges must deal with this inhomogeneity. We discuss this problem later after developing some understanding of the functioning of individual subnets.

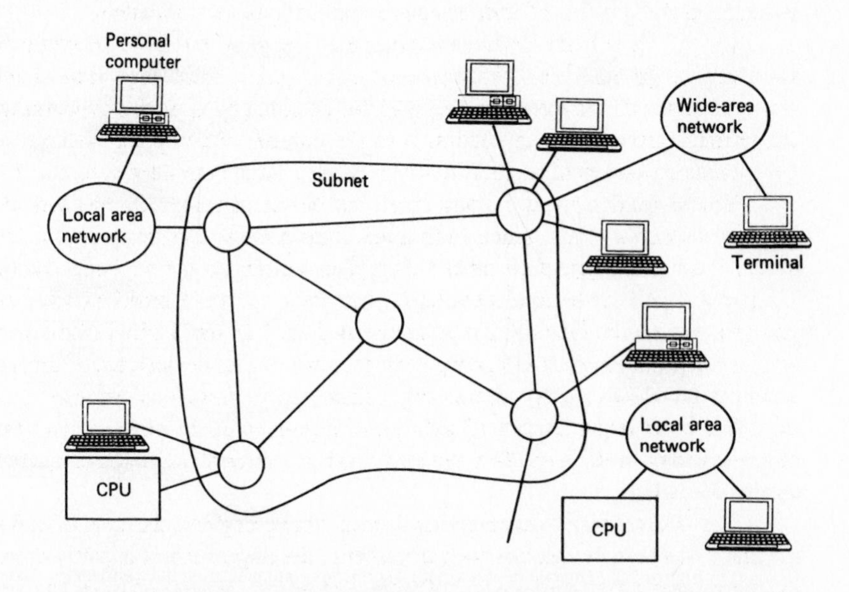

Figure 1.4 Network of interconnected networks. Individual wide area networks (WANs) and local networks (LANs) are connected via bridges and gateways.

In the future, it is likely that data networks, the voice network, and perhaps cable TV networks will be far more integrated than they are today. Data can be sent over the voice network today, as explained more fully in Section 2.2, and many of the links in data networks are leased from the voice network. Similarly, voice can be sent over data networks. What is envisioned for the future, however, is a single integrated network, called an integrated services digital network (ISDN), as ubiquitous as the present voice network. In this vision, offices and homes will each have an access point into the ISDN that will handle voice, current data applications, and new applications, all with far greater convenience and less expense than is currently possible. ISDN is currently available in some places, but it is not yet very convenient or inexpensive. Another possibility for the future is called broadband ISDN. Here the links will carry far greater data rates than ISDN and the network will carry video as well as voice and data. We discuss the pros and cons of this in Section 2.10.

1.1.1 Technological and Economic Background

Before data networks are examined, a brief overview will be given, first, of the technological and economic factors that have led to network development, and second, of the applications that require networks. The major driving force in the rapid advances in computers, communication, and data networks has been solid-state technology and in particular the development of very large scale integration (VLSI). In computers, this has led to faster, less expensive processors; faster, larger, less expensive primary memory; and faster, larger, less expensive bulk storage. The result has been the lowering of the cost of computation by roughly a factor of 2 every two years. This has led to a rapid increase in the number of cost effective applications for computers.

On the other hand, with the development of more and more powerful microprocessor chips, there has been a shift in cost effectiveness from large time-shared computer facilities to small but increasingly powerful personal computers and workstations. Thus, the primary growth in computation is in the number of computer systems rather than in the increasing power of a small number of very large computer systems.

This evolution toward many small but powerful computers has had several effects on data networks. First, since individual organizations use many computers, there is a need for them to share each other's data, thus leading to the network structures of Figs. 1.3 and 1.4. (If, instead, the evolution had been toward a small number of ever more powerful computer systems, the structure of Fig. 1.2 would still predominate.) Second, since subnet nodes are small computers, the cost of a given amount of processing within a subnet has been rapidly decreasing. Third, as workstations become more powerful, they deal with larger blocks of data (*e.g.*, high-resolution graphics) and the data rates of present-day wide area data networks are insufficient to transmit such blocks with acceptable delay.

The discussion of computational costs above neglects the cost of software. While the art of software design has been improving, the improvement is partly counterbalanced by the increasing cost of good software engineers. When software can be replicated, however, the cost per unit goes down inversely with the number of replicas. Thus, even though software is a major cost of a new computer system, the increasing market decreases its unit cost. Each advance in solid-state technology decreases cost and increases the performance of computer systems; this leads to an increase in market, thus generating decreased unit software costs, leading, in a feedback loop, to further increases in market. Each new application, however, requires new specialized software which is initially expensive (until a market develops) and which requires a user learning curve. Thus, it is difficult to forecast the details of the growth of both the computer market and the data network market.

1.1.2 Communication Technology

The communication links for wide area networks are usually leased from the facilities of the voice telephone network. In Section 2.2 we explain how these physical links are used to transmit a fixed-rate stream of binary digits. The rate at which a link transmits binary digits is usually referred to as the data rate, capacity, or speed of the link, and these data rates come in standard sizes. Early networks typically used link data rates of 2.4, 4.8, 9.6, and 56 kilobits/sec, whereas newer networks often use 64 kilobits/sec, 1.5 megabits/sec, and even 45 megabits/sec. There are major economies of scale associated with higher link speeds; for example, the cost of a 1.5 megabit/sec link is about six times that of a 64 kilobit/sec link, but the data rate is 24 times higher. This makes it economically advantageous to concentrate network traffic on a relatively small set of high-speed links. (This effect is seen in Fig. 1.2 with the use of multiplexers or concentrators to share communication costs.)

One result of sharing high-speed (*i.e.*, high data rate) communication links is that the cost of sending data from one point to another increases less than linearly with the geographic separation of the points. This occurs because a user with a long communication path can share one or more high-speed links with other users (thus achieving low cost per unit data) over the bulk of the path, and use low-speed links (which have high cost per unit data) only for local access to the high-speed links.

Estimating the cost of transmission facilities is highly specialized and complex. The cost of a communication link depends on whether one owns the facility or leases it; with leasing, the cost depends on the current competitive and regulatory situation. The details of communication cost will be ignored in what follows, but there are several overall effects of these costs that are important.

First, for wide area data networks, cost has until recently been dominated by transmission costs. Thus, it has been desirable to use the communication links efficiently, perhaps at added computational costs. As will be shown in Section 1.3, the sporadic nature of most data communication, along with the high cost of idle communication links, led to the development of packet data networks.

Second, because of the gradual maturing of optical fiber technology, transmission costs, particularly for high data rate links, are dropping at an accelerating rate which is expected to continue well into the future. The capacity of a single optical fiber using today's technology is 10^9 to 10^{10} bits/sec, and in the future this could rise to 10^{14} or more. In contrast, all the voice and data traffic in the United States amounts to about

10^{12} bits/sec. Optical fiber is becoming widespread in use and is expected to be the dominant mode of transmission in the future. One consequence of this is that network costs are not expected to be dominated by transmission costs in the future. Another consequence is that network link capacities will increase dramatically; as discussed later, this will change the nature of network applications.

Third, for local area networks, the cost of a network has never been dominated by transmission costs. Coaxial cable and even a twisted pair of wires can achieve relatively high-speed communication at modest cost in a small geographic area. The use of such media and the desire to avoid relatively expensive switching have led to a local area network technology in which many nodes share a common high-speed communication medium on a shared multiaccess basis. This type of network structure is discussed in Chapter 4.

1.1.3 Applications of Data Networks

With the proliferation of computers referred to above, it is not difficult to imagine a growing need for data communication. A brief description of several applications requiring communication will help in understanding the basic problems that arise with data networks.

First, there are many applications centered on remote accessing of central storage facilities and of data bases. One common example is that of a local area network in which a number of workstations without disk storage use one or more common file servers to access files. Other examples are the information services and financial services available to personal computer users. More sophisticated examples, requiring many interactions between the remote site and the data base and its associated programs, include remote computerized medical diagnoses and remote computer-aided education. In some of these examples, there is a cost trade-off between maintaining the data base wherever it might be required and the communication cost of remotely accessing it as required. In other examples, in which the data base is rapidly changing, there is no alternative to communication between the remote sites and the central data base.

Next, there are many applications involving the remote updating of data bases, perhaps in addition to accessing the data. Airline reservation systems, automatic teller machines, inventory control systems, automated order entry systems, and word processing with a set of geographically distributed authors provide a number of examples. Weather tracking systems and military early warning systems are larger-scale examples. In general, for applications of this type, there are many geographically separated points at which data enter the system and often many geographically separated points at which outputs are required. Whether the inputs are processed and stored at one point (as in Figs. 1.1 and 1.2) or processed and stored at many points (as in Fig. 1.3), there is a need for a network to collect the inputs and disseminate the outputs. In any data base with multiple users there is a problem maintaining consistency (*e.g.*, two users of an airline reservation system might sell the same seat on some flight). In geographically distributed systems these problems are particularly acute because of the networking delays.

The communication requirements for accessing files and data bases have been increasing rapidly in recent years. Part of the reason for this is just the natural growth of an expanding field. Another reason is that workstations are increasingly graphics oriented, and transmitting a high-resolution image requires millions of bits. Another somewhat related reason is that the link capacities available in local area networks have been much larger than those in wide area networks. As workstation users get used to sending images and large files over local area nets, they expect to do the same over wide area networks. Thus the need for increased link capacity for wide area networks is particularly pressing.

Another popular application is electronic mail between the human users of a network. Such mail can be printed, read, filed, forwarded to other individuals, perhaps with added comments, or read by the addressee at different locations. It is clear that such a service has many advantages over postal mail in terms of delivery speed and flexibility.

In comparison with facsimile, which has become very popular in recent years, electronic mail is more economical, has the flexibility advantages above, and is in principle more convenient for data already stored in a computer. Facsimile is far more convenient for data in hard-copy form (since the hard copy is fed directly into the facsimile machine). It appears clear, however, that the recent popularity of facsimile is due to the fact that it is relatively hassle-free, especially for the occasional or uninitiated user. Unfortunately, electronic mail, and more generally computer communication, despite all the cant about user friendliness, is full of hassles and pitfalls for the occasional or uninitiated user.

There is a similar comparison of electronic mail with voice telephone service. Voice service, in conjunction with an answering machine or voice mail service, in principle has most of the flexibility of electronic mail except for the ability to print a permanent record of a message. Voice, of course, has the additional advantage of immediate two-way interaction and of nonlinguistic communication via inflection and tone. Voice communication is more expensive, but requires only a telephone rather than a telephone plus computer.

As a final application, one might want to use a remote computer system for some computational task. This could happen as a means of load sharing if the local computer is overutilized. It could also arise if there is no local computer, if the local computer is inoperational, or the remote computer is better suited to the given task. Important special cases of the latter are very large problems that require supercomputers. These problems frequently require massive amounts of communication, particularly when the output is in high resolution graphic form. Present-day networks, with their limited link speeds, are often inadequate for these tasks. There are also "real-time" computational tasks in which the computer system must respond to inputs within some maximum delay. If such a task is too large for the local computer, it might be handled by a remote supercomputer or by a number of remote computers working together. Present-day networks are also often inadequate for the communication needs of these tasks.

It will be noted that all the applications above could be satisfied by a network with centralized computer facilities as in Fig. 1.1 or 1.2. To see this, simply visualize moving all the large computers, data bases, and subnet nodes in the network of Fig. 1.3 to one centralized location, maintaining links between all the nodes previously connected. The central facilities would then be connected by short communication lines rather than long, but aside from some changes in propagation delays, the overall network would be unchanged. Such a geographically centralized but logically distributed structure would both allow for shared memory between the computers and for centralized repair. Why, then, are data networks with geographically distributed computational and data base facilities growing so quickly in importance? One major reason is the cost and delay of communication. With distributed computers, many computational tasks can be handled locally. Even for remote tasks, communication costs can often be reduced significantly by some local processing. Another reason is that organizations often acquire computers for local automation tasks, and only after this local automation takes place does the need for remote interactions arise. Finally, organizations often wish to have control of their own computer systems rather than be overly dependent on the pricing policies, software changes, and potential security violations of a computer utility shared with many organizations.

Another advantage often claimed for a network with distributed computational facilities is increased reliability. For the centralized system in Fig. 1.2 there is some truth to this claim, since the failure of a communication link could isolate a set of sites from all access to computation. For the geographically centralized but logically distributed network, especially if there are several disjoint paths between each pair of sites, the failure of a communication link is less critical and the question of reliability becomes more complex. If all the large computers and data bases in a network were centralized, the network could be destroyed by a catastrophe at the central site. Aside from this possibility, however, a central site can be more carefully protected and repairs can be made more quickly and easily than with distributed computational sites. Other than these effects, there appears to be no reason why geographically distributed computational facilities are inherently more or less reliable than geographically centralized (but logically distributed) facilities. At any rate, the main focus in what follows will be on networks as in Figs. 1.3 and 1.4, where the communication subnet is properly viewed as the center of the entire network.

1.2 MESSAGES AND SWITCHING

1.2.1 Messages and Packets

A message in a data network corresponds roughly to the everyday English usage of the word. For example, in an airline reservation system, we would regard a request for a reservation, including date, flight number, passenger names, and so on, as a message. In an electronic mail system, a message would be a single document from one user to another. If that same document is then forwarded to several other users, we would sometimes want to regard this forwarding as several new messages and sometimes as forwarding of the same message, depending on the context. In a file transfer system, a message would usually be regarded as a file. In an image transmission system (*i.e.*, pictures, figures, diagrams, etc.), we would regard a message as an image. In an application requiring interactive communication between two or more users, a message would be one unit of communication from one user to another. Thus, in an interactive transaction, user 1 might send a message to user 2, user 2 might reply with a message to 1, who might then send another message to 2, and so forth until the completion of the overall transaction. The important characteristic of a message is that from the standpoint of the network users, it is a single unit of communication. If a recipient receives only part of a message, it is usually worthless.

It is sometimes necessary to make a distinction between a message and the representation of the message. Both in a subnet and in a computer, a message is usually represented as a string of binary symbols, 0 or 1. For brevity, a binary symbol will be referred to as a *bit.* When a message goes from sender to recipient, there can be several transformations on the string of bits used to represent the message. Such transformations are sometimes desirable for the sake of data compression and sometimes for the sake of facilitating the communication of the message through the network. A brief description of these two purposes follows.

The purpose of data compression is to reduce the length of the bit string representing the message. From the standpoint of information theory, a message is regarded as one of a collection of possible messages, with a probability distribution on the likelihood of different messages. Such probabilities can only be crudely estimated, either a priori or adaptively. The idea, then, is to assign shorter bit strings to more probable messages and longer bit strings to less probable messages, thus reducing the expected length of the representation. For example, with text, one can represent common letters in the alphabet (or common words in the dictionary) with a small number of bits and represent unusual letters or words with more bits. As another example, in an airline reservation system, the common messages have a very tightly constrained format (date, flight number, names, etc.) and thus can be very compactly represented, with longer strings for unusual types of situations. Data compression will be discussed more in Chapter 2 in the context of compressing control overhead. Data compression will not be treated in general here, since this topic is separable from that of data networks, and is properly studied in its own right, with applications both to storage and point-to-point communication.

Transforming message representations to facilitate communication, on the other hand, is a central topic for data networks. In subsequent chapters, there are many examples in which various kinds of control overhead must be added to messages to ensure reliable communication, to route the message to the correct destination, to control congestion, and so on. It will also be shown that transmitting very long messages as units in a subnet is harmful in several ways, including delay, buffer management, and congestion control. Thus, messages represented by long strings of bits are usually broken into shorter bit strings called *packets.* These packets can then be transmitted through the subnet as individual entities and reassembled into messages at the destination.

The purpose of a subnet, then, is to receive packets at the nodes from sites outside the subnet, then transmit these packets over some path of communication links and other nodes, and finally deliver them to the destination sites. The subnet must somehow obtain information about where the packet is going, but the meaning of the corresponding message is of no concern within the subnet. To the subnet, a packet is simply a string of bits that must be sent through the subnet reliably and quickly. We return to this issue in Section 1.3.

1.2.2 Sessions

Messages between two users usually occur as a sequence in some larger transaction; such a message sequence (or, equivalently, the larger transaction) is called a *session.* For example, updating a data base usually requires an interchange of several messages. Writing a program at a terminal for a remote computer usually requires many messages over a considerable time period. Typically, a setup procedure (similar to setting up a call in a voice network) is required to initiate a session between two users, and in this case a session is frequently called a *connection.* In other networks, no such setup is required and each message is treated independently; this is called a *connectionless* service. The reasons for these alternatives are discussed later.

From the standpoint of network users, the messages within a session are typically triggered by particular events. From the standpoint of the subnet, however, these message initiation times are somewhat arbitrary and unpredictable. It is often reasonable, for subnet purposes, to model the sequence of times at which messages or packets arrive for a given session as a random process. For simplicity, these arrivals will usually be modeled as occurring at random points in time, independently of each other and of the arrivals for other sessions. This type of arrival process is called a *Poisson* process and is defined and discussed in Section 3.3. This model is not entirely realistic for many types of sessions and ignores the interaction between the messages flowing in the two directions for a session. However, such simple models provide insight into the major trade-offs involved in network design, and these trade-offs are often obscured in more realistic and complex models.

Sometimes it will be more convenient to model message arrivals within a session by an on/off flow model. In such a model, a message is characterized by a sequence of bits flowing into the subnet at a given rate. Successive message arrivals are separated by random durations in which no flow enters the network. Such a model is appropriate, for example, for voice sessions and for real-time monitoring types of applications. When voice is digitized (see Section 2.2), there is no need to transmit when the voice is silent, so these silence periods correspond to the gaps in an on/off flow model. One might think that there is little fundamental difference between a model using point arrivals for messages and a model using on/off flow. The output from point message arrivals, followed by an access line of fixed rate, looks very much like an on/off flow (except for the possibilitity that one message might arrive while another is still being sent on the access line). The major difference between these models, however, is in the question of delay. For sessions naturally modeled by point message arrivals (*e.g.*, data base queries), one is usually interested in delay from message arrival to the delivery of the entire message (since the recipient will process the entire message as a unit). For sessions naturally modeled by flow (such as digitized voice), the concept of a message is somewhat artificial and one is usually interested in the delay experienced by the individual bits within the flow. It appears that the on/off flow model is growing in importance and is particularly appropriate for ISDN and broadband ISDN networks. Part of the reason for this growth is the prevalence of voice in ISDN and voice and video in broadband ISDN. Another reason, which will be more clear later, is that very long messages, which will be prevalent with ISDN, are probably better treated in the subnet as flows than as point arrivals.

To put this question of modeling message arrivals for a session in a more pragmatic way, note that networks, particularly wide area networks built around a subnet as in Fig. 1.3, generally handle multiple applications. Since the design and implementation of a subnet is a time-consuming process, and since applications are rapidly changing and expanding, subnets must be designed to handle a wide variety of applications, some of which are unknown and most of which are subject to change. Any complex model of message arrivals for sessions is likely to be invalid by the time the network is used. This point of view, that subnets must be designed to work independently of the fine details of applications, is discussed further in Section 1.3.

At this point we have a conceptual view, or model, of the function of a subnet. It will provide communication for a slowly varying set of sessions; within each session, messages of some random length distribution arrive at random times according to some random process. Since we will largely ignore the interaction between the two directions of message flow for a session, we shall usually model a two-way session as two one-way sessions, one corresponding to the message flow in one direction and the other in the opposite direction. In what follows we use the word *session* for such one-way sessions. In matters such as session initiation and end-to-end acknowledgment, distinctions are made between two-way and one-way sessions.

In principle a session could involve messages between more than two users. For example, one user could broadcast a sequence of messages to each of some set of other users, or the messages of each user in the set could be broadcast to each of the other users. Such sessions might become important in the future, especially for broadband ISDN, with applications such as video conferencing and television broadcast. We will not discuss such applications in any detail, but instead will simply model multiuser sessions as a multiplicity of one-way two-user sessions.

Although the detailed characteristics of different kinds of applications will not be examined, there are some gross characteristics of sessions that must be kept in mind. The most important are listed:

1. *Message arrival rate and variability of arrivals.* Typical arrival rates for sessions vary from zero to more than enough to saturate the network. Simple models for the variability of arrivals include Poisson arrivals, deterministic arrivals (*i.e.*, a fixed time interval from each message to the next message), and uniformly distributed arrivals (*i.e.*, the time interval between successive messages has a uniform probability density between some minimum and maximum interval).
2. *Session holding time.* Sometimes (as with electronic mail) a session is initiated for a single message. Other sessions last for a working day or even permanently.
3. *Expected message length and length distribution.* Typical message lengths vary roughly from a few bits to 10^9 bits, with file transfer applications at the high end and interactive sessions from a terminal to a computer at the low end. Simple models for length distribution include an exponentially decaying probability density, a uniform

probability density between some minimum and maximum, and fixed length. As mentioned above, long messages are becoming much more common because of graphics and long file transfers.

4. *Allowable delay*. The allowable expected delay varies from about 10 msec for some real-time control applications to 1 sec or less for interactive terminal to computer applications, to several minutes or more for some file transfer applications. In other applications, there is a maximum allowable delay (in contrast to expected delay). For example, with packetized voice, fixed-length segments of the incoming voice waveform are encoded into packets at the source. At the destination, these packets must be reconverted into waveform segments with some fixed overall delay; any packet not received by this time is simply discarded. As described above, delay is sometimes of interest on a message basis and sometimes, in the flow model, on a bit basis.
5. *Reliability*. For some applications, all messages must be delivered error-free. For example, in banking applications, in transmission of computer programs, or in file transfers, a single bit error in a message can have serious consequences. In other applications, such as electronic mail, all messages must be delivered, but an occasional bit error in a message can usually be visually corrected by the reader. Finally, in other applications, both occasional bit errors and occasional loss of entire packets or messages are allowable. For example, in distributed sensor systems, messages are sometimes noisy when transmitted, and occasional lost messages are soon replaced with more up-to-date messages. For packetized voice, the occasional loss (or late delivery) of a packet or an occasional bit error simply increases the noisiness of the received voice signal. It should be noted, however, that the use of data compression for packetized voice and other applications greatly increases the need for error-free communication.
6. *Message and packet ordering*. The packets within a message must either be maintained in the correct order going through the network or restored to the correct order at some point. For many applications (such as updating data bases), messages must also be delivered in the correct order, whereas for other applications, message order is unimportant. The question of where to handle reliability and message ordering (*i.e.*, at the external sites or within the subnet or both) is an important design issue. This is discussed in Section 2.8.

In keeping all these characteristics in mind, it is often helpful to focus on four types of applications which lie somewhat at the extreme points and which do not interact very well together in subnets. One is interactive terminal to computer sessions, in which messages are short, the message rate is low, the delay requirement is moderately stringent, and the need for reliability is high. Another is file transfer sessions, in which the messages are very long, the message arrival rate is typically low, the delay requirement is very relaxed, and the need for reliability is very high. The third is high-resolution graphics, in which the messages are again long, sometimes up to 10^9 bits, the delay requirement is stringent, and the arrival rate is low. The fourth is packetized voice. Here the concept of a message is not very useful, but the packets are short, the packet arrival rate is high, the maximum delay requirement is stringent, and the need for reliability is rather low. A network that can handle all these applications together will probably not have too much difficulty with the other applications of interest.

1.2.3 Circuit Switching and Store-and-Forward Switching

There are two general approaches, known as circuit switching and store-and-forward switching, that can be used within a subnet to transmit the traffic for the various sessions. A brief overview will be given of the circuit switching approach, followed by the reason why this approach leads to inefficient utilization of the communication channels for many types of sessions. Next, an overview of the store-and-forward approach will be given, showing how it overcomes the above inefficiency.

For the circuit switching approach, when a session s is initiated, it is allocated a given transmission rate r_s in bits per second (this could be different in the two directions of a two-way session, but we focus on a one-way session here). A path is then created from the transmitting site through the subnet and to the destination site. Each communication link on this path then allocates a portion r_s of its total transmission capacity in the given direction for that session. This allocation of transmission rates to different sessions on a communication link is usually done by time-division multiplexing (TDM) or frequency-division multiplexing (FDM), but the details of that are explained in Section 2.1. What is important is that the sum of the rates for all the sessions using a link cannot exceed the total capacity of the link. Thus, if a communication link is fully allocated to existing sessions, a new session cannot use that link. If no path can be found using links with at least r_s bits/sec of unused rate, the new session must be rejected (*i.e.*, given a busy signal). The other important point is that once the session has been successfully initiated, it has a guaranteed transmission rate r_s through the network. The nodes then simply take the incoming bit stream for a given session off the incoming link and switch it to the allocated portion of the outgoing link. This type of switching is quite similar to the well-developed technology for switching in the telephone network. In the telephone network, however, each session is allocated the same transmission rate, whereas in a data network, the required transmission rates are different and vary over a wide range.

Circuit switching is rarely used for data networks. In the past, the reason for this has had nothing to do with the potential complexity of the switching, but rather, as we now explain, has been because of very inefficient use of the links. Typical data sessions tend to have short bursts of high activity followed by lengthy inactive periods; circuit switching wastes the allocated rate during these inactive periods. For a more quantitative view, let λ be the message arrival rate for a given session s. More precisely, $1/\lambda$ is the expected interarrival time between messages of s. Let $\overline{X}$ be the expected transmission time of a message over a given link in the path; that is, if $\overline{L}$ is the expected length (in bits) of messages from s, and r_s is the bit rate allocated to s, then $\overline{X} = \overline{L}/r_s$. Figure 1.5 illustrates these arrivals and transmission times.

Note from the figure that the fraction of time in which session s's portion of the link is actually transmitting messages is rather small; that portion of the link is otherwise

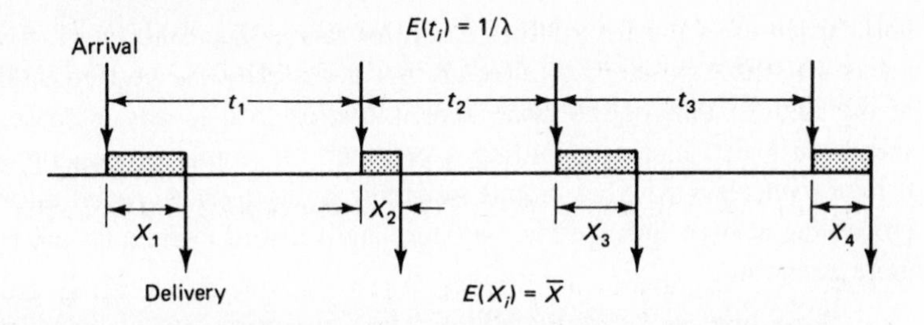

Figure 1.5 Link utilization. The expected transmission time of a message is $\overline{X}$. The expected interarrival period is $1/\lambda$. Thus, the link is used at most $\lambda\overline{X}$ of the time.

idle. It is intuitively plausible, since $1/\lambda$ is the expected interarrival time and $\overline{X}$ is the expected busy time between arrivals, that the ratio of $\overline{X}$ to $1/\lambda$ (*i.e.*, $\lambda\overline{X}$) is the fraction of time in which the portion of the link allocated to s is busy. This argument is made precise in Chapter 3. Our conclusion then is that if $\lambda\overline{X} \ll 1$, session s's portion of the link is idle most of the time (*i.e.*, inefficiently utilized).

To complete our argument about the inefficiency of circuit switching for data networks, we must relate $\overline{X}$ to the allowable expected delay T from message arrival at the source to delivery at the destination. Since $\overline{X}$ is the expected time until the last bit of the message has been sent on the first link, we must have $\overline{X} + P \leq T$, where P is the propagation delay through the network. Thus $\lambda\overline{X} < \lambda T$. If $\lambda T \ll 1$ (*i.e.*, the allowable delay is small relative to the message interarrival rate), the utilization $\lambda\overline{X}$ for the session is correspondingly small. In summary, the bit rate r_s allocated to a session must be large enough to allow message transmission within the required delay, and when $\lambda T \ll 1$, this implies inefficient utilization of the link. Sessions for which $\lambda T \ll 1$ are usually referred to as *bursty* sessions.

For many of the interactive terminal sessions carried by data networks, λT is on the order of 0.01 or less. Thus, with circuit switching, that fraction of a link allocated to such sessions is utilized at most 1% of the time. The conclusion we draw from this is that if link costs are a dominant part of the cost of a network and if bursty sessions require a dominant fraction of link capacity using circuit switching, then circuit switching is an unattractive choice for data networks. Up to the present, both the assumptions above have been valid, and for this reason, data networks have not used circuit switching. The argument above has ignored propagation delays, switching delays in the nodes, and queueing delays. (Queueing delay arises when a message from session s arrives while another message from s is in transmission.) Since these delays must be added to the link transmission time $\overline{X}$ in meeting the delay requirement T, $\overline{X}$ must often be substantially smaller than T, making circuit switching even more inefficient. While propagation and switching delays are often negligible, queueing delay is not, as shown in Chapter 3, particularly when λT is close to or exceeds 1.

In the future, it appears that link costs will become less important in the overall cost of a network. Also, with optical fiber, the marginal cost of link capacity is quite small, so that the wasted capacity of circuit switching will become less important. Finally, it appears that bursty interactive terminal traffic will grow considerably more slowly than link capacities in the future (the reason for this is discussed later). Thus circuit switching is a feasible possibility (although not necessarily the best possibility) for networks of the future. Part of the issue here is that as link speeds increase, node processing speed must also increase, putting a premium on simple processing within the subnet. It is not yet clear whether circuit switching or store-and-forward allows simpler subnet processing at high link speeds, but store-and-forward techniques are currently receiving more attention.

In the store-and-forward approach to subnet design, each session is initiated without necessarily making any reserved allocation of transmission rate for the session. Similarly, there is no conventional multiplexing of the communication links. Rather, one packet or message at a time is transmitted on a communication link, using the full transmission rate of the link. The link is shared between the different sessions using that link, but the sharing is done on an as needed basis (*i.e.*, demand basis) rather than a fixed allocation basis. Thus, when a packet or message arrives at a switching node on its path to the destination site, it waits in a queue for its turn to be transmitted on the next link in its path.

Store-and-forward switching has the advantage over circuit switching that each communication link is fully utilized whenever it has any traffic to send. In Chapter 3, when queueing is studied, it will be shown that using communication links on a demand basis often markedly decreases the delay in the network relative to the circuit switching approach. Store-and-forward switching, however, has the disadvantage that the queueing delays in the nodes are hard to control. The packets queued at a node come from inputs at many different sites, and thus there is a need for control mechanisms to slow down those inputs when the queueing delay is excessive, or even worse, when the buffering capacity at the node is about to be exceeded. There is a feedback delay associated with any such control mechanism. First, the overloaded node must somehow send the offending inputs some control information (through the links of the network) telling them to slow down. Second, a considerable number of packets might already be in the subnet heading for the given node. This is the general topic of flow control and is discussed in Chapter 6. The reader should be aware, however, that this problem is caused by the store-and-forward approach and is largely nonexistent in the circuit switching approach.

There is a considerable taxonomy associated with store-and-forward switching. *Message switching* is store-and-forward switching in which messages are sent as unit entities rather than being segmented into packets. If message switching were to be used, there would have to be a maximum message size, which essentially would mean that the user would have to packetize messages rather than having packetization done elsewhere. *Packet switching* is store-and-forward switching in which messages are broken into packets, and from the discussion above, we see that store-and-forward switching and packet switching are essentially synonymous. *Virtual circuit routing* is store-and-forward switching in which a particular path is set up when a session is initiated and maintained during the life of the session. This is like circuit switching in the sense of using a fixed path, but it is virtual in the sense that the capacity of each link is shared by the sessions using that link on a demand basis rather than by fixed allocations. *Dynamic routing* (or *datagram routing*) is store-and-forward switching in which each packet finds its own path through the network according to the current information available at the nodes visited. Virtual circuit routing is generally used in practice, although there are many interesting intermediate positions between virtual circuit routing and dynamic routing. The general issue of routing is treated in Chapter 5.

1.3 LAYERING

Layering, or layered architecture, is a form of hierarchical modularity that is central to data network design. The concept of modularity (although perhaps not the name) is as old as engineering. In what follows, the word *module* is used to refer either to a device or to a process within some computer system. What is important is that the module performs a given function in support of the overall function of the system. Such a function is often called the *service* provided by the module. The designers of a module will be intensely aware of the internal details and operation of that module. Someone who uses that module as a component in a larger system, however, will treat the module as a "black box." That is, the user will be uninterested in the internal workings of the module and will be concerned only with the inputs, the outputs, and, most important, the functional relation of outputs to inputs (*i.e.*, the service provided). Thus, a black box is a module viewed in terms of its input–output description. It can be used with other black boxes to construct a more complex module, which again will be viewed at higher levels as a bigger black box.

This approach to design leads naturally to a hierarchy of modules in which a module appears as a black box at one layer of the hierarchy, but appears as a system of lower-layer black boxes at the next lower layer of the hierarchy (see Fig. 1.6). At the overall system level (*i.e.*, at the highest layer of the hierarchy), one sees a small collection of top-layer modules, each viewed as black boxes providing some clear-cut service. At the next layer down, each top-layer module is viewed as a subsystem of lower-layer black boxes, and so forth, down to the lowest layer of the hierarchy. As shown in Fig. 1.6, each layer might contain not only black boxes made up of lower-layer modules but also simple modules that do not require division into yet simpler modules.

As an example of this hierarchical viewpoint, a computer system could be viewed as a set of processor modules, a set of memory modules, and a bus module. A processor module could, in turn, be viewed as a control unit, an arithmetic unit, an instruction fetching unit, and an input–output unit. Similarly, the arithmetic unit could be broken into adders, accumulators, and so on.

In most cases, a user of a black box does not need to know the detailed response of outputs to inputs. For example, precisely when an output changes in response to an input is not important as long as the output has changed by the time it is to be used. Thus, modules (*i.e.*, black boxes) can be specified in terms of tolerances rather than exact descriptions. This leads to standardized modules, which leads, in turn, to the possibility of using many identical, previously designed (*i.e.*, off-the-shelf) modules in the same system. In addition, such standardized modules can easily be replaced with new, functionally equivalent modules that are cheaper or more reliable.

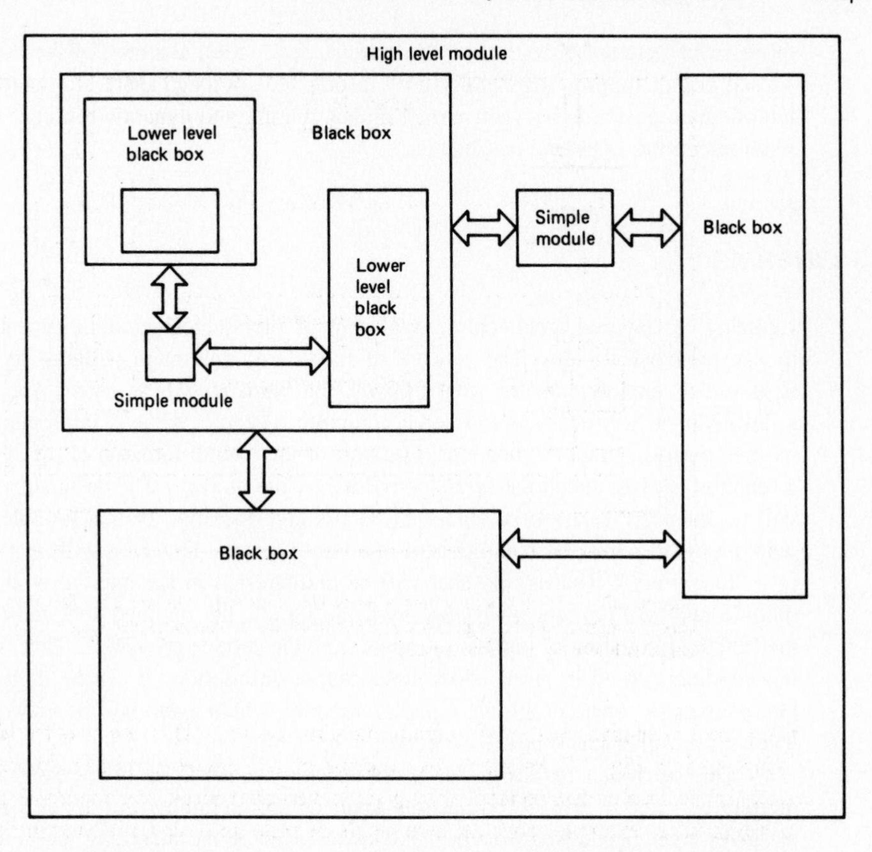

Figure 1.6 Hierarchy of nested black boxes. Each black box (except that at the lowest level) contains black boxes at a lower level, plus perhaps other modules.

All of these advantages of modularity (*i.e.*, simplicity of design; understandability; and standard, interchangeable, widely available modules) provide the motivation for a layered architecture in data networks. A layered architecture can be regarded as a hierarchy of nested modules or black boxes, as described above. Each given layer in the hierarchy regards the next lower layer as one or more black boxes which provide a specified service to the given higher layer.

What is unusual about the layered architecture for data networks is that the black boxes at the various layers are in fact distributed black boxes. The bottom layer of the hierarchy consists of the physical communication links, and at each higher layer, each black box consists of a lower-layer black box communication system plus a set of simple modules, one at each end of the lower-layer communication system. The simple modules associated with a black box at a given layer are called *peer processes* or *peer modules* (see Fig. 1.7).

In the simplest case, a black box consists of two peer processes, one at each of two nodes, and a lower-layer black box communication system connecting the two peer processes. One process communicates with its peer at the other node by placing a message

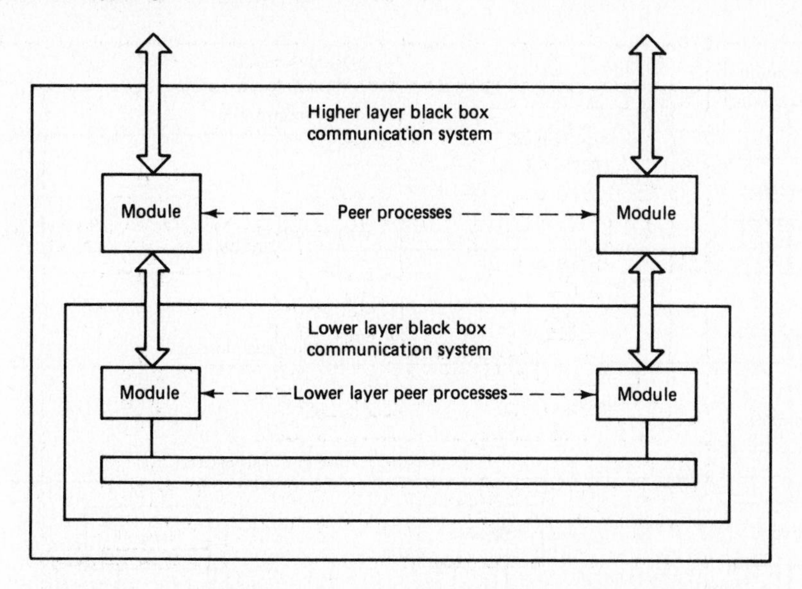

Figure 1.7 Peer processes within a black box communication system. The peer processes communicate through a lower-layer black box communication system that itself contains lower-layer peer processes.

into the lower-layer black box communication system. This lower-layer black box, as illustrated in Fig. 1.7, might in fact consist of two lower-layer peer processes, one at each of the two nodes, connected by a yet lower-layer black box communication system. As a familiar example, consider two heads of state who have no common language for communication. One head of state can then send a message to the peer head of state by a local translator, who communicates in a common language to a peer translator, who then delivers the message in the language of the peer head of state.

Note that there are two quite separate aspects to the communication between a module, say at layer n, and its layer n peer at another node. The first is the protocol (or distributed algorithm) that the peer modules use in exchanging messages or bit strings so as to provide the required functions or service to the next higher layer. The second is the specification of the precise interface between the layer n module at one node and the layer $n-1$ module at the same node through which the messages above are actually exchanged. The first aspect above is more important (and more interesting) for a conceptual understanding of the operation of a layered architecture, but the second is also vital in the actual design and standardization of the system. In terms of the previous example of communication between heads of state, the first aspect has to do with the negotiation between the heads of state, whereas the second has to do with each head of state ensuring that the translator can actually translate the messages faithfully.

Figure 1.8 illustrates such a layered architecture. The layers are those of the reference model of open systems interconnection (OSI) developed as an international standard for data networks by the International Standards Organization (ISO). Many existing networks, including SNA, DECNET, ARPANET, and TYMNET, have somewhat

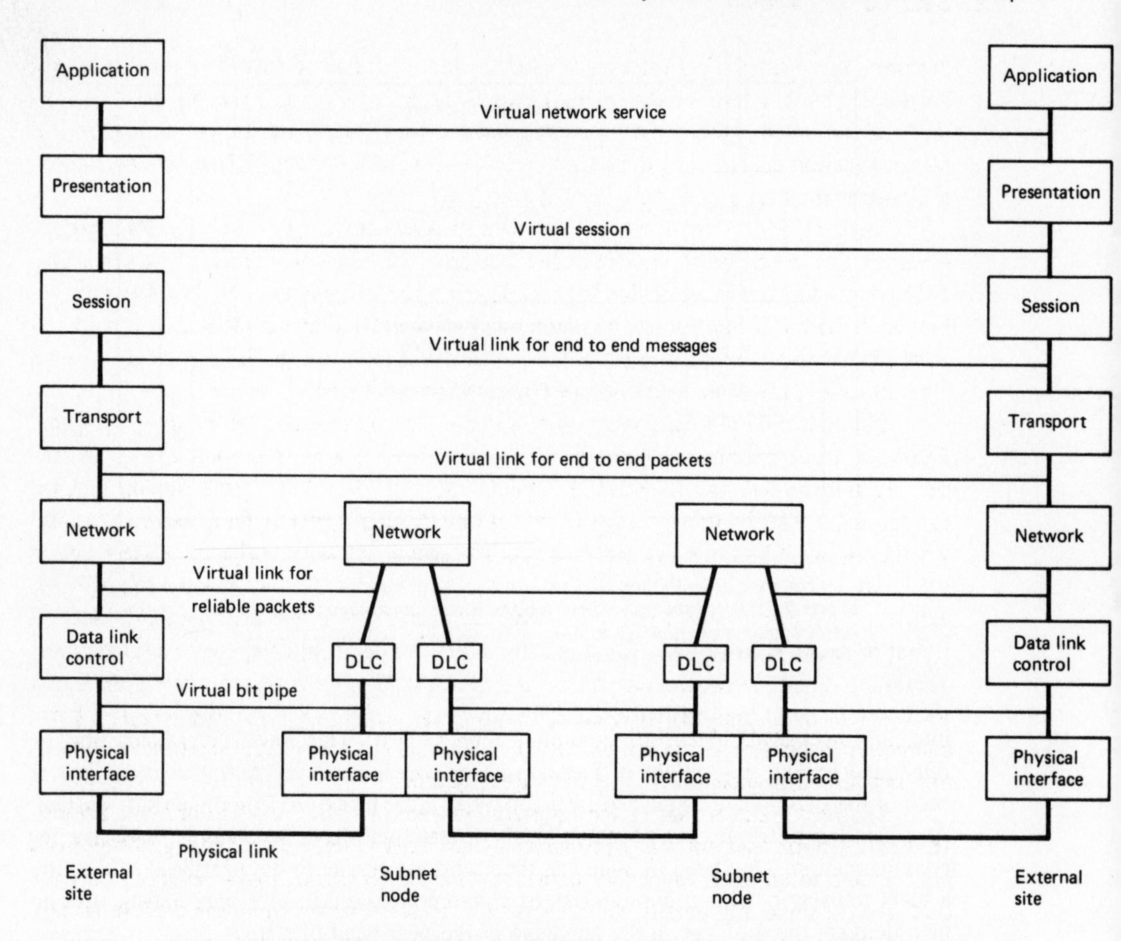

Figure 1.8 Seven-layer OSI network architecture. Each layer presents a virtual communication link with given properties to the next-higher layer.

different layers than this proposed standard. However, the OSI layers have a relatively clean structure that helps in understanding the concept of layering. Some of the variations used by these other networks are discussed later.

1.3.1 The Physical Layer

The function of the *physical layer* is to provide a virtual link for transmitting a sequence of bits between any pair of nodes (or any node and external site) joined by a physical communication channel. Such a virtual link is called a *virtual bit pipe*. To achieve this function, there is a physical interface module on each side of the communication channel whose function is to map the incoming bits from the next higher layer [*i.e.*, the data link control (DLC) layer] into signals appropriate for the channel, and at the receiving end, to map the signals back into bits. The physical interface module that performs these mapping functions is often called a *modem* (digital data *mo*dulator and *dem*odulator). The term *modem* is used broadly here to refer to any module that performs the function above, whether or not modulation is involved; for example, if the physical communication channel is a digital link (see Section 2.2), there is nothing for the modem to do other than interface with the DLC module.

Modems and communication channels are discussed in Section 2.2. The modem designer must be aware of the detailed characteristics of the communication channel (and different modems must be designed for different types of channels). To the higher layers, however, the black box formed by the modem–channel–modem combination appears as a bit pipe with the complexities of the physical channel hidden. Even viewed as a bit pipe, however, there are a few issues that must be discussed.

The first issue has to do with the timing of the bit sequence entering the bit pipe. There are three common situations. The first is that of a *synchronous* bit pipe where bits are transmitted and received at regular intervals (*i.e.*, 1 bit per t second interval for some t). The higher-layer DLC module must supply bits at this synchronous rate whether or not it has any real data to send. The second situation is that of an *intermittent synchronous* bit pipe where the DLC module supplies bits at a synchronous rate when it has data to send and stops sending bits when there are no data to send. The third situation is that of *asynchronous characters*, usually used with personal computers and low-speed terminals. Here, keyboard characters and various control characters are mapped into fixed-length bit strings (usually, eight-bit strings according to a standard mapping from characters to bit strings known as ASCII code), and the individual character bit strings are transmitted asynchronously as they are generated.

The next issue is that of the interface between the DLC module and the modem. One would think that not many problems should exist in delivering a string of bits from one module to another, especially if they are physically close. Unfortunately, there are a number of annoying details about such an interface. For example, the module on one end of the interface might be temporarily inoperable, and when both become operable, some initialization is required to start the flow of bits. Also, for synchronous operation, one side or the other must provide timing. To make matters worse, many different manufacturers provide the modules on either end, so there is a need for standardizing the interface. In fact, there are many such standards, so many that one applauds the effort but questions the success. Two of the better known are RS-232-C and the physical layer of X.21.

The RS-232-C interface approaches the problem by providing a separate wire between the two modules for each type of control signal that might be required. These wires from the modules are joined in a standard 25-pin connector (although usually many fewer wires are required). In communication jargon, the interface is between a DCE (data communication equipment), which is the modem in this case, and a DTE (data terminal equipment), which is the DLC layer and higher layers in this case.

As an example of the interface use, suppose that the DTE wants to start sending data (either on initialization or with a new data sequence in intermittent synchronous transmission). The DTE then sends a signal to the DCE on a "request-to-send" wire. The DCE replies with a signal on the "clear-to-send" wire. The DCE also sends a signal on the "DCE-ready" wire whenever it is operational and a signal on the "carrier detect" wire whenever it appears that the opposite modem and channel are operational. If the DTE receives all these signals (which are just level voltages), it starts to send data over the interface on the DTE-to-DCE data wire.

This interchange is a very simple example of a *protocol* or *distributed algorithm*. Each module performs operations based both on its own state and on the information received from the other module. Many less trivial protocols are developed in subsequent chapters. There are many other details in RS-232-C operation but no other new concepts.

It is sometimes helpful when focusing on the interface between the DLC module and the modem to view the wires between the modules as a physical channel and to view the DLC and modem as peer processes executing the interface protocol. To avoid confusion between the DLC module's major function as a peer process with the opposite DLC module and its lower-level function of interfacing with the modem, an extra dummy module is sometimes created (see Fig. 1.9) which exercises the interface protocol with the modem.

The X.21 physical layer interface is similar in function to RS-232-C, but it uses a smaller set of wires (eight wires are used, although there is a 15-pin connector) and is intended as an interface to a digital communication link. The idea is to avoid using a separate wire for each possible signal by doubling up on the use of wires by digital logic in the modules. The X.21 physical layer is used as the physical layer for the X.25 protocol, which is discussed in Chapter 2.

It should be clear from the above that there is a great conceptual advantage in removing the question of modem and modem interfaces from the higher-level aspects of networks. Note that this has already been done, in essence, in previous sections in referring to the number of bits per second that could be transmitted over communication links. It should also be noted, however, that modems cannot be totally segregated from network issues. For example, is it better to have a modem that transmits R bits/sec with an error rate of 10^{-6} or a modem that transmits $2R$ bits/sec with an error rate of 10^{-4}? This cannot be answered without some knowledge of how errors are eliminated at higher

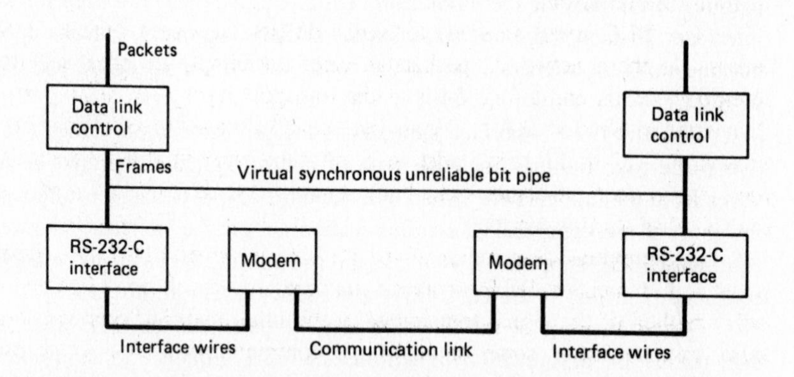

Figure 1.9 Layering with the interface between the DLC and the modem viewed as an interface over a physical medium consisting of a set of wires.

layers of the architecture. Conversely, decisions on how and where to eliminate errors at higher layers should depend on the error rate and error characteristics at the physical layer.

1.3.2 The Data Link Control Layer

The second layer in Fig. 1.8 is the *data link control (DLC) layer*. Each point-to-point communication link (*i.e.*, the two-way virtual bit pipe provided by layer 1) has data link control modules (as peer processes) at each end of the link. The customary purpose of data link control is to convert the unreliable bit pipe at layer 1 into a higher-level, virtual communication link for sending packets asynchronously but error-free in both directions over the link. From the standpoint of the DLC layer, a packet is simply a string of bits that comes from the next higher layer.

The communication at this layer is asynchronous in two ways. First, there is a variable delay between the entrance of a packet into the DLC module at one end of the link and its exit from the other end. This variability is due both to the need to correct the errors that occur at the physical layer and to the variable length of the packets. Second, the time between subsequent entries of packets into the DLC module at one end of the link is also variable. The latter variability is caused both because higher layers might have no packets to send at a given time and also because the DLC is unable to accept new packets when too many old packets are being retransmitted due to transmission errors.

Data link control is discussed in detail in Chapter 2. In essence, the sending DLC module places some overhead control bits called a *header* at the beginning of each packet and some more overhead bits called a *trailer* at the end of each packet, resulting in a longer string of bits called a *frame*. Some of these overhead bits determine if errors have occurred in the transmitted frames, some request retransmissions when errors occur, and some delineate the beginning and ending of frames. The algorithms (or protocols) for accomplishing these tasks are distributed between the peer DLC modules at the two ends of each link and are somewhat complex because the control bits themselves are subject to transmission errors.

The DLC layers in some networks do not retransmit packets in the presence of errors. In these networks, packets in error are simply dropped and retransmission is attempted on an end-to-end basis at the transport layer. The relative merits of this are discussed in Section 2.8.2. Typically, the DLC layer ensures that packets leave the receiving DLC in the same order in which they enter the transmitting DLC, but not all data link control strategies ensure this feature; the relative merits of ordering are also discussed in Section 2.8.2.

Our previous description of the physical layer and DLC was based on point-to-point communication links for which the received waveform at one end of the link is a noisy replica of the signal transmitted at the other end. In some networks, particularly local area networks, some or all of the communication takes place over multiaccess links. For these links, the signal received at one node is a function of the signals from a multiplicity of transmitting nodes, and the signal transmitted from one node might be heard at a multiplicity of other nodes. This situation arises in satellite communication, radio communication, and communication over cables, optical fibers, and telephone lines with multiple taps. Multiaccess communication is treated in Chapter 4.

The MAC sublayer The appropriate layers for multiaccess communication are somewhat different from those in networks of point-to-point links. There is still the need for a DLC layer to provide a virtual error-free packet link to higher layers, and there is still the need for a physical layer to provide a bit pipe. However, there is also a need for an intermediate layer to manage the multiaccess link so that frames can be sent by each node without constant interference from the other nodes. This is called *medium access control* (MAC). It is usually considered as the lower sublayer of layer 2 with the conventional DLC considered as the higher sublayer. Figure 1.10 illustrates the relationship between these layers. The service provided by the MAC to the DLC is that of an intermittent synchronous bit pipe. The function of the MAC sublayer is to allocate the multiaccess channel so that each node can successfully transmit its frames without undue interference from the other nodes; see Chapter 4 for various ways of accomplishing this function.

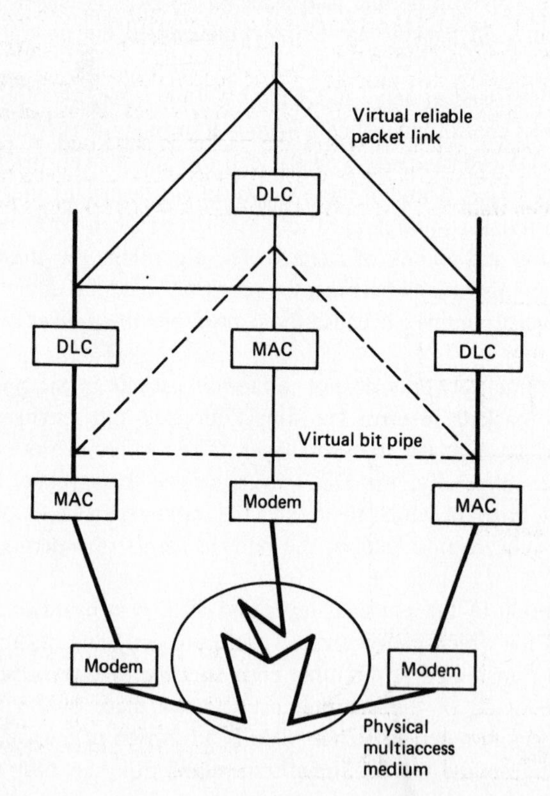

Figure 1.10 Layering for a multiaccess channel. The physical medium is accessed by all three users, each of whom hears the transmitted signals of the others. The DLC sublayer sees virtual point-to-point bit pipes below it. The MAC sublayer sees a multiaccess bit pipe, and the modems access the actual channel.

1.3.3 The Network Layer

The third layer in Fig. 1.8 is the *network layer*. There is one network layer process associated with each node and with each external site of the network. All these processes are peer processes and all work together in implementing routing and flow control for the network. When a frame enters a node or site from a communication link, the bits in that frame pass through the physical layer to the DLC layer. The DLC layer determines where the frame begins and ends, and if the frame is accepted as correct, the DLC strips off the DLC header and trailer from the frame and passes the resulting packet up to the network layer module (see Fig. 1.11). A packet consists of two parts, a packet header followed by the packet body (and thus a frame at the DLC layer contains first the DLC header, next the packet header, next the packet body, and then the DLC trailer). The network layer module uses the packet header of an incoming packet, along with stored information at the module, to accomplish its routing and flow control functions. Part of the principle of layering is that the DLC layer does not look at the packet header or packet body in performing its service function, which is to deliver the packet reliably to the network layer at the next node. Similarly, the network layer does not use any of the information in the DLC header or trailer in performing its functions of routing and flow control. The reason for this separation is to allow improvements, modifications, and replacements in the internal operation of one layer without forcing the other to be modified.

Newly generated messages from users at an external site are processed by the higher layers, broken into packet-sized pieces if need be, and passed down from the transport layer module to the network module. These packet-sized pieces constitute the packet body at the network layer. The transport layer also provides additional information about how to handle the packet (such as where the packet is supposed to go), but this information is passed to the network layer as a set of parameters in accordance with the interfacing protocol between transport and network layer. The network layer module uses

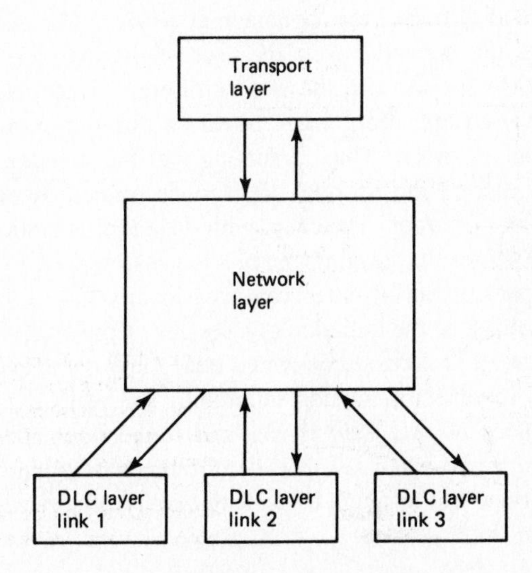

Figure 1.11 The network layer at a node or site can receive packets from a DLC layer for each incoming link and (in the case of a site) from the transport layer. It can send these packets out to the same set of modules.

these parameters, along with its own stored information, to generate the packet header in accordance with the protocol between peer network layer modules.

Along with the transit packets arriving at the network layer module from the lower layer, and the new packets arriving from the higher layer, the network layer can generate its own control packets. These control packets might be specific to a given session, dealing with initiating or tearing down the session, or might have a global function, informing other nodes of link congestion, link failures, and so on.

For networks using virtual circuit routing (*i.e.*, in which the route for a session is fixed over the life of the session), the routing function at the network layer module consists of two parts. The first is to select a route when the virtual circuit is being initiated, and the second is to ensure that each packet of the session follows the assigned route. The selection of a route could be carried out in a distributed way by all the nodes, or could be carried out by the source node or by some central node entrusted with this task. No matter how the job is allocated between the different nodes, however, there is need for considerable communication, via network control packets, concerning the operating characteristics and level of traffic and delay throughout the network. This subject is treated in considerable depth in Chapter 5. Ensuring that each packet follows the assigned route is accomplished by placing enough information in the packet header for the network layer module at each node of the route to be able to forward the packet on the correct outgoing link (or to pass the packet body up to the transport layer when the destination is reached). Ways of doing this are discussed in Section 2.8.

Datagram networks, on the other hand, do not have a connection phase in which a route for a session is determined. Rather, each packet is routed individually. This appears to be a very natural and simple approach, but as Chapter 5 shows, the dynamics of the traffic patterns in a network and the lack of timely knowledge about those patterns at the nodes make this much more difficult than one would think. Most wide area networks use virtual circuits for this reason.

It is necessary here to make a distinction between virtual circuit or datagram *operation* at the network layer and virtual circuit or datagram *service*. The discussion above concerned the *operation* of the network layer; the user of the network layer (usually the transport layer) is concerned only with the *service* offered. Since successive packets of a session, using datagram operation, might travel on different routes, they might appear at the destination out of order. Thus (assuming that the network layer module at the destination does not reorder the packets), the service offered by such a network layer allows packets to get out of order. Typically, with datagram operation, packets are sometimes dropped also. As a result, datagram *service* is usually taken to mean that the network layer can deliver packets out of order, can occasionally fail to deliver packets, and requires no connection phase at the initiation of a session. Conversely, virtual circuit *service* is taken to mean that all packets are delivered once, only once, and in order, but that a connection phase is required on session initiation. We will often use the term *connectionless service* in place of *datagram service* and *connection-oriented service* in place of *virtual circuit service*. We shall see that the difference between connectionless and connection-oriented service has as much to do with quality of service, flow control, and error recovery as it does with routing.

The other major function of the network layer, along with routing, is flow control, or congestion control. Some authors make a distinction between flow control and congestion control, viewing the first as avoiding sending data faster than the final destination can absorb it, and the second as avoiding congestion within the subnet. Actually, if the destination cannot absorb packets as fast as they are sent, those packets will remain in the subnet and cause congestion there. Similarly, if a link in the subnet is congested (*i.e.*, many packets are buffered in an adjacent node waiting for transmission on the link), then there are a number of mechanisms that cause the congestion to spread. Thus congestion is a global issue that involves both the subnet and the external sites, and at least at a conceptual level, it is preferable to treat it as a single problem.

Fundamentally, congestion occurs when the users of the network collectively demand more resources than the network (including the destination sites) has to offer. Good routing can help to alleviate this problem by spreading the sessions out over the available subnet resources. Good buffer management at the nodes can also help. Ultimately, however, the network layer must be able to control the flow of packets into the network, and this is what is meant by flow control (and why we use the term *flow control* in place of *congestion control*).

The control of packet flow into the network must be done in such a way as to prevent congestion and also to provide equitable service to the various sessions. Note that with connection-oriented service, it is possible for a session to negotiate its requirements from the network as a compromise between user desires and network utilization. Thus in some sense the network can guarantee the service as negotiated. With connectionless service, there is no such opportunity for negotiation, and equitable service between users does not have much meaning. This is another reason for the prevalence of connection-oriented service in wide area networks. In Chapter 6 we develop various distributed algorithms for performing the flow control function. As with routing, flow control requires considerable exchange of information between the nodes. Some of this exchange occurs through the packet headers, and some through control packets.

One might hope that the high link capacities that will be available in the future will make it possible to operate networks economically with low utilization, thus making flow control unnecessary. Unfortunately, this view appears overly simplistic. As link capacities increase, access rates into networks will also increase. Thus, even if the aggregate requirements for network service are small relative to the available capacity, a single malfunctioning user could dump enough data into the network quickly to cause serious congestion; if the network plays no regulatory role, this could easily lead to very chaotic service for other users.

The discussion of routing and flow control above has been oriented primarily toward wide area networks. Most local area networks can be viewed as using a single multiaccess channel for communication, and consequently any node is capable of receiving any packet. Thus routing is not a major problem for local area networks. There is a possibility of congestion in local area networks, but this must be dealt with in the MAC sublayer. Thus, in a sense, the major functions of the network layer are accomplished in the MAC sublayer, and the network layer is not of great importance in local area networks. For this reason, the arguments for virtual circuit operation and connection oriented service in the network layer do not apply to local area networks, and connectionless service is common there.

The network layer is conceptually the most complex of the layered hierarchy since *all* the peer processes at this layer must work together. For the lower layers (except for the MAC sublayer for multiaccess), the peer processes are paired, one at each side of a communication link. For the higher layers, the peer processes are again paired, one at each end of a session. Thus, the network layer and the MAC sublayer are the only layers in which the overall algorithms are distributed between many geographically separated processes.

Acquiring the ability to design and understand such distributed algorithms is one of the basic objectives of this book. Chapter 2 covers the simpler forms of distributed algorithms involving just two peer processes at opposite ends of a link. In Chapter 4 we treat distributed algorithms involving many peer processes in the context of the MAC sublayer, and Chapters 5 and 6 deal with distributed algorithms involving many peer processes at the network layer.

When the network layer and lower layers at all nodes and sites are regarded as one black box, a packet entering the network layer from the next higher layer at a site reappears at some later time at the interface between the network layer and the next higher layer at the destination site. Thus, the network layer appears as a virtual, packet-carrying, end-to-end link from origin site to destination site. Depending on the design of the network layer, this virtual link might be reliable, delivering every packet, once and only once, without errors, or might be unreliable, failing to deliver some packets and delivering some packets with errors. The higher layers then might have to recover from these errors. The network layer might also deliver all packets for each session in order or might deliver them out of order. The relative merits of these alternatives are discussed further in Section 2.8.

The internet sublayer Despite all efforts at standardization, different networks use different algorithms for routing and flow control at the network layer. We have seen some of the reasons for this variety in our discussion of wide area versus local area networks. Since these network layer algorithms are distributed and require close coordination between the various nodes, it is not surprising that one cannot simply connect different subnetworks together. The accepted solution to this problem is to create a new sublayer called the *internet sublayer*. This is usually regarded as being the top part of the network layer. Several subnets can be combined by creating special nodes called gateways between them. A gateway connecting two subnets will interface with each subnet through a network layer module appropriate for that subnet. From the standpoint of the subnet, then, a gateway looks like an external site.

Each gateway will have an internet sublayer module sitting on top of the network layer modules for the individual subnets. When a packet arrives at a gateway from one subnet, the corresponding network layer module passes the packet body and subsidiary information about the packet to the internet module (which thus acts like a transport layer module to the network layer module). This packet body and subsidiary information is then passed down to the other network layer module for forwarding on through the other subnet.

The internet modules also must play a role in routing and flow control. There is not a great deal of understanding in the field yet as to the appropriate ways for the internet sublayer and the various network layers to work together on routing and flow control. From a practical standpoint, the problem is exacerbated by the fact that the network layers for the subnets are usually in place, designed without the intention of later being used in a network of networks. Thus the internet layer must of necessity be somewhat ad hoc.

When combining local area networks, where routing and flow control are exercised at the MAC sublayer, it is often possible to replace the gateway between subnets with a bridge. Bridges interface different subnets at the DLC layer rather than at the network layer; for local area networks, this is possible because the routing and flow control are done in the MAC sublayer. In Chapter 5 we discuss gateways and bridges in greater detail, particularly with respect to routing.

1.3.4 The Transport Layer

The fourth layer in Fig. 1.8 is the *transport layer*. Here, for each virtual end-to-end link provided by the network layer (or internet sublayer), there is a pair of peer processes, one at each end of the virtual end-to-end link. The transport layer has a number of functions, not all of which are necessarily required in any given network.

First, the transport layer breaks messages into packets at the transmitting end and reassembles packets into messages at the receiving end. This reassembly function is relatively simple if the transport layer process has plenty of buffer space available, but can be quite tricky if there is limited buffer space that must be shared between many virtual end-to-end links. If the network layer delivers packets out of order, this reassembly problem becomes even more difficult.

Second, the transport layer might multiplex several low-rate sessions, all from the same source site and all going to the same destination site, into one session at the network layer. Since the subnet sees only one session in this case, the number of sessions in the subnet and the attendant overhead is reduced. Often this is carried to the extreme in which all sessions with a common source site and common destination site are multiplexed into the same session. In this case, the addressing at the network layer need only specify the source and destination sites; the process within the source site and destination site are then specified in a transport layer header.

Third, the transport layer might split one high-rate session into multiple sessions at the network layer. This might be desirable if the flow control at the network layer is incapable of providing higher-rate service to some sessions than others, but clearly a better solution to this problem would be for the network layer to adjust the rate to the session requirement.

Fourth, if the network layer is unreliable, the transport layer might be required to achieve reliable end-to-end communication for those sessions requiring it. Even when the network layer is designed to provide reliable communication, the transport layer has to be involved when one or the other end site fails or when the network becomes disconnected due to communication link failures. These failure issues are discussed further in Section 2.8 and in Chapters 5 and 6.

Fifth, end-to-end flow control is often done at the transport layer. There is little difference between end-to-end flow control at the transport layer and network layer (or internet sublayer if it exists). End-to-end flow control at the transport layer is common in practice but makes an integrated approach to avoiding congestion somewhat difficult. This is discussed further in Section 2.9.4 and in Chapter 6.

A header is usually required at the transport layer; this transport header, combined with the data being transported, serves as the packet body passed on to the network layer. Thus the actual body of data is encapsulated in a sequence of headers with the lowest layers on the outside (see Fig. 1.12). At the destination, these layer headers are peeled off in passing up through the various layers. In ISO terminology, the body of data shown in the figure is referred to as a transport service data unit (T-SDU). This data unit, along with the transport header, is referred to as a transport protocol data unit (T-PDU). This unit is also the body of the packet at the network layer, which is sometimes referred to as a network service data unit (N-SDU). Similarly, the packet body plus packet header is referred to as a network protocol data unit (N-PDU). Similarly, each layer in the hierarchy has an SDU, as the unit coming in from the higher layer, and a PDU as the unit going down to the next-lower layer. It is difficult to know where to take a stand against acronymitis in the network field, but we will continue to use the more descriptive terminology of messages, packets, and frames.

1.3.5 The Session Layer

The *session layer* is the next layer above the transport layer in the OSI hierarchy of Fig. 1.8. One function of the session layer is akin to the directory assistance service in the telephone network. That is, if a user wants an available service in the network

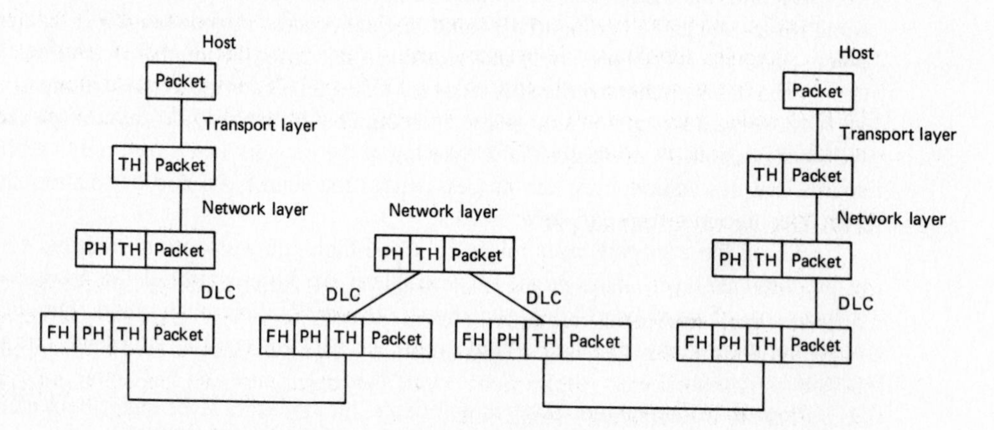

Figure 1.12 Illustration of various headers on a frame. Note that each layer looks only at its own header.

but does not know where to access that service, this layer provides the transport layer with the information needed to establish the session. For example, this layer would be an appropriate place to achieve load sharing between many processors that are sharing computational tasks within a network.

The session layer also deals with *access rights* in setting up sessions. For example, if a corporation uses a public network to exchange records between branch offices, those records should not be accessible to unauthorized users. Similarly, when a user accesses a service, the session layer helps deal with the question of who pays for the service.

In essence, the session layer handles the interactions between the two end points in setting up a session, whereas the network layer handles the subnet aspects of setting up a session. The way that session initiation is divided between session layer, transport layer, and network layer varies from network to network, and many networks do not have these three layers as separate entities.

1.3.6 The Presentation Layer

The major functions of the *presentation layer* are data encryption, data compression, and code conversion. The need for encryption in military organizations is obvious, but in addition, corporations and individual users often must send messages that should only be read by the intended recipient. Although data networks should be designed to prevent messages from getting to the wrong recipients, one must expect occasional malfunctions both at the external sites and within the subnet; this leads to the need for encryption of critical messages.

The desirability of data compression in reducing the number of bits to be communicated has already been mentioned. This function could be performed at any of the layers, but there is some advantage in compressing the data for each session separately, in the sense that different sessions have different types of redundancy in their messages. In particular, data compression must be done (if at all) before encryption, since encrypted data will not have any easily detectable redundancy.

Finally, code conversion is sometimes necessary because of incompatible terminals, printers, graphics terminals, file systems, and so on. For example, some terminals use the ASCII code to represent characters as 8-bit bytes, whereas other terminals use the EBCDIC code. Messages using one code must be converted to the other code to be readable by a terminal using the other code.

1.3.7 The Application Layer

The *application layer* is simply what is left over after the other layers have performed their functions. Each application requires its own software (*i.e.*, peer processes) to perform the desired application. The lower layers perform those parts of the overall task that are required for many different applications, while the application layer does that part of the task specific to the particular application.

At this point, the merits of a layered approach should be clear, but there is some question about which functions should be performed at each layer. Many networks omit the session and presentation layers, and as we have seen, the lower layers are now divided into sublayers. An even more serious issue is that in an effort to achieve agreement on the standards, a number of alternatives have been introduced which allow major functions to be either performed or not at various layers. For example, error recovery is sometimes done at the DLC layer and sometimes not, and because of this, the higher layers cannot necessarily count on reliable packet transfer. Thus, even within the class of networks that conform to the OSI reference model, there is considerable variation in the services offered by the various layers. Many of the existing networks described later do not conform to the OSI reference model, and thus introduce even more variation in layer services. Broadband ISDN networks, for example, do routing and flow control at the physical layer (in a desire to simplify switch design), thus making the network look like an end-to-end bit pipe from the origin to destination.

Even with all these problems, there is a strong desire for standardization of the interfaces and functions provided by each layer, even if the standard is slightly inappropriate. This desire is particularly strong for international networks and particularly strong among smaller equipment manufacturers who must design equipment to operate correctly with other manufacturers' equipment. On the other hand, standardization is an impediment to innovation, and this impediment is particularly important in a new field, such as data networks, that is not yet well understood. Fortunately, there is a constant evolution of network standards. For example, the asynchronous transfer node (ATM) protocol of broadband ISDN circumvents the ISO network layer standard by performing the function of the network layer at the physical layer (see Section 2.10).

One particular difficulty with the seven layers is that each message must pass through seven processes before even entering the subnet, and all of this might generate considerable delay. This text neither argues for this particular set of layers nor proposes another set. Rather, the objective is to create an increased level of understanding of the functions that must be accomplished, with the hope that this will contribute to standardization. The existing networks examined in subsequent chapters do not, in fact, have layers that quite correspond to the OSI layers.

1.4 A SIMPLE DISTRIBUTED ALGORITHM PROBLEM

All of the layers discussed in Section 1.3 have much in common. All contain peer processes, one at each of two or more geographically separated points, and the peer processes communicate via the communication facility provided by the next lower layer. The peer processes in a given layer have a common objective (*i.e.*, task or function) to be performed jointly, and that objective is achieved by some combination of processing and interchanging information. The algorithm to achieve the objective is a *distributed algorithm* or a *protocol*. The distributed algorithm is broken down into a set of *local algorithms*, one of which is performed by each peer process. The local algorithm performed by one process in a set of peers consists of carrying out various operations on the available data, and at various points in the algorithm, either sending data to one or more other peer processes or reading (or waiting for) data sent by another peer process.

In the simplest distributed algorithm, the order in which operations are carried out by the various local algorithms is completely determined. For example, one local algorithm might perform several operations and then reliably send some data to the other local algorithm, which then carries out some operations and returns some data. Only one local algorithm operates at a time and the distributed algorithm is similar to a centralized algorithm that performs all operations sequentially at one location.

In more complex cases, several local algorithms might operate concurrently, but each still waits at predetermined points in the local algorithm for predetermined messages from specific other local algorithms. In this case, the overall distributed algorithm still operates in a deterministic fashion (given the input data to the peer processes), but the lockstep ordering of operations between different local algorithms is removed.

In the most complex case (which is of most interest), the order in which a local algorithm performs its operations depends on the order in which data arrive (either from the next higher layer or from a peer process). Also, if the underlying communication facility is unreliable, data sent by a peer process might never arrive or might arrive with errors.

Most people are very familiar with the situation above, since people must often perform tasks requiring interacting with others, often with unreliable communication. In these situations, however, people deal with problems as they arise rather than thinking through all possible eventualities ahead of time, as must be done with a distributed algorithm.

To gain some familiarity with distributed algorithms, a very simple problem is presented, involving unreliable communication, which in fact has no solution. Analogous situations arise frequently in the study of data networks, so it is well to understand such a problem in its simplest context.

There are three armies, two colored magenta and one lavender. The lavender army separates the two magenta armies, and if the magenta armies can attack simultaneously, they win, but if they attack separately, the lavender army wins. The only communication between the magenta armies is by sending messengers through the lavender army lines, but there is a possibility that any such messenger will be captured, causing the message to go undelivered (see Fig. 1.13). The magenta armies would like to synchronize their attack at some given time, but each is unwilling to attack unless assured with certainty that the other will also attack. Thus, the first army might send a message saying, "Let's attack on Saturday at noon; please acknowledge if you agree."

The second army, hearing such a message, might send a return message saying, "We agree; send an acknowledgment if you receive our message." It is not hard to see that this strategy leads to an infinite sequence of messages, with the last army to send a message being unwilling to attack until obtaining a commitment from the other side.

What is more surprising is that no strategy exists for allowing the two armies to synchronize. To see this, assume that each army is initially in state 0 and stays in this state if it receives no messages. If an army commits itself to attack, it goes to state 1, but it will not go to state 1 unless it is certain that the other army will go to state 1. We also assume that an army can change state only at the time that it receives a message (this assumption in essence prevents side information other than messages from synchronizing the armies). Now consider any ordering in which the two armies receive messages. The

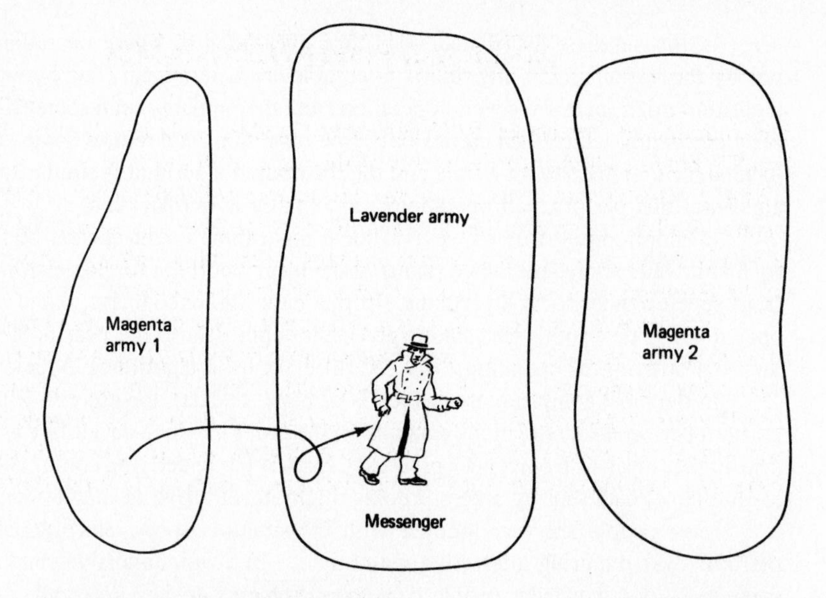

Figure 1.13 A messenger carries a message through enemy lines from magenta army 1 to magenta army 2. If the messenger is caught, the message is undelivered. Magenta army 1 is unaware of capture and magenta army 2 is unaware of existence of message.

first army to receive a message cannot go to state 1, since it has no assurance that any message will be received by the other army. The second army to receive a message also cannot go to state 1 since it is not assured that the other side will receive subsequent messages, and even if it knows that the other side received a first message, it knows that the other side is not currently in state 1. Proceeding in this way (or more formally by induction), neither army can ever go to state 1.

What is surprising about this argument is the difficulty in convincing oneself that it is correct. The difficulty does not lie with the induction argument, but rather with the question of whether the model fairly represents the situation described. It appears that the problem is that we are not used to dealing with distributed questions in a precise way; classical engineering problems do not deal with situations in which distributed decisions based on distributed information must be made.

If the conditions above are relaxed somewhat so as to require only a high probability of simultaneous attack, the problem can be solved. The first army simply decides to attack at a certain time and sends many messengers simultaneously to the other side. The first army is then assured with high probability that the second army will get the message, and the second army is assured that the first army will attack.

Fortunately, most of the problems of communication between peer processes that are experienced in data networks do not require this simultaneous agreement. Typically, what is required is for one process to enter a given state with the assurance that the peer process will *eventually* enter a corresponding state. The first process might be required to wait for a confirmation of this eventuality, but the deadlock situation of the three-army problem, in which neither process can act until after the other has acted, is avoided.

NOTES AND SUGGESTED READING

The introductory textbooks by Tanenbaum [Tan88], Stallings [Sta85], and Schwartz [Sch87] provide alternative treatments of the material in this chapter. Tanenbaum's text is highly readable and contains several chapters on the higher levels of the OSI architecture. Stalling's text contains a wealth of practical detail on current network practice. Schwartz's text also includes several chapters on circuit switching. Some perspectives on the historical evolution of data networks are given in [Gre84].

New developments in technology and applications are critically important in both network design and use. There are frequent articles in the *IEEE Spectrum*, *IEEE Communications Magazine*, and *IEEE Computer* that monitor these new developments. *Silicon Dreams: Information, Man and Machine* by Lucky [Luc90] provides an excellent overview of these areas. A good reference on layered architecture is [Gre82], and some interesting commentary on future standardization of layers is given in [Gre86].

PROBLEMS

1.1. A high quality image requires a spatial resolution of about 0.002 inch, which means that about 500 pixels (*i.e.* samples) per inch are needed in a digital representation. Assuming 24 bits per pixel for a color image of size 8.5 by 11 inches, find the total number of bits required for such an image representation.

1.2. **(a)** Suppose a city of one million inhabitants builds a data network. Suppose that each inhabitant, during the busiest hour of the day, is involved in an average of 4 transactions per hour that use this network (such as withdrawing money from a cash machine, buying some item in a store and thus generating an inventory control message, etc.). Suppose that each transaction, on the average, causes 4 packets of 1000 bits each to be sent. What is the aggregate average number of bits per second carried by the network? How many 64 kbit/sec voice telephone links are required to carry this traffic assuming that each packet travels over an average of 3 links?

(b) Suppose that the inhabitants use their telephones an average of 10% of the time during the busy hour. How many voice telephone links are required for this, assuming that all calls are within the city and travel over an average of three links?

1.3. Suppose packets can get dropped or arbitrarily delayed inside a packet network. Suppose two users are communicating in a session and want to terminate the session. We would like a protocol that exchanges packets in such a way that both users know that they can terminate with the knowledge that no further packets will arrive from the other user. Can such a protocol be designed? What is the relation between this problem and the three-army problem of Section 1.10?

2

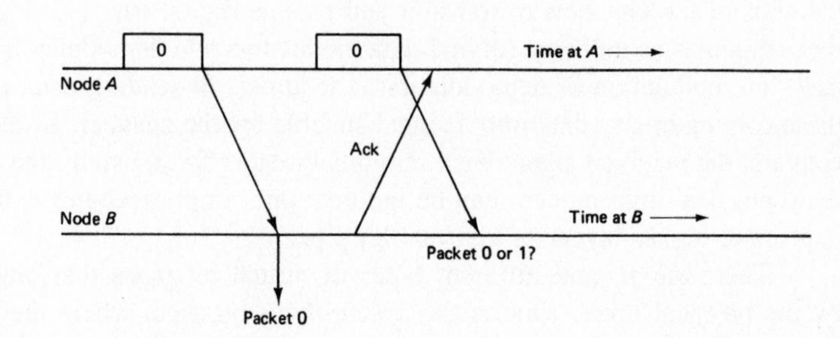

Point-to-Point Protocols and Links

2.1 INTRODUCTION

This chapter first provides an introduction to the physical communication links that constitute the building blocks of data networks. The major focus of the chapter is then data link control (*i.e.*, the point-to-point protocols needed to control the passage of data over a communication link). Finally, a number of point-to-point protocols at the network, transport, and physical layers are discussed. There are many similarities between the point-to-point protocols at these different layers, and it is desirable to discuss them together before addressing the more complex network-wide protocols for routing, flow control, and multiaccess control.

The treatment of physical links in Section 2.2 is a brief introduction to a very large topic. The reason for the brevity is not that the subject lacks importance or inherent interest, but rather, that a thorough understanding requires a background in linear system theory, random processes, and modern communication theory. In this section we provide a sufficient overview for those lacking this background and provide a review and perspective for those with more background.

In dealing with the physical layer in Section 2.2, we discuss both the actual communication channels used by the network and whatever interface modules are required at the ends of the channels to transmit and receive digital data (see Fig 2.1). We refer to these modules as modems (digital data modulators and demodulators), although in many cases no modulation or demodulation is required. In sending data, the modem converts the incoming binary data into a signal suitable for the channel. In receiving, the modem converts the received signal back into binary data. To an extent, the combination of modem, physical link, modem can be ignored; one simply recognizes that the combination appears to higher layers as a virtual bit pipe.

There are several different types of virtual bit pipes that might be implemented by the physical layer. One is the *synchronous* bit pipe, where the sending side of the data link control (DLC) module supplies bits to the sending side of the modem at a synchronous rate (*i.e.*, one bit each T seconds for some fixed T). If the DLC module temporarily has no data to send, it must continue to send dummy bits, called *idle fill*, until it again has data. The receiving modem recovers these bits synchronously (with delay and occasional errors) and releases the bits, including idle fill, to the corresponding DLC module.

The *intermittent synchronous* bit pipe is another form of bit pipe in which the sending DLC module supplies bits synchronously to the modem when it has data to send, but supplies nothing when it has no data. The sending modem sends no signal during these idle intervals, and the receiving modem detects the idle intervals and releases nothing to the receiving DLC module. This somewhat complicates the receiving modem, since it must distinguish between 0, 1, and idle in each time interval, and it must also regain synchronization at the end of an idle period. In Chapter 4 it will be seen that the capability to transmit nothing is very important for multiaccess channels.

A final form of virtual bit pipe is the *asynchronous character* pipe, where the bits within a character are sent at a fixed rate, but successive characters can be separated by

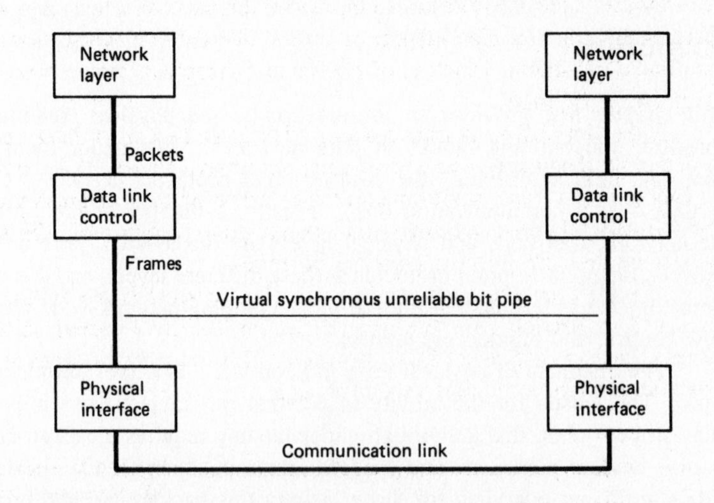

Figure 2.1 Data link control (DLC) layer with interfaces to adjacent layers.

variable delays, subject to a given minimum. As will be seen in the next section, this is used only when high data rates are not an important consideration.

In Sections 2.3 to 2.7 we treat the DLC layer, which is the primary focus of this chapter. For each point-to-point link, there are two DLC peer modules, one module at each end of the link. For traffic in a given direction, the sending DLC module receives packets from the network layer module at that node. The peer DLC modules employ a distributed algorithm, or protocol, to transfer these packets to the receiving DLC and thence to the network layer module at that node. In most cases, the objective is to deliver the packets in the order of arrival, with neither repetitions nor errors. In accomplishing this task, the DLC modules make use of the virtual bit pipe (with errors) provided by the physical layer.

One major problem in accomplishing the objective above is that of correcting the bit errors that occur on the virtual bit pipe. This objective is generally accomplished by a technique called ARQ (*a*utomatic *r*epeat re*q*uest). In this technique, errors are first detected at the receiving DLC module and then repetitions are requested from the transmitting DLC module. Both the detection of errors and the requests for retransmission require a certain amount of control overhead to be communicated on the bit pipe. This overhead is provided by adding a given number of bits (called a *header*) to the front of each packet and adding an additional number of bits (called a *trailer*) to the rear (see Fig. 2.2). A packet, extended by this header and trailer, is called a *frame*. From the standpoint of the DLC, a packet is simply a string of bits provided as a unit by the network layer; examples arise later where single "packets," in this sense, from the network layer contain pieces of information from many different sessions, but this makes no difference to the DLC layer.

Section 2.3 treats the problem of error detection and shows how a set of redundant bits in the trailer can be used for error detection. Section 2.4 then deals with retransmission requests. This is not as easy as it appears, first because the requests must be embedded into the data traveling in the opposite direction, and second because the opposite direction is also subject to errors. This provides our first exposure to a real distributed algorithm, which is of particular conceptual interest since it must operate in the presence of errors.

In Section 2.5 we discuss framing. The issue here is for the receiving DLC module to detect the beginning and end of each successive frame. For a synchronous bit pipe, the bits within a frame must contain the information to distinguish the end of the frame; also the idle fill between frames must be uniquely recognizable. This problem becomes even more interesting in the presence of errors.

A widely accepted standard of data link control is the HDLC protocol. This is discussed in Section 2.6, followed in Section 2.7 by a general discussion of how to initialize and to disconnect DLC protocols. This topic might appear to be trivial, but on closer inspection, it requires careful thought.

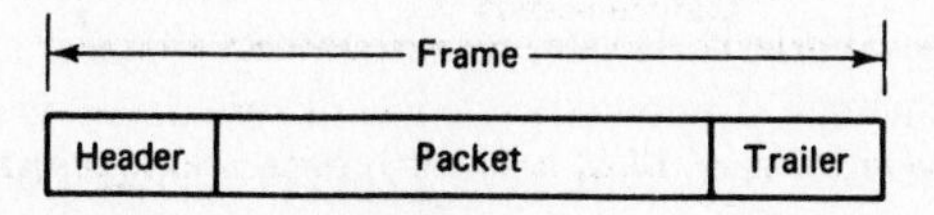

Figure 2.2 Frame structure. A packet from the network layer is extended in the DLC layer with control bits in front and in back of the packet.

Section 2.8 then treats a number of network layer issues, starting with addressing. End-to-end error recovery, which can be done at either the network or transport layer, is discussed next, along with a general discussion of why error recovery should or should not be done at more than one layer. The section ends with a discussion first of the X.25 network layer standard and next of the Internet Protocol (IP). IP was originally developed to connect the many local area networks in academic and research institutions to the ARPANET and is now a defacto standard for the internet sublayer.

A discussion of the transport layer is then presented in Section 2.9. This focuses on the transport protocol, TCP, used on top of IP. The combined use of TCP and IP (usually called TCP/IP) gives us an opportunity to explore some of the practical consequences of particular choices of protocols.

The chapter ends with an introduction to broadband integrated service data networks. In these networks, the physical layer is being implemented as a packet switching network called ATM, an abbreviation for "asynchronous transfer mode." It is interesting to compare how ATM performs its various functions with the traditional layers of conventional data networks.

2.2 THE PHYSICAL LAYER: CHANNELS AND MODEMS

As discussed in Section 2.1, the virtual channel seen by the data link control (DLC) layer has the function of transporting bits or characters from the DLC module at one end of the link to the module at the other end (see Fig. 2.1). In this section we survey how communication channels can be used to accomplish this function. We focus on point-to-point channels (*i.e.*, channels that connect just two nodes), and postpone consideration of multiaccess channels (*i.e.*, channels connecting more than two nodes) to Chapter 4. We also focus on communication in one direction, thus ignoring any potential interference between simultaneous transmission in both directions.

There are two broad classes of point-to-point channels: digital channels and analog channels. From a black box point of view, a digital channel is simply a bit pipe, with a bit stream as input and output. An analog channel, on the other hand, accepts a waveform (*i.e.*, an arbitrary function of time) as input and produces a waveform as output. We discuss analog channels first since digital channels are usually implemented on top of an underlying analog structure.

A module is required at the input of an analog channel to map the digital data from the DLC module into the waveform sent over the channel. Similarly, a module is required at the receiver to map the received waveform back into digital data. These modules will be referred to as modems (digital data *mo*dulator and *dem*odulator). The term *modem* is used broadly here, not necessarily implying any modulation but simply referring to the required mapping operations.

Let $s(t)$ denote the analog channel input as a function of time; $s(t)$ could represent a voltage or current waveform. Similarly, let $r(t)$ represent the voltage or current waveform at the output of the analog channel. The output $r(t)$ is a distorted, delayed, and attenuated version of $s(t)$, and our objective is to gain some intuitive appreciation of how to map the digital data into $s(t)$ so as to minimize the deleterious effects of this distortion. Note that a black box viewpoint of the physical channel is taken here, considering the input–output relation rather than the internal details of the analog channel. For example, if an ordinary voice telephone circuit (usually called a voice-grade circuit) is used as the analog channel, the physical path of the channel will typically pass through multiple switches, multiplexers, demultiplexers, modulators, and demodulators. For dial-up lines, this path will change on each subsequent call, although the specification of tolerances on the input–output characterization remain unchanged.

2.2.1 Filtering

One of the most important distorting effects on most analog channels is linear time-invariant filtering. Filtering occurs not only from filters inserted by the channel designer but also from the inherent behavior of the propagation medium. One effect of filtering is to "smooth out" the transmitted signal $s(t)$. Figure 2.3 shows two examples in which $s(t)$ is first a single rectangular pulse and then a sequence of rectangular pulses. The defining properties of linear time-invariant filters are as follows:

1. If input $s(t)$ yields output $r(t)$, then for any τ, input $s(t-\tau)$ yields output $r(t-\tau)$.
2. If $s(t)$ yields $r(t)$, then for any real number α, $\alpha s(t)$ yields $\alpha r(t)$.
3. If $s_1(t)$ yields $r_1(t)$ and $s_2(t)$ yields $r_2(t)$, then $s_1(t)+s_2(t)$ yields $r_1(t)+r_2(t)$.

The first property above expresses the time invariance; namely, if an input is delayed by τ, the corresponding output remains the same except for the delay of τ. The second and third properties express the linearity. That is, if an input is scaled by α, the corresponding output remains the same except for being scaled by α. Also, the

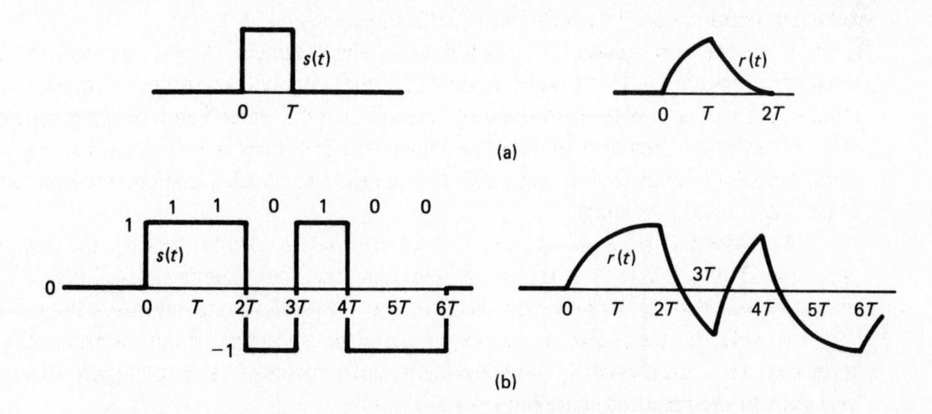

Figure 2.3 Relation of input and output waveforms for a communication channel with filtering. Part (a) shows the response $r(t)$ to an input $s(t)$ consisting of a rectangular pulse, and part (b) shows the response to a sequence of pulses. Part (b) also illustrates the NRZ code in which a sequence of binary inputs (1 1 0 1 0 0) is mapped into rectangular pulses. The duration of each pulse is equal to the time between binary inputs.

output due to the sum of two inputs is simply the sum of the corresponding outputs. As an example of these properties, the output for Fig. 2.3(b) can be calculated from the output in Fig. 2.3(a). In particular, the response to a negative-amplitude rectangular pulse is the negative of that from a positive-amplitude pulse (property 2), the output from a delayed pulse is delayed from the original output (property 1), and the output from the sum of the pulses is the sum of the outputs for the original pulses (property 3).

Figure 2.3 also illustrates a simple way to map incoming bits into an analog waveform $s(t)$. The virtual channel accepts a new bit from the DLC module each T seconds; the bit value 1 is mapped into a rectangular pulse of amplitude +1, and the value 0 into amplitude -1. Thus, Fig. 2.3(b) represents $s(t)$ for the data sequence 110100. This mapping scheme is often referred to as a nonreturn to zero (NRZ) code. The notation NRZ arises because of an alternative scheme in which the pulses in $s(t)$ have a shorter duration than the bit time T, resulting in $s(t)$ returning to 0 level for some interval between each pulse. The merits of these schemes will become clearer later.

Next, consider changing the rate at which bits enter the virtual channel. Figure 2.4 shows the effect of increasing the rate by a factor of 4 (*i.e.*, the signaling interval is reduced from T to $T/4$). It can be seen that output $r(t)$ is more distorted than before. The problem is that the response to a single pulse lasts much longer than a pulse time, so that the output at a given t depends significantly on the polarity of several input pulses; this phenomenon is called *intersymbol interference*.

From a more general viewpoint, suppose that $h(t)$ is the channel output corresponding to an infinitesimally narrow pulse of unit area at time 0. We call $h(t)$ the impulse response of the channel. Think of an arbitrary input waveform $s(t)$ as a superposition of very narrow pulses as shown in Fig. 2.5. The pulse from $s(\tau)$ to $s(\tau + \delta)$ can be viewed as a small impulse of area $\delta s(\tau)$ at time τ; this gives rise to the output $\delta s(\tau)h(t - \tau)$ at

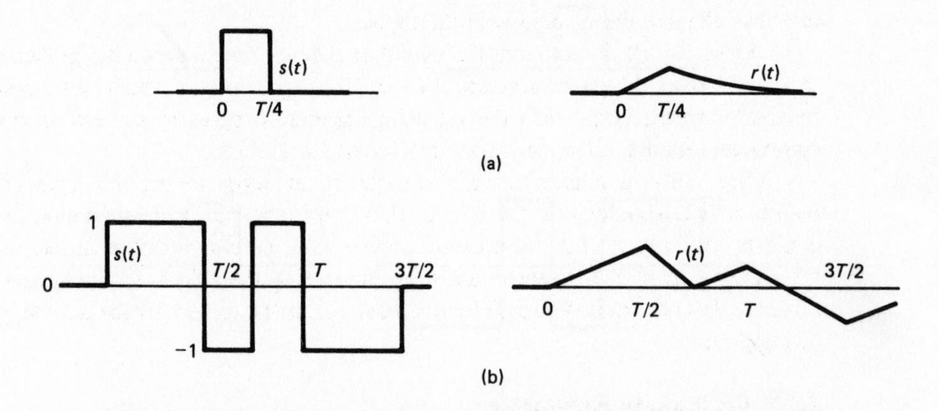

Figure 2.4 Relation of input and output waveforms for the same channel as in Fig. 2.3. Here the binary digits enter at 4 times the rate of Fig. 2.3, and the rectangular pulses last one-fourth as long. Note that the output $r(t)$ is more distorted and more attenuated than that in Fig. 2.3.

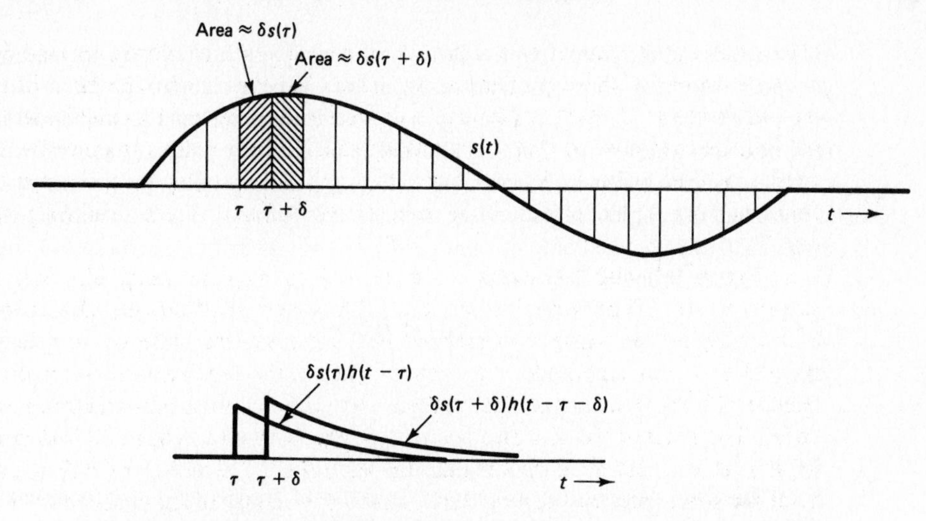

Figure 2.5 Graphical interpretation of the convolution equation. Input $s(t)$ is viewed as the superposition of narrow pulses of width δ. Each such pulse yields an output $\delta s(\tau)h(t-\tau)$. The overall output is the sum of these pulse responses.

time t. Adding the responses at time t from each of these input pulses and going to the limit $\delta \rightarrow 0$ shows that

$$r(t) = \int_{-\infty}^{+\infty} s(\tau)h(t-\tau)d\tau \tag{2.1}$$

This formula is called the *convolution integral*, and $r(t)$ is referred to as the convolution of $s(t)$ and $h(t)$. Note that this formula asserts that the filtering aspects of a channel are completely characterized by the impulse response $h(t)$; given $h(t)$, the output $r(t)$ for any input $s(t)$ can be determined. For the example in Fig. 2.3, it has been assumed that $h(t)$ is 0 for $t < 0$ and $\alpha e^{-\alpha t}$ for $t \geq 0$, where $\alpha = 2/T$. Given this, it is easy to calculate the responses in the figure.

From Eq. (2.1), note that the output at a given time t depends significantly on the input $s(\tau)$ over the interval where $h(t-\tau)$ is significantly nonzero. As a result, if this interval is much larger than the signaling interval T between successive input pulses, significant amounts of intersymbol interference will occur.

Physically, a channel cannot respond to an input before the input occurs, and therefore $h(t)$ should be 0 for $t < 0$; thus, the upper limit of integration in Eq. (2.1) could be taken as t. It is often useful, however, to employ a different time reference at the receiver than at the transmitter, thus eliminating the effect of propagation delay. In this case, $h(t)$ could be nonzero for $t < 0$, which is the reason for the infinite upper limit of integration.

2.2.2 Frequency Response

To gain more insight into the effects of filtering, it is necessary to look at the frequency domain as well as the time domain; that is, one wants to find the effect of filtering on sinusoids of different frequencies. It is convenient analytically to take a broader viewpoint here and allow the channel input $s(t)$ to be a complex function of t; that is, $s(t) = \text{Re}[s(t)] + j\ \text{Im}[s(t)]$, where $j = \sqrt{-1}$. The actual input to a channel is always real, of course. However, if $s(t)$ is allowed to be complex in Eq. (2.1), $r(t)$ will also be complex, but the output corresponding to $\text{Re}[s(t)]$ is simply $\text{Re}[r(t)]$ [assuming that $h(t)$ is real], and the output corresponding to $\text{Im}[s(t)]$ is $\text{Im}[r(t)]$. For a given frequency f, let $s(\tau)$ in Eq. (2.1) be the complex sinusoid $e^{j2\pi f\tau} = \cos(2\pi f\tau) + j\sin(2\pi f\tau)$. Integrating Eq. (2.1) (see Problem 2.3) yields

$$r(t) = H(f)e^{j2\pi ft} \tag{2.2}$$

where

$$H(f) = \int_{-\infty}^{\infty} h(\tau)e^{-j2\pi f\tau}d\tau \tag{2.3}$$

Thus, the response to a complex sinusoid of frequency f is a complex sinusoid of the same frequency, scaled by the factor $H(f)$. $H(f)$ is a complex function of the frequency f and is called the *frequency response* of the channel. It is defined for both positive and negative f. Let $|H(f)|$ be the magnitude and $\angle H(f)$ the phase of $H(f)$ [*i.e.*, $H(f) = |H(f)|e^{j\angle H(f)}$]. The response $r_1(t)$ to the real sinusoid $\cos(2\pi ft)$ is given by the real part of Eq. (2.2), or

$$r_1(t) = |H(f)|\cos[2\pi ft + \angle H(f)] \tag{2.4}$$

Thus, a real sinusoidal input at frequency f gives rise to a real sinusoidal output at the same freqency; $|H(f)|$ gives the amplitude and $\angle H(f)$ the phase of that output relative to the phase of the input. As an example of the frequency response, if $h(t) = \alpha e^{-\alpha t}$ for $t \geq 0$, then integrating Eq. (2.3) yields

$$H(f) = \frac{\alpha}{\alpha + j2\pi f} \tag{2.5}$$

Equation (2.3) maps an arbitrary time function $h(t)$ into a frequency function $H(f)$; mathematically, $H(f)$ is the *Fourier transform* of $h(t)$. It can be shown that the time function $h(t)$ can be recovered from $H(f)$ by the *inverse Fourier transform*

$$h(t) = \int_{-\infty}^{\infty} H(f)e^{j2\pi ft}df \tag{2.6}$$

Equation (2.6) has an interesting interpretation; it says that an (essentially) arbitrary function of time $h(t)$ can be represented as a superposition of an infinite set of infinitesimal complex sinusoids, where the amount of each sinusoid per unit frequency is $H(f)$, as given by Eq. (2.3). Thus, the channel input $s(t)$ (at least over any finite interval) can also be represented as a frequency function by

$$S(f) = \int_{-\infty}^{\infty} s(t)e^{-j2\pi ft}dt \tag{2.7}$$

$$s(t) = \int_{-\infty}^{\infty} S(f)e^{j2\pi ft}df \tag{2.8}$$

The channel output $r(t)$ and its frequency function $R(f)$ are related in the same way. Finally, since Eq. (2.2) expresses the output for a unit complex sinusoid at frequency f, and since $s(t)$ is a superposition of complex sinusoids as given by Eq. (2.8), it follows from the linearity of the channel that the response to $s(t)$ is

$$r(t) = \int_{-\infty}^{\infty} H(f)S(f)e^{j2\pi ft}df \tag{2.9}$$

Since $r(t)$ is also the inverse Fourier transform of $R(f)$, the input–output relation in terms of frequency is simply

$$R(f) = H(f)S(f) \tag{2.10}$$

Thus, the convolution of $h(t)$ and $s(t)$ in the time domain corresponds to the multiplication of $H(f)$ and $S(f)$ in the frequency domain. One sees from Eq. (2.10) why the frequency domain provides insight into filtering.

If $H(f) = 1$ over some range of frequencies and $S(f)$ is nonzero only over those frequencies, then $R(f) = S(f)$, so that $r(t) = s(t)$. One is not usually so fortunate as to have $H(f) = 1$ over a desired range of frequencies, but one could filter the received signal by an additional filter of frequency response $H^{-1}(f)$ over the desired range; this additional filtering would yield a final filtered output satisfying $r(t) = s(t)$. Such filters can be closely approximated in practice subject to additional delay [*i.e.*, satisfying $r(t) = s(t - \tau)$, for some delay τ]. Such filters can even be made to adapt automatically to the actual channel response $H(f)$; filters of this type are usually implemented digitally, operating on a sampled version of the channel output, and are called *adaptive equalizers*. These filters can and often are used to overcome the effects of intersymbol interference.

The question now arises as to what frequency range should be used by the signal $s(t)$; it appears at this point that any frequency range could be used as long as $H(f)$ is nonzero. The difficulty with this argument is that the additive noise that is always present on a channel has been ignored. Noise is added to the signal at various points along the propagation path, including at the receiver. Thus, the noise is not filtered in the channel in the same way as the signal. If $|H(f)|$ is very small in some interval of frequencies, the signal is greatly attenuated at those frequencies, but typically the noise is not attenuated. Thus, if the received signal is filtered at the receiver by $H^{-1}(f)$, the signal is restored to its proper level, but the noise is greatly amplified.

The conclusion from this argument (assuming that the noise is uniformly distributed over the frequency band) is that the digital data should be mapped into $s(t)$ in such a way that $|S(f)|$ is large over those frequencies where $|H(f)|$ is large and $|S(f)|$ is small (ideally zero) elsewhere. The cutoff point between large and small depends on the noise level and signal level and is not of great relevance here, particularly since the argument above is qualitative and oversimplified. Coming back to the example where $h(t) = \alpha e^{-\alpha t}$ for $t \geq 0$, the frequency response was given in Eq. (2.5) as $H(f) = \alpha/(\alpha + j2\pi f)$. It is seen that $|H(f)|$ is approximately 1 for small f and decreases as $1/f$ for large f. We shall not attempt to calculate $S(f)$ for the NRZ code of Fig. 2.3, partly because $s(t)$ and $S(f)$ depend on the particular data sequence being encoded, but the effect of changing the signaling interval T can easily be found. If T is decreased, then $s(t)$ is compressed in time, and this correspondingly expands $S(f)$ in frequency (see Problem 2.5). The equalization at the receiver would then be required either to equalize $H(f)$ over a broader band of frequencies, thus increasing the noise, or to allow more intersymbol interference.

2.2.3 The Sampling Theorem

A more precise way of looking at the question of signaling rates comes from the sampling theorem. This theorem states that if a waveform $s(t)$ is low-pass limited to frequencies at most W [*i.e.*, $S(f) = 0$, for $|f| > W$], then [assuming that $S(f)$ does not contain an impulse at $f = W$], $s(t)$ is completely determined by its values each $1/(2W)$ seconds; in particular,

$$s(t) = \sum_{i=-\infty}^{\infty} s\left(\frac{i}{2W}\right) \frac{\sin[2\pi W(t - i/(2W))]}{2\pi W[t - i/(2W)]} \tag{2.11}$$

Also, for any choice of sample values at intervals 1/(2W), there is a low-pass waveform, given by Eq. (2.11), with those sample values. Figure 2.6 illustrates this result.

The impact of this result, for our purposes, is that incoming digital data can be mapped into sample values at a spacing of $1/(2W)$ and used to create a waveform of the given sample values that is limited to $|f| \leq W$. If this waveform is then passed through an ideal low-pass filter with $H(f) = 1$ for $|f| \leq W$ and $H(f) = 0$ elsewhere, the received waveform will be identical to the transmitted waveform (in the absence of noise); thus, its samples can be used to recreate the original digital data.

The NRZ code can be viewed as mapping incoming bits into sample values of $s(t)$, but the samples are sent as rectangular pulses rather than the ideal $(\sin x)/x$ pulse shape of Eq. (2.11). In a sense, the pulse shape used at the transmitter is not critically important

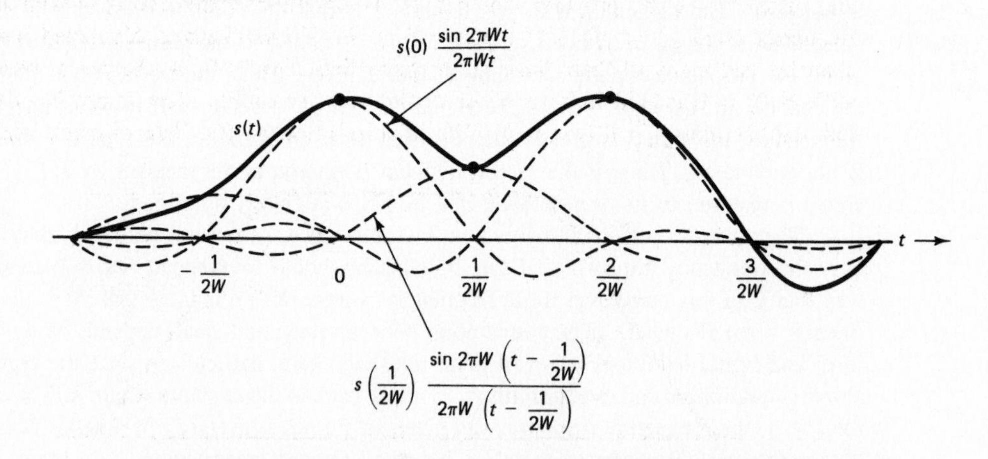

Figure 2.6 Sampling theorem, showing a function $s(t)$ that is low-pass limited to frequencies at most W. The function is represented as a superposition of $(\sin x)/x$ functions. For each sample, there is one such function, centered at the sample and with a scale factor equal to the sample value.

since the pulse shape can be viewed as arising from filtering sample impulses. If the combination of the pulse-shaping filter at the input with the channel filter and equalizer filter has the $(\sin x)/x$ shape of Eq. (2.11) (or, equivalently, a combined frequency response equal to the ideal low-pass filter response), the output samples will re-create the input samples.

The ideal low-pass filter and the $(\sin x)/x$ pulse shape suggested by the sampling theorem are nice theoretically but not very practical. What is required in practice is a more realistic filter for which the sample values at the output replicate those at the input (*i.e.*, no intersymbol interference) while the high-frequency noise is filtered out.

An elegant solution to this problem was given in a classic paper by Nyquist [Nyq28]. He showed that intersymbol interference is avoided by using a filter $H'(f)$ with odd symmetry at the band edge; that is, $H'(f+W) = 1 - H'(f-W)$ for $|f| \leq W$, and $H'(f) = 0$ for $|f| > 2W$ (see Fig. 2.7). The filter $H'(f)$ here is the composite of the pulse-shaping filter at the transmitter, the channel filter, and the equalizer filter. In practice, such a filter usually cuts off rather sharply around $f = W$ to avoid high-frequency noise, but ideal filtering is not required, thereby allowing considerable design flexibility.

It is important to recognize that the sampling theorem specifies the number of samples per second that can be utilized on a low-pass channel, but it does not specify how many bits can be mapped into one sample. For example, two bits per sample could be achieved by the mapping $11 \rightarrow 3, 10 \rightarrow 1, 00 \rightarrow -1$, and $01 \rightarrow -3$. As discussed later, it is the noise that limits the number of bits per sample.

2.2.4 Bandpass Channels

So far, we have considered only low-pass channels for which $|H(f)|$ is large only for a frequency band around $f = 0$. Most physical channels do not fall into this category, and instead have the property that $|(H(f)|$ is significantly nonzero only within some frequency band $f_1 \leq |f| \leq f_2$, where $f_1 > 0$. These channels are called bandpass channels and many of them have the property that $H(0) = 0$. A channel or waveform with $H(0) = 0$ is said to have no dc component. From Eq. (2.3), it can be seen that this implies that $\int_{-\infty}^{\infty} h(t)dt = 0$. The impulse response for these channels fluctuates

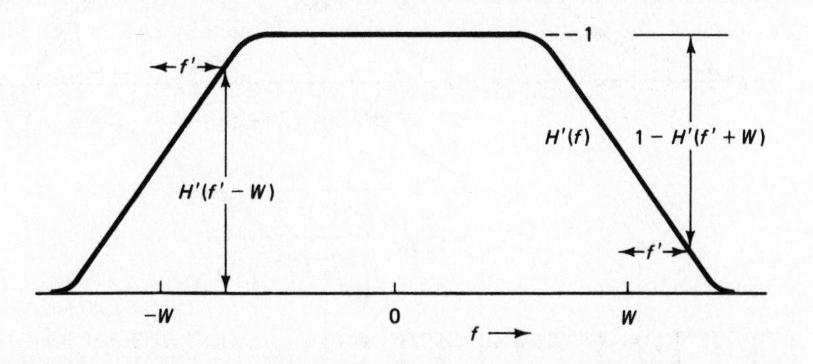

Figure 2.7 Frequency response $H'(f)$ that satisfies the Nyquist criterion for no intersymbol interference. Note that the response has odd symmetry around the point $f = W$.

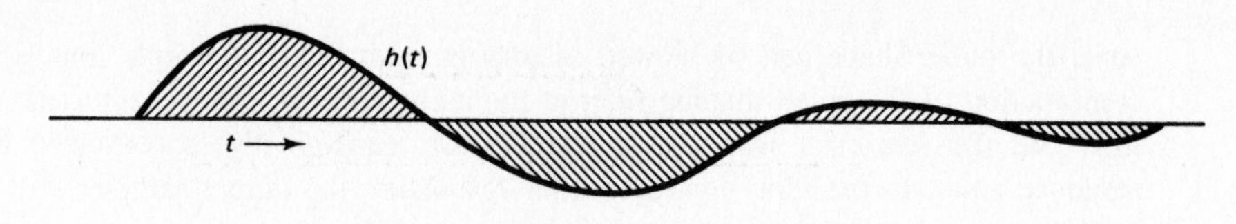

Figure 2.8 Impulse response $h(t)$ for which $H(f) = 0$ for $f = 0$. Note that the area over which $h(t)$ is positive is equal to that over which it is negative.

around 0, as illustrated in Fig. 2.8; this phenomenon is often called ringing. This type of impulse response suggests that the NRZ code is not very promising for a bandpass channel.

To avoid the foregoing problems, most modems for bandpass channels either directly encode digital data into signals with no dc component or else use modulation techniques. The best known direct encoding of this type is Manchester coding (see Fig. 2.9). As seen from the figure, the signals used to represent 0 and 1 each have no dc component and also have a transition (either 1 to -1 or -1 to 1) in the middle of each signaling interval. These transitions simplify the problem of timing recovery at the receiver (note that with the NRZ code, timing recovery could be difficult for a long run of 0's or 1's even in the absence of filtering and noise). The price of eliminating dc components in this way is the use of a much broader band of frequencies than required. Manchester coding is widely used in practice, particularly in the Ethernet system and the corresponding IEEE 802.3 standard described in Section 4.5. There are many other direct encodings, all of a somewhat ad hoc nature, with varying frequency characteristics and no dc component.

2.2.5 Modulation

One of the simplest modulation techniques is amplitude modulation (AM). Here a signal waveform $s(t)$ (called a baseband signal) is generated from the digital data as before, say by the NRZ code. This is then multiplied by a sinusoidal carrier, say $\cos(2\pi f_0 t)$, to generate a modulated signal $s(t)\cos(2\pi f_0 t)$. It is shown in Problem 2.6 that the frequency representation of this modulated signal is $[S(f - f_0) + S(f + f_0)]/2$ (see Fig. 2.10). At the receiver, the modulated signal is again multiplied by $\cos(2\pi f_0 t)$, yielding a received signal

$$
\begin{aligned}
r(t) &= s(t)\cos^2(2\pi f_0 t) \\
&= \frac{s(t)}{2} + \frac{s(t)\cos(4\pi f_0 t)}{2}
\end{aligned}
\tag{2.12}
$$

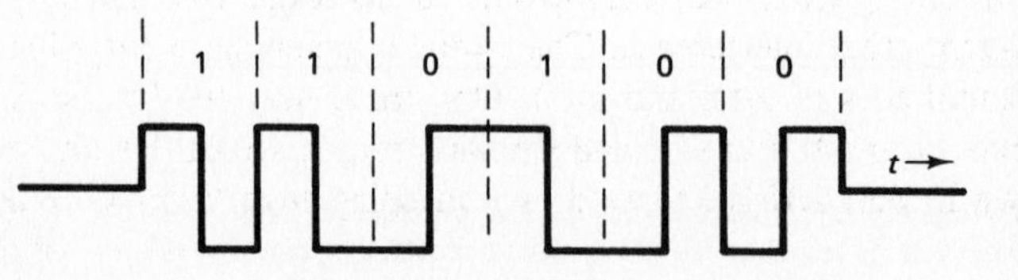

Figure 2.9 Manchester coding. A binary 1 is mapped into a positive pulse followed by a negative pulse, and a binary 0 is mapped into a negative pulse followed by a positive pulse. Note the transition in the middle of each signal interval.

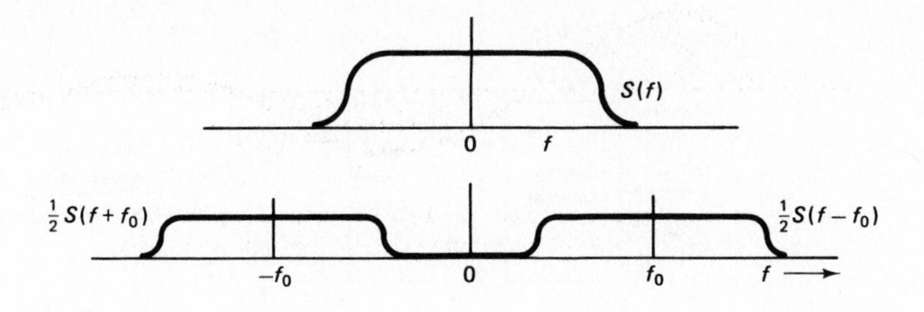

Figure 2.10 Amplitude modulation. The frequency characteristic of the waveform $s(t)$ is shifted up and down by f_0 in frequency.

The high-frequency component, at $2f_0$, is then filtered out at the receiver, leaving the demodulated signal $s(t)/2$, which is then converted back to digital data. In practice, the incoming bits are mapped into shorter pulses than those of NRZ and then filtered somewhat to remove high-frequency components. It is interesting to note that Manchester coding can be viewed as AM in which the NRZ code is multiplied by a square wave. Although a square wave is not a sinusoid, it can be represented as a sinusoid plus odd harmonics; the harmonics serve no useful function and are often partly filtered out by the channel.

AM is rather sensitive to the receiver knowing the correct phase of the carrier. For example, if the modulated signal $s(t)\cos(2\pi f_0 t)$ were multiplied by $\sin(2\pi f_0 t)$, the low-frequency demodulated waveform would disappear. This suggests, however, the possibility of transmitting twice as many bits by a technique known as quadrature amplitude modulation (QAM). Here, the incoming bits are mapped into two baseband signals, $s_1(t)$ and $s_2(t)$. Then $s_1(t)$ is multiplied by $\cos(2\pi f_0 t)$ and $s_2(t)$ by $\sin(2\pi f_0 t)$; the sum of these products forms the transmitted QAM signal (see Fig. 2.11). The received waveform is separately multiplied by $\cos(2\pi f_0 t)$ and $\sin(2\pi f_0 t)$ The first multiplication (after filtering out the high-frequency component) yields $s_1(t)/2$, and the second yields $s_2(t)/2$.

QAM is widely used in high-speed modems for voice-grade telephone channels. These channels have a useful bandwidth from about 500 to 2900 Hz (hertz, *i.e.*, cycles per second), and the carrier frequency is typically about 1700 Hz. The signals $s_1(t)$ and $s_2(t)$ are usually low-pass limited to 1200 Hz, thus allowing 2400 sample inputs per second for each baseband waveform. Adaptive equalization is usually used at the receiver (either before or after demodulation) to compensate for variations in the channel frequency response.

The next question is how to map the incoming bits into the low-pass waveforms $s_1(t)$ and $s_2(t)$. The pulse shape used in the mapping is not of central importance; it will be filtered anyway, and the equalizer will compensate for its shape automatically in attempting to eliminate intersymbol interference. Thus, what is important is mapping the input bits into sample amplitudes of $s_1(t)$ and $s_2(t)$. One could map one bit into a sample of s_1, mapping 1 into +1 and 0 into -1, and similarly map a second bit into a sample of s_2. This is shown in Fig. 2.12(a), viewed as a mapping from two bits into two amplitudes.

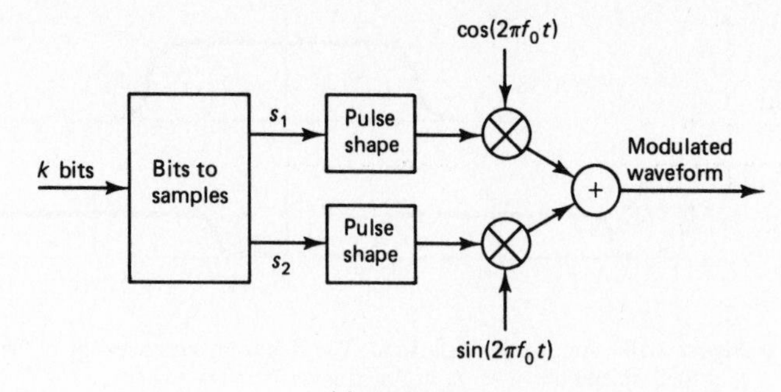

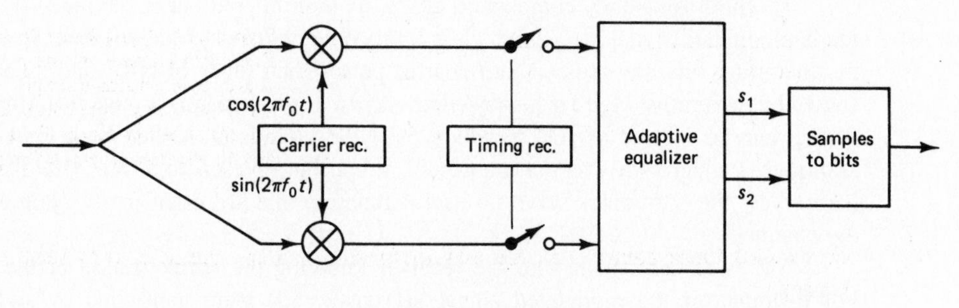

Figure 2.11 Quadrature amplitude modulation. (a) Each sample period, k bits enter the modulator, are converted into quadrature amplitudes, are then modulated by sine and cosine functions, respectively, and are then added. (b) The reverse operations take place at the demodulator.

Similarly, for any given integer k, one can map k bits into two amplitudes. Each of the 2^k combinations of k bits map into a different amplitude pair. This set of 2^k amplitude pairs in a mapping is called a signal constellation (see Fig. 2.12 for a number of examples). The first two constellations in Fig. 2.12 are also referred to as phase-shift keying (PSK), since they can be viewed as simply changing the phase of the carrier rather than changing its amplitude. For voice-grade modems, where $W = 2400$ Hz, $k = 2$ [as in Fig. 2.12(a)] yields 4800 bits per second (bps), whereas $k = 4$ [as in Fig. 2.12(d)] yields 9600 bps.

To reduce the likelihood of noise causing one constellation point to be mistaken for another, it is desirable to place the signal points as far from each other as possible subject to a constraint on signal power (*i.e.*, on the mean-squared distance from origin to constellation point, taken over all constellation points). It can be shown that the constellation in Fig. 2.12(c) is preferable in this sense to that in Fig. 2.12(b) for $k = 3$.

Modems for voice-grade circuits are also available at rates of 14,400 and 19,200 bps. These use correspondingly larger signal constellations and typically also use error-correction coding. It is reasonable to ask at this point what kind of limit is imposed on data rate by channels subject to bandwidth constraints and noise. Shannon [Sha48] has

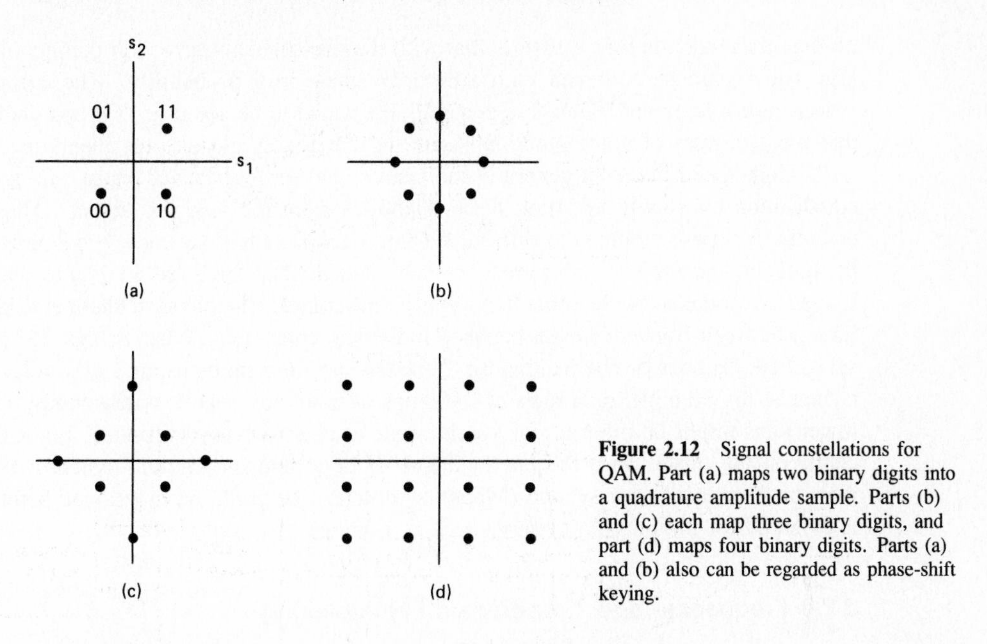

Figure 2.12 Signal constellations for QAM. Part (a) maps two binary digits into a quadrature amplitude sample. Parts (b) and (c) each map three binary digits, and part (d) maps four binary digits. Parts (a) and (b) also can be regarded as phase-shift keying.

shown that the capacity (*i.e.*, the maximum reliable data rate in bps) of such a channel is given by

$$C = W \log_2 \left(1 + \frac{S}{N_0 W}\right) \tag{2.13}$$

where W is the available bandwidth of the channel (essentially 2400 Hz for voice-grade telephone), S is the allowable signal power as seen by the receiver, and $N_0 W$ is the noise power within the bandwidth W seen by the receiver (*i.e.*, N_0 is the noise power per unit bandwidth, assumed uniformly distributed over W).

This is a deep and subtle result, and the machinery has not been developed here even to state it precisely, let alone to justify its validity. The description of QAM, however, at least provides some intuitive justification for the form of the result. The available sampling rate $1/T$ is equal to W by the sampling theorem [recall that W is twice the low-pass bandwidth of $s_1(t)$ and $s_2(t)$]. Since k bits are transmitted per sample, Wk bps can be sent. Next note that some minimum area must be provided around each point in the signal constellation to allow for noise, and this area is proportional to $N_0 W$. Also, the total area occupied by the entire constellation is proportional to the signal power S. Thus, 2^k, the number of points in the constellation, should be proportional to $S/(N_0 W)$, making k proportional to $\log_2[S/(N_0 W)]$.

Communication engineers usually express the signal-to-noise ratio [*i.e.*, $S/(N_0 W)$] in terms of decibels (dB), where the number of dB is defined as $10 \log_{10}[S/(N_0 W)]$. Thus, k, the number of bits per quadrature sample, is proportional to the signal-to-noise ratio in dB. The additional 1 in Eq. (2.13) is harder to explain; note, however, that it is required to keep the capacity positive when the signal-to-noise ratio is very small. The intuitive argument above would make it appear that the error probability depends on the spacing of points in the constellation and thus on k. The beauty of Shannon's theorem, on the other hand, is that it asserts that with the use of error-correction coding, any rate less than C can be achieved with arbitrarily small error probability. The capacity of voice-grade telephone channels is generally estimated to be about 25,000 bps, indicating that the data rates of voice-grade modems are remarkably close to the theoretical limit.

High-speed modems generally maintain signal timing, carrier phase, and adaptive equalization by slowly updating these quantities from the received signal. Thus, it is essential for these modems to operate synchronously; each T seconds, k bits must enter the modem, and the DLC unit must provide idle fill when there are no data to send. For low-speed modems, on the other hand, it is permissible for the physical channel to become idle in between frames or even between individual characters. What is regarded as high speed here depends on the bandwidth of the channel. For modems on 3 kHz voice-grade channels, for example, data rates of 2400 bps or more are usually synchronous, whereas lower rates might be intermittent synchronous or character asynchronous. For a coaxial cable, on the other hand, the bandwidth is so large that intermittent synchronous data rates of 10 megabits per second (Mbps) or more can be used. As mentioned before, this is essential for multiaccess systems such as Ethernet.

2.2.6 Frequency- and Time-Division Multiplexing

In previous subsections, bandwidth constraints imposed by filtering on the physical channel and by noise have been discussed. Often, however, a physical channel is shared by multiple signals each constrained to a different portion of the available bandwidth; this is called frequency-division multiplexing (FDM). The most common examples of this are AM, FM, and TV broadcasting, in which each station uses a different frequency band. As another common example, voice-grade channels are often frequency multiplexed together in the telephone network. In these examples, each multiplexed signal is constrained to its own allocated frequency band to avoid interference (sometimes called crosstalk) with the other signals. The modulation techniques of the preceding subsection are equally applicable whether a bandwidth constraint is imposed by FDM or by channel filtering.

FDM can be viewed as a technique for splitting a big channel into many little channels. Suppose that a physical channel has a usable bandwidth of W hertz and we wish to split it into m equal FDM subchannels. Then, each subchannel has W/m hertz available, and thus W/m available quadrature samples per second for sending data (in practice, the bandwidth per subchannel must be somewhat smaller to provide guard bands between the subchannels, but that is ignored here for simplicity). In terms of Eq. (2.13), each subchannel is allotted $1/m$ of the overall available signal power and is subject to $1/m$ of the overall noise, so each subchannel has $1/m$ of the total channel capacity.

Time-division multiplexing (TDM) is an alternative technique for splitting a big channel into many little channels. Here, one modem would be used for the overall bandwidth W. Given m equal rate streams of binary data to be transmitted, the m bit streams would be multiplexed together into one bit stream. Typically, this is done by sending the data in successive frames. Each frame contains m slots, one for each bit stream to be multiplexed; a slot contains a fixed number of bits, sometimes 1, sometimes 8, and sometimes more. Typically, each frame also contains extra bits to help the receiver maintain frame synchronization. For example, T1 carrier, which is in widespread telephone network use in the United States and Japan, multiplexes 24 data streams into slots of eight bits each with one extra bit per frame for synchronization. The overall data rate is 1.544 Mbps, with 64,000 bps for each of the multiplexed streams. In Europe, there is a similar system with 32 data streams and an overall data rate of 2.048 Mbps.

One way to look at FDM and TDM is that in each case one selects W quadrature samples per second (or more generally, $2W$ individual samples) for transmission; in one case, the samples are distributed in frequency and in the other they are distributed in time.

2.2.7 Other Channel Impairments

The effects of filtering and noise have been discussed in preceding subsections. There are a number of other possible impairments on physical channels. Sometimes there are multiple stages of modulation and demodulation internal to the physical channel. These can cause small amounts of phase jitter and carrier frequency offset in the received waveform. Amplification of the signals, within the physical channel and in the modems, can cause nonnegligible nonlinear distortion. Impulses of noise can occur due to lightning and switching effects. Repair personnel can short out or disconnect the channel for periods of time. Finally, there can be crosstalk from other frequency bands and from nearby wires.

To a certain extent, these effects can all be considered as extra noise sources. However, these noise sources have different statistical properties than the noise assumed by Shannon's theorem. (Technically, that noise is known as additive white Gaussian noise.) The principal difference is that the errors caused by these extra noise sources tend to occur in bursts of arbitrarily long length. As a result, one cannot, in practice, achieve the arbitrarily low error probabilities promised by Shannon's theorem. One must also use error detection and retransmission at the DLC layer; this is treated in Sections 2.3 and 2.4.

2.2.8 Digital Channels

In many cases, physical channels are designed to carry digital data directly, and the DLC unit can interface almost directly to the digital channel rather than to a modem mapping digital data into analog signals. To a certain extent, this is simply the question of whether the channel supplier or the channel user supplies the modem. This is an oversimplified view, however. A channel designed to carry digital data directly (such as the T1 carrier mentioned earlier) is often capable of achieving higher data rates at lower error probabilities than one carrying analog signals.

A major reason for this improved performance is the type of repeater used for digital channels. Repeaters are basically amplifiers inserted at various points in the propagation path to overcome attenuation. For a channel designed for analog signals, both the signal and the noise must be amplified at each repeater. Thus, the noise at the final receiver is an accumulation of noise over each stage of the path. For a digital channel, on the other hand, the digital signal can be recovered at each repeater. This means that the noise does not accumulate from stage to stage, and an error occurs only if the noise in a single stage is sufficient to cause an error. In effect, the noise is largely suppressed at each stage. Because of this noise suppression, and also because of the low cost of digital processing, it is increasingly common to use digital channels (such as the T1 carrier system) for the transmission of analog signals such as voice. Analog signals are sampled (typically at 8000 samples per second for voice) and then quantized (typically at eight bits per sample) to provide a digital input for a digital channel.

The telephone network can be separated into two parts—the local loops, which go from subscribers to local offices, and the internal network connecting local offices, central offices, and toll switches. Over the years, the internal network has become mostly digital, using TDM systems such as the T1 carrier just described and a higher-speed carrier system called T3, at 44.736 Mbps, that will multiplex 28 T1 signals.

There are two consequences to this increasing digitization. First, the telephone companies are anxious to digitize the local loops, thus allowing 64 kilobit per second (kbps) digital service over a single voice circuit. An added benefit of digitizing the local loops is that more than a single voice circuit could be provided to each subscriber. Such an extension of the voice network in which both voice and data are simultaneously available in an integrated way is called an *integrated services digital network* (ISDN). The other consequence is that data network designers can now lease T1 lines (called DS1 service) and T3 lines (called DS3 service) at modest cost and thus construct networks with much higher link capacities than are available on older networks. These higher-speed links are making it possible for wide area networks to send data at speeds comparable to local area nets.

With the increasing use of optical fiber, still higher link speeds are being standardized. *SONET* (Synchronous Optical Network) is the name for a standard family of interfaces for high-speed optical links. These start at 51.84 Mbps (called STS-1), and have various higher speeds of the form n times 51.84 Mbps (called STS-n) for $n = 1, 3, 9, 12, 18, 24, 36, 48$. Like the T1 and T3 line speeds, each of these speeds has a 125 μs frame structure which can be broken down into a very large number of 64 kbps voice circuits, each with one byte per frame. What is ingenious and interesting about these standards is the way that the links carrying these frames can be joined together, in the presence of slight clock drifts, without losing or gaining bits for any of the voice circuits. What is relevant for our purposes, however, is that 1.5 and 45 Mbps link speeds are now economically available, and much higher speeds will be economically available in the future. The other relevant fact is that optical fibers tend to be almost noise-free, with error rates of less than 10^{-10}.

ISDN As outlined briefly above, an integrated services digital network (ISDN) is a telephone network in which both the internal network and local loops are digital. The "integrated service" part of the name refers to the inherent capability of such a network to carry both voice and data in an integrated fashion. A telephone network with analog voice lines for local loops also allows both data and voice to be carried, but not easily at the same time and certainly not with much speed or convenience. There has been considerable effort to standardize the service provided by ISDN. Two standard types of service are available. The first, called *basic service*, provides two 64 kbps channels plus one 16 kbps channel to the user. Each 64 kbps channel can be used as an ordinary voice channel or as a point-to-point data channel. The 16 kbps channel is connected to the internal signaling network of the telephone system (*i.e.*, the network used internally for setting up and disconnecting calls, managing the network, etc.). Thus, the 16 kbps channel can be used to set up and control the two 64 kbps channels, but it could also be used for various low-data-rate services such as alarm and security systems.

In ISDN jargon, the 64 kbps channels are called B channels and the 16 kbps channel is called a D channel; the overall basic service is thus referred to as $2B + D$. With these facilities, a subscriber could conduct two telephone conversations simultaneously, or have one data session (at 64 kbps) plus one voice conversation. The latter capability appears to be rather useful to someone who works at home with a terminal but does not want to be cut off from incoming telephone calls. Most terminal users, of course, would be quite happy, after getting used to 300 bps terminals, to have the data capabilities of the 16 kbps D channel, and it is likely that subscribers will be offered a cheaper service consisting of just one B and one D channel.

The 64 kbps channels can be used in two ways: to set up a direct connection to some destination (thus using ISDN as a circuit switched network), and as an access line into a node of a packet switching network. On the one hand, a 64 kbps access into a packet switched network appears almost extravagant to one used to conventional networks. On the other hand, if one is trying to transfer a high-resolution graphics image of 10^9 bits, one must wait for over 4 hours using a 64 kbps access line. This illustrates a rather peculiar phenomenon. There are a very large number of applications of data networks that can be accomplished comfortably at very low data rates, but there are others (usually involving images) that require very high data rates.

Optical fiber will gradually be introduced into the local loops of the telephone network, and this will allow ISDN to operate at much higher data rates than those described above. Broadband ISDN (BISDN) is the name given to such high-speed ISDN networks. There has already been considerable standardization on broadband ISDN, including a standard user access rate of 155 Mbps (*i.e.*, the SONET STS-3 rate). These high data rates will allow for high-resolution TV as well as for fast image transmission, high-speed interconnection of supercomputers, and video conferencing. There are many interesting questions about how to build data networks handling such high data rates; one such strategy, called asynchronous transfer mode (ATM), is discussed briefly in Section 2.10.

The $2B + D$ basic service above is appropriate (or even lavish, depending on one's point of view) for the home or a very small office, but is inappropriate for larger offices used to using a PBX with a number of outgoing telephone lines. What ISDN (as proposed) offers here is something called *primary service* as opposed to basic service. This consists of 24 channels at 64 kbps each (in the United States and Japan) or 31 channels at 64 kbps each (in Europe). One of these channels is designated as the D channel and the others as B channels. The D channel is the one used for signaling and call setup and has a higher data rate here than in the basic service to handle the higher traffic levels. Subscribers can also obtain higher rate channels than the 64 kbps

B channels above. These higher rate channels are called H channels and come in 384, 1536, and 1920 kbps flavors. Such higher rate channels partially satisfy the need for the high data rate file transfers discussed above.

One of the interesting technical issues with ISDN is how to provide the required data rates over local loops consisting of a twisted wire pair. The approach being taken, for basic service, is to use time-division multiplexing of the B and D channels together with extra bits for synchronization and frequency shaping, thus avoiding a dc component. The most interesting part of this is that data must travel over the local loop in both directions, and that a node trying to receive the attenuated signal from the other end is frustrated by the higher-level signal being generated at the local end. For voice circuits, this problem has been solved in the past by a type of circuit called a hybrid which isolates the signal going out on the line from that coming in. At these high data rates, however, this circuit is not adequate and adaptive echo cancelers are required to get rid of the remaining echo from the local transmitted signal at the receiver.

2.2.9 Propagation Media for Physical Channels

The most common media for physical channels are twisted pair (*i.e.*, two wires twisted around each other so as to partially cancel out the effects of electromagnetic radiation from other sources), coaxial cable, optical fiber, radio, microwave, and satellite. For the first three, the propagated signal power decays exponentially with distance (*i.e.*, the attenuation in dB is linear with distance). Because of the attenuation, repeaters are used every few kilometers or so. The rate of attenuation varies with frequency, and thus as repeaters are spaced more closely, the useful frequency band increases, yielding a trade-off between data rate and the cost of repeaters. Despite this trade-off, it is helpful to have a ballpark estimate of typical data rates for channels using these media.

Twisted pair is widely used in the telephone network between subscribers and local stations and is increasingly used for data. One Mbps is a typical data rate for paths on the order of 1 km or less. Coaxial cable is widely used for local area networks, cable TV, and high-speed point-to-point links. Typical data rates are from 10 to several hundred Mbps. For optical fiber, data rates of 1000 Mbps or more are possible. Optical fiber is growing rapidly in importance, and the major problems lie in the generation, reception, amplification, and switching of such massive amounts of data.

Radio, microwave, and satellite channels use electromagnetic propagation in open space. The attenuation with distance is typically much slower than with wire channels, so repeaters can either be eliminated or spaced much farther apart than for wire lines. Frequencies below 1000 MHz are usually referred to as radio frequencies, and higher frequencies are referred to as microwave.

Radio frequencies are further divided at 30 MHz. Above 30 MHz, the ionosphere is transparent to electromagnetic waves, whereas below 30 MHz, the waves can be reflected by the ionosphere. Thus, above 30 MHz, propagation is on line-of-sight paths. The antennas for such propagation are frequently placed on towers or hills to increase the length of these line-of-sight paths, but the length of a path is still somewhat limited and repeaters are often necessary. This frequency range, from 30 to 1000 MHz, is used for UHF and VHF TV broadcast, for FM broadcast, and many specialized applications. Packet radio networks, discussed in Section 4.6, use this frequency band. Typical data rates in this band are highly variable; the DARPA (U.S. Department of Defense Advanced Research Projects Agency) packet radio network, for example, uses 100,000 and 400,000 bps.

Below 30 MHz, long-distance propagation beyond line-of-sight is possible by reflection from the ionosphere. Ironically, the 3 to 30 MHz band is called the high-frequency (HF) band, the terminology coming from the early days of radio. This band is very noisy, heavily used (*e.g.*, by ham radio), and subject to fading. Fading can be viewed as a channel filtering phenomenon with the frequency response changing relatively rapidly in time; this is caused by time-varying multiple propagation paths from source to destination. Typical data rates in this band are 2400 bps and less.

Microwave links (above 1000 MHz) must use line-of-sight paths. The antennas (usually highly directional dishes) yield typical path lengths of 10 to 200 km. Longer paths than this can be achieved by the use of repeaters. These links can carry 1000 Mbps or so and are usually multiplexed between long-distance telephony, TV program distribution, and data.

Satellite links use microwave frequencies with a satellite as a repeater. They have similar data rates and uses as microwave links. One satellite repeater can receive signals from many ground stations and broadcast back in another frequency band to all those ground stations. The satellite can contain multiple antenna beams, allowing it to act as a switch for multiple microwave links. In Chapter 4, multiaccess techniques for sharing individual frequency bands between different ground stations are studied.

This section has provided a brief introduction to physical channels and their use in data transmission. A link in a subnet might use any of these physical channels, or might share such a channel on a TDM or FDM basis with many other uses. Despite the complexity of this subject (which we have barely touched), these links can be regarded simply as unreliable bit pipes by higher layers.

2.3 ERROR DETECTION

The subject of the next four sections is data link control. This section treats the detection of transmission errors, and the next section treats retransmission requests. Assume initially that the receiving data link control (DLC) module knows where frames begin and end. The problem then is to determine which of those frames contain errors. From the layering viewpoint, the packets entering the DLC are arbitrary bit strings (*i.e.*, the function of the DLC layer is to provide error-free packets to the next layer up, no matter what the packet bit strings are). Thus, at the receiving DLC, any bit string is acceptable as a packet and errors cannot be detected by analysis of the packet itself. Note that a transformation on packets of K bits into some other representation of length K cannot help; there are 2^K possible packets and all possible bit strings of length K must be used to represent all possible packets. The conclusion is that extra bits must be appended to a packet to detect errors.

2.3.1 Single Parity Checks

The simplest example of error detection is to append a single bit, called a *parity check*, to a string of data bits. This parity check bit has the value 1 if the number of 1's in the bit string is odd, and has the value 0 otherwise (see Fig. 2.13). In other words, the parity check bit is the sum, modulo 2, of the bit values in the original bit string (k modulo j, for integer k and positive integer j, is the integer m, $0 \leq m < j$, such that $k - m$ is divisible by j).

In the ASCII character code, characters are mapped into strings of seven bits and then a parity check is appended as an eighth bit. One can also visualize appending a parity check to the end of a packet, but it will soon be apparent that this is not a sufficiently reliable way to detect errors.

Note that the total number of 1's in an encoded string (*i.e.*, the original bit string plus the appended parity check) is always even. If an encoded string is transmitted and a single error occurs in transmission, then, whether a 1 is changed to a 0 or a 0 to a 1, the resulting number of 1's in the string is odd and the error can be detected at the receiver. Note that the receiver cannot tell which bit is in error, nor how many errors occurred; it simply knows that errors occurred because of the odd number of 1's.

It is rather remarkable that for bit strings of any length, a single parity check enables the detection of any single error in the encoded string. Unfortunately, two errors in an encoded string always leave the number of 1's even so that the errors cannot be detected. In general, any odd number of errors are detected and any even number are undetected.

Despite the appealing simplicity of the single parity check, it is inadequate for reliable detection of errors; in many situations, it only detects errors in about half of the encoded strings where errors occur. There are two reasons for this poor behavior. The first is that many modems map several bits into a single sample of the physical channel input (see Section 2.2.5), and an error in the reception of such a sample typically causes several bit errors. The second reason is that many kinds of noise, such as lightning and temporarily broken connections, cause long bursts of errors. For both these reasons, when one or more errors occur in an encoded string, an even number of errors is almost as likely as an odd number and a single parity check is ineffective.

2.3.2 Horizontal and Vertical Parity Checks

Another simple and intuitive approach to error detection is to arrange a string of data bits in a two-dimensional array (see Fig. 2.14) with one parity check for each row and one for each column. The parity check in the lower right corner can be viewed as a parity check on the row parity checks, on the column parity checks, or on the data array. If an even number of errors are confined to a single row, each of them can be detected by the

s_1	s_2	s_3	s_4	s_5	s_6	s_7	c
1	0	1	1	0	0	0	1

Figure 2.13 Single parity check. Final bit c is the modulo 2 sum of s_1 to s_k, where $k = 7$ here.

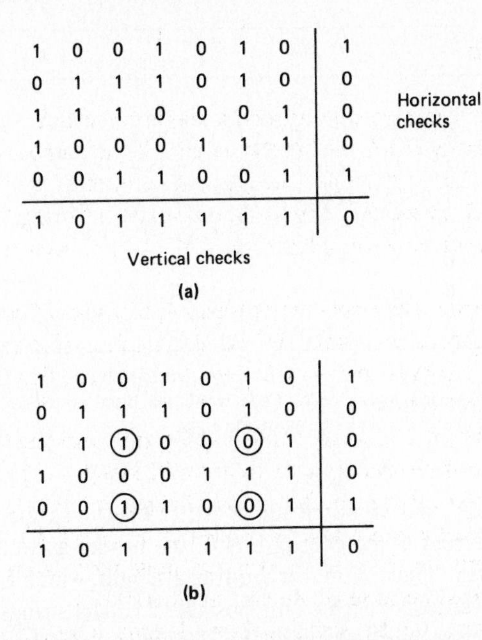

Figure 2.14 Horizontal and vertical parity checks. Each horizontal parity check checks its own row, and each column parity check checks its own column. Note, in part (b), that if each circled bit is changed, all parity checks are still satisfied.

corresponding column parity checks; similarly, errors in a single column can be detected by the row parity checks. Unfortunately, any pattern of four errors confined to two rows and two columns [*i.e.*, forming a rectangle as indicated in Fig. 2.14(b)] is undetectable.

The most common use of this scheme is where the input is a string of ASCII encoded characters. Each encoded character can be visualized as a row in the array of Fig. 2.14; the row parity check is then simply the last bit of the encoded character. The column parity checks can be trivially computed by software or hardware. The major weakness of this scheme is that it can fail to detect rather short bursts of errors (*e.g.*, two adjacent bits in each of two adjacent rows). Since adjacent errors are quite likely in practice, the likelihood of such failures is undesirably high.

2.3.3 Parity Check Codes

The nicest feature about horizontal and vertical parity checks is that the underlying idea generalizes immediately to arbitrary parity check codes. The underlying idea is to start with a bit string (the array of data bits in Fig. 2.14) and to generate parity checks on various subsets of the bits (the rows and columns in Fig. 2.14). The transformation from the string of data bits to the string of data bits and parity checks is called a *parity check code* or *linear code*. An example of a parity check code (other than the horizontal and vertical case) is given in Fig. 2.15. A parity check code is defined by the particular collection of subsets used to generate parity checks. Note that the word *code* refers to the transformation itself; we refer to an encoded bit string (data plus parity checks) as a *code word*.

Let K be the length of the data string for a given parity check code and let L be the number of parity checks. For the frame structure in Fig. 2.2, one can view the data

s_1	s_2	s_3	c_1	c_2	c_3	c_4	
1	0	0	1	1	1	0	
0	1	0	0	1	1	1	$c_1 = s_1 + s_3$
0	0	1	1	1	0	1	$c_2 = s_1 + s_2 + s_3$
1	1	0	1	0	0	1	$c_3 = s_1 + s_2$
1	0	1	0	0	1	1	$c_4 = s_2 + s_3$
1	1	1	0	1	0	0	
0	0	0	0	0	0	0	
0	1	1	1	0	1	0	

Figure 2.15 Example of a parity check code. Code words are listed on the left, and the rule for generating the parity checks is given on the right.

string as the header and packet, and view the set of parity checks as the trailer. Note that it is important to detect errors in the control bits of the header as well as to detect errors in the packets themselves. Thus, $K + L$ is the frame length, which for the present is regarded as fixed. For a given code, each of the possible 2^K data strings of length K is mapped into a frame (*i.e.*, code word) of length $K + L$. In an error-detection system, the frame is transmitted and the receiving DLC module determines if each of the parity checks is still the modulo 2 sum of the corresponding subset of data bits. If so, the frame is regarded by the receiver as error-free, and if not, the presence of errors is detected. If errors on the link convert one code word into another, the frame is regarded by the receiver as error-free, and undetectable errors are said to have occurred in the frame.

Given any particular code, one would like to be able to predict the probability of undetectable errors in a frame for a particular link. Unfortunately, this is very difficult. First, errors on links tend to be dependent and to occur in bursts; there are no good models for the length or intensity of these bursts, which vary widely among links of the same type. Second, for any reasonable code, the frequency of undetectable errors is very small and is thus both difficult to measure experimentally and dependent on rare, difficult-to-model events. The literature contains many calculations of the probability of undetectable errors, but these calculations are usually based on the assumption of independent errors; the results are usually orders of magnitude away from reality.

As a result of these difficulties, the effectiveness of a code for error detection is usually measured by three parameters: (1) the minimum distance of the code, (2) the burst-detecting capability, and (3) the probability that a completely random string will be accepted as error-free. The minimum distance of a code is defined as the smallest number of errors that can convert one code word into another. As we have seen, the minimum distance of a code using a single parity check is 2, and the minimum distance of a code with horizontal and vertical parity checks is 4.

The length of a burst of errors in a frame is the number of bits from the first error to the last, inclusive. The burst-detecting capability of a code is defined as the largest integer B such that a code can detect all bursts of length B or less. The burst-detecting capability of the single parity check code is 1, whereas the burst-detecting capability of a code with horizontal and vertical parity checks is 1 plus the length of a row (assuming that rows are sent one after the other).

By a completely random string of length $K + L$ is meant that each such string is received with probability 2^{-K-L}. Since there are 2^K code words, the probability of an undetected error is the probability that the random string is one of the code words; this occurs with probability 2^{-L} (the possibility that the received random string happens to be the same as the transmitted frame is ignored). This is usually a good estimate of the probability of undetectable errors given that both the minimum distance and the burst-detecting capability of the code are greatly exceeded by the set of errors on a received frame.

Parity check codes can be used for error correction rather than just for error detection. For example, with horizontal and vertical parity checks, any single error can be corrected simply by finding the row and column with odd parity. It is shown in Problem 2.10 that a code with minimum distance d can be used to correct any combination of fewer than $d/2$ errors. Parity check codes (and convolutional codes, which are closely related but lack the frame structure) are widely used for error correction at the physical layer. The modern approach to error correction is to view it as part of the modulation and demodulation process, with the objective of creating a virtual bit pipe with relatively low error rate.

Error correction is generally not used at the DLC layer, since the performance of an error correction code is heavily dependent on the physical characteristics of the channel. One needs error detection at the DLC layer, however, to detect rare residual errors from long noisy periods.

2.3.4 Cyclic Redundancy Checks

The parity check codes used for error detection in most DLCs today are cyclic redundancy check (CRC) codes. The parity check bits are called the CRC. Again, let L be the length of the CRC (*i.e.*, the number of check bits) and let K be the length of the string of data bits (*i.e.*, the header and packet of a frame). It is convenient to denote the data bits as $s_{K-1}, s_{K-2}, \ldots, s_1, s_0$, and to represent the string as a polynomial $s(D)$ with coefficients $s_{K-1}, \ldots, s_0$,

$$s(D) = s_{K-1}D^{K-1} + s_{K-2}D^{K-2} + \cdots + s_0 \tag{2.14}$$

The powers of the indeterminate D can be thought of as keeping track of which bit is which; high-order terms are viewed as being transmitted first. The CRC is represented as another polynomial,

$$c(D) = c_{L-1}D^{L-1} + \cdots + c_1D + c_0 \tag{2.15}$$

The entire frame of transmitted information and CRC can then be represented as $x(D) = s(D)D^L + c(D)$, that is, as

$$x(D) = s_{K-1}D^{L+K-1} + \cdots + s_0D^L + c_{L-1}D^{L-1} + \cdots + c_0 \tag{2.16}$$

The CRC polynomial $c(D)$ is a function of the information polynomial $s(D)$, defined in terms of a *generator polynomial* $g(D)$; this is a polynomial of degree L with binary coefficients that specifies the particular CRC code to be used.

$$g(D) = D^L + g_{L-1}D^{L-1} + \cdots + g_1 D + 1 \tag{2.17}$$

For a given $g(D)$, the mapping from the information polynomial to the CRC polynomial $c(D)$ is given by

$$c(D) = \text{Remainder}\left[\frac{s(D)D^L}{g(D)}\right] \tag{2.18}$$

The polynomial division above is just ordinary long division of one polynomial by another, except that the coefficients are restricted to be binary and the arithmetic on coefficients is performed modulo 2. Thus, for example, $(1+1)$ modulo $2 = 0$ and $(0-1)$ modulo $2 = 1$. Note that subtraction using modulo 2 arithmetic is the same as addition. As an example of the operation in Eq. (2.18),

$$\begin{array}{r l}
 & D^2 + D \\
D^3 + D^2 + 1 \,\big) & D^5 + \qquad D^3 \\
 & D^5 + D^4 + \qquad D^2 \\
\hline
 & \quad D^4 + D^3 + D^2 \\
 & \quad D^4 + D^3 + \qquad D \\
\hline
 & \qquad\qquad D^2 + D = \text{Remainder}
\end{array}$$

Since $g(D)$ is a polynomial of degree at most L, the remainder is of degree at most $L-1$. If the degree of $c(D)$ is less than $L-1$, the corresponding leading coefficients $c_{L-1}, \ldots$, in Eq. (2.18) are taken as zero.

This long division can be implemented easily in hardware by the feedback shift register circuit shown in Fig. 2.16. By comparing the circuit with the long division above, it can be seen that the successive contents of the shift register cells are just the coefficients of the partial remainders in the long division. In practice, CRCs are usually calculated by VLSI chips, which often perform the other functions of DLC as well.

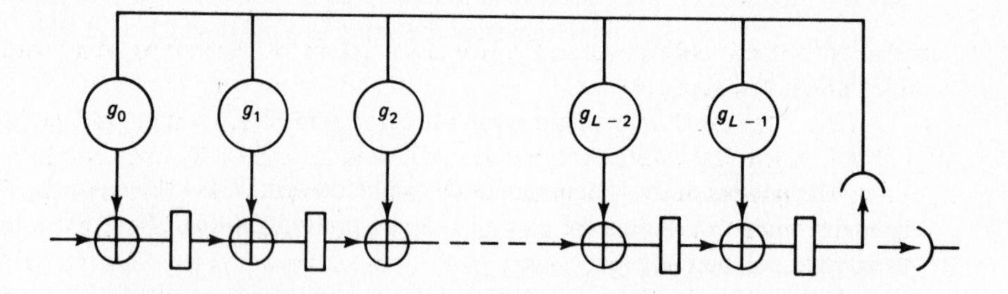

Figure 2.16 Shift register circuit for dividing polynomials and finding the remainder. Each rectangle indicates a storage cell for a single bit and the preceding circles denote modulo 2 adders. The large circles at the top indicate multiplication by the value of g_i. Initially, the register is loaded with the first L bits of $s(D)$ with s_{K-1} at the right. On each clock pulse, a new bit of $s(D)$ comes in at the left and the register reads in the corresponding modulo 2 sum of feedback plus the contents of the previous stage. After K shifts, the switch at the right moves to the horizontal position and the CRC is read out.

Let $z(D)$ be the quotient resulting from dividing $s(D)D^L$ by $g(D)$. Then, $c(D)$ can be represented as

$$s(D)D^L = g(D)z(D) + c(D) \tag{2.19}$$

Subtracting $c(D)$ (modulo 2) from both sides of this equation and recognizing that modulo 2 subtraction and addition are the same, we obtain

$$x(D) = s(D)D^L + c(D) = g(D)z(D) \tag{2.20}$$

Thus, all code words are divisible by $g(D)$, and all polynomials divisible by $g(D)$ are code words. It has not yet been shown that the mapping from $s(D)$ to $x(D)$ corresponds to a parity check code. This is demonstrated in Problem 2.15 but is not necessary for the subsequent development.

Now suppose that $x(D)$ is transmitted and that the received sequence is represented by a polynomial $y(D)$, where $x(D)$ and $y(D)$ differ because of the errors on the communication link. If the error sequence is represented as a polynomial $e(D)$, then $y(D) = x(D) + e(D)$, where, as throughout this section, + means modulo 2 addition; each error in the frame corresponds to a nonzero coefficient in $e(D)$ [*i.e.*, a coefficient in which $y(D)$ and $x(D)$ differ]. At the receiver, Remainder$[y(D)/g(D)]$ can be calculated by essentially the same circuit as that above. Since it has been shown that $x(D)$ is divisible by $g(D)$,

$$\text{Remainder}\left[\frac{y(D)}{g(D)}\right] = \text{Remainder}\left[\frac{e(D)}{g(D)}\right] \tag{2.21}$$

If no errors occur, then $e(D) = 0$ and the remainder above will be 0. The rule followed by the receiver is to decide that the frame is error- free if this remainder is 0 and to decide that there are errors otherwise. When errors actually occur [*i.e.*, $e(D) \neq 0$], the receiver fails to detect the errors only if this remainder is 0; this occurs only if $e(D)$ is itself some code word. In other words, $e(D) \neq 0$ is undetectable if and only if

$$e(D) = g(D)z(D) \tag{2.22}$$

for some nonzero polynomial $z(D)$. We now explore the conditions under which undetected errors can occur.

First, suppose that a single error occurs, say $e_i = 1$, so that $e(D) = D^i$. Since $g(D)$ has at least two nonzero terms (*i.e.*, D^L and 1), $g(D)z(D)$ must also have at least two nonzero terms for any nonzero $z(D)$ (see Problem 2.13). Thus $g(D)z(D)$ cannot equal D^i; since this is true for all i, all single errors are detectable. By the same type of argument, since the highest-order and lowest-order terms in $g(D)$ (*i.e.*, D^L and 1, respectively) differ by L, the highest-order and lowest-order terms in $g(D)z(D)$ differ by at least L for all nonzero $z(D)$. Thus, if $e(D)$ is a code word, the burst length of the errors is at least $L + 1$ (the +1 arises from the definition of burst length as the number of positions from the first error to the last error *inclusive*).

Next, suppose that a double error occurs, say in positions i and j, so that

$$e(D) = D^i + D^j = D^j(D^{i-j} + 1), \quad i > j \tag{2.23}$$

From the argument above, D^j is not divisible by $g(D)$ or by any factor of $g(D)$; thus, $e(D)$ fails to be detected only if $D^{i-j}+1$ is divisible by $g(D)$. For any binary polynomial $g(D)$ of degree L, there is some smallest n for which $D^n + 1$ is divisible by $g(D)$. It is known from the theory of finite fields that this smallest n can be no larger than $2^L - 1$; moreover, for all $L > 0$, there are special L-degree polynomials, called *primitive polynomials*, such that this smallest n is equal to $2^L - 1$. Thus, if $g(D)$ is chosen to be such a primitive polynomial of degree L, and if the frame length is restricted to be at most $2^L - 1$, then $D^{i-j} + 1$ cannot be divisible by $g(D)$; thus, all double errors are detected.

In practice, the generator polynomial $g(D)$ is usually chosen to be the product of a primitive polynomial of degree $L-1$ times the polynomial $D+1$. It is shown in Problem 2.14 that a polynomial $e(D)$ is divisible by $D+1$ if and only if $e(D)$ contains an even number of nonzero coefficients. This ensures that all odd numbers of errors are detected, and the primitive polynomial ensures that all double errors are detected (as long as the block length is less than 2^{L-1}). Thus, any code of this form has a minimum distance of at least 4, a burst-detecting capability of at least L, and a probability of failing to detect errors in completely random strings of 2^{-L}. There are two standard CRCs with length $L = 16$ (denoted CRC-16 and CRC-CCITT). Each of these CRCs is the product of $D+1$ times a primitive $(L-1)$- degree polynomial, and thus both have the foregoing properties. There is also a standard CRC with $L = 32$. It is a 32-degree primitive polynomial, and has been shown to have a minimum distance of 5 for block lengths less than 3007 and 4 for block lengths less than 12,145 [FKL86]. These polynomials are as follows:

$$g(D) = D^{16} + D^{15} + D^2 + 1 \qquad \text{for CRC-16}$$

$$g(D = D^{16} + D^{12} + D^5 + 1 \qquad \text{for CRC-CCITT}$$

$$g(D) = D^{32} + D^{26} + D^{23} + D^{22} + D^{16} + D^{12} + D^{11} + D^{10} + D^8 + D^7 + D^5 + D^4 + D^2 + D^1 + 1$$

2.4 ARQ: RETRANSMISSION STRATEGIES

The general concept of *a*utomatic *r*epeat re*q*uest (ARQ) is to detect frames with errors at the receiving DLC module and then to request the transmitting DLC module to repeat the information in those erroneous frames. Error detection was discussed in the preceding section, and the problem of requesting retransmissions is treated in this section. There are two quite different aspects of retransmission algorithms or protocols. The first is that of correctness: Does the protocol succeed in releasing each packet, once and only once, without errors, from the receiving DLC? The second is that of efficiency: How much of the bit-transmitting capability of the bit pipe is wasted by unnecessary waiting and by sending unnecessary retransmissions? First, several classes of protocols are developed and shown to be correct (in a sense to be defined more precisely later). Later, the effect that the various parameters in these classes have on efficiency is considered.

Recall from Fig. 2.2 that packets enter the DLC layer from the network layer. The DLC module appends a header and trailer to each packet to form a frame, and the frames are transmitted on the virtual bit pipe (*i.e.*, are sent to the physical layer for transmission). When errors are detected in a frame, a new frame containing the old packet is transmitted. Thus, the first transmitted frame might contain the first packet, the next frame the second packet, the third frame a repetition of the first packet, and so forth. When a packet is repeated, the frame header and trailer might or might not be the same as in the earlier version.

Since framing will not be discussed until the next section, we continue to assume that the receiving DLC knows when frames start and end; thus a CRC (or any other technique) may be used for detecting errors. We also assume, somewhat unrealistically, that *all* frames containing transmission errors are detected. The reason for this is that we want to prove that ARQ works correctly except when errors are undetected. This is the best that can be hoped for, since error detection cannot work with perfect reliability and bounded delay; in particular, any code word can be changed into another code word by some string of transmission errors. This can cause erroneous data to leave the DLC or, perhaps worse, can cause some control bits to be changed. In what follows, we refer to frames without transmission errors as *error-free frames* and those with transmission errors as *error frames*. We are assuming, then, that the receiver can always disinguish error frames from error-free frames.

Finally, we need some assumptions about the bit pipes over which these frames are traveling. The reason for these assumptions will be clearer when framing is studied; in effect, these assumptions will allow us to relax the assumption that the receiving DLC has framing information. Assume first that each transmitted frame is delayed by an arbitrary and variable time before arriving at the receiver, and assume that some frames might be "lost" and never arrive. Those frames that arrive, however, are assumed to arrive in the same order as transmitted, with or without errors. Figure 2.17 illustrates this behavior.

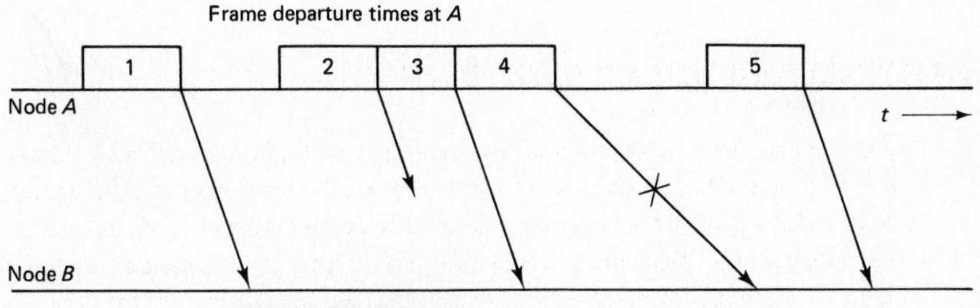

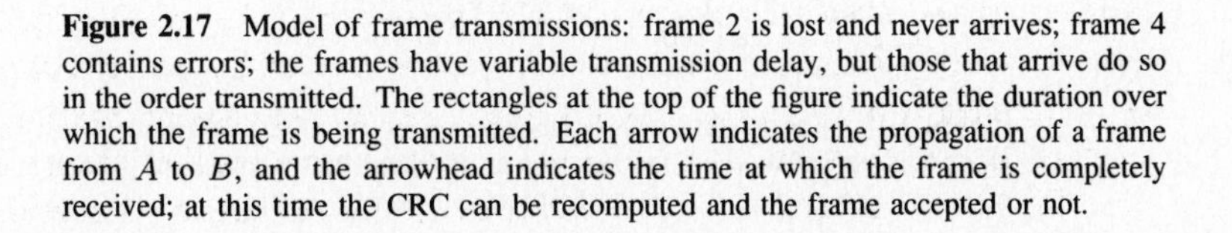

Figure 2.17 Model of frame transmissions: frame 2 is lost and never arrives; frame 4 contains errors; the frames have variable transmission delay, but those that arrive do so in the order transmitted. The rectangles at the top of the figure indicate the duration over which the frame is being transmitted. Each arrow indicates the propagation of a frame from A to B, and the arrowhead indicates the time at which the frame is completely received; at this time the CRC can be recomputed and the frame accepted or not.

2.4.1 Stop-and-Wait ARQ

The simplest type of retransmission protocol is called *stop-and-wait*. The basic idea is to ensure that each packet has been received correctly before initiating transmission of the next packet. Thus, in transmitting packets from point A to B, the first packet is transmitted in the first frame, and then the sending DLC waits. If the frame is error-free, B sends an acknowledgment (called an ack) back to A; if the frame is an error frame, B sends a negative acknowledgment (called a nak) back to A. Since errors can occur from B to A as well as from A to B, the ack or nak is protected with a CRC.

If an error-free frame is received at B, and the corresponding ack frame to A is error-free, then A can start to send the next packet in a new frame. Alternatively, detected errors can occur either in the transmission of the frame or the return ack or nak, and in either case A resends the old packet in a new frame. Finally, if either the frame or the return ack or nak is lost, A must eventually time-out and resend the old packet.

Figure 2.18 illustrates a potential malfunction in such a strategy. Since delays are arbitrary, it is possible for node A to time-out and resend a packet when the first transmission and/or the corresponding ack is abnormally delayed. If B receives both transmissions of the given packet correctly, B has no way of knowing whether the second transmission is a new packet or a repetition of the old packet. One might think that B could simply compare the packets to resolve this issue, but as far as the DLC layer is concerned, packets are arbitrary bit strings and the first and second packets could be identical; it would be a violation of the principle of layering for the DLC layer to rely on higher layers to ensure that successive packets are different.

The simplest solution to this problem is for the sending DLC module (at A) to use a sequence number in the frame header to identify successive packets. Unfortunately, even the use of sequence numbers from A to B is not quite enough to ensure correct operation. The problem is that acks can get lost on the return channel, and thus when B gets the same packet correctly twice in a row, it has to send a new ack for the second reception (see Fig. 2.19). After transmitting the packet twice but receiving only one ack,

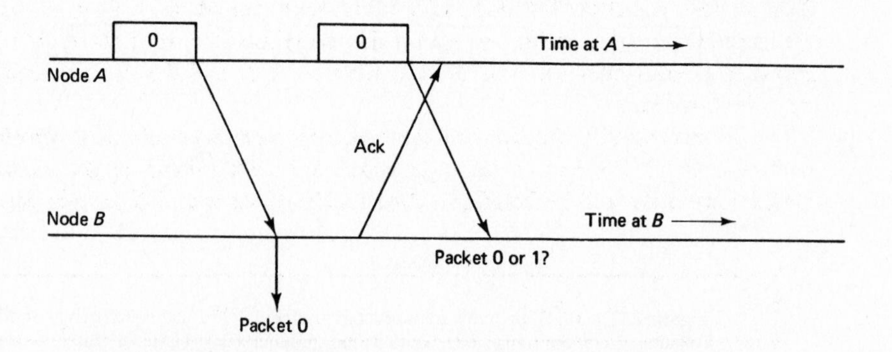

Figure 2.18 The trouble with unnumbered packets. If the transmitter at A times-out and sends packet 0 twice, the receiver at B cannot tell whether the second frame is a retransmission of packet 0 or the first transmission of packet 1.

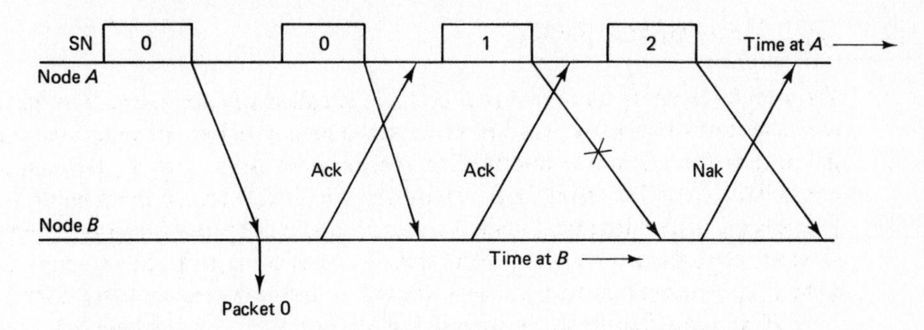

Figure 2.19 The trouble with unnumbered acks. If the transmitter at A times-out and sends packet 0 twice, node B can use the sequence numbers to recognize that packet 0 is being repeated. It must send an ack for both copies, however, and (since acks can be lost) the transmitter cannot tell whether the second ack is for packet 0 or 1.

node A could transmit the next packet in sequence, and then on receiving the second ack, could interpret that as an ack for the new packet, leading to a potential failure of the system.

To avoid this type of problem, the receiving DLC (at B), instead of returning ack or nak on the reverse link, returns the number of the next packet awaited. This provides all the information of the ack/nak, but avoids ambiguities about which frame is being acked. An equivalent convention would be to return the number of the packet just accepted, but this is not customary. Node B can request this next awaited packet upon the receipt of each packet, at periodic intervals, or at an arbitrary selection of times. In many applications, there is another stream of data from B to A, and in this case, the frames from B to A carrying requests for new A to B packets must be interspersed with data frames carrying data from B to A. It is also possible to "piggyback" these requests for new packets into the headers of the data frames from B to A (see Fig. 2.20), but as shown in Problem 2.38, this is often counterproductive for stop-and-wait ARQ. Aside from its effect on the timing of the requests from node B to A, the traffic from B to A does not affect the stop-and-wait strategy from A to B; thus, in what follows, we ignore this reverse traffic (except for the recognition that requests might be delayed). Figure 2.21 illustrates the flow of data from A to B and the flow of requests in the opposite direction.

We next specify the stop-and-wait strategy more precisely and then show that it works correctly. The correctness might appear to be self-evident, but the methodology of the demonstration will be helpful in understanding subsequent distributed algorithms. It

	SN	RN		Packet	CRC

Figure 2.20 The header of a frame contains a field carrying the sequence number, SN, of the packet being transmitted. If piggybacking is being used, it also contains a field carrying the request number, RN, of the next packet awaited in the opposite direction.

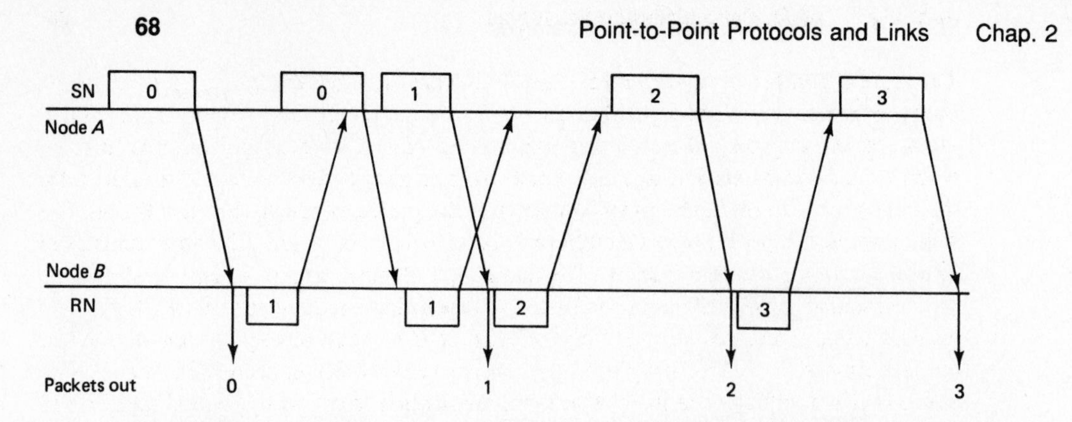

Figure 2.21 Example of use of sequence and request numbers for stop-and-wait transmission from A to B. Note that packet 0 gets repeated, presumably because node A times-out too soon. Note also that node A delays repeating packet 1 on the second request for it. This has no effect on the correctness of the protocol, but avoids unnecessary retransmissions.

will be seen that what is specified is not a single algorithm, but rather a class of algorithms in which the timing of frame transmissions is left unspecified. Showing that all algorithms in such a class are correct then allows one to specify the timing as a compromise between simplicity and efficiency without worrying further about correctness.

Assume that when the strategy is first started on the link, nodes A and B are correctly initialized in the sense that no frames are in transit on the link and that the receiver at B is looking for a frame with the same sequence number as the first frame to be transmitted from A. It makes no difference what this initial sequence number SN is as long as A and B agree on it, so we assume that $SN = 0$, since this is the conventional initial value.

The algorithm at node A for A-to-B transmission:

1. Set the integer variable SN to 0.
2. Accept a packet from the next higher layer at A; if no packet is available, wait until it is; assign number SN to the new packet.
3. Transmit the SNth packet in a frame containing SN in the sequence number field.
4. If an error-free frame is received from B containing a request number RN greater than SN, increase SN to RN and go to step 2. If no such frame is received within some finite delay, go to step 3.

The algorithm at node B for A-to-B transmission:

1. Set the integer variable RN to 0 and then repeat steps 2 and 3 forever.
2. Whenever an error-free frame is received from A containing a sequence number SN equal to RN, release the received packet to the higher layer and increment RN.
3. At arbitrary times, but within bounded delay after receiving any error-free data frame from A, transmit a frame to A containing RN in the request number field.

There are a number of conventional ways to handle the arbitrary delays between subsequent transmissions in the algorithm above. The usual procedure for node A (recall that we are discussing only the stop-and-wait strategy for A-to-B traffic) is to set a timer when a frame transmission begins. If the timer expires before receipt of a request for the next packet from B, the timer is reset and the packet is resent in a new frame. If a request for a new packet is received from B before the timer expires, the timer is disabled until A transmits the next packet. The same type of timer control can be used at node B. Alternatively, node B could send a frame containing a request for the awaited packet each time that it receives a frame from A. Also, if B is piggybacking its request numbers on data frames, it could simply send the current value of *RN* in each such frame. Note that this is not sufficient in itself, since communication from A to B would then halt in the absence of traffic from B to A; thus, node B must send nondata frames containing the *RN* values when there is no B-to-A data traffic. The important point is that whatever timing strategy is being used, it must guarantee that the intervening interval between repetitions in each direction is bounded.

Correctness of stop and wait We now go through an informal proof that this class of algorithms is correct in the sense that a never-ending stream of packets can be accepted from the higher layer at A and delivered to the higher layer at B in order and without repetitions or deletions. We continue to assume that all error frames are detected by the CRC. We also assume that there is some $q > 0$ such that each frame is received error-free with probability at least q. Finally, we recall the assumption that the link is initially empty, that the first packet from A has $SN = 0$, and that node B is initially awaiting the packet with $SN = 0$.

Proofs of this type usually break into two parts, characterized as *safety* and *liveness*. An algorithm is safe if it never produces an incorrect result, which in this case means never releasing a packet out of the correct order to the higher layer at B. An algorithm is live if it can continue forever to produce results (*i.e.*, if it can never enter a deadlock condition from which no further progress is possible). In this case liveness means the capability to continue forever to accept new packets at A and release them at B.

The safety property is self-evident for this algorithm; that is, node B is initially awaiting packet 0, and the only packet that can be released is packet 0. Subsequently (using induction if one wants to be formal), node B has released all the packets in order, up to, but not including, the current value of RN; packet RN is the only packet that it can next accept and release. When an error-free frame containing packet RN is received and released, the value of RN is incremented and the situation above is repeated with the new value of RN.

To see that the liveness property is satisfied, assume that node A first starts to transmit a given packet i at time t_1 (see Fig. 2.22). Let t_2 be the time at which this packet is received error-free and released to the higher layer at node B; let $t_2 = \infty$ if this event never occurs. Similarly, let t_3 be the time at which the sequence number at A is increased to $i + 1$, and let $t_3 = \infty$ if this never occurs. We will show that $t_1 < t_2 < t_3$ and that t_3 is finite. This is sufficient to demonstrate liveness, since using the argument repeatedly for $i = 0$, then $i = 1$, and so on, shows that each packet is transmitted with

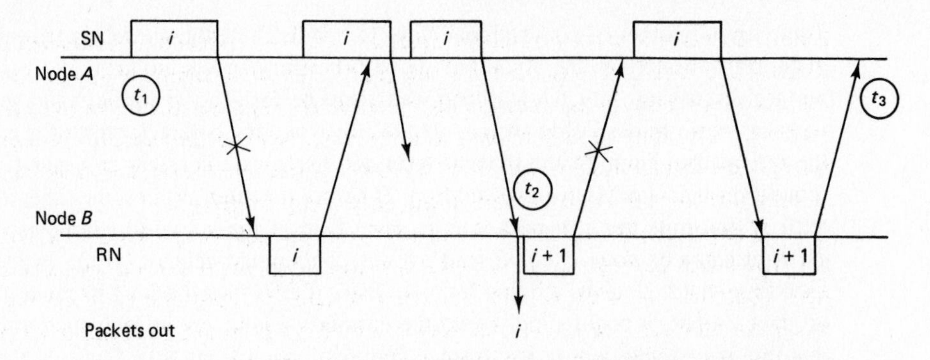

Figure 2.22 Times t_1, t_2, and t_3 when a packet is first placed in a frame for transmission at A, first received and released to the higher layer at B, and first acknowledged at A.

finite delay. Note that we cannot guarantee that the higher layer at A will always supply packets within finite delay, so that the notion of liveness here can only require finite delay given available packets to send.

Let $RN(t)$ be the value of the variable RN at node B as a function of time t and let $SN(t)$ be the corresponding value of SN at node A. It is seen directly from the algorithm statement that $SN(t)$ and $RN(t)$ are nondecreasing in t. Also, since $SN(t)$ is the largest request number received from B up to time t, $SN(t) \leq RN(t)$. By assumption, packet i has never been transmitted before time t_1, so (using the safety property) $RN(t_1) \leq i$. Since $SN(t_1) = i$, it follows that $SN(t_1) = RN(t_1) = i$. By definition of t_2 and t_3, $RN(t)$ is incremented to $i + 1$ at t_2 and $SN(t)$ is incremented to $i + 1$ at t_3. Using the fact that $SN(t) \leq RN(t)$, it follows that $t_2 < t_3$.

We have seen that node A transmits packet i repeatedly, with finite delay between successive transmissions, from t_1 until it is first received error-free at t_2. Since there is a probability $q > 0$ that each retransmission is received correctly, and retransmissions occur within finite intervals, an error-free reception eventually occurs and t_2 is finite. Node B then transmits frames carrying $RN = i+1$ from time t_2 until received error-free at time t_3. Since node A is also transmitting frames in this interval, the delay between subsequent transmissions from B is finite, and, since $q > 0$, t_3 eventually occurs; thus the interval from t_1 to t_3 is finite and the algorithm is live.

One trouble with the stop-and-wait strategy developed above is that the sequence and request numbers become arbitrarily large with increasing time. Although one could simply use a very large field for sending these numbers, it is nicer to send these numbers modulo some integer. Given our assumption that frames travel in order on the link, it turns out that a modulus of 2 is sufficient.

To understand why it is sufficient to send sequence numbers modulo 2, we first look more carefully at what happens in Fig. 2.22 when ordinary integers are used. Note that after node B receives packet i at time t_2, the subsequent frames received at B must all have sequence numbers i or greater (since frames sent before t_1 cannot arrive after t_2). Similarly, while B is waiting for packet $i + 1$, no packet greater than $i + 1$ can be sent [since $SN(t) \leq RN(t)$]. Thus, in the interval while $RN(t) = i + 1$, the received frames all carry sequence numbers equal to i or $i+1$. Sending the sequence number modulo 2 is sufficient to resolve this ambiguity. The same argument applies for all i. By the same argument, in the interval t_1 to t_3, while $SN(t)$ is equal to i, the request numbers received at A must be either i or $i+1$, so again sending *RN* modulo 2 is sufficient to resolve the ambiguity. Finally, since *SN* and *RN* need be transmitted only modulo 2, it is sufficient to keep track of them at the nodes only modulo 2.

Using modulo 2 values for *SN* and *RN*, we can view nodes A and B as each having two states (for purposes of A to B traffic), corresponding to the binary value of SN at node A and RN at node B. Thus, A starts in state 0; a transition to state 1 occurs upon receipt of an error-free request for packet 1 modulo 2. Note that A has to keep track of more information than just this state, such as the contents of the current packet and the time until time-out, but the binary state above is all that is of concern here.

Node B similarly is regarded as having two possible states, 0 and 1, corresponding to the number modulo 2 of the packet currently awaited. When a packet of the desired number modulo 2 is received, the DLC at B releases that packet to the higher layer and changes state, awaiting the next packet (see Fig. 2.23). The combined state of A and B is then initially (0,0); when the first packet is received error-free, the state of B changes to 1, yielding a combined state (0, 1). When A receives the new *RN* value (*i.e.*, 1), the state of A changes to 1 and the combined state becomes (1,1). Note that there is a fixed sequence for these combined states, (0,0), (0,1), (1,1), (1,0), (0,0), and so on, and that A and B alternate in changing states. It is interesting that at the instant of the transition from (0,0) to (0,1), B knows the combined state, but subsequently, up until the transition later from (1,1) to (1,0), it does not know the combined state (*i.e.*, B never knows that A has received the ack information until the next packet is received). Similarly, A knows the combined state at the instant of the transition from (0,1) to (1,1) and of the transition from (1,0) to (0,0). The combined state is always unknown to either A or B, and is frequently unknown to both. The situation here is very similar to that in the three-army problem discussed in Section 1.4. Here, however, information is transmitted even though the combined state information is never jointly known, whereas in the three-army problem, it is impossible to coordinate an attack because of the impossibility of obtaining this joint knowledge.

The stop-and-wait strategy is not very useful for modern data networks because of its highly inefficient use of communication links. In particular, it should be possible to do something else while waiting for an ack. There are three common strategies for extending the basic idea of stop-and-wait ARQ so as to achieve higher efficiency: go back n ARQ, selective repeat ARQ, and finally, the ARPANET ARQ.

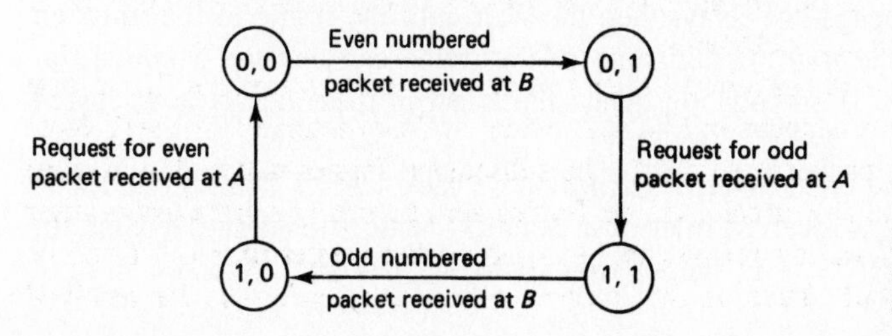

Figure 2.23 State transition diagram for stop-and-wait ARQ. The state is (*SN* mod 2 at *A*, *RN* mod 2 at *B*).

2.4.2 Go Back *n* ARQ

Go back n ARQ is the most widely used type of ARQ protocol; it appears in the various standard DLC protocols, such as HDLC, SDLC, ADCCP, and LAPB. It is not elucidating to know the meaning of these acronyms, but in fact, these standards are almost the same. They are discussed in Section 2.6, and some of their differences are mentioned there. Go back n is also the basis of most error recovery procedures at the transport layer.

The basic idea of go back n is very simple. Incoming packets to a transmitting DLC module for a link from A to B are numbered sequentially, and this sequence number *SN* is sent in the header of the frame containing the packet. In contrast to stop-and-wait ARQ, several successive packets can be sent without waiting for the next packet to be requested. The receiving DLC at B operates in essentially the same way as stop-and-wait ARQ. It accepts packets only in the correct order and sends request numbers RN back to A; the effect of a given request RN is to acknowledge all packets prior to RN and to request transmission of packet RN.

The go back number $n \geq 1$ in a go back n protocol is a parameter that determines how many successive packets can be sent in the absence of a request for a new packet. Specifically, node A is not allowed to send packet $i+n$ before i has been acknowledged (*i.e.*, before $i+1$ has been requested). Thus, if i is the most recently received request from node B, there is a "window" of n packets, from i to $i+n-1$, that the transmitter is allowed to send. As successively higher-numbered requests are received from B, this window slides upward; thus go back n protocols are often called *sliding window* ARQ protocols.

Figure 2.24 illustrates the operation of go back 7 ARQ when piggybacking of request numbers is being used and when there are no errors and a constant supply of traffic. Although the figure portrays data traffic in both directions, the flow of sequence numbers is shown in one direction and request numbers in the other. Note that when the first frame from A (containing packet 0) is received at B, node B is already in the middle of transmitting its second frame. The piggybacked request number at node B is traditionally in the frame header, and the frame is traditionally completely assembled before transmission starts. Thus when packet 0 is received from A, it is too late for node B to request packet 1 in the second frame, so that $RN = 1$ does not appear until the third frame from B. When this frame is completely received at A, the window at A "slides up" from $[0, 6]$ to $[1, 7]$.

Note that even in the absence of transmission errors, there are several sources of delay between the time that a packet is first assembled into a frame at A and the time when A receives an acknowledgment of the packet. First there is the transmission time of the frame, then the propagation delay, then the wait until the frame in transmission at B is completed, then the transmission time of the frame carrying the acknowledgment, and finally, the B-to-A propagation delay; the effect of these delays is discussed later.

Figure 2.25 shows the effect of error frames on go back 4 ARQ. The second frame from A, carrying packet 1, is received in error at node B. Node B continues to look for packet 1 and to transmit $RN = 1$ in frames from B to A. Packets 2, 3, and 4 from A

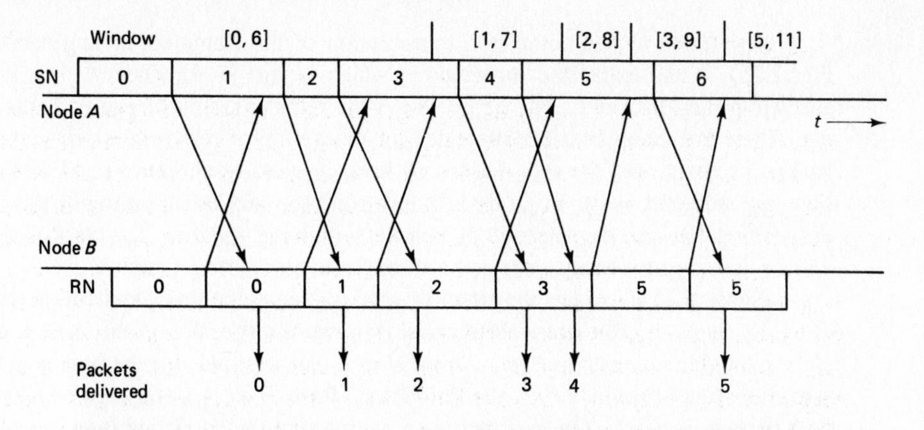

Figure 2.24 Example of go back 7 protocol for A-to-B traffic. Both nodes are sending data, but only the sequence numbers are shown for the frames transmitted from A and only the request numbers are shown for the frames from B. When packet 0 is completely received at B, it is delivered to the higher layer, as indicated at the lower left of the figure. At this point, node B wants to request packet 1, so it sends $RN = 1$ in the next outgoing frame. When that outgoing frame is completely received at A, node A updates its window from [0,6] to [1,7]. Note that when packets 3 and 4 are both received at B during the same frame transmission at B, node B awaits packet 5 and uses $RN = 5$ in the next frame from B to acknowledge both packets 3 and 4.

all arrive at B in error-free frames but are not accepted since node B is looking only for packet 1. One might think that it would be more efficient for node B to buffer packets 2, 3, and 4, thus avoiding the necessity for A to retransmit them after packet 1 is finally retransmitted. Such a buffering strategy is indeed possible and is called *selective repeat* ARQ; this is discussed in Section 2.4.3, but, by definition, go back n does not include this possibility.

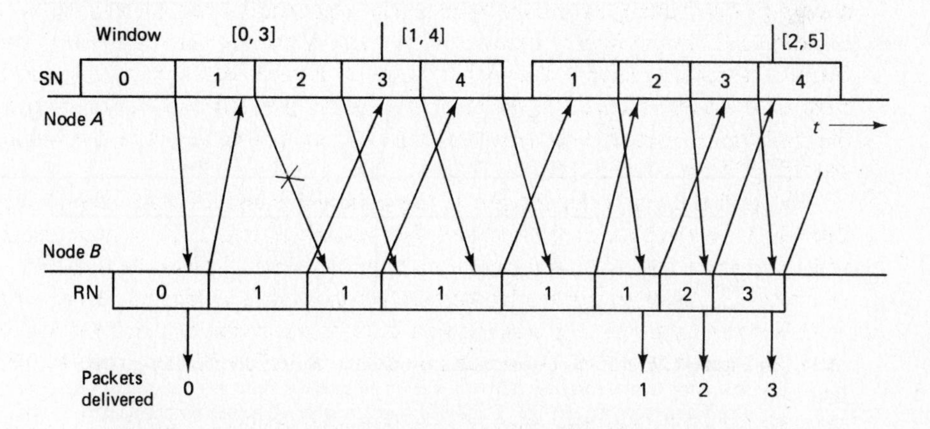

Figure 2.25 Effect of a transmission error on go back 4. Packet 1 is received in error at B, and node B continues to request packet 1 in each reverse frame until node A transmits its entire window, times-out, and goes back to packet 1.

After node A has completed transmission of the frame containing packet 4 (see Fig. 2.25), it has exhausted its window, which is still $[1, 4]$; node A then goes back and retransmits packet 1. The figure shows the retransmission of packet 1 after a time-out. There are many possible strategies for the timing of retransmissions within the go back n protocol, and for this reason, go back n is really an entire class of algorithms (just like stop and wait). In go back n, however, not only is the timing of transmissions unspecified, but also the selection of a packet within the window. The class of algorithms will be specified precisely after going through several more examples.

Figure 2.26 illustrates the effect of error frames in the reverse direction (from node B to A). Such an error frame need not slow down transmission in the A-to-B direction, since a subsequent error-free frame from B to A can acknowledge the packet and perhaps some subsequent packets (*i.e.*, the third frame from B to A in the figure acknowledges both packets 1 and 2). On the other hand, with a small window and long frames from B to A, an error frame from B to A can delay acknowledgments until after all the packets in the window are transmitted, thus causing A to either wait or to go back. Note that when the delayed acknowledgments get through, node A can jump forward again (from packet 3 to 5 in Fig. 2.26).

Finally, Fig. 2.27 shows that retransmissions can occur even in the absence of any transmission errors. This happens particularly in the case of short frames in one direction and long frames in the other. We discuss the impact of this phenomenon on the choice of window size and frame length later.

Rules followed by transmitter and receiver in go back *n* We now specify the precise rules followed by the class of go back n protocols and then, in the next

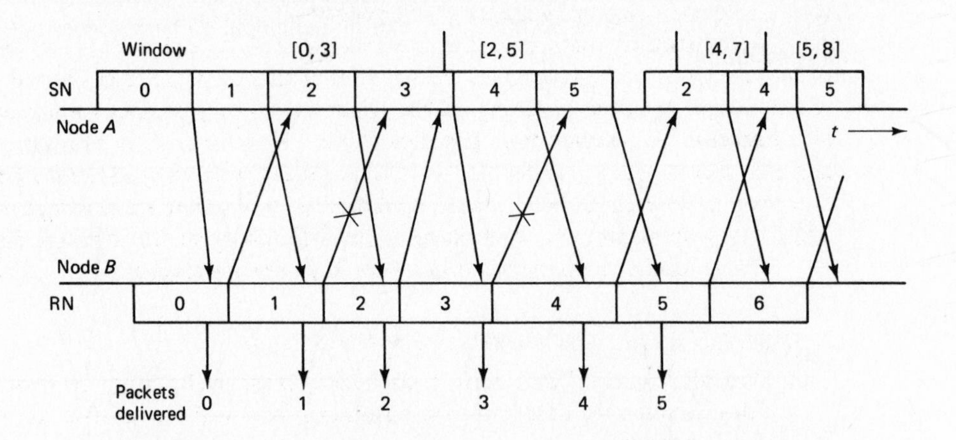

Figure 2.26 Effect of transmission errors in the reverse direction for go back 4. The first error frame, carrying $RN = 1$, causes no problem since it is followed by an error-free frame carrying $RN = 2$ and this frame reaches A before packet number 3, (*i.e.*, the last packet in the current window at A) has completed transmission and before a time-out occurs. The second error frame, carrying $RN = 3$, causes retransmissions since the following reverse frame is delayed until after node A sends its entire window and times-out. This causes A to go back and retransmit packet 2.

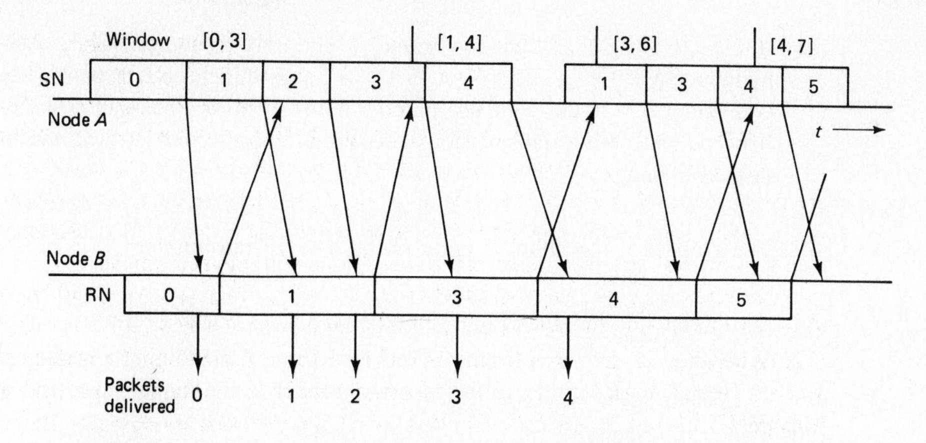

Figure 2.27 Effect of delayed feedback for go back 4. The frames in the B-to-A direction are much longer than those in the A-to-B direction, thus delaying the request numbers from getting back to A. The request for packet 1 arrives in time to allow packet 4 to be sent, but after sending packet 4, node A times-out and goes back to packet 1.

subsection, demonstrate that the algorithms in the class work correctly. We assume here that the sequence numbers and request numbers are integers that can increase without bound. The more practical case in which SN and RN are taken modulo some integer m is considered later.

The rules given here do not treat the initialization of the protocol. We simply assume that initially there are no frames in transit on the link, that node A starts with the transmission of packet number 0, and that node B is initially awaiting packet number 0. How to achieve such an initialization is discussed in Section 2.7.

The transmitter uses two integer variables, SN_{min} and SN_{max}, to keep track of its operations. SN_{min} denotes the smallest-numbered packet that has not yet been acknowledged (*i.e.*, the lower end of the window). SN_{max} denotes the number of the next packet to be accepted from the higher layer. Thus the DLC layer is attempting to transmit packets SN_{min} to $SN_{max} - 1$. Conceptually we can visualize the DLC layer as storing these packets, but it makes no difference where they are stored physically as long as they are maintained somewhere at the node for potential retransmission.

The go back n algorithm at node A for A-to-B transmission:

1. Set the integer variables SN_{min} and SN_{max} to 0.
2. Do steps 3, 4, and 5 repeatedly in any order. There can be an arbitrary but bounded delay between the time when the conditions for a step are satisfied and when the step is executed.
3. If $SN_{max} < SN_{min} + n$, and if a packet is available from the higher layer, accept a new packet into the DLC, assign number SN_{max} to it, and increment SN_{max}. send this packet
4. If an error-free frame is received from B containing a request number RN greater than SN_{min}, increase SN_{min} to RN.

5. If $SN_{min} < SN_{max}$ and no frame is currently in transmission, choose some number SN, $SN_{min} \leq SN < SN_{max}$; transmit the SNth packet in a frame containing SN in the sequence number field. At most a bounded delay is allowed between successive transmissions of packet SN_{min} over intervals when SN_{min} does not change.

The go back n algorithm at node B for A-to-B transmission:

1. Set the integer variable RN to 0 and repeat steps 2 and 3 forever.
2. Whenever an error-free frame is received from A containing a sequence number SN equal to RN, release the received packet to the higher layer and increment RN.
3. At arbitrary times, but within bounded delay after receiving any error-free data frame from A, transmit a frame to A containing RN in the request number field.

There are many conventional ways of handling the timing and ordering of the various operations in the algorithm above. Perhaps the simplest is for node A to set a timer whenever a packet is transmitted. If the timer expires before that packet is acknowledged (*i.e.*, before SN_{min} increases beyond that packet number), the packet is retransmitted. Sometimes when this approach is used, the transmitter, after going back and retransmitting SN_{min}, simply retransmits subsequent packets in order up to $SN_{max} - 1$, whether or not subsequent request numbers are received. For example, at the right-hand edge of Fig. 2.26, the transmitter might have followed packet 3 with 4 rather than 5. In terms of the algorithm as stated, this corresponds to the transmitter delaying the execution of step 4 while in the process of retransmitting a window of packets. Another possibility is for node A to cycle back whenever all the available packets in the window have been transmitted. Also, A might respond to a specific request from B for retransmission; such extra communication between A and B can be considered as part of this class of protocols in the sense that it simply guides the available choices within the algorithm.

Perhaps the simplest approach to timing at node B is to piggyback the current value of RN in each data frame going from B to A. When there are no data currently going from B to A, a nondata frame containing RN should be sent from B to A whenever a data frame is received from A.

Correctness of go back *n* We first demonstrate the correctness of this class of algorithms under our current assumptions that SN and RN are integers; we then show that correctness is maintained if SN and RN are integers modulo m, for m strictly greater than the go back number n. The correctness demonstration when SN and RN are integers is almost the same as the demonstration in Section 2.4.1 for stop and wait. In particular, we start by assuming that all frames with transmission errors are detected by the CRC, that there is some $q > 0$ such that each frame is received error-free with probability at least q, and that the system is correctly initialized in the sense that there are no frames on the link and that nodes A and B both start at step 1 of their respective algorithms.

The safety property of the go back n algorithm is exactly the same as for stop and wait. In particular, node B releases packets to the higher layer in order, using the variable RN to track the next packet awaited. To verify the liveness property, assume that i is the value of SN_{min} at node A at a given time t_1 (see Fig. 2.28). Let t_2 be the time at which packet i is received error-free and released to the higher layer at node B; let $t_2 = \infty$ if this event never occurs. Similarly, let t_3 be the time at which SN_{min} is increased beyond i and let $t_3 = \infty$ if this never occurs. We will show that t_3 is finite and that $t_1 < t_3$ and $t_2 < t_3$. This is sufficient to demonstrate liveness, since using the argument for each successive value of SN_{min} shows that each packet is transmitted with finite delay.

Let $RN(t)$ be the value of the variable RN at node B as a function of time t and let $SN_{min}(t)$ be the corresponding value of SN_{min} at node A. It is seen directly from the algorithm statement that $SN_{min}(t)$ and $RN(t)$ are nondecreasing in t. Also, since $SN_{min}(t)$ is the largest request number (if any) received from B up to time t, $SN_{min}(t) \leq RN(t)$. By definition of t_2 and t_3, $RN(t)$ is incremented to $i+1$ at t_2 and $SN_{min}(t)$ is increased beyond i at t_3. Using the fact that $SN_{min}(t) \leq RN(t)$, it follows that $t_2 < t_3$. Note that it is possible that $t_2 < t_1$, since packet i might have been received error-free and released at B before time t_1 and even before SN_{min} became equal to i.

From the algorithm statement, node A transmits packet i repeatedly, with finite delay between successive transmissions, from t_1 until t_3. If $t_1 < t_2$, then $RN(t) = i$ for $t_1 \leq t \leq t_2$, so the first error-free reception of packet i after t_1 will be accepted and released to the higher layer at B. Since $t_2 < t_3$, node A will retransmit packet i until this happens. Since there is a probability $q > 0$ that each retransmission is received correctly, and retransmissions occur within finite intervals, the time from t_1 to t_2 is finite. Node

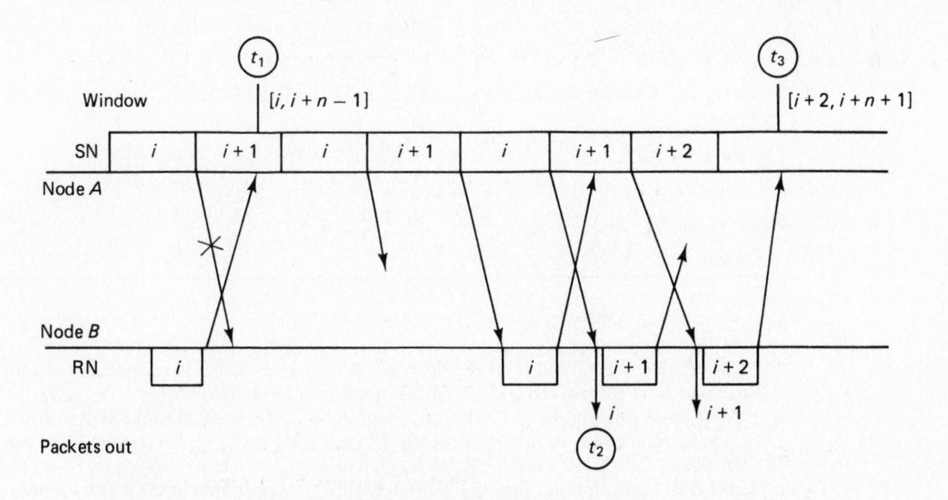

Figure 2.28 SN_{min} is i at time t_1. Packet i is then released to the higher layer at B at time t_2 and the window is increased at A at time t_3.

B (whether $t_1 < t_2$, or vice versa) transmits frames carrying $RN \geq i+1$ from time t_2 until some such frame is received error-free at A at time t_3. Since node A is also transmitting frames in this interval, the delay between subsequent transmissions from B is finite, and, since $q > 0$, the interval from t_2 to t_3 is finite. Thus the interval from t_1 to t_3 is finite and the algorithm is live.

It will be observed that no assumption was made in the demonstration above about the frames traveling in order on the links, and the algorithm above operates correctly even when frames get out of order. When we look at error recovery at the transport layer, the role of a link will be played by a subnetwork and the role of a frame will be played by a packet. For datagram networks, packets can get out of order, so this generality will be useful.

Go back *n* with modulus *m* > *n*. It will now be shown that if the sequence number SN and the request number RN are sent modulo m, for some m strictly greater than the go back number n, the correctness of go back n is maintained as long as we reimpose the condition that frames do not get out of order on the links. To demonstrate this correctness, we first look more carefully at the ordering of events when ordinary integers are used for SN and RN.

Consider the transmission of an arbitrary frame from node A to B. Suppose that the frame is generated at time t_1 and received at t_2 (see Fig. 2.29). The sequence number SN of the frame must lie in node A's window at time t_1, so

$$SN_{min}(t_1) \leq SN \leq SN_{min}(t_1) + n - 1 \qquad (2.24)$$

Also, as shown in the figure,

$$SN_{min}(t_1) \leq RN(t_2) \leq SN_{min}(t_1) + n \qquad (2.25)$$

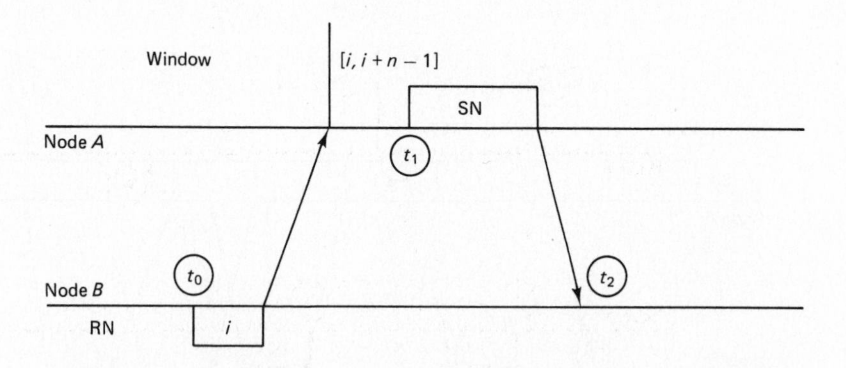

Figure 2.29 Let t_1 and t_2 be the times at which a given frame is generated at A and received at B respectively. The sequence number in the frame satisfies $SN_{min}(t_1) \leq SN \leq SN_{min}(t_1) + n - 1$. The value i of $SN_{min}(t_1)$ is equal to the last received value of RN, which is $RN(t_0) \leq RN(t_2)$. Thus $SN_{min}(t_1) \leq RN(t_2)$, which is the left side of Eq. (2.25). Conversely, no frame with sequence number $SN_{min}(t_1)+n$ can have been sent before t_1 since this value is beyond the upper limit of the window. Since frames travel in order on the link, no frame with this number has arrived at B before t_2, and $RN(t_2) \leq SN_{min}(t_1) + n$, which is the right side of Eq. (2.25).

We see from Eqs. (2.24) and (2.25) that SN and $RN(t_2)$ are both contained in the interval from $SN_{min}(t_1)$ to $SN_{min}(t_1) + n$, and thus must satisfy

$$|RN(t_2) - SN| \leq n \tag{2.26}$$

Now suppose that when packet number SN is sent, the accompanying sequence number is sent modulo m, and let sn denote SN mod m. Step 3 of the algorithm at node B must then be modified to: If an error-free frame is received from A containing a sequence number sn equal to RN mod m, release the received packet to the higher layer and increment RN. Since $m > n$ by assumption, we see from Eq. (2.26) that $sn = RN$ mod m will be satisfied if and only if the packet number SN is equal to RN; thus, the algorithm still works correctly.

Next consider the ordering of arriving request numbers (using ordinary integers) relative to the window at node A. From Fig. 2.30, we see that

$$SN_{min}(t_2) \leq RN \leq SN_{min}(t_2) + n \tag{2.27}$$

Now suppose that RN is sent modulo m, and let $rn = RN$ mod m. Step 4 of the algorithm at node A must then be modified to: If an error-free frame is received from B containing $rn \neq SN_{min}$ mod m, then increment SN_{min} until $rn = SN_{min}$ mod m. Because of the range of RN in Eq. (2.27), we see that this new rule is equivalent to the old rule, and it is sufficient to send request numbers modulo m. At this point, however, we see that it is unnecessary for SN_{min}, SN_{max}, and RN to be saved at nodes A and B as ordinary integers; everything can be numbered modulo m, and the algorithm has been demonstrated to work correctly for $m > n$.

For completeness, we restate the algorithm for operation modulo m; since all numbers are taken modulo m, we use capital letters for these numbers.

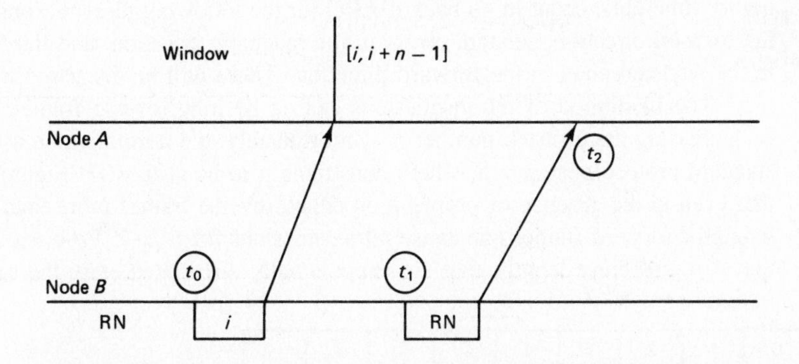

Figure 2.30 Let t_1 and t_2 be the times at which a given frame with request number RN is generated at B and received at A, respectively. Let $SN_{min}(t_2) = i$ be the lower edge of the window at t_2 and let t_0 be the generation time of the frame from B that caused the window at A to move to $[i, i + n - 1]$. Since the frames travel in order on the link, $t_0 < t_1$, so $i \leq RN$ and thus $SN_{min}(t_2) \leq RN$. Similarly, node A cannot have sent packet $i + n$ before t_2, so it certainly cannot have been received before t_1. Thus $RN \leq SN_{min}(t_2) + n$.

The go back n algorithm at node A for modulo m operation, $m > n$:

1. Set the modulo m variables SN_{min} and SN_{max} to 0.
2. Do steps 3, 4, and 5 repeatedly in any order. There can be an arbitrary but bounded delay between the time when the conditions for a step are satisfied and when the step is executed.
3. If $(SN_{max} - SN_{min})$ mod $m < n$, and if a packet is available from the higher layer, accept a new packet into the DLC, assign number SN_{max} to it, and increment SN_{max} to $(SN_{max} + 1)$ mod m.
4. If an error-free frame is received from B containing a request number RN, and $(RN - SN_{min})$ mod $m \leq (SN_{max} - SN_{min})$ mod m, set SN_{min} to equal RN.
5. If $SN_{min} \neq SN_{max}$ and no frame is currently in transmission, choose some number SN such that $(SN - SN_{min})$ mod $m < (SN_{max} - SN_{min})$ mod m ; transmit packet SN in a frame containing SN in the sequence number field.

The go back n algorithm at node B for modulo m operation, $m > n$:

1. Set the modulo m variable RN to 0.
2. Whenever an error-free frame is received from A containing a sequence number SN equal to RN, release the received packet to the higher layer and increment RN to $(RN + 1)$ mod m.
3. At arbitrary times, but within bounded delay after receiving any error-free data frame from A, transmit a frame to A containing RN in the request number field.

Efficiency of go back *n* implementations Retransmissions, or delays waiting for time-outs, occur in go back n ARQ for the following three reasons: first, errors in the forward direction, second, errors in the feedback direction, and third, longer frames in the feedback than in the forward direction. These will be discussed in reverse order.

The likelihood of retransmissions caused by long reverse frames can be reduced by increasing the go back number n. Unfortunately, the normal value of modulus in the standard protocols is $m = 8$, which constrains n to be at most 7. Figure 2.31 illustrates that even in the absence of propagation delay, reverse frames more than three times the length of forward frames can cause retransmissions for $n = 7$. Problem 2.23 also shows that if frames have lengths that are exponentially distributed, with the same distribution

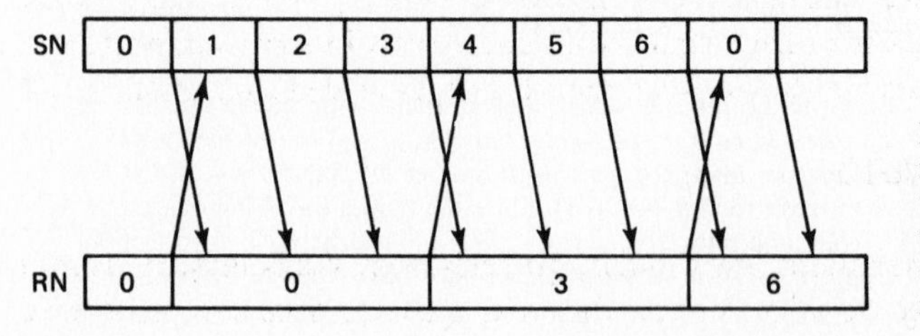

Figure 2.31 Go back 7 ARQ with long frames in the reverse direction. Note that the ack for packet 1 has not arrived at the sending side by the time packet 6 finishes transmission, thereby causing a retransmission of packet 0.

in each direction, the probability p that a frame is not acked by the time the window is exhausted is given by

$$p = (1+n)2^{-n} \tag{2.28}$$

For $n = 7$, p is equal to 1/16. Frames normally have a maximum permissible length (and a minimum length because of the control overhead), so in practice p is somewhat smaller. However, links sometimes carry longer frames in one direction than in the other, and in this case, there might be considerable waste in link utilization for $n = 7$. When propagation delay is large relative to frame length (as, for example, on high-speed links and satellite links), this loss of utilization can be quite serious. Fortunately, the standard protocols have an alternative choice of modulus as $m = 128$.

When errors occur in a reverse frame, the acknowledgment information is postponed for an additional reverse frame. This loss of line utilization in one direction due to errors in the other can also be avoided by choosing n sufficiently large.

Finally, consider the effect of errors in the forward direction. If n is large enough to avoid retransmissions or delays due to large propagation delay and long frames or errors in the reverse direction, and if the sending DLC waits to exhaust its window of n packets before retransmitting, a large number of packets are retransmitted for each forward error. The customary solution to this problem is the use of time-outs. In its simplest version, if a packet is not acknowledged within a fixed time-out period after transmission, it is retransmitted. This time-out should be chosen long enough to include round-trip propagation and processing delay plus transmission time for two maximum-length packets in the reverse direction (one for the frame in transit when a packet is received, and one to carry the new RN). In a more sophisticated version, the sending DLC, with knowledge of propagation and processing delays, can determine which reverse frame should carry the ack for a given packet; it can go back if that frame is error free and fails to deliver the ack.

In a more fundamental sense, increasing link utilization and decreasing delay is achieved by going back quickly when a forward error occurs, but avoiding retransmissions caused by long frames and errors in the reverse direction. One possibility here is for the receiving DLC to send back a short supervisory frame upon receiving a frame in error. This allows the sending side to go back much sooner than if RN were simply piggybacked on a longer reverse data frame. Another approach is to insert RN in the trailer of the reverse frame, inserting it before the CRC and inserting it at the last moment, after the packet part of the frame has been sent. This cannot be done in the standard DLC protocols, but would have the effect of reducing the feedback delay by almost one frame length.

It is not particularly difficult to invent new ways of reducing both feedback delay and control overhead in ARQ strategies. One should be aware, however, that it is not trivial to ensure the correctness of such strategies. Also, except in special applications, improvements must be substantial to outweigh the advantages of standardization.

2.4.3 Selective Repeat ARQ

Even if unnecessary retransmissions are avoided, go back n protocols must retransmit at least one round-trip-delay worth of frames when a single error occurs in an awaited packet. In many situations, the probability of one or more errors in a frame is 10^{-4} or less, and in this case, retransmitting many packets for each frame in error has little effect on efficiency. There are some communication links, however, for which small error probabilities per frame are very difficult to achieve, even with error correction in the modems. For other links (*e.g.*, high-speed links and satellite links), the number of frames transmitted in a round- trip delay time is very large. In both these cases, selective repeat ARQ can be used to increase efficiency.

The basic idea of selective repeat ARQ for data on a link from A to B is to accept out-of-order packets and to request retransmissions from A only for those packets that are not correctly received. There is still a go back number, or window size, n, specifying how far A can get ahead of RN, the lowest-numbered packet not yet correctly received at B.

Note that whatever ARQ protocol is used, only error-free frames can deliver packets to B, and thus, if p is the probability of frame error, the expected number η of packets delivered to B per frame from A to B is bounded by

$$\eta \leq 1 - p \tag{2.29}$$

As discussed later, this bound can be approached (in principle) by selective repeat ARQ; thus $1 - p$ is sometimes called the throughput of ideal selective repeat. Ideal go back n ARQ can similarly be defined as a protocol that retransmits the packets in one round-trip delay each time that a frame carrying the packet awaited by the receiving DLC is corrupted by errors. The throughput of this ideal is shown in Problem 2.26 to be

$$\eta \leq \frac{1 - p}{1 + p\beta} \tag{2.30}$$

where β is the expected number of frames in a round-trip delay interval. This indicates that the increase in throughput available with selective repeat is significant only when $p\beta$ is appreciable relative to 1.

The selective repeat algorithm at node A, for traffic from A to B, is the same as the go back n algorithm, except that (for reasons soon to be made clear) the modulus m must satisfy $m \geq 2n$. At node B, the variable RN has the same significance as in go back n; namely, it is the lowest-numbered packet (or the lowest-numbered packet modulo m) not yet correctly received. In the selective repeat algorithm, node B accepts packets anywhere in the range RN to $RN + n - 1$. The value of RN must be sent back to A as in go back n, either piggybacked on data frames or sent in separate frames. Usually, as discussed later, the feedback from node B to A includes not only the value of RN but also additional information about which packets beyond RN have been correctly received. In principle, the DLC layer still releases packets to the higher layer in order, so that the accepted out-of-order packets are saved until the earlier packets are accepted and released.

We now assume again that frames do not get out of order on the links and proceed to see why a larger modulus is required for selective repeat than for go back n. Assume that a frame is received at node B at some given time t_2 and that the frame was generated at node A at time t_1. If SN and RN are considered as integers, Eqs. (2.24) and (2.25) are still valid, and we can conclude from them that the sequence number SN in the received frame must satisfy

$$RN(t_2) - n \leq SN \leq RN(t_2) + n - 1 \tag{2.31}$$

If sequence numbers are sent mod m, and if packets are accepted in the range $RN(t_2)$ to $RN(t_2) + n - 1$, it is necessary for node B to distinguish values of SN in the entire range of Eq. (2.31). This means that the modulus m must satisfy

$$m \geq 2n, \qquad \text{for selective repeat} \tag{2.32}$$

With this change, the correctness of this class of protocols follows as before. The real issue with selective repeat, however, is using it efficiently to achieve throughputs relatively close to the ideal $1 - p$. Note first that using RN alone to provide acknowledgment information is not very efficient, since if several frame errors occur in one round-trip delay period, node A does not find out about the second frame error until one round-trip delay after the first error is retransmitted. There are several ways of providing the additional acknowledgment information required by A. One is for B to send back the lowest j packet numbers that it has not yet received; j should be larger than $p\beta$ (the expected number of frame errors in a round-trip delay time), but is limited by the overhead required by the feedback. Another possibility is to send RN plus an additional k bits (for some constant k), one bit giving the ack/nak status of each of the k packets after RN.

Assume now that the return frames carry sufficient information for A to determine, after an expected delay of β frames, whether or not a packet was successfully received. The typical algorithm for A is then to repeat packets as soon as it is clear that the previous transmission contained errors; if A discovers multiple errors simultaneously, it retransmits them in the order of their packet numbers. When there are no requested retransmissions, A continues to send new packets, up to $SN_{max} - 1$. At this limit, node A can wait for some time-out or immediately go back to SN_{min} to transmit successive unacknowledged packets.

Node A acts like an ideal selective repeat system until it is forced to wait or go back from $SN_{max} - 1$. When this happens, however, the packet numbered SN_{min} must have been transmitted unsuccessfully about n/β times. Thus, by making n large enough, the probability of such a go back can be reduced to negligible value. There are two difficulties with very large values of n (assuming that packets must be reordered at the receiving DLC). The first is that storage must be provided at B for all of the accepted packets beyond RN. The second is that the large number of stored packets are all delayed waiting for RN.

The amount of storage provided can be reduced to $n - \beta$ without much harm, since whenever a go back from $SN_{max} - 1$ occurs, node A can send no new packets beyond $SN_{max} - 1$ for a round-trip delay; thus, it might as well resend the yet-unacknowledged packets from to $SN_{max} - \beta$ to $SN_{max} - 1$; this means that B need not save these packets. The value of n can also be reduced, without increasing the probability of go back, by retransmitting a packet several times in succession if the previous transmission contained errors. For example, Fig. 2.32 compares double retransmissions with single retransmissions for $n = 2\beta + 2$. Single retransmissions fail with probability p and cause β extra retrans-

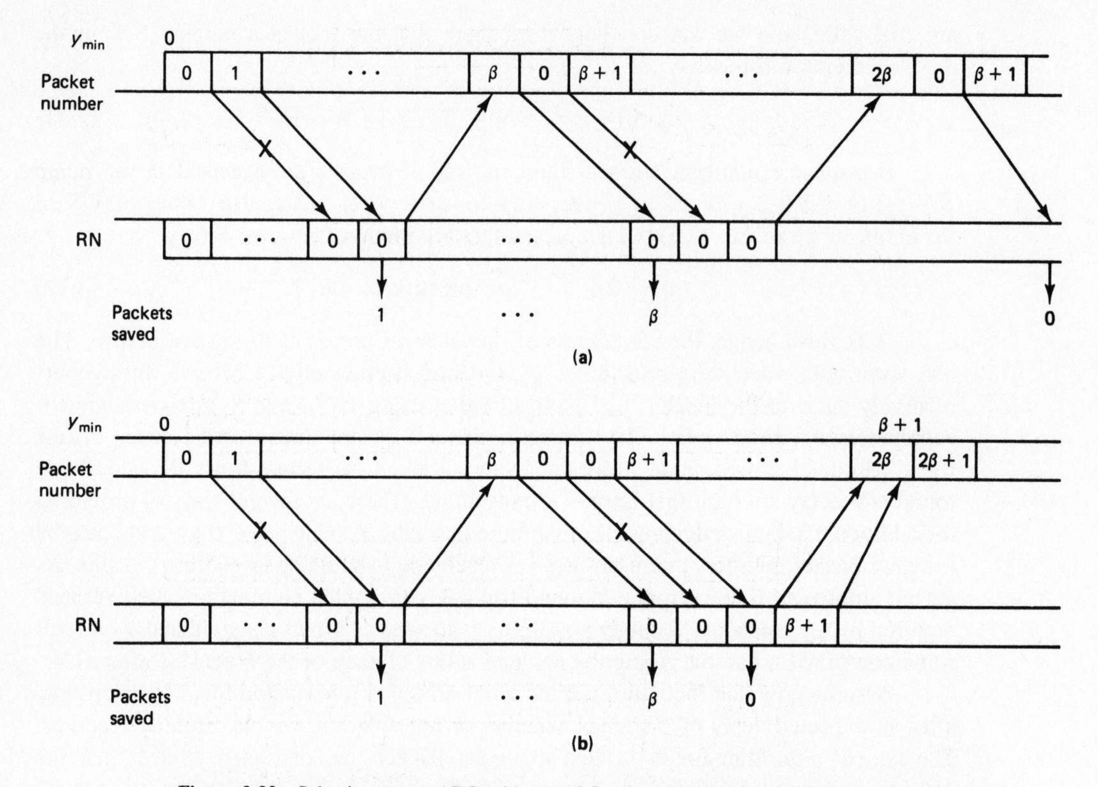

Figure 2.32 Selective repeat ARQ with $n = 2\beta + 2$ and receiver storage for $\beta + 1$ packets. (a) Note the wasted transmissions if a given packet (0) is transmitted twice with errors. (b) Note that this problem is cured, at the cost of one extra frame, if the second transmission of packet 0 is doubled. Feedback contains not only RN but additional information on accepted packets.

missions; double retransmissions rarely fail, but always require one extra retransmission. Thus, double retransmissions can increase throughput if $pn > 1$, but might be desirable in other cases to reduce the variability of delay. (See [Wel82] for further analysis.)

2.4.4 ARPANET ARQ

The ARPANET achieves efficiency by using eight stop-and-wait strategies in parallel, multiplexing the bit pipe between the eight. That is, each incoming packet is assigned to one of eight virtual channels, assuming that one of the eight is idle; if all the virtual channels are busy, the incoming packet waits outside the DLC (see Fig. 2.33). The busy virtual channels are multiplexed on the bit pipe in the sense that frames for the different virtual channels are sent one after the other on the link. The particular order in which frames are sent is not very important, but a simple approach is to send them in round-robin order. If a virtual channel's turn for transmission comes up before an ack has been received for that virtual channel, the packet is sent again, so that the multiplexing removes the need for any time-outs. (The actual ARPANET protocol, however, does use

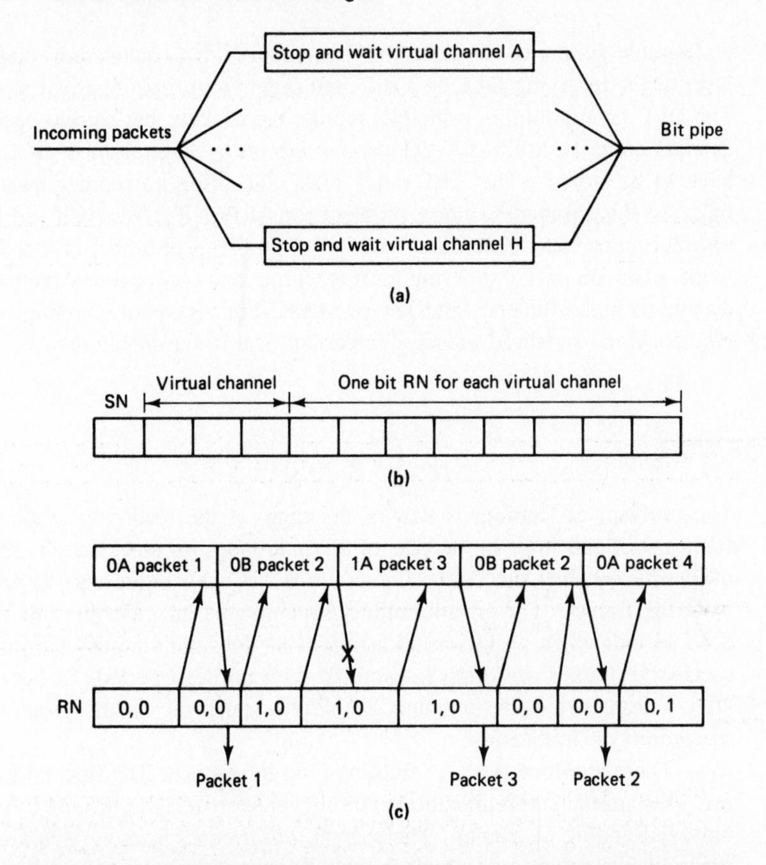

Figure 2.33 ARPANET ARQ. (a) Eight multiplexed, stop-and-wait virtual channels. (b) Bits in the header for ARQ control. (c) Operation of multiplexed stop and wait for two virtual channels. Top-to-bottom frames show SN and the channel number, and bottom-to-top frames show RN for both channels. The third frame from bottom to top acks packet 1 on the A channel.

time-outs.) When an ack is received for a frame on a given virtual channel, that virtual channel becomes idle and can accept a new packet from the higher layer.

Somewhat more overhead is required here than in the basic stop-and-wait protocol. In particular, each frame carries both the virtual channel number (requiring three bits) and the sequence number modulo 2 (*i.e.*, one bit) of the packet on that virtual channel. The acknowledgment information is piggybacked onto the frames going in the opposite direction. Each such frame, in fact, carries information for all eight virtual channels. In particular, an eight-bit field in the header of each return frame gives the number modulo 2 of the awaited packet for each virtual channel.

One of the desirable features of this strategy is that the ack information is repeated so often (*i.e.*, for all virtual channels in each return frame) that relatively few retransmissions are required because of transmission errors in the reverse direction. Typically, only one retransmission is required for each frame in error in the forward direction. The undesirable feature of the ARPANET protocol is that packets are released to the higher layer at the receiving DLC in a different order from that of arrival at the sending DLC. The DLC layer could, in principle, reorder the packets, but since a packet on one virtual channel could be arbitrarily delayed, an arbitrarily large number of later packets might have to be stored. The ARPANET makes no effort to reorder packets on individual links, so this protocol is not a problem for ARPANET. We shall see later that the lack of ordering on links generates a number of problems at higher layers. Most modern networks maintain packet ordering for this reason, and consequently do not use this protocol despite its high efficiency and low overhead. For very poor communication links, where efficiency and overhead are very important, it is a reasonable choice.

2.5 FRAMING

The problem of framing is that of deciding, at the receiving DLC, where successive frames start and stop. In the case of a synchronous bit pipe, there is sometimes a period of idle fill between successive frames, so that it is also necessary to separate the idle fill from the frames. For an intermittent synchronous bit pipe, the idle fill is replaced by dead periods when no bits at all arrive. This does not simplify the problem since, first, successive frames are often transmitted with no dead periods in between, and second, after a dead period, the modems at the physical layer usually require some idle fill to reacquire synchronization.

There are three types of framing used in practice. The first, *character-based framing*, uses special communication control characters for idle fill and to indicate the beginning and ending of frames. The second, *bit-oriented framing with flags*, uses a special string of bits called a flag both for idle fill and to indicate the beginning and ending of frames. The third, *length counts*, gives the frame length in a field of the header. The following three subsections explain these three techniques, and the third also gives a more fundamental view of the problem. These subsections, except for a few comments, ignore the possibility of errors on the bit pipe. Section 2.5.4 then treats the joint problems of ARQ and framing in the presence of errors. Finally, Section 2.5.5 explains the trade-offs involved in the choice of frame length.

2.5.1 Character-Based Framing

Character codes such as ASCII generally provide binary representations not only for keyboard characters and terminal control characters, but also for various communication control characters. In ASCII, all these binary representations are seven bits long, usually with an extra parity bit which might or might not be stripped off in communication [since a cyclic redundancy check (CRC) can be used more effectively to detect errors in frames].

SYN (synchronous idle) is one of these communication control characters; a string of SYN characters provides idle fill between frames when a sending DLC has no data to send but a synchronous modem requires bits. SYN can also be used within frames, sometimes for synchronization of older modems, and sometimes to bridge delays in supplying data characters. STX (start of text) and ETX (end of text) are two other communication control characters used to indicate the beginning and end of a frame, as shown in Fig. 2.34.

The character-oriented communication protocols used in practice, such as the IBM binary synchronous communication system (known as Bisynch or BSC), are far more complex than this, but our purpose here is simply to illustrate that framing presents no insurmountable problems. There is a slight problem in the example above in that either the header or the CRC might, through chance, contain a communication control character. Since these always appear in known positions after STX or ETX, this causes no problem for the receiver. If the packet to be transmitted is an arbitrary binary string, however, rather than a string of ASCII keyboard characters, serious problems arise; the packet might contain the ETX character, for example, which could be interpreted as ending the frame. Character-oriented protocols use a special mode of transmission, called *transparent mode*, to send such data.

The transparent mode uses a special control character called DLE (data link escape). A DLE character is inserted before the STX character to indicate the start of a frame in transparent mode. It is also inserted before intentional uses of communication control characters within such a frame. The DLE is not inserted before the possible appearances of these characters as part of the binary data. There is still a problem if the DLE character itself appears in the data, and this is solved by inserting an extra DLE before each appearance of DLE in the data proper. The receiving DLC then strips off one DLE from each arriving pair of DLEs, and interprets each STX or ETX preceded by an unpaired DLE as an actual start or stop of a frame. Thus, for example, DLE ETX (preceded by something other than DLE) would be interpreted as an end of frame, whereas DLE DLE ETX (preceded by something other than DLE) would be interpreted as an appearance of the bits corresponding to DLE ETX within the binary data.

With this type of protocol, the frame structure would appear as shown in Fig. 2.35. This frame structure is used in the ARPANET. It has two disadvantages, the first of which is the somewhat excessive use of framing overhead (*i.e.,* DLE STX precedes each frame, DLE ETX follows each frame, and two SYN characters separate each frame for a total of six framing characters per frame). The second disadvantage is that each frame must consist of an integral number of characters.

Let us briefly consider what happens in this protocol in the presence of errors. The CRC checks the header and packet of a frame, and thus will normally detect errors

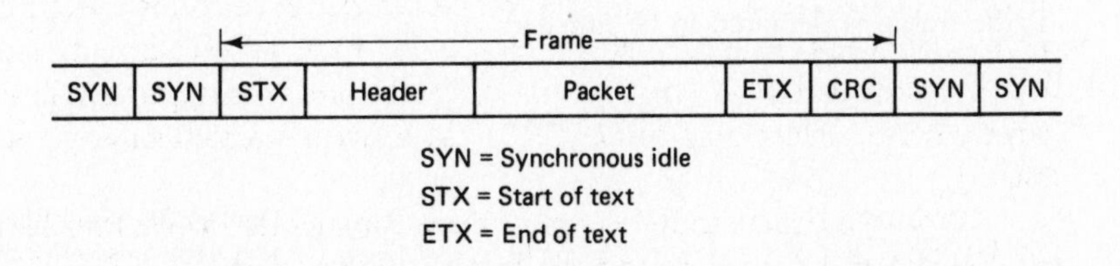

Figure 2.34 Simplified frame structure with character-based framing.

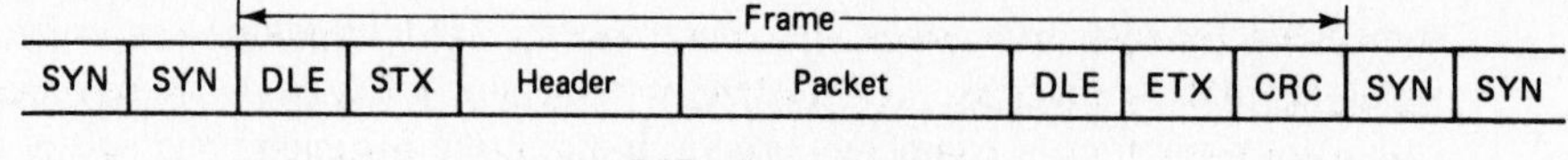

Figure 2.35 Character-based framing in a transparent mode as used in the ARPANET.

there. If an error occurs in the DLE ETX ending a frame, however, the receiver will not detect the end of the frame and thus will not check the CRC; this is why in the preceding section we assumed that frames could get lost. A similar problem is that errors can cause the appearance of DLE ETX in the data itself; the receiver would interpret this as the end of the frame and interpret the following bits as a CRC. Thus, we have an essentially random set of bits interpreted as a CRC, and the preceding data will be accepted as a packet with probability 2^{-L}, where L is the length of the CRC. The same problems occur in the bit-oriented framing protocols to be studied next and are discussed in greater detail in Section 2.5.4.

2.5.2 Bit-Oriented Framing: Flags

In the transparent mode for character-based framing, the special character pair DLE ETX indicated the end of a frame and was avoided within the frame by doubling each DLE character. Here we look at another approach, using a *flag* at the end of the frame. A flag is simply a known bit string, such as DLE ETX, that indicates the end of a frame. Similar to the technique of doubling DLEs, a technique called *bit stuffing* is used to avoid confusion between possible appearances of the flag as a bit string within the frame and the actual flag indicating the end of the frame. One important difference between bit-oriented and character based framing is that a bit-oriented frame can have any length (subject to some minimum and maximum) rather than being restricted to an integral number of characters. Thus, we must guard against appearances of the flag bit pattern starting in any bit position rather than just on character boundaries.

In practice, the flag is the bit string 01^60, where the notation 1^j means a string of j 1's. The rule used for bit stuffing is to insert (stuff) a 0 into the data string of the frame proper after each successive appearance of five 1's (see Fig. 2.36). Thus, the frame, after stuffing, never contains more than five consecutive 1's, and the flag at the end of the frame is uniquely recognizable. At the receiving DLC, the first 0 after each string of five consecutive 1's is deleted; if, instead, a string of five 1's is followed by a 1, the frame is declared to be finished.

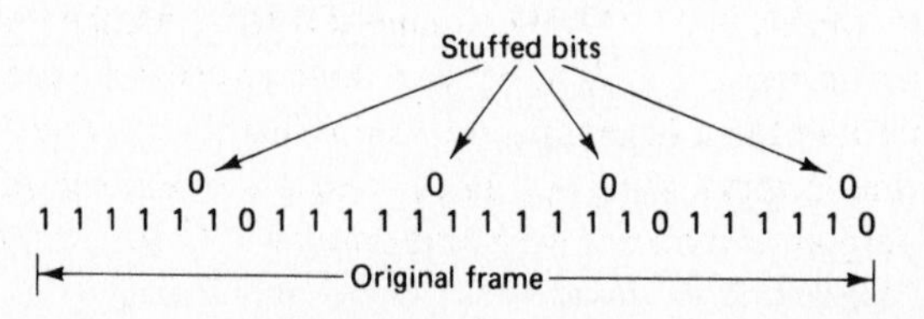

Figure 2.36 Bit stuffing. A 0 is stuffed after each consecutive five 1's in the original frame. A flag, 01111110, without stuffing, is sent at the end of the frame.

Bit stuffing has a number of purposes beyond eliminating flags within the frame. Standard DLCs have an abort capability in which a frame can be aborted by sending seven or more 1's in a row; in addition, a link is regarded as idle if 15 or more 1's in a row are received. What this means is that 01^6 is really the string denoting the end of a frame. If 01^6 is followed by a 0, it is the flag, indicating normal frame termination; if followed by a 1, it indicates abnormal termination. Bit stuffing is best viewed as preventing the appearance of 01^6 within the frame. One other minor purpose of bit stuffing is to break up long strings of 1's, which cause some older modems to lose synchronization.

It is easy to see that the bit stuffing rule above avoids the appearance of 01^6 within the frame, but it is less clear that so much bit stuffing is necessary. For example, consider the first stuffed bit in Fig. 2.36. Since the frame starts with six 1's (following a distinguishable flag), this could not be logically mistaken for 01^6. Thus, stuffing is not logically necessary after five 1's at the beginning of the frame (provided that the receiver's rule for deleting stuffed bits is changed accordingly).

The second stuffed bit in the figure is clearly necessary to avoid the appearance of 01^6. From a strictly logical standpoint, the third stuffed bit could be eliminated (except for the synchronization problem in older modems). Problem 2.31 shows how the receiver rule could be appropriately modified. Note that the reduction in overhead, by modifying the stuffing rules as above, is almost negligible; the purpose here is to understand the rules rather than to suggest changes.

The fourth stuffed bit in the figure is definitely required, although the reason is somewhat subtle. The original string 01^50 surrounding this stuffed bit could not be misinterpreted as 01^6, but the receiving DLC needs a rule to eliminate stuffed bits; it cannot distinguish a stuffed 0 following 01^5 from a data 0 following 01^5. Problem 2.32 develops this argument in detail.

There is nothing particularly magical about the string 01^6 as the bit string used to signal the termination of a frame (except for its use in preventing long strings of 1's on the link), and in fact, any bit string could be used with bit stuffing after the next-to-last bit of the string (see Problem 2.33). Such strings, with bit stuffing, are often useful in data communication to signal the occurrence of some rare event.

Consider the overhead incurred by using a flag to indicate the end of a frame. Assume that a frame (before bit stuffing and flag addition) consists of independent, identically distributed, random binary variables, with equal probability of 0 or 1. Assume for increased generality that the terminating signal for a frame is 01^j for some j (with 01^j0 being the flag and 01^{j+1} indicating abnormal termination); thus, $j = 6$ for the standard flag. An insertion will occur at (*i.e.*, immediately following) the ith bit of the original frame (for $i \geq j$) if the string from $i - j + 1$ to i is 01^{j-1}; the probability of this is 2^{-j}. An insertion will also occur (for $i \geq 2j - 1$) if the string from $i - 2j + 2$ to i is 01^{2j-2}; the probability of this is 2^{-2j+1}. We ignore this term and the probability of insertions due to yet longer strings of 1's: first, because these probabilities are practically negligible, and second, because these insertions are used primarily to avoid long strings of 1's rather than to provide framing. Bit $j - 1$ in the frame is somewhat different than the other bits, since an insertion here occurs with probability 2^{-j+1} (*i.e.*, if the first $j - 1$ bits of the frame are all 1's).

Recall that the expected value of a sum of random variables is equal to the sum of the expected values (whether or not the variables are independent). Thus, the expected number of insertions in a frame of original length K is the sum, over i, of the expected number of insertions at each bit i of the frame. The expected number of insertions at a given bit, however, is just the probability of insertion there. Thus, the expected number of insertions in a string of length $K \geq j - 1$ is

$$(K - j + 3)2^{-j}$$

Taking the expected value of this over frame lengths K (with the assumption that all frames are longer than $j - 1$) and adding the $j + 1$ bits in the termination string, the expected overhead for framing becomes

$$E\{OV\} = (E\{K\} - j + 3)2^{-j} + j + 1 \tag{2.33}$$

Since $E\{K\}$ is typically very much larger than j, we have the approximation and upper bound (for $j \geq 3$)

$$E\{OV\} \leq E\{K\}2^{-j} + j + 1 \tag{2.34}$$

One additional bit is needed to distinguish a normal from an abnormal end of frame.

It will be interesting to find the integer value of j that minimizes this expression for a given value of expected frame length. As j increases from 1, the quantity on the right-hand side of Eq. (2.34) first decreases and then increases. Thus, the minimizing j is the smallest integer j for which the right-hand side is less than the same quantity with j increased by 1, that is, the smallest j for which

$$E\{K\}2^{-j} + j + 1 < E\{K\}2^{-j-1} + j + 2 \tag{2.35}$$

This inequality simplifies to $E\{K\}2^{-j-1} < 1$, and the smallest j that satisfies this is

$$j = \lfloor \log_2 E\{K\} \rfloor \tag{2.36}$$

where $\lfloor x \rfloor$ means the integer part of x. It is shown in Problem 2.34 that for this optimal value of j,

$$E\{OV\} \leq \log_2 E\{K\} + 2 \tag{2.37}$$

For example, with an expected frame length of 1000 bits, the optimal j is 9 and the expected framing overhead is less than 12 bits. For the standard flag, with $j = 6$, the expected overhead is about 23 bits (hardly cause for wanting to change the standard).

2.5.3 Length Fields

The basic problem in framing is to inform the receiving DLC where each idle fill string ends and where each frame ends. In principle, the problem of determining the end of an idle fill string is trivial; idle fill is represented by some fixed string (*e.g.*, repeated SYN characters or repeated flags) and idle fill stops whenever this fixed pattern is broken. In principle, one bit inverted from the pattern is sufficient, although in practice, idle fill is usually stopped at a boundary between flags or SYN characters.

Since a frame consists of an arbitrary and unknown bit string, it is somewhat harder to indicate where it ends. A simple alternative to flags or special characters is to include a length field in the frame header. DECNET, for example, uses this framing technique. Assuming no transmission errors, the receiving DLC simply reads this length in the header and then knows where the frame ends. If the length is represented by ordinary binary numbers, the number of bits in the length field has to be at least $\lfloor \log_2 K_{\max} \rfloor + 1$, where $K_{\max}$ is the maximum frame size. This is the overhead required for framing in this technique; comparing it with Eq. (2.37) for flags, we see that the two techniques require similar overhead.

Could any other method of encoding frame lengths require a smaller expected number of bits? This question is answered by information theory. Given any probability assignment $P(K)$ on frame lengths, the source coding theorem of information theory states that the minimum expected number of bits that can encode such a length is at least the entropy of that distribution, given by

$$H = \sum_K P(K) \log_2 \frac{1}{P(K)} \tag{2.38}$$

According to the theorem, at least this many bits of framing overhead, on the average, must be sent over the link per frame for the receiver to know where each frame ends. If $P(K) = 1/K_{\max}$, for $1 \le K \le K_{\max}$, then H is easily calculated to be $\log_2 K_{\max}$. Similarly, for a geometric distribution on lengths, with given $E\{K\}$, the entropy of the length distribution is approximately $\log_2 E\{K\} + \log_2 e$, for large $E\{K\}$. This is about 1/2 bit below the expression in Eq. (2.37). Thus, for the geometric distribution on lengths, the overhead using flags for framing is essentially minimum. The geometric distribution has an interesting extremal property; it can be shown to have the largest entropy of any probability distribution over the positive integers with given $E\{K\}$ (*i.e.*, it requires more bits than any other distribution).

The general idea of source coding is to map the more likely values of K into short bit strings and less likely values into long bit strings; more precisely, one would like to map a given K into about $\log_2[1/P(K)]$ bits. If one does this for a geometric distribution, one gets an interesting encoding known as the unary–binary encoding. In particular, for a given j, the frame length K is represented as

$$K = i2^j + r; \quad 0 \le r < 2^j \tag{2.39}$$

The encoding for K is then i 0's followed by a 1 (this is known as a unary encoding of i) followed by the ordinary binary encoding of r (using j bits). For example, if $j = 2$ and $K = 7$, K is represented by $i = 1$, $r = 3$, which encodes into 0111 (where 01 is the unary encoding of $i = 1$ and 11 is the binary encoding of $r = 3$). Note that different values of K are encoded into different numbers of bits, but the end of the encoding can always be recognized as occurring j bits after the first 1.

In general, with this encoding, a given K maps into a bit string of length $\lfloor K/2^j \rfloor + 1 + j$. If the integer value above is neglected and the expected value over K is taken, then

$$E\{OV\} = E\{K\}2^{-j} + 1 + j \tag{2.40}$$

Note that this is the same as the flag overhead in Eq. (2.34). This is again minimized by choosing $j = \lfloor \log_2 E\{K\} \rfloor$. Thus, this unary–binary length encoding and flag framing both require essentially the minimum possible framing overhead for the geometric distribution, and no more overhead for any other distribution of given $E\{K\}$.

2.5.4 Framing with Errors

Several peculiar problems arise when errors corrupt the framing information on the communication link. First, consider the flag technique. If an error occurs in the flag at the end of a frame, the receiver will not detect the end of frame and will not check the cyclic redundancy check (CRC). In this case, when the next flag is detected, the receiver assumes the CRC to be in the position preceding the flag. This perceived CRC might be the actual CRC for the following frame, but the receiver interprets it as checking what was transmitted as two frames. Alternatively, if some idle fill follows the frame in which the flag was lost, the perceived CRC could include the error-corrupted flag. In any case, the perceived CRC is essentially a random bit string in relation to the perceived preceding frame, and the receiver fails to detect the errors with a probability essentially 2^{-L}, where L is the length of the CRC.

An alternative scenario is for an error within the frame to change a bit string into the flag, as shown for the flag 01^60:

$$0\,1\,0\,0\,1\,1\,0\,1\,1\,1\,0\,0\,1\,\ldots \qquad \text{(sent)}$$

$$0\,1\,0\,0\,1\,1\,1\,1\,1\,1\,0\,0\,1\,\ldots \qquad \text{(received)}$$

It is shown in Problem 2.35 that the probability of this happening somewhere in a frame of K independent equiprobable binary digits is approximately $(1/32)Kp$, where p is the probability of a bit error. In this scenario, as before, the bits before the perceived flag are interpreted by the receiver as a CRC, and the probability of accepting a false frame, given this occurrence, is 2^{-L}. This problem is often called the data sensitivity problem of DLC, since even though the CRC is capable of detecting any combination of three or fewer errors, a single error that creates or destroys a flag, plus a special combination of data bits to satisfy the perceived preceding CRC, causes an undetectable error.

If a length field in the header provides framing, an error in this length field again causes the receiver to look for the CRC in the wrong place, and again an incorrect frame is accepted with probability about 2^{-L}. The probability of such an error is smaller using a length count than using a flag (since errors can create false flags within the frame); however, after an error occurs in a length field, the receiver does not know where to look for any subsequent frames. Thus, if a length field is used for framing, some synchronizing string must be used at the beginning of a frame whenever the sending DLC goes back to retransmit. (Alternatively, synchronization could be used at the start of every frame, but this would make the length field almost redundant.)

There are several partial solutions to these problems, but none are without disadvantages. DECNET uses a fixed-length header for each frame and places the length of the frame in that header; in addition, the header has its own CRC. Thus, if an error occurs in the length field of the header, the receiver can detect it by the header CRC, which is in a known position. One difficulty with this strategy is that the transmitter must still resynchronize after such an error, since even though the error is detected, the receiver will not know when the next frame starts. The other difficulty is that two CRCs must be used in place of one, which is somewhat inefficient.

A similar approach is to put the length field of one frame into the trailer of the preceding frame. This avoids the inefficiency of the DECNET approach, but still requires a special synchronizing sequence after each detected error. This also requires a special header frame to be sent whenever the length of the next frame is unknown when a given frame is transmitted.

Another approach, for any framing technique, is to use a longer CRC. This at least reduces the probability of falsely accepting a frame if framing errors occur. It appears that this is the most likely alternative to be adopted in practice; a standard 32 bit CRC exists as an option in standard DLCs.

A final approach is to regard framing as being at a higher layer than ARQ. In such a system, packets would be separated by flags, and the resulting sequence of packets and flags would be divided into fixed-length frames. Thus, frame boundaries and packet boundaries would bear no relation. If a packet ended in the middle of a frame and no further packets were available, the frame would be completed with idle fill. These frames would then enter the ARQ system, and because of the fixed-length frames, the CRC would always be in a known place. One disadvantage of this strategy is delay; a packet could not be accepted until the entire frame containing the end of the packet was accepted. This extra delay would occur on each link of the packet's path.

2.5.5 Maximum Frame Size

The choice of a maximum frame length, or maximum packet length, in a data network depends on many factors. We first discuss these factors for networks with variable packet lengths and then discuss some additional factors for networks using fixed packet lengths. Most existing packet networks use variable packet lengths, but the planners of broadband ISDN are attempting to standardize on a form of packet switching called asynchronous transfer mode (ATM) which uses very short frames (called cells in ATM) with a fixed length of 53 bytes. The essential reason for the fixed-length is to simplify the hardware for high-speed switching. There are also some advantages of fixed-length frames for multiaccess systems, as discussed in Chapter 4.

Variable frame length Assume that each frame contains a fixed number V of overhead bits, including frame header and trailer, and let K_{max} denote the maximum length of a packet. Assume, for the time being, that each message is broken up into as many maximum-length packets as possible, with the last packet containing what is left over. That is, a message of length M would be broken into $\lceil M/K_{max} \rceil$ packets, where $\lceil x \rceil$ is the smallest integer greater than or equal to x. The first $\lceil M/K_{max} \rceil - 1$ of these packets each contain K_{max} bits and the final packet contains between 1 and K_{max} bits.

The total number of bits in the resulting frames is then

$$\text{total bits} = M + \left\lceil \frac{M}{K_{max}} \right\rceil V \tag{2.41}$$

We see from this that as K_{max} decreases, the number of frames increases and thus the total overhead in the message, $\lceil M/K_{max} \rceil V$, increases. In the limit of very long messages, a fraction $V/(V + K_{max})$ of the transmitted bits are overhead bits. For shorter messages, the fraction of overhead bits is typically somewhat larger because of the reduced length of the final packet.

A closely related factor is that the nodes and external sites must do a certain amount of processing on a frame basis; as the maximum frame length decreases, the number of frames, and thus this processing load, increase. With the enormous increase in data rates available from optical fiber, it will become increasingly difficult to carry out this processing for small frame lengths. In summary, transmission and processing overhead both argue for a large maximum frame size.

We next discuss the many factors that argue for small frame size. The first of these other factors is the pipelining effect illustrated in Fig. 2.37. Assume that a packet must be completely received over one link before starting transmission over the next. If an entire message is sent as one packet, the delay in the network is the sum of the message transmission times over each link. If the message is broken into several packets, however, the earlier packets may proceed along the path while the later packets are still being transmitted on the first link, thus reducing overall message delay.

Since delay could be reduced considerably by starting the transmission of a packet on one link before receiving it completely on the preceding link, we should understand why this is not customarily done. First, if the DLC is using some form of ARQ, the CRC must be checked before releasing the packet to the next link. This same argument holds even if a CRC is used to discard packets in error rather than to retransmit them. Finally, if the links on a path have different data rates, the timing required to supply incoming bits to the outgoing link becomes very awkward; the interface between the DLC layer and the network layer also becomes timing dependent.

Let us investigate the combined effect of overhead and pipelining on message delay, assuming that each packet is completely received over one link before starting transmission on the next. Suppose that a message of length M is broken into maximum-length packets, with a final packet of typically shorter length. Suppose that the message must be transmitted over j equal-capacity links and that the network is lightly loaded, so that waiting at the nodes for other traffic can be ignored. Also, ignore errors on the links (which will be discussed later), and ignore propagation delays (which are independent of maximum packet length). The total time T required to transmit the message to the destination is the time it takes the first packet to travel over the first $j - 1$ links, plus the time it takes the entire message to travel over the final link (*i.e.*, when a nonfinal frame finishes traversing a link, the next frame is always ready to start traversing the link). Let C be the capacity of each link in bits per second, so that TC is the number of bit transmission times required for message delivery. Then, assuming that $M \geq K_{max}$,

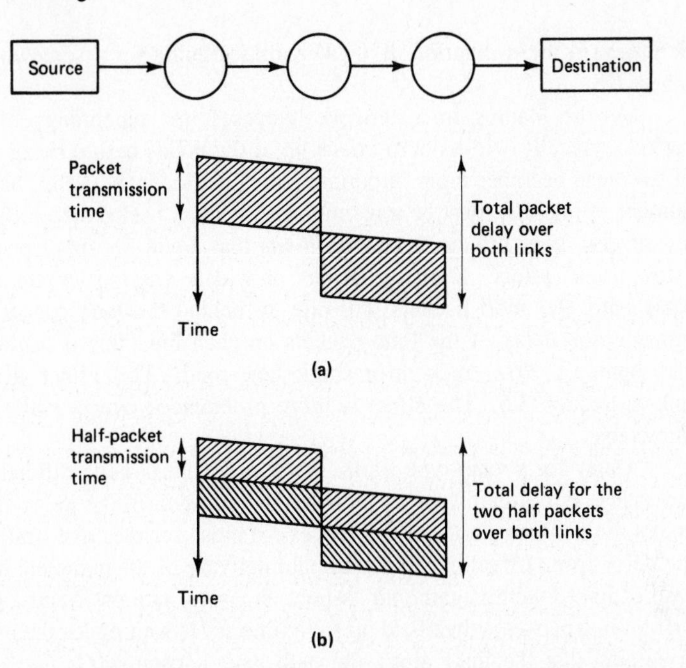

Figure 2.37 Decreasing delay by shortening packets to take advantage of pipelining. (a) The total packet delay over two empty links equals twice the packet transmission time on a link plus the overall propagation delay. (b) When each packet is split in two, a pipelining effect occurs. The total delay for the two half packets equals 1.5 times the original packet transmission time on a link plus the overall propagation delay.

$$TC = (K_{max} + V)(j - 1) + M + \left\lceil \frac{M}{K_{max}} \right\rceil V \tag{2.42}$$

In order to find the expected value of this over message lengths M, we make the approximation that $E\{\lceil M/K_{max} \rceil\} = E\{M/K_{max}\} + \frac{1}{2}$ (this is reasonable if the distribution of M is reasonably uniform over spans of K_{max} bits). Then

$$E\{TC\} \approx (K_{max} + V)(j - 1) + E\{M\} + \frac{E\{M\}V}{K_{max}} + \frac{V}{2} \tag{2.43}$$

We can differentiate this with respect to K_{max} (ignoring the integer constraint) to find the value of K_{max} that minimizes $E\{TC\}$. The result is

$$K_{max} \approx \sqrt{\frac{E\{M\}V}{j - 1}} \tag{2.44}$$

This shows the trade-off between overhead and pipelining. As the overhead V increases, K_{max} should be increased, and as the path length j increases, K_{max} should be reduced. As a practical detail, recall that delay is often less important for file transfers than for other messages, so file transfers should probably be weighted less than other messages in the estimation of $E\{M\}$, thus arguing for a somewhat smaller K_{max} than otherwise.

As the loading in a network increases, the pipelining effect remains, although packets typically will have to queue up at the nodes before being forwarded. The effect of overhead becomes more important at heavier loads, however, because of the increased number of bits that must be transmitted with small packet sizes. On the other hand, there are several other effects at heavier loads that argue for small packet sizes. One is the "slow truck" effect. If many packets of widely varying lengths are traveling over the same path, the short packets will pile up behind the long packets because of the high transmission delay of the long packets on each link; this is analogous to the pileup of cars behind a slow truck on a single-lane road. This effect is analyzed for a single link in Section 3.5. The effect is more pronounced over a path, but is mathematically intractable.

Delay for stream-type traffic (such as voice) is quite different from delay for data messages. For stream-type traffic, one is interested in the delay from when a given bit enters the network until that bit leaves, whereas for message traffic, one is interested in the delay from arrival of the message to delivery of the complete message. Consider the case of light loading again and assume an arrival rate of R and a packet length K. The first bit in a packet is then held up for a time K/R waiting for the packet to be assembled. Assuming that the links along the path have capacities $C_1, C_2, \ldots$, each exceeding R, and assuming V bits of framing overhead, a given packet is delayed by $(K+V)/C_i$ on the ith link. When a given packet is completely received at the last node of the network, the first bit of the packet can be released immediately, yielding a total delay

$$T = \frac{K}{R} + (K+V)\sum_i \frac{1}{C_i} \tag{2.45}$$

Assuming that the received data stream is played out at rate R, all received bits have the same delay, which is thus given by Eq. (2.45). We have tacitly assumed in deriving this equation that $(K+V)/C_i \leq K/R_i$ for each link i (*i.e.*, that each link can transmit frames as fast as they are generated). If this is violated, the queueing delay becomes infinite, even with no other traffic in the network. We see then from Eq. (2.45) that T decreases as K decreases until $(K+V)/C_i = K/R$ for some link, and this yields the minimum possible delay. Packet lengths for stream traffic are usually chosen much larger than this minimum because of the other traffic that is expected on the links. As link speeds increase, however, the dominant delay term in Eq. (2.45) is K/R, which is unaffected by other traffic. For 64 kbps voice traffic, for example, packets usually contain on the order of 500 bits or less, since the delay from the K/R term starts to become objectionable for longer lengths.

Note that under light loading, the delay of a session (either message or stream) is controlled by the packet length of that session. Under heavy-loading conditions, however, the use of long packets by some users generally increases delay for all users. Thus a maximum packet length should be set by the network rather than left to the users.

Several effects of high variability in frame lengths on go back n ARQ systems were discussed in Section 2.4.2. High variability either increases the number of packets that must be retransmitted or increases waiting time. This again argues for small maximum packet size. Finally, there is the effect of transmission errors. Large frames have a somewhat higher probability of error than small frames (although since errors are usually correlated, this effect is not as pronounced as one would think). For most links in use (with the notable exception of radio links), the probability of error on reasonable-sized frames is on the order of 10^{-4} or less, so that this effect is typically less important than the other effects discussed. Unfortunately, there are many analyses of optimal maximum frame length in the literature that focus only on this effect. Thus, these analyses are relevant only in those special cases where error probabilities are very high.

In practice, typical maximum frame lengths for wide area networks are on the order of 1 to a few thousand bits. Local area networks usually have much longer maximum frame lengths, since the path usually consists of a single multiaccess link. Also, delay and congestion are typically less important there and long frames allow most messages to be sent as a single packet.

Fixed frame length When all frames (and thus all packets) are required to be the same length, message lengths are not necessarily an integer multiple of the packet length, and the last packet of a message will have to contain some extra bits, called fill, to bring it up to the required length. Distinguishing fill from data at the end of a packet is conceptually the same problem as determining the end of a frame for variable-length frames. One effect of the fill in the last packet is an additional loss of efficiency, especially if the fixed packet length is much longer than many messages. Problem 2.39 uses this effect to repeat the optimization of Eq. (2.44), and as expected, the resulting optimal packet length is somewhat smaller with fixed frame lengths than with variable frame lengths. A considerably more important practical effect comes from the need to achieve a small delay for stream-type traffic. As was pointed out earlier, 64 kbps voice traffic must use packets on the order or 500 bits or less, and this requirement, for fixed frame length, forces *all* packets to have such short lengths. This is the primary reason why ATM (see Section 2.10) uses 53 byte frames even though very much longer frames would be desirable for most of the other traffic.

2.6 STANDARD DLCs

There are a number of similar standards for data link control, namely HDLC, ADCCP, LAPB, and SDLC. These standards (like most standards in the network field) are universally known by their acronyms, and the words for which the acronyms stand are virtually as cryptic as the acronyms themselves. HDLC was developed by the International Standards Organization (ISO), ADCCP by the American National Standards Institute (ANSI), LAPB by the International Consultative Committee on Telegraphy and Telephony (CCITT), and SDLC by IBM. HDLC and ADCCP are virtually identical and are described here. They have a wide variety of different options and modes of operation. LAPB is the DLC layer for X.25, which is the primary standard for connecting an external site to a subnet; LAPB is, in essence, a restricted subset of the HDLC options and modes, as will be described later. Similarly, SDLC, which was the precursor of HDLC and ADCCP, is essentially another subset of options and modes.

HDLC and ADCCP are designed for a variety of link types, including both multi-access or point-to-point links and both full-duplex links (*i.e.*, both stations can transmit at once) or half-duplex links (*i.e.*, only one station can transmit at a time). There are three possible *modes* of operation, one of which is selected when the link is initialized for use.

The first mode, the *normal response mode* (NRM), is for a master–slave type of link use such as is typical between a computer and one or more peripherals (the word "normal" is of historical significance only). There is a primary or master station (the computer) and one or more secondary or slave stations. The secondary stations can send frames only in response to polling commands from the primary station. This mode is reasonable in multiaccess situations where the polling commands ensure that only one station transmits at a time (see Chapter 4 for a general discussion of multiaccess communication). NRM is also reasonable for half-duplex links or for communication with peripherals incapable of acting in other modes.

The second mode, the *asynchronous response mode* (ARM), is also a master–slave mode in which the secondary nodes are not so tightly restricted. It is not widely used and will not be discussed further.

The third mode, the *asynchronous balanced mode* (ABM), is for full-duplex point-to-point links between stations that take equal responsibility for link control. This is the mode of greatest interest for point-to-point links in data networks. It is also used, in slightly modified form, for multiaccess communication in local area networks. LAPB uses only this mode, whereas SDLC uses only the first two modes.

We first describe the frame structure common to all three modes and all four standards. Then the operation of the normal response mode (NRM) and the asynchronous balanced mode (ABM) are described in more detail. Although ABM is our major interest, the peculiarities of ABM will be understood more easily once NRM is understood.

Figure 2.38 shows the frame structure. The flag, 01^60, is as described in Section 2.5.2, with bit stuffing used within the frame (the flag is not considered as part of the frame proper). One or more flags must separate each pair of frames. The CRC uses the CRC-CCITT generator polynomial $g(D) = D^{16} + D^{12} + D^5 + 1$ as explained in Section 2.3.4; it checks the entire frame (not including the flag). One modification is that the first 16 bits of the frame are inverted in the calculation of the CRC (although not inverted in transmission). Also, the remainder is inverted to form the CRC. At the receiving DLC, the first 16 bits of the frame are similarly inverted for recalculation of the remainder, which is then inverted for comparison with the received CRC. One reason for this is to avoid having an all-zero string (which is common if the link or modem is malfunctioning) satisfy the CRC. Another reason is to avoid satisfying the CRC if a few

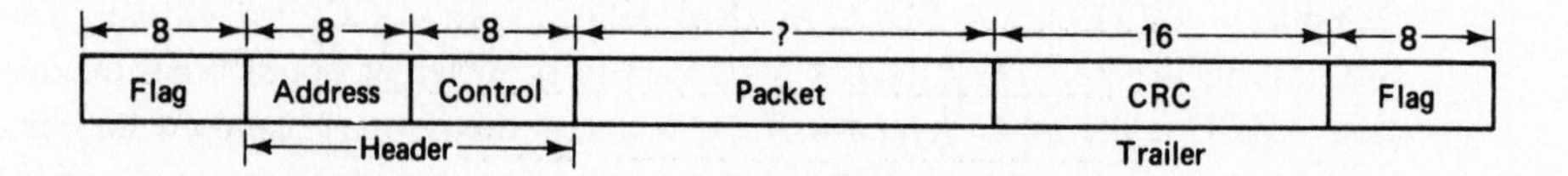

Figure 2.38 Frame structure of standard DLCs. The address, control, and CRC fields can each be optionally extended in length.

zeros are added or deleted from the beginning or end of the frame. As an option, the 32-bit CRC discussed in Section 2.3.4 can be used. This optional 32-bit CRC is used in the IEEE 802 standard for local area networks.

The address field normally consists of one byte (eight bits). For the normal response mode (NRM), the address is always that of the secondary station; that is, when the primary sends a frame to a secondary, it uses the secondary's address, and when the secondary responds, it uses its own address. Note that this address is not intended as a destination for a session, but simply to distinguish stations for multiaccess use. For point-to-point links, the address field has no natural function, although we shall see later that it has some rather peculiar uses in ABM. There is an option in which the address field is extendable to an arbitrary number of 1 or more bytes; the first bit in a byte is 1 if that byte is the last address byte, and otherwise it is 0.

The layout of the control field is shown in Fig. 2.39. There are three different frame formats: *information*, *supervisory*, and *unnumbered*. The information frames are the actual packet-carrying frames; they normally use go back n ARQ with a modulus of 8. As shown in the figure, the sequence and request numbers SN and RN are part of the control field. The supervisory frames are used to return ARQ information (*i.e.*, RN) either when there are no data packets to send or when a speedy ack or nak is needed. Finally, the unnumbered frames are used for the initiation and termination of a link and for sending various kinds of supplementary control information.

There is an option under which the control field is two bytes long and in which SN and RN are each seven bits long and each use a modulus of 128. Aside from this lengthening of SN and RN, this option is essentially the same as the normal single-byte control field, which is assumed in the following description.

The first bit of the control field distinguishes the information format from the supervisory and unnumbered formats. Similarly, for noninformation formats, the second bit distinguishes supervisory from unnumbered formats. The fifth bit of the control field is the *poll final* (P/F) bit. For the normal response mode, the primary sends a 1 in this bit position to poll the secondary. The secondary must then respond with one or more frames, and sends a 1 in this bit position to denote the final frame of the response. This polling and response function of the P/F bit in NRM is used for supervisory and unnumbered frames as well as information frames. The P/F bit is used in a rather strange way for the asynchronous balanced mode (ABM), as described later.

There are four types of supervisory frames: *receive-ready* (RR), *receive-not-ready* (RNR), *reject* (REJ), and *selective-reject* (SREJ). The type is encoded in the third and

1	2	3	4	5	6	7	8	
0	SN			P/F	RN			Information
1	0	type		P/F	RN			Supervisory
1	1	type		P/F	type			Unnumbered

Figure 2.39 Structure of the control field for the information, supervisory, and unnumbered frame formats.

fourth bits of the control field. All of these have the basic function of returning ARQ information (*i.e.*, all packets with numbers less than RN are acknowledged) and consequently they contain neither a sequence number nor an information field. RR is the normal response if there is no data packet on which to piggyback RN. RNR means in addition that the station is temporarily unable to accept further packets; this allows a rudimentary form of link flow control—that is, if the buffers at one end of the link are filling up with received packets, the other end of the link can be inhibited from further transmission.

The REJ supervisory frame is sent to indicate either a received frame in error or a received frame with other than the awaited sequence number. Its purpose is to improve the efficiency of go back n ARQ, as explained in Section 2.4.2. Similarly, SREJ provides a primitive form of selective repeat. It acks all packets before RN and requests retransmission of RN itself followed by new packets; one cannot provide supplementary information about multiple required retransmissions. This facility can be useful, however, on satellite links, where there is a very small probability of more than one frame error in a round-trip delay. The implementation of REJ and SREJ is optional, whereas RR and RNR are required.

The unnumbered frames carry no sequence numbers and are used for initializing the link for communication, disconnecting the link, and special control purposes. There are five bits in the control field available for distinguishing different types of unnumbered frames, but not all are used. Six of these types will be referred to generically as *set mode* commands. There is one such command for each mode (NRM, ARM, and ABM), and one for each mode using the extended control field. This command is sent by a primary (for NRM or ARM) and by either node (for ABM) to initialize communication on a link. Upon sending or receiving this command, a station sets its sequence and request numbers to zero. The recipient of the command must acknowledge it with an unnumbered frame, either an *unnumbered ack* which agrees to start using the link, or a *disconnected mode* which refuses to start using the link. Similarly, there is a disconnect command which is used to stop using the link; this also requires an unnumbered ack in response.

Figure 2.40 shows the operation of the NRM mode for a primary and two secondaries. Note that the primary initiates communication with each separately and disconnects from each separately. Note also that the P/F bit is set to 1 on each unnumbered frame that requires a response. After a given unnumbered frame with $P = 1$ is sent to a particular secondary, no new unnumbered frame with $P = 1$ may be sent to that secondary until the given frame is acknowledged by a frame with $F = 1$; naturally the given frame may be repeated after a time-out waiting for the ack. This alternation of unnumbered frames with acknowledgments is similar to the stop-and-wait protocol. Section 2.7 explains this similarity carefully and shows that the acknowledgment strategy here contains the same defect as stop-and-wait strategies without numbered acknowledgments.

In the asynchronous balanced mode, the decision was made, in designing these protocols, to have each station act as both a primary and a secondary (thus replacing a simple problem with a more familiar but more difficult problem). When a station sends a primary frame (in its role as a primary station), it uses the other station's address in the address field, and the P/F bit is interpreted as a poll bit. When a station sends

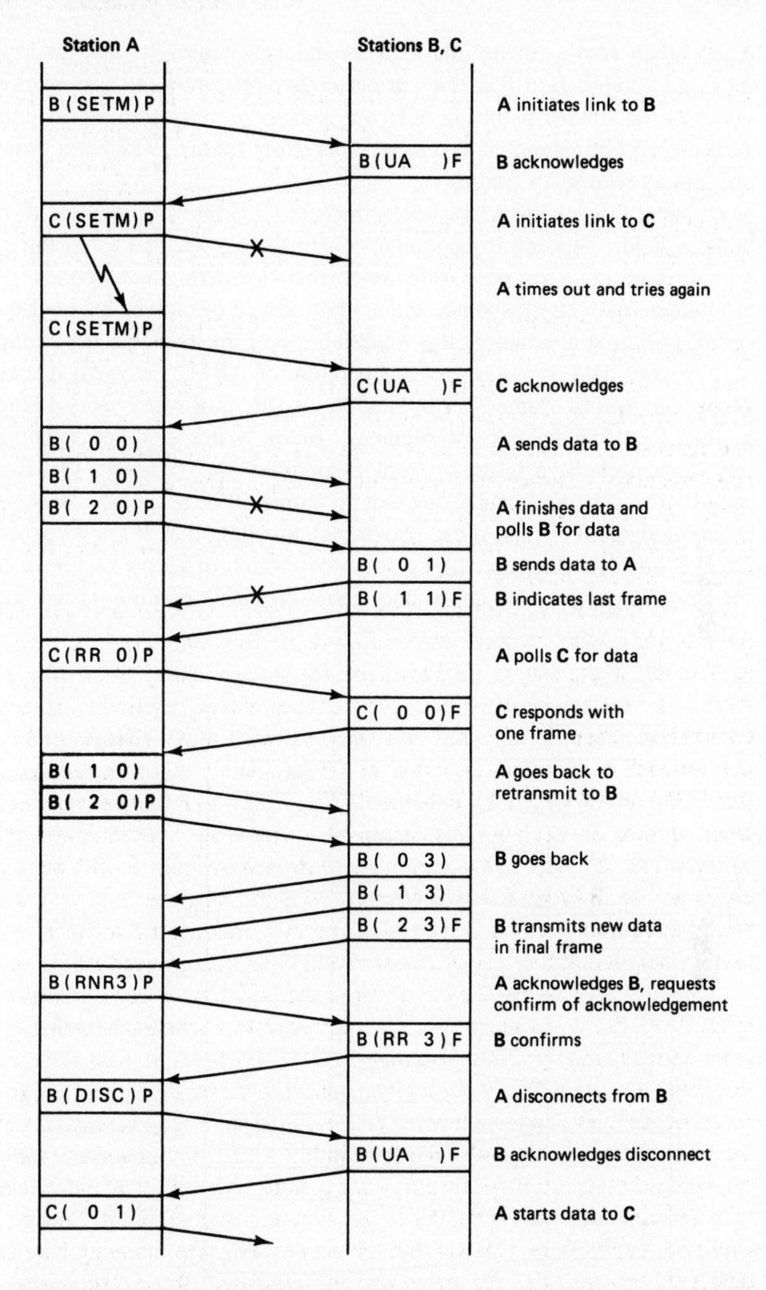

Figure 2.40 Normal response mode (NRM) operation of primary A with two secondaries, B and C. SETM refers to the command setting, the NRM mode with $SN = 0$ and $RN = 0$ at both stations. The address of each frame is given first, followed by (SN,RN) for information frames, (type,RN) for supervisory frames, and (type) for unnumbered frames; this is followed by a P/F bit if set to 1. Note that A uses different SN and RN values for B and C.

a secondary frame (in its role as a secondary), it uses its own address, and the P/F bit is interpreted as a final bit. In customary operation, all information frames and all unnumbered frames requiring acks are sent as primary frames; all unnumbered frames carrying acks or responses, and most supervisory frames (which ack primary data frames), are sent as secondary frames.

What this means in practice is that the P/F bit is either P or F, depending on the address field. This bit, in conjunction with the address field, has three possible values: $P = 1$, requiring a response from the other station acting in a secondary role; $F = 1$, providing the secondary response, and neither. Given the somewhat inadequate performance of this poll/final alternation, this modified use of the P/F bit was an unfortunate choice.

Figure 2.41 shows a typical operation of ABM. The most interesting part of this is the next-to-last frame sent by station A; this is a supervisory frame sent with $P = 1$, thus requiring an acknowledgment. Since A has signified unwillingness to accept any more packets, a response from B signifies that B knows that packets up to three inclusive have been accepted and that no more will be accepted. Thus, upon receiving the acknowledgment of this, A can disconnect, knowing that B knows exactly which packets from B have been accepted; A also knows which of its packets have been accepted, so the disconnect leaves no uncertainty about the final condition of any packets at either A

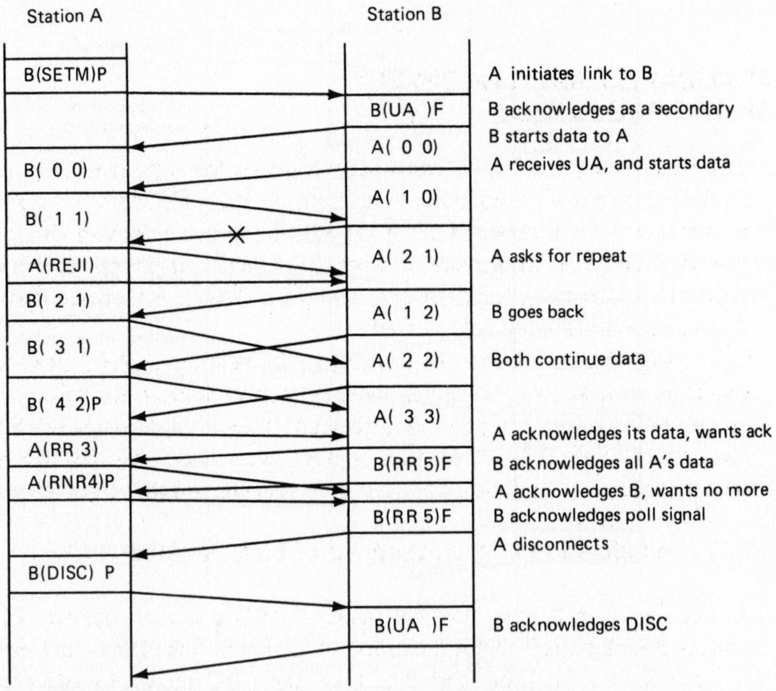

Figure 2.41 Asynchronous balanced mode (ABM) of operation. SETM refers to the set ABM command, which initiates the link with SN and RN at both sides equal to 0. Note the use of addresses in different types of frames.

or B. Note that this "ideal" disconnect requires some previous agreement (not specified by the standard) between A and B; in particular, A must send RNR, with $P = 1$ and get an ack before disconnecting—this lets B know that A did not go back to accepting packets again temporarily before the disconnect. This "ideal" disconnect does not work, of course, if the line is being disconnected because of a malfunction; in this case, the idea of an orderly sequence of RNR, ack, disconnect, and ack is unrealistic.

Another type of unnumbered frame is "frame reject." This is sent in response to a received frame with a valid CRC but otherwise impossible value (such as exceeding the maximum frame length, being shorter than the minimum frame length, having an RN value not between $SN_{\min}$ and $SN_{\max} - 1$, or having an impossible control field). Thus, either an undetectable error has occurred or one of the stations has malfunctioned. Such an error is outside the realm of what the ARQ is designed for, and it should be reported at each station. The usual response to a "frame reject" is for the opposite station either to reinitialize the link with a set mode command or send an unnumbered "reset" command. The first resets RN and SN at each station to 0, and the second sets SN to 0 at the station that sent the faulty frame and RN to 0 at the receiving station. Both stations will again start to send packets (starting from the first packet not acked), but because of the resetting, some packets might arrive more than once and out of order. This is not a fault of the protocol, but a recognition that node malfunctions and undetectable errors cannot be dealt with.

2.7 INITIALIZATION AND DISCONNECT FOR ARQ PROTOCOLS

The discussion of ARQ in Section 2.4 assumed throughout that the protocol was correctly initialized with no frames traveling on the link and with sequence and request numbers equal to 0 at both ends of the link. Initialization is a general problem for protocols involving communication between two nodes, and arises, for example, in transport protocols and many special-purpose protocols. We shall soon see that initialization and disconnection are intimately related.

One might think that the problem of initializing an ARQ protocol is trivial—one might imagine simply initializing the DLC at each node when the link is first physically connected and then using the ARQ protocol forever. Unfortunately, if the link fails or if one or both nodes fail, initialization is more complicated. We first discuss the problem in the presence of link failures alone and then consider the case in which nodes can fail also.

2.7.1 Initialization in the Presence of Link Failures

It was shown in Section 2.4 that standard ARQ protocols operate correctly (assuming correct initialization) in the presence of arbitrary link delays and errors. Thus, even if the link fails for some period of time, the ARQ protocol could continue attempting packet transmission, and those packets not received correctly before the failure would be received after the failure. The difficulty with this point of view is that when a link fails for a protracted period of time, it is necessary for the transport and/or network layers to take over the problem of reliable end-to-end delivery, using alternative paths for those packets that were not delivered over the failed link. When the failed link eventually returns to operation, the higher layers set up new paths through the link, thus presenting the link with a completely new stream of packets.

We see then that the operation of a link protocol over time is properly viewed as being segmented into an alternation of up and down periods. The DLC must be properly initialized at the beginning of each up period, whereas at the end of an up period, there may be packets still in the process of communication that have been accepted into the DLC for transmission at one node but not yet received and released at the opposite node. Stated more mathematically, correct operation of a DLC means that in each up period, the string of packets released to the higher layer at one node must be a prefix of the string of packets accepted for transmission from the higher layer at the opposite node.

2.7.2 Master–Slave Protocol for Link Initialization

The careful reader will have observed that there is a problem in defining an up period carefully; in particular, the nodes at opposite ends of a link might not agree at any given instant about whether the link is up or down. Rather than resolving this issue in the abstract, we present a simple protocol for initializing and disconnecting the link. The protocol is a master–slave protocol where one node, say A, is in charge of determining when the link should be up or down. Node A then informs node B at the other end of the link whenever it determines that the link has changed from down to up or up to down.

The sequence of messages (for this protocol) that go from A to B is then simply an alternating sequence of initialize and disconnect messages (see Fig. 2.42). In this discussion we ignore the data messages that are sent during up periods and simply focus on the messages of the initialization protocol. We view these messages as using a stop-and-wait protocol, using modulo 2 sequence numbering. Thus, the initialize messages (denoted INIT) correspond to frames with $SN = 1$, and the disconnect messages (denoted DISC) correspond to frames with $SN = 0$. The acknowledgments sent by node B will be denoted ACKI and ACKD. In terms of stop-and-wait ARQ, ACKI corresponds to $RN = 0$ [*i.e.*, the INIT message ($SN = 1$) has been correctly received and DISC ($SN = 0$) can be sent when desired by A].

When node A sends INIT, it waits for the acknowledgment ACKI and continues to send INIT on successive time-outs until receiving ACKI. Node A is then initialized (as described below) and can send and receive data packets until it decides to disconnect, at which time it sends DISC (repeatedly if need be) until receiving ACKD. The decision to disconnect (after A is initialized and up) corresponds to the acceptance of a new packet from the higher layer in ordinary stop and wait. The decision to initialize (after A has received ACKD and is down) has the same interpretation. Note that these decisions to try to initialize or disconnect are not part of the initialization protocol, but rather should be viewed either as coming from a higher layer or as coming from measurements of the link. The decisions to actually send the messages INIT or DISC are part of the protocol, however, and rely on having already received an acknowledgment for the previous message.

From our analysis of the stop and wait protocol, we know that each message from A is accepted once and only once in order. There is also a strict alternation of events between A and B (*i.e.*, B accepts INIT, A accepts ACKI, B accepts DISC, A accepts ACKD, etc.). Now assume that A regards the link as up from the time that it receives ACKI from B until it sends DISC to B (see Fig. 2.42). Similarly, B regards the link as up from the time that it receives INIT from A until it receives DISC from A. Each node sends and accepts data packets only when it regards the link as up. Assuming that frames stay in order on the links, we see that the link from A to B must be free of data when a DISC message reaches B, and it must stay free until the next up period starts at A. Similarly, the link from B to A must be free of data when ACKD reaches A and must stay free until the next up period starts at B (see Fig. 2.43). It follows that the link is free of data in both directions from the time when an ACKD message reaches A until the next INIT message reaches B. Assuming that A initializes its ARQ protocol on sending the INIT message and that B initializes on receiving the INIT message, we see that the ARQ is correctly initialized.

A peculiar feature of this protocol is that once node A starts to initialize, there is no way to abort the initialization before receiving ACKI from node B (see Problem 2.40). Naturally, node A can start to disconnect as soon as ACKI is received, thus sending no data, but A must continue to send INIT until receiving ACKI. The protocol could be complicated to include a message to abort an attempted initialization, but the example described in Fig. 2.44 should make the reader wary of ad hoc additions to protocols. A similar issue is that node B might wish to refuse to initialize (which would be handled by the "disconnected mode" message in HDLC). The cleanest way to handle this is for node B to acknowledge A's INIT message, but to use an extra control field to request A to disconnect immediately without sending data.

The initialization protocol used in the normal response mode of HDLC (see Fig. 2.40) behaves in almost the same way as the initialization protocol above. The SETM unnumbered frame corresponds to INIT, the DISC unnumbered frame corresponds to

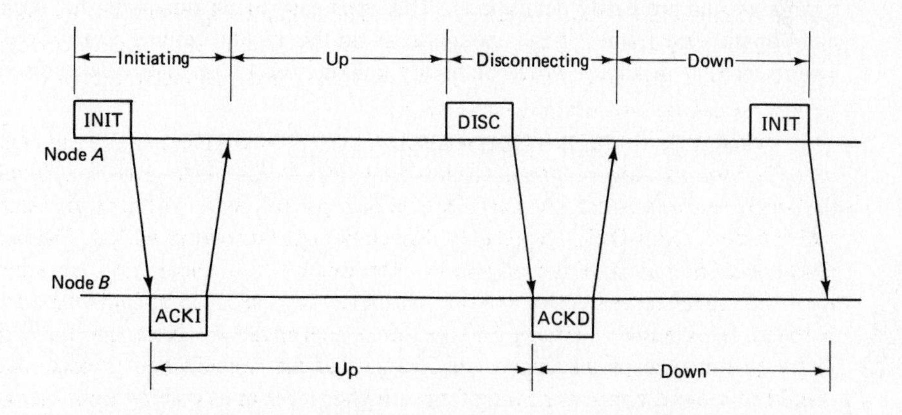

Figure 2.42 Initialization and disconnection protocol. Data are transmitted only in up periods and the ARQ is initialized at the beginning of an up period. Note that node A cycles between periods of initiating, being up, disconnecting, and being down.

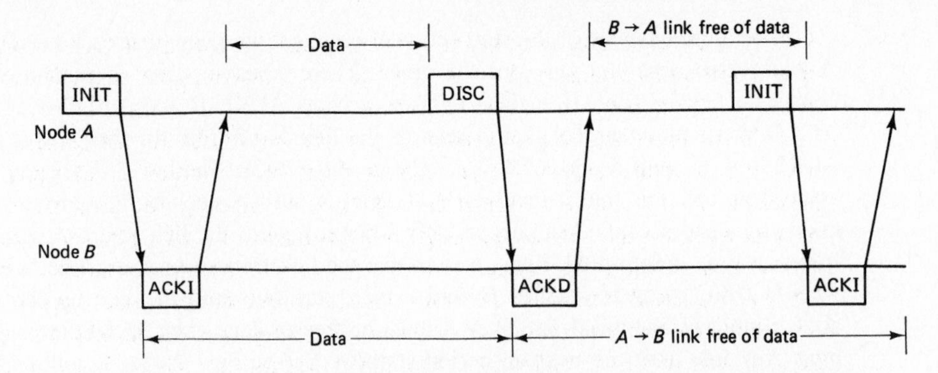

Figure 2.43 Illustration of periods when link is free of data in each direction. Note that the link is free in both directions both when A sends INIT and when B receives it.

DISC, and the ACK unnumbered frame corresponds to both ACKI and ACKD; that is, the unnumbered acks in HDLC fail to indicate whether they are acking initialize or disconnect frames. Figure 2.44 illustrates how this can cause incorrect operation. Node A times-out twice trying to initialize a link between A and B. Finally, on receiving an ack, it initializes the ARQ and sends data packet D with $SN = 0$. After timing out waiting for an acknowledgment, node A decides that the link is not operational, sends a disconnect, and the higher layers send D over some other path. An ack from one of the earlier SETM messages then arrives at A and is viewed as an ack to DISC, so A tries to initialize the link again. The third old ack then arrives, A sends a new data packet D', again initialized with $SN = 0$; D' is lost on the link, but an ack for the old packet D arrives; packet D' then never reaches node B and the ARQ system has failed.

This conceptual error in HDLC does not seem to have caused many problems over the years. The reason for this appears to be that in most networks, link delays are not arbitrary and are easily determined. Thus it is easy to set time-outs (for retransmission of unnumbered frames) large enough to avoid this type of failure. Even if the time-outs were set very poorly, it would be highly unlikely for a node to decide to disconnect and

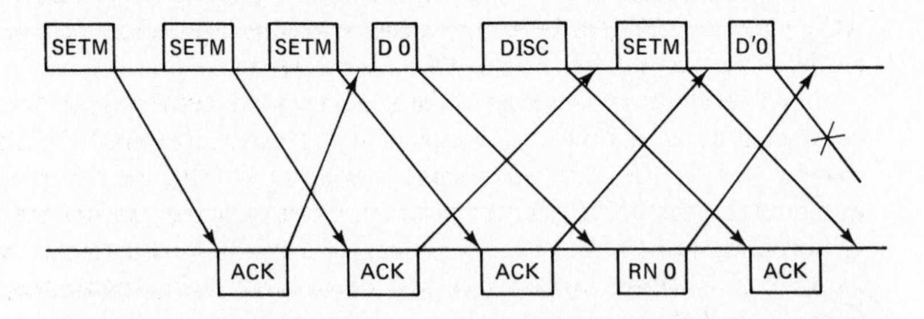

Figure 2.44 Example of HDLC in which node A attempts to initialize the link three times, due to delayed acknowledgments from B. The delayed acks are falsely interpreted as being acks for a subsequent disconnect and initialization. This in turn causes a failure in the data packets.

then to try to reinitialize over a time span comparable to the link delay. Despite the lack of practical problems caused by this error, however, this is a good example of the danger of designing protocols in an ad hoc fashion.

In the initialization and disconnect protocol just discussed, the process of disconnecting is a critical part of the protocol, and can be viewed as the first part of a new initialization of the link. In fact, if the link fails, the message DISC cannot be sent over the link until the link has recovered; it is this message, however, that guarantees that the next INIT message is not an earlier copy of an earlier attempt to initialize. As seen in Section 2.6, initialization and disconnection are often used on a multidrop line to set up and terminate communication with a given secondary node in the absence of any link failures, and in these applications, the DISC message is really a signal to disconnect rather than part of the initialization process.

Some readers may have noticed that the stop-and-wait protocol that we are using for initialization itself requires initialization. This is true, but this protocol does not require reinitialization on each link failure; it requires initialization only once when the link is physically set up between the two nodes. Recall that the ARQ protocol requires reinitialization on each link failure because of the need to break the packet stream between successive up periods on the link.

2.7.3 A Balanced Protocol for Link Initialization

It is often preferable to have an initialization protocol that is the same at both ends of the link. At the very least, there is a conceptual simplification in not having to decide which node is master and ensuring that each node uses the appropriate protocol. The essential idea of a balanced initialization protocol is to use two master–slave protocols, with node A playing the master for one and node B playing the master for the other. The only new requirement is to synchronize the two protocols so that they have corresponding up and down periods.

To simplify this synchronization somewhat, we assume that each INIT or DISC message from a node (acting as master) also contains a piggybacked ACKI or ACKD for the master–slave protocol in the opposite direction. Thus a single received message can affect both master–slave protocols, and we regard the ACK as being acted on first. ACK messages can also be sent as stand alone messages when there is no need to send an INIT or DISC but an INIT or DISC is received.

We recall that in the master–slave initialization protocol from A to B, the link was considered up at A from the receipt of ACKI until an external decision to disconnect (see Fig. 2.42). The link was regarded as up at B from the receipt of INIT until the receipt of the next DISC. For the balanced protocol, a node regards the link as *up* if it is up according to both the $A \rightarrow B$ protocol and the $B \rightarrow A$ protocol. A node initializes its ARQ at the beginning of an up period and sends and acknowledges data during this period. Similarly, a node regards the link as down if it is down according to both the $A \rightarrow B$ and $B \rightarrow A$ protocols.

If the link is regarded as up at node A (according to both master–slave protocols), either an external decision at A or a DISC message from B causes A to leave the up period and start to disconnect by sending a DISC message. As a special case of this, if a DISC message from B carries a piggybacked ACKI, the ACKI can cause node A to regard the link as up, in which case the DISC message immediately ends the up period (see Problem 2.40). Similarly, if the link is regarded as down at node A, either an external decision or an INIT message from B causes A to leave the down period and start to initialize by sending an INIT message. Again, if an INIT message from B is accompanied by a piggybacked ACKD, the ACKD can cause A to regard the link as down, and the INIT message immediately ends the down period.

Figure 2.45 illustrates the operation of this balanced protocol. Each of the master–slave protocols operate as described in the preceding section, and in particular, each node, in its role as slave, acks each INIT and DISC message from the other node; also each node, in its role as master, continues to retransmit INIT and DISC messages until acknowledged. The synchronizing rules of the preceding paragraph simply limit when external decisions can be made about starting to initialize and disconnect. For this reason, the safety of the balanced protocol follows immediately from the correctness of the master–slave protocols. The safety of the protocol can be verified, using the same assumptions as in Section 2.4, simply by verifying that when one node starts to initialize (or disconnect), the other node will eventually start to do the same (if it did not start earlier), and both nodes will eventually reach the up state (or the down state in the case of disconnect). See Problem 2.40 for details.

This protocol has the same peculiar problems as the master-slave protocol in that a node might be sending INIT messages with no intention of transferring data; this could occur either because of a link failure while in the process of initializing or because of sending an INIT message in response to the opposite node sending an INIT message. These problems are best treated by using extra fields for control information and recognizing that the protocol here is primarily designed for synchronizing the two nodes.

The initialization part of this balanced protocol (*i.e.*, sending INIT, waiting for both ACKI and INIT from the other side, and responding to INIT with ACKI) is often

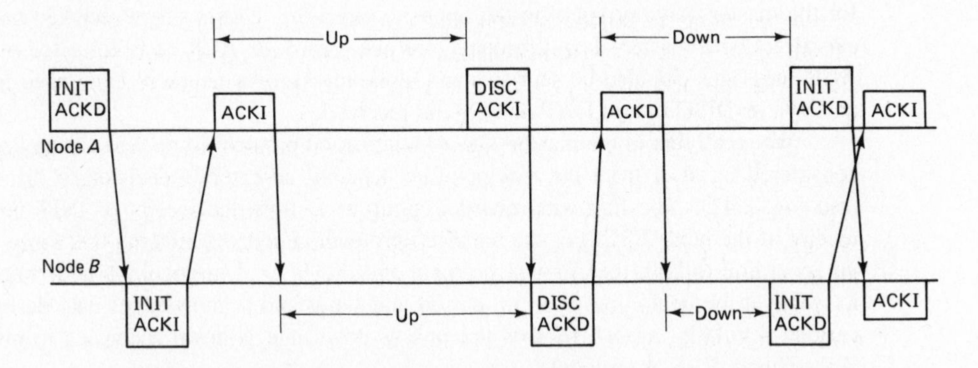

Figure 2.45 Example of balanced initialization and disconnect protocol. Either node can start to initialize from the down state or start to disconnect from the up state and the other node will follow.

called a three-way handshake. The reason for this terminology is that the ACKI from the second node is usually piggybacked on the INIT from the second node, making a three-message interchange. It is always possible that both sides start to initialize independently, however, and then all four messages must be sent separately (as seen on the right-hand side of Fig. 2.45). We shall see a closely related type of three-way handshake in transport protocols, but the problem is more complex there because of the possibility of packets traveling out of order.

2.7.4 Link Initialization in the Presence of Node Failures

From the viewpoint of a network, a node failure is similar to (but, as we shall see, not the same as) the failures of all the adjacent links to that node; in both cases, no communication is possible passing through that node or going to or from that node. For our purposes, a node is modeled as either working correctly or being in a failure state; in the failure state, no input is accepted, no output is produced, and no operations occur (except for a possible loss of state information). We view each node as alternating in time between correct operation and the failure state (typically, with very long periods between failures). Thus the question of interest here is that of reinitialization after a node failure. Another interesting class of problems, which we do not address here, is that of intermittent faulty operation, or even malicious operation, at a node. This is really a network layer problem and is treated in [Per88].

If a node maintains its state information during a failure period, we view the data at the node as being lost or retransmitted over other paths. Thus when the node returns to correct operation, the adjacent links must be reinitialized and the problem is really the same as if all the adjacent links had failed. If a node loses its state information during a failure period, the problem of reinitialization is much more difficult than that of adjacent link failures.

To illustrate the problem of initialization after loss of state information, consider the master–slave initialization protocol of Section 2.7.2 and suppose that node A, the master node, fails and recovers several times in rapid succession. Assume also that when A recovers, the initialization protocol starts in the down state (*i.e.*, in the state for which ACKD was the last received message); Problem 2.41 looks at the opposite case where it starts in the up state. Each time node A recovers, it receives a new sequence of packets to transmit to node B after initializing the (A, B) link.

Figure 2.46 illustrates what can happen if the time between failure and recovery is on the order of the round-trip delay. Node B has no way of knowing that A has failed (other than the messages of the initialization protocol, which are supposed to initialize the link after a failure). In the figure, A sends just the INIT message and then fails. When it recovers, it sends INIT again, and receives the ack from the previous INIT; this completes the initialization and allows A to send data packet D with sequence number 0. Node A then fails again, and on recovery gets the responses from B from the previous period when A was functioning.

The failure to initialize correctly here is not due to the particular protocol and is not due to the particular starting state. In fact, incorrect operation can occur for any

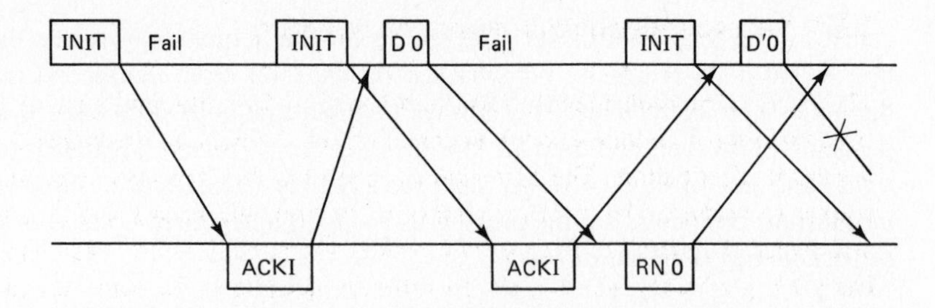

Figure 2.46 Example of a sequence of node failures, with loss of state, that causes a failure in correct initialization. Each time the node recovers from a failure, it assumes that ACKD was the last protocol message received, thus allowing acknowledgments from earlier initialization periods to get confused with later periods.

protocol and any starting state, assuming that there is no upper bound on link delay (see [Spi89] and [LMF88]). Problem 2.41 illustrates this phenomenon.

There are several solutions to the problem of initialization after a node failure. The first is to provide nodes with a small amount of nonvolatile storage. For example, the initialization protocols in this section can operate correctly with only a single bit of nonvolatile storage per adjacent link. If the links have a maximum propagation delay, another solution is simply to use sufficiently long time-outs in the initialization protocol to avoid the problems above (see [Spi89]). Yet another type of solution, which we explore later in terms of transport protocols, is to use a pseudo-random-number generator to avoid confusion between different periods of correct operation. We shall see that such protocols are very robust, although they have the disadvantage that there is a small probability of incorrect operation.

2.8 POINT-TO-POINT PROTOCOLS AT THE NETWORK LAYER

The major conceptual issues at the network layer, namely routing and flow control, involve algorithms that are distributed between all the nodes of the network; these issues are treated in Chapters 5 and 6, respectively. There are a number of smaller issues, however, that simply involve interactions between a pair of nodes. One of these issues is the transfer of packets between adjacent nodes or sites. At the network layer, this involves being able to distinguish packets of one session from another and being able to distinguish different packets within the same session. These issues are of particular concern between a source–destination site and the adjacent subnet node because of the need for standardization between users of the network and the subnet.

In the following subsections, we first treat the problem of addressing and session identification from a general viewpoint and give several examples to illustrate the range of possibilities. We then discuss packet numbering with a brief description of how this relates to flow control and to error recovery at the network layer. Finally, we discuss the X.25 network layer standard.

2.8.1 Session Identification and Addressing

The issue of session identification is relevant at the network layer. A packet received by a node must contain enough information for the node to determine how to forward it toward its destination. The brute-force approach to this problem is for the header of each packet to contain identification numbers for both the source and destination sites and additional identification numbers to identify the session within each site. This approach has great generality, since it allows different packets of the same session to travel over different paths in the subnet; it also requires considerable overhead.

If the network uses virtual circuits, far less overhead is required in the packet headers. With virtual circuits, each (one way) session using the network at a given time has a given path through the network, and thus there is a certain set of sessions using each link. It is helpful to visualize each link as being shared by a set of "virtual channels," distinguished by numbers. When a new session is set up, a path is established by assigning, on each link of the path, one unused virtual channel to that session. Each node then maintains a table mapping each busy incoming virtual channel on each link onto the corresponding outgoing virtual channel and link for the corresponding session (see Fig. 2.47). For example, if a given session has a path from node 3 to 5 to 8, using virtual channel 13 on link (3,5) and virtual channel 7 on link (5,8), then node 5 would map [link (3,5), virtual channel 13] to [link (5,8), virtual channel 7], and node 8 would map [link (5,8), virtual channel 7] to the access link to the appropriate destination site with a virtual channel number for that link. The nice thing about these virtual channels is that each link has its own set of virtual channels, so that no coordination is required between the links with respect to numbering.

With this virtual channel strategy, the only thing that must be transmitted in the packet header on a link is an encoding of the virtual channel number. The simplest way

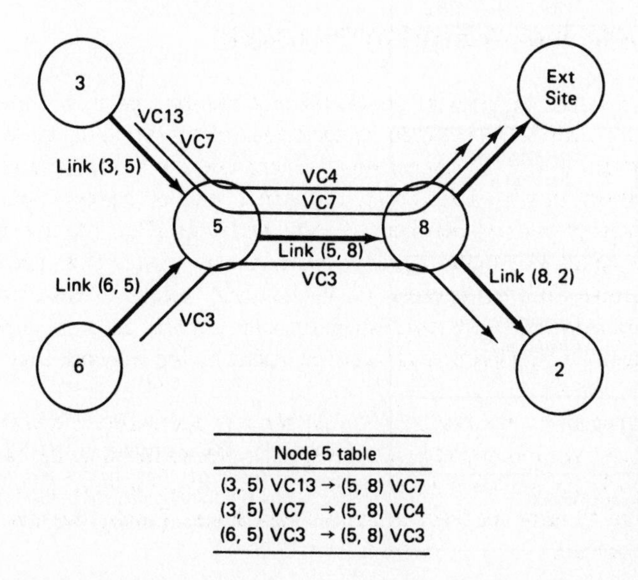

Node 5 table
(3, 5) VC13 → (5, 8) VC7
(3, 5) VC7 → (5, 8) VC4
(6, 5) VC3 → (5, 8) VC3

Figure 2.47 Use of virtual channel (VC) numbers to establish virtual circuits in a network. Each packet carries its VC number on each link and the node changes that number (by a table representing the virtual circuit) for the next link on the path.

to do this, of course, is to represent the virtual channel number as a binary number in a fixed field of the header. Because of the limited capacity of the link, there is some natural upper bound on the number of sessions a link can handle, so there is no problem representing the virtual channel number by a fixed number of binary digits.

Later, other ways of representing virtual channel numbers will be discussed, but for now note that any method of distinguishing sessions must do something equivalent to representing virtual channel numbers. In other words, a given link carries data for a number of different sessions, and the transmitted bits themselves must distinguish which data belong to which session. The packet header might supply a great deal of additional information, such as source and destination identities, but the bare minimum required is to distinguish the sessions of a given link. We now look at how this representation is handled in TYMNET.

Session identification in TYMNET TYMNET is a network developed in 1970 primarily to provide access for a time-sharing computer service, but expanded rapidly to allow users to communicate with their own computers. Since the terminals of that period usually transmitted only one character at a time and usually waited for an echo of that character to come back before transmitting the next character, the protocols were designed to work well with very short messages consisting of one or several characters. Although the network never received much academic publicity, many of the ideas first implemented there are now standard network practice. The frames are very short, with a maximum length of 66 bytes; the frames have the format shown in Fig. 2.48.

The frame header of 16 bits is broken into five bits for a synchronization string, five bits for a word count, three bits for *SN*, and three bits for *RN*. The word count gives the number of 16 bit words in the frame, not counting the header and trailer. Since

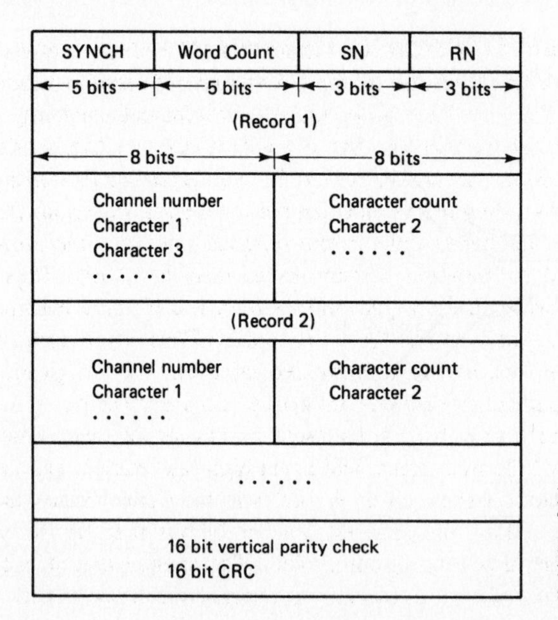

Figure 2.48 Frame structure in TYMNET. The first 16 bits is a frame header, and the last 32 bits is a frame trailer. Note that data from multiple sessions are included in a single frame.

the nodes were implemented with 16 bit minicomputers, all the fields of the frame are organized in 16 bit (two byte) increments. Note that the 5 bit word count can represent up to 31 words of two bytes each, or 62 bytes. Thus, restricting the frames to be multiples of 16 bits serves to reduce (slightly) the overhead required for framing. The SN and RN fields provide conventional go back n ARQ with a modulus of $m = 8$. The last four bytes (the frame trailer) provide error detection, using both a 16 bit CRC and 16 vertical parity checks. Thus, somewhat more error detection is provided than with the standard 16 bit CRC, but slightly less than with a 32 bit CRC.

Each record within a frame corresponds to data from a different session. A record starts with a byte giving the virtual channel number, which identifies the session. The next byte tells how many characters of data are contained in the record. If the number is odd, the last byte of the record is just a fill character, so that the next record can start on a word boundary. A record here can be thought of as a very short packet, so that the channel number and character count correspond to the packet header. Thus, the packets are so short that a number of packets are normally grouped into one frame so as to avoid excessive frame overhead. These packets are somewhat different from the usual concept of a packet in that they do not necessarily contain the same number of characters from one link to the next on their path.

This type of strategy is usually referred to as *statistical multiplexing*. This means that the link is multiplexed between the virtual channels, but multiplexed on a demand basis. Conventional packet switching, in which there is just one packet per frame, is another form of statistical multiplexing, but conventionally, one refers to that as packet switching rather than statistical multiplexing. When a frame is assembled for transmission at a node, the sessions are served in round-robin order, taking some number of characters from each session that has characters waiting.

Session identification in the Codex networks The Codex networks use quite a similar statistical multiplexing strategy, again with multiple sessions transmitting characters within the same frame. Here, however, a more efficient encoding is used to represent the virtual channel number and the amount of data for that session. In assembling a frame, the sessions are again served in sequential order, starting with virtual channel 1. Instead of sending the virtual channel number, however, an encoding is sent of the gap (*i.e.*, the number of idle virtual channels) between the number of the session being served and the previous session served (see Fig. 2.49). Thus, when all sessions have characters to send, these gaps are all zero, and if many but not all of the sessions have characters, most of the gaps are small. Thus, by encoding small gaps into a small number of bits, and large gaps into more bits, very high efficiency is maintained when large numbers of sessions are active. On the other hand, if only a small number of sessions are active, but have a large amount of data each, then only a small number of records will be in a frame, and again high efficiency is maintained.

Because of implementation issues, all of the fields in the Codex frames are multiples of four bit "nibbles." Thus, the smallest number of bits that can be used to represent a gap for identifying a session is one nibble. The method that identifies the number of characters in a record uses a special start record character, 0000, to start

Unary-binary gap encoding:
$0 \rightarrow 10$
$1 \rightarrow 11$
$2 \rightarrow 010$
$3 \rightarrow 011$
$4 \rightarrow 0010$
......

Start record character: 0000

Sample encoding of data:
$(0000)(10)(x_1)(x_2)(x_3)(x_4)(0000)(10)(x_1)(x_2)$
$(0000)(11)(x_1)(0000)(0010)(x_1)(x_2)(x_3)\ldots$

Figure 2.49 Efficient encoding for virtual channels and lengths of records. In the example shown, there are four characters x_1, x_2, x_3, x_4 for VC1, two characters for VC2, one for VC4, and three for VC9. Each record starts with 0000 to indicate the start of new record (where 0000 cannot occur in data characters); this is followed by the encoding of the gap between virtual channels with data. The Codex protocols are similar, but gaps are encoded into multiples of four bits. Parentheses are for visual clarity.

each new record. The characters for each session first go through a data compressor that maps highly probable characters into one nibble, less likely characters into two nibbles, and least likely characters into three nibbles. This is done adaptively, thus automatically using the current character statistics for the session. The 0000 nibble is reserved to indicate the start of a new record. Thus, the "packet header" consists of first a nibble to indicate a new packet (record), and next one or more nibbles representing the new session number. Under heavy loading, the packet header is typically just one byte long.

This strategy of encoding the difference between successive virtual channel numbers can also be used in conventional packet switched networks. There are two advantages in this. First, the packet overhead is reduced under heavy loading, and second, serving the sessions on a link in round-robin order tends to promote fairness in the service that each session receives.

2.8.2 Packet Numbering, Window Flow Control, and Error Recovery

In datagram networks, packets might be dropped and might get out of order relative to other packets in the same session. Thus, to achieve reliable communication, it is necessary to provide some way to identify successive packets in a session. The most natural and usual way of doing this is to number successive packets modulo 2^k for some k (*i.e.*, to provide a k-bit sequence number). This sequence number is placed in the packet header if the network layer modules at the source and destination have the responsibility for reordering the packets and retransmitting dropped packets. In many networks, for reasons we discuss later, the transport layer takes the responsibility for reordering and retransmission, and in that case the sequence number above is placed in the transport header. Sometimes sequence numbers are used at the network layer, at the internet sublayer, and at the transport layer.

For virtual circuit networks, there is also a need to number the packets in a session, but the reason is less obvious than for datagram networks. In our discussion of data link control, we saw a number of ways in which packets might get lost or arrive at the destination containing errors. First, a sequence of channel errors might (with very low probability) lead to an incorrect frame that still satisfies the CRC at the DLC layer. Second, errors made while data packets pass through a node are not checked by the CRC at the DLC layer. Third, if a link on the session path fails, there is uncertainty about how many packets were successfully transmitted before the failure. Finally, if a node fails, the packets stored at that node are typically lost. The error events described above are presumably quite rare, but for some sessions, correct transmission is important enough that error recovery at the network or transport layer is necessary.

Error recovery From a conceptual standpoint, error recovery at either the network or transport layer is quite similar to ARQ at the DLC layer. Each packet within a given session is numbered modulo $m = 2^k$ at the source site, and this sequence number SN is sent in the packet header at the network or transport layer. The destination site receives these packets and at various intervals sends acknowledgment packets back to the source site containing a request number RN equal (mod m) to the lowest-numbered yet-unreceived packet for the session. The receiver can operate either on a go back n basis or on a selective repeat basis.

It is important to understand both the similarity and the difference between the problem of end-to-end error recovery and that of error detection and retransmission at the DLC layer. End-to-end error recovery involves two sites and a subnet in between, whereas at the DLC layer, there are two nodes with a link in between. The sequence numbering for end-to-end error recovery involves only the packets (or bytes, messages, etc.) of a given session, whereas at the DLC layer, all packets using the link are sequentially numbered. At the DLC layer, we assumed that frames could get lost on the link, and similarly here we are allowing for the possibility of packets being thrown away or lost. At the DLC layer, we assumed variable delay for frame transmission, and here there is certainly widely variable delay. (Note that variable delay on a link is somewhat unnatural, but we assumed this so as to avoid worrying about processing delays.) The only real difference (and it is an important difference) is that frames must stay in order on a link, whereas packets might arrive out of order in a network.

Recall that datagram networks typically make no attempt to keep packets of the same session on the same path, so that out-of-order packets pose a generic problem for datagram networks. One would think that virtual circuit networks would be guaranteed to keep packets of the same session in order, but there are some subtle problems even here concerned with setting up new paths. If a link fails and a new path is set up for a session, the last packets on the old path might arrive out of order relative to the first packets on the new path. This problem can be avoided by a well-designed network layer, but for simplicity, we simply assume here that we are dealing with a datagram network and that packets can get out of order arbitrarily on the network.

Since packets can get lost, packets must sometimes be retransmitted end-to-end, and since packets can be delayed arbitrarily, a packet can be retransmitted while the earlier copy of the packet is lurking in the network; that earlier copy can then remain in the network until the packet numbering wraps around, and the packet can then cause an uncorrectable error. This scenario is perhaps highly unlikely, but it serves to show that end-to-end error recovery at the network or transport layer cannot be guaranteed to work correctly if packets are sometimes lost, sometimes get out of order, and sometimes become arbitrarily delayed.

There are three possible ways out of the dilemma described above: first, insist that the network layer use virtual circuits and maintain ordering even if the path for a session is changed; second, make the modulus for numbering packets so large that errors of the foregoing type are acceptably unlikely; and third, create a mechanism whereby packets are destroyed after a maximum lifetime in the network, using a modulus large enough that wraparound cannot occur in that time. It is shown in Problem 2.43 that the required modulus, using selective repeat, must be twice the window size plus the maximum number of packets of a session that can be transmitted while a given packet is alive. In Section 2.9 we discuss how the transport protocol TCP uses a combination of the last two approaches described above.

End-to-end error recovery also requires some sort of end-to-end check sequence such as a supplementary CRC to detect errors either made at the subnet nodes or undetected by the CRC at the link layer. In TCP, for example, this check sequence is 16 bits long and is the one's complement of the one's complement sum of the 16 bit words making up the packet. This type of check sequence is not as good at detecting errors as a CRC, but is easier to implement in software for typical processors. One interesting problem that occurs with end-to-end check sequences is that packets frequently change somewhat when passing from node to node. The primary example of this is the changing virtual channel numbers in the virtual channel addressing scheme discussed in Section 2.8.1. One certainly does not want to recompute the end-to-end check sequence at each node, since this would fail in the objective of detecting node errors.

The solution to this problem is to calculate the end-to-end check sequence before placing values in those fields that can vary from node to node. At the destination site, the check sequence is recomputed, again ignoring the values in those varying fields. This solution leaves the varying fields unprotected except by the link layer. For the example of virtual channel addressing, however, there is a good way to restore this protection. Both end sites for the session know the universal address of the other (from the setup of the session). Thus these addresses can be included in the check sequence, thus protecting against packets going to the wrong destination. Other parameters known to both end sites can also be included in the check sequence. As one example, we discussed the large end-to-end sequence number fields required to number packets in datagram networks. Only a few of the least significant bits are required to reorder the packets over a window, whereas the rest of the bits are required to ensure that very old packets (and very old acknowledgments) get rejected. By transmitting only the least significant bits of the sequence and request numbers, and including the more significant bits in the check sequence, one achieves greater efficiency at no cost in reliability.

Flow control The use of either go back n or selective repeat ARQ for end-to-end error recovery provides a rather effective form of flow control as an added bonus. If we ignore errors and retransmissions, we see that with a window size of n, at most n packets for a given session can be outstanding in the network at a given time. A new packet j can be sent only after an acknowledgment is received for packet $j - n$. The effect of this is that as congestion builds up and delay increases, acknowledgments are delayed and the source is slowed down. This is called end-to-end window flow control and is analyzed in Chapter 6. If the destination wishes to receive packets less rapidly, it can simply delay sending the current request number RN; thus this type of flow control can be effective in reducing congestion both in the subnet and at the destination site.

In the discussion of data link control we did not mention this flow control aspect of ARQ. The reason is that a link simply carries a fixed number of bits per second and there is no reason to limit this flow (other than the ability of the receiving node to handle the incoming data). Recall that standard DLCs provide a receive-not-ready supervisory packet as a way to protect the receiving node from too many incoming packets. This mechanism allows speedy acknowledgments, thus preventing unnecessary retransmissions due to time-outs, but also permits the receiving node to exert some flow control over the sending node.

Combining end-to-end acknowledgments with flow control has some limitations. If an acknowledgment does not arrive because of an error event, the packet should be retransmitted, whereas if the ack is delayed due to network congestion, the packet should not be retransmitted, since it would only add to the congestion. Since delays in the subnet are highly variable, it is difficult for time-outs to distinguish between these cases. To make matters worse, the destination might be overloaded and postpone sending acks as a way of preventing further transmissions from the source; this again might cause unnecessary retransmissions due to time-outs.

One partial solution to these problems is to provide the destination with a way to slow down the source without delaying acknowledgments. One way of doing this is by a technique such as the receive-not-ready supervisory frames in the standard data link controls. Such a frame both provides a speedy ack and prevents further transmission. This same technique can be used at the network layer (and is used in the X.25 network layer standard to be discussed shortly). A *permit* scheme is a more refined technique for the same objective. The destination sends two feedback numbers back to the source rather than one. The first, the usual RN, provides the ack function, and the second, a permit, tells the source how many additional packets the destination is prepared to receive. Thus if the permit number is j, the source is permitted to send packets from number RN to $RN + j - 1$ inclusive. In effect, this allows the destination to change the window size with each acknowledgment to the source at will. It allows the destination to send permits for only as many packets as will fit in the available buffers. This permit scheme could be provided in DLCs as well, but it is less useful there because the buffering requirements at a node are relatively modest because of the fixed number of links at a node.

Error recovery at the transport layer versus the network layer We have discussed end-to-end error recovery and its relationship to end-to-end window flow control. We now consider whether error recovery belongs at the transport layer or the network layer. From a practical standpoint, the answer now and in the near future is that error recovery belongs at the transport layer. The ISO transport layer standard known as

TP4 and the de facto standard TCP both provide for error recovery at the transport layer. Part of the reason for this is the prevalence of internetworking and the large number of networks that do not provide reliable service. The Internet Protocol, which is designed to work with TCP and which connects many universities and research labs, is itself an unreliable protocol, as will be discussed shortly.

The major disadvantage of having unreliable network and internet protocols and error recovery at the transport layer is that the transport layer has no way to distinguish between acknowledgments that are slow because of congestion and acknowledgments that will never arrive because the packet was thrown away. This problem is particularly severe for datagram networks. As a result, when the network becomes congested, packets start to be retransmitted, leading to still greater congestion, more dropped packets, and thus unstable behavior. There are a number of ad hoc approaches to reducing the severity of this problem, but it appears that error recovery at the transport layer will always lead to problems unless such recovery is required only rarely . This in turn will require almost reliable operation in the internet layer and in the various network layers.

2.8.3 The X.25 Network Layer Standard

The X.25 standard was developed by CCITT to provide a standard interface between external sites and subnet nodes. Recall that the physical layer, X.21, of this standard was discussed briefly in Section 2.2, and the DLC layer, LAPB, was discussed in Section 2.6. In X.25 terminology, the external site is referred to as DTE (data terminal equipment) and the subnet node as DCE (data communication equipment), but we shall continue to refer to them as sites and nodes.

The structure of an X.25 data packet is shown in Fig. 2.50. Recall that the DLC layer adds a frame header and trailer to this packet structure before transmission on the link. The third byte of the packet header is very similar to the control byte of the standard DLCs (except that SN and RN here refer to sequence numbers within the session, as discussed above). The control bit, C, is 0 for data packets and 1 for control packets (to be discussed later). This control bit is different from the control bit in the frame header; at the DLC layer, all packets, data and control, are treated indistinguishably, using 0 as the DLC layer control bit. Finally, the "more" bit, M, in the packet header is 1 for nonfinal packets in a message and 0 for the final packet.

The last half of the first byte and the second byte of the packet header is a 12-bit virtual channel number, as discussed in Section 2.7. This number can be different at the two ends of the virtual circuit for a given session. Many subnets use this same packet structure within the subnet, and the virtual channel number can then change from link to link, as described in Section 2.7.

1	2	3	4	5	6	7	8	
Q	D	Modulus		Virtual channel				Byte 1
Virtual channel								Byte 2
RN			M	SN			C=0	Byte 3

Figure 2.50 Packet header for data in X.25. For a control packet the C bit is 1.

The first bit, Q, in the packet header (if used at all) has the value 1 for control packets at the transport and higher layers and has the value 0 for control packets at the network layer and for data packets. The next bit, D, indicates whether the acknowledgment (RN) has end-to-end significance or just link significance. If $D = 1$, then RN signifies that the destination site has received all packets for the session previous to RN, that RN is awaited, and that the source is free to send a window of packets from RN on. If $D = 0$, RN signifies that the local subnet node or site has received packets previous to RN, that RN is awaited, and the other site or node is free to send a window of packets from RN on. It is possible for some sessions to use end-to-end significance and others, single-link significance. As discussed above, end-to-end acks are very helpful in providing error recovery. Some sessions, however, do not require error recovery, and some provide it at a higher layer, for example, by using a query–response interaction between the sites. For single-link significance, the acknowledgment function of RN is redundant (the DLC layer guarantees that the packet is received). Thus, RN is used strictly for flow control, and the receiving node delays sending RN until it is prepared to accept a window of packets beyond RN.

The third and fourth bits of the header indicate whether the modulus is 8 or 128. If the modulus is 128, the third byte is extended to two bytes to provide seven bits for SN and seven for RN.

The control packets in X.25 are quite similar to the supervisory frames and unnumbered frames in the standard DLCs. The third byte of the packet header indicates the type of control packet, with the final bit (control bit) always 1 to distinguish it from a data packet. The virtual channel number is the same as in a data packet, so that these control packets always refer to a particular session. There are receive-ready, receive-not-ready, and reject control packets, each of which carry RN; these are used for acknowledgments in the same way as in DLCs. The receive-not-ready packet provides some ability to acknowledge while asking for no more packets, but there is no ready-made facility for permits.

One interesting type of control packet is the call-request packet. After the three-byte header, this packet contains an address field, containing the length of the calling address and called address in the first byte, followed by the addresses themselves. This is used to set up a virtual circuit for a new session in the network. Note that the virtual channel number is of no help in finding the destination; it simply specifies the virtual channel number to be used after the virtual circuit is established. The subnet node, on receiving a call-request packet, sets up a path to the destination and forwards the packet to the destination site. If the recipient at the destination site is willing and able to take part in the session, it sends back a call-accepted packet; when this reaches the originating site, the session is established.

Both the call-request and call-accepted packet contain a facilities field; this contains data such as the maximum packet length for the session, the window size, who is going to pay for the session, and so on. The default maximum packet length is 128 bytes, and the default window size is 2. If the call-request packet specifies different values for maximum packet length or window size, the recipient can specify values closer to the default values, which must then be used. This is a sensible procedure for the external sites, but it is somewhat awkward for the subnet. The subnet can specify a maximum packet length and maximum window size to be used by the sessions using the subnet, but it is also desirable to have the ability to reduce window sizes under heavy loading, and this cannot be done naturally with X.25.

If the subnet is unable to set up the virtual circuit (because of heavy loading, for example), or if the recipient is unwilling to accept the call, a clear-request control packet goes back to the initiating site, explaining why the session could not be established.

2.8.4 The Internet Protocol

The Internet is a collection of thousands of networks interconnected by gateways. It was developed under the U.S. Department of Defense Advanced Research Projects Agency (DARPA) support as an effort to connect the myriad of local area networks at U.S. universities to the ARPANET. It now provides almost universal data access within the academic and research communities. The protocol developed to interconnect the various networks was called the Internet Protocol, usually referred to as IP. The backbone functions have since been taken over by the U.S. National Science Foundation (NSF), and the current topology is quite chaotic and still growing rapidly.

The individual networks comprising the Internet are joined together by gateways. A gateway appears as an external site to an adjacent network, but appears as a node to the Internet as a whole. In terms of layering, the internet sublayer is viewed as sitting on top of the various network layers. The internet sublayer packets are called datagrams, and each datagram can pass through a path of gateways and individual networks on the way to its final destination. When a gateway sends a datagram into an individual network, the datagram becomes the body of the packet at the network layer (see Fig. 2.51). Thus the individual network sends the datagram on to the next gateway or to the final destination (if on that network) in the same way that it handles other packets. Assuming that the individual network succeeds in getting the datagram to the next gateway or destination, the individual network looks like a link to the internet layer.

One of the goals of the Internet project was to make it as easy as possible for networks to join the Internet. Arbitrary networks were allowed to join the Internet, even if they allowed datagrams to be delayed arbitrarily, lost, duplicated, corrupted by noise, or reversed in order. This lack of reliability underneath was one of the reasons for making IP a datagram protocol; the problem of error recovery was left for the transport layer. Note that if an individual network happens to be a virtual circuit network, this poses no problem. Most virtual circuit networks can also handle datagrams, and alternatively, permanent virtual circuits could be established between gateways and sites on a given network to carry the internet datagrams between those gateways and sites. It would also have been possible to develop a virtual circuit internet protocol. This would have involved using error recovery at each gateway, thus making the individual networks appear like virtual circuit networks. We discuss this later.

Two transport layer protocols were developed along with IP. The first, the transmission control protocol, known as TCP, uses error recovery and reordering to present

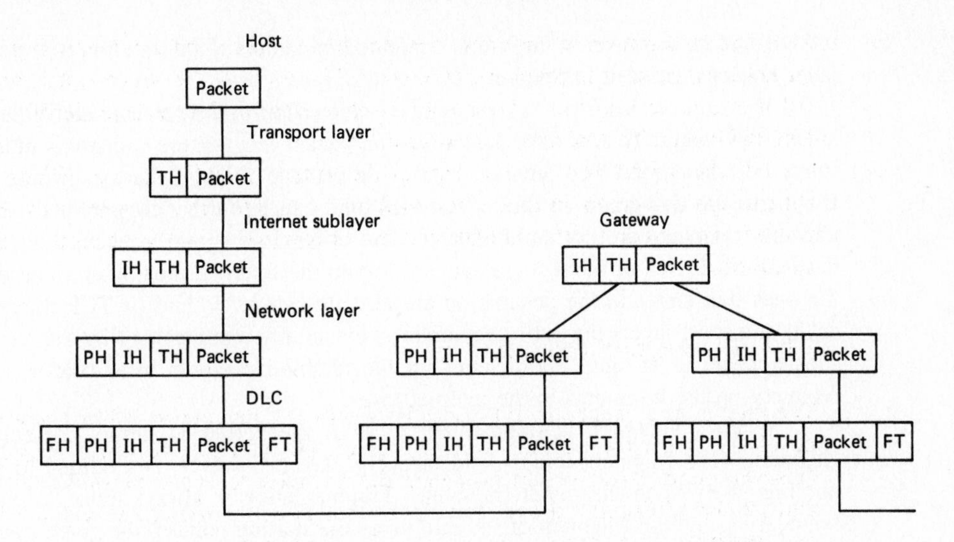

Figure 2.51 Illustration of a packet as it passes through a gateway. Within any given network, only the packet header and frame header are used, whereas at a gateway, the internet header is used to select the next network and to construct a new packet header appropriate to the new network.

the user with reliable virtual circuit service. The second, the user datagram protocol, simply deals with the question of where to send datagrams within a given destination.

The primary functions of IP are, first, to route datagrams through the Internet; second, to provide addressing information to identify the source and destination sites; and third, to fragment datagrams into smaller pieces when necessary because of size restrictions in the various networks. We shall not discuss routing here, since that is a topic for Chapter 5. We first discuss addressing and then fragmentation.

Since datagrams are being used, it is necessary for each datagram to contain identifications of the source and destination sites that are universally recognizable throughout the Internet (*i.e.*, the virtual channel addressing scheme of Section 2.8.1 is not applicable to datagram networks). Assigning and keeping track of these addresses is difficult because of the size and rapidly changing nature of the Internet. The addressing scheme used for IP is hierarchical in the sense that each site is identified first by a network identifier and then by an identifier for the site within the network. One peculiar consequence of this is that a site that directly accesses several networks has several addresses, and the address used determines which of those networks a datagram will arrive on.

The addresses in IP are all 32 bits in length, with a variable number of bits allocated to the network ID and the rest to the site (or host as it is called in IP terminology). In particular, the networks are divided into three classes, A, B, and C, according to size. The largest networks, class A, have a 7 bit network ID and a 24 bit host ID. Class B networks have 14 bits for network ID and 16 for host, and class C have 22 bits for network and 8 for host. The remaining one or two bits out of the 32 bit field are used to encode which class is being used. These addresses indicate only the source and destination sites. The intended process or user at the site is contained in additional information in the transport layer header discussed in Section 2.9.

The original length of a datagram is selected primarily for user convenience, but since individual networks have limitations on packet lengths, the datagrams must sometimes be fragmented into smaller pieces for passage through these networks. These fragments are datagrams in their own right (and can be further fragmented if need be) and the fragments are not combined into the original datagram until reaching the final destination. If one or more fragments are lost in the Internet, then, after some time-out, the ones that arrive at the destination are also thrown away, and (if TCP is being used at the transport layer) the entire datagram is eventually retransmitted by the source site. This appears to be quite inefficient, but the alternative would have been to do error recovery on the fragments in the internet layer.

When a datagram is fragmented, most of the header is simply copied to each fragment. The fragment length field (see Fig. 2.52), however, is modified to give the number of bytes in the given fragment. Fragmentation is always done on eight byte boundaries, and the fragment offset field gives the starting point of the given fragment in eight byte increments. One of the three flag bits in the header is used to indicate whether or not a fragment is the last fragment of the datagram. The final piece of information needed at the destination to reconstruct a datagram from its fragments is the datagram ID. The source site numbers successive datagrams, and this sequence number is used as the datagram ID; this is copied to each fragment and allows the destination to know which fragments are part of the same datagram. It is not difficult to see that the combination of fragment length, offset, last fragment bit, and datagram ID provided for each fragment allows the original datagram to be reconstructed.

We now give a quick summary of what the other fields in the IP datagram header do. The field "Vers" tells which version of the protocol is being used. This is useful when a new version of the protocol is issued, since it is virtually impossible to synchronize a changeover from one version to another. The field "HL" gives the length of the header (including options) in 32 bit increments. "Service" allows users to specify what kind of service they would like, but there is no assurance that the gateways pay any attention to this.

1	4	8	16		32
Vers	HL	Service	Fragment length		
Datagram Id			Flag	Fragment offset	
TTL		Protocol	Header checksum		
Source address (net and node)					
Destination address (net and node)					
Options (if any)					
Data					

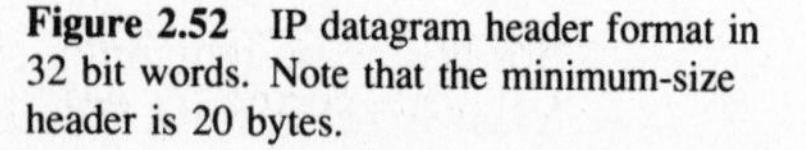

Figure 2.52 IP datagram header format in 32 bit words. Note that the minimum-size header is 20 bytes.

The time to live field, "TTL," is decremented each time the datagram reaches a new gateway, and the datagram is thrown away when the count reaches zero. This is supposed to prevent old datagrams from existing in the Internet too long, but unfortunately, there is no mechanism to prevent a datagram from remaining in one of the individual networks for an arbitrarily long period. The field "protocol" indicates which higher-layer protocol is to operate on the datagram. This is necessary since the information about where the datagram is to go at the destination site is in the transport header, and the format of this header depends on which transport protocol is being used. Finally, the header check sum checks on the header. This seems a little peculiar for a datagram, but as a practical matter, one does not want datagrams in a large internet going to some random destination in the event of an error.

2.9 THE TRANSPORT LAYER

The transport layer is the fourth layer in the OSI layered architecture and has the functions, (1) of breaking messages into packets, (2) of performing error recovery if the lower layers are not adequately error free, (3) of doing flow control if not done adequately at the network layer, and (4) of multiplexing and demultiplexing sessions together. The first function, breaking messages into packets, is relatively straightforward, aside from the question of selecting the maximum packet length for a network. As discussed in the preceding subsection, even the selection of packet size is less important for TCP combined with IP, since IP fragments packets when necessary; thus, the transport layer packets going from TCP to IP can really be viewed as messages with a transport layer header.

We have already discussed the general problem of end-to-end error recovery and the reasons why it is normally done at the transport layer. Rather than discussing the other functions of the transport layer in general, we shall discuss them in terms of TCP and the ISO standards.

2.9.1 Transport Layer Standards

One has a wide choice of standard transport protocols. The International Standards Organization (ISO), in its OSI layered architecture, has five different classes of transport protocol (denoted TP), numbered 0 to 4. TP class 0 is a relatively simple protocol that splits messages into packets and reassembles them at the destination but provides neither for multiplexing of sessions nor for error recovery. Class 1 allows for some error recovery but relies on the network layer and consistency checks for detection of the need for retransmission. Class 2 provides multiplexing but no error recovery, and class 3 provides the error recovery of class 1 and the multiplexing of class 2. Finally, class 4 provides multiplexing plus a more complete form of error recovery using both detection of errors and retransmission. The class 4 protocol is designed for use with datagram networks and other networks that do not adequately deliver each packet once, only once, correctly, and in order.

The U.S. National Institute for Standards and Technology has also developed a U.S. Federal Information Processing Standard (FIPS) for the transport layer; this is essentially the same as TP classes 2 and 4 of the OSI standard except for some additional options.

The U.S. Department of Defense has standardized TCP (transmission control protocol). As discussed in the preceding subsection, this protocol was originally developed for the ARPANET and was designed to operate on top of IP, thus allowing internetworking between the ARPANET and other networks. TCP was developed before the OSI TP protocols, and TP class 4 uses many of the features of TCP. In what follows, we discuss TCP and then contrast it briefly with TP class 4.

2.9.2 Addressing and Multiplexing in TCP

In Section 2.8.4 we discussed addressing in the Internet and pointed out that the IP header must contain sufficient information to route the given packet through the Internet, namely a network ID and a host ID. One of the basic tenets of layering is that the header (and perhaps trailer) of a packet at a given layer contains all the information necessary for the module at that layer to do its job; the body of the packet at that layer (which contains the headers for the higher layers) should not be consulted by the module at the given layer. When a packet is passed from TCP to IP at a given node, there are extra parameters passed along with the packet that provide IP with addressing information. Similarly, when the datagram arrives at its destination, the IP header is stripped off, but the addressing information and other header information can be passed up as parameters.

In TCP, the entire address of a source (or destination) is called a *socket*. It is organized hierarchically as, first, network ID, then host ID, and finally, the user or process ID within that node. The user or process ID within a host is called a *port* in TCP (*i.e.*, the transport layer is viewed as having many ports accessing the next higher layer, one for each session). The port ID, for both source and destination, is included in the transport header, whereas the network and node IDs appear in the IP header. This means that all sessions going from a given source host to a given destination host will normally have the same source and destination address in the IP header and will be distinguished only in the transport header. Thus all sessions between the same pair of hosts could be viewed as being multiplexed together at the transport layer into a common lower-layer session. The fact that IP is a datagram protocol is irrelevant from this standpoint; TCP could also operate with an X.25 form of network layer, and then the network layer would have to set up a session and map the host address into an address in the X.25 packet header.

It is possible to have multiple logical addresses for the same host in TCP. Thus one could avoid multiplexing by using a separate address in the IP header for each transport layer session. The general point to be learned from this, however, is that transport headers, internet headers, and network headers all have address fields. Sometimes there is redundancy between the information in these various fields, but the information in the network header must be sufficient for routing and flow control in an individual network. Similarly, the internet header plus the network header must be sufficient for routing between gateways. Finally, the transport header must be sufficient to distinguish all sessions multiplexed together.

The size of the address fields at different layers in the various standards is somewhat bizarre. TCP uses 16 bit port numbers, and TP class 4 similarly uses 16 bits for a reference number that performs the same role. IP uses 32 bit addresses with 8, 16, or 24 bits representing the node and the rest representing the component network (along with an encoding of whether 8, 16, or 24 bits represent the node). On the other hand, X.25 uses only 12 bits as a virtual circuit number. The use of virtual circuits allows part of this efficiency at the network layer, but one still wonders why the standard network layer is so stingy with address bits and the transport layer and internet sublayer are so profligate. This same concern for efficiency at the network (and DLC) layer and lack of concern at higher layers runs through all the standard protocols. One might have hoped that with the enormous flexibility in these protocols, some flexibility allowing a little efficiency would have been included (such as variable-length address fields), but the designers of transport protocols and internet protocols seem dedicated to the principle that bigger is better. The large size of address fields at the transport and IP levels and the small size at the network layer sometimes inhibit maintaining session identities (and thus providing different types of service, such as delay and error recovery to different sessions) at the network layer.

2.9.3 Error Recovery in TCP

Error recovery in TCP follows the general pattern described in Section 2.8.2; the strategy is selective repeat ARQ. The transport layer packets each contain both a 32 bit sequence number, SN, and a 32 bit request number, RN, in the packet header. The number SN is used in a rather peculiar way in TCP, since it counts successive bytes in the session data rather than successive packets. In particular, SN is the number of the first data byte in the given transport layer packet. Thus if a given transport layer packet has a sequence number $SN = m$ and contains n bytes in the packet body (*i.e.*, the part of the packet labeled "data" in Fig. 2.53), then the next transport layer packet for that session has $SN = m + n$. There appears to be no good reason for this, but it does not do any harm, other than adding complexity and overhead. One might think that this type of numbering would make it easier to fragment packets into smaller pieces for internetworking, but as we have seen, IP has its own mechanisms for handling fragmentation. The request number RN, as one might expect, denotes the next desired

Source port			Destination port
Sequence number			
Request number			
Data offset	Reserved	Control	Window
Check sum			Urgent pointer
Options (if any)			
Data			

Figure 2.53 TCP header format in 32 bit words; note that the minimum-sized header is 20 bytes.

packet in the opposite direction by giving the number of the first data byte in that packet. As with most ARQ schemes, the RN field allows acknowledgments to be piggybacked on data for the given session in the reverse direction, and in the absence of such data, a packet with just the header can carry the RN information.

As discussed earlier, errors are detected in TCP by a 16 bit check sum calculated from both the transport layer packet and some parameters (such as source and destination address and packet length) that are passed to the next lower layer. These addresses are also checked in IP with a header check sum, but TCP is supposed to work with other lower layers, so this provides worthwhile extra protection.

From the description above, it should be clear that error recovery for packets within an established connection at the transport layer is not a logically complicated matter . The same principles apply as at the DLC layer, and the only new feature is that of enlarging the modulus to allow for out-of-order packet arrivals. There are several complications, however, that remain to be discussed.

The first such complication is that of retransmission time-outs. At the DLC layer, retransmission time-outs provide an interesting issue to think through in understanding how distributed algorithms work, but the issue is not very critical. At the transport layer, the issue is quite serious, since high congestion in the network causes large delays, which trigger retransmissions from the transport layer, which further increase delays. When the congestion becomes bad enough, packets will also be discarded in the network. This triggers more retranmissions, and eventually throughput is reduced to almost zero. One advantage of doing error recovery at the network layer is that it is possible to provide better estimates on delay, and thus avoid retransmission under high-delay conditions.

The final complication is the problem of error recovery when setting up and tearing down connections. What makes this problem tricky is that one does not want to save information about connections after they have been torn down. Thus there is the possibility of a packet from an old connection entering a node after a new connection has been set up between the same processes. There is a similar possibility of confusion between the packets used to initiate or tear down a connection, and finally there is a possibility of confusion when one node crashes and loses track of connections that still exist at other nodes.

In TCP, the major mechanism used to deal with the initiation and disconnection of sessions is the synchronization of the sequence number and request number at the initiation of a session. Each node has a clock counter (not synchronized between nodes) that increments once each 4 μsec. When a new connection is attempted, this clock counter is used as the initial sequence number for the first packet in an initiation handshake. The opposite node responds by acknowledging this sequence number in its request number (as usual) and using its own clock counter for its own initial sequence number. Since the clocks are incrementing faster than the sequence numbers of any established connection can increment (at least for data nets at today's speeds) and since the 32 bit clock counter wraps around less than once in 4 hours, this mechanism solves most of the problems noted above. With this type of protocol, the responding node cannot start to send data (beyond this initial packet and its potential repetitions) until it receives a second packet from the initiating node acking the responding node's first packet. Because of this third required packet, this type of initiation is called a three-way handshake. We discussed this type of handshake in Section 2.7. What is new here is that avoiding confusion between old and new connections is achieved by a pseudorandom choice of sequence numbers rather than by using an alternation of connection and disconnection. The strategy here has the disadvantage of potential malfunction under arbitrary delays, but the use of a 32 bit number makes this possibility remote.

The major type of problem for this kind of handshake is that it is possible for one node to believe that a given connection is open while the other node believes that it is closed. This can happen due to one node crashing, the network disconnecting temporarily, or incomplete handshakes on opening or closing. If the node that believes the connection is open sends a packet, and then the other node starts to initiate a connection, it is possible for the new connection to get set up and then for the old packet to arrive. If the old connection had been in existence for a very long time, it is possible for both the sequence and request numbers to be in the appropriate windows, causing an error. This type of situation appears very unlikely, especially considering the large amount of redundancy in the rest of the header.

For completeness, the format of the header in TCP is shown below as a sequence of 32 bit words. We have discussed the source port, destination port, sequence number, request number, and checksum. The data offset gives the number of 32 bit words in the header, the urgent pointer has to do with high-priority data and need not concern us, and the options also need not concern us. Each of the six control bits has a particular function. For example, the fifth control bit is the SYN bit, used to indicate an initiation packet that is setting the sequence number field. The second control bit is the ACK bit, used to distinguish the first initiation packet (which does not carry an ACK) from the second and subsequent packets, which do carry an ACK.

2.9.4 Flow Control in TCP/IP

TCP uses a permit scheme to allow the destination to control the flow of data from the source host. This is implemented by a 16 bit field called a *window* stating how many bytes beyond the request number can be accepted for the given connection by the given node. This provides some rationale for TCP to count bytes (or some fixed-length unit of data), since there is such high potential variability in datagram lengths. For a given size of window, the level of congestion in the Internet (or network in the case of no internetworking) controls how fast data flow for the given connection. Thus this scheme both protects the destination from being saturated and slows down the sources in the event of congestion.

It is also possible for a source to monitor the round-trip delay from datagram transmission to acknowledgment receipt. This can provide an explicit measure of congestion and there are many strategies that use such measurements to constrain data flow into the Internet or subnet. There are some inherent difficulties with strategies that use such measurements to decrease source flow. The first is that such strategies are subject to instability, since there is a delay between the onset of congestion and its measurement. This is made worse by throwing away packets under congestion. The next problem is that "good citizen" sessions that cut back under congestion are treated unfairly relative to sessions that do not cut back. Since the Internet uses datagrams, the Internet cannot distinguish "good citizen" sessions from bad, and thus datagrams are thrown away without regard to session.

It is important to recognize that these are inherent problems associated with any datagram network. The network resources are shared among many sessions, and if the resources are to be shared fairly, the network has to have some individual control over the sessions. The session that a datagram belongs to, in principle, is not known at the network layer (note from Fig. 2.52 that the IP protocol has no header field to indicate session number), and thus the datagram network or internet cannot monitor individual sessions. A corollary of this is that datagram networks cannot provide guaranteed types of service to given sessions and cannot negotiate data rate. A datagram network might give different classes of service to different kinds of datagrams, but the choice of service class comes from the transport layer and is not controlled by the network. If some sessions place very heavy demands on the network, all sessions suffer.

TCP/IP has been enormously successful in bringing wide area networking to a broad set of users. The goal of bringing in a wide variety of networks precluded the possibility of any effective network or internet-oriented flow control, and thus precluded the possibility of guaranteed classes of service. How to provide such guarantees and still maintain internetworking among a wide variety of networks is a challenge for the future.

2.9.5 TP Class 4

The ISO transport protocol, TP class 4, is very similar to TCP, with the following major differences. First, sequence numbers and request numbers count packets rather than bytes (in the jargon of the standards, packets at the transport level are called transport protocol data units or TPDUs). Next, acknowledgments are not piggybacked on packets going the other way, but rather, separate acknowledgment packets are sent (using the same selective repeat strategy as before). These ack packets are sent periodically after some time-out even with no data coming the other way, and this serves as a way to terminate connections at one node that are already terminated at the opposite node.

Another difference is that sequence numbers are not initiated from a clock counter, but rather start from 0. The mechanism used in TP class 4 to avoid confusion between old and new connections is a destination reference number. This is similar to the destination port number in TCP, but here the reference number maps into the port number and can essentially be chosen randomly or sequentially to avoid confusion between old and new connections.

2.10 BROADBAND ISDN AND THE ASYNCHRONOUS TRANSFER MODE

In Sections 2.8 and 2.9 we have described some of the point-to-point aspects of network layer and transport layer protocols. In this section we describe a new physical layer protocol called asynchronous transfer mode (ATM). From a conceptual viewpoint, ATM performs the functions of a network layer protocol, and thus we must discuss why network layer functions might be useful at the physical layer. Since ATM was developed for use in broadband ISDN systems, however, we start with some background on broadband ISDN.

A broadband integrated services digital network (broadband ISDN) is a network designed to carry data, voice, images, and video. The applications for such networks are expected to expand rapidly after such networks are available. Recall that in our discussion of ISDN in Section 2.2, we pointed out that a high-resolution image represented by 10^9 bits would require over 4 hours for transmission on a 64 kbit/sec circuit and would require 11 minutes on a 1.5 Mbit/sec circuit. Broadband ISDN is an effort to provide, among other things, data rates that are high enough to comfortably handle image data in the future. The planned access rate for broadband ISDN is 150 Mbit/sec, which is adequate for image traffic and also allows for the interconnection of high-speed local area networks (see Section 4.5.5). This access rate also allows video broadcast traffic, video conferencing, and many potential new applications.

The technology is currently available for building such networks, and the potential for new applications is clearly very exciting. However, the evolution from the networks of today to a full-fledged broadband ISDN will be very difficult, and it is debatable whether this is the right time to standardize the architectural details of such a network. Some of the arguments for rapid standardization are as follows. There are current needs for these high data rates, including image transmission, accessing high-speed local area networks, and accessing supercomputers; the rapid development of such a network would meet these needs and also encourage new applications. Many people also argue that the cost of networks is dominated by the "local loop" (*i.e.*, the access link to the user); optical fiber will increasingly be installed here, and economy dictates combining all the telecommunication services to any given user on a single fiber rather than having multiple networks with separate fibers. The difficulty with this argument is that it is not difficult to multiplex the access to several networks on a single fiber. Another popular argument for the rapid deployment of broadband ISDN is that there are economies of scale in having a single network provide all telecommunication services.

One of the arguments against rapid standardization is that there are also diseconomies of scale involved with a single all-purpose network. These diseconomies come from trying to force very different types of communication requirements, such as video and conventional low-speed data, into a common mold. Another argument is that there is no large-scale engineering experience with such networks and little knowledge of how the potential applications will evolve. This is exacerbated by the rapid development of optical fiber technology, which is rapidly changing the range of choices available in the backbone of a broadband network. Finally, the evolution of broadband ISDN is central to the future of the telecommunication industry, and thus political, legal, and regulatory issues, as well as corporate competitive positioning, will undoubtedly play a large and uncertain role in the evolution and structure of future networks. In what follows we look only at technological issues, but the reader should be aware that this is only part of the story.

Broadband ISDN can be contrasted both with data networks and with the voice network. In contrast with data networks, there is an increase of over three orders of magnitude in the link speeds envisioned for broadband ISDN. Because of this, broadband

ISDN designers are more concerned with computational complexity and less concerned with efficient link utilization than are designers of conventional data networks. Because of the speed constraints, the protocols in broadband ISDN should be simple and amenable to both parallel and VLSI implementation.

Another important contrast with data networks is that individual user requirements are critically important in broadband ISDN. Part of the reason for this is that voice and video normally require both a fixed bit rate and fixed delay, whereas packet-switched data networks generally have variable delays and often use flow control to reduce data rates. In the past, data applications that required fixed rate and/or fixed delay usually avoided data networks and instead used dedicated leased lines from the voice network. In other words, traditional data networks provide a limited range of service, which users can take or leave. Broadband ISDN, on the other hand, is intended to supply all necessary telecommunication services, and thus must provide those services, since there will be no additional voice network on which to lease lines.

Next consider the contrast of broadband ISDN with the voice network. All voice sessions on a voice network are fundamentally the same. Except on the local loops, voice is usually sent digitally at 64 kbits/sec, with one eight bit sample each 125 μsec. In Section 2.2 we described briefly how these sessions are time-division multiplexed together. The important point here is that all of these multiplexing systems, including the SONET standard, use a 125 μsec frame broken up into a large number of one byte slots. Switching is accomplished at a node by mapping the slots in the incoming frames into the appropriate slots of the outgoing frames. In contrast, broadband ISDN must cope with potential user rates from less than 1 bit/sec to hundreds of megabits per second. It must also cope with both bursty traffic and constant-rate traffic.

Despite the enormous difference between broadband ISDN and the voice network, broadband ISDN will probably evolve out of the present-day voice network if and when it is built. The reason for this is that the bulk of transmission and switching facilities currently exist within the current voice network, and current wide area data networks typically lease their transmission facilities from the voice network. Because of this, the early thinking about broadband ISDN focused on using the underlying synchronous 125-μsec frame structure of the voice network to provide the physical layer transmission.

Within the frame structure of the voice network, it is possible to group multiple slots per frame into a single circuit whose rate is a multiple of 64 kbits/sec. These higher-rate circuits can be leased either for dedicated data channels or for use as links in data networks. They come in standard multiples of 64 kbits/sec (*e.g.*, DS1 service is 24 times 64 kbits/sec and DS3 service is 28 times the DS1 speed). These higher-rate circuits could also be used in broadband ISDN for high-rate services such as video. In this sense, broadband ISDN was initially visualized as an evolution from (narrowband) ISDN, maintaining the emphasis both on 64 kbit/sec circuit switched sessions and on various standard multiples of the 64 kbit/sec speed. The early objective for broadband ISDN was simply to increase data rates to allow for video, high-resolution images, and interconnection of high-speed local area networks.

As interest in broadband ISDN increased and as standardization efforts started, an increasing number of people questioned the underlying reliance on the synchronous frame structure of the voice network and proposed using an asynchronous packet structure for multiplexing at the physical layer. As this debate quickened, use of the conventional synchronous frame structure became known as STM (synchronous transfer mode), and the use of a packet structure became known as ATM (asynchronous transfer mode). Fundamentally, therefore, the choice between STM and ATM is the choice between circuit switching and packet switching.

As discussed in Section 1.2.3, conventional data networks use packet switching principally because of the burstiness of the data, but burstiness is far less important for broadband ISDN. In the first place, voice traffic totally outweighs conventional data traffic and will probably continue to do so (see Problem 1.2). Next, circuit switching is already used in the voice network, is ideal for voice traffic, and is well suited for video traffic. High-resolution images would not fit well into circuit switching, but could be handled (along with conventional data) on a packet network using the transmission resources of a circuit-switched net as a physical layer.

A more important aspect of the choice between STM and ATM is the question of the flexibility of STM for handling a wide variety of different data rates. STM would require each circuit-switched application to use some standard multiple of 64 kbits/sec. This would lead to great switching complexity if the number of standard rates were large, and to great inefficiency otherwise. In addition, the proponents of ATM argue that the use of a small set of standard rates would inhibit the development of new applications and the development of improved data compression techniques.

This lack of flexibility in STM is particularly troublesome at the local loop. Here, given a small set of sessions, each would have to be assigned one of the standard rates. Any attempt to multiplex several sessions (particularly bursty data sessions) on one standard rate, or to use several standard rates for one session, would lead to considerable complexity. In comparison with these problems of matching sessions to standard rates, the ATM solution of simply packetizing all the data looks very simple.

While the relative merits of STM and ATM were being debated, there was great activity in developing high-speed packet-switching techniques. These techniques were highly parallelizable and amenable to VLSI implementation. As a result of both the flexibility of ATM for local access and the promise of simple implementation, the CCITT study group on broadband ISDN selected ATM for standardization.

Before describing ATM in more detail, we should understand how it fits into the OSI seven-layer architecture. In the original view of ISDN, the synchronous frame structure of STM was quite properly viewed as the physical layer. Thus, when ATM replaced STM, it was reasonable to consider ATM as the physical layer for data transmission. This had the added advantage of relieving ATM from any responsibility to conform to the OSI standard network layer. Another consequence of using ATM in place of STM, however, is that there is no longer a need for a packet-switched network on top of ATM; ATM deals with bursty data directly as a packet-switching technique in its own right. We will see later that ATM, as a packet-switching layer, has an *adaptation layer* on top of it. This adaptation layer is much like the transport layer of OSI; it breaks up the incoming messages or bit streams into fixed-length packets (called cells in ATMese), and it is the responsibility of the adaptation layer, using the ATM layer, to reconstitute the messages or data stream at the other end. What comes in could be OSI layer 2 frames; the adaptation layer will view this as a user message, break it into "cells," and reconstitute the user message (*i.e.*, the frame) at the destination. Thus, as far as the OSI layer 2 is concerned, the broadband ISDN looks like a bit pipe. Naturally, there is no need for a user to go through all the OSI layers, since data can be presented directly to the broadband ISDN.

2.10.1 Asynchronous Transfer Mode (ATM)

ATM packets (or cells) all have the same length, 53 bytes, consisting of 5 bytes of header information followed by 48 bytes of data (see Fig. 2.54). The reason for choosing a fixed length was that implementation of fast packet switches seems to be somewhat simpler if packets arrive synchronously at the switch. One might question being so restrictive in a long-term standard because of a short-term implementation issue, but the standards committee clearly felt some pressure to develop a standard that would allow early implementation.

The major reason for choosing 48 bytes of data for this standard length was because of the packetization delay of voice. Standard 64 kbit/sec voice is generated by an eight bit sample of the voice waveform each 125 μsec. Thus, the time required to collect 48 bytes of voice data in a cell is 6 msec. Even if a cell is transmitted with no delay, the reconstruction of the analog voice signal at the destination must include this 6 msec delay. The problem now is that this reconstructed voice can be partially reflected at the destination and be added to the voice in the return direction. There is another 6 msec packetization delay for this echo on the return path, and thus an overall delay of 12 msec in the returning echo. A delayed echo is quite objectionable in speech, so it should either be kept small or be avoided by echo cancellation. In the absence of echo cancellation everywhere, it is desirable to keep the delay small.

There are several reasons why a larger cell length would have been desirable. One is that there is a fixed amount of computation that must be done on each cell in an ATM switch, and thus, the computational load on an ATM switch is inversely proportional to the amount of data in a cell. The other relates to the communication inefficiency due to the cell header (and as we shall see later, the header in the adaptation layer). The length of 48 bytes was a compromise between voice packetization delay and these reasons for wanting larger cells.

The format of the header for ATM cells has two slightly different forms, one for the access from user to subnet and the other within the subnet (see Fig. 2.55). The only difference between the header at the user–subnet interface and that inside the network

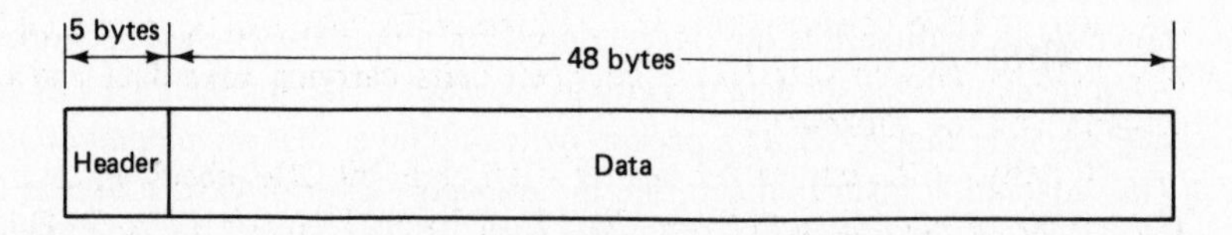

Figure 2.54 Packet (cell) format in ATM. Note that the cell size is fixed.

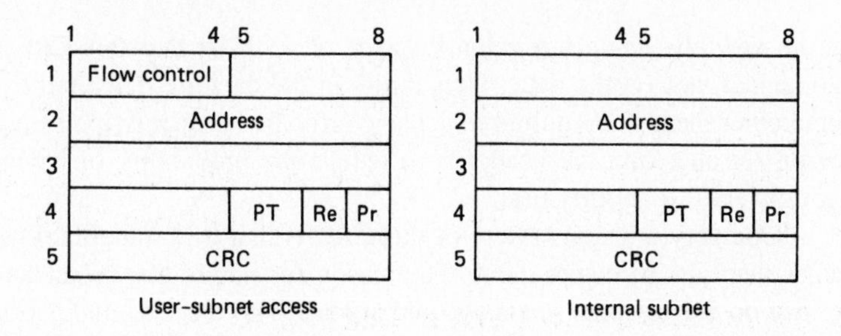

Figure 2.55 ATM header format for the user–subnet interface and for the internal subnet. PT stands for payload type, Re for reserved, and Pr for priority bit.

is a four bit "generic flow control" field provided at the user–subnet interface. This is intended to allow the user to mediate between the demands of different applications at the user site. The use of this field is not standardized and is used only to help the user to statistically multiplex the cells from different applications onto the access link to the network. The question of how to accomplish flow control within the broadband ISDN network is discussed in Section 2.10.3, but there is no field in the ATM header (other than the priority bit) to help accomplish this function.

The contents of the address field in the header can change as the cell travels from link to link, and aside from a detail to be discussed later, this address is used in the same way as the virtual channel numbers described for addressing in Section 2.8. That is, ATM uses virtual circuits, and when a virtual circuit is set up for a session, a virtual channel number is assigned to that session on each link along the session's path. Since the address field contains 24 bits for the user–network interface, over 16 million sessions can access the network from one user site. Since the "user" might be a corporation or a local area network, for example, this is not quite as extravagant as it sounds. Within the network, the address field contains 28 bits, allowing for over 268 million sessions to share a link.

There is an additional quirk about addressing in ATM. The address field is divided into two subfields, called the virtual channel identifier (VCI) and the virtual path identifier (VPI) in ATMese. The reason for this is to allow sessions sharing the same path (or at least sharing the same path over some portion of their entire paths) to be assigned the same virtual path identifier and to be switched together. For example, two office buildings of a given corporation might require a large number of virtual circuits between the two offices, but they could all use the same virtual path identifiers and be switched together within the network. The VCI subfield contains 16 bits, and the VPI subfield contains 8 bits at the user–subnet interface and 12 bits within the subnet.

After the address field comes the "payload type" field, the "reserved field," which is not used, and the "priority bit" field, which is discussed in Section 2.10.3. The payload type field is used to distinguish between cells carrying user data and cells containing network control information.

Finally, at the end of the header is an eight bit CRC checking on the ATM header. The generator polynomial of this CRC is $g(D) = D^8 + D^2 + D + 1$. This is the product of a primitive polynomial of degree 7 times $D + 1$, and, as explained in Section 2.3.4, the header thus has a minimum distance of 4. Note that this CRC checks only on the cell header, not on the data. Thus errors in the data are delivered to the destination and some other means is required to achieve error-free transmission of user information. The reason for checking the header is to reduce the probability of having data for one user get delivered to another user.

One very unusual feature of this CRC is that it is sometimes used to *correct* errors rather than just to detect them. The reason for this is that broadband ISDN is intended for use on optical fiber channels, and at least with current modulation techniques, errors are very improbable (the error rate is less than 10^{-9}) and more or less independent. Thus single error correction might actually remove most header errors. In an attempt to guard against bursts of errors due to equipment problems, the receiving node switches between single error correction and pure error detection as illustrated in Fig. 2.56. After either correcting an error or detecting more than a single error, the decoder switches to an error-detecting mode and stays there until the next header is received without detected errors. This is quite a novel approach to improving the reliability of optical channels in a simple way. The only problem is that optical technology is still rather fluid, and this approach is very dependent both on a very low channel error probability and on an extremely low probability of multiple errors.

We next investigate how ATM interfaces to the physical channel underneath. Since ATM performs the functions of a network layer, we would expect to find some sort of data link control layer beneath it. However, in this case neither ARQ nor framing is required, so there is no need for DLC. There is no need for ARQ for two reasons: first, the error probability is very low on the links, and second, a large part of the traffic is expected to be voice or video, where retransmission is inappropriate. Thus, error recovery is more appropriate on an end-to-end basis as required. In North America and Japan, ATM will use SONET as a lower layer and there are control bytes within the SONET frame structure that can specify the beginning of an ATM cell within the SONET frame. This of course explains why the user access rate for ATM was standardized at the SONET STS-3 rate. Since all cells are of the same length, framing information is not needed on each cell. Some European countries are not planning to use SONET as a lower layer to ATM, so they will require some other means of framing.

Since framing information is not provided on each cell, it is necessary to be careful about what to send when no data are available. One solution is to send "unassigned cells" which carry no data but fill up a cell slot. A special reserved virtual circuit number is used to identify this type of cell so that it can be discarded at each ATM switching point.

The use of SONET underneath ATM has a number of other advantages. One advantage is that very-high-data-rate applications could have the possibility of using SONET directly rather than being broken down into tiny ATM cells that must be individually

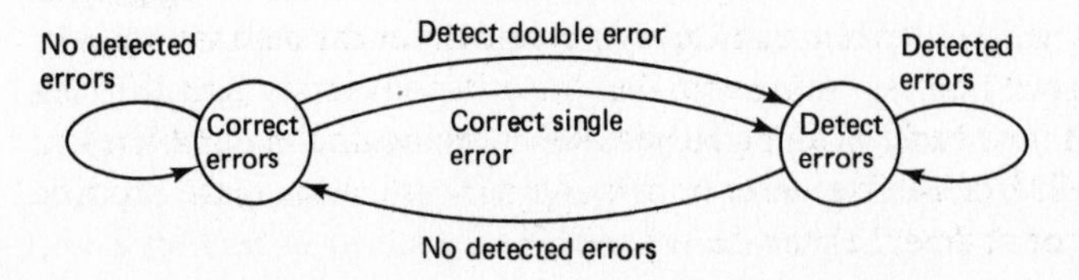

Figure 2.56 Error correction and detection on the headers of ATM cells. A single error is corrected only if the preceding cell header contained no detected errors.

switched at each node. Another is that SONET circuits carrying ATM traffic could be switched directly through intermediate nodes rather than being broken down at each switch into ATM cells. If one takes the ATM user access rate of 150 mbits/sec seriously, one can see that backbone links might be required to carry enormous data rates, and switches that operate on 53 bytes at a time are not attractive. If one looks further into the future and visualizes optical fibers carrying 10^{12} or 10^{13} bits/sec, the thought of ATM switches at every node is somewhat bizarre.

2.10.2 The Adaptation Layer

The adaptation layer of broadband ISDN sits on top of the ATM layer and has the function of breaking incoming source data into small enough segments to fit into the data portions of ATM cells. In the sense that ATM is performing the functions of a network layer, the adaptation layer performs the function of a transport layer. Thus the adaptation layer is normally used only on entry and exit from the network. Since the adaptation layer interfaces with incoming voice, video, and message data, its functions must depend on the type of input. The adaptation layer is not yet well defined within the standards bodies and is in a state of flux. Thus we give only a brief introduction to the way in which it accomplishes its functions.

Different types of input are separated into the following four major classes within the adaptation layer:

Class 1: Constant-bit-rate traffic. Examples of this are 64 kbit/sec voice, fixed-rate video, and leased lines for private data networks.

Class 2: Variable-bit-rate packetized data that must be delivered with fixed delay. Examples of this are packetized voice or video; the packetization allows for data compression, but the fixed delay is necessary to reconstruct the actual voice or video signal at the receiver.

Class 3: Connection-oriented data. Examples of this include all the conventional applications of data networks described in Section 1.1.3 (assuming that a connection is set up before transferring data). This class includes the signaling used within the subnet for network control and establishment of connections. Finally, it includes very-low-rate stream-type data.

Class 4: Connectionless data. Examples include the conventional applications of data networks in cases where no connection is set up (*i.e.*, datagram traffic).

For classes 2, 3, and 4, the source data coming into the adaptation layer are in the form of frames or messages. The adaptation layer must break these frames into segments that fit into the ATM cells and provide enough extra information for the destination to be able to reconstruct the original frames. Along with this, the different classes (and different applications within the classes) have differing requirements for treating the problems of dropped or misdirected ATM cells. We describe the approach taken for class 3 traffic and then briefly describe the differences for the other classes.

Class 3 (connection-oriented) traffic The adaptation layer is split into two sublayers, called the convergence sublayer and the segmentation and reassembly sublayer. The first treats the user frames as units and is concerned with flow control and error recovery. The second treats the segments of a frame as units and is concerned with reassemblying the segments into frames. As usual, peer communication between nodes at each sublayer takes place via headers and trailers as illustrated in Fig. 2.57.

Current plans call for a two byte header and two byte trailer in the segmentation and reassembly sublayer. The header consists of the following fields:

Segment type (2 bits). This distinguishes whether a segment is the beginning segment of a convergence sublayer frame, an intermediate segment, a final segment, or a segment that contains an entire frame.

Sequence number (4 bits). This numbers segments within a convergence sublayer frame; this helps check on dropped or misdirected cells.

Reserved (10 bits). This might be used to multiplex multiple-user sessions on a single virtual circuit.

The trailer consists of the following fields:

Length indicator (6 bits). This tells how many bytes from the convergence sublayer frame are contained in the given segment. This is useful primarily for the last segment of a frame.

CRC (10 bits). The CRC polynomial is $D^{10} + D^9 + D^5 + D^4 + D + 1$. This is a primitive polynomial times $D+1$ and has the properties given in Section 2.3.4. It is used to check on the entire segment, including the header and the length indicator of the trailer.

It should be clear that the header and trailer above provide enough data to reconstruct the convergence sublayer frames from the received segments. The sequence

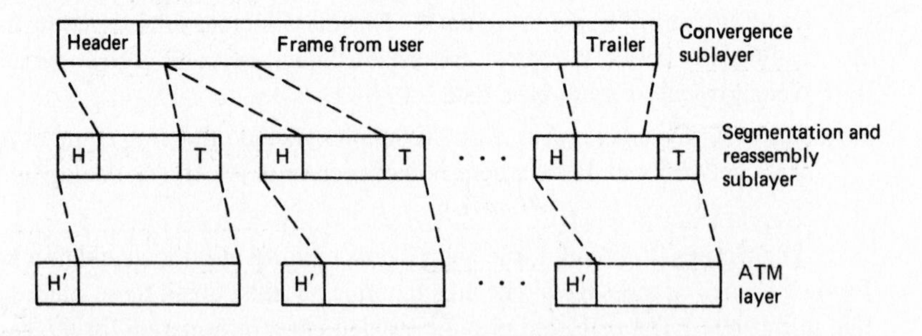

Figure 2.57 Headers and trailers at the convergence sublayer and segmentation and reassembly sublayer of the adaptation layer. Note that the user frame and its convergence sublayer header and trailer need not completely fill the data area at the segmentation and reassembly sublayer. The data plus header and trailer at the segmentation and reassembly sublayer must be exactly 48 bytes in length, however, to fit into the ATM data area.

number and CRC also provide some protection against bit errors and dropped or misdirected cells, but the approach is ad hoc.

The header and trailer for the convergence sublayer have not yet been defined. It is surprising that a CRC is used at the segmentation and reassembly sublayer rather than at the convergence sublayer. The longer block length at the convergence sublayer would have allowed greater error detection (or correction) power and higher efficiency. It would also have made it more convenient to use a CRC only when required by the user.

Class 4 (connectionless) traffic The headers and trailers at the segmentation and reassembly sublayer are expected to be the same for class 4 and class 3 traffic. The major new problem is that of connectionless traffic in a network where the cells are routed by virtual circuits. As discussed in Section 2.10.3, the most promising approach to this traffic is to view the routing as taking place at a higher layer than the adaptation layer; essentially this means viewing the datagram traffic as a higher-layer network using ATM virtual circuits as links (see Fig. 2.58). Thus, whereas connection-oriented traffic enters the adaptation layer only at origin and destination, connectionless traffic passes through the adaptation layer at each node used as a datagram switch. Between any two datagram switches, the traffic would use permanent virtual circuits assigned to carry all the connectionless traffic between the two switches. An interesting question is whether to view the switching of connectionless traffic as taking place in the convergence sublayer or strictly at a higher layer.

Class 1 and 2 traffic Class 1 traffic is constant-bit-rate traffic without any framing structure on the input to the adaptation layer. Thus a user data stream simply has to be broken into segments. The current plan is to use a one byte header and no trailer at the segmentation and reassembly sublayer (thus providing 47 bytes of user data per ATM cell). The header would consist of a four bit segment number plus a four bit CRC on the segment number. The reason for this is that occasional errors in this type

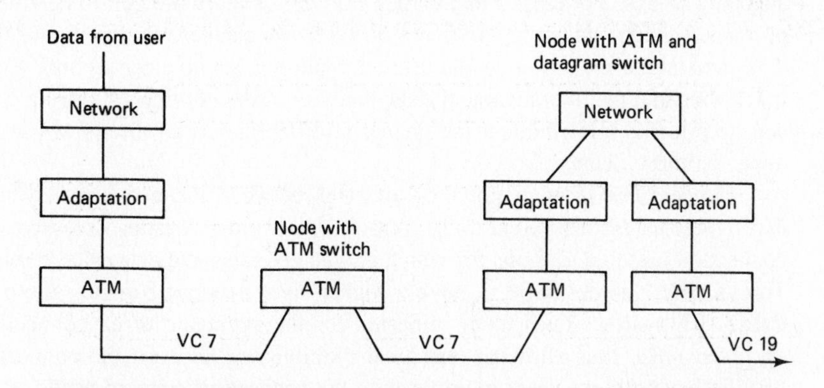

Figure 2.58 Datagram traffic can use an ATM network by being routed from one datagram switch to another. The virtual link between neighboring datagram switches is a virtual circuit in the ATM network dedicated to the aggregate of all the datagram traffic traveling between the two datagram switches.

of data are permissible, but to maintain framing, it is important to know when data have been dropped or misinserted, and the segment numbers provide this protection. Class 2 traffic has the same resilience to occasional errors as class 1 and the same need to know about dropped cells. At the same time, it has an incoming frame structure, requiring some of the features of the class 3 protocol for reassembling the frames.

2.10.3 Congestion

One would think that congestion could not be a problem in a network with the enormous link capacities of a broadband ISDN network. Unfortunately, this first impression is naive. Video and image applications can use large data rates, and marginal increases in quality will always appear to be possible by further increases in rate. Also, given a broadband network, there will be vested interests in ensuring that it is reasonably loaded, and reasonable loading will typically lead in time to heavy loading.

There are three mechanisms in broadband ISDN for limiting congestion. The first is for the network and user to agree to the required rate of the user at connection setup time. This agreement is expected to involve other quantities, such as the burstiness of the source and the quality of service required. The second mechanism is for the network to monitor each connection (within the convergence sublayer of the adaptation layer) to ensure that the user is complying with the agreed-upon rate and burstiness. Finally, the third mechanism is the priority bit in the ATM header (see Fig. 2.55). We explain briefly how these mechanisms work and interact. Chapter 6 develops the background required for a more thorough understanding of these mechanisms.

The ability to negotiate an agreed-upon rate between user and network is one of the primary reasons for a connection-oriented network. Without such negotiation, it would be impossible for the network to guarantee data rates and limited delay to users. It is easy to determine the effect of a new constant-bit-rate user, since such a session simply uses part of the available capacity on each link of the path without otherwise contributing to queueing delays. For class 2 and class 3 traffic, it is more difficult to determine the effect of a new session, since the burstiness of the traffic can cause queueing delays which can delay existing users as well as the user attempting to set up a connection. Relatively little is known about the interactions of delay between users of widely differing characteristics, and even less is known about the impact of different service disciplines at the switching nodes on these delays.

Since there is no negotiation of rates between the network and connectionless users, these users must be carefully controlled to avoid generating excessive delays for the connection-oriented sessions for which guaranteed rates and delays have been negotiated. This is why it is desirable to have a higher-layer network handling the connectionless traffic. This allows rates to be allocated for the aggregate of all connectionless traffic on given links, thus allowing statistical multiplexing between the connectionless users without having those users interfere with the connection-oriented traffic.

Along with negotiating rates when virtual circuits are established, it is necessary for the network to monitor and regulate the incoming traffic to ensure that it satisfies the agreed-upon limits. There are a number of mechanisms, such as the leaky bucket mechanism, for monitoring and regulating incoming traffic, and these are discussed in Chapter 6. The problem of interest here is what to do with traffic that exceeds the agreed-upon limits. The simplest approach is simply to discard such traffic, but this appears to be foolish when the network is lightly loaded. Another approach is to allow the excessive traffic into the network but to mark it so that it can be discarded later at any congested node.

The priority bit in the ATM header could be used to mark cells that exceed the negotiated rate limits of a user. This priority bit could also be set by a user to indicate lower-priority traffic that could be dropped by the network. Such a capability is useful in some video compression and voice compression schemes. It appears at first that using a single bit to carry all priority information in such a large and diverse network severely limits flexibility in applying flow control to the network. For example, suppose that the priority bit were used both by users to send relatively unimportant data and by the network to mark excessive data from a user. Then it would be possible for the network to be flooded by marked but unimportant video traffic; this could cause the network to discard important bursty traffic that was marked by the network because of exceeding the negotiated burstiness.

One possibility for an increased differentiation of priorities comes from the fact that individual users could negotiate priority while setting up a connection, and the subnet nodes could use these user priorities along with the priority bit in making discard decisions. It is also possible in principle to establish priorities concerning delay while setting up a connection, thus enabling the subnet switches to serve those virtual circuits requiring small delay before other circuits. Designing switches that are simple and yet account flexibly for both delay and discarding requirements is clearly a challenge, but the lack of information in the cell header does not seem to be a critical drawback.

Connectionless traffic poses a particular challenge to ATM networks. In the preceding section it was suggested that this traffic should be routed through a packet network sitting on top of the adaptation layer (or perhaps within the convergence sublayer). One benefit of this is that rate could be allocated to virtual circuits used as links in that higher-layer net. Assuming buffering in these higher-layer packet switches, the aggregate connectionless traffic could stay within its allocated limit and thus not interfere with any of the other classes of users. On the other hand, since the connectionless traffic is aggregated on these virtual circuits, individual connectionless users would obtain the benefits of statistical multiplexing. One could also adjust the rates of the virtual circuits over time to provide yet better service for connectionless traffic over periods of light loading from other classes.

SUMMARY

In this chapter, an introductory survey of the physical layer, a fairly complete treatment of the data link control layer, and a few closely related topics at other layers were presented. From a conceptual standpoint, one important issue has been distributed algorithms between two stations in the presence of errors and arbitrary delays. This has been developed in the context of ARQ, framing, and error recovery at higher layers. The correctness proof of the go back n class of protocols is important here since it shows how to reason carefully about such algorithms. There are many ad hoc algorithms used in data networks that contain subtle bugs, and it is important for designers to learn how to think about such algorithms in a more fundamental and structured way.

Another conceptual issue has been the encoding of control information. This was developed in the context of framing and session identification. Although the emphasis here was on efficiency of representation, the most important issue is to focus on the information actually conveyed; for example, it is important to recognize that flags with bit stuffing, length count fields, and special communication control characters all convey the same information but in quite different ways.

From a more practical standpoint, discussions were given on how modems work, how CRCs work (and fail), how ARQ works and how it can be made to work more efficiently with carefully chosen time-outs and control feedback, how various kinds of framing techniques work (and fail), how session identification can be accomplished, and how error recovery at different layers can be effected. Many existing networks have been used to illustrate these ideas.

Various standards have been discussed, but not in enough detail for one actually to implement them. Unfortunately, if one wishes to implement a protocol in conformance with a standard, one must read the documentation of the standard, which is long and tedious. The purpose here was to give some conceptual understanding of the standard so as to see what the standard accomplishes and how. There are many special-purpose networks in which standard protocols are inappropriate, and it is helpful to know why.

NOTES, SOURCES, AND SUGGESTED READING

Section 2.2. For more information on linear filters and their representation in the time and frequency domain, see any standard undergraduate text (*e.g.*, [Sie86]) on linear systems. For an extensive and up-to-date treatment of modulation and demodulation of digital signals, see [Pro83]. Also, [Qur85] provides a complete and beautifully organized treatment of adaptive equalization. For the information-theoretic aspects of this chapter (*i.e.*, Shannon's theorem, entropy, error detection and correction coding, and source coding), see [Gal68]. Finally, for an elementary treatment of the topics in this section oriented toward current practice, see [Sta85].

Section 2.3. Treatments of parity check codes and cyclic codes (CRC) can be found in [Bla83], [ClC81], and [Gal68].

Sections 2.4 and 2.5. A classic early paper on ARQ and framing is [Gra72]. An entire subdiscipline has developed to prove the correctness of point-to-point protocols. Typically (see, *e.g.*, [BoS82]) those approaches use a combination of exhaustive computer search plus analytic tools to reduce the search space. More information on highly efficient ARQ and framing is given in [Gal81]. Selective repeat and the advantages of sending packets multiple times after errors is discussed in [Wel82]. A more complete analysis of choice of frame lengths is given in [Alt86].

Sections 2.6 and 2.7. An excellent description of the standard DLCs is given in [Car80]. Initialization and disconnection are treated in [BaS83] and [Spi89].

Sections 2.8 and 2.9. TYMNET is discussed in [Rin76] and [Tym81]. The Codex networks are discussed in [HuS86]. See [Ryb80] for more discussion of the X.25 standard and [Com88] for discussion of TCP/IP.

Section 2.10. Basic tutorials on ATM and SONET are given in [Min89] and [BaC89], respectively. The CCITT and T1S1 working documents on ATM are surprisingly readable for an in-depth treatment.

PROBLEMS

2.1 Suppose that the output in Fig. 2.3(a) is $1 - e^{-2t/T}$ for $0 \leq t \leq T$, $(e^2 - 1)e^{-2t/T}$ for $t > T$, and zero for $t < 0$. Find the output in Fig. 2.3(b) using the linearity and time invariance of the filter.

2.2 Let $s(t) = 1$, for $0 \leq t \leq T$, and $s(t) = 0$ elsewhere. Let $h(t) = \alpha e^{-\alpha t}$ for $t \geq 0$ and $h(t) = 0$ for $t < 0$. Use the convolution equation to find the output when $s(t)$ is passed through a filter of impulse response $h(t)$.

2.3 Integrate Eq. (2.1) for $s(\tau) = e^{j2\pi f\tau}$ to get Eqs. (2.2) and (2.3). *Hint*: Transform the variable of integration by $\tau' = t - \tau$.

2.4 Suppose that a channel has the ideal low-pass frequency response $H(f) = 1$ for $-f_0 \leq f \leq f_0$ and $H(f) = 0$ elsewhere. Find the impulse response of the channel. [Use the inverse Fourier transform, Eq. (2.6).]

2.5 Let $s(t)$ be a given waveform and let $s_1(t) = s(\beta t)$ for some given positive β. Sketch the relationship between $s(t)$ and $s_1(t)$ for $\beta = 2$. Let $S(f)$ be the Fourier transform of $s(t)$. Find $S_1(f)$, the Fourier transform of $s_1(t)$, in terms of $S(f)$. Sketch the relationship for $\beta = 2$ [assume $S(f)$ real for your sketch].

2.6 Let $S(f)$ be the Fourier transform of $s(t)$.

(a) Show that the Fourier transform of $s(t)\cos(2\pi f_0 t)$ is $[S(f - f_0) + S(f + f_0)]/2$.

(b) Find the Fourier transform of $s(t)\cos^2(2\pi f_0 t)$ in terms of $S(f)$.

2.7 Suppose that the expected frame length on a link is 1000 bits and the standard deviation is 500 bits.

(a) Find the expected time to transmit a frame on a 9600-bps link and on a 50,000-bps link.

(b) Find the expected time and standard deviation of the time required to transmit 10^6 frames on each of the links above (assume that all frame lengths are statistically independent of each other).

(c) The *rate* at which frames can be transmitted is generally defined as the reciprocal of the expected transmission time. Using the result in part (b), discuss whether this definition is reasonable.

2.8 Show that the final parity check in a horizontal and vertical parity check code, if taken as the modulo 2 sum of all the data bits, is equal to the modulo 2 sum of the horizontal parity checks and also equal to the modulo 2 sum of the vertical parity checks.

2.9 **(a)** Find an example of a pattern of six errors that cannot be detected by the use of horizontal and vertical parity checks. *Hint*: Each row with errors and each column with errors will contain exactly two errors.

(b) Find the number of different patterns of four errors that will not be detected by a horizontal and vertical parity check code; assume that the array is K bits per row and J bits per column.

2.10 Suppose that a parity check code has minimum distance d. Define the distance between a code word and a received string (of the same length) as the number of bit positions in which the two are different. Show that if the distance between one code word and a given string is less than $d/2$, the distance between any other code word and the given string must exceed $d/2$. Show that if a decoder maps a given received string into a code word at smallest distance from the string, all combinations of fewer than $d/2$ errors will be corrected.

2.11 Consider a parity check code with three data bits and four parity checks. Suppose that three of the code words are 1001011, 0101101, and 0011110. Find the rule for generating each of the parity checks and find the set of all eight code words. What is the minimum distance of this code?

2.12 Let $g(D) = D^4 + D^2 + D + 1$, and let $s(D) = D^3 + D + 1$. Find the remainder when $D^4 s(D)$ is divided by $g(D)$, using modulo 2 arithmetic.

2.13 Let $g(D) = D^L + g_{L-1}D^{L-1} + \cdots + g_1 D + 1$ (assume that $L > 0$). Let $z(D)$ be a nonzero polynomial with highest and lowest order terms of degree j and i, respectively; that is, $z(D) = D^j + z_{j-1}D^{j-1} + \cdots + D^i$ [or $z(D) = D^j$, if $i = j$]. Show that $g(D)z(D)$ has at least two nonzero terms. *Hint*: Look at the coefficients of D^{L+j} and of D^i.

2.14 Show that if $g(D)$ contains the factor $1 + D$, then all error sequences with an odd number of errors are detected. *Hint*: Recall that a nonzero error polynomial $e(D)$ is detected unless $e(D) = g(D)z(D)$ for some polynomial $z(D)$. Look at what happens if 1 is substituted for D in this equation.

2.15 For a given generator polynomial $g(D)$ of degree L and given data length K, let $c^{(i)}(D) = c^{(i)}_{L-1}D^{L-1} + \cdots + c^{(i)}_1 D + c^{(i)}_0$ be the CRC resulting from the data string with a single 1 in position i [*i.e.*, $s(D) = D^i$ for $0 \leq i \leq K - 1$].

(a) For an arbitrary data polynomial $s(D)$, show that the CRC polynomial is $c(D) = \sum_{i=0}^{K-1} s_i c^{(i)}(D)$ (using modulo 2 arithmetic).

(b) Letting $c(D) = c_{L-1}D^{L-1} + \cdots + c_1 D + c_0$, show that

$$c_j = \sum_{i=0}^{K-1} s_i c^{(i)}_j; \qquad 0 \leq j < L$$

This shows that each c_j is a parity check and that a cyclic redundancy check code is a parity check code.

2.16 **(a)** Consider a stop-and-wait ARQ strategy in which, rather than using a sequence number for successive packets, the sending DLC sends the number of times the given packet has been retransmitted. Thus, the format of the transmitted frames is | j | packet | CRC |, where j is 0 the first time a packet is transmitted, 1 on the first retransmission, and so on. The receiving DLC returns an ack or nak (without any request number) for each frame it receives. Show by example that this strategy does not work correctly no matter what rule the receiving DLC uses for accepting packets. (Use the assumptions preceding Section 2.4.1.)

(b) Redo part (a) with the following two changes: (1) the number j above is 0 on the initial transmission of a packet and 1 on *all* subsequent retransmissions of that packet, and (2) frames have a fixed delay in each direction, are always recognized as frames, and the sending DLC never times-out, simply waiting for an ack or either a nak or error.

2.17 **(a)** Let T_t be the expected transmission time for sending a data frame. Let T_f be the feedback transmisson time for sending an ack or nak frame. Let T_d be the expected propagation and processing delay in one direction (the same in each direction). Find the expected time T between successive frame transmissions in a stop-and-wait system (assume that there are no other delays and that no frames get lost).

(b) Let p_t be the probability of frame error in a data frame (independent of frame length), and p_f the probability of error in a feedback ack or nak frame. Find the probability q that a data packet is correctly received and acked on a given transmission. Show that $1/q$ is the expected number of times a packet must be transmitted for a stop-and-wait system (assume independent errors on all frames).

(c) Combining parts (a) and (b), find the expected time required per packet. Evaluate for $T_t = 1$, $T_f = T_d = 0.1$, and $p_t = p_f = 10^{-3}$.

2.18 Redraw Figs. 2.25 and 2.26 with the same frame lengths and the same set of frames in error, but focusing on the packets from B to A. That is, show *SN* and the window for node B, and show RN and packets out for node A (assume that all frames are carrying packets).

2.19 Give an example in which go back n will deadlock if receiving DLCs ignore the request number RN in each frame not carrying the awaited packet. *Hint*: Construct a pattern of delays leading to a situation where RN at node B is n plus SN_{min} at node A, and similarly for A and B reversed.

2.20 Give an example in which go back n ARQ fails if the modulus m is equal to n. For $m > n$, give an example where the right hand inequality in Eq. (2.24) is satisfied with equality and an example where the right hand inequality in Eq. (2.25) is satisfied with equality.

2.21 Assume that at time t, node A has a given SN_{min} as the smallest packet number not yet acknowledged and $SN_{max} - 1$ as the largest number yet sent. Let T_m be the minimum packet transmission time on a link and T_d be the propagation delay. Show that RN at node B must lie in the range from SN_{min} to SN_{max} for times in the range from $t - T_m - T_d$ to $t + T_m + T_d$.

2.22 **(a)** Let T_{min} be the minimum transmission time for data frames and T_d be the propagation and processing delay in each direction. Find the maximum allowable value T_{max} for frame transmission time such that a go back n ARQ system (of given n) will never have to go back or wait in the absence of transmission errors or lost frames. *Hint*: Look at Fig. 2.31.

(b) Redo part (a) with the added possibility that isolated errors can occur in the feedback direction.

2.23 Assume that frame transmission times τ are exponentially distributed with, for example, probability density $p(\tau) = e^{-\tau}$, for $\tau \geq 0$. Assume that a new frame always starts in each direction as soon as the previous one in that direction terminates. Assume that transmission times are all independent and that processing and propagation delays are negligible. Show that the probability q that a given frame is not acked before $n - 1$ additional frames have been sent (*i.e.*, the window is exhausted) is $q = (1 + n)2^{-n}$. *Hint*: Note that the set of times at which frame transmissions terminate at either node is a Poisson process and that a point in this process is equally likely to be a termination at A or B, with independence between successive terminations.

2.24 Show that if an isolated error in the feedback direction occurs on the ack of the given packet, then q in Problem 2.23 becomes $q = [2 + n + (n+1)n/2]2^{-n-1}$.

2.25 Assume that frame transmission times τ have an Erlang distribution with probability density $p(\tau) = \tau e^{-\tau}$. Find the corresponding value of q in Problem 2.23. *Hint*: The termination times at A can be regarded as alternative points in a Poisson process and the same is true for B.

2.26 Let γ be the expected number of transmitted frames from A to B per successfully accepted packet at B. Let β be the expected number of transmitted frames from A to B between the transmission of a given frame and the reception of feedback about that frame (including the frame in transmission when the feedback arrives). Let p be the probability that a frame arriving at B contains errors (with successive frames assumed independent). Assume that A is always busy transmitting frames, that n is large enough that A never goes back in the absence of feedback, and that A always goes back on the next frame after hearing that the awaited frame contained errors. Show that γ satisfies $\gamma = 1 + p(\beta + \gamma)$. Define the efficiency η as $1/\gamma$, and find η as a function of β and p.

2.27 Assume that a selective repeat system has the property that the sending DLC always finds out whether a given transmitted packet was successfully received during or before the βth frame transmission following the frame in which the given packet was transmitted. Give a careful demonstration that the sending DLC need never save more than $\beta + 1$ unacknowledged packets.

2.28 Find the efficiency of the ARPANET ARQ system under the assumptions that all eight virtual channels are always busy sending packets and that feedback always arrives about the fate of a given packet during or before the seventh frame transmission following the frame carrying the given packet.

2.29 Consider a generalized selective repeat ARQ system that operates in the following way: as in conventional go back n and selective repeat, let RN be the smallest numbered packet not yet correctly received, and let SN_{min} be the transmitter's estimate of RN. Let y_{top} be the largest-numbered packet correctly received and accepted (thus, for go back n, y_{top} would be $RN - 1$, whereas for conventional selective repeat, y_{top} could be as large as $RN + n - 1$). The rule at the transmitter is the same as for conventional go back n or selective repeat: The number z of the packet transmitted must satisfy

$$SN_{min} \le z \le SN_{min} + n - 1$$

The rule at the receiver, for some given positive number k, is that a correctly received packet with number z is accepted if

$$RN \le z \le y_{\text{top}} + k \tag{2.46}$$

(a) Assume initially that z (rather than $SN = z \bmod m$) is sent with each packet and that RN (rather than $RN \bmod m$) is sent with each packet in the reverse direction. Show that for each received packet at the receiver,

$$z \le RN + n - 1 \tag{2.47}$$

$$z \ge y_{\text{top}} - n + 1$$

(b) Now assume that $SN = z \bmod m$ and RN is sent mod m. When z is received Eq. (2.47) is satisfied, but erroneous operation will result if $z - m$ or $z + m$ lie in the window specified by Eq. (2.46). How large need m be to guarantee that $z + m > y_{\text{top}} + k$ [*i.e.*, that $z + m$ is outside the window of Eq. (2.46)]?

(c) Is the value of m determined in part (b) large enough to ensure that $z - m < RN$?

(d) How large need m be (as a function of n and k) to ensure correct operation?

(e) Interpret the special cases $k = 1$ and $k = n$.

2.30 (a) Apply the bit stuffing rules of Section 2.5.2 to the following frame:
0110111110011111101011111 1111101111010

(b) Suppose that the following string of bits is received:
011111101111110110011111100 11111011111011
000111111010111110
Remove the stuffed bits and show where the actual flags are.

2.31 Suppose that the bit stuffing rule of Section 2.5.2 is modified to stuff a 0 only after the appearance of 01^5 in the original data. Carefully describe how the destuffing rule at the receiver must be modified to work with this change. Show how your rule would destuff the following string:
0110111110111111011111010 1111110
If your destuffing rule is correct, you should remove only two 0's and find only one actual flag.

2.32 Bit stuffing must avoid the appearance of 01^6 within the transmitted frame, so we accept as given that an original string 01^6 will always be converted into 01^501. Use this, plus the necessity to destuff correctly at the receiver, to show that a 0 must always be stuffed after 01^5. *Hint*: Consider the data string $01^501x_1x_2\ldots$. If a 0 is not stuffed after 01^5, the receiver cannot distinguish this from $01^6x_1x_2\ldots$ after stuffing, so stuffing is required in this case. Extend this argument to $01^50^k1x_1x_2\ldots$ for any $k > 1$.

2.33 Suppose that the string 0101 is used as the bit string to indicate the end of a frame and the bit stuffing rule is to insert a 0 after each appearance of 010 in the original data; thus, 010101 would be modified by stuffing to 01001001. In addition, if the frame proper ends in 01, a 0 would be stuffed after the first 0 in the actual terminating string 0101. Show how the string 11011010010101011101 would be modified by this rule. Describe the destuffing rule required at the receiver. How would the string 11010001001001100101 be destuffed?

2.34 Let $A = E\{K\}2^{-j} + j + 1$ be the upper bound on overhead in Eq. (2.34), and let $j = \lfloor \log_2 E\{K\} \rfloor$. Show that A satisfies

$$1.914\ldots + \log_2 E\{K\} \leq A \leq 2 + \log_2 E\{K\}$$

Find the analytic expression for 1.914.... *Hint*: Define γ as $\log_2 E\{K\} - j$ and find the minimum and maximum of $A - \log_2 E\{K\}$ as γ varies between 0 and 1.

2.35 Show that the probability that errors will cause a flag 01^60 to appear within a transmitted frame is approximately $(1/32)Kp$, where K is the number of bits in the frame proper before stuffing and p is the probability of bit error. Assume that p is very small and that the probability of multiple errors in a frame is negligible. Assume that the bits of the original frame are IID (independent, identically distributed) with equal probability of 0 and 1. *Hint*: This is trickier than it appears, since the bits after stuffing are no longer IID. First, find the probability that a stuffed bit in error causes a flag (not an abort) to appear and use the analysis of Section 2.5.2 to approximate the expected number of stuffed bits as $K2^{-6}$. Then find the probability of a flag appearing due to errors in the original bits of the frame.

2.36 A certain virtual circuit network is designed for the case where all sessions carry heavy, almost steady traffic. It is decided to serve the various sessions using each given link in round-robin order, first sending one packet from the first session, then one from the second, and so on, up to the last, and then starting over with the first, thus avoiding the necessity of sending a session number with each packet. To handle the unlikely possibility that a session has no packet to send on a link at its turn, we place a k-bit flag, 01^{k-1}, in front of the information packet for the next session that has a packet. This flag is followed by a unary encoding of the number of sessions in order with nothing to send, 1 for one session, 01 for two, 001 for three, and so on. Thus, for example, if a packet for the second session has just been sent, sessions 3 and 4 have no packets, and session 5 has the packet $\overline{x}_5$, we send $01^{k-1}01\overline{x}_5$ as the next packet. In addition, an insertion is used if an information packet happens to start with 01^{k-2}. Otherwise, insertions are not required in information packets, and information packets carry their own length information.

(a) Assume that p is the probability that a session has no packet when it is that session's turn; this event is assumed independent of other sessions and of past history. Find the expected number of overhead bits per transmitted packet for flags, unary code, and insertions. Ignore, as negligible, any problems of what to do if no session has a packet to send.

(b) Suppose that the packets above, as modified by flags, unary code, and insertions, are the inputs to a data link control system using ARQ for error control and flags to indicate the end of packets. Explain if any problems arise due to the use of flags both on the DLC system and in the higher-level scheme for identifying packets with sessions.

2.37 Consider a stop-and-wait data link control protocol. Each transmitted frame contains the sequence number, modulo 2, of the contained packet. When a frame arrives at the receiver, the sequence number of the packet is compared with the packet number awaited by the receiver; if the numbers are the same modulo 2 and the CRC is satisfied, the packet is accepted and the awaited packet number is incremented. An ack containing the new awaited packet number modulo 2 is sent back to the transmitter. If a frame is received with correct CRC but the wrong sequence number, an ack (containing the awaited packet number modulo 2) is also sent back to the transmitter. The new twist here is that frames can go out of order on the channel. More precisely, there is a known maximum delay T on the channel. If a packet is transmitted at some time t, it is either not received at all or is received at an arbitrary time in the interval from t to $t + T$, independently of other transmissions. The return channel for the acks behaves the same way with the same maximum delay T. Assume that the receiver sends acks instantaneously upon receiving packets with valid CRCs.

(a) Describe rules for the transmitter to follow so as to ensure that each packet is eventually accepted, once and in order, by the receiver. You are not to put any extra protocol information in the packets nor to change the receiver's operation. You should try to leave as much freedom as possible to the transmitter's operation subject to the restriction of correct operation.

(b) Explain why it is impossible to achieve correct operation if there is no bound on delay.

2.38 Consider a stop-and-wait system with two-way traffic between nodes A and B. Assume that data frames are all of the same length and require D seconds for transmission. Acknowledgment frames require R seconds for transmission and there is a propagation delay P on the link. Assume that A and B both have an unending sequence of packets to send, assume that no transmission errors occur, and assume that the time-out interval at which a node resends a previously transmitted packet is very large. Finally, assume that each node sends new data packets and acknowledgments as fast as possible, subject to the rules of the stop-and-wait protocol.

(a) Show that the rate at which packets are transmitted in each direction is $(D+R+2P)^{-1}$. Show that this is true whether or not the starting time between A and B is synchronized.

(b) Assume, in addition, that whenever a node has both an acknowledgment and a data packet to send, the acknowledgment is piggybacked onto the data frame (which still requires D seconds). Assume that node B does not start transmitting data until the instant that it receives the first frame from A (*i.e.*, B's first frame contains a piggybacked ack). Show that all subsequent frames in each direction are data frames with a piggybacked ack. Show that the rate in each direction is now $(2D+2P)^{-1}$. Note that if $D > R$, this rate is less than that in part (a).

2.39 **(a)** Consider a data link control using fixed-length packets of K bits each. Each message is broken into packets using fill in the last packet of the message as required; for example, a message of 150 bits is broken into two packets of 100 bits each, using 50 bits of fill. Modify Eq. (2.42) to include fill for fixed-length packets.

(b) Find $E\{TC\}$ using the approximation $E\{\lceil M/K \rceil\} \approx E\{M/K\}+1/2$; find the value of K that minimizes this approximation to $E\{TC\}$.

(c) Note that for $j = 1$ the value of K_{max} in Eq. (2.44) is infinite, whereas the minimizing fixed length K is finite. Explain why this is reasonable.

2.40 **(a)** Assume that nodes A and B both regard the link between them as up according to the balanced initialization and disconnect protocol of Section 2.7.3. Assume that at time t node A starts to disconnect (*i.e.*, transmits DISC). Show, using the same assumptions as in Section 2.4, that each node eventually regards the link as down.

(b) Give an example for the situation in part (a), where node A regards the link as down and then starts to initialize the link before node B regards the link as down; check your proof for part (a) in light of this example. Show that B regards the link as down before A again regards the link as up.

(c) Now suppose that node A regards the link as up and node B is initializing the link but has not yet received ACKI from A. Suppose that A starts to disconnect. Show that B eventually regards the link as up and then starts to disconnect and that after that A and B both eventually regard the link as down.

2.41 **(a)** Create an example of incorrect operation, analogous to that of Fig. 2.46, for the master–slave initialization protocol under the assumption that a node, after recovering from a node failure, starts in the up state and initializes by first sending DISC and then INIT. *Hint:* Use the same structure as Fig. 2.46; that is, assume that after each failure, node A receives the messages from B that were generated by B in response to the messages in the earlier period at A.

(b) Now suppose that an arbitrary protocol is being used for initialization. It is known that node A, on recovery from a failure, will send message x. Node B, on recovering from failure and receiving message x, will send message y, and node A, upon receiving y under these circumstances, will be correctly initialized. Construct a sequence of failures and delays such that the protocol fails, assuming only that the nodes operate as above under the above conditions. *Hint:* Follow the hint in part (a), but also assume a sequence of failures at B, with B receiving responses in one period to messages in the preceding period.

2.42 The balanced initialization protocol of Section 2.7.3 piggybacks acks on each INIT or DISC message. Explain why it is a bad idea to include also either an INIT or DISC message every time that an ACKI or ACKD must be sent.

2.43 Consider a datagram network and assume that M is the maximum number of packets that can be sent by a session while a given packet (in either direction) still exists within the network. Assume that selective repeat ARQ is used for the session with a window size of n. Show that the modulus m must satisfy

$$m \geq 2n + M$$

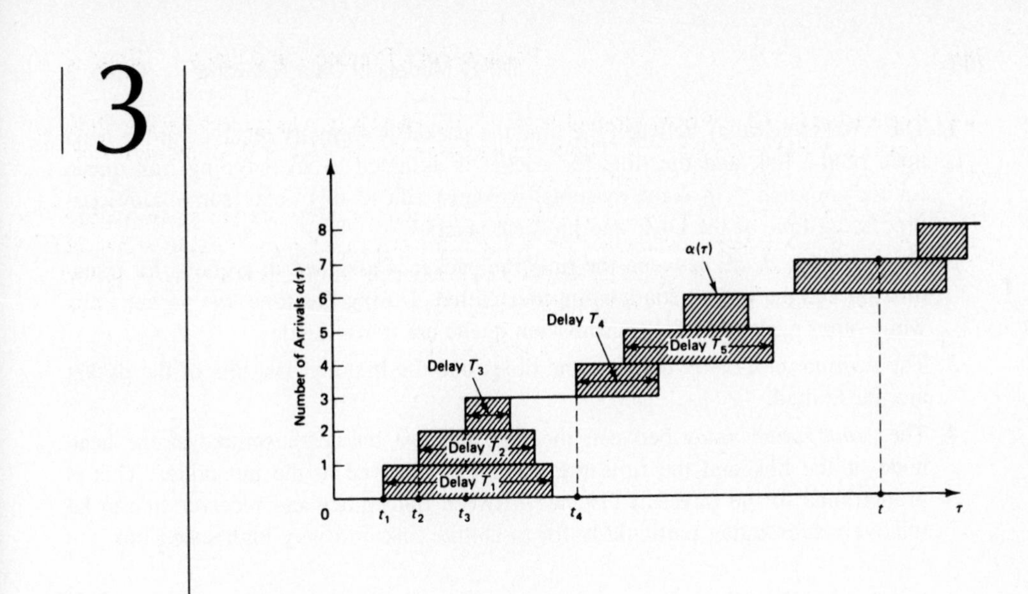

3 Delay Models in Data Networks

3.1 INTRODUCTION

One of the most important performance measures of a data network is the average delay required to deliver a packet from origin to destination. Furthermore, delay considerations strongly influence the choice and performance of network algorithms, such as routing and flow control. For these reasons, it is important to understand the nature and mechanism of delay, and the manner in which it depends on the characteristics of the network.

Queueing theory is the primary methodological framework for analyzing network delay. Its use often requires simplifying assumptions since, unfortunately, more realistic assumptions make meaningful analysis extremely difficult. For this reason, it is sometimes impossible to obtain accurate quantitative delay predictions on the basis of queueing models. Nevertheless, these models often provide a basis for adequate delay approximations, as well as valuable qualitative results and worthwhile insights.

In what follows, we will mostly focus on packet delay within the communication subnet (*i.e.*, the network layer). This delay is the sum of delays on each subnet link traversed by the packet. Each link delay in turn consists of four components.

1. The *processing delay* between the time the packet is correctly received at the head node of the link and the time the packet is assigned to an outgoing link queue for transmission. (In some systems, we must add to this delay some additional processing time at the DLC and physical layers.)
2. The *queueing delay* between the time the packet is assigned to a queue for transmission and the time it starts being transmitted. During this time, the packet waits while other packets in the transmission queue are transmitted.
3. The *transmission delay* between the times that the first and last bits of the packet are transmitted.
4. The *propagation delay* between the time the last bit is transmitted at the head node of the link and the time the last bit is received at the tail node. This is proportional to the physical distance between transmitter and receiver; it can be relatively substantial, particularly for a satellite link or a very high speed link.

This accounting neglects the possibility that a packet may require retransmission on a link due to transmission errors or various other causes. For most links in practice, other than multiaccess links to be considered in Chapter 4, retransmissions are rare and will be neglected. The propagation delay depends on the physical characteristics of the link and is independent of the traffic carried by the link. The processing delay is also independent of the amount of traffic handled by the corresponding node if computation power is not a limiting resource. This will be assumed in our discussion. Otherwise, a separate processing queue must be introduced prior to the transmission queues. Most of our subsequent analysis focuses on the queueing and transmission delays. We first consider a single transmission line and analyze some classical queueing models. We then take up the network case and discuss the type of approximations involved in deriving analytical delay models.

While our primary emphasis is on packet-switched network models, some of the models developed are useful in a circuit-switched network context. Indeed, queueing theory was developed extensively in response to the need for performance models in telephony.

3.1.1 Multiplexing of Traffic on a Communication Link

The communication link considered is viewed as a bit pipe over which a given number of bits per second can be transmitted. This number is called the *transmission capacity* of the link. It depends on both the physical channel and the interface (*e.g.*, modems), and is simply the rate at which the interface accepts bits. The link capacity may serve several traffic streams (*e.g.*, virtual circuits or groups of virtual circuits) multiplexed on the link. The manner of allocation of capacity among these traffic streams has a profound effect on packet delay.

In the most common scheme, *statistical multiplexing*, the packets of all traffic streams are merged into a single queue and transmitted on a first-come first-serve basis. A variation of this scheme, which has roughly the same average delay per packet, maintains a separate queue for each traffic stream and serves the queues in sequence one packet at a time. However, if the queue of a traffic stream is empty, the next traffic stream is served and no communication resource is wasted. Since the entire transmission capacity C (bits/sec) is allocated to a single packet at a time, it takes L/C seconds to transmit a packet that is L bits long.

In *time-division* (TDM) *and frequency-division multiplexing* (FDM) with m traffic streams, the link capacity is essentially subdivided into m portions—one per traffic stream. In FDM, the channel bandwidth W is subdivided into m channels each with bandwidth W/m (actually slightly less because of the need for guard bands between channels). The transmission capacity of each channel is roughly C/m, where C is the capacity that would be obtained if the entire bandwidth were allocated to a single channel. The transmission time of a packet that is L bits long is Lm/C, or m times larger than in the corresponding statistical multiplexing scheme. In TDM, allocation is done by dividing the time axis into slots of fixed length (*e.g.*, one bit or one byte long, or perhaps one packet long for fixed length packets). Again, conceptually, we may view the communication link as consisting of m separate links with capacity C/m. In the case where the slots are short relative to packet length, we may again regard the transmission time of a packet L bits long as Lm/C. In the case where the slots are of packet length, the transmission time of an L bit packet is L/C, but there is a wait of $(m - 1)$ packet transmission times between packets of the same stream.

One of the themes that will emerge from our queueing analysis is that statistical multiplexing has smaller average delay per packet than either TDM or FDM. This is particularly true when the traffic streams multiplexed have a relatively low duty cycle. The main reason for the poor delay performance of TDM and FDM is that communication resources are wasted when allocated to a traffic stream with a momentarily empty queue, while other traffic streams have packets waiting in their queue. For a traffic analogy, consider an m-lane highway and two cases. In one case, cars are not allowed to cross over to other lanes (this corresponds to TDM or FDM), while in the other case, cars can change lanes (this corresponds roughly to statistical multiplexing). Restricting crossover increases travel time for the same reason that the delay characteristics of TDM or FDM are poor: namely, some system resources (highway lanes or communication channels) may not be utilized, while others are momentarily stressed.

Under certain circumstances, TDM or FDM may have an advantage. Suppose that each traffic stream has a "regular" character (*i.e.*, all packets arrive sufficiently apart so that no packet has to wait while the preceding packet is transmitted.) If these traffic streams are merged into a single queue, it can be shown that the average delay per packet will decrease, but the variance of waiting time in queue will generally become positive (for an illustration, see Prob. 3.7). Therefore, if maintaining a small variability of delay is more important than decreasing delay, it may be preferable to use TDM or FDM. Another advantage of TDM and FDM is that there is no need to include identification of the traffic stream on each packet, thereby saving some overhead and simplifying packet processing at the nodes. Note also that when overhead is negligible, one can afford to make packets very small, thereby reducing delay through pipelining (cf. Fig. 2.37).

3.2 QUEUEING MODELS—LITTLE'S THEOREM

We consider queueing systems where customers arrive at random times to obtain service. In the context of a data network, customers represent packets assigned to a communication link for transmission. Service time corresponds to the packet transmission time and is equal to L/C, where L is the packet length in bits and C is the link transmission capacity in bits/sec. In this chapter it is convenient to ignore the layer 2 distinction between packets and frames; thus packet lengths are taken to include frame headers and trailers. In a somewhat different context (which we will not emphasize very much), customers represent ongoing conversations (or virtual circuits) between points in a network and service time corresponds to the duration of a conversation. In a related context, customers represent active calls in a telephone or circuit switched network and again service time corresponds to the duration of the call.

We shall be typically interested in estimating quantities such as:

1. The average number of customers in the system (*i.e.*, the "typical" number of customers either waiting in queue or undergoing service)
2. The average delay per customer (*i.e.*, the "typical" time a customer spends waiting in queue plus the service time).

These quantities will be estimated in terms of known information such as:

1. The customer arrival rate (*i.e.*, the "typical" number of customers entering the system per unit time)
2. The customer service rate (*i.e.*, the "typical" number of customers the system serves per unit time when it is constantly busy)

In many cases the customer arrival and service rates are not sufficient to determine the delay characteristics of the system. For example, if customers tend to arrive in groups, the average customer delay will tend to be larger than when their arrival times are regularly spaced apart. Thus to predict average delay, we will typically need more detailed (statistical) information about the customer interarrival and service times. In this section, however, we will largely ignore the availability of such information and see how far we can go without it.

3.2.1 Little's Theorem

We proceed to clarify the meaning of the terms "average" and "typical" that we used somewhat liberally above in connection with the number of customers in the system, the customer delay, and so on. In doing so we will derive an important result known as *Little's Theorem.*

Suppose that we observe a sample history of the system from time $t = 0$ to the indefinite future and we record the values of various quantities of interest as time progresses. In particular, let

$$N(t) = \text{Number of customers in the system at time } t$$

$$\alpha(t) = \text{Number of customers who arrived in the interval } [0, t]$$

$$T_i = \text{Time spent in the system by the } i^{\text{th}} \text{ arriving customer}$$

Our intuitive notion of the "typical" number of customers in the system observed up to time t is

$$N_t = \frac{1}{t} \int_0^t N(\tau)\, d\tau$$

which we call the *time average of* $N(\tau)$ *up to time* t. Naturally, N_t changes with the time t, but in many systems of interest, N_t tends to a steady-state N as t increases, that is,

$$N = \lim_{t \to \infty} N_t$$

In this case, we call N the *steady-state time average* (or simply time average) of $N(\tau)$. It is also natural to view

$$\lambda_t = \frac{\alpha(t)}{t}$$

as the *time average arrival rate* over the interval $[0, t]$. The *steady-state arrival rate* is defined as

$$\lambda = \lim_{t \to \infty} \lambda_t$$

(assuming that the limit exists). The *time average of the customer delay up to time* t is similarly defined as

$$T_t = \frac{\sum_{i=0}^{\alpha(t)} T_i}{\alpha(t)} \tag{3.1}$$

that is, the average time spent in the system per customer up to time t. The *steady-state time average customer delay* is defined as

$$T = \lim_{t \to \infty} T_t$$

(assuming that the limit exists).

It turns out that the quantities N, λ, and T above are related by a simple formula that makes it possible to determine one given the other. This result, known as Little's Theorem, has the form

$$N = \lambda T$$

Little's Theorem expresses the natural idea that crowded systems (large N) are associated with long customer delays (large T) and reversely. For example, on a rainy day, traffic on a rush hour moves slower than average (large T), while the streets are more crowded (large N). Similarly, a fast-food restaurant (small T) needs a smaller waiting room (small N) than a regular restaurant for the same customer arrival rate.

The theorem is really an accounting identity and its derivation is very simple. We will give a graphical proof under some simplifying assumptions. Suppose that the system is initially empty [$N(0) = 0$] and that customers depart from the system in the order they arrive. Then the number of arrivals $\alpha(t)$ and departures $\beta(t)$ up to time t form a staircase graph as shown in Fig. 3.1. The difference $\alpha(t) - \beta(t)$ is the number in the system $N(t)$ at time t. The shaded area between the graphs of $\alpha(\tau)$ and $\beta(\tau)$ can be expressed as

$$\int_0^t N(\tau)\,d\tau$$

and if t is any time for which the system is empty [$N(t) = 0$], the shaded area is also equal to

$$\sum_{i=1}^{\alpha(t)} T_i$$

Dividing both expressions above with t, we obtain

$$\frac{1}{t}\int_0^t N(\tau)\,d\tau = \frac{1}{t}\sum_{i=1}^{\alpha(t)} T_i = \frac{\alpha(t)}{t}\,\frac{\sum_{i=1}^{\alpha(t)} T_i}{\alpha(t)}$$

or equivalently,

$$N_t = \lambda_t T_t \tag{3.2}$$

Little's Theorem is obtained assuming that

$$N_t \to N,\ \lambda_t \to \lambda,\ T_t \to T$$

and that the system becomes empty infinitely often at arbitrarily large times. With a minor modification in the preceding argument, the latter assumption becomes unnecessary. To see this, note that the shaded area in Fig. 3.1 lies between $\sum_{i=1}^{\alpha(t)} T_i$ and $\sum_{i=1}^{\beta(t)} T_i$, so we obtain

$$\frac{\beta(t)}{t}\,\frac{\sum_{i=1}^{\beta(t)} T_i}{\beta(t)} \le N_t \le \lambda_t T_t$$

Assuming that $N_t \to N$, $\lambda_t \to \lambda$, $T_t \to T$, and that the departure rate $\beta(t)/t$ up to time t tends to the steady-state arrival rate λ, we obtain Little's Theorem.

The simplifying assumptions used in the preceding graphical proof can be relaxed considerably, and one can construct an analytical proof that requires only that the limits $\lambda = \lim_{t\to\infty} \alpha(t)/t$, $\delta = \lim_{t\to\infty} \beta(t)/t$, and $T = \lim_{t\to\infty} T_t$ exist, and that $\lambda = \delta$. In particular, it is not necessary that customers are served in the order they arrive, and that the system is initially empty (see Problem 3.41). Figure 3.2 explains why the order of customer service is not essential for the validity of Little's Theorem.

3.2.2 Probabilistic Form of Little's Theorem

Little's Theorem admits also a probabilistic interpretation provided that we can replace time averages with statistical or ensemble averages, as we now discuss. Our preceding

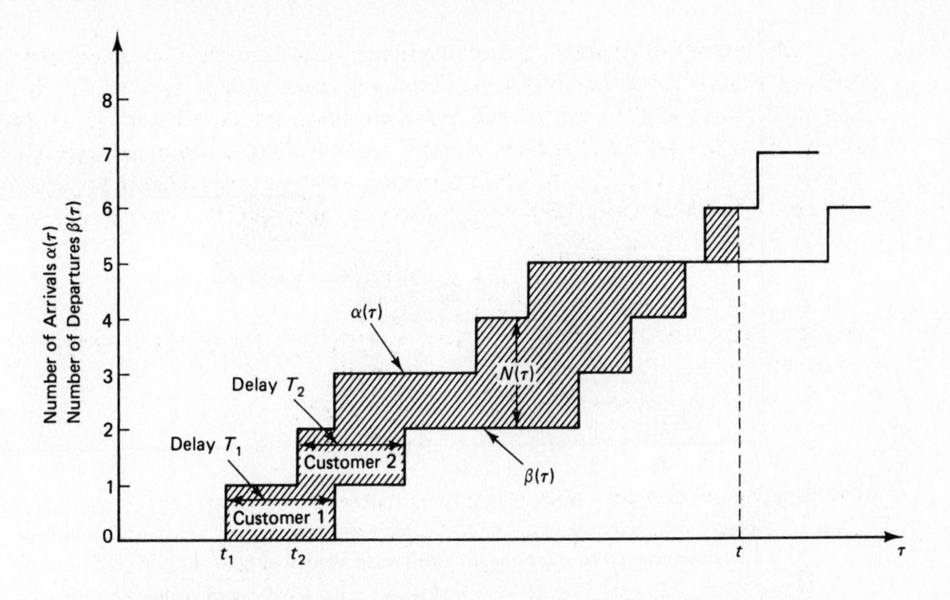

Figure 3.1 Proof of Little's Theorem. If the system is empty at time t $[N(t) = 0]$, the shaded area can be expressed both as $\int_0^t N(\tau)\,d\tau$ and as $\sum_{i=1}^{\alpha(t)} T_i$. Dividing both expressions by t, equating them, and taking the limit as $t \to \infty$ gives Little's Theorem. If $N(t) > 0$, we have

$$\sum_{i=1}^{\beta(t)} T_i \leq \int_0^t N(\tau)\,d\tau \leq \sum_{i=1}^{\alpha(t)} T_i$$

and assuming that the departure rate $\beta(t)/t$ up to time t tends to the steady-state arrival rate λ, the same argument applies.

analysis deals with a single sample function; now we will look at the probabilities of many sample functions and other events.

We first need to clarify the meaning of an ensemble average. Let us denote

$$p_n(t) = \text{Probability of } n \text{ customers in the system at time } t$$
$$\text{(waiting in queue or under service)}$$

In a typical situation we are given the initial probabilities $p_n(0)$ at time 0, together with enough statistical information to determine, at least in principle, the probabilities $p_n(t)$ for all times t. For example, the probability distribution of the time between two successive arrivals (the interarrival time), and the probability distribution of the customers' service time at various parts of the queueing system may be given. Then the average number in the system at time t is given by

$$\overline{N}(t) = \sum_{n=0}^{\infty} n p_n(t)$$

Note that both $\overline{N}(t)$ and $p_n(t)$ depend on t as well as the initial probability distribution

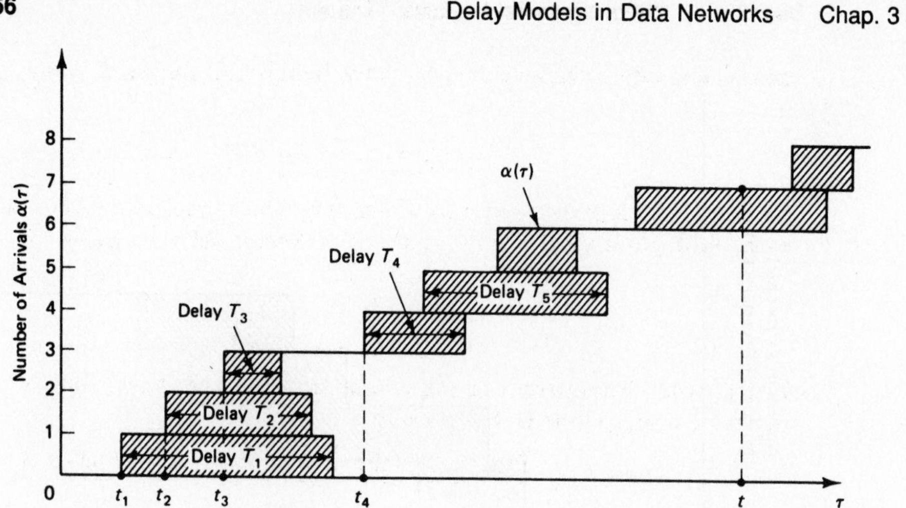

Figure 3.2 Informal justification of Little's Theorem without assuming first-in first-out customer service. The shaded area can be expressed both as $\int_0^t N(\tau)\,d\tau$ and as $\sum_{i\in D(t)} T_i + \sum_{i\in \overline{D}(t)}(t-t_i)$, where $D(t)$ is the set of customers that have departed up to time t, $\overline{D}(t)$ is the set of customers that are still in the system at time t, and t_i is the time of arrival of the ith customer. Dividing both expressions by t, equating them, and taking the limit as $t \to \infty$ gives Little's Theorem.

$\{p_0(0),\ p_1(0),\ldots\}$. However, the queueing systems that we will consider typically reach a steady-state in the sense that for some p_n (independent of the initial distribution), we have

$$\lim_{t\to\infty} p_n(t) = p_n\,, \quad n = 0, 1, \ldots$$

The average number in the system at steady-state is given by

$$\overline{N} = \sum_{n=0}^{\infty} np_n$$

and we typically have

$$\overline{N} = \lim_{t\to\infty} \overline{N}(t)$$

Regarding average delay per customer, we are typically given enough statistical information to determine in principle the probability distribution of delay of each individual customer (*i.e.*, the first, second, etc.). From this, we can determine the average delay of each customer. The average delay of the kth customer, denoted $\overline{T}_k$, typically converges as $k \to \infty$ to a steady-state value

$$\overline{T} = \lim_{k\to\infty} \overline{T}_k$$

To make the connection with time averages, we note that almost every system of interest to us is *ergodic* in the sense that the time average, $N = \lim_{t\to\infty} N_t$, of a sample function is, with probability 1, equal to the steady-state average $\overline{N} = \lim_{t\to\infty} \overline{N}(t)$, that is,

$$N = \lim_{t\to\infty} N_t = \lim_{t\to\infty} \overline{N}(t) = \overline{N}$$

Similarly, for the systems of interest to us, the time average of customer delay T is also equal (with probability 1) to the steady-state average delay $\overline{T}$, that is,

$$T = \lim_{k\to\infty} \frac{1}{k} \sum_{i=1}^{k} T_i = \lim_{k\to\infty} \overline{T}_k = \overline{T}$$

Under these circumstances, Little's formula, $N = \lambda T$, holds with N and T being stochastic averages and with λ given by

$$\lambda = \lim_{t\to\infty} \frac{\text{Expected number of arrivals in the interval } [0,t]}{t}$$

The equality of long term time and ensemble averages of various stochastic processes will often be accepted in this chapter on intuitive grounds. This equality can often be shown by appealing to general results from the theory of Markov chains (see Appendix A, at the end of this chapter, which states these results without proof). In other cases, this equality, though highly plausible, requires a specialized mathematical proof. Such a proof is typically straightforward for an expert in stochastic processes but requires background that is beyond what is assumed in this book. In what follows we will generally use the time average notation T and N in place of the ensemble average notation $\overline{T}$ and $\overline{N}$, respectively, implicitly assuming the equality of the corresponding time and ensemble averages.

3.2.3 Applications of Little's Theorem

The significance of Little's Theorem is due in large measure to its generality. It holds for almost every queueing system that reaches a steady-state. The system need not consist of just a single queue. Indeed, with appropriate interpretation of the terms N, λ, and T, the theorem holds for many complex arrival–departure systems. The following examples illustrate its broad applicability.

Example 3.1

If λ is the arrival rate in a transmission line, N_Q is the average number of packets waiting in queue (but not under transmission), and W is the average time spent by a packet waiting in queue (not including the transmission time), Little's Theorem gives

$$N_Q = \lambda W$$

Furthermore, if $\overline{X}$ is the average transmission time, then Little's Theorem gives the average number of packets under transmission as

$$\rho = \lambda \overline{X}$$

Since at most one packet can be under transmission, ρ is also the line's *utilization factor*, (*i.e.*, the proportion of time that the line is busy transmitting a packet).

Example 3.2

Consider a network of transmission lines where packets arrive at n different nodes with corresponding rates $\lambda_1, \ldots, \lambda_n$. If N is the average total number of packets inside the network, then (regardless of the packet length distribution and method for routing packets) the average delay per packet is

$$T = \frac{N}{\sum_{i=1}^{n} \lambda_i}$$

Furthermore, Little's Theorem also yields $N_i = \lambda_i T_i$, where N_i and T_i are the average number in the system and average delay of packets arriving at node i, respectively.

Example 3.3

A packet arrives at a transmission line every K seconds with the first packet arriving at time 0. All packets have equal length and require αK seconds for transmission where $\alpha < 1$. The processing and propagation delay per packet is P seconds. The arrival rate here is $\lambda = 1/K$. Because packets arrive at a regular rate (equal interarrival times), there is no delay for queueing, so the time T a packet spends in the system (including the propagation delay) is

$$T = \alpha K + P$$

According to Little's Theorem, we have

$$N = \lambda T = \alpha + \frac{P}{K}$$

Here the number in the system $N(t)$ is a deterministic function of time. Its form is shown in Fig. 3.3 for the case where $K < \alpha K + P < 2K$, and it can be seen that $N(t)$ does not converge to any value (the system never reaches statistical equilibrium). However, Little's Theorem holds with N viewed as a time average.

Example 3.4

Consider a window flow control system (as described in Section 2.8.1) with a window of size W for each session. Since the number of packets in the system per session is always no more than W, Little's Theorem asserts that the arrival rate λ of packets into the system for each session, and the average packet delay are related by $W \geq \lambda T$. Thus, if congestion builds up in the network and T increases, λ must eventually decrease. Next, suppose that the network is congested and capable of delivering only λ packets per unit time for each session. Assuming that acknowledgment delays are negligible relative to the forward packet delays, we have $W \simeq \lambda T$. Then, increasing the window size W for all sessions merely serves to increase the delay T without appreciably changing λ.

Example 3.5

Consider a queueing system with K servers, and with room for at most $N \geq K$ customers (either in queue or in service). The system is always full; we assume that it starts with N customers and that a departing customer is immediately replaced by a new customer. (Queueing systems of this type are called *closed* and are discussed in detail in Section 3.8.) Suppose that the average customer service time is $\overline{X}$. We want to find the average

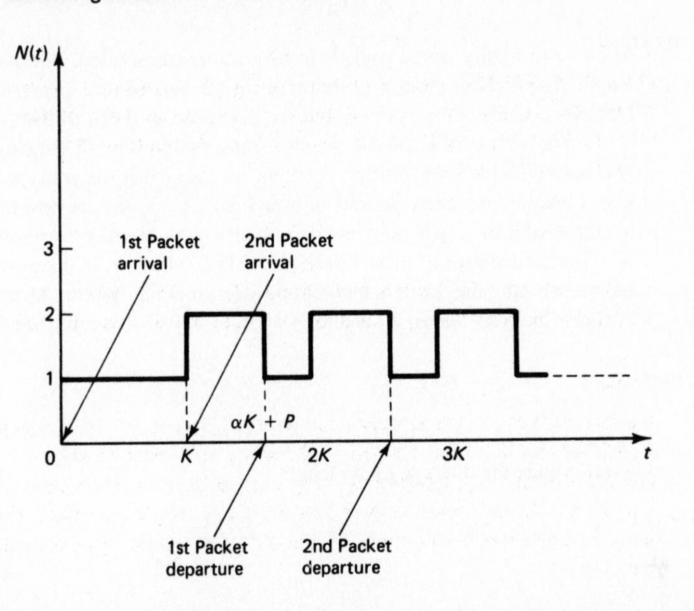

Figure 3.3 The number in the system in Example 3.3, $N(t)$, is deterministic and does not converge as $t \to \infty$. Little's Theorem holds with N, λ, and T viewed as time averages.

customer time in the system T. We apply Little's Theorem twice, first for the entire system, obtaining $N = \lambda T$, and then for the service portion of the system, obtaining $K = \lambda \overline{X}$ (since all servers are constantly busy). By eliminating λ in these two relations we have

$$T = \frac{N\overline{X}}{K}$$

Consider also the same system but under different customer arrival assumptions. In particular, assume that customers arrive at a rate λ but are blocked (and lost) from the system if they find the system full. Then the number of servers that are busy may be less than K. Let $\overline{K}$ be the average number of busy servers and let β be the proportion of customers that are blocked from entering the system. Applying Little's Theorem to the service portion of the system, we obtain

$$\overline{K} = (1 - \beta)\lambda\overline{X}$$

from which

$$\beta = 1 - \frac{\overline{K}}{\lambda\overline{X}}$$

Since $\overline{K} \leq K$, we obtain a lower bound on the blocking probability, namely,

$$\beta \geq 1 - \frac{K}{\lambda\overline{X}}$$

Example 3.6

A transmission line serves m packet streams, also called users, in round-robin cycles. In each cycle, some packets of user 1 are transmitted, followed by some packets of user 2, and so on, until finally, some packets of user m are transmitted. An overhead period of average length A_i precedes the transmission of the packets of user i in each cycle. Systems of this type are called *polling systems* and are discussed in detail in Section 3.5.2.

The arrival rate and the average transmission time of the packets of user i are λ_i and $\overline{X}_i$, respectively. From Little's theorem we know that the fraction of time the transmission line is busy transmitting packets of user i is $\lambda_i \overline{X}_i$. Consider now the time intervals used for overhead of user i. We can view these intervals as "packets" with average transmission time A_i. The arrival rate of these "packets" is $1/L$, where L is the average cycle length, and as before, we may use Little's theorem to assert that the fraction of time used for transmission of these "packets" is A/L, where $A = A_1 + A_2 + \cdots + A_m$. Therefore, we have

$$1 = \frac{A}{L} + \sum_{i=1}^{m} \lambda_i \overline{X}_i$$

which yields the average cycle length

$$L = \frac{A}{1 - \sum_{i=1}^{m} \lambda_i \overline{X}_i}$$

Example 3.7 Estimating Throughput in a Time-Sharing System

Little's Theorem can sometimes be used to provide bounds on the attainable system throughput λ. In particular, known bounds on N and T can be translated into throughput bounds via $\lambda = N/T$. As an example, consider a time-sharing computer system with N terminals. A user logs into the system through a terminal, and after an initial reflection period of average length R, submits a job that requires an average processing time P at the computer. Jobs queue up inside the computer and are served by a single CPU according to some unspecified priority or time-sharing rule.

We would like to get estimates of the throughput sustainable by the system (in jobs per unit time), and corresponding estimates of the average delay of a user. Since we are interested in maximum attainable throughput, we assume that there is always a user ready to take the place of a departing user, so the number of users in the system is always N. For this reason, it is appropriate to adopt a model whereby a departing user immediately reenters the system as shown in Fig. 3.4.

Applying Little's Theorem to the portion of the system between the entry to the terminals and the exit of the system (points A and C in Fig. 3.4), we have

$$\lambda = \frac{N}{T} \tag{3.3}$$

where T is the average time a user spends in the system. We have

$$T = R + D \tag{3.4}$$

where D is the average delay between the time a job is submitted to the computer and the time its execution is completed. Since D can vary between P (case where the user's job does not have to wait for other jobs to be completed) and NP (case where the user's job has to wait for the jobs of all the other users; compare with Example 3.5), we have

$$R + P \leq T \leq R + NP \tag{3.5}$$

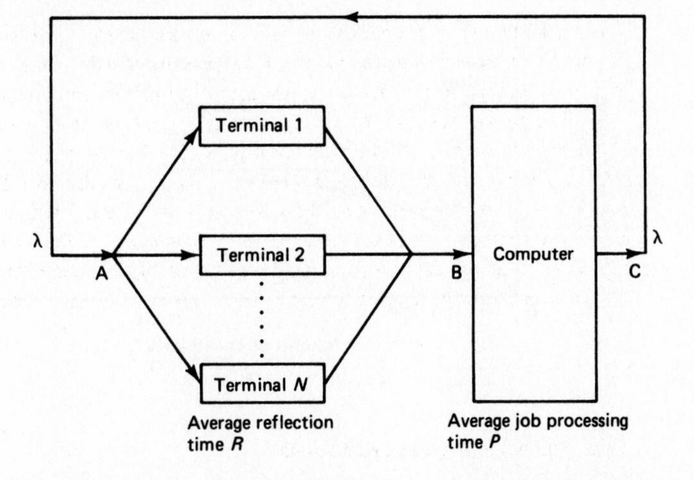

Figure 3.4 N terminals connected with a time-sharing computer system. To estimate maximum attainable throughput, we assume that a departing user immediately reenters the system or, equivalently, is immediately replaced by a new user.

Combining this relation with $\lambda = N/T$ [cf. Eq. (3.3)], we obtain

$$\frac{N}{R+NP} \leq \lambda \leq \frac{N}{R+P} \tag{3.6}$$

The throughput λ is also bounded above by the processing capacity of the computer. In particular, since the execution time of a job is P units on the average, it follows that the computer cannot process in the long run more than $1/P$ jobs per unit time, that is,

$$\lambda \leq \frac{1}{P} \tag{3.7}$$

(This conclusion can also be reached by applying Little's Theorem between the entry and exit points of the computer's CPU.)

By combining the preceding two relations, we obtain the bounds

$$\frac{N}{R+NP} \leq \lambda \leq \min\left\{\frac{1}{P}, \frac{N}{R+P}\right\} \tag{3.8}$$

for the throughput λ. By using $T = N/\lambda$, we also obtain bounds for the average user delay when the system is fully loaded:

$$\max\{NP,\ R+P\} \leq T \leq R+NP \tag{3.9}$$

These relations are illustrated in Fig. 3.5.

It can be seen that as the number of terminals N increases, the throughput approaches the maximum $1/P$, while the average user delay rises essentially in direct proportion with N. The number of terminals becomes a throughput bottleneck when $N < 1 + R/P$, in which case the computer resource stays idle for a substantial portion of the time while all users are engaged in reflection. In contrast, the limited processing power of the computer becomes the bottleneck when $N > 1 + R/P$. It is interesting to note that while the exact maximum attainable throughput depends on system parameters, such as the statistics of the reflection and processing times, and the manner in which jobs are served by the CPU, the

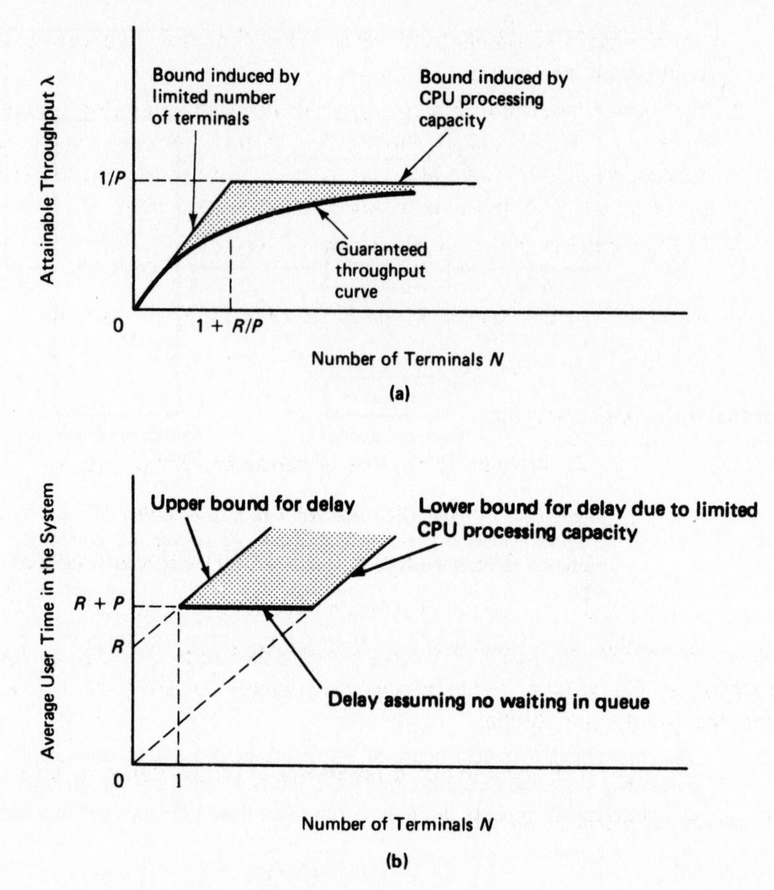

Figure 3.5 Bounds on throughput and average user delay in a time-sharing system. (a) Bounds on attainable throughput [Eq. (3.8)]. (b) Bounds on average user time in a fully loaded system [Eq. (3.9)]. The time increases essentially in proportion with the number of terminals N.

bounds obtained are independent of these parameters. We owe this convenient situation to the generality of Little's Theorem.

3.3 THE *M/M/1* QUEUEING SYSTEM

The $M/M/1$ queueing system consists of a single queueing station with a single server (in a communication context, a single transmission line). Customers arrive according to a Poisson process with rate λ, and the probability distribution of the service time is exponential with mean $1/\mu$ sec. We will explain the meaning of these terms shortly. The name $M/M/1$ reflects standard queueing theory nomenclature whereby:

1. The first letter indicates the nature of the arrival process [*e.g.*, M stands for memoryless, which here means a Poisson process (*i.e.*, exponentially distributed inter-

arrival times), G stands for a general distribution of interarrival times, D stands for deterministic interarrival times].

2. The second letter indicates the nature of the probability distribution of the service times (*e.g.*, M, G, and D stand for exponential, general, and deterministic distributions, respectively). In all cases, successive interarrival times and service times are assumed to be statistically independent of each other.
3. The last number indicates the number of servers.

We have already established, via Little's Theorem, the relations

$$N = \lambda T, \qquad N_Q = \lambda W$$

between the basic quantities,

$$N = \text{Average number of customers in the system}$$

$$T = \text{Average customer time in the system}$$

$$N_Q = \text{Average number of customers waiting in queue}$$

$$W = \text{Average customer waiting time in queue}$$

However, N, T, N_Q, and W cannot be specified further unless we know something more about the statistics of the system. Given these statistics, we will be able to derive the steady-state probabilities

$$p_n = \text{Probability of } n \text{ customers in the system, } n = 0, 1, \ldots$$

From these probabilities, we can get

$$N = \sum_{n=0}^{\infty} n p_n$$

and using Little's Theorem,

$$T = \frac{N}{\lambda}$$

Similar formulas exist for N_Q and W. Appendix B provides a summary of the results for the $M/M/1$ system and the other major systems analyzed later.

The analysis of the $M/M/1$ system as well as several other related systems, such as the $M/M/m$ or $M/M/\infty$ systems, is based on the theory of Markov chains summarized in Appendix A. An alternative approach is to use simple graphical arguments based on the concept of mean residual time introduced in Section 3.5. This approach does not require that the service times are exponentially distributed (*i.e.*, it applies to the $M/G/1$ system). The price paid for this generality is that the characterization of the steady-state probabilities is more complicated than for the $M/M/1$ system. The reader wishing to circumvent the Markov chain analysis may start directly with the $M/G/1$ system in Section 3.5 after a reading of the preliminary facts on the Poisson process given in Sections 3.3.1 and 3.3.2.

3.3.1 Main Results

We first introduce our assumptions on the arrival and service statistics of the $M/M/1$ system.

Arrival statistics—the Poisson process. In the $M/M/1$ system, customers arrive according to a Poisson process which we now define:

A stochastic process $\{A(t) \mid t \geq 0\}$ taking nonnegative integer values is said to be a *Poisson process* with rate λ if

1. $A(t)$ is a counting process that represents the total number of arrivals that have occurred from 0 to time t [*i.e.*, $A(0) = 0$], and for $s < t$, $A(t) - A(s)$ equals the numbers of arrivals in the interval $(s, t]$.
2. The numbers of arrivals that occur in disjoint time intervals are independent.
3. The number of arrivals in any interval of length τ is Poisson distributed with parameter $\lambda\tau$. That is, for all t, $\tau > 0$,

$$P\{A(t+\tau) - A(t) = n\} = e^{-\lambda\tau}\frac{(\lambda\tau)^n}{n!}, \qquad n = 0, 1, \ldots \tag{3.10}$$

The average number of arrivals within an interval of length τ is $\lambda\tau$ (based on the mean of the Poisson distribution). This leads to the interpretation of λ as an arrival rate (average number of arrivals per unit time).

We list some of the properties of the Poisson process that will be of interest:

1. Interarrival times are independent and exponentially distributed with parameter λ; that is, if t_n denotes the time of the nth arrival, the intervals $\tau_n = t_{n+1} - t_n$ have the probability distribution

$$P\{\tau_n \leq s\} = 1 - e^{-\lambda s}, \qquad s \geq 0 \tag{3.11}$$

and are mutually independent. [The corresponding probability density function is $p(\tau_n) = \lambda e^{-\lambda\tau_n}$. The mean and variance of τ_n are $1/\lambda$ and $1/\lambda^2$, respectively.] For a proof of this property, see [Ros83], p. 35.

2. For every $t \geq 0$ and $\delta \geq 0$,

$$P\{A(t+\delta) - A(t) = 0\} = 1 - \lambda\delta + o(\delta) \tag{3.12}$$

$$P\{A(t+\delta) - A(t) = 1\} = \lambda\delta + o(\delta) \tag{3.13}$$

$$P\{A(t+\delta) - A(t) \geq 2\} = o(\delta) \tag{3.14}$$

where we generically denote by $o(\delta)$ a function of δ such that

$$\lim_{\delta \to 0} \frac{o(\delta)}{\delta} = 0$$

These equations can be verified by expanding the Poisson distribution on the number of arrivals in an interval of length δ [Eq. (3.10)] in a Taylor series [or equivalently, by writing $e^{-\lambda\delta} = 1 - \lambda\delta + (\lambda\delta)^2/2 - \cdots$].

3. If two or more independent Poisson processes $A_1, \ldots, A_k$ are merged into a single process $A = A_1 + A_2 + \cdots + A_k$, the latter process is Poisson with a rate equal to the sum of the rates of its components (see Problem 3.10).
4. If a Poisson process is split into two other processes by independently assigning each arrival to the first (second) of these processes with probability p ($1 - p$, respectively), the two arrival processes thus obtained are Poisson (see Problem 3.11). (For this it is essential that the assignment of each arrival be independent of the assignment of other arrivals. If, for example, the assignment is done by alternation, with even-numbered arrivals assigned to one process and odd-numbered arrivals assigned to the other, the two generated processes are not Poisson. This will prove to be significant in the context of data networks; see Example 3.17 in Section 3.6.)

A Poisson process is generally considered to be a good model for the aggregate traffic of a large number of similar and independent users. In particular, suppose that we merge n independent and identically distributed packet arrival processes. Each process has arrival rate λ/n, so that the aggregate process has arrival rate λ. The interarrival times τ between packets of the same process have a given distribution $F(s) = P\{\tau \le s\}$ and are independent [$F(s)$ need not be an exponential distribution]. Then under relatively mild conditions on F [*e.g.*, $F(0) = 0$, $dF(0)/ds > 0$], the aggregate arrival process can be approximated well by a Poisson process with rate λ as $n \to \infty$ (see [KaT75], p. 221).

Service statistics. Our assumption regarding the service process is that *the customer service times have an exponential distribution with parameter* μ, that is, if s_n is the service time of the nth customer,

$$P\{s_n \le s\} = 1 - e^{-\mu s}, \qquad s \ge 0$$

[The probability density function of s_n is $p(s_n) = \mu e^{-\mu s_n}$, and its mean and variance are $1/\mu$ and $1/\mu^2$, respectively.] Furthermore, *the service times s_n are mutually independent and also independent of all interarrival times.* The parameter μ is called the *service rate* and represents the rate (in customers served per unit time) at which the server operates when busy. In the context of a packet transmission system, the independence of interarrival and service times implies, among other things, that the length of an arriving packet does not affect the arrival time of the next packet. It will be seen in Section 3.6 that this condition is often violated in practice, particularly when the arriving packets have just departed from another queue.

An important fact regarding the exponential distribution is its *memoryless* character, which can be expressed as

$$P\{\tau_n > r + t \mid \tau_n > t\} = P\{\tau_n > r\}, \qquad \text{for } r, t \ge 0$$

$$P\{s_n > r + t \mid s_n > t\} = P\{s_n > r\}, \qquad \text{for } r, t \ge 0$$

for the interarrival and service times τ_n and s_n, respectively. This means that the additional time needed to complete a customer's service in progress is independent of when the service started. Similarly, the time up to the next arrival is independent of when the previous arrival occurred. Verification of the memoryless property follows from the calculation

$$P\{\tau_n > r+t \mid \tau_n > t\} = \frac{P\{\tau_n > r+t\}}{P\{\tau_n > t\}} = \frac{e^{-\lambda(r+t)}}{e^{-\lambda t}} = e^{-\lambda r} = P\{\tau_n > r\}$$

Markov chain formulation. An important consequence of the memoryless property is that it allows the use of the theory of Markov chains. Indeed, this property, together with our earlier independence assumptions on interarrival and service times, imply that once we know the number $N(t)$ of customers in the system at time t, the times at which customers will arrive or complete service in the future are independent of the arrival times of the customers presently in the system and of how much service the customer currently in service (if any) has already received. This means that the future numbers of customers depend on past numbers only through the present number; that is, $\{N(t) \mid t \geq 0\}$ *is a continuous-time Markov chain.*

We could analyze the process $N(t)$ in terms of continuous-time Markov chain methodology; most of the queueing literature follows this line of analysis (see also Problem 3.12). It is sufficient, however, for our purposes in this section to use the simpler theory of discrete-time Markov chains (briefly summarized in Appendix A).

Let us focus attention at the times

$$0,\ \delta,\ 2\delta, \ldots, k\delta, \ldots$$

where δ is a small positive number. We denote

$$N_k = \text{ Number of customers in the system at time } k\delta$$

Since $N_k = N(k\delta)$ and, as discussed, $N(t)$ is a continuous-time Markov chain, we see that $\{N_k \mid k = 0, 1, \ldots\}$ is a discrete-time Markov chain with steady-state occupancy probabilities equal to those of the continuous chain. Let P_{ij} denote the corresponding transition probabilities

$$P_{ij} = P\{N_{k+1} = j \mid N_k = i\}$$

Note that P_{ij} depends on δ, but to keep notation simple, we do not show this dependence. By using Eqs. (3.12) through (3.14), one can show that

$$P_{oo} = 1 - \lambda\delta + o(\delta) \tag{3.15}$$

$$P_{ii} = 1 - \lambda\delta - \mu\delta + o(\delta), \qquad i \geq 1 \tag{3.16}$$

$$P_{i,i+1} = \lambda\delta + o(\delta), \qquad i \geq 0 \tag{3.17}$$

$$P_{i,i-1} = \mu\delta + o(\delta), \qquad i \geq 1 \tag{3.18}$$

$$P_{ij} = o(\delta), \qquad i \text{ and } j \neq i,\ i+1,\ i-1$$

To see how these equations are verified, note that when at a state $i \geq 1$, the probability of 0 arrivals and 0 departures in a δ-interval $I_k = (k\delta, (k+1)\delta]$ is $(e^{-\lambda\delta})(e^{-\mu\delta})$; this is because the number of arrivals and the number of departures are Poisson distributed and independent of each other. Expanding this in a power series in δ,

$$P\{0 \text{ customers arrive and } 0 \text{ depart in } I_k\} = 1 - \lambda\delta - \mu\delta + o(\delta) \tag{3.19}$$

The probability of 0 arrivals and 1 departure in the interval I_k is $e^{-\lambda\delta}(1 - e^{-\mu\delta})$ if $i = 1$ (since $1 - e^{-\mu\delta}$ is the probability that the customer in service will complete its service within I_k), and $e^{-\lambda\delta}(\mu\delta e^{-\mu\delta})$ if $i > 1$ (since $\mu\delta e^{-\mu\delta}$ is the probability that within the interval I_k, the customer in service will complete its service while the subsequent customer will not). In both cases we have

$$P\{0 \text{ customers arrive and } 1 \text{ departs in } I_k\} = \mu\delta + o(\delta)$$

Similarly, the probability of 1 arrival and 0 departures in I_k is $(\lambda\delta e^{-\lambda\delta})e^{-\mu\delta}$, so

$$P\{1 \text{ customer arrives and } 0 \text{ depart in } I_k\} = \lambda\delta + o(\delta)$$

These probabilities add up to 1 plus $o(\delta)$. Thus, the probability of more than one arrival or departure is negligible for δ small. It follows that for $i \geq 1$, P_{ii}, which is the probability of an equal number of arrivals and departures in I_k, is within $o(\delta)$ of the value in Eq. (3.19); this verifies Eq. (3.16). Equations (3.15), (3.17), and (3.18) are verified in the same way.

The state transition diagram for the Markov chain $\{N_k\}$ is shown in Fig. 3.6, where we have omitted the terms $o(\delta)$.

Derivation of the stationary distribution. Consider now the steady-state probabilities

$$p_n = \lim_{k\to\infty} P\{N_k = n\} = \lim_{t\to\infty} P\{N(t) = n\}$$

Note that during any time interval, the total number of transitions from state n to $n+1$ must differ from the total number of transitions from $n+1$ to n by at most 1. Thus asymptotically, the frequency of transitions from n to $n+1$ is equal to the frequency of transitions from $n+1$ to n. Equivalently, the probability that the system is in state n and makes a transition to $n+1$ in the next transition interval is the same as the probability that the system is in state $n+1$ and makes a transition to n, that is,

$$p_n\lambda\delta + o(\delta) = p_{n+1}\mu\delta + o(\delta)$$

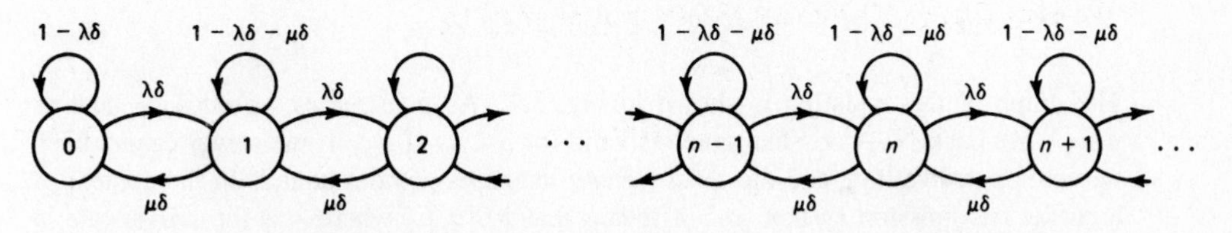

Figure 3.6 Discrete-time Markov chain for the $M/M/1$ system. The state n corresponds to n customers in the system. Transition probabilities shown are correct up to an $o(\delta)$ term.

By taking the limit in this equation as $\delta \to 0$, we obtain

$$p_n \lambda = p_{n+1} \mu \tag{3.20}$$

(The preceding equations are called *global balance equations*, corresponding to the set of states $\{0, 1, \ldots, n\}$ and $\{n+1, n+2, \ldots\}$. See Appendix A for a more general statement of these equations and for an interpretation that parallels the argument given above.) These equations can also be written as

$$p_{n+1} = \rho p_n, \qquad n = 0, 1, \ldots$$

where

$$\rho = \frac{\lambda}{\mu}$$

It follows that

$$p_{n+1} = \rho^{n+1} p_0, \quad n = 0, 1, \ldots \tag{3.21}$$

If $\rho < 1$ (service rate exceeds arrival rate), the probabilities p_n are all positive and add up to unity, so

$$1 = \sum_{n=0}^{\infty} p_n = \sum_{n=0}^{\infty} \rho^n p_0 = \frac{p_0}{1-\rho} \tag{3.22}$$

Combining the last two equations, we finally obtain

$$p_n = \rho^n (1-\rho), \qquad n = 0, 1, \ldots \tag{3.23}$$

We can now calculate the average number of customers in the system in steady-state:

$$\begin{aligned} N = \lim_{t \to \infty} E\{N(t)\} &= \sum_{n=0}^{\infty} n p_n = \sum_{n=0}^{\infty} n \rho^n (1-\rho) \\ &= \rho(1-\rho) \sum_{n=0}^{\infty} n \rho^{n-1} = \rho(1-\rho) \frac{\partial}{\partial \rho} \left(\sum_{n=0}^{\infty} \rho^n \right) \\ &= \rho(1-\rho) \frac{\partial}{\partial \rho} \left(\frac{1}{1-\rho} \right) = \rho(1-\rho) \frac{1}{(1-\rho)^2} \end{aligned}$$

and finally, using $\rho = \lambda/\mu$, we have

$$N = \frac{\rho}{1-\rho} = \frac{\lambda}{\mu - \lambda} \tag{3.24}$$

The graph of this equation is shown in Fig. 3.7. As ρ increases, so does N, and as $\rho \to 1$, we have $N \to \infty$. The graph is valid for $\rho < 1$. If $\rho > 1$, the server cannot keep up with the arrival rate and the queue length increases without bound. In the context of a packet transmission system, $\rho > 1$ means that $\lambda L > C$, where λ is the arrival rate in packets/sec, L is the average packet length in bits, and C is the transmission capacity in bits/sec.

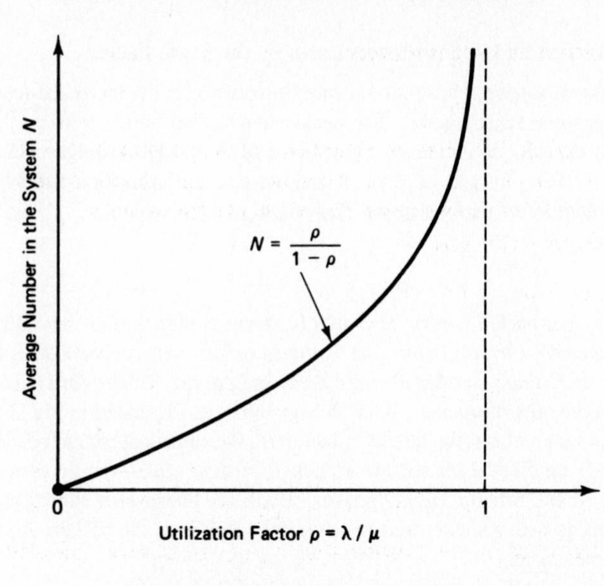

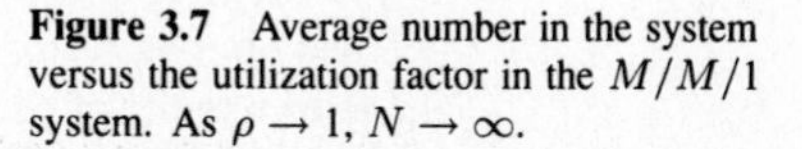
Figure 3.7 Average number in the system versus the utilization factor in the $M/M/1$ system. As $\rho \to 1$, $N \to \infty$.

The average delay per customer (waiting time in queue plus service time) is given by Little's Theorem,

$$T = \frac{N}{\lambda} = \frac{\rho}{\lambda(1-\rho)} \tag{3.25}$$

Using $\rho = \lambda/\mu$, this becomes

$$T = \frac{1}{\mu - \lambda} \tag{3.26}$$

We note that it is actually possible to show that the customer delay is exponentially distributed in steady-state [see Problem 3.11(b)].

The average waiting time in queue, W, is the average delay T less the average service time $1/\mu$, so

$$W = \frac{1}{\mu - \lambda} - \frac{1}{\mu} = \frac{\rho}{\mu - \lambda}$$

By Little's Theorem, the average number of customers in queue is

$$N_Q = \lambda W = \frac{\rho^2}{1-\rho}$$

A very useful interpretation is to view the quantity ρ as the *utilization factor* of the queueing system, (*i.e.*, the long-term proportion of time the server is busy). We showed this earlier in a broader context by using Little's Theorem (Example 3.1). Based on this interpretation, it follows that $\rho = 1 - p_0$, where p_0 is the probability of having no customers in the system, and we obtain an alternative verification of the formula derived for p_0 [Eq. (3.22)].

We illustrate these results by means of some examples from data networks.

Example 3.8 Increasing the Arrival and Transmission Rates by the Same Factor

Consider a packet transmission system whose arrival rate (in packets/sec) is increased from λ to $K\lambda$, where $K > 1$ is some scalar factor. The packet length distribution remains the same but the transmission capacity is increased by a factor of K, so the average packet transmission time is now $1/(K\mu)$ instead of $1/\mu$. It follows that the utilization factor ρ, and therefore the average number of packets in the system, remain the same:

$$N = \frac{\rho}{1-\rho} = \frac{\lambda}{\mu - \lambda}$$

However, the average delay per packet is now $T = N/(K\lambda)$ and is therefore decreased by a factor of K. In other words, *a transmission line K times as fast will accommodate K times as many packets/sec at K times smaller average delay per packet.* This result is quite general, even applying to networks of queues. What is happening, as illustrated in Fig. 3.8, is that by increasing arrival rate and service rate by a factor K, the statistical characteristics of the queueing process are unaffected except for a change in time scale—the process is speeded up by a factor K. Thus, when a packet arrives, it will see ahead of it statistically the same number of packets as with a slower transmission line. However, the packets ahead of it will be moving K times faster.

Example 3.9 Statistical Multiplexing Compared with Time- and Frequency-Division Multiplexing

Assume that m statistically identical and independent Poisson packet streams each with an arrival rate of λ/m packets/sec are transmitted over a communication line. The packet lengths for all streams are independent and exponentially distributed. The average transmission time is $1/\mu$. If the streams are merged into a single Poisson stream, with rate λ, as in statistical multiplexing, the average delay per packet is

$$T = \frac{1}{\mu - \lambda}$$

If, instead, the transmission capacity is divided into m equal portions, one per packet stream as in time- and frequency-division multiplexing, each portion behaves like an $M/M/1$ queue with arrival rate λ/m and average service rate μ/m. Therefore, the average delay per packet is

$$T = \frac{m}{\mu - \lambda}$$

that is, m times larger than for statistical multiplexing.

The preceding argument indicates that multiplexing a large number of traffic streams on separate channels in a transmission line performs very poorly in terms of delay. The performance is even poorer if the capacity of the channels is not allocated in direct proportion to the arrival rates of the corresponding streams—something that cannot be done (at least in the scheme considered here) if these arrival rates change over time. This is precisely why data networks, which most of the time serve many low duty cycle traffic streams, are typically organized on the basis of some form of statistical multiplexing. An argument in favor of time- and frequency-division multiplexing arises when each traffic stream is "regular" (as opposed to Poisson) in the sense that no packet arrives while another is transmitted, and thus there is no waiting in queue if that stream is transmitted on a dedicated transmission line. If several streams of this type are statistically multiplexed on a single transmission line, the

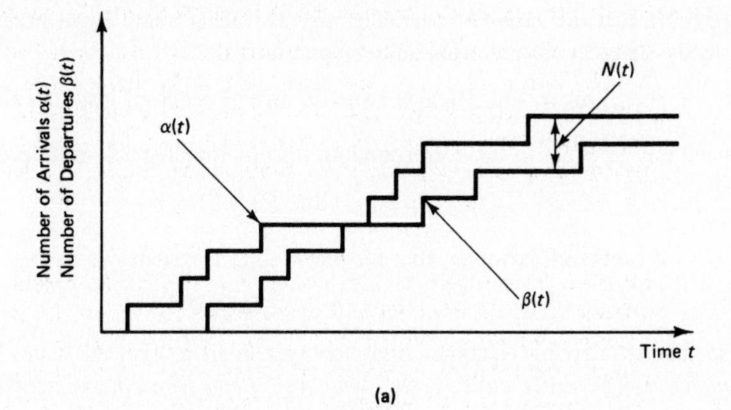

(a)

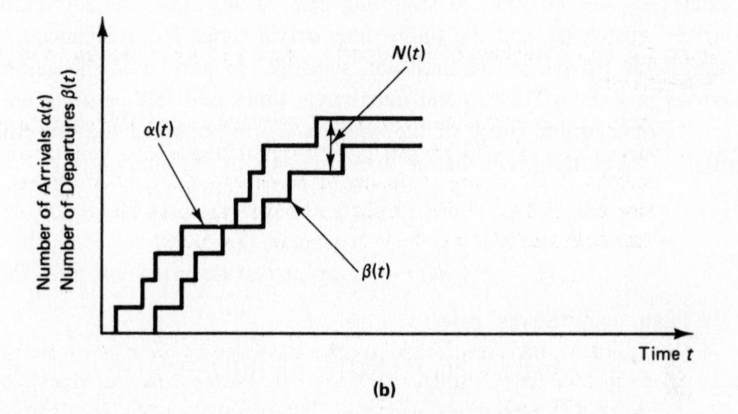

(b)

Figure 3.8 Increasing the arrival rate and the service rate by the same factor (see Example 3.8). (a) Sample paths of number of arrivals $\alpha(t)$ and departures $\beta(t)$ in the original system. (b) Corresponding sample paths of number of arrivals $\alpha(t)$ and departures $\beta(t)$ in the "speeded up" system, where the arrival rate and the service rate have been increased by a factor of 2. The average number in the system is the same as before, but the average delay is reduced by a factor of 2 since customers are moving twice as fast.

average delay per packet will decrease, but the average waiting time in queue will become positive and the variance of delay will also become positive. Thus in telephony, where each traffic stream is a voice conversation that is regular in the sense above and small variability of delay is critical, time- and frequency-division multiplexing are still used widely.

3.3.2 Occupancy Distribution upon Arrival

In our subsequent development, there are several situations where we will need a probabilistic characterization of a queueing system as seen by an arriving customer. It is possible that the times of customer arrivals are in some sense nontypical, so that the steady-state occupancy probabilities upon arrival,

$$a_n = \lim_{t\to\infty} P\{N(t) = n \mid \text{an arrival occurred just after time } t\} \tag{3.27}$$

need not be equal to the corresponding unconditional steady-state probabilities,

$$p_n = \lim_{t\to\infty} P\{N(t) = n\} \tag{3.28}$$

It turns out, however, that for the $M/M/1$ system, we have

$$p_n = a_n, \qquad n = 0, 1, \ldots \tag{3.29}$$

so that an arriving customer finds the system in a "typical" state. Indeed, *this holds under very general conditions for queueing systems with Poisson arrivals regardless of the distribution of the service times*. The only additional requirement we need is that future arrivals are independent of the current number in the system. More precisely, *we assume that for every time t and increment $\delta > 0$, the number of arrivals in the interval $(t, t+\delta)$ is independent of the number in the system at time t*. Given the Poisson hypothesis, essentially this amounts to assuming that, at any time, the service times of previously arrived customers and the future interarrival times are independent—something that is reasonable for packet transmission systems. In particular, the assumption holds if the arrival process is Poisson and interarrival times and service times are independent.

For a formal proof of the equality $a_n = p_n$ under the preceding assumption, let $A(t, t+\delta)$ be the event that an arrival occurs in the interval $(t, t+\delta)$. Let

$$p_n(t) = P\{N(t) = n\} \tag{3.30}$$

$$a_n(t) = P\{N(t) = n \mid \text{an arrival occurred just after time } t\} \tag{3.31}$$

We have, using Bayes' rule,

$$\begin{aligned} a_n(t) &= \lim_{\delta\to 0} P\{N(t) = n \mid A(t, t+\delta)\} \\ &= \lim_{\delta\to 0} \frac{P\{N(t) = n,\ A(t, t+\delta)\}}{P\{A(t, t+\delta)\}} \\ &= \lim_{\delta\to 0} \frac{P\{A(t, t+\delta) \mid N(t) = n\}\ P\{N(t) = n\}}{P\{A(t, t+\delta)\}} \end{aligned} \tag{3.32}$$

By assumption, the event $A(t, t+\delta)$ is independent of the number in the system at time t. Therefore,

$$P\{A(t, t+\delta) \mid N(t) = n\} = P\{A(t, t+\delta)\}$$

and we obtain from Eq. (3.32)

$$a_n(t) = P\{N(t) = n\} = p_n(t)$$

Taking the limit as $t \to \infty$, we obtain $a_n = p_n$.

As an example of what can happen if the arrival process is not Poisson, suppose that interarrival times are independent and uniformly distributed between 2 and 4 sec, while customer service times are all equal to 1 sec. Then an arriving customer always finds an empty system. On the other hand, the average number in the system as seen by an outside observer looking at a system at a random time is $1/3$. (The time in the system of each customer is 1 sec, so by Little's Theorem, N is equal to the arrival rate λ, which is $1/3$ since the expected time between arrivals is 3.)

For a similar example where the arrival process is Poisson but the service times of customers in the system and the future arrival times are correlated, consider a packet transmission system where packets arrive according to a Poisson process. The transmission time of the nth packet equals one half the interarrival time between packets n and $n+1$. Upon arrival, a packet finds the system empty. However, the average number in the system, as seen by an outside observer, is easily seen to be $1/2$.

3.3.3 Occupancy Distribution upon Departure

Let us consider the distribution of the number of customers in the system just after a departure has occurred, that is, the probabilities

$$d_n(t) = P\{N(t) = n \mid \text{a departure occurred just before time } t\}$$

The corresponding steady-state values are denoted

$$d_n = \lim_{t \to \infty} d_n(t), \qquad n = 0, 1, \ldots$$

It turns out that

$$d_n = a_n, \qquad n = 0, 1, \ldots$$

under very general assumptions—the only requirement essentially is that the system reaches a steady-state with all n having positive steady-state probabilities, and that $N(t)$ changes in unit increments. [These assumptions certainly hold for a stable $M/M/1$ system ($\rho < 1$), but they also hold for most stable single-queue systems of interest.] For any sample path of the system and for every n, the number in the system will be n infinitely often (with probability 1). This means that for each time the number in the system increases from n to $n+1$ due to an arrival, there will be a corresponding future decrease from $n+1$ to n due to a departure. Therefore, in the long run, the frequency of transitions from n to $n+1$ out of transitions from any k to $k+1$ equals the frequency of transitions from $n+1$ to n out of transitions from any $k+1$ to k, which implies that $d_n = a_n$. Therefore, *in steady-state, the system appears statistically identical to an arriving and a departing customer. When arrivals are Poisson,* we saw earlier that $a_n = p_n$; so, in this case, *both an arriving and a departing customer in steady-state see a system that is statistically identical to the one seen by an observer looking at the system at an arbitrary time.*

3.4 THE $M/M/m$, $M/M/\infty$, $M/M/m/m$, AND OTHER MARKOV SYSTEMS

We consider now a number of queueing systems that are similar to $M/M/1$ in that the arrival process is Poisson and the service times are independent, exponentially distributed, and independent of the interarrival times. Because of these assumptions, these systems can be modeled with continuous- or discrete-time Markov chains. From the corresponding state transition diagram, we can derive a set of equations that can be solved for the steady-state occupancy probabilities. Application of Little's Theorem then yields the average delay per customer.

3.4.1 $M/M/m$: The m-Server Case

The $M/M/m$ queueing system is identical to the $M/M/1$ system except that there are m servers (or channels of a transmission line in a data communication context). A customer at the head of the queue is routed to any server that is available. The corresponding state transition diagram is shown in Fig. 3.9.

By writing down the global balance equations for the steady-state probabilities p_n and taking $\delta \to 0$, we obtain

$$\begin{aligned} \lambda p_{n-1} &= n\,\mu\, p_n, \qquad n \le m \\ \lambda p_{n-1} &= m\,\mu\, p_n, \qquad n > m \end{aligned} \tag{3.33}$$

From these equations we obtain

$$p_n = \begin{cases} p_0 \dfrac{(m\rho)^n}{n!}, & n \le m \\ \\ p_0 \dfrac{m^m \rho^n}{m!}, & n > m \end{cases} \tag{3.34}$$

where ρ is given by

$$\rho = \frac{\lambda}{m\mu} < 1$$

We can calculate p_0 using Eq. (3.34) and the condition $\sum_{n=0}^{\infty} p_n = 1$. We obtain

$$p_0 = \left[1 + \sum_{n=1}^{m-1} \frac{(m\rho)^n}{n!} + \sum_{n=m}^{\infty} \frac{(m\rho)^n}{m!} \frac{1}{m^{n-m}}\right]^{-1}$$

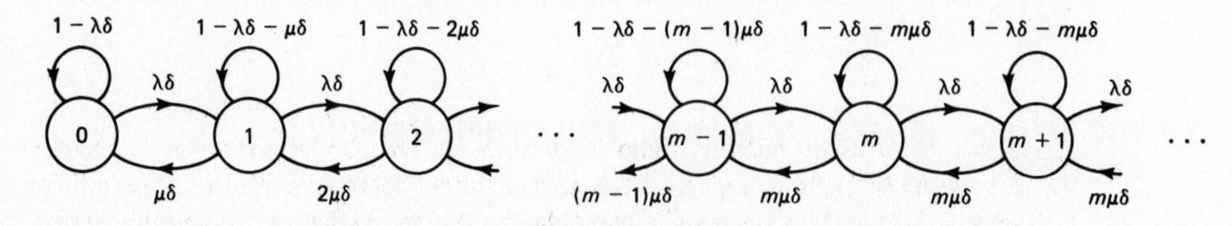

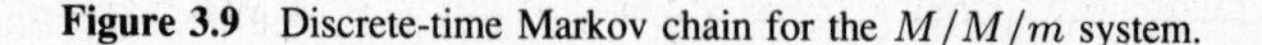

Figure 3.9 Discrete-time Markov chain for the $M/M/m$ system.

and finally,

$$p_0 = \left[\sum_{n=0}^{m-1} \frac{(m\rho)^n}{n!} + \frac{(m\rho)^m}{m!(1-\rho)}\right]^{-1} \tag{3.35}$$

The probability that an arrival will find all servers busy and will be forced to wait in queue is an important measure of performance of the $M/M/m$ system. Since an arriving customer finds the system in "typical" state (see Section 3.3.2), we have

$$P\{\text{Queueing}\} = \sum_{n=m}^{\infty} p_n = \sum_{n=m}^{\infty} \frac{p_0 m^m \rho^n}{m!} = \frac{p_0(m\rho)^m}{m!} \sum_{n=m}^{\infty} \rho^{n-m}$$

and, finally,

$$P_Q \triangleq P\{\text{Queueing}\} = \frac{p_0(m\rho)^m}{m!(1-\rho)} \tag{3.36}$$

where p_0 is given by Eq. (3.35). This is known as the *Erlang C formula*, honoring Denmark's A. K. Erlang, the foremost pioneer of queueing theory. This equation is often used in telephony (and more generally in circuit switching systems) to estimate the probability of a call request finding all of the m circuits of a transmission line busy. In an $M/M/m$ model it is assumed that such a call request "remains in queue," that is, continuously attempts to find a free circuit. The alternative model where such a call departs from the system and never returns is discussed in the context of the $M/M/m/m$ system in Section 3.4.3.

The expected number of customers waiting in queue (not in service) is given by

$$N_Q = \sum_{n=0}^{\infty} n p_{m+n}$$

Using the expression for p_{m+n} [Eq. (3.34)], we obtain

$$N_Q = \sum_{n=0}^{\infty} n p_0 \frac{m^m \rho^{m+n}}{m!} = \frac{p_0(m\rho)^m}{m!} \sum_{n=0}^{\infty} n\rho^n$$

Using the Erlang C formula of Eq. (3.36) to express p_0 in terms of P_Q, and the equation $(1-\rho)\sum_{n=0}^{\infty} n\rho^n = \rho/(1-\rho)$ encountered in the $M/M/1$ system analysis, we finally obtain

$$N_Q = P_Q \frac{\rho}{1-\rho} \tag{3.37}$$

Note that

$$\frac{N_Q}{P_Q} = \frac{\rho}{1-\rho}$$

represents the expected number found in queue by an arriving customer conditioned on the fact that he is forced to wait in queue, and is independent of the number of servers for a given $\rho = \lambda/m\mu$. This suggests in particular that as long as there are customers waiting in queue, the queue size of the $M/M/m$ system behaves identically as in an $M/M/1$

system with service rate $m\mu$—the aggregate rate of the m servers. Some thought shows that indeed this is true in view of the memoryless property of the exponential distribution.

Using Little's Theorem and the expression (3.37) for N_Q, we obtain the average time W a customer has to wait in queue:

$$W = \frac{N_Q}{\lambda} = \frac{\rho P_Q}{\lambda(1-\rho)}$$

The average delay per customer is, therefore,

$$T = \frac{1}{\mu} + W = \frac{1}{\mu} + \frac{\rho P_Q}{\lambda(1-\rho)}$$

and using $\rho = \lambda/m\mu$, we obtain

$$T = \frac{1}{\mu} + W = \frac{1}{\mu} + \frac{P_Q}{m\mu - \lambda} \tag{3.38}$$

Using Little's Theorem again, the average number of customers in the system is

$$N = \lambda T = \frac{\lambda}{\mu} + \frac{\lambda P_Q}{m\mu - \lambda}$$

and using $\rho = \lambda/m\mu$, we finally obtain

$$N = m\rho + \frac{\rho P_Q}{1-\rho}$$

Example 3.10 Using One vs. Using Multiple Channels in Statistical Multiplexing

Consider a communication link serving m independent Poisson traffic streams with overall rate λ. Suppose that the link is divided into m separate channels with one channel assigned to each traffic stream. However, if a traffic stream has no packet awaiting transmission, its corresponding channel is used to transmit a packet of another traffic stream. The transmission times of packets on each of the channels are exponentially distributed with mean $1/\mu$. The system can be modeled by the same Markov chain as the $M/M/m$ queue. Let us compare the average delays per packet of this system, and an $M/M/1$ system with the same arrival rate λ and service rate $m\mu$ (statistical multiplexing with one channel having m times larger capacity). In the former case, the average delay per packet is given by the $M/M/m$ average delay expression (3.38)

$$T = \frac{1}{\mu} + \frac{P_Q}{m\mu - \lambda}$$

while in the latter case, the average delay per packet is

$$\hat{T} = \frac{1}{m\mu} + \frac{\hat{P}_Q}{m\mu - \lambda}$$

where P_Q and $\hat{P}_Q$ denote the queueing probability in each case. When $\rho \ll 1$ (lightly loaded system) we have $P_Q \cong 0$, $\hat{P}_Q \cong 0$, and

$$\frac{T}{\hat{T}} \cong m$$

When ρ is only slightly less than 1, we have $P_Q \cong 1$, $\hat{P}_Q \cong 1$, $1/\mu \ll 1/(m\mu - \lambda)$, and

$$\frac{T}{\hat{T}} \cong 1$$

Therefore, for a light load, statistical multiplexing with m channels produces a delay almost m times larger than the delay of statistical multiplexing with the m channels combined in one (about the same as time- and frequency-division multiplexing). For a heavy load, the ratio of the two delays is close to 1.

3.4.2 $M/M/\infty$: The Infinite-Server Case

In the limiting case where $m = \infty$ in the $M/M/m$ system, we obtain from the global balance equations (3.33)

$$\lambda p_{n-1} = n\mu p_n, \qquad n = 1, 2, \ldots$$

so

$$p_n = p_0 \left(\frac{\lambda}{\mu}\right)^n \frac{1}{n!}, \qquad n = 1, 2, \ldots$$

From the condition $\sum_{n=0}^{\infty} p_n = 1$, we obtain

$$p_0 = \left[1 + \sum_{n=1}^{\infty} \left(\frac{\lambda}{\mu}\right)^n \frac{1}{n!}\right]^{-1} = e^{-\lambda/\mu}$$

so finally,

$$p_n = \left(\frac{\lambda}{\mu}\right)^n \frac{e^{-\lambda/\mu}}{n!}, \qquad n = 0, 1, \ldots$$

Therefore, in steady-state, *the number in the system is Poisson distributed with parameter* λ/μ. The average number in the system is

$$N = \frac{\lambda}{\mu}$$

By Little's Theorem, the average delay is N/λ or

$$T = \frac{1}{\mu}$$

This last equation can also be obtained simply by arguing that in an $M/M/\infty$ system, there is no waiting in queue, so T equals the average service time $1/\mu$. It can be shown that the number in the system is Poisson distributed even if the service time distribution is not exponential (*i.e.*, in the $M/G/\infty$ system; see Problem 3.47).

Example 3.11 The Quasistatic Assumption

It is often convenient to assume that the external packet traffic entering a subnet node and destined for some other subnet node can be modeled by a stationary stochastic process that has a constant bit arrival rate (average bits/sec). This approximates a situation where the arrival rate changes slowly with time and constitutes what we refer to as the quasistatic assumption.

When there are only a few active sessions (*i.e.*, user pairs) for the given origin–destination pair, this assumption is seriously violated since the addition or termination of a single session can change the total bit arrival rate by a substantial factor. When, however, there are many active sessions, each with a bit arrival rate that is small relative to the total, it seems plausible that the quasistatic assumption is approximately valid. The reason is that session additions are statistically counterbalanced by session terminations, with variations in the total rate being relatively small. For an analytical substantiation, assume that sessions are generated according to a Poisson process with rate λ, and terminate after a time which is exponentially distributed with mean $1/\mu$. Then the number of active sessions n evolves like the number of customers in an $M/M/\infty$ system (*i.e.*, is Poisson distributed with parameter λ/μ in steady-state). In particular, the mean and standard deviation of n are

$$N = E\{n\} = \frac{\lambda}{\mu}$$

$$\sigma_n = \left[E\left\{(n-N)^2\right\}\right]^{1/2} = \left(\frac{\lambda}{\mu}\right)^{1/2}$$

Suppose the ith active session generates traffic according to a stationary stochastic process having a bit arrival rate γ_i bits/sec. Assume that the rates γ_i are independent random variables with common mean $E\{\gamma_i\} = \Gamma$ and second moment $s_\gamma^2 = E\{\gamma_i^2\}$. Then the total bit arrival rate for n active sessions is the random variable $f = \sum_{i=1}^{n} \gamma_i$, which has mean

$$F = E\{f\} = \frac{\lambda}{\mu}\Gamma$$

The standard deviation of f, denoted σ_f, can be obtained by writing

$$\sigma_f^2 = E\left\{\left(\sum_{i=1}^{n}\gamma_i\right)^2\right\} - F^2$$

and carrying out the corresponding calculations (Problem 3.28). The result is

$$\sigma_f = \left(\frac{\lambda}{\mu}\right)^{1/2} s_\gamma$$

Therefore, we have

$$\frac{\sigma_f}{F} = \left(\frac{s_\gamma}{\Gamma}\right)\left(\frac{\mu}{\lambda}\right)^{1/2}$$

Suppose now that the average bit rate Γ of a session is small relative to the total F; that is, a "many-small-sessions assumption" holds. Then, since $\Gamma/F = \mu/\lambda$, we have that μ/λ is small. If we reasonably assume that s_γ/Γ has a moderate value, it follows from the equation above that σ_f/F is small. Therefore, the total arrival rate f is approximately constant, thereby justifying the quasistatic assumption.

3.4.3 *M/M/m/m*: The *m*-Server Loss System

Consider a system which is identical to the $M/M/m$ system except that if an arrival finds all m servers busy, it does not enter the system and is lost instead; the last m in the

$M/M/m/m$ notation indicates the limit on the number of customers in the system. This model is in wide use in telephony (and also, more generally, in circuit switched networks). In this context, customers in the system correspond to active telephone conversations and the m servers represent a single transmission line consisting of m circuits. The average service time $1/\mu$ is the average duration of a telephone conversation. The principal quantity of interest here is the *blocking probability*, that is, the steady-state probability that all circuits are busy, in which case an arriving call is refused service. Note that in an $M/M/m/m$-based model, the assumption is that blocked calls are lost (not reattempted). This is in contrast with an $M/M/m$-based model, where the assumption is that blocked calls continuously reattempt admission into service. In data networks, the $M/M/m/m$ system can be used as a model where arrivals correspond to requests for virtual circuit connections between two nodes and the maximum number of virtual circuits allowed is m.

The corresponding state transition diagram is shown in Fig. 3.10. We have

$$\lambda p_{n-1} = n\mu p_n\,, \qquad n = 1, 2, \ldots, m$$

so

$$p_n = p_0 \left(\frac{\lambda}{\mu}\right)^n \frac{1}{n!}, \qquad n = 1, 2, \ldots, m$$

Solving for p_0 in the equation $\sum_{n=0}^{m} p_n = 1$, we obtain

$$p_0 = \left[\sum_{n=0}^{m} \left(\frac{\lambda}{\mu}\right)^n \frac{1}{n!}\right]^{-1}$$

The probability that an arrival will find all m servers busy and will therefore be lost is

$$p_m = \frac{(\lambda/\mu)^m/m!}{\sum_{n=0}^{m}(\lambda/\mu)^n/n!}$$

This equation is known as the *Erlang B formula* and finds wide use in evaluating the blocking probability of telephone systems. It can be shown to hold even if the service time has mean $1/\mu$ but arbitrary probability distribution (*i.e.*, for an $M/G/m/m$ system; see [Ros83], p. 170).

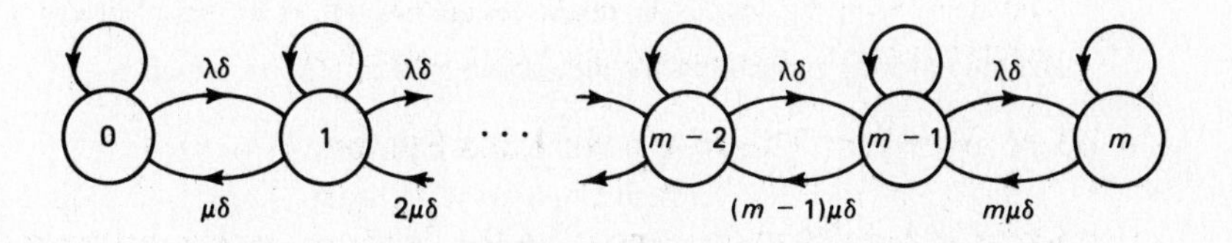

Figure 3.10 Discrete-time Markov chain for the $M/M/m/m$ system.

3.4.4 Multidimensional Markov Chains—Applications in Circuit Switching

We have considered so far queueing systems with a single type of customer where the state can be described by the number of customers in the system. In some important systems there are several classes of customers, each with its own statistical characteristics for arrival and service, which cannot be lumped into a single class for the purpose of analysis. Here are some examples:

Example 3.12 Two Session Classes in a Circuit Switching System

Consider a transmission line consisting of m independent circuits of equal capacity. There are two types of sessions arriving with Poisson rates λ_1 and λ_2, respectively. A session is blocked and lost for the system if all circuits are busy upon arrival, and is otherwise routed to any free circuit. The durations (or holding times) of the sessions of the two types are exponentially distributed with means $1/\mu_1$ and $1/\mu_2$. We are interested in finding the steady-state blocking probability for this system.

We first note that if $\mu_1 = \mu_2$, the two session types are indistinguishable for queueing purposes and the system can be modeled by an $M/M/m/m$ queue with arrival rate $\lambda_1 + \lambda_2$ and state equal to the total number of busy circuits. The desired blocking probability p_m is then given by the Erlang B formula of the preceding subsection. If, however, $\mu_1 \neq \mu_2$, then the total number of busy circuits does not fully specify the future statistical behavior of the queue; the number of each session type is also important since the duration of a session depends statistically on its type. Thus, the appropriate Markov chain model involves the two-dimensional state (n_1, n_2), where n_i is the number of circuits occupied by a session of type i, for $i = 1, 2$. The transition probability diagram for this chain is shown in Fig. 3.11. Generally, for multidimensional chains one may write the global balance equations for the stationary distribution

$$P(n_1, n_2), \qquad n_1 \geq 0,\ n_2 \geq 0,\ n_1 + n_2 \leq m$$

and try to solve them numerically. For this example, however, a closed-form expression is possible. We will demonstrate this shortly, once we develop the appropriate methodology.

Example 3.13 Two-Class System with Preferential Treatment for One Class

Consider the system of the preceding example with the difference that there is a limit $k < m$ on the number of circuits that can be used by sessions of the second type, so there are always $m - k$ circuits for use by sessions of the first type. The corresponding two-dimensional Markov chain is shown in Fig. 3.12. Note that here we should distinguish between the blocking probability for the first type of session, which is

$$\sum_{\{(n_1,n_2)|m-k\leq n_1\leq m, n_2=m-n_1\}} P(n_1, n_2)$$

and the blocking probability for the second type of session, which is

$$\sum_{\{(n_1,n_2)|0\leq n_1\leq m, n_2=\min\{k,m-n_1\}\}} P(n_1, n_2)$$

Again, it turns out that there is a closed-form expression for $P(n_1, n_2)$, as will be seen shortly.

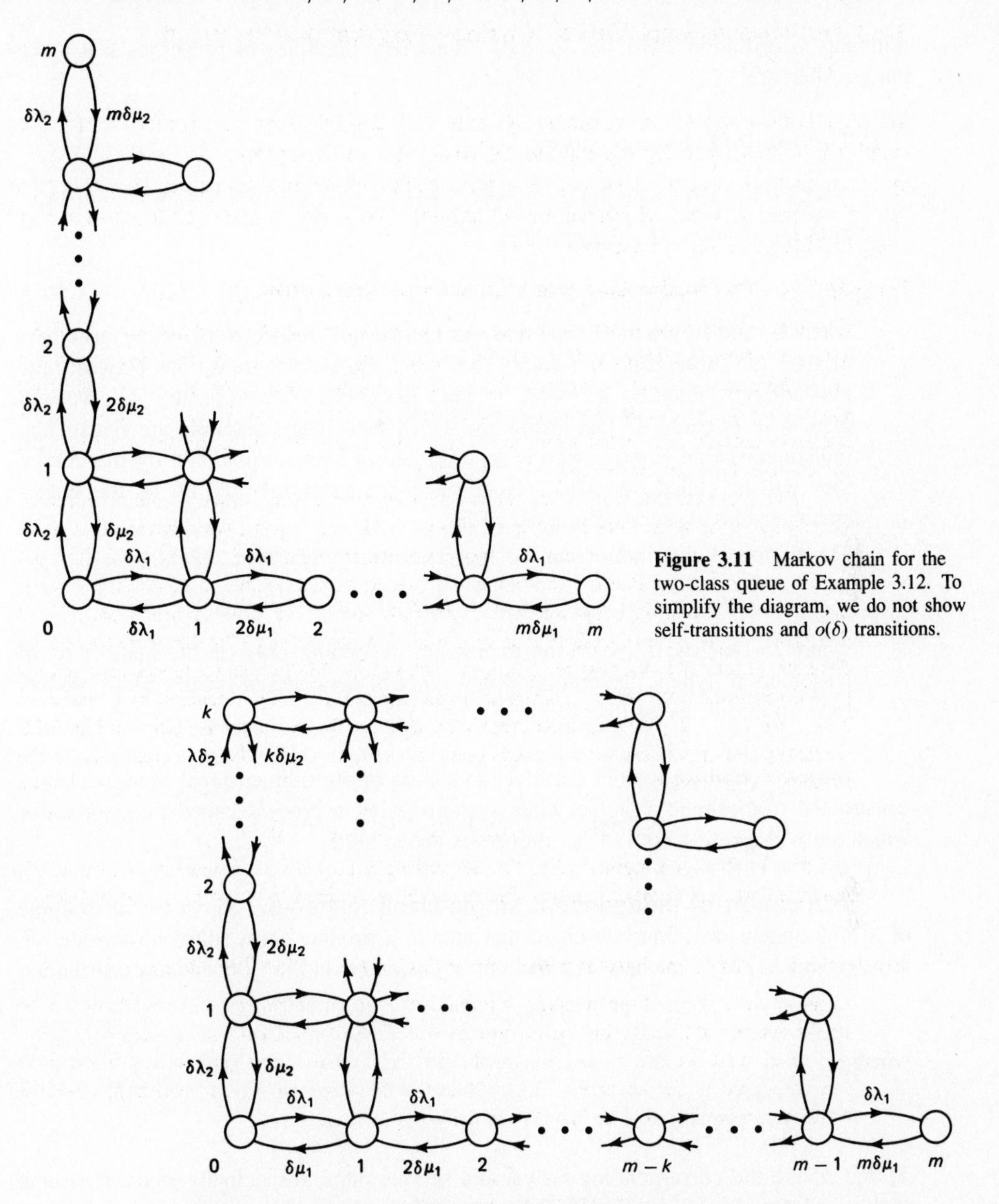

Figure 3.11 Markov chain for the two-class queue of Example 3.12. To simplify the diagram, we do not show self-transitions and $o(\delta)$ transitions.

Figure 3.12 Markov chain for the two class queue with preferential treatment for one class (cf. Example 3.13). Self-transitions and $o(\delta)$ transitions are not shown.

Multidimensional Markov chains usually involve K customer types. Their states are of the form $(n_1, n_2, \ldots, n_K)$, where n_i is the number of customers of type i in the system. Such chains are usually harder to analyze than their one-dimensional counterparts, but in many interesting special cases one can obtain a closed-form solution for the stationary distribution $P(n_1, n_2, \ldots, n_K)$. Important examples of properties that make this possible are:

1. The *detailed balance equations*

$$\lambda_i P(n_1, \ldots, n_{i-1}, n_i, n_{i+1}, \ldots, n_K) = \mu_i P(n_1, \ldots, n_{i-1}, n_i + 1, n_{i+1}, \ldots, n_K)$$

hold for all pairs of adjacent states

$$(n_1, \ldots, n_{i-1}, n_i, n_{i+1}, \ldots, n_K) \quad \text{and} \quad (n_1, \ldots, n_{i-1}, n_i + 1, n_{i+1}, \ldots, n_K)$$

where λ_i and μ_i are the arrival rate and service rate, respectively, of the customers of type i. These equations imply that the frequency of transitions between any two adjacent states is the same in both directions (see Appendix A). We will explain in Section 3.7 that chains for which these equations hold are statistically indistinguishable when looked in forward and in reverse time, and for this reason they will be called *reversible*. Note that these equations hold for all the single-customer class systems discussed so far.

2. The stationary distribution can be expressed in *product form*, that is,

$$P(n_1, n_2, \ldots, n_K) = P_1(n_1)P_2(n_2)\cdots P_K(n_K)$$

where for each i, $P_i(n_i)$ is an expression depending only on the number n_i of customers of type i. Several important types of networks of queues admit product form solutions, as will be seen in Section 3.8.

In this section we restrict ourselves to a class of multidimensional Markov chains, constructed from single-customer class systems using a process called *truncation*, for which we will see that both of the properties above hold.

Truncation of independent single-class systems. For a trivial example of a multidimensional Markov chain that admits a product form solution, consider K independent $M/M/1$ queues. The number of customers in the ith queue has distribution

$$P_i(n_i) = \rho_i^{n_i}(1 - \rho_i)$$

where

$$\rho_i = \frac{\lambda_i}{\mu_i}$$

λ_i and μ_i are the corresponding arrival and service rates, respectively, and we assume that $\rho_i < 1$ for all i. Since the K queues are independent, we have the product form

$$P(n_1, n_2, \ldots, n_K) = P_1(n_1)P_2(n_2)\cdots P_K(n_K)$$

Note that the reasoning above would also apply if each of the $M/M/1$ queues were replaced by a birth–death type of queue (two successive states can differ only by a single unit: for example, the $M/M/m$, $M/M/\infty$, and $M/M/m/m$ queues). The only requirement is that the queues are independent.

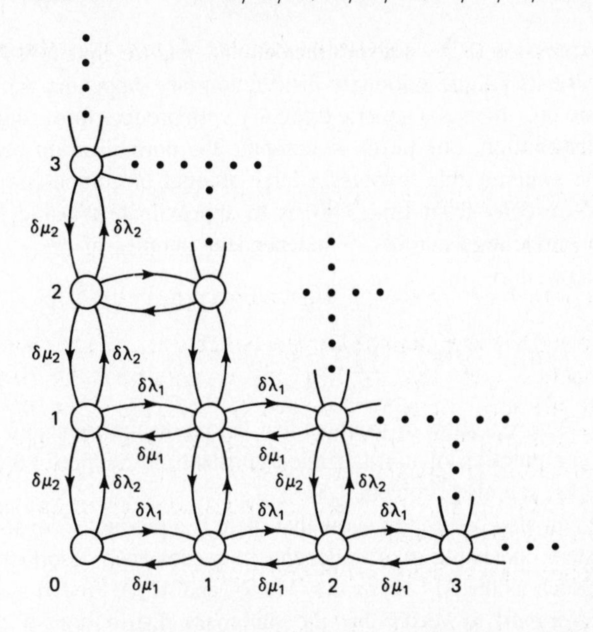

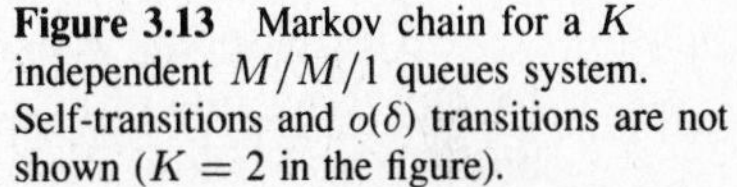

Figure 3.13 Markov chain for a K independent $M/M/1$ queues system. Self-transitions and $o(\delta)$ transitions are not shown ($K = 2$ in the figure).

Consider now the transition probability diagram of the K independent $M/M/1$ queues system, shown in Fig. 3.13. A *truncation* of this system is a Markov chain having the same transition probability diagram with the only difference that some states have been eliminated, while transitions between all other pairs of states, together with their corresponding transition probabilities, have been left unchanged except for $O(\delta)$ terms (see Fig. 3.14). We require that the truncation is an irreducible Markov chain, that is, all states communicate with each other (see Appendix A).

We claim that the stationary distribution of this truncated system has the product form

$$P(n_1, n_2, \ldots, n_K) = \frac{\rho_1^{n_1} \rho_2^{n_2} \cdots \rho_K^{n_K}}{G} \tag{3.39}$$

where G is a normalization constant guaranteeing that $P(n_1, n_2, \ldots, n_K)$ is a probability distribution, that is,

$$G = \sum_{(n_1, n_2, \ldots, n_K) \in S} \rho_1^{n_1} \rho_2^{n_2} \cdots \rho_K^{n_K} \tag{3.40}$$

where S is the set of states of the truncated system.

To show this, we consider the detailed balance equations

$$\lambda_i P(n_1, \ldots, n_{i-1}, n_i, n_{i+1}, \ldots, n_K) = \mu_i P(n_1, \ldots, n_{i-1}, n_i + 1, n_{i+1}, \ldots, n_K)$$

By substituting the probabilities (3.39) in these equations, we obtain

$$\lambda_i \frac{\rho_1^{n_1} \cdots \rho_{i-1}^{n_{i-1}} \rho_i^{n_i} \rho_{i+1}^{n_{i+1}} \cdots \rho_K^{n_K}}{G} = \mu_i \frac{\rho_1^{n_1} \cdots \rho_{i-1}^{n_{i-1}} \rho_i^{n_i+1} \rho_{i+1}^{n_{i+1}} \cdots \rho_K^{n_K}}{G}$$

which holds as an identity in view of the definition $\rho_i = \lambda_i/\mu_i$. Therefore, the probability distribution given by the expression (3.39) satisfies the detailed balance equations for the truncated chain, so it must be its unique stationary distribution (see Appendix A).

It should be noted here that there is a generic difficulty with product form solutions. To obtain the stationary distribution, one needs to compute the normalization constant G of Eq. (3.40). For some systems, this involves a large amount of computation. An alternative to computing G directly from Eq. (3.40) is to approximate it using Monte Carlo simulation. Here, a fairly large number of independent samples of $(n_1, \ldots, n_K)$ are generated using the distribution

$$P(n_1, \ldots, n_K) = \prod_{i=1}^{K} (1 - \rho_i)\rho_i^{n_i}$$

and G is approximated by the proportion of samples that belong to the truncated space S. We will return to the computation of normalization constants in Section 3.8 in the context of queueing networks; see also Problem 3.51.

The reasoning above can also be used to show that there is a product form solution for any truncation of a system consisting of K independent queues each described by a birth–death Markov chain, such as the $M/M/m$, $M/M/\infty$, and $M/M/m/m$ systems. For example, it is straightforward to verify that the stationary distribution of the K independent $M/M/\infty$ queues system is given by

$$P(n_1, n_2, \ldots, n_K) = \frac{\frac{\rho_1^{n_1}}{n_1!} \frac{\rho_2^{n_2}}{n_2!} \cdots \frac{\rho_K^{n_K}}{n_K!}}{G}$$

where G is a normalization constant,

$$G = \sum_{(n_1, n_2, \ldots, n_K) \in S} \frac{\rho_1^{n_1}}{n_1!} \frac{\rho_2^{n_2}}{n_2!} \cdots \frac{\rho_K^{n_K}}{n_K!}$$

and S is the set of states of the truncated chain.

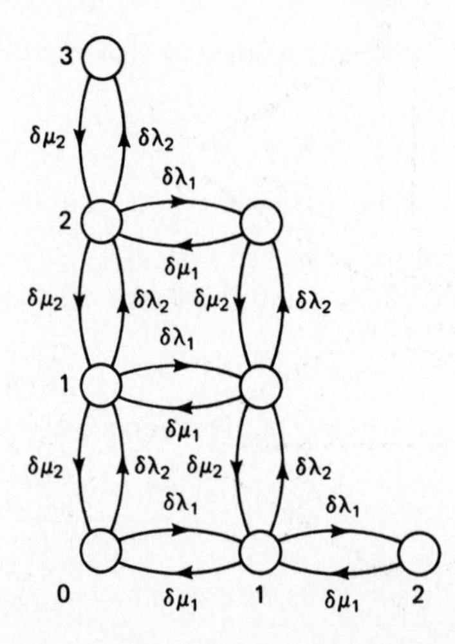

Figure 3.14 Example of a Markov chain which is a truncation of the K independent $M/M/1$ queues system ($K = 2$ in the figure).

Blocking probabilities for circuit switching systems. Using the product form solution (3.39), it is straightforward to write closed-form expressions for the blocking probabilities of the circuit switching systems with two session classes of Examples 3.12 and 3.13. The two-dimensional chains of these examples are truncations of the two independent $M/M/\infty$ queues system (Figs. 3.11 to 3.13). Thus, in the case of Example 3.13, the blocking probability for the first type of session is

$$\sum_{\{(n_1,n_2)|m-k\le n_1\le m, n_2=m-n_1\}} P(n_1,n_2)$$

$$= \frac{\sum_{n_1=k}^{m} \frac{\rho_1^{n_1}}{n_1!}\frac{\rho_2^{m-n_1}}{m-n_1!}}{\sum_{n_1=0}^{k}\sum_{n_2=0}^{k} \frac{\rho_1^{n_1}}{n_1!}\frac{\rho_2^{n_2}}{n_2!} + \sum_{n_1=k+1}^{m}\sum_{n_2=0}^{m-n_1} \frac{\rho_1^{n_1}}{n_1!}\frac{\rho_2^{n_2}}{n_2!}}$$

The blocking probability for the second type of session is

$$\sum_{\{(n_1,n_2)|0\le n_1\le m, n_2=\min\{k,m-n_1\}\}} P(n_1,n_2)$$

$$= \frac{\sum_{n_1=0}^{k} \frac{\rho_1^{n_1}}{n_1!}\frac{\rho_2^{k-n_1}}{m-n_1!} + \sum_{n_1=k+1}^{m} \frac{\rho_1^{n_1}}{n_1!}\frac{\rho_2^{m-n_1}}{m-n_1!}}{\sum_{n_1=0}^{k}\sum_{n_2=0}^{k} \frac{\rho_1^{n_1}}{n_1!}\frac{\rho_2^{n_2}}{n_2!} + \sum_{n_1=k+1}^{m}\sum_{n_2=0}^{m-n_1} \frac{\rho_1^{n_1}}{n_1!}\frac{\rho_2^{n_2}}{n_2!}}$$

The following important example illustrates the wide applicability of product form solutions in circuit switching networks.

Example 3.14 Circuit Switching Networks with Fixed Routing

Consider a network of transmission lines shared by sessions of K different types (see Fig. 3.15). Sessions of type i arrive according to a Poisson rate λ_i and have an exponentially distributed holding time with mean $1/\mu_i$.

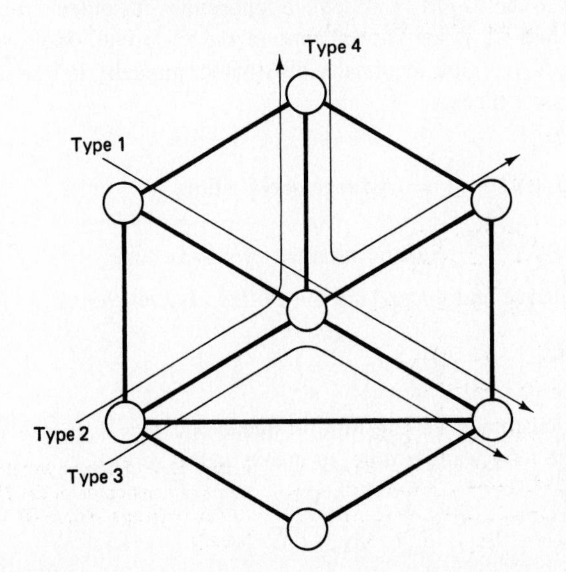

Figure 3.15 Model of a circuit switching network. There are K different session types. All sessions of the same type go over the same path and reserve the same amount of transmission capacity on each link of their path. A session is blocked if some link on its path is loaded to the point that it cannot accommodate the transmission capacity of the session.

We assume that all sessions of a given type i traverse the same set of network links (fixed routing) and reserve a fixed amount b_i of transmission capacity at each link. Thus if C_j is the transmission capacity of a link j and $I(j)$ is the set of session types using this link, we must have

$$\sum_{i \in I(j)} b_i n_i \leq C_j$$

where n_i is the number of sessions of type i in the network. A session of a given type m is blocked from entering the network (and is assumed lost to the system) if upon arrival it finds that it cannot be accommodated due to insufficient link capacity, that is,

$$b_m + \sum_{i \in I(j)} b_i n_i > C_j$$

for some link j that the session must traverse.

The quality of service of this system may be described by the blocking probabilities for the different session types. To obtain these probabilities, we model the system as a truncation of the K independent $M/M/\infty$ queues system. The truncated chain is the same as for the latter system, except that all states $(n_1, n_2, \ldots, n_K)$ for which the inequality $\sum_{i \in I(j)} b_i n_i \leq C_j$ is violated for some link j have been eliminated. The stationary distribution has a product form, which yields the desired blocking probabilities.

A remarkable fact about the product form solution of this example is that it is valid for a broad class of holding time distributions that includes the exponential as a special case (see [BLL84] and [Kau81]).

3.5 THE $M/G/1$ SYSTEM

Consider a single-server queueing system where customers arrive according to a Poisson process with rate λ, but the customer service times have a general distribution—not necessarily exponential as in the $M/M/1$ system. Suppose that customers are served in the order they arrive and that X_i is the service time of the ith arrival. We assume that the random variables $(X_1, X_2, \ldots)$ are identically distributed, mutually independent, and independent of the interarrival times.

Let

$$\overline{X} = E\{X\} = \frac{1}{\mu} = \text{ Average service time}$$

$$\overline{X^2} = E\{X^2\} = \text{ Second moment of service time}$$

Our objective is to derive and understand the *Pollaczek–Khinchin (P-K) formula:*

$$W = \frac{\lambda \overline{X^2}}{2(1-\rho)} \tag{3.41}$$

where W is the expected customer waiting time in queue and $\rho = \lambda/\mu = \lambda\overline{X}$. Given the P-K formula (3.41), the total waiting time, in queue and in service, is

$$T = \overline{X} + \frac{\lambda \overline{X^2}}{2(1-\rho)} \tag{3.42}$$

Applying Little's formula to W and T, we get the expected number of customers in the queue N_Q and the expected number in the system N:

$$N_Q = \frac{\lambda^2 \overline{X^2}}{2(1-\rho)} \tag{3.43}$$

$$N = \rho + \frac{\lambda^2 \overline{X^2}}{2(1-\rho)} \tag{3.44}$$

For example, when service times are exponentially distributed, as in the $M/M/1$ system, we have $\overline{X^2} = 2/\mu^2$, and the P-K formula (3.41) reduces to the equation (see Section 3.3.2)

$$W = \frac{\rho}{\mu(1-\rho)} \qquad (M/M/1)$$

When service times are identical for all customers (the $M/D/1$ system, where D means deterministic), we have $\overline{X^2} = 1/\mu^2$, and

$$W = \frac{\rho}{2\mu(1-\rho)} \qquad (M/D/1) \tag{3.45}$$

Since the $M/D/1$ case yields the minimum possible value of $\overline{X^2}$ for given μ, it follows that the values of W, T, N_Q, and N for an $M/D/1$ queue are lower bounds to the corresponding quantities for an $M/G/1$ queue of the same λ and μ. It is interesting to note that W and N_Q for the $M/D/1$ queue are exactly one half their values for the $M/M/1$ queue of the same λ and μ. The values of T and N for $M/D/1$, on the other hand, range from the same as $M/M/1$ for small ρ to one half of $M/M/1$ as ρ approaches 1. The reason is that the expected service time is the same in the two cases, and for ρ small, most of the waiting occurs in service, whereas for ρ large, most of the waiting occurs in the queue.

We provide a proof of the Pollaczek–Khinchin formula based on the concept of the *mean residual service time*. This same concept will prove useful in a number of subsequent developments. One example is $M/G/1$ queues with priorities. Another is reservation systems where part of the service time is occupied with sending packets (*i.e.*, serving customers), and part with sending control information or making reservations for sending the packets.

Denote

W_i = Waiting time in queue of the ith customer

R_i = Residual service time seen by the ith customer. By this we mean that if customer j is already being served when i arrives, R_i is the remaining time until customer j's service time is complete. If no customer is in service (*i.e.*, the system is empty when i arrives), then R_i is zero

X_i = Service time of the ith customer

N_i = Number of customers found waiting in queue by the ith customer upon arrival

We have

$$W_i = R_i + \sum_{j=i-N_i}^{i-1} X_j$$

By taking expectations and using the independence of the random variables N_i and $X_{i-1}, \ldots, X_{i-N_i}$, we have

$$E\{W_i\} = E\{R_i\} + E\left\{\sum_{j=i-N_i}^{i-1} E\{X_j \mid N_i\}\right\} = E\{R_i\} + \overline{X}E\{N_i\}$$

Taking the limit as $i \to \infty$, we obtain

$$W = R + \frac{1}{\mu}N_Q \tag{3.46}$$

where

$R =$ Mean residual time, defined as $R = \lim_{i\to\infty} E\{R_i\}$.

In Eq. (3.46) (and throughout this section) all long-term average quantities should be viewed as limits when time or customer index increases to infinity. Thus, W, R, and N_Q are limits (as $i \to \infty$) of the average waiting time, residual time, and number found in queue, respectively, corresponding to the ith customer. We assume that these limits exist, and this is true of almost all systems of interest to us provided that $\lambda < \mu$. Note that in the waiting time equation (3.46), the average number in queue N_Q and the mean residual time R as seen by an arriving customer are also equal to the average number in queue and mean residual time seen by an outside observer at a random time. This is due to the Poisson character of the arrival process, which implies that the occupancy distribution upon arrival is typical (see Section 3.3.2).

By Little's Theorem, we have

$$N_Q = \lambda W$$

and by substitution in the waiting time formula (3.46), we obtain

$$W = R + \rho W \tag{3.47}$$

where $\rho = \lambda/\mu$ is the utilization factor; so, finally,

$$W = \frac{R}{1-\rho} \tag{3.48}$$

We can calculate R by a graphical argument. In Fig. 3.16 we plot the residual service time $r(\tau)$ (*i.e.*, the remaining time for completion of the customer in service at time τ) as a function of τ. Note that when a new service of duration X begins, $r(\tau)$ starts at X and decays linearly for X time units. Consider a time t for which $r(t) = 0$. The time average of $r(\tau)$ in the interval $[0, t]$ is

$$\frac{1}{t}\int_0^t r(\tau)\,d\tau = \frac{1}{t}\sum_{i=1}^{M(t)} \frac{1}{2}X_i^2 \tag{3.49}$$

where $M(t)$ is the number of service completions within $[0,t]$, and X_i is the service time of the ith customer. We can also write this equation as

$$\frac{1}{t}\int_0^t r(\tau)\,d\tau = \frac{1}{2}\frac{M(t)}{t}\frac{\sum_{i=1}^{M(t)} X_i^2}{M(t)} \tag{3.50}$$

and assuming the limits below exist, we obtain

$$\lim_{t\to\infty}\frac{1}{t}\int_0^t r(\tau)\,d\tau = \frac{1}{2}\lim_{t\to\infty}\frac{M(t)}{t}\cdot\lim_{t\to\infty}\frac{\sum_{i=1}^{M(t)} X_i^2}{M(t)} \tag{3.51}$$

The two limits on the right are the time averages of the departure rate (which equals the arrival rate) and the second moment of the service time, respectively, while the limit on the left is the time average of the residual time. Assuming that time averages can be replaced by ensemble averages, we obtain

$$R = \frac{1}{2}\lambda\overline{X^2} \tag{3.52}$$

The P-K formula,

$$W = \frac{\lambda\overline{X^2}}{2(1-\rho)} \tag{3.53}$$

now follows by substituting the expression obtained for R [cf. Eq. (3.52)] into the formula $W = R/(1-\rho)$ [cf. Eq. (3.48)].

Note that our derivation was based on two assumptions:

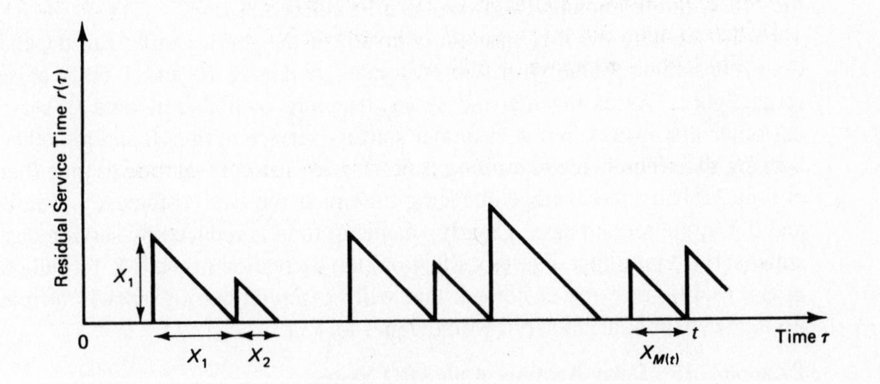

Figure 3.16 Derivation of the mean residual service time. During period $[0,t]$, the time average of the residual service time $r(\tau)$ is

$$\frac{1}{t}\int_0^t r(\tau)\;d\tau = \frac{1}{t}\sum_{i=1}^{M(t)}\frac{1}{2}X_i^2 = \frac{1}{2}\frac{M(t)}{t}\frac{\sum_{i=1}^{M(t)} X_i^2}{M(t)}$$

where X_i is the service time of the ith customer, and $M(t)$ is the number of service completions in $[0,t]$. Taking the limit as $t\to\infty$ and equating time and ensemble averages, we obtain the mean residual time $R = (1/2)\lambda\overline{X^2}$.

1. The existence of the steady-state averages W, R, and N_Q
2. The equality (with probability one) of the long-term time averages appearing in Eq. (3.51) with the corresponding ensemble averages

These assumptions can be justified by careful applications of the law of large numbers, but the details are beyond the scope of this book. However, these are natural assumptions for the systems of interest to us, and we will base similar derivations on graphical arguments and interchange of time averages with ensemble averages without further discussion.

One curious feature of the P-K formula (3.53) is that an $M/G/1$ queue can have $\rho < 1$ but infinite W if the second moment $\overline{X^2}$ is ∞. What is happening in this case is that a small fraction of customers have incredibly long service times. When one of these customers is served, an incredible number of arrivals are queued and delayed by a significant fraction of that long service time. Thus, the contribution to W is proportional to the square of the service time, leading to an infinite W if $\overline{X^2}$ is infinite.

The derivation of the P-K formula above assumed that customers were served in order of arrival, that is, that the number of customers served between the ith arrival and service is just the number in queue at the ith arrival. It turns out, however, that this formula is valid for any order of servicing customers as long as the order is determined independently of the required service time. To see this, suppose that the ith and j^{th} customers are both in the queue and that they exchange places. The expected queueing time of customer i (over the service times of the customers in queue) will then be exchanged with that for customer j, but the average, over all customers, is unchanged. Since any service order can be considered as a sequence of reversals in queue position, the P-K formula remains valid (see also Problem 3.32).

To see why the P-K formula is invalid if the service order can depend on service time, consider a queue with two customers requiring 10 and 1 units of service time, respectively. Assuming that the server becomes available at time 0, serving the first customer first results in one customer starting service at time 0 and the other at time 10. Serving the second customer first results in one customer starting at time 0 and the other at time 1. Thus, the average queueing time over the two customers is 5 in the first case and 0.5 in the second case. Clearly, queueing time is reduced by serving customers with small service time first. For this situation, the derivation of the P-K formula breaks down at Eq. (3.46) since the customers that will be served before a newly arriving customer no longer have a mean service time equal to $1/\mu$.

Example 3.15 Delay Analysis of an ARQ System

Consider a go back n ARQ system such as the one discussed in Section 2.4. Assume that packets are transmitted in frames that are one time unit long, and there is a maximum wait for an acknowledgment of $n - 1$ frames before a packet is retransmitted (see Fig. 3.17). In this system packets are retransmitted for two reasons:

1. A given packet transmitted in frame i might be rejected at the receiver due to errors, in which case the transmitter will transmit packets in frames $i+1, i+2, \ldots, i+n-1$, (if any are available), and then go back to retransmit the given packet in frame $i+n$.

2. A packet transmitted in frame i might be accepted at the receiver, but the corresponding acknowledgment (in the form of the receive number) might not have arrived at the transmitter by the time the transmission of packet $i+n-1$ is completed. This can happen due to errors in the return channel, large propagation delays, long return frames relative to the size of the goback number n, or a combination thereof.

We will assume (somewhat unrealistically) that retransmissions occur only due to reason 1, and that a packet is rejected at the receiver with probability p independently of other packets.

Consider the case where packets arrive at the transmitter according to a Poisson process with rate λ. It follows that the time interval between start of the first transmission of a given packet after the last transmission of the previous packet and end of the last transmission of the given packet is $1+kn$ time units with probability $(1-p)p^k$. (This corresponds to k retransmissions following the last transmission of the previous packet; see Fig. 3.17.) Thus, the transmitter's queue behaves like an $M/G/1$ queue with service time distribution given by

$$P\{X = 1 + kn\} = (1-p)p^k, \quad k = 0, 1, \ldots$$

The first two moments of the service time are

$$\overline{X} = \sum_{k=0}^{\infty}(1+kn)(1-p)p^k = (1-p)\left(\sum_{k=0}^{\infty}p^k + n\sum_{k=0}^{\infty}kp^k\right)$$

$$\overline{X^2} = \sum_{k=0}^{\infty}(1+kn)^2(1-p)p^k = (1-p)\left(\sum_{k=0}^{\infty}p^k + 2n\sum_{k=0}^{\infty}kp^k + n^2\sum_{k=0}^{\infty}k^2p^k\right)$$

We now note that

$$\sum_{k=0}^{\infty}p^k = \frac{1}{1-p}, \qquad \sum_{k=0}^{\infty}kp^k = \frac{p}{(1-p)^2}, \qquad \sum_{k=0}^{\infty}k^2p^k = \frac{p+p^2}{(1-p)^3}$$

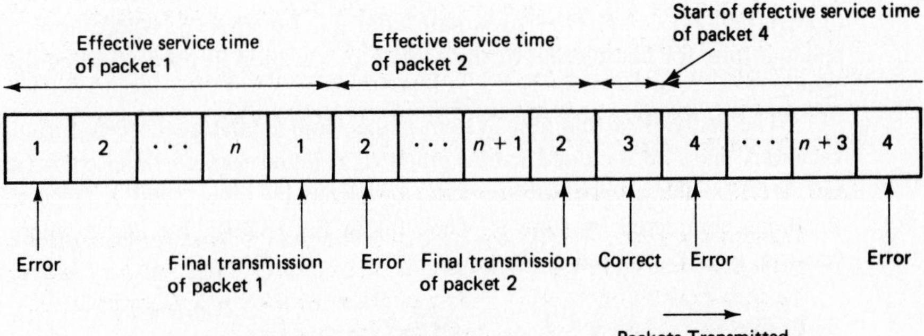

Figure 3.17 Illustration of the effective service times of packets in the ARQ system of Example 3.15. For example, packet 2 has an effective service time of $n+1$ because there was an error in the first attempt to transmit it following the last transmission of packet 1, but no error in the second attempt.

(The first sum is the usual geometric series sum, while the other two sums are obtained by differentiating the first sum twice.) Using these formulas in the equations for $\overline{X}$ and $\overline{X^2}$ above, we obtain

$$\overline{X} = 1 + \frac{np}{1-p}$$

$$\overline{X^2} = 1 + \frac{2np}{1-p} + \frac{n^2(p+p^2)}{(1-p)^2}$$

The P-K formula gives the average packet time in queue and in the system (up to the end of the last transmission):

$$W = \frac{\lambda \overline{X^2}}{2(1-\lambda\overline{X})}$$

$$T = \overline{X} + W$$

3.5.1 *M/G/1* Queues with Vacations

Suppose that at the end of each busy period, the server goes on "vacation" for some random interval of time. Thus, a new arrival to an idle system, rather than going into service immediately, waits for the end of the vacation period (see Fig. 3.18). If the system is still idle at the completion of a vacation, a new vacation starts immediately. For data networks, vacations correspond to the transmission of various kinds of control and record-keeping packets when there is a lull in the data traffic; other applications will become apparent later.

Let $V_1, V_2, \ldots$ be the durations of the successive vacations taken by the server. We assume that $V_1, V_2, \ldots$ are independent and identically distributed (IID) random variables, also independent of the customer interarrival times and service times. As before, the arrivals are Poisson and the service times are IID with a general distribution. A new arrival to the system has to wait in the queue for the completion of the current service or vacation and then for the service of all the customers waiting before it. Thus, the waiting time formula $W = R/(1-\rho)$ is still valid [cf. Eq. (3.48)], where now R is the mean residual time for completion of the service *or* vacation in process when the ith customer arrives.

The analysis of this new system is the same as that of the P-K formula except that vacations must be included in the graph of residual service times $r(\tau)$ (see Fig. 3.19). Let $M(t)$ be the number of services completed by time t and $L(t)$ be the number of vacations completed by time t. Then [as in Eq. (3.49)], for any t where a service or vacation is just completed, we have

$$\frac{1}{t}\int_0^t r(\tau)\,d\tau = \frac{1}{t}\sum_{i=1}^{M(t)} \frac{1}{2}X_i^2 + \frac{1}{t}\sum_{i=1}^{L(t)} \frac{1}{2}V_i^2 = \frac{M(t)}{t}\,\frac{\sum_{i=1}^{M(t)} \frac{1}{2}X_i^2}{M(t)} + \frac{L(t)}{t}\,\frac{\sum_{i=1}^{L(t)} \frac{1}{2}V_i^2}{L(t)} \tag{3.54}$$

As before, assuming that a steady-state exists, $M(t)/t$ approaches λ with increasing t, and the first term on the right side of Eq. (3.54) approaches $\lambda\overline{X^2}/2$ as in the derivation

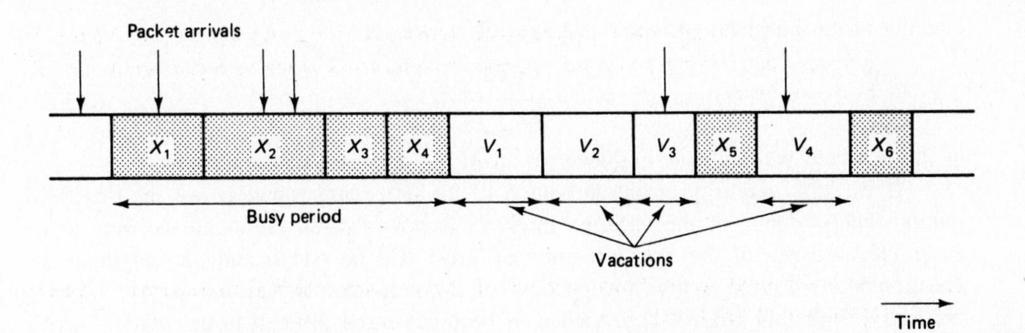

Figure 3.18 $M/G/1$ system with vacations. At the end of a busy period, the server goes on vacation for time V with first and second moments $\overline{V}$ and $\overline{V^2}$, respectively. If the system is empty at the end of a vacation, the server takes a new vacation. An arriving customer to an empty system must wait until the end of the current vacation to get service.

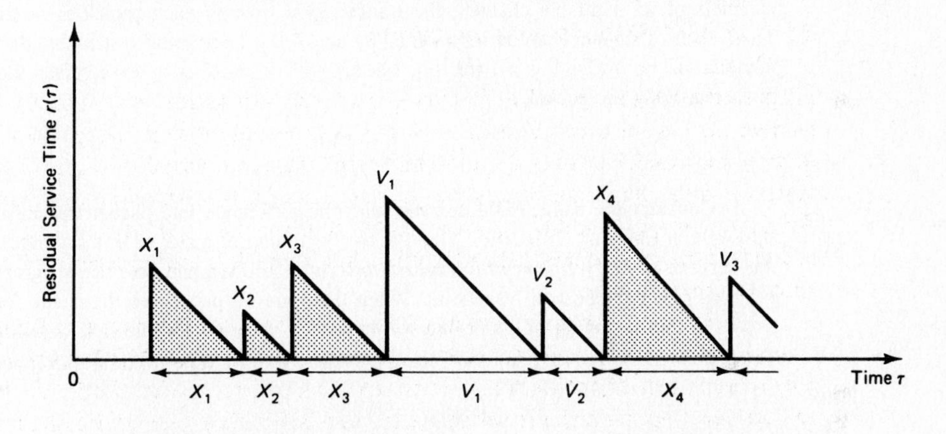

Figure 3.19 Residual service times for an $M/G/1$ system with vacations. Busy periods alternate with vacation periods. If $M(t)$ and $L(t)$ are the numbers of services and vacations completed by time t, respectively, and t is a time of completion of a service or a vacation, we have

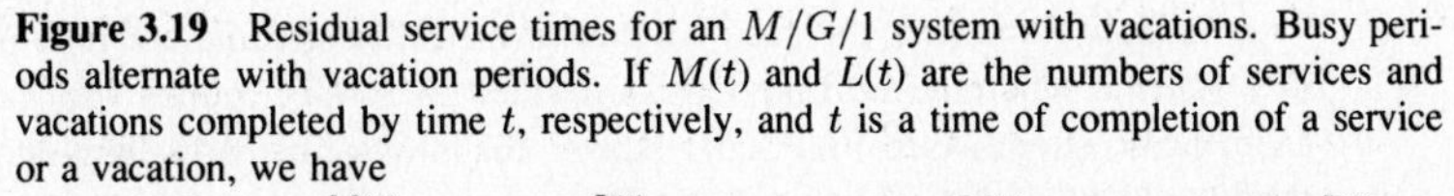

$$\frac{1}{t}\int_0^t r(\tau)\,d\tau = \frac{1}{t}\sum_{i=1}^{M(t)}\frac{1}{2}X_i^2 + \frac{1}{t}\sum_{i=1}^{L(t)}\frac{1}{2}V_i^2 = \frac{M(t)}{t}\frac{\sum_{i=1}^{M(t)}\frac{1}{2}X_i^2}{M(t)} + \frac{L(t)}{t}\frac{\sum_{i=1}^{L(t)}\frac{1}{2}V_i^2}{L(t)}$$

Taking limit as $t \to \infty$ and arguing that $M(t)/t \to \lambda$ and $L(t)/t \to (1-\rho)/\overline{V}$, we obtain the mean residual time $R = \frac{\lambda\overline{X^2}}{2} + \frac{(1-\rho)\overline{V^2}}{2\overline{V}}$.

of the P-K formula [cf. Eq. (3.52)]. For the second term, note that as $t \to \infty$, the fraction of time spent serving customers approaches ρ, and thus the fraction of time occupied with vacations is $1-\rho$. Assuming that time averages can be replaced by ensemble averages, we have $t(1-\rho)/L(t) \to \overline{V}$ with increasing t, and thus the second term in Eq. (3.54) approaches $(1-\rho)\overline{V^2}/(2\overline{V})$, where $\overline{V}$ and $\overline{V^2}$ are the first and second moments of the vacation interval, respectively. Combining this with $W = R/(1-\rho)$, and assuming equality of the time and ensemble averages of R, we get

$$W = \frac{\lambda \overline{X^2}}{2(1-\rho)} + \frac{\overline{V^2}}{2\overline{V}} \tag{3.55}$$

as the expected waiting time in queue for an $M/G/1$ system with vacations.

If we look carefully at the derivation of the preceding equation, we see that the mutual independence of the vacation intervals is not required (although the time and ensemble averages of the vacation intervals must still be equal) and the length of a vacation interval need not be independent of the customer arrival and service times. Naturally, with this kind of dependence, it becomes more difficult to calculate $\overline{V}$ and $\overline{V^2}$, as these quantities might be functions of the underlying $M/G/1$ process.

Example 3.16 Frequency- and Time-Division Multiplexing on a Slot Basis

We have m traffic streams of equal-length packets arriving according to a Poisson process with rate λ/m each. If the traffic streams are frequency-division multiplexed on m subchannels of an available channel, the transmission time of each packet is m time units. Then, each subchannel can be represented by an $M/D/1$ queueing system and the $M/D/1$ formula $W = \rho/(2\mu(1-\rho))$ [cf. Eq. (3.45)] with $\rho = \lambda$, $\mu = 1/m$, gives the average queueing delay per packet,

$$W_{\text{FDM}} = \frac{\lambda m}{2(1-\lambda)} \tag{3.56}$$

Consider the same FDM scheme with the difference that packet transmissions can start only at times $m, 2m, 3m, \ldots$ (*i.e.*, at the beginning of a slot of m time units). We call this scheme *slotted frequency-division multiplexing* (SFDM), and note that it can be viewed as an $M/D/1$ queue with vacations. When there are no packets in the queue for a given stream at the beginning of a slot, the server takes a vacation for one slot, or m time units. Thus, $\overline{V} = m$, $\overline{V^2} = m^2$, and the vacation system waiting time formula (3.55) becomes

$$W_{\text{SFDM}} = W_{\text{FDM}} + \frac{m}{2} \tag{3.57}$$

Finally, consider the case where the m traffic streams are time-division multiplexed in a scheme whereby the time axis is divided in m-slot frames with one slot dedicated to each traffic stream (see Fig. 3.20). Each slot is one time unit long and can carry a single packet. Then, if we compare this TDM scheme with the SFDM scheme, we see that the queue for a given stream in TDM is precisely the same as the queue for SFDM, and

$$W_{\text{TDM}} = W_{\text{SFDM}} = W_{\text{FDM}} + \frac{m}{2} = \frac{m}{2(1-\lambda)} \tag{3.58}$$

If we now look at the total delay for TDM, we get a different picture, since the service time is 1 unit of time rather than m units as in SFDM. By adding the service times to the queueing delays, we obtain

$$T_{\text{FDM}} = m + \frac{\lambda m}{2(1-\lambda)}$$

$$T_{\text{SFDM}} = T_{\text{FDM}} + \frac{m}{2}$$

$$T_{\text{TDM}} = 1 + \frac{m}{2(1-\lambda)} = T_{\text{FDM}} - \left(\frac{m}{2} - 1\right) \tag{3.59}$$

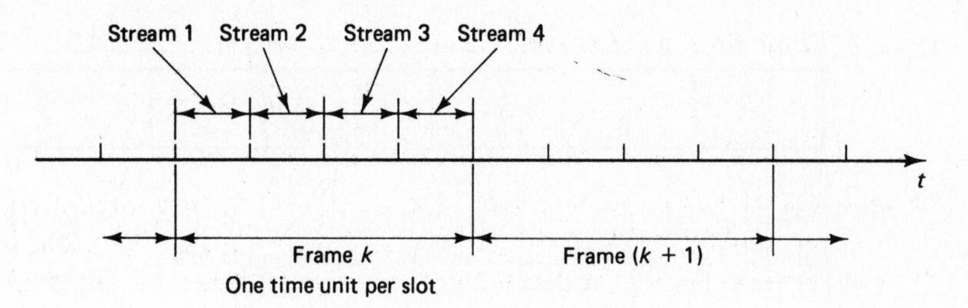

Figure 3.20 TDM with $m = 4$ traffic streams.

Thus, the customer's average total delay is more favorable in TDM than in FDM (assuming that $m > 2$). The longer average waiting time in queue for TDM is more than compensated by the faster service time. Contrast this with the Example 3.9, which treats TDM with slots that are a very small portion of the packet size. Problem 3.33 outlines an alternative approach for deriving the TDM average delay.

3.5.2 Reservations and Polling

Organizing transmissions from several packet streams into a statistical multiplexing system requires some form of scheduling. In some cases, this scheduling is naturally and easily accomplished; in other cases, however, some form of reservation or polling system is required.

Situations of this type arise often in multiaccess channels, which will be treated extensively in Chapter 4. For a typical example, consider a communication channel that can be accessed by several spatially separated users; however, only one user can transmit successfully on the channel at any one time. The communication resource of the channel can be divided over time into a portion used for packet transmissions and another portion used for reservation or polling messages that coordinate the packet transmissions. In other words, the time axis is divided into *data intervals*, where actual data are transmitted, and *reservation intervals*, used for scheduling future data. For uniform presentation, we use the term "reservation" even though "polling" may be more appropriate to the practical situation.

We will consider m traffic streams (also called users) and assume that each data interval contains packets of a *single* user. Reservations for these packets are made in the immediately preceding reservation interval. All users are taken up in cyclic order (see Fig. 3.21). There are several versions of this system differing in the rule for deciding which packets are transmitted during the data interval of each user. In the *gated* system, the rule is that only those packets that arrived prior to the user's preceding reservation interval are transmitted. By contrast, in the *exhaustive* system, the rule is that all available packets of a user are transmitted during the corresponding data interval, including those that arrived in this data interval or the preceding reservation interval. An intermediate version, which we call the *partially gated* system, results when the packets transmitted in a user's data interval are those that arrived up to the time this data interval began (and the corresponding reservation interval ended). A typical example of such reservation systems

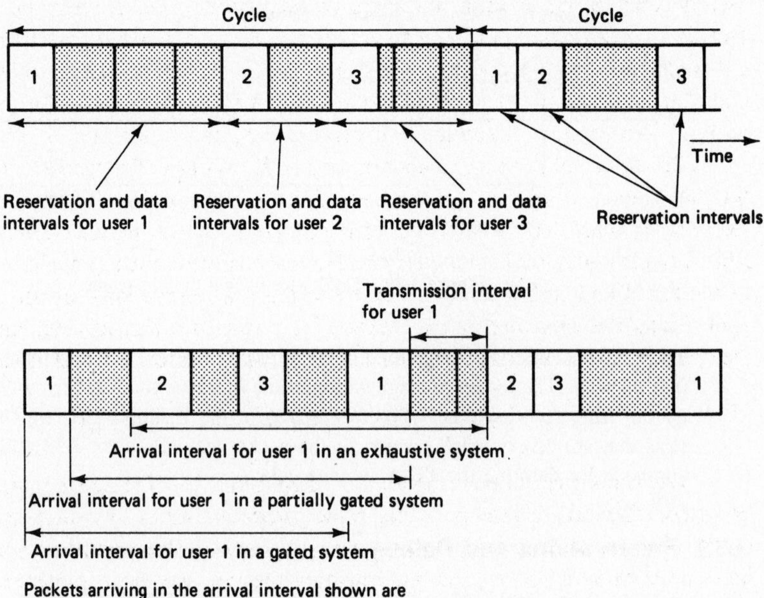

Figure 3.21 Reservation or polling system with three users. In the exhaustive version, a packet of a user that arrives during the user's reservation or data interval is transmitted in the same data interval. In the partially gated version, a packet of a user arriving during the user's data interval must wait for an entire cycle and be transmitted during the next data interval of the user. In the fully gated version, packets arriving during the user's reservation interval must also wait for an entire cycle. The figure shows, for the three systems, the association between the interval in which a packet arrives and the interval in which the packet is transmitted.

is one of the most common local area networks, the token ring. The users are connected by cable in a unidirectional loop. Each user transmits the current packet backlog, then gives the opportunity to a neighbor to transmit, and the process is repeated. (A more detailed description of the token ring is given in Chapter 4.)

We assume that the arrival processes of all users are independent Poisson with rate λ/m, and that the first and second moments of the packet transmission times are $\overline{X} = 1/\mu$ and $\overline{X^2}$, respectively. The utilization factor is $\rho = \lambda/\mu$. Interarrival times and transmission times are, as usual, assumed independent. While we assume that all users have identical arrival and service statistics, we allow the reservation intervals of different users to have different statistics.

Single-user system. Our general line of analysis of reservation systems can be better understood in terms of the special case where $m = 1$, so we consider this case first. We may also view this as a system where all users share reservation and data intervals. Let V_ℓ be the duration of the ℓth reservation interval and assume that successive reservation intervals are independent and identically distributed random variables with first and second moments $\overline{V}$ and $\overline{V^2}$, respectively. We consider a gated system and assume that the reservation intervals are statistically independent of the arrival times and service durations. Finally, for convenience of exposition, we assume that packets are transmitted in the order of their arrival. As in our derivation of the P-K formula, expected delays and queue lengths are independent of service order as long as service order is independent of service requirement (*i.e.*, packet length).

Consider the ith data packet arriving at the system. This packet must wait in queue for the residual time R_i until the end of the current packet transmission or reservation interval. It must also wait for the transmission of the N_i packets currently in the queue (this includes both packets for which reservations were already made in the last reservation interval and earlier arrivals waiting to make a reservation). Finally, the packet must wait during the next reservation interval $V_{\ell(i)}$, say, in which its reservation will be made (see Fig. 3.22). Thus, the expected queueing delay for the ith packet is given by

$$E\{W_i\} = E\{R_i\} + \frac{E\{N_i\}}{\mu} + E\{V_{\ell(i)}\} \tag{3.60}$$

The similarity of this reservation system to the $M/G/1$ queue with vacations should be noted. The only difference is that in the gated reservation system, a reservation interval starts when all packets that arrived prior to the start of the preceding reservation interval have been served, whereas in the vacation system, a vacation interval starts when all arrivals up to the current time have been served and the system is empty. (Thus in the gated reservation system, every packet has to wait in queue for a full reservation interval, while in the vacation system, only the packets that find the system empty upon arrival have to wait for part of a vacation interval. Note that the exhaustive version of this reservation system is equivalent to the vacation system.) The time-average mean residual time for the two systems is the same (the calculation based on Fig. 3.19 still applies) and is given by $\lambda\overline{X^2}/2 + (1-\rho)\overline{V^2}/2\overline{V}$. The value of $\lim_{i\to\infty} E\{N_i\}/\mu$ is ρW in both systems, and finally the value of $\lim_{i\to\infty} E\{V_{\ell(i)}\}$ is just $\overline{V}$. Thus, from

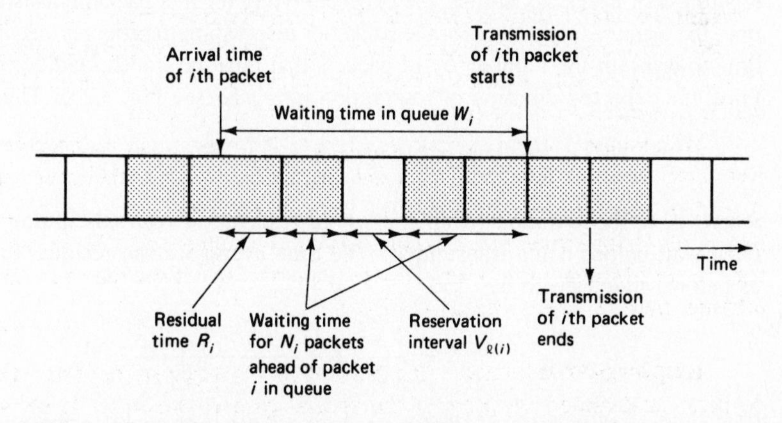

Figure 3.22 Calculation of the average waiting time in the single-user gated system. The expected waiting time $E\{W_i\}$ of the ith packet is

$$E\{W_i\} = E\{R_i\} + \frac{E\{N_i\}}{\mu} + E\{V_{\ell(i)}\}$$

Eq. (3.60) the expected time in queue for the single-user reservation system is

$$W = \frac{\lambda \overline{X^2}}{2(1-\rho)} + \frac{\overline{V^2}}{2\overline{V}} + \frac{\overline{V}}{1-\rho} \quad \text{(single user, gated)} \tag{3.61}$$

In the common situation where the reservation interval is a constant A, this simplifies to

$$W = \frac{\lambda \overline{X^2}}{2(1-\rho)} + \frac{A}{2}\left(\frac{3-\rho}{1-\rho}\right) \tag{3.62}$$

There is an interesting paradox associated with the waiting time formula (3.61). We have seen that a fraction $1-\rho$ of time is used on reservations. Since there is one reservation interval of mean duration $\overline{V}$ per cycle, we can conclude that the expected cycle length must be $\overline{V}/(1-\rho)$ (see also Example 3.6). The mean queueing delay in Eq. (3.61) can be an arbitrarily large multiple of this mean cycle length, which seems paradoxical since each packet is transmitted on the cycle following its arrival. The explanation of this is that more packets tend to arrive in long cycles than in short cycles, and thus mean cycle length is not representative of the cycle lengths seen by arriving packets; this is the same phenomenon that makes the mean residual service time used in the P-K formula derivation larger than one might think (see also Problem 3.31).

Multiuser system. Suppose that the system has m users, each with independent Poisson arrivals of rate λ/m. Again $\overline{X}$ and $\overline{X^2}$ are the first two moments of the service time for each user's packets. We denote by $\overline{V_i}$ and $\overline{V_i^2}$, respectively, the first two moments of the reservation intervals of user i. The service times and reservation intervals are all independent. We number the users $0, 1, \ldots, m-1$ and assume that the ℓth reservation interval is used to make reservations for user $\ell \bmod m$ and the subsequent (ℓ^{th}) data interval is used to send the packets corresponding to those reservations.

Consider the ith packet arrival into the system (counting packets in order of arrival, regardless of user). As before, the expected delay for this packet consists of three terms: first, the mean residual time for the packet or reservation in progress; second, the expected time to transmit the number N_i of packets that must be transmitted before packet i; and third, the expected duration of reservation intervals (see Fig. 3.23). Thus,

$$E\{W_i\} = E\{R_i\} + \frac{E\{N_i\}}{\mu} + E\{Y_i\} \tag{3.63}$$

where Y_i is the duration of all the whole reservation intervals during which packet i must wait before being transmitted. The time average mean residual time is calculated as before, and is given by

$$R = \frac{\lambda \overline{X^2}}{2} + \frac{(1-\rho)\sum_{\ell=0}^{m-1} \overline{V_\ell^2}}{2\sum_{\ell=0}^{m-1} \overline{V_\ell}} \tag{3.64}$$

The number of packets N_i that i must wait for is not equal to the number already in queue, but the order of serving packets is independent of packet service time; thus, each packet served before i still has a mean transmission time $1/\mu$ as indicated in Eq. (3.63) and by Little's formula, the value of $\lim_{i\to\infty} E\{N_i\}/\mu$ is ρW as before.

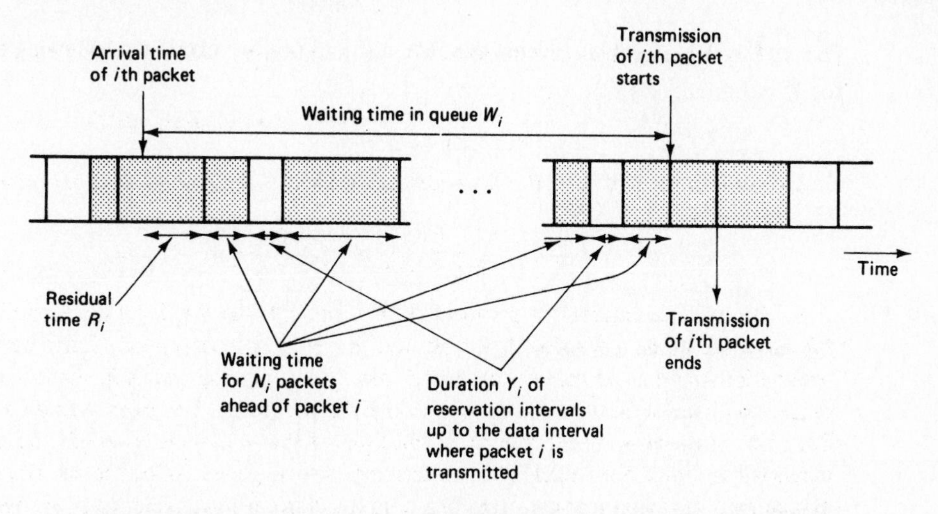

Figure 3.23 Calculation of the average waiting time in the multiuser system. The expected waiting time $E\{W_i\}$ of the ith packet is

$$E\{W_i\} = E\{R_i\} + \frac{E\{N_i\}}{\mu} + E\{Y_i\}$$

Letting $Y = \lim_{i\to\infty} E\{Y_i\}$, we can thus write the steady-state version of Eq. (3.63):

$$W = R + \rho W + Y$$

or, equivalently,

$$W = \frac{R+Y}{1-\rho} \tag{3.65}$$

We first calculate Y for an exhaustive system. Denote

$$\alpha_{\ell j} = E\{Y_i \mid \text{packet } i \text{ arrives in user } \ell\text{'s reservation or data interval and belongs to user } (\ell + j) \bmod m\}$$

We have

$$\alpha_{\ell j} = \begin{cases} 0, & j = 0 \\ \overline{V}_{(\ell+1) \bmod m} + \cdots + \overline{V}_{(\ell+j) \bmod m}, & j > 0 \end{cases}$$

Since packet i belongs to any user with equal probability $1/m$, we have

$E\{Y_i \mid$ packet i arrives in user ℓ's reservation or data interval$\}$

$$= \frac{1}{m} \sum_{j=1}^{m-1} \alpha_{\ell j} = \sum_{j=1}^{m-1} \frac{m-j}{m} \overline{V}_{(\ell+j) \bmod m} \tag{3.66}$$

Since all users have equal data rate, the data intervals of all users have equal average length in steady-state. Therefore, in steady-state, a packet will arrive during user ℓ's data interval with probability ρ/m, and during user ℓ's reservation interval with probability

$(1-\rho)\overline{V}_\ell / \left(\sum_{k=0}^{m-1} \overline{V}_k\right)$. Using this fact in Eq. (3.66), we obtain the following equation for $Y = \lim_{i\to\infty} E\{Y_i\}$:

$$Y = \sum_{\ell=0}^{m-1} \left(\frac{\rho}{m} + \frac{(1-\rho)\overline{V}_\ell}{\sum_{k=0}^{m-1} \overline{V}_k} \right) \sum_{j=1}^{m-1} \frac{m-j}{m} \overline{V}_{(\ell+j) \bmod m}$$

$$= \frac{\rho}{m} \sum_{j=1}^{m-1} \frac{m-j}{m} \left(\sum_{\ell=0}^{m-1} \overline{V}_\ell \right) + \frac{1-\rho}{\sum_{k=0}^{m-1} \overline{V}_k} \sum_{\ell=0}^{m-1} \sum_{j=1}^{m-1} \frac{m-j}{m} \overline{V}_\ell \overline{V}_{(\ell+j) \bmod m} \tag{3.67}$$

The last sum above can be written

$$\sum_{\ell=0}^{m-1} \sum_{j=1}^{m-1} \frac{m-j}{m} \overline{V}_\ell \overline{V}_{(\ell+j) \bmod m} = \frac{1}{2} \left[\left(\sum_{\ell=0}^{m-1} \overline{V}_\ell \right)^2 - \sum_{\ell=0}^{m-1} \overline{V}_\ell^2 \right]$$

(To see this, note that the right side above is the sum of all possible products $\overline{V}_\ell \overline{V}_{\ell'}$ for $\ell \neq \ell'$. The left side is the sum of all possible terms $(j/m)\overline{V}_\ell \overline{V}_{\ell'}$ and $[(m-j)/m]\overline{V}_\ell \overline{V}_{\ell'}$, where $j = |\ell - \ell'|$ and $\ell \neq \ell'$.) Using this expression, and denoting

$$\overline{V} = \frac{1}{m} \sum_{\ell=0}^{m-1} \overline{V}_\ell$$

as the reservation interval averaged over all users, we can write Eq. (3.67) as

$$Y = \frac{\rho\overline{V}(m-1)}{2} + \frac{(1-\rho)m\overline{V}}{2} - \frac{(1-\rho)\sum_{\ell=0}^{m-1} \overline{V}_\ell^2}{2m\overline{V}}$$

$$= \frac{(m-\rho)\overline{V}}{2} - \frac{(1-\rho)\sum_{\ell=0}^{m-1} \overline{V}_\ell^2}{2m\overline{V}} \tag{3.68}$$

Combining Eqs. (3.64), (3.65), and (3.68), we obtain

$$W = \frac{\lambda\overline{X^2}}{2(1-\rho)} + \frac{(m-\rho)\overline{V}}{2(1-\rho)} + \frac{\sum_{\ell=0}^{m-1} \left(\overline{V_\ell^2} - \overline{V}_\ell^2 \right)}{2m\overline{V}}$$

Denoting

$$\sigma_V^2 = \frac{\sum_{\ell=0}^{m-1} \left(\overline{V_\ell^2} - \overline{V}_\ell^2 \right)}{m}$$

as the variance of the reservation intervals averaged over all users, we finally obtain

$$W = \frac{\lambda\overline{X^2}}{2(1-\rho)} + \frac{(m-\rho)\overline{V}}{2(1-\rho)} + \frac{\sigma_V^2}{2\overline{V}} \qquad \text{(exhaustive)} \tag{3.69}$$

The partially gated system is the same as the exhaustive except that if a packet of a user arrives during a user's own data interval (an event of probability ρ/m in steady-state), it is delayed by an additional $m\overline{V}$, the average sum of reservation intervals in a cycle. Thus, Y is increased by $\rho\overline{V}$ in the preceding calculation, and we obtain

$$W = \frac{\lambda\overline{X^2}}{2(1-\rho)} + \frac{(m+\rho)\overline{V}}{2(1-\rho)} + \frac{\sigma_V^2}{2\overline{V}} \quad \text{(partially gated)} \tag{3.70}$$

Consider, finally, the fully gated system. This is the same as the partially gated system except that if a packet of a user arrives during a user's own reservation interval [an event of probability $(1-\rho)/m$ in steady-state], it is delayed by an additional $m\overline{V}$. This increases Y by an additional $(1-\rho)\overline{V}$ and results in the equation

$$W = \frac{\lambda\overline{X^2}}{2(1-\rho)} + \frac{(m+2-\rho)\overline{V}}{2(1-\rho)} + \frac{\sigma_V^2}{2\overline{V}} \quad \text{(gated)} \tag{3.71}$$

In comparing these results with the single-user system, consider the case where the reservation interval is a constant A/m. Thus, A is the overhead or reservation time for an entire cycle of reservations for each user, which is usually the appropriate parameter to compare with A in the single-user waiting-time formula (3.62). We then have ($\overline{V} = A/m,\ \sigma_V^2 = 0$)

$$W = \frac{\lambda\overline{X^2}}{2(1-\rho)} + \frac{A}{2}\left(\frac{1-\rho/m}{1-\rho}\right) \quad \text{(exhaustive)} \tag{3.72}$$

$$W = \frac{\lambda\overline{X^2}}{2(1-\rho)} + \frac{A}{2}\left(\frac{1+\rho/m}{1-\rho}\right) \quad \text{(partially gated)} \tag{3.73}$$

$$W = \frac{\lambda\overline{X^2}}{2(1-\rho)} + \frac{A}{2}\left(\frac{1+(2-\rho)/m}{1-\rho}\right) \quad \text{(gated)} \tag{3.74}$$

It can be seen that delay is somewhat reduced in the multiuser case; essentially, packets are delayed by roughly the same amount until the reservation time in all cases, but delay is quite small after the reservation in the multiuser case.

Limited service systems. We now consider a variation of the multiuser system whereby, in each user's data interval, *only the first* packet of the user waiting in queue (if any) is transmitted (rather than *all* waiting packets). We concentrate on the gated and partially gated versions of this system, since an exhaustive version does not make sense. As before, we have

$$E\{W_i\} = E\{R_i\} + \frac{E\{N_i\}}{\mu} + E\{Y_i\}$$

and by taking the limit as $i \to \infty$, we obtain

$$W = R + \rho W + Y \tag{3.75}$$

Here R is given by Eq. (3.64) as before. To calculate the new formula for Y for the partially gated system, we argue as follows. A packet arriving during user ℓ's data or reservation interval will belong to any one of the users with equal probability $1/m$. Therefore, in steady-state, the expected number of packets waiting in the queue of the user that owns the arriving packet, averaged over all users, is $\lim_{i\to\infty} E\{N_i\}/m = \lambda W/m$.

Each of these packets causes an extra cycle of reservations $m\overline{V}$, so Y is increased by an amount $\lambda W\overline{V}$. Using this fact in Eq. (3.75), we see that

$$W = \frac{R + \widetilde{Y}}{1 - \rho - \lambda\overline{V}}$$

where $\widetilde{Y}$ is the value of Y obtained earlier for the partially gated system without the single-packet-per-data-interval restriction. Equivalently, we see from Eq. (3.65) that *the single-packet-per-data-interval restriction results in an increase of the average waiting time for the partially gated system by a factor*

$$\frac{1 - \rho}{1 - \rho - \lambda\overline{V}}$$

Using this fact in Eq. (3.70), we obtain

$$W = \frac{\lambda\overline{X^2}}{2(1 - \rho - \lambda\overline{V})} + \frac{(m + \rho)\overline{V}}{2(1 - \rho - \lambda\overline{V})} + \frac{\sigma_V^2(1 - \rho)}{2\overline{V}(1 - \rho - \lambda\overline{V})}$$

(limited service, partially gated) (3.76)

Consider now the gated version. Y_i is the same as for the partially gated system except for an additional cycle of reservation intervals of average length $m\overline{V}$ associated with the event where packet i arrives during the reservation interval of its owner, and the subsequent data interval is empty. It is easily verified (Problem 3.34) that the latter event occurs with steady-state probability $(1 - \rho - \lambda\overline{V})/m$. Therefore, for the gated system Y equals the corresponding value for the partially gated system plus $(1 - \rho - \lambda\overline{V})\overline{V}$. This adds $\overline{V}$ to the value of W for the partially gated system, and the average waiting time now is

$$W = \frac{\lambda\overline{X^2}}{2(1 - \rho - \lambda\overline{V})} + \frac{(m + 2 - \rho - 2\lambda\overline{V})\overline{V}}{2(1 - \rho - \lambda\overline{V})} + \frac{\sigma_V^2(1 - \rho)}{2\overline{V}(1 - \rho - \lambda\overline{V})}$$

(limited service, gated) (3.77)

Note that it is not enough that $\rho = \lambda/\mu < 1$ for W to be bounded; rather, $\rho + \lambda\overline{V} < 1$ is required or, equivalently,

$$\lambda\left(\frac{1}{\mu} + \overline{V}\right) < 1$$

This is due to the fact that each packet requires a separate reservation interval of average length $\overline{V}$, thereby effectively increasing the average transmission time from $1/\mu$ to $1/\mu + \overline{V}$.

As a final remark, consider the case of a very large number of users m and a very small average reservation interval $\overline{V}$. An examination of the equation given for the average waiting time W of every multiuser system considered so far shows that as $m \to \infty$, $\overline{V} \to 0$, $\sigma_V^2/\overline{V} \to 0$, and $m\overline{V} \to A$, where A is a constant, we have

$$W \to \frac{\lambda\overline{X^2}}{2(1 - \rho)} + \frac{A}{2(1 - \rho)}$$

It can be shown (see Example 3.6) that $A/(1-\rho)$ is the average length of a cycle (m successive reservation and data intervals). Thus, W approaches the $M/G/1$ average waiting time plus one half the average cycle length.

3.5.3 Priority Queueing

Consider the $M/G/1$ system with the difference that arriving customers are divided into n different priority classes. Class 1 has the highest priority, class 2 has the second highest, and so on. The arrival rate and the first two moments of service time of each class k are denoted λ_k, $\overline{X}_k = 1/\mu_k$, and $\overline{X_k^2}$, respectively. The arrival processes of all classes are assumed independent, Poisson, and independent of the service times.

Nonpreemptive priority. We first consider the nonpreemptive priority rule whereby a customer undergoing service is allowed to complete service without interruption even if a customer of higher priority arrives in the meantime. A separate queue is maintained for each priority class. When the server becomes free, the first customer of the highest nonempty priority queue enters service. This priority rule is one of the most appropriate for modeling packet transmission systems.

We will develop an equation for average delay for each priority class, which is similar to the P-K formula and admits a similar derivation. Denote

$$N_Q^k = \text{Average number in queue for priority } k$$

$$W_k = \text{Average queueing time for priority } k$$

$$\rho_k = \lambda_k/\mu_k = \text{System utilization for priority } k$$

$$R = \text{Mean residual service time}$$

We assume that the overall system utilization is less than 1, that is,

$$\rho_1 + \rho_2 + \cdots + \rho_n < 1$$

When this assumption is not satisfied, there will be some priority class k such that the average delay of customers of priority k and lower will be infinite while the average delay of customers of priority higher than k will be finite. Problem 3.39 takes a closer look at this situation.

As in the derivation of the P-K formula given earlier, we have for the highest-priority class,

$$W_1 = R + \frac{1}{\mu_1} N_Q^1$$

Eliminating N_Q^1 from this equation using Little's Theorem,

$$N_Q^1 = \lambda_1 W_1$$

we obtain

$$W_1 = R + \rho_1 W_1$$

and finally,

$$W_1 = \frac{R}{1 - \rho_1} \tag{3.78}$$

For the second priority class, we have a similar expression for the queueing delay W_2 except that we have to count the additional queueing delay due to customers of higher priority that arrive while a customer is waiting in queue. This is the meaning of the last term in the formula

$$W_2 = R + \frac{1}{\mu_1} N_Q^1 + \frac{1}{\mu_2} N_Q^2 + \frac{1}{\mu_1} \lambda_1 W_2$$

Using Little's Theorem ($N_Q^k = \lambda_k W_k$), we obtain

$$W_2 = R + \rho_1 W_1 + \rho_2 W_2 + \rho_1 W_2$$

which yields

$$W_2 = \frac{R + \rho_1 W_1}{1 - \rho_1 - \rho_2}$$

Using the expression $W_1 = R/(1 - \rho_1)$ obtained earlier, we finally have

$$W_2 = \frac{R}{(1 - \rho_1)(1 - \rho_1 - \rho_2)}$$

The derivation is similar for all priority classes $k > 1$. The formula for the waiting time in queue is

$$W_k = \frac{R}{(1 - \rho_1 - \cdots - \rho_{k-1})(1 - \rho_1 - \cdots - \rho_k)} \tag{3.79}$$

The average delay per customer of class k is

$$T_k = \frac{1}{\mu_k} + W_k \tag{3.80}$$

The mean residual service time R can be derived as for the P-K formula (compare with Fig. 3.16). We have

$$R = \frac{1}{2} \sum_{i=1}^{n} \lambda_i \overline{X_i^2} \tag{3.81}$$

The average waiting time in queue and the average delay per customer for each class is obtained by combining Eqs. (3.79) to (3.81):

$$W_k = \frac{\sum_{i=1}^{n} \lambda_i \overline{X_i^2}}{2(1 - \rho_1 - \cdots - \rho_{k-1})(1 - \rho_1 - \cdots - \rho_k)} \tag{3.82}$$

$$T_k = \frac{1}{\mu_k} + W_k \tag{3.83}$$

The analysis given above does not extend easily to the case of multiple servers, primarily because there is no simple formula for the mean residual time R. If, however, the service times of all priority classes are identically and exponentially distributed, there is a convenient characterization of R. Equation (3.79) then yields a closed-form expression for the average waiting times W_k (see Problem 3.38).

Note that it is possible to affect the average delay per customer by choosing the priority classes appropriately. It is generally true that average delay tends to be reduced when customers with short service times are given higher priority. (For an example from common experience, consider the supermarket practice of having special checkout counters for customers with few items. A similar situation can be seen in copying machine waiting lines, where people often give priority to others who need to make just a few copies.) For an analytical substantiation, consider a nonpreemptive system and two customer classes A and B, with respective arrival and service rates λ_A, μ_A, and λ_B, μ_B. A straightforward calculation using the formulas above shows that if $\mu_A > \mu_B$, then the average delay per customer (averaged over both classes)

$$T = \frac{\lambda_A T_A + \lambda_B T_B}{\lambda_A + \lambda_B}$$

is smaller when A is given priority over B than when B is given priority over A. For related results, see Problem 3.40.

Preemptive resume priority. One of the features of the nonpreemptive priority rule is that the average delay of a priority class depends on the arrival rate of lower-priority classes. This is evident from Eq. (3.82), which gives the average waiting times W_k, and is due to the fact that a high-priority customer must wait for a lower-priority customer already in service. This dependence is not present in the *preemptive resume priority discipline*, whereby service of a customer is interrupted when a higher-priority customer arrives and is resumed from the point of interruption once all customers of higher priority have been served.

As an example of an (approximation to) such a system, consider a transmission line serving several Poisson packet streams of different priorities. The packets of each stream are subdivided into many small "subpackets" (*e.g.*, ATM cells), which in the absence of packets of higher priority, are contiguously transmitted on the line. The transmission of the subpackets of a given packet is halted when a packet of higher priority arrives and is resumed when no subpackets of higher priority packets are left in the system.

As we consider the calculation of T_k, the average time in the system of priority k customers, we should keep in mind that the presence of customers of priorities $k+1$ through n does not affect this calculation. Therefore, we can treat each priority class as if it were the lowest in the system. The system time T_k consists of three terms:

1. The customer's average service time $1/\mu_k$.
2. The average time required, upon arrival of a priority k customer, to service customers of priority 1 to k already in the system (*i.e.*, the average unfinished work corresponding to priorities 1 through k). It can be seen that this time is equal to the average waiting time in the corresponding, ordinary $M/G/1$ system (without priorities), where the customers of priorities $k+1$ through n are neglected, that is

[cf. Eq. (3.48)],

$$\frac{R_k}{1-\rho_1-\cdots-\rho_k}$$

where R_k is the mean residual time

$$R_k = \frac{\sum_{i=1}^{k} \lambda_i \overline{X_i^2}}{2} \tag{3.84}$$

The reason is that at all times, the unfinished work (sum of remaining service times of all customers in the system) of an $M/G/1$-type system is independent of the priority discipline of the system. This is true for any system where the server is always busy while the system is nonempty, and customers leave the system only after receiving their required service. (An example of a system that does not have this property is the vacation system of Section 3.5.1.)

3. The average waiting time for customers of priorities 1 through $k-1$ who arrive while the customer of class k is in the system. This term is

$$\sum_{i=1}^{k-1} \frac{1}{\mu_i} \lambda_i T_k = \sum_{i=1}^{k-1} \rho_i T_k$$

for $k > 1$, and is zero for $k = 1$.

Collecting the three terms above, we obtain the equation

$$T_k = \frac{1}{\mu_k} + \frac{R_k}{1-\rho_1-\cdots-\rho_k} + \left(\sum_{i=1}^{k-1} \rho_i\right) T_k \tag{3.85}$$

The final result is, for $k = 1$,

$$T_1 = \frac{(1/\mu_1)(1-\rho_1)+R_1}{1-\rho_1} \tag{3.86}$$

and for $k > 1$,

$$T_k = \frac{(1/\mu_k)(1-\rho_1-\cdots-\rho_k)+R_k}{(1-\rho_1-\cdots-\rho_{k-1})(1-\rho_1-\cdots-\rho_k)} \tag{3.87}$$

where R_k is given by Eq. (3.84). As for the nonpreemptive system, there is no easy extension of this formula to the case of multiple servers unless the service times of all priority classes are identically and exponentially distributed (see Problem 3.38).

3.5.4 An Upper Bound for the $G/G/1$ System

Consider the $G/G/1$ system, which is the same as $M/G/1$ except that the interarrival times have a general rather than exponential distribution. We continue to assume that the interarrival times and service times are all independent. We want to show that the average waiting time in queue satisfies

$$W \le \frac{\lambda(\sigma_a^2+\sigma_b^2)}{2(1-\rho)} \tag{3.88}$$

where

$$\sigma_a^2 = \text{Variance of the interarrival times}$$

$$\sigma_b^2 = \text{Variance of the service times}$$

$$\lambda = \text{Average interarrival time}$$

$$\rho = \text{Utilization factor } \lambda/\mu, \text{ where } 1/\mu \text{ is the average service time}$$

The upper bound (3.88) becomes exact asymptotically as $\rho \to 1$, that is, as the system becomes heavily loaded.

Let us denote

$$W_k = \text{Waiting time of the } k^{\text{th}} \text{ customer}$$

$$X_k = \text{Service time of the } k^{\text{th}} \text{ customer}$$

$$\tau_k = \text{Interarrival time between the } k^{\text{th}} \text{ and } (k+1)^{\text{st}} \text{ customer}$$

From Fig. 3.24 we see that

$$W_{k+1} = \max\{0, W_k + X_k - \tau_k\} \tag{3.89}$$

To simplify the analysis, we will use the following notation for any random variable Y:

$$Y^+ = \max\{0, Y\}, \qquad Y^- = -\min\{0, Y\}$$

$$\overline{Y} = E\{Y\}, \qquad \sigma_Y^2 = E\{Y^2 - \overline{Y}^2\}$$

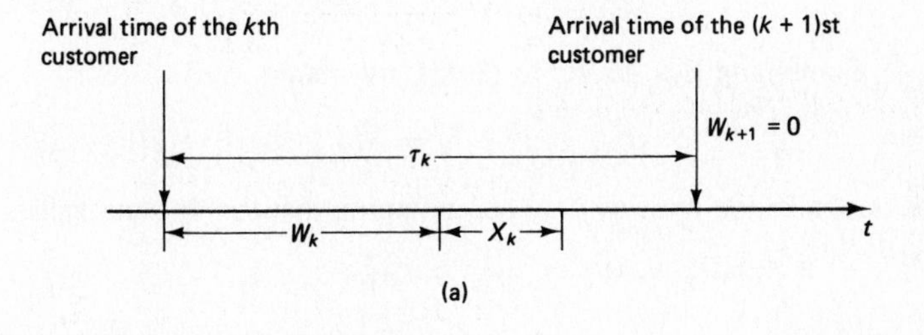

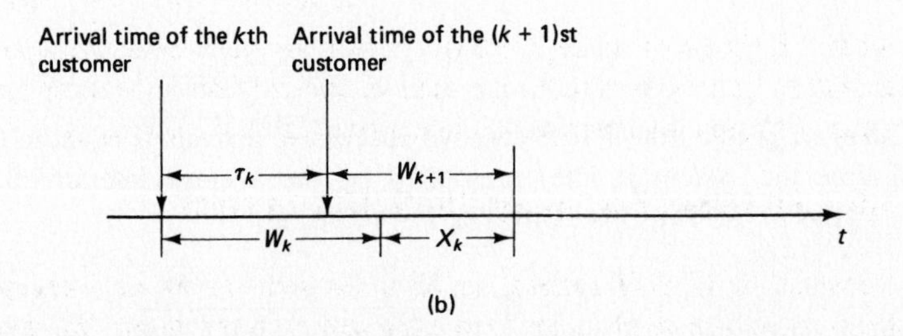

Figure 3.24 Expressing the waiting time W_{k+1} of the $(k+1)^{\text{st}}$ customer in terms of the waiting time W_k, the service time X_k, and the interarrival time τ_k of the kth customer. If the kth customer has departed before the $(k+1)^{\text{st}}$ customer's arrival, which is equivalent to $W_k + X_k - \tau_k \leq 0$, then $W_{k+1} = 0$ [case (a)]. Otherwise, we have $W_{k+1} = W_k + X_k - \tau_k$ [case (b)].

Note that we have

$$Y = Y^+ - Y^-, \qquad Y^+ \cdot Y^- = 0$$

from which we see that

$$\overline{Y} = \overline{Y^+} - \overline{Y^-}, \qquad \sigma_Y^2 = \sigma_{Y^+}^2 + \sigma_{Y^-}^2 + 2\overline{Y^+} \cdot \overline{Y^-} \tag{3.90}$$

Let us now write the expression (3.89) as

$$W_{k+1} = (W_k + V_k)^+ \tag{3.91}$$

where

$$V_k = X_k - \tau_k \tag{3.92}$$

Let us also denote

$$I_k = (W_k + V_k)^- \tag{3.93}$$

From Fig. 3.24 we see that I_k is the length of the idle period between the arrival of the kth and the arrival of the $(k+1)^{\text{st}}$ customer.

We have, using Eq. (3.90),

$$\sigma_{(W_k+V_k)}^2 = \sigma_{(W_k+V_k)^+}^2 + \sigma_{(W_k+V_k)^-}^2 + 2\overline{(W_k + V_k)^+} \cdot \overline{(W_k + V_k)^-} \tag{3.94}$$

Since W_k and V_k are independent, we also have

$$\sigma_{(W_k+V_k)}^2 = \sigma_{W_k}^2 + \sigma_{V_k}^2 = \sigma_{W_k}^2 + \sigma_a^2 + \sigma_b^2 \tag{3.95}$$

Combining Eqs. (3.91) to (3.95), we obtain

$$\sigma_{W_k}^2 + \sigma_a^2 + \sigma_b^2 = \sigma_{W_{k+1}}^2 + \sigma_{I_k}^2 + 2\overline{W}_{k+1}\overline{I}_k$$

We now take the limit as $k \to \infty$, assuming that steady-state values exist, that is,

$$\overline{W}_k \to W, \qquad \sigma_{W_k}^2 \to \sigma_W^2, \qquad \overline{I}_k \to I, \qquad \sigma_{I_k}^2 \to \sigma_I^2$$

We obtain

$$W = \frac{\sigma_a^2 + \sigma_b^2}{2I} - \frac{\sigma_I^2}{2I} \tag{3.96}$$

The average idle time I between two successive arrivals is equal to $(1-\rho)$ (the fraction of time the system is idle) multiplied by the average interarrival time $1/\lambda$, that is, $I = (1-\rho)/\lambda$. Thus, we can write Eq. (3.96) as

$$W = \frac{\lambda(\sigma_a^2 + \sigma_b^2)}{2(1-\rho)} - \frac{\lambda\sigma_I^2}{2(1-\rho)} \tag{3.97}$$

Since $\sigma_I^2 \geq 0$, we obtain the inequality

$$W \leq \frac{\lambda(\sigma_a^2 + \sigma_b^2)}{2(1-\rho)} \tag{3.98}$$

which is the desired result. Note that as the system becomes more heavily loaded, the average idle time I_k tends to diminish and so does the variance σ_I^2, thereby making the upper bound increasingly accurate.

As an example, consider the $M/G/1$ queue. By the Pollatcheck–Khinchin formula we have

$$W = \frac{\lambda \overline{X^2}}{2(1-\rho)} = \frac{\lambda(\sigma_b^2 + 1/\mu^2)}{2(1-\rho)} \tag{3.99}$$

Since for the Poisson arrival process with rate λ we have $\sigma_a^2 = 1/\lambda^2$, by comparing Eqs. (3.97) and (3.99) we see that

$$\sigma_I^2 = \frac{1}{\lambda^2} - \frac{1}{\mu^2}$$

Thus the term that was neglected to derive the upper bound (3.98) for W is equal to

$$\frac{\lambda \sigma_I^2}{2(1-\rho)} = \frac{\lambda}{2(1-\lambda/\mu)}\left(\frac{1}{\lambda^2} - \frac{1}{\mu^2}\right) = \frac{1}{2}\left(\frac{1}{\lambda} + \frac{1}{\mu}\right)$$

This term is always less than the average interarrival time $1/\lambda$ when $\rho < 1$. As $\rho \to 1$, it approaches $1/\mu$ and is negligible relative to the upper bound of Eq. (3.98).

We finally note that several other bounds and approximations for the $G/G/1$ queue have been obtained. A particularly simple improvement to the one we have given here is

$$W \leq \frac{\lambda(\sigma_a^2 + \sigma_b^2)}{2(1-\rho)} - \frac{\lambda(1-\rho)\sigma_a^2}{2}$$

Its derivation is outlined in Problem 3.48.

3.6 NETWORKS OF TRANSMISSION LINES

In a data network, there are many transmission queues that interact in the sense that a traffic stream departing from one queue enters one or more other queues, perhaps after merging with portions of other traffic streams departing from yet other queues. Analytically, this has the unfortunate effect of complicating the character of the arrival processes at downstream queues. The difficulty is that the packet interarrival times become strongly correlated with packet lengths once packets have traveled beyond their entry queue. As a result it is impossible to carry out a precise and effective analysis comparable to the one for the $M/M/1$ and $M/G/1$ systems.

As an illustration of the phenomena that complicate the analysis, consider two transmission lines of equal capacity in tandem, as shown in Fig. 3.25. Assume that Poisson arrivals of rate λ packets/sec enter the first queue, and that all packets have *equal* length. Therefore, the first queue is $M/D/1$ and the average packet delay there is given by the Pollaczek–Khinchin formula. However, at the second queue the interarrival times must be greater than or equal to $1/\mu$ (the packet transmission time). Furthermore, because the packet transmission times are equal at both queues, each packet arriving at the second queue will complete transmission at or before the time the next packet arrives, so there is *no waiting at the second queue.* Therefore, a delay model based on Poisson assumptions is totally inappropriate for the second queue.

Consider next the case of the two tandem transmission lines where packet lengths are exponentially distributed and are independent of each other as well as of the interarrival times at the first queue. Then the first queue is $M/M/1$. The second queue, however, *cannot* be modeled as $M/M/1$. The reason is, again, that *the interarrival times at the second queue are strongly correlated with the packet lengths.* In particular, the interarrival time of two packets at the second queue is greater than or equal to the transmission time of the second packet at the first queue (see Fig. 3.26). As a result, long packets will typically wait less time at the second queue than short packets, since their transmission at the first queue takes longer, thereby giving the second queue more time to empty out. For a traffic analogy, consider a slow truck traveling on a busy narrow street together with several faster cars. The truck will typically see empty space ahead of it while being closely followed by the faster cars.

As an indication of the difficulty of analyzing queueing network problems involving dependent interarrival and service times, no analytical solution is known for even the simple tandem queueing problem of Fig. 3.25 involving Poisson arrivals and exponentially distributed service times. In the real situation where packet lengths and interarrival times are correlated, a simulation has shown that under heavy traffic conditions, average delay per packet is smaller than in the idealized situation where there is no such correlation. The reverse is true under light traffic conditions. It is not known whether and in what form this result can be extended to more general networks.

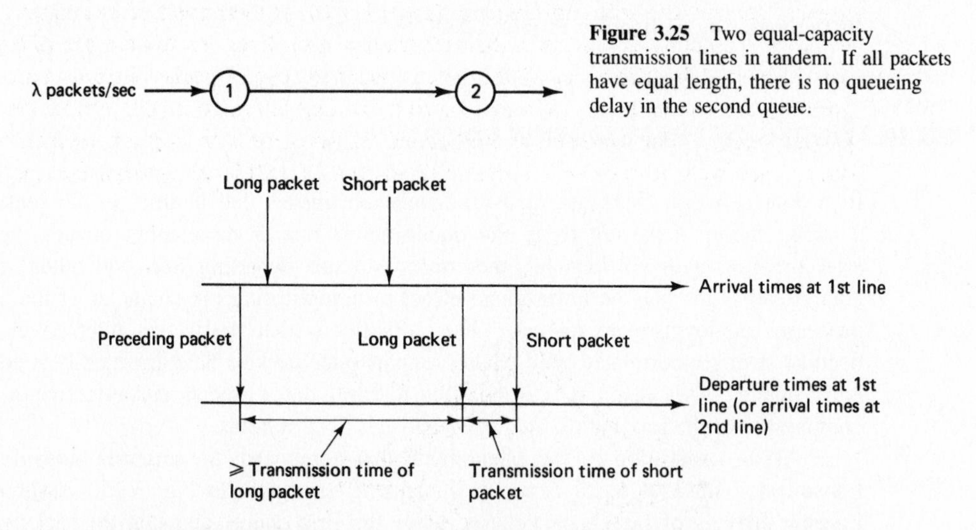

Figure 3.25 Two equal-capacity transmission lines in tandem. If all packets have equal length, there is no queueing delay in the second queue.

Figure 3.26 Timing diagram of packet arrivals and departures completions in a system of two transmission lines in tandem. The interarrival time of two packets at the second queue is greater or equal to the transmission time of the second packet. (It is greater if and only if the second packet finds the first queue empty upon arrival.) Hence the interarrival times at the second queue are correlated with the packet lengths.

3.6.1 The Kleinrock Independence Approximation

We now formulate a framework for approximation of average delay per packet in data networks. Consider a network of communication links as shown in Fig. 3.27. Assume that there are several packet streams, each following a unique path that consists of a sequence of links through the network. Let x_s, in packets/sec, be the arrival rate of the packet stream s. Then the total arrival rate at link (i, j) is

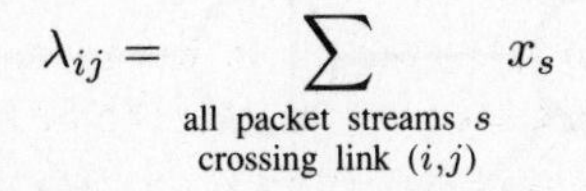

$$\lambda_{ij} = \sum_{\substack{\text{all packet streams } s \\ \text{crossing link } (i,j)}} x_s$$

The preceding network model is well suited for virtual circuit networks, with each packet stream modeling a separate virtual circuit. For datagram networks, it is sometimes necessary to use a more general model that allows bifurcation of the traffic of a packet stream. Here there are again several packet streams, each having a unique origin and destination. However, there may be several paths followed by the packets of a stream (see Fig. 3.28). Assume that no packets travel in a loop, let x_s denote the arrival rate of packet stream s, and let $f_{ij}(s)$ denote the fraction of the packets of stream s that go through link (i, j). Then the total arrival rate at link (i, j) is

$$\lambda_{ij} = \sum_{\substack{\text{all packet streams } s \\ \text{crossing link } (i,j)}} f_{ij}(s) x_s$$

We have seen from the special case of two tandem queues that even if the packet streams are Poisson with independent packet lengths at their point of entry into the network, this property is lost after the first transmission line. To resolve the dilemma, it was suggested by Kleinrock [Kle64] that merging several packet streams on a transmission line has an effect akin to restoring the independence of interarrival times and packet lengths. For example, if the second transmission line in the preceding tandem queue case were to receive a substantial amount of additional external Poisson traffic,

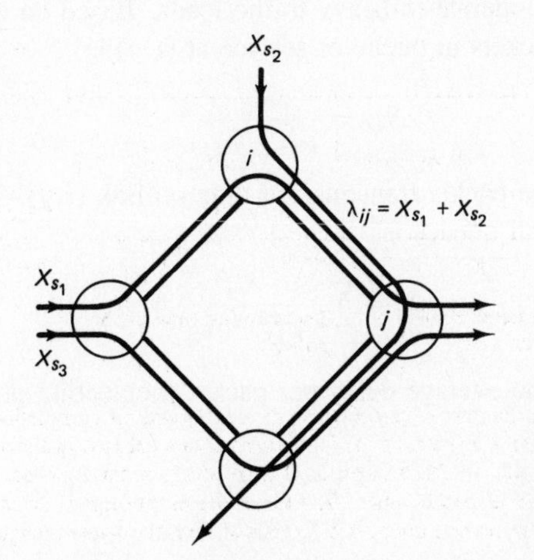

Figure 3.27 Model suitable for virtual circuit networks. There are several packet streams, each using a single path. The total arrival rate λ_{ij} at a link (i, j) is equal to the sum of the arrival rates x_s of all packet streams s traversing the link.

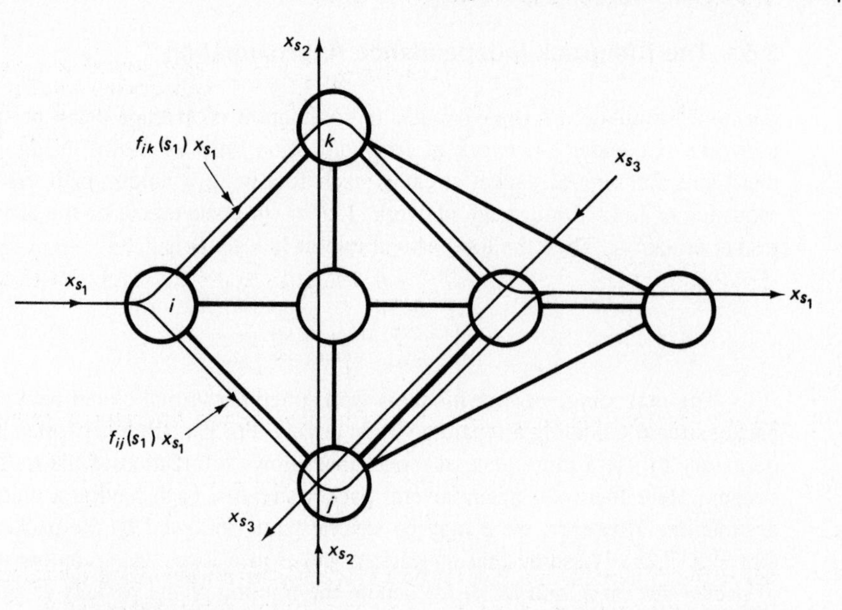

Figure 3.28 Model suitable for datagram networks. There are several packet streams, each associated with a unique origin–destination pair. However, packets of the same stream may follow one of several paths. The total arrival rate λ_{ij} at a link (i, j) is equal to the sum of the fractions $f_{ij}(s)x_s$ of the arrival rates of all packet streams s traversing the link.

the dependence of interarrival and service times displayed in Fig. 3.26 would be weakened considerably. It was concluded that it is often appropriate to adopt an $M/M/1$ queueing model for each communication link regardless of the interaction of traffic on this link with traffic on other links. (See also the discussion preceding Jackson's theorem in Section 3.8.) This is known as the *Kleinrock independence approximation* and seems to be a reasonably good approximation for systems involving Poisson stream arrivals at the entry points, packet lengths that are nearly exponentially distributed, a densely connected network, and moderate-to-heavy traffic loads. Based on this $M/M/1$ model, the average number of packets in queue or service at (i, j) is

$$N_{ij} = \frac{\lambda_{ij}}{\mu_{ij} - \lambda_{ij}} \tag{3.100}$$

where $1/\mu_{ij}$ is the average packet transmission time on link (i, j). The average number of packets summed over all queues is

$$N = \sum_{(i,j)} \frac{\lambda_{ij}}{\mu_{ij} - \lambda_{ij}} \tag{3.101}$$

so by Little's Theorem, the average delay per packet (neglecting processing and propagation delays) is

$$T = \frac{1}{\gamma} \sum_{(i,j)} \frac{\lambda_{ij}}{\mu_{ij} - \lambda_{ij}} \tag{3.102}$$

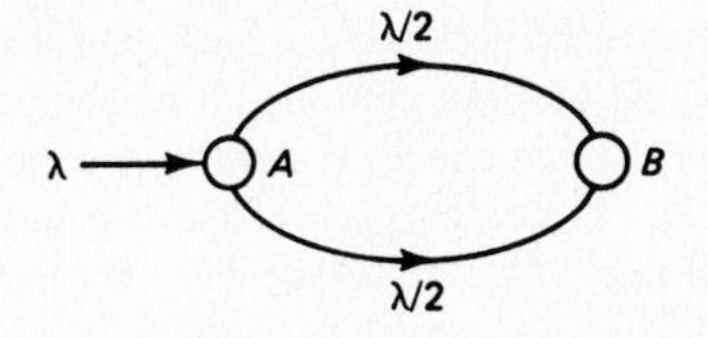

Figure 3.29 Poisson process with rate λ divided among two links. If division is done by randomization, each link behaves like an $M/M/1$ queue. If division is done by metering, the whole system behaves like an $M/M/2$ queue.

where $\gamma = \sum_s x_s$ is the total arrival rate in the system. If the average processing and propagation delay d_{ij} at link (i,j) is not negligible, this formula should be adjusted to

$$T = \frac{1}{\gamma} \sum_{(i,j)} \left(\frac{\lambda_{ij}}{\mu_{ij} - \lambda_{ij}} + \lambda_{ij} d_{ij} \right) \tag{3.103}$$

Finally, the average delay per packet of a traffic stream traversing a path p is given by

$$T_p = \sum_{\substack{\text{all } (i,j) \\ \text{on path } p}} \left(\frac{\lambda_{ij}}{\mu_{ij}(\mu_{ij} - \lambda_{ij})} + \frac{1}{\mu_{ij}} + d_{ij} \right) \tag{3.104}$$

where the three terms in the sum above represent average waiting time in queue, average transmission time, and processing and propagation delay, respectively.

In many networks, the assumption of exponentially distributed packet lengths is not appropriate. Given a different type of probability distribution of the packet lengths, one may keep the approximation of independence between queues but use the P-K formula for average number in the system in place of the $M/M/1$ formula (3.100). Equations (3.101) to (3.104) for average delay would then be modified in an obvious way.

For virtual circuit networks (cf. Fig. 3.27), the main approximation involved in the $M/M/1$ formula (3.101) is due to the correlation of the packet lengths and the packet interarrival times at the various queues in the network. If somehow this correlation was not present (*e.g.*, if a packet upon departure from a transmission line was assigned a new length drawn from an exponential distribution), then the average number of packets in the system would be given indeed by the formula

$$N = \sum_{(i,j)} \frac{\lambda_{ij}}{\mu_{ij} - \lambda_{ij}}$$

This fact (by no means obvious) is a consequence of Jackson's Theorem, which will be discussed in Section 3.8.

In datagram networks that involve multiple path routing for some origin–destination pairs (cf. Fig. 3.28), the accuracy of the $M/M/1$ approximation deteriorates for another reason, which is best illustrated by an example.

Example 3.17

Suppose that node A sends traffic to node B along two links with service rate μ in the network of Fig. 3.29. Packets arrive at A according to a Poisson process with rate λ packets/sec. Packet transmission times are exponentially distributed and independent of interarrival times as in the $M/M/1$ system. Assume that the arriving traffic is to be divided equally among the two links. However, how should this division be implemented? Consider the following possibilities.

1. *Randomization.* Here each packet is assigned upon arrival at A to one of the two links based on the outcome of a fair coin flip. It is then possible to show that the arrival process on each of the two queues is Poisson and independent of the packet lengths (see Problem 3.11). Therefore, each of the two queues behaves like an $M/M/1$ queue with arrival rate $\lambda/2$ and average delay per packet

$$T_R = \frac{1}{\mu - \lambda/2} = \frac{2}{2\mu - \lambda} \tag{3.105}$$

which is consistent with the Kleinrock independence approximation.

2. *Metering.* Here each arriving packet is assigned to a queue that currently has the smallest total backlog in bits and will therefore empty out first. An equivalent system maintains a common queue for the two links and routes the packet at the head of the queue to the link that becomes idle first. This works like an $M/M/2$ system with arrival rate λ and with each link playing the role of a server. Using the result of Section 3.4.1, the average delay per packet can be calculated to be

$$T_M = \frac{2}{(2\mu - \lambda)(1 + \rho)} \tag{3.106}$$

where $\rho = \lambda/2\mu$.

Comparing the average delay expressions (3.105) and (3.106), we see that metering performs better than randomization in terms of delay by a factor $1/(1+\rho)$. This is basically the same advantage that statistical multiplexing with multiple channels holds over time-division multiplexing as discussed in Example 3.10. Generally, it is preferable to use some form of metering rather than randomization when dividing traffic among alternative routes. However, in contrast with randomization, metering destroys the Poisson character of the arrival process at the point of division. In our example, when metering is used, the interarrival times at each link are neither exponentially distributed nor independent of preceding packet lengths. Therefore, the use of metering (which is recommended for performance reasons) tends to degrade the accuracy of the $M/M/1$ approximation.

We finally mention an alternative approach for approximating average delay in a network of transmission lines. This approach uses $G/G/1$ approximations in place of $M/M/1$ or $M/G/1$ approximations. The key idea is that given the first two moments of the interarrival and service times of each of the external packet streams, one may approximate reasonably well the first two moments of the interarrival and service times of the total packet arrival stream at each queue (see [Whi83a], [Whi83b], and the references quoted there). Then the average delay at each queue can be estimated using $G/G/1$ bounds and approximations of the type discussed in Section 3.5.4.

3.7 TIME REVERSIBILITY—BURKE'S THEOREM

The analysis of the $M/M/1$, $M/M/m$, $M/M/\infty$, and $M/M/m/m$ systems was based on the equality of the steady-state frequency of transitions from j to $j+1$, that is, $p_j P_{j(j+1)}$, with the steady-state frequency of transitions from $j+1$ to j, that is, $p_{j+1} P_{(j+1)j}$. These relations, called *detailed balance equations*, are valid for any Markov chain with integer states in which transitions can occur only between neighboring states (*i.e.*, from j to $j-1$, j, or $j+1$); these Markov chains are called *birth–death* processes. The detailed balance equations lead to an important property called time reversibility, as we now explain.

Consider an irreducible, aperiodic, discrete-time Markov chain $X_n, X_{n+1}, \ldots$ having transition probabilities P_{ij} and stationary distribution $\{p_j \mid j \geq 0\}$ with $p_j > 0$ for all j. Suppose that the chain is in steady-state, that is,

$$P\{X_n = j\} = p_j\,, \qquad \text{for all } n$$

(This occurs if the initial state is chosen according to the stationary distribution, and is equivalent to imagining that the process began at time $-\infty$.)

Suppose that we trace the sequence of states going backward in time. That is, starting at some n, consider the sequence of states $X_n, X_{n-1}, \ldots$. This sequence is itself a Markov chain, as seen by the following calculation:

$$\begin{aligned}
&P\{X_m = j \mid X_{m+1} = i,\ X_{m+2} = i_2, \ldots, X_{m+k} = i_k\} \\
&= \frac{P\{X_m = j, X_{m+1} = i,\ X_{m+2} = i_2, \ldots, X_{m+k} = i_k\}}{P\{X_{m+1} = i,\ X_{m+2} = i_2, \ldots, X_{m+k} = i_k\}} \\
&= \frac{P\{X_m = j, X_{m+1} = i\}\, P\{X_{m+2} = i_2, \ldots, X_{m+k} = i_k \mid X_m = j, X_{m+1} = i\}}{P\{X_{m+1} = i\} P\{X_{m+2} = i_2, \ldots, X_{m+k} = i_k \mid X_{m+1} = i\}} \\
&= \frac{P\{X_m = j, X_{m+1} = i\}}{P\{X_{m+1} = i\}} \\
&= \frac{P\{X_m = j\}\, P\{X_{m+1} = i \mid X_m = j\}}{P\{X_{m+1} = i\}} \\
&= \frac{p_j P_{ji}}{p_i}
\end{aligned}$$

where the third equality follows from the Markov property of the chain $X_n, X_{n+1}, \ldots$ Thus, conditional on the state at time $m+1$, the state at time m is independent of that at times $m+2$, $m+3, \ldots$. The backward transition probabilities are given by

$$P^*_{ij} = P\{X_m = j \mid X_{m+1} = i\} = \frac{p_j P_{ji}}{p_i}, \qquad i, j \geq 0 \tag{3.107}$$

If $P^*_{ij} = P_{ij}$ for all i, j (*i.e.*, the transition probabilities of the forward and reversed chain are identical), we say that the chain is *time reversible*.

We list some properties of the reversed chain:

1. The reversed chain is irreducible, aperiodic, and has the same stationary distribution as the forward chain. [This property can be shown either by elementary reasoning using the definition of the reversed chain, or by verifying the equality $p_j = \sum_{i=0}^{\infty} p_i P^*_{ij}$ using Eq. (3.107).] The intuitive idea here is that the reversed chain corresponds to the same process, looked at in the reversed time direction. Thus, if the steady-state probabilities are viewed as proportions of time the process visits the states, then the steady-state occupancy distributions of the forward and the reverse chains are equal. Note that in view of this equality, the form of the transition probabilities of the reversed chain $P^*_{ij} = p_j P_{ji}/p_i$ [cf. Eq. (3.107)] can be intuitively explained. It expresses the fact that (with probability 1) the proportion of transitions from j to i out of all transitions in the forward chain (which is $p_j P_{ji}$) equals the proportion of transitions from i to j out of all transitions in the reversed chain (which is $p_i P^*_{ij}$).

2. If we can find positive numbers p_i, $i \geq 0$, summing to unity and such that the scalars

$$P^*_{ij} = \frac{p_j P_{ji}}{p_i}, \qquad i, j \geq 0 \tag{3.108}$$

form a transition probability matrix, that is,

$$\sum_{j=0}^{\infty} P^*_{ij} = 1, \qquad i = 0, 1, \ldots$$

then $\{p_i \mid i \geq 0\}$ is the stationary distribution and P^*_{ij} are the transition probabilities of the reversed chain. [To see this, note that by multiplying with p_i Eq. (3.108) and adding over j, we obtain

$$\sum_{j=0}^{\infty} p_j P_{ji} = p_i \sum_{j=0}^{\infty} P^*_{ij} = p_i$$

which is the global balance equation and implies that $\{p_i \mid i \geq 0\}$ is the stationary distribution.] This property, which holds regardless of whether the chain is time reversible, is useful if through an intelligent guess, we can verify Eq. (3.108), thereby obtaining both the p_j and P^*_{ij}; for examples of such applications, see Section 3.8.

3. A chain is time reversible if and only if the detailed balance equations hold:

$$p_i P_{ij} = p_j P_{ji}, \qquad i, j \geq 0$$

This follows from the equality $p_i P^*_{ij} = p_j P_{ji}$ [cf. Eq. (3.107)] and the definition of time reversibility. In other words, a system is time reversible if in a typical system history, transitions from i to j occur with the same frequency as transitions from j to i (and therefore also with the same frequency as transitions from i to j when this system history is reversed in time). In particular, the chains corresponding to the queueing systems $M/M/1$, $M/M/m$, $M/M/\infty$, and $M/M/m/m$ discussed in Sections 3.3 and 3.4 are time reversible (in the limit as $\delta \to 0$). More generally, chains corresponding to birth–death processes ($P_{ij} = 0$ if $|i - j| > 1$) are time reversible. Figure 3.30 gives some additional examples of reversible and nonreversible systems.

The idea of time reversibility extends in a straightforward manner to irreducible continuous-time Markov chains. The corresponding analysis can be carried out either directly or by discretizing time in intervals of length δ, considering the corresponding

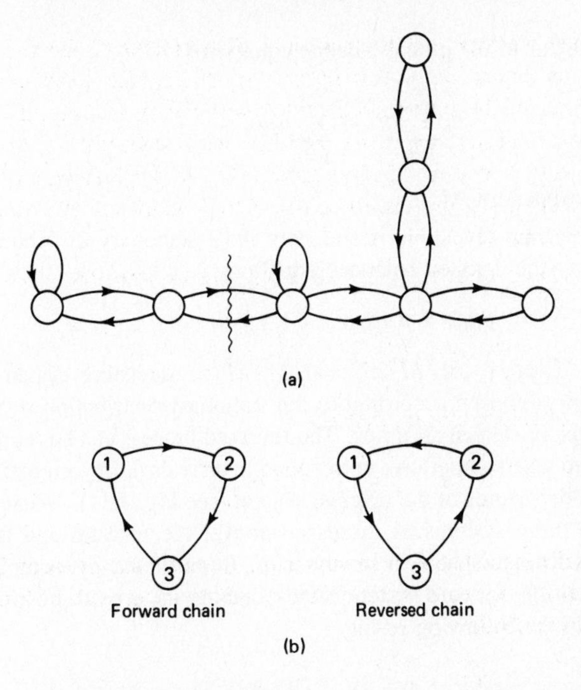

Figure 3.30 (a) Example of a time reversible chain. To see this, note that by splitting the state space in two subsets as shown we obtain global balance equations which are identical with the detailed balance equations. (b) Example of a chain which is not time reversible. The states in the forward and the reversed systems move in the clockwise and counterclockwise directions, respectively.

discrete-time chain, and passing back to the continuous chain by taking the limit as $\delta \to 0$. All results regarding the reversed chain carry over almost verbatim from their discrete-time counterparts by replacing transition probabilities with transition rates. In particular, if the continuous-time chain has transition rates q_{ij} and a stationary distribution $\{p_j \mid j \geq 0\}$ with $p_j > 0$ for all j, then:

1. The reversed chain is a continuous-time Markov chain with the same stationary distribution as the forward chain and with transition rates

$$q_{ij}^* = \frac{p_j q_{ji}}{p_i}, \qquad i, j \geq 0 \tag{3.109}$$

2. If we can find positive numbers p_i, $i \geq 0$, summing to unity and such that the scalars

$$q_{ij}^* = \frac{p_j q_{ji}}{p_i}, \qquad i, j \geq 0 \tag{3.110}$$

satisfy for all $i \geq 0$

$$\sum_{j=0}^{\infty} q_{ij} = \sum_{j=0}^{\infty} q_{ij}^* \tag{3.111}$$

then $\{p_i \mid i \geq 0\}$ is the stationary distribution of both the forward and the reversed chain, and q_{ij}^* are the transition rates of the reversed chain. The relation $\sum_{j=0}^{\infty} q_{ij} = \sum_{j=0}^{\infty} q_{ij}^*$ equates, for every state i, the total rate out of i in the forward and the reversed chains, and by taking into account also the relation $q_{ij}^* = p_j q_{ji}/p_i$, it can be seen to be equivalent to the global balance equation

$$p_i \sum_{j=0}^{\infty} q_{ij} = \sum_{j=0}^{\infty} p_j q_{ji}$$

[cf. Eq. (3A.10) of Appendix A].

3. The forward chain is time reversible if and only if its stationary distribution and transition rates satisfy the detailed balanced equations

$$p_i q_{ij} = p_j q_{ji}, \qquad i, j \geq 0$$

Consider now the $M/M/1$, $M/M/m$, and $M/M/\infty$ queueing systems. We assume that the initial state is chosen according to the stationary distribution so that the queueing systems are in steady-state at all times. The reversed process can be represented by another queueing system where departures correspond to arrivals of the original system and arrivals correspond to departures of the original system (see Fig. 3.31). Because time reversibility holds for all these systems as discussed above, the forward and reversed systems are statistically indistinguishable in steady-state. In particular by using the fact that the departure process of the forward system corresponds to the arrival process of the reversed system, we obtain the following result:

Burke's Theorem. Consider an $M/M/1$, $M/M/m$, or $M/M/\infty$ system with arrival rate λ. Suppose that the system starts in steady-state. Then the following hold true:

(a) The departure process is Poisson with rate λ.

(b) At each time t, the number of customers in the system is independent of the sequence of departure times prior to t.

Proof: (a) This follows from the fact that the forward and reversed systems are statistically indistinguishable in steady-state, and the departure process in the forward system is the arrival process in the reversed system.

(b) As shown in Fig. 3.32, for a fixed time t, the departures prior to t in the forward process are also the arrivals after t in the reversed process. The arrival process in the reversed system is independent Poisson, so the future arrival process does not depend on the current number in the system, which in forward system terms means that the past departure process does not depend on the current number in the system. **Q.E.D.**

Note that part (b) of Burke's Theorem is quite counterintuitive. One would expect that a recent stream of closely spaced departures suggests a busy system with an atypically large number of customers in queue. Yet Burke's Theorem shows that this is not so. Note, however, that Burke's Theorem says nothing about the state of the system *before* a stream of closely spaced departures. Such a state would tend to have abnormally many customers in queue, in accordance with intuition.

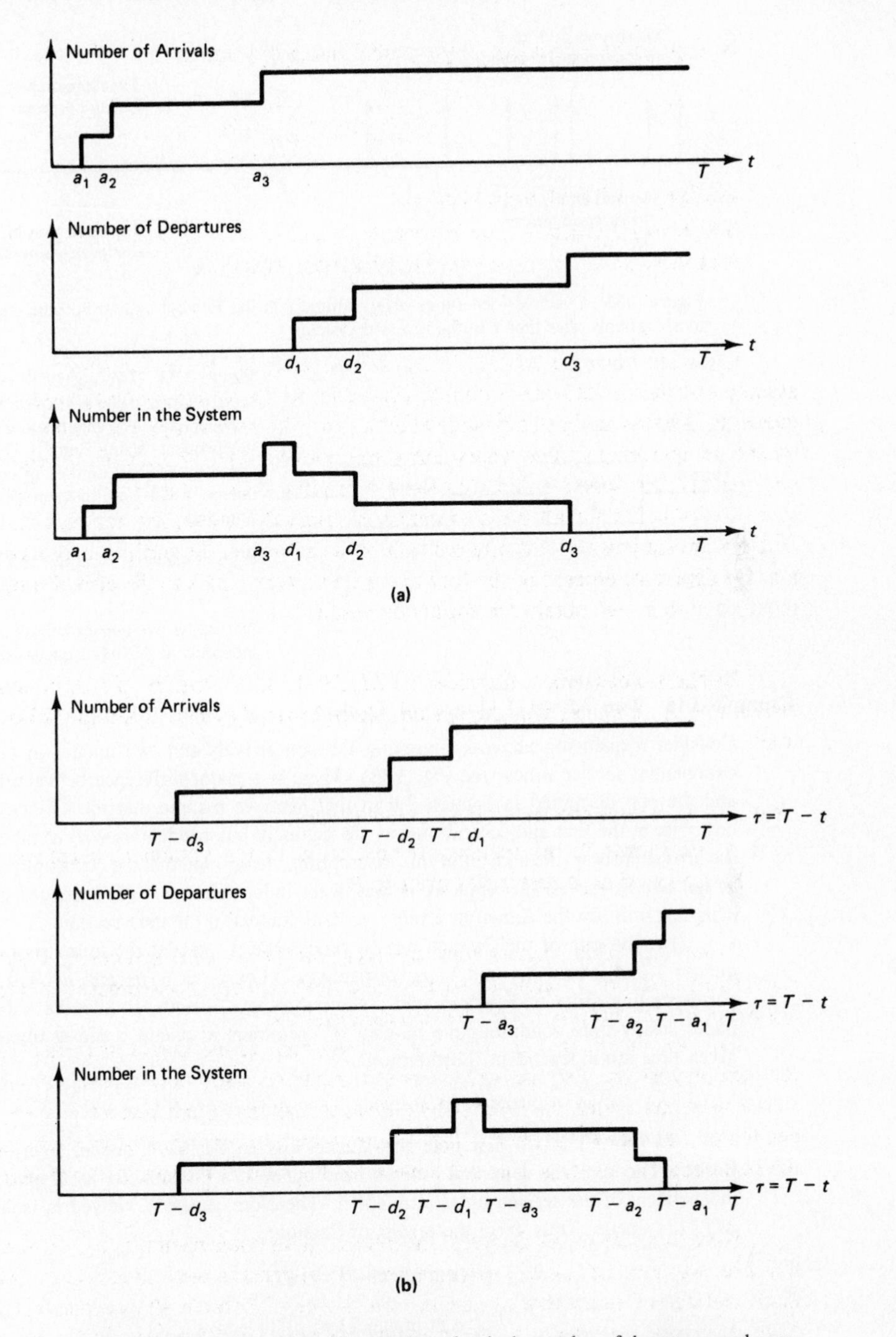

Figure 3.31 (a) Forward system number of arrivals, number of departures, and occupancy during $[0, T]$. (b) Reversed system number of arrivals, number of departures, and occupancy during $[0, T]$.

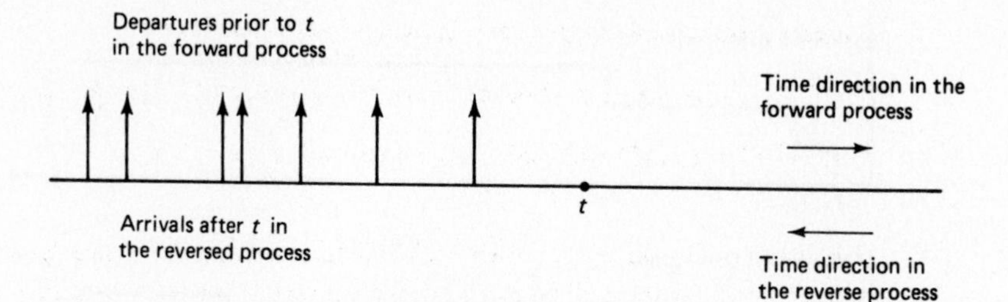

Figure 3.32 Customer departures *prior* to time t in the forward system become customer arrivals *after* time t in the reversed system.

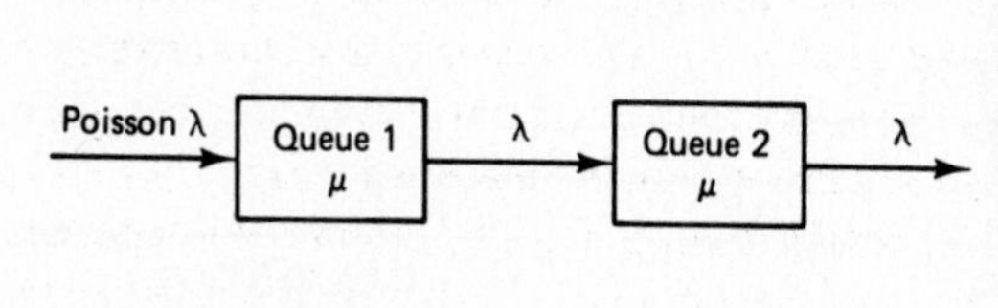

Figure 3.33 Two queues in tandem. The service times at the two queues are exponentially distributed and mutually independent. Using Burke's Theorem, we can show that the number of customers in queues 1 and 2 are independent at a given time and

$$P\{n \text{ at queue 1}, m \text{ at queue 2}\} = \rho_1^n(1-\rho_1)\rho_2^m(1-\rho_2)$$

that is, the two queues behave as if they are independent $M/M/1$ queues in isolation.

Example 3.18 Two $M/M/1$ Queues in Tandem

Consider a queueing network involving Poisson arrivals and two queues in tandem with exponential service times (see Fig. 3.33). There is a major difference between this system and the one discussed in Section 3.6 in that here we assume that the service times of a customer at the first and second queues are mutually independent as well as independent of the arrival process. As a result of this assumption, we will see that the occupancy distribution in the two queues is the same as if they were independent $M/M/1$ queues in isolation. This fact will also be shown in a more general context in the next section.

Let the rate of the Poisson arrival process be λ, and let the mean service times at queues 1 and 2 be $1/\mu_1$ and $1/\mu_2$, respectively. Let $\rho_1 = \lambda/\mu_1$ and $\rho_2 = \lambda/\mu_2$ be the corresponding utilization factors, and assume that $\rho_1 < 1$ and $\rho_2 < 1$. We will show that under steady-state conditions the number of customers at queue 1 and at queue 2 at any given time are independent. Furthermore,

$$P\{n \text{ at queue 1}, m \text{ at queue 2}\} = \rho_1^n(1-\rho_1)\rho_2^m(1-\rho_2) \tag{3.112}$$

To prove this we first note that queue 1 is an $M/M/1$ queue, so by part (a) of Burke's Theorem, the departure process from queue 1 is Poisson. By assumption, it is also independent of the service times at queue 2. Therefore, queue 2, viewed in isolation, is an $M/M/1$ queue. Thus, from the results of Section 3.1,

$$\begin{aligned} P\{n \text{ at queue 1}\} &= \rho_1^n(1-\rho_1) \\ P\{m \text{ at queue 2}\} &= \rho_2^m(1-\rho_2) \end{aligned} \tag{3.113}$$

From part (b) of Burke's Theorem it follows that the number of customers presently in queue 1 is independent of the sequence of earlier arrivals at queue 2 and therefore also of the number of customers presently in queue 2. This implies that

$$P\{n \text{ at queue } 1,\, m \text{ at queue } 2\} = P\{n \text{ at queue } 1\} \cdot P\{m \text{ at queue } 2\}$$

and using Eq. (3.113) we obtain the desired product form (3.112).

We note that, by part (a) of Burke's Theorem, the arrival and the departure processes at both queues of the preceding example are Poisson. This fact can be similarly shown for a much broader class of queueing networks with Poisson arrivals and independent, exponentially distributed service times. We call such networks *acyclic* and define them as follows. We say that queue j is a *downstream neighbor* of queue i if there is a positive probability that a departing customer from queue i will next enter queue j. We say that queue j *lies downstream* of queue i if there is a sequence of queues starting from i and ending at j such that each queue after i in the sequence is a downstream neighbor of its predecessor. A queueing network is called acyclic if it is impossible to find two queues i and j such that j lies downstream of i, and i lies downstream of j. Having an acyclic network is essential for the Poisson character of the arrival and departure processes at each queue to be maintained (see Section 3.8). However, the product form (3.112) of the occupancy distribution generalizes in a natural way to networks that are not acyclic, as we show in the next section.

3.8 NETWORKS OF QUEUES—JACKSON'S THEOREM

As discussed in Section 3.6, the main difficulty with analysis of networks of transmission lines is that the packet interarrival times after traversing the first queue are correlated with their lengths. It turns out that if somehow this correlation were eliminated (which is the premise of the Kleinrock independence approximation) and randomization is used to divide traffic among different routes, then the average number of packets in the system can be derived as if each queue in the network were $M/M/1$. This is an important result known as Jackson's Theorem. In this section we derive a simple version of this theorem and some of its extensions.

Consider a network of K first-come first-serve, single-server queues in which customers arrive from outside the network at each queue i in accordance with independent Poisson processes at rate r_i. We allow the possibility that $r_i = 0$, in which case there are no external arrivals at queue i, but we require that $r_i > 0$ for at least one i. Once a customer is served at queue i, it proceeds to join each queue j with probability P_{ij} or to exit the network with probability $1 - \sum_{j=1}^{K} P_{ij}$.

The routing probabilities P_{ij} together with the external input rates r_j can be used to determine the total arrival rate of customers λ_j at each queue j, that is, the sum of r_j and the arrival rate of customers coming from other queues. Calculating λ_j is fairly easy when the network is of the acyclic type discussed at the end of Section 3.7. If there is a positive probability that a customer may visit the same queue twice, a more complex computation is necessary, based on the equations

$$\lambda_j = r_j + \sum_{i=1}^{K} \lambda_i P_{ij}, \qquad j = 1, \ldots, K \tag{3.114}$$

These equations represent a linear system in which the rates λ_j, $j = 1, \ldots, K$, constitute a set of K unknowns. To guarantee that they can be solved uniquely to yield λ_j, $j = 1, \ldots, K$ in terms of r_j, P_{ij}, $i, j = 1, \ldots, K$, we make a fairly natural assumption that essentially asserts that each customer will eventually exit the system with probability 1. This assumption is that for every queue i_1, there is a queue i with $1 - \sum_{j=1}^{K} P_{ij} > 0$ and a sequence $i_1, i_2, \ldots, i_k, i$ such that $P_{i_1 i_2} > 0, \ldots, P_{i_k i} > 0$.*

The service times of customers at the jth queue are assumed exponentially distributed with mean $1/\mu_j$ and are assumed mutually independent and independent of the arrival process at the queue. The utilization factor of each queue is denoted

$$\rho_j = \frac{\lambda_j}{\mu_j}, \qquad j = 1, \ldots, K \tag{3.115}$$

and we assume that $\rho_j < 1$ for all j.

In order to model a packet network such as the one considered in Section 3.6 within the framework described above, it is necessary to accept several simplifying conditions in addition to assuming Poisson arrivals and exponentially distributed packet lengths. The first is the independence of packet lengths and interarrival times discussed earlier. The second is relevant to datagram networks, and has to do with the assumption that bifurcation of traffic at a network node can be modeled reasonably well by a randomization process whereby each departing packet from queue i joins queue j with probability P_{ij}—this need not be true, as discussed in Section 3.6. Still a packet network differs from the model of this section because it involves several traffic streams which may have different routing probabilities at each node, and which maintain their identity as they travel along different routes (see the virtual circuit and datagram network models of Figs. 3.27 and 3.28). This difficulty can be partially addressed by using an extension of Jackson's Theorem that applies to a network with multiple classes of customers. Within this more general framework, we can model traffic streams corresponding to different origin–destination pairs as different classes of customers. If all traffic streams have the same average packet length, it turns out that Jackson's Theorem as stated below is valid assuming the simplifying conditions mentioned earlier; see the analysis in the next subsection.

*For a brief explanation aimed at the advanced reader, consider the Markov chain with states $0, 1, \ldots, K$ and transition probabilities from states $i \neq 0$ to states $j \neq 0$ equal to P_{ij}, and transition probabilities to state 0 equal to $P_{00} = 1$, $P_{i0} = 1 - \sum_{j=1}^{K} P_{ij}$ for $i \neq 0$. (Thus state 0 is an absorbing state that corresponds to exit of a customer from the system.) Let P be the $K \times K$ matrix with elements P_{ij}. The sum of the ith row elements of the matrix P^m (P to the mth power) is the probability that the Markov chain has not arrived at state 0 after m transitions starting from state i. Our hypothesis on P_{ij} implies that the chain will eventually (with probability 1) arrive at state 0 regardless of the initial state. It follows that $\lim_{m \to \infty} P^m = 0$, so unity is not an eigenvalue of P. Therefore, $I - P$ is nonsingular, where I is the identity matrix, from which it can be seen that the system of equations (3.114) has a unique solution.

For an analysis, we view the system as a continuous-time Markov chain in which the state n is the vector $(n_1, n_2, \ldots, n_K)$, where n_i denotes the number of customers at queue i. At a given state

$$n = (n_1, n_2, \ldots, \ldots, n_K)$$

the possible successor states correspond to a single customer arrival and/or departure. In particular, the transition from n to state

$$n(j^+) = (n_1, \ldots, n_{j-1}, n_j + 1, n_{j+1}, \ldots, n_K)$$

corresponding to an external arrival at queue j, has transition rate

$$q_{nn(j^+)} = r_j$$

The transition from n to state

$$n(j^-) = (n_1, \ldots, n_{j-1}, n_j - 1, n_{j+1}, \ldots, n_K)$$

corresponding to a departure from queue j to the outside, has transition rate

$$q_{nn(j^-)} = \mu_j \left(1 - \sum_i P_{ji}\right)$$

The transition from n to state

$$n(i^+, j^-) = (n_1, \ldots, n_{i-1}, n_i + 1, n_{i+1}, \ldots, n_{j-1}, n_j - 1, n_{j+1}, \ldots, n_K)$$

corresponding to a customer moving from queue j to queue i, has transition rate

$$q_{nn(i^+, j^-)} = \mu_j P_{ji}$$

Let $P(n_1, \ldots, n_K)$ denote the stationary distribution of the chain. We have:

Jackson's Theorem. Assuming that $\rho_j < 1$, $j = 1, \ldots, K$, we have for all $n_1, \ldots, n_K \geq 0$,

$$P(n) = P_1(n_1)P_2(n_2)\cdots P_K(n_K) \tag{3.116}$$

where $n = (n_1, \ldots, n_K)$ and

$$P_j(n_j) = \rho_j^{n_j}(1 - \rho_j), \qquad n_j \geq 0 \tag{3.117}$$

Proof: In our proof we will assume that $\lambda_j > 0$ for all j. There is no loss of generality in doing so because every queue j with $\lambda_j = 0$ is empty in steady-state, so we have $P_j(0) = 1$ and $P_j(n_j) = 0$ for $n_j > 0$, and queue j can be ignored in deriving the stationary distribution of Eqs. (3.116) and (3.117). It can be verified that the condition $\lambda_j > 0$ for all j together with the assumption made earlier to guarantee the uniqueness of solution of Eq. (3.114) imply that the Markov chain with states $n = (n_1, \ldots, n_K)$ describing the system is irreducible; we leave the proof of this for the reader. We will use a technique outlined in Section 3.7 whereby we guess at the transition rates of the reversed process and verify that, together with the probability distribution of Eqs. (3.116) and (3.117), they satisfy the total departure rate equation (3.111). (The Markov chain is not time reversible here. Nonetheless, the use of the reversed process is both analytically convenient and conceptually useful.)

For any two state vectors n and n', let $q_{nn'}$ be the corresponding transition rate. Jackson's Theorem will be proved if the rates $q^*_{nn'}$ defined for all n, n' by the equation

$$q^*_{nn'} = \frac{P(n')q_{n'n}}{P(n)} \tag{3.118}$$

satisfy, for all n, the total rate equation

$$\sum_m q_{nm} = \sum_m q^*_{nm} \tag{3.119}$$

which as mentioned in Section 3.7, is equivalent to the global balance equations.

For transitions between states n, $n(j^+)$, and $n(j^-)$, we have

$$q_{nn(j^+)} = r_j \tag{3.120}$$

$$q_{nn(j^-)} = \mu_j \left(1 - \sum_i P_{ji}\right) \tag{3.121}$$

The rates $q^*_{nn(j^+)}$ and $q^*_{nn(j^-)}$ are defined by Eqs. (3.118), (3.120), and (3.121). Using the fact $P\big(n(j^+)\big) = \rho_j P(n) = \lambda_j P(n)/\mu_j$ [cf. Eqs. (3.115)–(3.117)], we obtain

$$q^*_{nn(j^+)} = \lambda_j \left(1 - \textstyle\sum_i P_{ji}\right) \tag{3.122}$$

$$q^*_{nn(j^-)} = \frac{\mu_j r_j}{\lambda_j} \tag{3.123}$$

Next consider transitions between states n and $n(i^+, j^-)$ corresponding to a customer moving from queue j to queue i. We have

$$q_{nn(i^+,j^-)} = \mu_j P_{ji} \tag{3.124}$$

and using the fact that $P\big(n(i^+, j^-)\big) = \rho_i P(n)/\rho_j = \lambda_i \mu_j P(n)/(\lambda_j \mu_i)$, we obtain $q^*_{n(i^+,j^-)n}$ from Eq. (3.118) as

$$q^*_{n(i^+,j^-)n} = \frac{\mu_i \lambda_j P_{ji}}{\lambda_i} \tag{3.125}$$

Since for all other types of pairs of state vectors n, n', we have

$$q_{nn'} = 0 \tag{3.126}$$

it follows from Eq. (3.118) that

$$q^*_{n'n} = 0 \tag{3.127}$$

There remains to verify that the rates q_{nm} and q^*_{nm} satisfy the total rate equation $\sum_m q_{nm} = \sum_m q^*_{nm}$. We have for the forward system, using Eqs. (3.120), (3.121), and (3.124),

$$\sum_m q_{nm} = \sum_{j=1}^{K} q_{nn(j^+)} + \sum_{\{(j,i)|n_j>0\}} q_{nn(i^+,j^-)} + \sum_{\{j|n_j>0\}} q_{nn(j^-)}$$

$$= \sum_{j=1}^{K} r_j + \sum_{\{(j,i)|n_j>0\}} \mu_j P_{ji} + \sum_{\{j|n_j>0\}} \mu_j \left(1 - \sum_{i=1}^{K} P_{ji}\right)$$

$$= \sum_{j=1}^{K} r_j + \sum_{\{j|n_j>0\}} \mu_j \tag{3.128}$$

Similarly, using Eqs. (3.122), (3.123), (3.125), and (3.114), we obtain for the reversed system

$$\sum_m q^*_{nm} = \sum_{j=1}^{K} q^*_{nn(j^+)} + \sum_{\{(j,i)|n_j>0\}} q^*_{nn(i^+,j^-)} + \sum_{\{j|n_j>0\}} q^*_{nn(j^-)}$$

$$= \sum_{j=1}^{K} \lambda_j \left(1 - \sum_{i=1}^{K} P_{ji}\right) + \sum_{\{(j,i)|n_j>0\}} \frac{\mu_j \lambda_i P_{ij}}{\lambda_j} + \sum_{\{j|n_j>0\}} \frac{\mu_j r_j}{\lambda_j}$$

$$= \sum_{j=1}^{K} \lambda_j \left(1 - \sum_{i=1}^{K} P_{ji}\right) + \sum_{\{j|n_j>0\}} \frac{\mu_j (r_j + \sum_{i=1}^{K} \lambda_i P_{ij})}{\lambda_j}$$

$$= \sum_{j=1}^{K} \lambda_j \left(1 - \sum_{i=1}^{K} P_{ji}\right) + \sum_{\{j|n_j>0\}} \mu_j \tag{3.129}$$

By writing Eq. (3.114) as $r_j = \lambda_j - \sum_{i=1}^{K} \lambda_i P_{ij}$ and adding over $j = 1, \ldots, K$, we obtain

$$\sum_{j=1}^{K} r_j = \sum_{j=1}^{K} \lambda_j \left(1 - \sum_{i=1}^{K} P_{ji}\right) \tag{3.130}$$

By combining the last three equations, we see that the total rate equation $\sum_m q_{nm} = \sum_m q^*_{nm}$ is satisfied. **Q.E.D.**

Note that the transition rates $q^*_{nn'}$ defined by Eqs. (3.122), (3.123), (3.125), and (3.127) are those of the reversed process. It can be seen that the reversed process corresponds to a network of queues where traffic arrives at queue i from outside the network according to a Poisson process with rate $\lambda_i \left(1 - \sum_j P_{ij}\right)$ [cf. Eq. (3.122)]. The routing probability from queue i to queue j in the reversed process is

$$\frac{\lambda_j P_{ji}}{r_i + \sum_k \lambda_k P_{ki}}$$

[cf. Eqs. (3.123) and (3.125)]. This is also the probability that an arriving customer at queue i just departed from queue j in the forward process. Note that the processes of departure out of the forward system are the exogenous arrival processes of the reversed system, which suggests that the processes of departure out of the system are independent

Poisson. Indeed, this can be proved by observing that the interarrival times in the reversed system are independent and exponentially distributed.

Example 3.19 Computer System with Feedback Loop for I/O

Consider a model of a computer CPU connected to an I/O device as shown in Fig. 3.34(a). Jobs enter the system according to a Poisson process with rate λ, and use the CPU for an exponentially distributed time interval with mean $1/\mu_1$. Upon exiting the CPU, a job with probability p_1 exits the system, and with probability p_2 $(= 1 - p_1)$ uses the I/O device for a time which is exponentially distributed with mean $1/\mu_2$. Upon exit from the I/O device, a job again joins the CPU queue. We assume that all service times, including successive service times of the same job at the CPU or the I/O device, are independent.

We first calculate the arrival rates λ_1 and λ_2 at the CPU and I/O device queues, respectively. We have (cf. Fig. 3.34)

$$\lambda_1 = \lambda + \lambda_2, \qquad \lambda_2 = p_2 \lambda_1$$

[These are Eqs. (3.114) specialized to this example.] By solving for λ_1 and λ_2 we obtain

$$\lambda_1 = \frac{\lambda}{p_1}, \qquad \lambda_2 = \frac{\lambda p_2}{p_1} \tag{3.131}$$

Let

$$\rho_1 = \frac{\lambda_1}{\mu_1}, \qquad \rho_2 = \frac{\lambda_2}{\mu_2} \tag{3.132}$$

The steady-state probability distribution of the system is given by Jackson's Theorem

$$P(n_1, n_2) = \rho_1^{n_1}(1 - \rho_1)\rho_2^{n_2}(1 - \rho_2)$$

The average number of jobs N_i in the ith queue is the same as for an $M/M/1$ system with utilization factor ρ_i, that is,

$$N_1 = \frac{\rho_1}{1 - \rho_1}, \qquad N_2 = \frac{\rho_2}{1 - \rho_2}$$

The total number in the system is

$$N = N_1 + N_2 = \frac{\rho_1}{1 - \rho_1} + \frac{\rho_2}{1 - \rho_2}$$

(a)

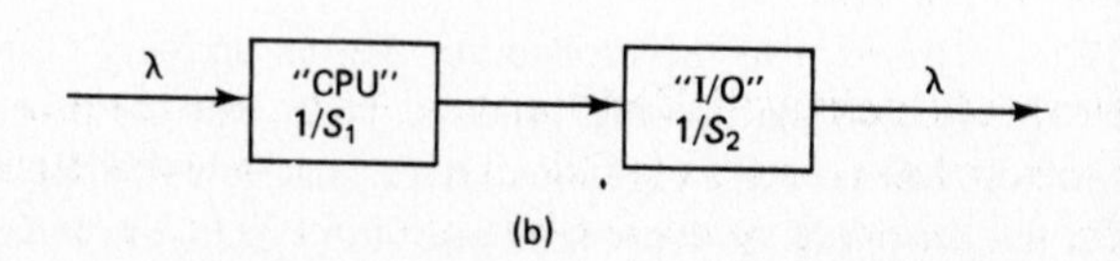

Figure 3.34 (a) Feedback model of a CPU and an I/O device (cf. Example 3.19). (b) "Equivalent" tandem model of a CPU and an I/O queue which has the same occupancy distribution as the feedback model.

and the average time in the system is

$$T = \frac{N}{\lambda} = \frac{\rho_1}{\lambda(1-\rho_1)} + \frac{\rho_2}{\lambda(1-\rho_2)}$$

Using Eqs. (3.131) and (3.132) we can write this relation as

$$T = \frac{\lambda_1/\mu_1}{\lambda(1-\lambda_1/\mu_1)} + \frac{\lambda_2/\mu_2}{\lambda(1-\lambda_2/\mu_2)} = \frac{\lambda/(\mu_1 p_1)}{\lambda\big(1-\lambda/(\mu_1 p_1)\big)} + \frac{\lambda p_2/(\mu_2 p_1)}{\lambda\big(1-\lambda p_2/(\mu_2 p_1)\big)}$$

$$= \frac{S_1}{1-\lambda S_1} + \frac{S_2}{1-\lambda S_2} \tag{3.133}$$

where

$$S_1 = \frac{1}{\mu_1 p_1}, \qquad S_2 = \frac{p_2}{\mu_2 p_1} \tag{3.134}$$

Since the utilization factor of the CPU queue is $\rho_1 = \lambda_1/\mu_1 = \lambda/(\mu_1 p_1)$, while the arrival rate of *new* job arrivals at the CPU (as opposed to feedback arrivals) is λ, we see from Little's Theorem that S_1 is the total CPU time a job requires on the average (this includes all visits of the job to the CPU). Similarly, S_2 is the total I/O time a job requires on the average.

An interesting interpretation of Eqs. (3.133) and (3.134) is that the average number of jobs and time in the system are the same as in an "equivalent" tandem model of CPU and I/O queues with service rates $1/S_1$ and $1/S_2$, respectively, as shown in Fig. 3.34(b). However, the probability density function of the time in the system is not the same in the feedback and tandem systems. To get some idea of this fact, suppose that $p_1 = p_2 = 1/2$ and that the CPU service rate is much faster than the I/O service rate ($\mu_1 >> \mu_2$). Then half the jobs in the feedback system do not require any I/O service and their average time in the system is much smaller than the average time of the other half. This is not so in the tandem system where the average job time in the "CPU" queue is very small and the system time is distributed approximately as in the "I/O" queue, that is, as in an $M/M/1$ queue with Poisson rate λ and service rate $1/S_1$.

Jackson's Theorem says in effect that the numbers of customers in the system's queues are distributed as if each queue is $M/M/1$ and is independent of the other queues [compare Eq. (3.117) and the corresponding equations in Section 3.3]. Despite this fact, *the total arrival process at each queue need not be Poisson*. As an example (see Fig. 3.35), suppose that there is a single queue with a service rate which is very large relative to the arrival rate from the outside. Suppose also that with probability p near unity, a customer upon completion of service is fed back into the queue. Hence, when an arrival occurs at the queue, there is a large probability of another arrival at the queue in a short time (namely, the feedback arrival), whereas at an arbitrary time point, there will be only a very slight chance of an arrival occurring shortly since λ is small. In other words, queue arrivals tend to occur in bursts triggered by the arrival of a single customer from the outside. Hence, the queue arrival process does not have independent interarrival times and cannot be Poisson.

Heuristic explanation of Jackson's Theorem. Our proof of Jackson's theorem is based on algebraic manipulation and gives little insight as to why this remarkable result holds. For this reason we provide a heuristic explanation for the case of the feed-

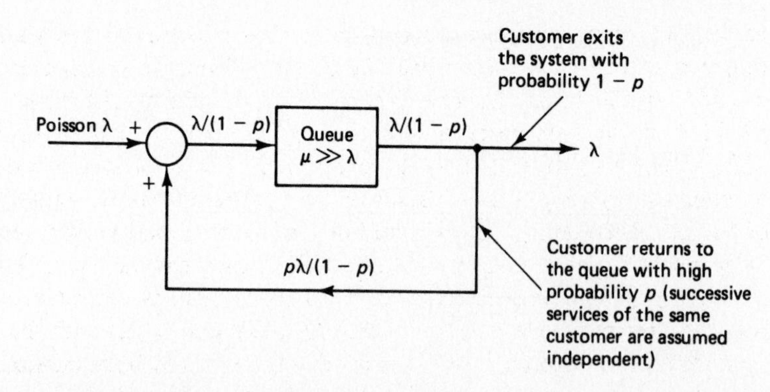

Figure 3.35 Example of a queue within a network where the external arrival process is Poisson but the total arrival process at the queue is not Poisson. An external arrival is typically processed fast (since μ is much larger than λ) and with high probability returns to the queue through the feedback loop. As a result, the total queue arrival process typically consists of bursts of arrivals, with each burst triggered by the arrival of a single customer from the outside.

back network of Fig. 3.35. This explanation can be generalized and made rigorous albeit at the expense of a great deal of technical complications (see [Wal83]).

Suppose that we introduce a delay Δ in the feedback loop of the single-queue network discussed above (see Fig. 3.36). Let us denote by $n(t)$ the number in the queue at time t, and by $f_\Delta(t)$ the content of the delay line at time t. The interpretation here is that $f_\Delta(t)$ is a function of time that specifies the customer output of the delay line in the subsequent Δ interval $(t, t+\Delta]$. Suppose that the initial distribution $n(0)$ of the queue state at time 0, is equal to the steady-state distribution of an $M/M/1$ queue, that is,

$$P\{n(0) = n\} = \rho^n(1-\rho) \tag{3.135}$$

where $\rho = \lambda/\big(\mu(1-p)\big)$ is the utilization factor. Suppose also that $f_\Delta(0)$ is a portion of a Poisson arrival process with rate λ. The customers in $f_\Delta(0)$ have service times that are independent, exponentially distributed with parameter μ. We assume that $n(0)$ and $f_\Delta(0)$ are independent. Then, the input to the queue over the interval $[0, \Delta)$ will

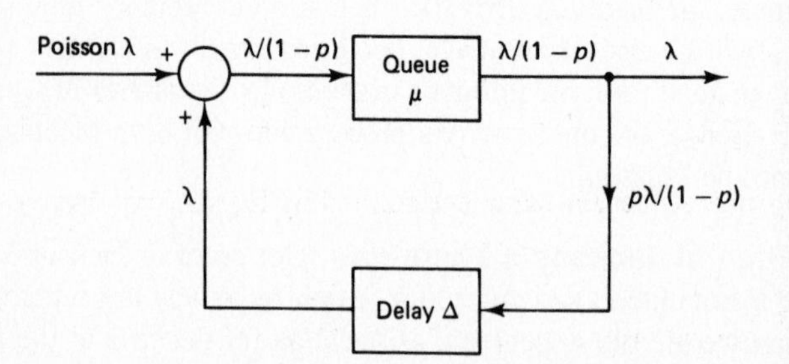

Figure 3.36 Heuristic explanation of Jackson's Theorem. Consider the introduction of an arbitrarily small positive delay Δ in the feedback loop of the network of Fig. 3.35. An occupancy distribution of the queue that equals the $M/M/1$ equilibrium, and a content of the delay line that is an independent Δ segment of a Poisson process form an equilibrium distribution of the overall system. Therefore, the $M/M/1$ equilibrium distribution is an equilibrium for the queue as suggested by Jackson's Theorem even though the total arrival process to the queue is not Poisson.

be the sum of two independent Poisson streams which are independent of the number in queue at time 0. It follows that the queue will behave in the interval $[0, \Delta)$ like an $M/M/1$ queue in equilibrium. Therefore, $n(\Delta)$ will be distributed according to the $M/M/1$ steady-state distribution of Eq. (3.135), and by part (b) of Burke's theorem, $n(\Delta)$ will be independent of the departure process from the queue in the interval $[0, \Delta)$, or, equivalently, of $f_\Delta(\Delta)$ —the delay line content at time Δ. Furthermore, by part (a) of Burke's Theorem, $f_\Delta(\Delta)$ will be Poisson. Thus, to summarize, we started out with independent initial conditions $n(0)$ and $f_\Delta(0)$ which had the equilibrium distribution of an $M/M/1$ queue and the statistics of a Poisson process, respectively, and Δ seconds later we obtained corresponding quantities $n(\Delta)$ and $f_\Delta(\Delta)$ with the same properties. Using the same reasoning, we can show that for all t which are multiples of Δ, $n(t)$ and $f_\Delta(t)$ have the same properties. It follows that the $M/M/1$ steady-state distribution of Eq. (3.135) is an equilibrium distribution for the queueing system for an arbitrary positive value of the feedback delay Δ, and this strongly suggests the validity of Jackson's Theorem. Note that this argument does not suggest that the feedback process, and therefore also the total arrival process to the queue, are Poisson. Indeed, it can be seen that successive Δ portions of the feedback arrival stream are correlated since, with probability p, a departing customer from the queue appears as an arrival Δ seconds later. Therefore, over the interval $[0, \infty)$, the feedback process is not Poisson. This is consistent with our earlier observations regarding the example of Fig. 3.35.

3.8.1 Extensions of Jackson's Theorem

There are a number of interesting extensions and variations of Jackson's Theorem, and in this and the next subsections we will describe a few of them.

State-dependent service rates. The model for Jackson's Theorem assumed so far requires that all queues have a single server. An extension to the multiserver case can be obtained by allowing the service rate at each queue to depend on the number of customers at that queue. Thus the model is the same as before but the service time at the jth queue is exponentially distributed with mean $1/\mu_j(m)$, where m is the number of customers in the jth queue just before the customer's departure (m includes the customer). The single-queue version of this model includes as special cases the $M/M/m$ and $M/M/\infty$ queues, and can be analyzed by means of a Markov chain (see Problem 3.16). The corresponding network of queues model can also be analyzed by means of a Markov chain, and is characterized by a product form structure for the stationary distribution.

Let us define

$$\rho_j(m) = \frac{\lambda_j}{\mu_j(m)}, \qquad j = 1, \ldots, K, \ m = 1, 2, \ldots \tag{3.136}$$

where λ_j is the total arrival rate at the jth queue determined by Eq. (3.114). Let us also define

$$\hat{P}_j(n_j) = \begin{cases} 1, & \text{if } n_j = 0 \\ \rho_j(1)\rho_j(2)\cdots\rho_j(n_j), & \text{if } n_j > 0 \end{cases} \tag{3.137}$$

We have:

Jackson's Theorem for State-Dependent Service Rates. We have for all states $n = (n_1, \ldots, n_K)$

$$P(n) = \frac{\hat{P}_1(n_1) \cdots \hat{P}_K(n_K)}{G} \tag{3.138}$$

assuming that $0 < G < \infty$, where the normalizing constant G is given by

$$G = \sum_{n_1=0}^{\infty} \cdots \sum_{n_K=0}^{\infty} \hat{P}_1(n_1) \cdots \hat{P}_K(n_K) \tag{3.139}$$

Proof: Note that the formula for G guarantees that $P(n)$ is a probability distribution, that is, the sum of all $P(n)$ is unity. Using this fact, the proof is obtained by repeating the steps of the earlier proof of Jackson's Theorem, substituting the state-dependent service rates $\mu_j(m)$ in place of the rates μ_j at the appropriate points, and is left for the reader. **Q.E.D.**

Multiple classes of customers. In many interesting networks of queues the routing probabilities P_{ij} are not the same for all customers. Typical examples arise in data networks where the transmission queue joined by a packet at each intermediate node depends on the packet's destination and possibly its origin. It is therefore necessary to distinguish between customers of different types or classes. We will show that the product form expressions derived so far remain valid provided that the service time distribution at each queue is the same for all customer classes.

Let the customer classes be $c = 1, 2, \ldots, C$, let $r_j(c)$ be the rate of the external Poisson arrival process of class c at queue j, and let $P_{ij}(c)$ be the routing probabilities of class c. The assumptions made for an open Jackson network with a single customer class are replicated for each customer class, so that the equations

$$\lambda_j(c) = r_j(c) + \sum_{i=1}^{K} \lambda_i(c) P_{ij}(c), \qquad j = 1, \ldots, K, \quad c = 1, 2, \ldots, C \tag{3.140}$$

can be solved uniquely to give the total arrival rate $\lambda_j(c)$ at each queue j and for each customer class c. We assume that the service times at queue j are exponentially distributed with a common mean $1/\mu_j(m)$ for all customer classes, which depends on m, the total number of customers in the queue. As earlier, customers are served on a first-come first-serve basis.

The state of each queue is characterized not just by the total number of customers present in the queue, but also by the class of the customers and the relative order of arrival of the customers of different classes. Thus, we define the *composition of the* j^{th} *queue* at a given time as

$$z_j = (c_1, c_2, \ldots, c_{n_j})$$

where n_j is the total number of customers in the queue and c_i is the class of the customer in the ith queue position.

The state of the queueing network at a given time is

$$z = (z_1, z_2, \ldots, z_K)$$

where z_j is the composition of the jth queue at that time. It can be viewed as the state of a Markov chain the transition probabilities of which can be described in terms of the given quantities $\lambda_j(c)$, $\mu_j(m)$, and $P_{ij}(c)$. To state the appropriate form of Jackson's Theorem, define

$$\hat{\rho}_j(c, m) = \frac{\lambda_j(c)}{\mu_j(m)}, \qquad j = 1, \ldots, K, \quad c = 1, 2, \ldots, C \tag{3.141}$$

$$\hat{P}_j(z_j) = \begin{cases} 1, & \text{if } n_j = 0 \\ \hat{\rho}_j(c_1, 1)\hat{\rho}_j(c_2, 2)\cdots\hat{\rho}_j(c_{n_j}, n_j), & \text{if } n_j > 0 \end{cases} \tag{3.142}$$

$$G = \sum_{(z_1, \ldots, z_K)} \prod_{j=1}^{K} \hat{P}_j(z_j) \tag{3.143}$$

The proof of the following theorem follows the same pattern as the corresponding proof for the single customer class case, and is left for the reader.

Jackson's Theorem for Multiple Classes of Customers. Assuming that $0 < G < \infty$, the steady-state probability $\hat{P}(z)$ of state $z = (z_1, z_2, \ldots, z_K)$ is given by

$$\hat{P}(z) = \frac{\hat{P}_1(z_1)\cdots\hat{P}_K(z_K)}{G} \tag{3.144}$$

The steady-state probability $P(n) = P(n_1, \ldots, n_K)$ of having a total of n_j customers at queue $j = 1, \ldots, K$ (irrespective of class) is given by

$$P(n) = \sum_{z \in Z(n)} \hat{P}(z)$$

where $Z(n)$ is the set of states for which there is a total of n_j customers in queue j. By adding the expression (3.144) over $z \in Z(n)$, it is straighforward to verify that when the service rate at each queue is the same for all customer classes and is independent of the queue size, we have

$$P(n) = \prod_{j=1}^{K} \rho_j^{n_j}(1 - \rho_j) \tag{3.145}$$

where

$$\rho_j = \frac{\sum_{c=1}^{C} \lambda_j(c)}{\mu_j} \tag{3.146}$$

and μ_j is the service rate at queue j. In other words, the expression for $P(n)$ is the same as when there is a single customer class with total arrival rate at each queue j equal to the sum of the arrival rates of all customer classes $\sum_{j=1}^{C} \lambda_j(c)$.

We note that when the service rates at the queues are state dependent (but identical for all classes) the steady-state probabilities $P(n)$ can be shown to be given by the (single class) formulas (3.136) to (3.139). (See the references cited at the end of the chapter.)

The following example addresses the first data network model discussed in Section 3.6 (cf. Fig. 3.27). A similar analysis can be used for the datagram network model of Fig. 3.28.

Example 3.20 Virtual Circuit Network

Consider the network of communication links discussed in Section 3.5 (cf. Fig. 3.27). There are several traffic streams (or virtual circuits) denoted $c = 1, 2, \ldots, C$. Virtual circuit c uses a path p_c and has a Poisson arrival rate x_c. The total arrival rate of each link (i, j) is

$$\lambda_{ij} = \sum_{\{c|(i,j) \text{ lies on the path } p_c\}} x_c$$

Assume that the transmission times of all packets at link (i, j) are exponentially distributed with mean $1/\mu_{ij}$, which is the same for all virtual circuits. Assume also that the transmission times of all packets are independent, including the transmission times of the same packet at two different links (this is the essence of the Kleinrock independence approximation). Then the multiple-class model of this subsection applies and based on Eq. (3.145), the average number of packets in the system, N, is the same as if each link were an $M/M/1$ queue in isolation, that is,

$$N = \sum_{(i,j)} \frac{\lambda_{ij}}{\mu_{ij} - \lambda_{ij}}$$

Example 3.20 shows how multiple customer classes can be used to model data network situations where the route used by a packet depends on its origin and destination. There is still an unrealistic assumption in this example, namely that the transmission times of the same packet at two different links are independent. Furthermore, the assumption that all packet transmission times are exponentially distributed with common mean is often violated in practice. For a more realistic model, we would like to be able to assume more general transmission time distributions (*e.g.*, deterministic transmission times). It turns out that the product form of Eqs. (3.141) to (3.144) holds even when the service time distributions belong to a broad class of "phase-type" distributions, which can approximate arbitrarily closely deterministic service times (see [GeP87] and [Wal88]). For this, however, we need to assume that the service discipline at each queue is either *processor sharing* or *last-come first-serve* instead of first-come first-serve. Processor sharing refers to a situation where all customers in the queue are simultaneously served at the same rate (which is μ/n when μ is the total service rate and n is the number of customers). Last-come first-serve refers to the situation where upon arrival at a queue, a customer goes immediately into service, replacing the customer who is in service at the time (if any) on a preemptive-resume basis. While processor sharing or last-come first-serve may not be reasonable models for most data networks, the validity of the product form expression (3.141) to (3.144) under a variety of different assumptions is reassuring. It suggests that product forms provide a good first approximation in many practical situations where their use cannot be rigorously justified. Current practice and experience seems to be supporting this view. We note, however, that for special types of priority disciplines, there are queueing networks that are unstable (some queue lengths grow indefinitely) even though the arrival rate is smaller than the service rate at each queue [KuS89]. We refer to the sources given at the end of the chapter for more details and discussion on the subject.

3.8.2 Closed Queueing Networks

Many interesting queueing problems involve a network of queues where the total number of customers is fixed because no customers are allowed to arrive or depart. Networks of this type are called *closed*, emphasizing the distinction from the earlier networks in this section which are called *open*. Examples 3.5 and 3.7 illustrate applications of closed networks. In both examples the fixed number of customers in the network depends on some limited resource, and the main purpose of analysis is to understand how the availability of this resource affects performance characteristics such as system throughput.

Closed networks can also be analyzed using Markov chains and it can be shown that the steady-state occupancy distribution has a product form under assumptions similar to those used earlier for open networks. For simplicity, we assume a single customer class, but extensions involving multiple customer classes are possible. Let M be the fixed number of customers in the system and let P_{ij} be the routing probability that a customer that departs from queue i will next visit queue j. Note that because no customer can exit the system, we have

$$\sum_{j=1}^{K} P_{ij} = 1, \qquad i = 1, \ldots, K$$

Let also $\mu_j(m)$ be the service rate at the jth queue when the number of customers at that queue is m.

An important difference from the open network case is that the total arrival rates, denoted $\lambda_j(M)$, at the queues $j = 1, \ldots K$ are not easily determined. We still have the equations

$$\lambda_j = \sum_{i=1}^{K} \lambda_i P_{ij}, \qquad j = 1, \ldots, K \tag{3.147}$$

obtained by setting to zero the external arrival rates r_j in Eq. (3.114). These equations do not have a unique solution anymore, but under some fairly natural assumptions, they determine the arrival rates $\lambda_j(M)$ up to a multiplicative constant. In particular, let us assume that the Markov chain with states $1, \ldots, K$ and transition probabilities P_{ij} is irreducible (see Appendix A). Then it can be shown that all solutions λ_j, $j = 1, \ldots, K$, of Eq. (3.147) are of the form*

*For a brief explanation, fix λ_1 at some positive value a and consider the system of equations $\lambda_j = aP_{1j} + \sum_{i=2}^{K} \lambda_i P_{ij}$, $j = 2, \ldots, K$. Because of the irreducibility assumption, this system has a unique solution. [See the explanation given in connection with the uniqueness of solution of the corresponding open network equation (3.114).] This unique solution is proportional to a and it can be shown to have positive elements.

$$\lambda_j = \alpha\overline{\lambda}_j, \qquad j = 1, \ldots, K$$

where α is a scalar and $\overline{\lambda}_j$, $j = 1, \ldots, K$ is a particular solution with $\overline{\lambda}_j > 0$ for all j. Thus the true arrival rates are given by

$$\lambda_j(M) = \alpha(M)\overline{\lambda}_j, \qquad j = 1, \ldots, K \tag{3.148}$$

where $\alpha(M)$ is the constant of proportionality corresponding to M. Note that while $\overline{\lambda}_j$ can be chosen to be independent of M, both $\alpha(M)$ and the true total arrival rates $\lambda_j(M)$ increase with M. In the case where the queue service rates μ_j are independent of the number of customers, $\alpha(M)$ tends asymptotically to the value that makes the maximum utilization factor $\max\{\lambda_1(M)/\mu_1, \ldots, \lambda_K(M)/\mu_K\}$ equal to one.

We now describe the form of Jackson's Theorem for closed networks. Let

$$\rho_j(m) = \frac{\overline{\lambda}_j}{\mu_j(m)}, \qquad j = 1, \ldots, K, \quad m = 1, 2, \ldots \tag{3.149}$$

where $\{\overline{\lambda}_j \mid j = 1, \ldots, K\}$ is some positive solution of the system of equations (3.147). Denote

$$\hat{P}_j(n_j) = \begin{cases} 1, & \text{if } n_j = 0 \\ \rho_j(1)\rho_j(2)\cdots\rho_j(n_j), & \text{if } n_j > 0 \end{cases} \tag{3.150}$$

$$G(M) = \sum_{\{(n_1,\ldots,n_K) \mid n_1+\cdots+n_K=M\}} \hat{P}_1(n_1)\cdots\hat{P}_K(n_K) \tag{3.151}$$

We have:

Jackson's Theorem for Closed Networks. Under the preceding assumptions, we have for all states $n = (n_1, \ldots, n_K)$ with $n_1 + \cdots + n_K = M$

$$P(n) = \frac{\hat{P}_1(n_1)\cdots\hat{P}_K(n_K)}{G(M)} \tag{3.152}$$

[Note that because all solutions of Eq. (3.147) are scalar multiples of each other, the expression (3.152) for the probabilities $P(n)$ is not affected by the choice of the solution as long as this solution is nonzero. Note also that $G(M)$ is a normalization constant that ensures that $P(n)$ is a probability distribution.]

Proof: The proof is similar to the proof of Jackson's Theorem for open networks. We consider state vectors n and n' of the form

$$n = (n_1, \ldots, n_i, \ldots, n_j, \ldots, n_K)$$

$$n' = (n_1, \ldots, n_{i-1}, n_i + 1, n_{i+1}, \ldots, n_{j-1}, n_j - 1, n_{j+1}, \ldots, n_K)$$

Let $q_{nn'}$ be the corresponding transition rate. Jackson's Theorem will be proved if the rates $q^*_{nn'}$ defined for all n, n' by the equation

$$q^*_{nn'} = \frac{P(n')q_{n'n}}{P(n)} \tag{3.153}$$

satisfy, for all states n, the total rate equation

$$\sum_m q_{nm} = \sum_m q^*_{nm} \tag{3.154}$$

Indeed, let us assume for the purpose of the proof that the particular solutions $\overline{\lambda}_j$ are taken to be equal to the true arrival rates $\lambda_j(M)$, and for convenience let us denote both $\overline{\lambda}_j$ and $\lambda_j(M)$ as λ_j. Then we have [cf. Eqs. (3.124) and (3.125)]

$$q_{nn'} = \mu_j(n_j)P_{ji}, \qquad q^*_{nn'} = \frac{\mu_j(n_j)\lambda_i P_{ij}}{\lambda_j}$$

and the total rate equation (3.154) is written as

$$\sum_{\{(j,i)|n_j>0\}} \mu_j(n_j)P_{ji} = \sum_{\{(j,i)|n_j>0\}} \frac{\mu_j(n_j)\lambda_i P_{ij}}{\lambda_j}$$

We have

$$\begin{aligned}\sum_{\{(j,i)|n_j>0\}} \mu_j(n_j)P_{ji} &= \sum_{i=1}^{K} \sum_{\{j|n_j>0\}} \mu_j(n_j)P_{ji} = \sum_{\{j|n_j>0\}} \mu_j(n_j) \sum_{i=1}^{K} P_{ji} \\ &= \sum_{\{j|n_j>0\}} \mu_j(n_j)\end{aligned} \tag{3.155}$$

We also have

$$\begin{aligned}\sum_{\{(j,i)|n_j>0\}} \frac{\mu_j(n_j)\lambda_i P_{ij}}{\lambda_j} &= \sum_{i=1}^{K} \sum_{\{j|n_j>0\}} \frac{\mu_j(n_j)\lambda_i P_{ij}}{\lambda_j} = \sum_{\{j|n_j>0\}} \frac{\mu_j(n_j)}{\lambda_j} \sum_{i=1}^{K} \lambda_i P_{ij} \\ &= \sum_{\{j|n_j>0\}} \mu_j(n_j)\end{aligned} \tag{3.156}$$

From Eqs. (3.155) and (3.156) we see that the total rate equation (3.154) holds. **Q.E.D.**

Example 3.21 Closed Computer System with Feedback Loop for I/O

Consider a model of a computer CPU connected to an I/O device as shown in Fig. 3.37. This is a similar model to the one discussed in Example 3.19. The difference is that here

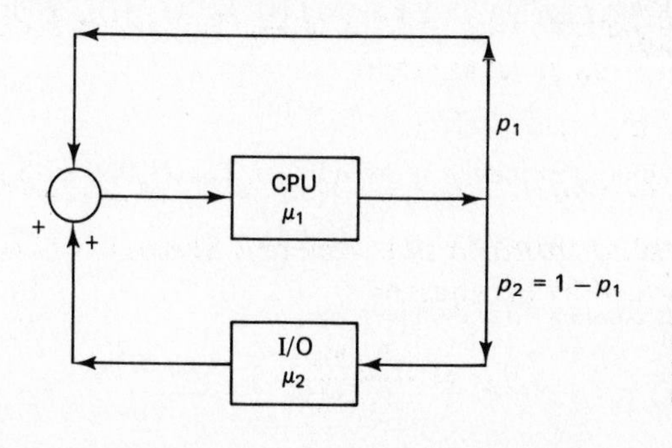

Figure 3.37 Closed network model of a feedback system of a CPU and an I/O device.

we have a closed network with each job reentering the CPU directly (with probability p_1) or after using the I/O device (with probability $p_2 = 1 - p_1$). There are M jobs in the system. We select

$$\overline{\lambda}_1 = \mu_1, \qquad \overline{\lambda}_2 = p_2\mu_1$$

as the particular solution of the system $\lambda_j = \sum_{i=1}^{2} \lambda_i P_{ij}$, $j = 1, 2$. With this choice we have

$$\rho_1 = 1, \qquad \rho_2 = \frac{p_2\mu_1}{\mu_2}$$

and the steady-state distribution of the system is given by [cf. Eqs. (3.149) to (3.151)]

$$P(M - n, n) = \frac{\rho_2^n}{G(M)}, \qquad n = 0, 1, \ldots, M$$

where the normalizing constant $G(M)$ is given by

$$G(M) = \sum_{n=0}^{M} \rho_2^n$$

The CPU utilization factor is given by

$$U(M) = 1 - P(0, M) = 1 - \frac{\rho_2^M}{G(M)} = \frac{G(M-1)}{G(M)}$$

and from Little's Theorem we obtain the arrival rate at the CPU as $\lambda_1(M) = U(M)\mu_1$. The expression above for the utilization factor $U(M)$ is a special case of a more general formula (see Problem 3.65).

Example 3.22 Throughput of a Time-Sharing System

Consider the time-sharing computer system with N terminals discussed in Example 3.7 [cf. Fig. 3.38(a)]. We will make detailed statistical assumptions on the times spent by jobs at the terminals and the CPU. We will consequently be able to obtain a closed-form expression for the throughput of the system in place of the upper and lower bounds obtained in Section 3.2.

In particular, let us assume that the reflection time of a job at a terminal is exponentially distributed with mean R and the processing time of a job at the CPU is exponentially distributed with mean P. All reflection and processing times are assumed independent. Then the terminal and CPU queues, viewed in isolation, can be modeled as an $M/M/\infty$ and an $M/M/1$ queue, respectively [see Fig. 3.38(b)]. Let $\overline{\lambda} = 1/P$ be the particular solution of the arrival rate equation for the system [cf. Eq. (3.147)]. With this choice we have

$$\rho_1 = \frac{R}{P}, \qquad \rho_2 = 1$$

The steady-state probability distribution is given by [cf. Eqs. (3.149) to (3.151)]

$$P(n, N - n) = \frac{(R/P)^n}{n!G(N)}$$

where the normalizing constant $G(N)$ is given by

$$G(N) = 1 + (R/P) + \frac{(R/P)^2}{2!} + \cdots + \frac{(R/P)^N}{N!}$$

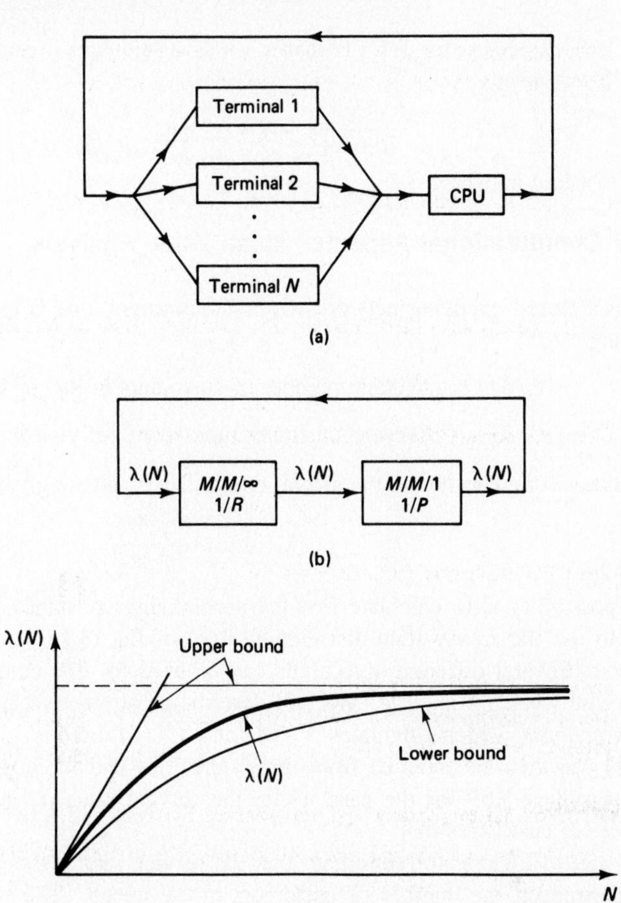

Figure 3.38 (a) Closed network model of a time-sharing system consisting of N terminals and a CPU. (b) Network of queues model of the system. There are at N jobs in the system at all times. (c) Throughput $\lambda(N)$ as a function of the number of terminals compared with the upper and lower bounds

$$\frac{N}{R+NP} \leq \lambda(N) \leq \min\left\{\frac{1}{P}, \frac{N}{R+P}\right\}$$

derived in Example 3.7.

The CPU utilization factor is

$$U(N) = 1 - \frac{P(N,0)}{G(N)} = \frac{(R/P)^N}{N!G(N)} = \frac{G(N-1)}{G(N)}$$

and by Little's Theorem, it is also equal to $\lambda(N)P$, where $\lambda(N)$ is the system throughput. Therefore, we have $\lambda(N) = U(N)/P$, or

$$\lambda(N) = \frac{1}{P}\frac{G(N-1)}{G(N)}$$

This expression for $\lambda(N)$ is shown in Fig. 3.38(c) and is contrasted with the upper and lower bounds

$$\frac{N}{R+NP} \leq \lambda(N) \leq \min\left\{\frac{1}{P}, \frac{N}{R+P}\right\} \tag{3.157}$$

obtained in Section 3.2.

3.8.3 Computational Aspects—Mean Value Analysis

Given a closed queueing network with M customers, one is typically interested in calculating

$$N_j(M) = \text{ Average number of customers in the } j^{\text{th}} \text{ queue}$$

$$T_j(M) = \text{ Average customer time spent per visit in the } j^{\text{th}} \text{ queue}$$

From these one can obtain the arrival rate at the jth queue given by Little's Theorem as

$$\lambda_j(M) = \frac{N_j(M)}{T_j(M)} \tag{3.158}$$

One possibility is to calculate first the normalizing constant $G(M)$ of Eq. (3.151) and then to use the steady-state distribution $P(n)$ of Eq. (3.152) to obtain all quantities of interest. Several different algorithms can be used for this computation, which is often nontrivial when M is large. We will describe an alternative approach, known as *mean value analysis*, which calculates $N_j(M)$ and $T_j(M)$ directly. The normalizing constant $G(M)$ can then be obtained from these quantities and the arrival rates of Eq. (3.158). [See Problem 3.65 for the case where the service rates $\mu_j(m)$ do not depend on the number of customers m.]

Let us assume for simplicity that the service rate at the jth queue is μ_j and does not depend on the number of customers in the queue. The main idea in mean value analysis is to start with the known quantities

$$T_j(0) = N_j(0) = 0, \qquad j = 1, \ldots, K \tag{3.159}$$

(corresponding to an empty system) and then calculate $T_j(1)$ and $N_j(1)$ (corresponding to one customer in the system), then calculate $T_j(2)$ and $N_j(2)$, and so on until the desired quantities $T_j(M)$ and $N_j(M)$ are obtained. This calculation is based on the equation (to be justified shortly)

$$T_j(s) = \frac{1}{\mu_j}\big(1 + N_j(s-1)\big), \qquad j = 1, \ldots, K, \quad s = 1, \ldots, M \tag{3.160}$$

which obtains $T_j(s)$ from $N_j(s-1)$ for all j. Then $N_j(s)$ is calculated from $T_j(s)$ for all j, using the equation (which is in effect Little's Theorem, as will be seen shortly)

$$N_j(s) = s\frac{\overline{\lambda}_j T_j(s)}{\sum_{i=1}^{K} \overline{\lambda}_i T_i(s)}, \qquad j = 1, \ldots, K, \quad s = 1, \ldots, M \tag{3.161}$$

where $\overline{\lambda}_j$, $j = 1, \ldots, K$ is a positive solution of the system of equations $\lambda_j = \sum_{i=1}^{K} \lambda_i P_{ij}$, $j = 1, \ldots, K$ [cf. Eq. (3.147)].

We proceed to derive Eqs. (3.160) and (3.161). Since we have for all j, $\lambda_j(s) = \alpha(s)\overline{\lambda}_j$ for some scalar $\alpha(s) > 0$, Eq. (3.161) can be written as

$$N_j(s) = s\frac{\lambda_j(s)T_j(s)}{\sum_{i=1}^K \lambda_i(s)T_i(s)}$$

and becomes evident once we observe that we have $\lambda_i(s)T_i(s) = N_i(s)$ for all i (by Little's Theorem) and $s = \sum_{i=1}^K N_i(s)$ [by the definition of $N_i(s)$].

To derive Eq. (3.160), we need an important result known as the *Arrival Theorem.* It states that the occupancy distribution found by a customer upon arrival at the jth queue is the same as the steady-state distribution of the jth queue in a closed network with the arriving customer removed. Thus, in an s-customer closed network, the average number of customers found upon arrival by a customer at the jth queue is equal to $N_j(s-1)$, the average number seen by a random observer in the $(s-1)$-customer closed network. This explains the form of Eq. (3.160).

An intuitive explanation of the Arrival Theorem is given in Problem 3.59. For an analytical justification, assume that the s-customer closed network is in steady-state at time t and let $x(t)$ denote the state at that time. For each state $n = (n_1, \ldots, n_K)$ with $n_i > 0$, we want to calculate

$$\alpha_{ij}(n) = P\{x(t) = n \mid \text{a customer moved from queue } i \text{ to queue } j \text{ just after time } t\} \tag{3.162}$$

Let us denote by $M_{ij}(t)$ the event of a customer move from queue i to queue j just after time t, and let us denote by $M_i(t)$ the event of a customer move from queue i just after time t. Then Eq. (3.162) can be written as

$$\begin{aligned}\alpha_{ij}(n) &= \frac{P\{x(t) = n, M_{ij}(t) \mid M_i(t)\}}{P\{M_{ij}(t) \mid M_i(t)\}} \\ &= \frac{P\{x(t) = n \mid M_i(t)\}P\{M_{ij}(t) \mid x(t) = n,\ M_i(t)\}}{P\{M_{ij}(t) \mid M_i(t)\}} \\ &= \frac{P(n)P_{ij}}{\sum_{\{n'=(n'_1,\ldots,n'_K)\mid n'_i>0\}} P(n')P_{ij}}\end{aligned}$$

and finally, using Eqs. (3.149) to (3.152) for the steady-state probabilities $P(n)$,

$$\alpha_{ij}(n) = \frac{\hat{P}_1(n_1)\cdots\hat{P}_K(n_K)}{\sum_{\{(n'_1,\ldots,n'_K)\mid n'_1+\cdots+n'_K=s,\ n'_i>0\}} \hat{P}_1(n'_1)\cdots\hat{P}_K(n'_K)} \tag{3.163}$$

The numerator and the denominator of this equation contain a common factor ρ_i because $n_i > 0$ in the numerator and $n'_i > 0$ in each term of the denominator. By dividing with ρ_i, and by using the expression (3.150) for $\hat{P}_j(n_j)$, we obtain

$$\alpha_{ij}(n) = \frac{\hat{P}_1(n_1)\cdots\hat{P}_{i-1}(n_{i-1})\hat{P}_i(n_i-1)\hat{P}_{i+1}(n_{i+1})\cdots\hat{P}_K(n_K)}{\sum_{\{(n'_1,\ldots,n'_K)\mid n'_1+\cdots+n'_K=s-1\}} \hat{P}_1(n'_1)\cdots\hat{P}_K(n'_K)}$$

Therefore, $\alpha_{ij}(n)$ is equal to the steady-state probability of state $(n_1, \ldots, n_{i-1}, n_i - 1, n_{i+1}, \ldots, n_K)$ in the $(s-1)$-customer closed network, as stated by the Arrival Theorem.

We note that the Arrival Theorem holds also in some cases where there are multiple classes of customers and where the queues have multiple servers. Mean value analysis can also be used in these cases, with Eq. (3.160) replaced by the appropriate formula.

Finally, a number of approximate methods based on mean value analysis have been proposed. As an example, suppose that an approximate relation of the form

$$N_j(M-1) = f_j\big(N_j(M)\big)$$

is hypothesized; for large M, one reasonable possibility is

$$N_j(M-1) = \frac{M-1}{M} N_j(M)$$

Then Eqs. (3.160) and (3.161) yield the system of nonlinear equations

$$T_j(M) = \frac{1}{\mu_j}\Big(1 + f_j\big(N_j(M)\big)\Big), \qquad j = 1, \ldots, K$$

$$N_j(M) = M \frac{\overline{\lambda}_j T_j(M)}{\sum_{i=1}^{K} \overline{\lambda}_i T_i(M)}, \qquad j = 1, \ldots, K$$

which can be solved by iterative methods to yield approximate values for $T_j(M)$ and $N_j(M)$.

SUMMARY

Queueing models provide qualitative insights on the performance of data networks, and quantitative predictions of average packet delay. An example of the former is the comparison of time-division and statistical multiplexing, while an example of the latter is the delay analysis of reservation systems.

To obtain tractable queueing models for data networks, it is frequently necessary to make simplifying assumptions. A prime example is the Kleinrock independence approximation discussed in Section 3.6. Delay predictions based on this approximation are adequate for many uses. A more accurate alternative is simulation, which, however, can be slow, expensive, and lacking in insight.

Little's Theorem is a simple but extremely useful result since it holds under very general conditions. To proceed beyond this theorem we assumed Poisson arrivals and independent interarrival and service times. This led to the $M/G/1$ system and its extensions in reservation and priority queueing systems. We analyzed a surprisingly large number of important delay models using simple graphical arguments. An alternative analysis was based on the use of Markov chain models and led to the derivation of the occupancy probability distribution of the $M/M/1$, $M/M/m$, and related systems.

Reversibility is an important notion that helps to prove and understand Jackson's Theorem and provides a taste of advanced queueing topics.

NOTES, SOURCES, AND SUGGESTED READING

Section 3.2. Little's Theorem was formalized in [Lit61]. Rigorous proofs under various assumptions are given in [Sti72] and [Sti74]. Several applications in finding performance bounds of computer systems are described in [StA85].

Section 3.3. For a general background on the Poisson process, Markov chains, and related topics, see [Ros80], [Ros83], and [KaT75]. Standard texts on queueing theory include [Coo81], [GrH85], [HeS82], [Kle75], and [Wol89]. A reference for the fact that Poisson arrivals see a typical occupancy distribution (Section 3.3.2) is [Wol82a].

Section 3.4. Queueing systems that admit analysis via Markov chain theory include those where the service times have an Erlang distribution; see [Kle76], Chap. 4. For extensions to more general models and computational methods, see [Kei79], [Neu81], [Haj82], and [Twe82]. For methods to calculate the blocking probability in circuit switching networks (Example 3.14), see [Kau81], [Kel86], [LeG89], [RoT90], and [TsR90].

Section 3.5. The P-K formula is often derived by using z-transforms; see [Kle75]. This derivation is not very insightful, but gives the probability distribution of the system occupancy (not just the mean that we obtained via our much simpler analysis). For more on delay analysis of ARQ systems, see [AnP86] and [ToW79].

The results on polling and reservation systems are fairly recent; see [Coo70], [Eis79], [FeA85], [FuC85], [IEE86], and [Kue79]. The original references that are closest to our analysis are [Has72] for unlimited service systems, [NoT78] for limited service systems, and [Hum78] for nonsymmetric polling systems. Reference [Tak86] is a monograph devoted to polling. There are two main reservation and polling systems considered in the literature: the symmetric case, where all users have identical arrival and reservation interval statistics, and the nonsymmetric case, where these statistics are user dependent. The former case admits simple expressions for the mean waiting times while the latter does not. We have considered the partially symmetric case, where all users have identical arrival statistics but different reservation interval statistics. The fact that simple expressions hold for this case has not been known earlier, and in this respect, our formulas are original. Our treatment in terms of simple graphical arguments is also original. Approximate formulas for nonsymmetric polling systems are given in [BoM86] and [IbC89]. The result of Problem 3.35 on limited service systems with shared reservation and data intervals is new.

An extensive treatment of priority queueing systems is [Jai68]. A simpler, less comprehensive exposition is given in [Kle75].

The material on the $G/G/1$ queue is due to [Kin62]. For further material, see Chapter 11 of [Wol89], [Whi83a], [Whi83b], and the references quoted there.

Section 3.6. Delay analysis for data networks in terms of $M/M/1$ approximations was introduced in [Kle64]. References [Wol82b] and [PiW82] study via analysis and simulation the behavior of two queues in tandem when the service times of a customer at the two queues are dependent. The special issue [IEE86] provides a view of recent work on the subject.

Section 3.7. The notion of reversibility was used in Markov chain analysis by Kolmogorov [Kol36], and was explored in depth in [Kel79] and [Wal88].

Section 3.8. There is an extensive literature on product form solutions of queueing networks following Jackson's original paper [Jac57]. The survey [DiK85] lists 314 references. There are also several books on the subject: [Kel79], [BrB80], [GeP87], [Wal88], and [CoG89]. The heuristic explanation of Jackson's theorem is due to [Wal83].

PROBLEMS

3.1 Customers arrive at a fast-food restaurant at a rate of five per minute and wait to receive their order for an average of 5 minutes. Customers eat in the restaurant with probability 0.5 and carry out their order without eating with probability 0.5. A meal requires an average of 20 minutes. What is the average number of customers in the restaurant? (Answer: 75.)

3.2 Two communication nodes 1 and 2 send files to another node 3. Files from 1 and 2 require on the average R_1 and R_2 time units for transmission, respectively. Node 3 processes a file of node i $(i = 1, 2)$ in an average of P_i time units and then requests another file from either node 1 or node 2 (the rule of choice is left unspecified). If λ_i is the throughput of node i in files sent per unit time, what is the region of all feasible throughput pairs (λ_1, λ_2) for this system?

3.3 A machine shop consists of N machines that occasionally fail and get repaired by one of the shop's m repairpersons. A machine will fail after an average of R time units following its previous repair and requires an average of P time units to get repaired. Obtain upper and lower bounds (functions of R, N, P, and m) on the number of machine failures per unit time and on the average time between repairs of the same machine.

3.4 The average time T a car spends in a certain traffic system is related to the average number of cars N in the system by a relation of the form $T = \alpha + \beta N^2$, where $\alpha > 0$, $\beta > 0$ are given scalars.

(a) What is the maximal car arrival rate λ^* that the system can sustain?

(b) When the car arrival rate is less than λ^*, what is the average time a car spends in the system assuming that the system reaches a statistical steady state? Is there a unique answer? Try to argue against the validity of the statistical steady-state assumption.

3.5 An absent-minded professor schedules two student appointments for the same time. The appointment durations are independent and exponentially distributed with mean 30 minutes. The first student arrives on time, but the second student arrives 5 minutes late. What is the expected time between the arrival of the first student and the departure of the second student? (Answer: 60.394 minutes.)

3.6 A person enters a bank and finds all of the four clerks busy serving customers. There are no other customers in the bank, so the person will start service as soon as one of the customers in service leaves. Customers have independent, identical, exponential distribution of service time.

(a) What is the probability that the person will be the last to leave the bank assuming that no other customers arrive?

(b) If the average service time is 1 minute, what is the average time the person will spend in the bank?

(c) Will the answer in part (a) change if there are some additional customers waiting in a common queue and customers begin service in the order of their arrival?

3.7 A communication line is divided in two identical channels each of which will serve a packet traffic stream where all packets have equal transmission time T and equal interarrival time $R > T$. Consider, alternatively, statistical multiplexing of the two traffic streams by combining the two channels into a single channel with transmission time $T/2$ for each packet. Show that the average system time of a packet will be decreased from T to something between $T/2$ and $3T/4$, while the variance of waiting time in queue will be increased from 0 to as much as $T^2/16$.

3.8 Consider a packet stream whereby packets arrive according to a Poisson process with rate 10 packets/sec. If the interarrival time between any two packets is less than the transmission time of the first to arrive, the two packets are said to collide. (This notion will be made more meaningful in Chapter 4 when we discuss multiaccess schemes.) Find the probabilities that a packet does not collide with either its predecessor or its successor, and that a packet does not collide with another packet assuming:

(a) All packets have a transmission time of 20 msec. (Answer: Both probabilities are equal to 0.67.)

(b) Packets have independent, exponentially distributed transmission times with mean 20 msec. (This part requires the $M/M/\infty$ results.) (Answer: The probability of no collision with predecessor or successor is 0.694. The probability of no collision is 0.682.)

3.9 A communication line capable of transmitting at a rate of 50 Kbits/sec will be used to accommodate 10 sessions each generating Poisson traffic at a rate 150 packets/min. Packet lengths are exponentially distributed with mean 1000 bits.

(a) For each session, find the average number of packets in queue, the average number in the system, and the average delay per packet when the line is allocated to the sessions by using:

(1) 10 equal-capacity time-division multiplexed channels. (Answer: $N_Q = 5$, $N = 10$, $T = 0.4$ sec.)

(2) Statistical multiplexing. (Answer: $N_Q = 0.5$, $N = 1$, $T = 0.04$ sec.)

(b) Repeat part (a) for the case where five of the sessions transmit at a rate of 250 packets/min while the other five transmit at a rate of 50 packets/min. (Answer: $N_Q = 21$, $N = 26$, $T = 1.038$ sec.)

3.10 This problem deals with some of the basic properties of the Poisson process.

(a) Derive Eqs. (3.11) to (3.14).

(b) Show that if the arrivals in two disjoint time intervals are independent and Poisson distributed with parameters $\lambda\tau_1$, $\lambda\tau_2$, then the number of arrivals in the union of the intervals is Poisson distributed with parameter $\lambda(\tau_1 + \tau_2)$. (This shows in particular that the Poisson distribution of the number of arrivals in any interval [cf. Eq. (3.10)] is consistent with the independence requirement in the definition of the Poisson process.) *Hint*: Verify the correctness of the following calculation, where N_1 and N_2 are the number of arrivals in the two disjoint intervals:

$$P\{N_1 + N_2 = n\} = \sum_{k=0}^{n} P\{N_1 = k\}P\{N_2 = n-k\}$$

$$= e^{-\lambda(\tau_1+\tau_2)} \sum_{k=0}^{n} \frac{(\lambda\tau_1)^k(\lambda\tau_2)^{n-k}}{k!(n-k)!}$$

$$= e^{-\lambda(\tau_1+\tau_2)} \frac{(\lambda\tau_1 + \lambda\tau_2)^n}{n!}$$

(c) Show that if k independent Poisson processes $A_1, \ldots, A_k$ are combined into a single process $A = A_1 + A_2 + \cdots + A_k$, then A is Poisson with rate λ equal to the sum of the rates $\lambda_1, \ldots \lambda_k$ of $A_1, \ldots, A_k$. Show also that the probability that the first arrival of the combined process comes from A_1 is λ_1/λ independently of the time of arrival. *Hint*: For $k = 2$ write

$$P\{A_1(t+\tau) + A_2(t+\tau) - A_1(t) - A_2(t) = n\}$$
$$= \sum_{m=0}^{n} P\{A_1(t+\tau) - A_1(t) = m\}P\{A_2(t+\tau) - A_2(t) = n-m\}$$

and continue as in the hint for part (b). Also write for any t

$$P\{1 \text{ arrival from } A_1 \text{ prior to } t \mid 1 \text{ occurred}\}$$
$$= \frac{P\{1 \text{ arrival from } A_1 \text{ prior to } t, 0 \text{ from } A_2\}}{P\{1 \text{ occurred}\}}$$
$$= \frac{\lambda_1 t e^{-\lambda_1 t} e^{-\lambda_2 t}}{\lambda t e^{-\lambda t}} = \frac{\lambda_1}{\lambda}$$

(d) Suppose we know that in an interval $[t_1, t_2]$ only one arrival of a Poisson process has occurred. Show that, conditional on this knowledge, the time of this arrival is uniformly distributed in $[t_1, t_2]$. *Hint*: Verify that if t is the time of arrival, we have for all $s \in [t_1, t_2]$,

$$P\{t < s \mid 1 \text{ arrival occurred in } [t_1, t_2]\}$$
$$= \frac{P\{1 \text{ arrival occurred in } [t_1, s), 0 \text{ arrivals occurred in } [s, t_2]\}}{P\{1 \text{ arrival occurred}\}}$$
$$= \frac{s - t_1}{t_2 - t_1}$$

3.11 Packets arrive at a transmission facility according to a Poisson process with rate λ. Each packet is independently routed with probability p to one of two transmission lines and with probability $(1-p)$ to the other.

(a) Show that the arrival processes at the two transmission lines are Poisson with rates λp and $\lambda(1-p)$, respectively. Furthermore, the two processes are independent. *Hint*: Let $N_1(t)$ and $N_2(t)$ be the number of arrivals in $[0, t]$ in lines 1 and 2, respectively. Verify the correctness of the following calculation:

$$P\{N_1(t) = n, N_2(t) = m\}$$

$$= P\{N_1(t) = n, N_2(t) = m \mid N(t) = n + m\} \frac{e^{-\lambda t}(\lambda t)^{n+m}}{(n+m)!}$$

$$= \binom{n+m}{n} p^n (1-p)^m \frac{e^{-\lambda t}(\lambda t)^{n+m}}{(n+m)!}$$

$$= \frac{e^{-\lambda tp}(\lambda tp)^n}{n!} \frac{e^{-\lambda t(1-p)}(\lambda t(1-p))^m}{m!}$$

$$P\{N_1(t) = n\} = \sum_{m=0}^{\infty} P\{N_1(t) = n, N_2(t) = m\} = \frac{e^{-\lambda tp}(\lambda tp)^n}{n!}$$

(b) Use the result of part (a) to show that the probability distribution of the customer delay in a (first-come first-serve) $M/M/1$ queue with arrival rate λ and service rate μ is exponential, that is, in steady-state we have

$$P\{T_i \geq \tau\} = e^{-(\mu-\lambda)\tau}$$

where T_i is the delay of the ith customer. *Hint*: Consider a Poisson process A with arrival rate μ, which is split into two processes, A_1 and A_2, by randomization according to a probability $\rho = \lambda/\mu$; that is, each arrival of A is an arrival of A_1 with probability ρ and an arrival of A_2 with probability $(1 - \rho)$, independently of other arrivals. Show that the interarrival times of A_2 have the same distribution as T_i.

3.12 Let τ_1 and τ_2 be two exponentially distributed, independent random variables with means $1/\lambda_1$ and $1/\lambda_2$. Show that the random variable $\min\{\tau_1, \tau_2\}$ is exponentially distributed with mean $1/(\lambda_1 + \lambda_2)$ and that $P\{\tau_1 < \tau_2\} = \lambda_1/(\lambda_1 + \lambda_2)$. Use these facts to show that the $M/M/1$ queue can be described by a continuous-time Markov chain with transition rates $q_{n(n+1)} = \lambda$, $q_{(n+1)n} = \mu$, $n = 0, 1, \ldots$. (See Appendix A for material on continuous-time Markov chains.)

3.13 Persons arrive at a taxi stand with room for W taxis according to a Poisson process with rate λ. A person boards a taxi upon arrival if one is available and otherwise waits in a line. Taxis arrive at the stand according to a Poisson process with rate μ. An arriving taxi that finds the stand full departs immediately; otherwise, it picks up a customer if at least one is waiting, or else joins the queue of waiting taxis.

(a) Use an $M/M/1$ queue formulation to obtain the steady-state distribution of the person's queue. What is the steady-state probability distribution of the taxi queue size when $W = 5$ and λ and μ are equal to 1 and 2 per minute, respectively? (Answer: Let p_i = Probability of i taxis waiting. Then $p_0 = 1/32$, $p_1 = 1/32$, $p_2 = 1/16$, $p_3 = 1/8$, $p_4 = 1/4$, $p_5 = 1/2$.)

(b) In the leaky bucket flow control scheme to be discussed in Chapter 6, packets arrive at a network entry point and must wait in a queue to obtain a permit before entering the network. Assume that permits are generated by a Poisson process with given rate and can be stored up to a given maximum number; permits generated while the maximum number of permits is available are discarded. Assume also that packets arrive according to a Poisson process with given rate. Show how to obtain the occupancy distribution of the queue of packets waiting for permits. *Hint*: This is the same system as the one of part (a).

(c) Consider the flow control system of part (b) with the difference that permits are not generated according to a Poisson process but are instead generated periodically at a given rate. (This is a more realistic assumption.) Formulate the problem of finding the occupancy distribution of the packet queue as an $M/D/1$ type of problem.

3.14 A communication node A receives Poisson packet traffic from two other nodes, 1 and 2, at rates λ_1 and λ_2, respectively, and transmits it, on a first-come first-serve basis, using a link with capacity C bits/sec. The two input streams are assumed independent and their packet lengths are identically and exponentially distributed with mean L bits. A packet from node 1 is always accepted by A. A packet from node 2 is accepted only if the number of packets in A (in queue or under transmission) is less than a given number $K > 0$; otherwise, it is assumed lost.

(a) What is the range of values of λ_1 and λ_2 for which the expected number of packets in A will stay bounded as time increases?

(b) For λ_1 and λ_2 in the range of part (a) find the steady-state probability of having n packets in $A\,(0 \leq n < \infty)$. Find the average time needed by a packet from source 1 to clear A once it enters A, and the average number of packets in A from source 1. Repeat for packets from source 2.

3.15 Consider a system that is identical to $M/M/1$ except that when the system empties out, service does not begin again until k customers are present in the system (k is given). Once service begins it proceeds normally until the system becomes empty again. Find the steady-state probabilities of the number in the system, the average number in the system, and the average delay per customer. [Answer: The average number in the system is $N = \rho/(1-\rho) + (k-1)/2$.]

3.16 *M/M/1-Like System with State-Dependent Arrival and Service Rate.* Consider a system which is the same as $M/M/1$ except that the rate λ_n and service rate μ_n when there are n customers in the system depend on n. Show that

$$p_{n+1} = (\rho_0 \cdots \rho_n)p_0$$

where $\rho_k = \lambda_k/\mu_{k+1}$ and

$$p_0 = \left[1 + \sum_{k=0}^{\infty}(\rho_0 \cdots \rho_k)\right]^{-1}$$

3.17 *Discrete-Time Version of the M/M/1 System.* Consider a queueing system where interarrival and service times are integer valued, so customer arrivals and departures occur at integer times. Let λ be the probability that an arrival occurs at any time k, and assume that at most one arrival can occur. Also let μ be the probability that a customer who was in service at time k will complete service at time $k+1$. Find the occupancy distribution p_n in terms of λ and μ.

3.18 Empty taxis pass by a street corner at a Poisson rate of 2 per minute and pick up a passenger if one is waiting there. Passengers arrive at the street corner at a Poisson rate of 1 per minute and wait for a taxi only if there are fewer than four persons waiting; otherwise, they leave and never return. Find the average waiting time of a passenger who joins the queue. (Answer: 13/15 min.)

3.19 A telephone company establishes a direct connection between two cities expecting Poisson traffic with rate 30 calls/min. The durations of calls are independent and exponentially distributed with mean 3 min. Interarrival times are independent of call durations. How many circuits should the company provide to ensure that an attempted call is blocked (because all circuits are busy) with probability less than 0.01? It is assumed that blocked calls are lost (*i.e.*, a blocked call is not attempted again).

3.20 A mail-order company receives calls at a Poisson rate of one per 2 min and the duration of the calls is exponentially distributed with mean 3 min. A caller who finds all telephone operators busy patiently waits until one becomes available. Write a computer program to determine how many operators the company should use so that the average waiting time of a customer is half a minute or less?

3.21 Consider the $M/M/1/m$ system which is the same as $M/M/1$ except that there can be no more than m customers in the system and customers arriving when the system is full are lost. Show that the steady-state occupancy probabilities are given by

$$p_n = \frac{\rho^n(1-\rho)}{1-\rho^{m+1}}, \qquad 0 \le n \le m$$

3.22 An athletic facility has five tennis courts. Players arrive at the courts at a Poisson rate of one pair per 10 min and use a court for an exponentially distributed time with mean 40 min.

(a) Suppose that a pair of players arrives and finds all courts busy and k other pairs waiting in queue. How long will they have to wait to get a court on the average?

(b) What is the average waiting time in queue for players who find all courts busy on arrival?

3.23 Consider an $M/M/\infty$ queue with servers numbered $1, 2, \ldots$ There is an additional restriction that upon arrival a customer will choose the lowest-numbered server that is idle at the time. Find the fraction of time that each server is busy. Will the answer change if the number of servers is finite? *Hint*: Argue that in steady-state the probability that all of the first m servers are busy is given by the Erlang B formula of the $M/M/m/m$ system. Find the total arrival rate to servers $(m+1)$ and higher, and from this, the arrival rate to each server.

3.24 $M/M/1$ *Shared Service System.* Consider a system which is the same as $M/M/1$ except that whenever there are n customers in the system they are all served simultaneously at an equal rate $1/n$ per unit time. Argue that the steady-state occupancy distribution is the same as for the $M/M/1$ system. *Note*: It can be shown that the steady-state occupancy distribution is the same as for $M/M/1$ even if the service time distribution is not exponential (*i.e.*, for an $M/G/1$ type of system) ([Ros83], p. 171).

3.25 *Blocking Probability for Single-Cell Radio Systems ([BaA81] and [BaA82]).* A cellular radiotelephone system serves a given geographical area with m radiotelephone channels connected to a single switching center. There are two types of calls: radio-to-radio calls, which occur with a Poisson rate λ_1 and require two radiochannels per call, and radio-to-nonradio calls, which occur with a Poisson rate λ_2 and require one radiochannel per call. The duration of all calls is exponentially distributed with mean $1/\mu$. Calls that cannot be accommodated by the system are blocked. Give formulas for the blocking probability of the two types of calls.

3.26 A facility of m identical machines is sharing a single repairperson. The time to repair a failed machine is exponentially distributed with mean $1/\lambda$. A machine, once operational, fails after a time that is exponentially distributed with mean $1/\mu$. All failure and repair times are independent. What is the steady-state proportion of time where there is no operational machine?

3.27 $M/M/2$ *System with Heterogeneous Servers.* Derive the stationary distribution of an $M/M/2$ system where the two servers have different service rates. A customer that arrives when the system is empty is routed to the faster server.

3.28 In Example 3.11, verify the formula $\sigma_f = (\lambda/\mu)^{1/2}s_\gamma$. *Hint*: Write

$$E\{f^2\} = E\left\{\left(\sum_{i=1}^{n}\gamma_i\right)^2\right\} = E\left\{E\left\{\left(\sum_{i=1}^{n}\gamma_i\right)^2 \mid n\right\}\right\},$$

and use the fact that n is Poisson distributed.

3.29 Customers arrive at a grocery store's checkout counter according to a Poisson process with rate 1 per minute. Each customer carries a number of items that is uniformly distibuted between 1 and 40. The store has two checkout counters, each capable of processing items at a rate of 15 per minute. To reduce the customer waiting time in queue, the store manager considers dedicating one of the two counters to customers with x items or less and dedicating the other counter to customers with more than x items. Write a small computer program to find the value of x that minimizes the average customer waiting time.

3.30 In the $M/G/1$ system, show that

$$P\{\text{the system is empty}\} = 1 - \lambda\overline{X}$$

$$\text{Average length of time between busy periods} = \frac{1}{\lambda}$$

$$\text{Average length of busy period} = \frac{\overline{X}}{1 - \lambda\overline{X}}$$

$$\text{Average number of customers served in a busy period} = \frac{1}{1 - \lambda\overline{X}}$$

3.31 Consider the following argument in the $M/G/1$ system: When a customer arrives, the probability that another customer is being served is $\lambda\overline{X}$. Since the served customer has mean service time $\overline{X}$, the average time to complete the service is $\overline{X}/2$. Therefore, the mean residual service time is $\lambda\overline{X}^2/2$. What is wrong with this argument?

3.32 *$M/G/1$ System with Arbitrary Order of Service.* Consider the $M/G/1$ system with the difference that customers are not served in the order they arrive. Instead, upon completion of a customer's service, one of the waiting customers in queue is chosen according to some rule, and is served next. Show that the P-K formula for the average waiting time in queue W remains valid provided that the relative order of arrival of the customer chosen is independent of the service times of the customers waiting in queue. *Hint*: Argue that the independence hypothesis above implies that at any time t, the number $N_Q(t)$ of customers waiting in queue is independent of the service times of these customers. Show that this in turn implies that $U = R + \rho W$, where R is the mean residual time and U is the average steady-state unfinished work in the system (total remaining service time of the customers in the system). Argue that U and R are independent of the order of customer service.

3.33 Show that Eq. (3.59) for the average delay of time-division multiplexing on a slot basis can be obtained as a special case of the results for the limited service reservation system. *Hint*: Consider the gated system with zero packet length.

3.34 Consider the limited service reservation system. Show that for both the gated and the partially gated versions:

(a) The steady-state probability of arrival of a packet during a reservation interval is $1 - \rho$.

(b) The steady-state probability of a reservation interval being followed by an empty data interval is $(1 - \rho - \lambda\overline{V})/(1 - \rho)$. *Hint:* If p is the required probability, argue that the ratio of the times used for data intervals and for reservation intervals is $(1 - p)\overline{X}/\overline{V}$.

3.35 *Limited Service Reservation System with Shared Reservation and Data Intervals.* Consider the gated version of the limited service reservation system with the difference that the m users share reservation and data intervals, (*i.e.*, all users make reservations in the same interval and transmit at most one packet each in the subsequent data interval). Show that

$$W = \frac{\lambda \overline{X^2}}{2(1-\rho-\lambda\overline{V}/m)} + \frac{(1-\rho)\overline{V^2}}{2(1-\rho-\lambda\overline{V}/m)\overline{V}} + \frac{(1-\rho\alpha-\lambda\overline{V}/m)\overline{V}}{1-\rho-\lambda\overline{V}/m}$$

where $\overline{V}$ and $\overline{V^2}$ are the first two moments of the reservation interval, and α satisfies

$$\frac{\overline{K}+(\hat{K}-1)(2\overline{K}-\hat{K})}{2m\overline{K}} - \frac{1}{2m} \leq \alpha \leq \frac{1}{2} - \frac{1}{2m}$$

where

$$\overline{K} = \frac{\lambda\overline{V}}{1-\rho}$$

is the average number of packets per data interval, and $\hat{K}$ is the smallest integer which is larger than $\overline{K}$. Verify that the formula for W becomes exact as $\rho \to 0$ (light load) and as $\rho \to 1 - \lambda\overline{V}/m$ (heavy load). *Hint:* Verify that

$$W = R + \lambda W + \left(1 + \frac{\lambda W}{m} - S\right)\overline{V}$$

where $S = \lim_{i\to\infty} E\{S_i\}$ and S_i is the number (0 or 1) of packets of the owner of packet i that will start transmission between the time of arrival of packet i and the end of the cycle in which packet i arrives. Try to obtain bounds for S by considering separately the cases where packet i arrives in a reservation and in a data interval.

3.36 Repeat part (a) of Problem 3.9 for the case where packet lengths are not exponentially distributed, but 10% of the packets are 100 bits long and the rest are 1500 bits long. Repeat the problem for the case where the short packets are given nonpreemptive priority over the long packets. (Answer: $N_Q = 0.791$, $N = 1.47$, $T = 0.588$ sec.)

3.37 Persons arrive at a Xerox machine according to a Poisson process with rate one per minute. The number of copies to be made by each person is uniformly distributed between 1 and 10. Each copy requires 3 sec. Find the average waiting time in queue when:

(a) Each person uses the machine on a first-come first-serve basis. (Answer: $W = 3.98$.)

(b) Persons with no more than 2 copies to make are given nonpreemptive priority over other persons.

3.38 *Priority Systems with Multiple Servers.* Consider the priority systems of Section 3.5.3 assuming that there are m servers and that all priority classes have exponentially distributed service times with common mean $1/\mu$.

(a) Consider the nonpreemptive system. Show that Eq. (3.79) yields the average queueing times with the mean residual time R given by

$$R = \frac{P_Q}{m\mu}$$

where P_Q is the steady-state probability of queueing given by the Erlang C formula of Eq. (3.36). [Here $\rho_i = \lambda_i/(m\mu)$ and $\rho = \sum_{i=1}^{n} \rho_i$.]

(b) Consider the preemptive resume system. Argue that $W_{(k)}$, defined as the average time in queue averaged over the first k priority classes, is the same as for an $M/M/m$ system with arrival rate $\lambda_1 + \cdots + \lambda_k$ and mean service time $1/\mu$. Use Little's Theorem to show that the average time in queue of a kth priority class customer can be obtained recursively from

$$W_1 = W_{(1)}$$

$$W_k = \frac{1}{\lambda_k}\left[W_{(k)}\sum_{i=1}^{k}\lambda_i - W_{(k-1)}\sum_{i=1}^{k-1}\lambda_i\right], \qquad k = 2, 3, \ldots, n$$

3.39 Consider the nonpreemptive priority queueing system of Section 3.5.3 for the case where the available capacity is sufficient to handle the highest-priority traffic but cannot handle the traffic of all priorities, that is,

$$\rho_1 < 1 < \rho_1 + \rho_2 + \cdots + \rho_n$$

Find the average delay per customer of each priority class. *Hint*: Determine the departure rate of the highest-priority class that will experience infinite average delay and the mean residual service time.

3.40 *Optimization of Class Ordering in a Nonpreemptive System.* Consider an n-class, nonpreemptive priority system:

(a) Show that the sum $\sum_{k=1}^{n} \rho_k W_k$ is independent of the priority order of classes, and in fact

$$\sum_{k=1}^{n} \rho_k W_k = \frac{R\rho}{1-\rho}$$

where $\rho = \rho_1 + \rho_2 + \cdots + \rho_n$. (This is known as the $M/G/1$ conservation law [Kle64].) *Hint:* Use Eq. (3.79). Alternatively, argue that $U = R + \sum_{k=1}^{n} \rho_k W_k$, where U is the average steady-state unfinished work in the system (total remaining service time of customers in the system), and U and R are independent of the priority order of the classes.

(b) Suppose that there is a cost c_k per unit time for each class k customer that waits in queue. Show that cost is minimized when classes are ordered so that

$$\frac{\overline{X}_1}{c_1} \le \frac{\overline{X}_2}{c_2} \le \cdots \le \frac{\overline{X}_n}{c_n}$$

Hint: Express the cost as $\sum_{k=1}^{n}(c_k/\overline{X}_k)(\rho_k W_k)$ and use part (a). Also use the fact that interchanging the order of any two adjacent classes leaves the waiting time of all other classes unchanged.

3.41 *Little's Theorem for Arbitrary Order of Service; Analytical Proof [Sti74].* Consider the analysis of Little's Theorem in Section 3.2 and the notation introduced there. We allow the possibility that the initial number in the system is positive [*i.e.*, $N(0) > 0$]. Assume that the time-average arrival and departure rates exist and are equal:

$$\lambda = \lim_{t\to\infty} \frac{\alpha(t)}{t} = \lim_{t\to\infty} \frac{\beta(t)}{t}$$

and that the following limit defining the time-average system time exists:

$$T = \lim_{k\to\infty} \frac{1}{N(0)+\alpha(t)}\left(\sum_{i\in D(t)} T_i + \sum_{i\in \overline{D}(t)} (t - t_i)\right)$$

where $D(t)$ is the set of customers departed by time t and $\overline{D}(t)$ is the set of customers that are in the system at time t. (For all customers that are initially in the system, the time T_i is counted starting at time 0.) Show that regardless of the order in which customers are served, Little's Theorem ($N = \lambda T$) holds with

$$N = \lim_{t\to\infty} \frac{1}{t} \int_0^t N(\tau)\, d\tau$$

Show also that

$$T = \lim_{k\to\infty} \frac{1}{k} \sum_{i=1}^{k} T_i$$

Hint: Take $t \to \infty$ below:

$$\frac{1}{t} \sum_{i\in D(t)} T_i \le \frac{1}{t} \int_0^t N(\tau)\, d\tau \le \frac{1}{t} \sum_{i\in D(t)\cup \overline{D}(t)} T_i$$

3.42 *A Generalization of Little's Theorem.* Consider an arrival–departure system with arrival rate λ, where entering customers are forced to pay money to the system according to some rule.

(a) Argue that the following identity holds:

Average rate at which the system earns $= \lambda \cdot$ (Average amount a customer pays)

(b) Show that Little's Theorem is a special case.

(c) Consider the $M/G/1$ system and the following cost rule: Each customer pays at a rate of y per unit time when its remaining service time is y, whether in queue or in service. Show that the formula in (a) can be written as

$$W = \lambda \left(\overline{X} W + \frac{\overline{X^2}}{2} \right)$$

which is the Pollaczek–Khinchin formula.

3.43 *$M/G/1$ Queue with Random-Sized Batch Arrivals.* Consider the $M/G/1$ system with the difference that customers are arriving in batches according to a Poisson process with rate λ. Each batch has n customers, where n has a given distribution and is independent of customer service times. Adapt the proof of Section 3.5 to show that the waiting time in queue is given by

$$W = \frac{\lambda \overline{n}\, \overline{X^2}}{2(1-\rho)} + \frac{\overline{X}(\overline{n^2} - \overline{n})}{2\overline{n}(1-\rho)}$$

Hint: Use the equation $W = R + \rho W + W_B$, where W_B is the average waiting time of a customer for other customers who arrived in the same batch.

3.44 *$M/G/1$ Queue with Overhead for Each Busy Period.* Consider the $M/G/1$ queue with the difference that the service of the first customer in each busy period requires an increment Δ over the ordinary service time of the customer. We assume that Δ has a given distribution and is independent of all other random variables in the model. Let $\rho = \lambda \overline{X}$ be the utilization factor. Show that

(a) $p_0 = P\{\text{the system is empty}\} = (1-\rho)/(1+\lambda\overline{\Delta})$.

(b) Average length of busy period $= (\overline{X} + \overline{\Delta})/(1-\rho)$.

(c) The average waiting time in queue is

$$W = \frac{\lambda\overline{X^2}}{2(1-\rho)} + \frac{\lambda[\overline{(X+\Delta)^2} - \overline{X^2}]}{2(1+\lambda\overline{\Delta})}$$

(d) Parts (a), (b), and (c) also hold in the case where Δ may depend on the interarrival and service time of the first customer in the corresponding busy period.

3.45 Consider a system that is identical to $M/G/1$ except that when the system empties out, service does not begin again until k customers are present in the system (k is given). Once service begins, it proceeds normally until the system becomes empty again. Show that:

(a) In steady-state:

$$P\{\text{system empty}\} = \frac{1-\rho}{k}$$

$$P\{\text{system nonempty and waiting}\} = \frac{(k-1)(1-\rho)}{k}$$

$$P\{\text{system nonempty and serving}\} = \rho$$

(b) The average length of a busy period is

$$\frac{\rho + k - 1}{\lambda(1-\rho)}$$

Verify that this average length is equal to the average time between arrival and start of service of the first customer in a busy period, plus k times the average length of a busy period for the corresponding $M/G/1$ system ($k = 1$).

(c) Suppose that we divide a busy period into a busy/waiting portion and a busy/serving portion. Show that the average number in the system during a busy/waiting portion is $k/2$ and the average number in the system during a busy/serving portion is

$$\frac{N_{M/G/1}}{\rho} + \frac{k-1}{2}$$

where $N_{M/G/1}$ is the average number in the system for the corresponding $M/G/1$ system ($k = 1$). *Hint*: Relate a busy/serving portion of a busy period with k independent busy periods of the corresponding $M/G/1$ system where $k = 1$.

(d) The average number in the system is

$$N_{M/G/1} + \frac{k-1}{2}$$

3.46 *Single-Vacation $M/G/1$ System.* Consider the $M/G/1$ system with the difference that each busy period is followed by a single vacation interval. Once this vacation is over, an arriving customer to an empty system starts service immediately. Assume that vacation intervals are independent, identically distributed, and independent of the customer interarrival and service times. Prove that the average waiting time in queue is

$$W = \frac{\lambda\overline{X^2}}{2(1-\rho)} + \frac{\overline{V^2}}{2I}$$

where I is the average length of an idle period, and show how to calculate I.

3.47 *The $M/G/\infty$ System.* Consider a queueing system with Poisson arrivals at rate λ. There are an infinite number of servers, so that each arrival starts service at an idle server immediately on arrival. Each server has a general service time distribution and $F_X(x) = P\{X \le x\}$

denotes the probability that a service starting at any given time τ is completed by time $\tau + x$ [$F_X(x) = 0$ for $x \leq 0$]. The servers have independent and identical service time distributions.

(a) For x and δ ($0 < \delta < x$) very small, find the probability that there was an arrival in the interval $[\tau - x, \ \tau - x + \delta]$ *and* that this arrival is still being served at time τ.

(b) Show that the mean service time for any arrival is given by

$$\overline{X} = \int_0^\infty [1 - F_X(x)]\, dx$$

Hint: Use a graphical argument.

(c) Use parts (a) and (b) to verify that the number in the system is Poisson distributed with mean $\lambda\overline{X}$.

3.48 *An Improved Bound for the* $G/G/1$ *Queue.*

(a) Let r be a nonnegative random variable and let x be a nonnegative scalar. Show that

$$\frac{\overline{\left(\max\{0, r - x\}\right)^2}}{\left(\overline{\max\{0, r - x\}}\right)^2} \geq \frac{\overline{r^2}}{(\overline{r})^2}$$

where overbar denotes expected value. *Hint*: Prove that the left-hand expression is monotonically nondecreasing as a function of x.

(b) Using the notation of Section 3.5.4, show that

$$\sigma_I^2 \geq (1 - \rho)^2 \sigma_a^2$$

and that

$$W \leq \frac{\lambda(\sigma_a^2 + \sigma_b^2)}{2(1 - \rho)} - \frac{\lambda(1 - \rho)\sigma_a^2}{2}$$

Hint: Use part (a) with r being the customer interarrival time and x equal to the time in the system [cf. Eq. (3.93)].

3.49 *Last-Come First-Serve* $M/G/1$ *System.* Consider an $M/G/1$ system with the difference that upon arrival at the queue, a customer goes immediately into service, replacing the customer who is in service at the time (if any) on a preemptive-resume basis. When a customer completes service, the customer most recently preempted resumes service. Show that:

(a) The expected length of a busy period, denoted $E\{B\}$, is the same as in the ordinary $M/G/1$ queue.

(b) Show that the expected time in the system of a customer is equal to $E\{B\}$. *Hint*: Argue that a customer who starts a busy period stays in the system for the entire duration of the busy period.

(c) Let C be the average time in the system of a customer requiring one unit of service time. Argue that the average time in the system of a customer requiring X units of service time is XC. *Hint*: Argue that a customer requiring two units of service time is "equivalent" to two customers with one unit service time each, and with the second customer arriving at the time that the first departs.

(d) Show that

$$C = \frac{E\{B\}}{E\{X\}} = \frac{1}{1 - \rho}$$

3.50 *Truncation of Queues.* This problem illustrates one way to use simple queues to obtain results about more complicated queues.

(a) Consider a continuous-time Markov chain with state space S, stationary distribution $\{p_j\}$, and transition rates q_{ij}. Suppose that we have a truncated version of this chain, that is, a new chain with space $\overline{S}$, which is a subset of S and has the same transition rates q_{ij} between states i and j of $\overline{S}$. Assume that for all $j \in \overline{S}$, we have

$$p_j \sum_{i \notin \overline{S}} q_{ji} = \sum_{i \notin \overline{S}} p_i q_{ij}$$

Show that if the truncated chain is irreducible, then its stationary distribution $\{\overline{p}_j\}$ satisfies $\overline{p}_j = p_j / \sum_{i \in \overline{S}} p_i$ for all $j \in \overline{S}$. (Note that $\overline{p}_j$ is the conditional probability for the state of the original chain to be j conditioned on the fact that it lies within $\overline{S}$.)

(b) Show that the condition of part (a) on the stationary distribution $\{p_j\}$ and the transition rates $\{q_{ij}\}$ is satisfied if the original chain is time reversible, and that in this case, the truncated chain is also time reversible.

(c) Consider two queues with independent Poisson arrivals and independent exponentially distributed service times. The arrival and service rates are denoted λ_i, μ_i, for $i = 1, 2$, respectively. The two queues share a waiting room with finite capacity B (including customers in service). Arriving customers that find the waiting room full are lost. Use part (b) to show that the system is reversible and that for $m + n \leq B$, the steady-state probabilities are

$$P\{m \text{ in queue } 1,\ n \text{ in queue } 2\} = \frac{\rho_1^m \rho_2^n}{G}$$

where $\rho_i = \lambda_i / \mu_i$, $i = 1, 2$, and G is a normalizing constant.

3.51 *Decomposition/Aggregation of Reversible Chains.* Consider a time reversible continuous-time Markov chain in equilibrium, with state space S, transition rates q_{ij}, and stationary probabilities p_j. Let $S = \cup_{k=1}^{K} S_k$ be a partition of S in mutually disjoint sets, and denote for all k and $j \in S_k$:

u_k = Probability of the state being in S_k (*i.e.*, $u_k = \sum_{j \in S_k} p_j$)

π_j = Probability of the state being equal to j conditioned on the fact that the state belongs to S_k (*i.e.*, $\pi_j = P\{X_n = j \mid X_n \in S_k\} = p_j / u_k$)

Assume that all states in S_k communicate with all other states in S_k.

(a) Show that $\{\pi_j \mid j \in S_k\}$ is the stationary distribution of the truncated chain with state space S_k (cf. Problem 3.50).

(b) Show that $\{u_k \mid k = 1, \ldots, K\}$ is the stationary distribution of the so-called *aggregate chain*, which is the Markov chain with states $k = 1, \ldots, K$ and transition rates

$$\tilde{q}_{km} = \sum_{j \in S_k,\ i \in S_m} \pi_j q_{ji}, \qquad k, m = 1, \ldots, K$$

Show also that the aggregate chain is reversible. (Note that the aggregate chain corresponds to a fictitious process; the actual process, corresponding to transitions between sets of states, need not be Markov.)

(c) Outline a divide-and-conquer solution method that first solves for the distributions of the truncated chains and then solves for the distribution of the aggregate chain. Apply this method to Examples 3.12 and 3.13.

(d) Suppose that the truncated chains are reversible but the original chain is not. Show that the results of parts (a) and (b) hold except that the aggregate chain need not be reversible.

3.52 *An Extension of Burke's Theorem.* Consider an $M/M/1$ system in steady state where customers are served in the order that they arrive. Show that given that a customer departs at time t, the arrival time of that customer is independent of the departure process prior to t. *Hint*: Consider a customer arriving at time t_1 and departing at time t_2. In reversed system terms, the arrival process is independent Poisson, so the arrival process to the left of t_2 is independent of the times spent in the system of customers that arrived at or to the right of t_2.

3.53 Consider the model of two queues in tandem of Section 3.7 and assume that customers are served at each queue in the order they arrive.

(a) Show that the times (including service) spent by a customer in queue 1 and in queue 2 are mutually independent, and independent of the departure process from queue 2 prior to the customer's departure from the system. *Hint:* By Burke's Theorem, the time spent by a customer in queue 1 is independent of the sequence of arrival times at queue 2 prior to the customer's arrival at queue 2. These arrival times (together with the corresponding independent service times) determine the time the customer spends at queue 2 as well as the departure process from queue 2 prior to the customer's departure from the system.

(b) Argue by example that the times a customer spends waiting *before entering service* at the two queues are *not* independent.

3.54 Use reversibility to characterize the departure process of the $M/M/1/m$ queue.

3.55 Consider the feedback model of a CPU and I/O device of Example 3.19 with the difference that the CPU consists of m identical parallel processors. The service time of a job at each parallel processor is exponentially distributed with mean $1/\mu_1$. Derive the stationary distribution of the system.

3.56 Consider the discrete-time approximation to the $M/M/1$ queue of Fig. 3.6. Let X_n be the state of the system at time $n\Delta$ and let D_n be a random variable taking on the value 1 if a departure occurs between $n\Delta$ and $(n+1)\Delta$, and the value 0 if no departure occurs. Assume that the system is in steady-state at time $n\Delta$. Answer the following without using reversibility.

(a) Find $P\{X_n = i, D_n = j\}$ for $i \geq 0$, $j = 0, 1$.

(b) Find $P\{D_n = 1\}$.

(c) Find $P\{X_n = i, D_n = 1\}$ for $i \geq 0$.

(d) Find $P\{X_{n+1} = i, D_n = 1\}$ and show that X_{n+1} is statistically independent of D_n. *Hint*: Use part (c); also show that $P\{X_{n+1} = i\} = P\{X_{n+1} = i \mid D_n = 1\}$ for all $i \geq 0$ is sufficient to show independence.

(e) Find $P\{X_{n+k} = i, D_{n+1} = j \mid D_n\}$ and show that the pair of variables (X_{n+1}, D_{n+1}) is statistically independent of D_n.

(f) For each $k > 1$, find $P\{X_{n+k} = i, D_{n+k} = j \mid D_{n+k-1}, D_{n+k-2}, \ldots, D_n\}$ and show that the pair (X_{n+k}, D_{n+k}) is statistically independent of $(D_{n+k-1}, D_{n+k-2}, \ldots, D_n)$. *Hint*: Use induction on k.

(g) Deduce a discrete-time analog to Burke's Theorem.

3.57 Consider the network in Fig. 3.39. There are four sessions: ACE, ADE, BCEF, and BDEF sending Poisson traffic at rates 100, 200, 500, and 600 packets/min, respectively. Packet lengths are exponentially distributed with mean 1000 bits. All transmission lines have capacity 50 kbits/sec, and there is a propagation delay of 2 msec on each line. Using the Kleinrock independence approximation, find the average number of packets in the system, the average delay per packet (regardless of session), and the average delay per packet of each session.

3.58 *Jackson Networks with a Limit on the Total Number of Customers.* Consider an open Jackson network as described in the beginning of Section 3.8, with the difference that all customers who arrive when there are a total of M customers in the network are blocked from entering and are lost for the system. Derive the stationary distribution. *Hint*: Convert the system into a closed network with M customers by introducing an additional queue $K+1$ with service rate equal to $\sum_{j=1}^{K} r_j$. A customer exiting queue $i \in \{1, \ldots, K\}$ enters queue $K+1$ with probability $1 - \sum_j P_{ij}$, and a customer exiting queue $K+1$ enters queue $i \in \{1, \ldots, K\}$ with probability $r_i / \sum_{j=1}^{K} r_j$.

3.59 Justify the Arrival Theorem for closed networks by inserting a very fast $M/M/1$ queue between every pair of queues. Argue that conditioning on a customer moving from one queue to another is essentially equivalent to conditioning on a single customer being in the fast $M/M/1$ queue that lies between the two queues.

3.60 Consider a closed Jackson network where the service time at each queue is independent of the number of customers at the queue. Suppose that for a given number of customers, the utilization factor of one of the queues, say queue 1, is strictly larger than the utilization factors of the other queues. Show that as the number of customers increases, the proportion of time that a customer spends in queue 1 approaches unity.

3.61 Consider a model of a computer CPU connected to m I/O devices as shown in Fig. 3.40. Jobs enter the system according to a Poisson process with rate λ, use the CPU and with

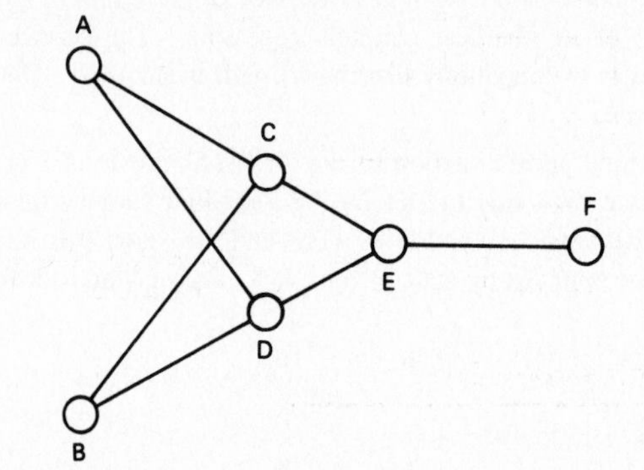

Figure 3.39 Network of transmission lines for Problem 3.57.

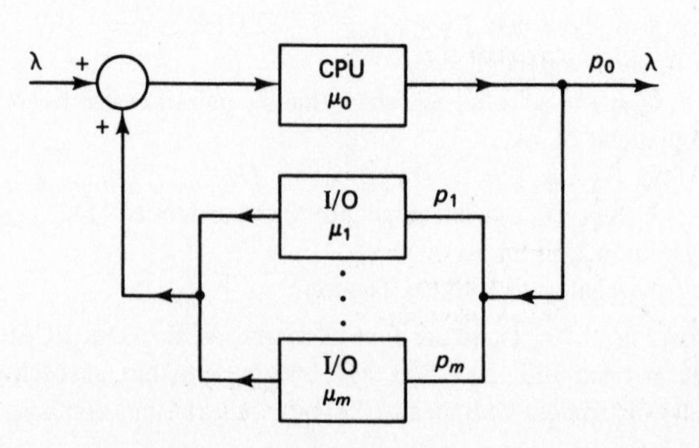

Figure 3.40 Model of a computer CPU connected to m I/O devices for Problem 3.61.

probability p_i, $i = 1, \ldots, m$, are routed to the ith I/O device, while with probability p_0 they exit the system. The service time of a job at the CPU (or the ith I/O device) is exponentially distributed with mean $1/\mu_0$ (or $1/\mu_i$, respectively). We assume that all job service times at all queues are independent (including the times of successive visits to the CPU and I/O devices of the same job). Find the occupancy distribution of the system and construct an "equivalent" system with $m + 1$ queues in tandem that has the same occupancy distribution.

3.62 Consider a closed version of the queueing system of Problem 3.61, shown in Fig. 3.41. There are M jobs in the system at all times. A job uses the CPU and with probability p_i, $i = 1, \ldots, m$, is routed to the ith I/O device. The service time of a job at the CPU (or the ith I/O device) is exponentially distributed with mean $1/\mu_0$ (or $1/\mu_i$, respectively). We assume that all job service times at all queues are independent (including the times of successive visits to the CPU and I/O devices of the same job). Find the arrival rate of jobs at the CPU and the occupancy distribution of the system.

3.63 *Bounds on the Throughput of a Closed Queueing Network.* Packets enter the network of transmission lines shown in Fig. 3.42 at point A and exit at point B. A packet is first transmitted on one of the lines $L_1, \ldots, L_K$, where it requires on the average a transmission time $\overline{X}$, and is then transmitted in line L_{K+1}, where it requires on the average a transmission time $\overline{Y}$. To effect flow control, a maximum of $N \geq K$ packets are admitted into the system.

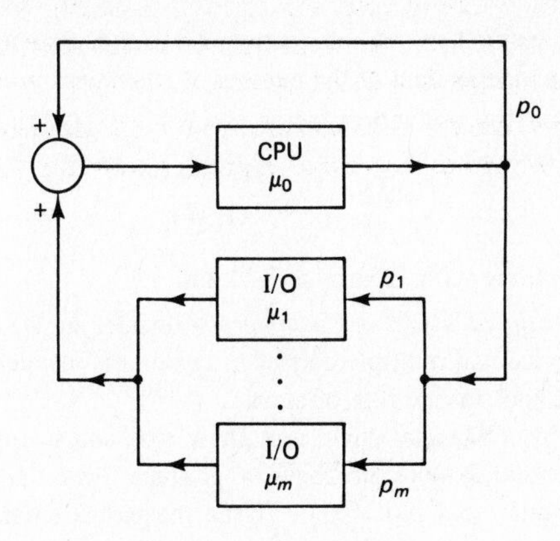

Figure 3.41 Closed queueing system for Problem 3.62.

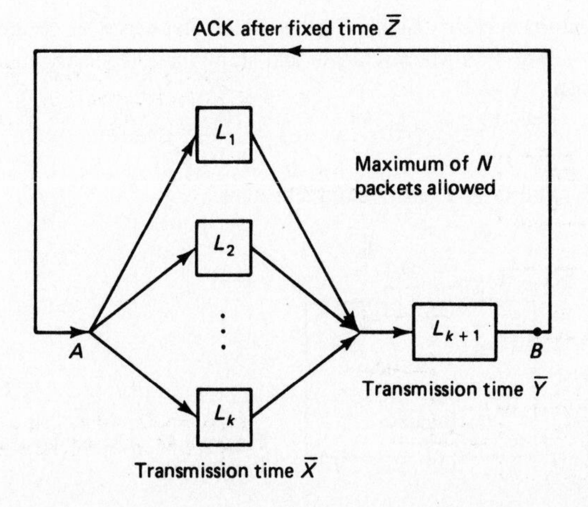

Figure 3.42 Closed queueing network for Problem 3.63.

Each time a packet exits the system at point B, an acknowledgment is sent back and reaches point A after a fixed time $\overline{Z}$. At that time, a new packet is allowed to enter the system. Use Little's Theorem to find upper and lower bounds for the system throughput under two circumstances:

(a) The method of routing a packet to one of the lines $L_1, \ldots, L_K$ is unspecified.

(b) The routing method is such that whenever one of the lines $L_1, \ldots, L_K$ is idle, there is no packet waiting at any of the other lines.

3.64 Consider the closed queueing network in Fig. 3.43. There are three customers who are doomed forever to cycle between queue 1 and queue 2. The service times at the queues are independent and exponentially distributed with mean μ_1 and μ_2. Assume that $\mu_2 < \mu_1$.

(a) The system can be represented by a four-state Markov chain. Find the transition rates of the chain.

(b) Find the steady-state probabilities of the states.

(c) Find the customer arrival rate at queue 1.

(d) Find the rate at which a customer cycles through the system.

(e) Show that the Markov chain is reversible. What does a departure from queue 1 in the forward process correspond to in the reversed process? Can the transitions of a single customer in the forward proces be associated with transitions of a single customer in the reverse process?

3.65 Consider the closed queueing network of Section 3.8.2 and assume that the service rate $\mu_j(m)$ at the jth queue is independent of the number of customers m in the queue $[\mu_j(m) = \mu_j$ for all $m]$. Show that the utilization factor $U_j(M) = \lambda_j(M)/\mu_j$ of the jth queue is given by

$$U_j(M) = \rho_j \frac{G(M-1)}{G(M)}$$

where $\rho_j = \overline{\lambda}_j/\mu_j$ (compare with Examples 3.21 and 3.22).

3.66 *$M/M/1$ System with Multiple Classes of Customers.* Consider an $M/M/1$-like system with first-come first-serve service and multiple classes of customers denoted $c = 1, 2, \ldots, C$. Let λ_i and μ_i be the arrival and service rate of class i.

(a) Model this system by a Markov chain and show that unless $\mu_1 = \mu_2 = \cdots = \mu_C$, its steady-state distribution does not have a product form. *Hint:* Consider a state $z = (c_1, c_2, \ldots, c_n)$ such that $\mu_{c_1} \neq \mu_{c_n}$. Write the global balance equations for state z.

(b) Suppose instead that the service discipline is last-come first-serve (as defined in Problem 3.49). Model the system by a Markov chain and show that the steady-state distribution has the product form

$$P(z) = P(c_1, c_2, \ldots, c_n) = \frac{\rho_{c_1}\rho_{c_2}\cdots\rho_{c_n}}{G}$$

where $\rho_c = \lambda_c/\mu_c$ and G is a normalizing constant.

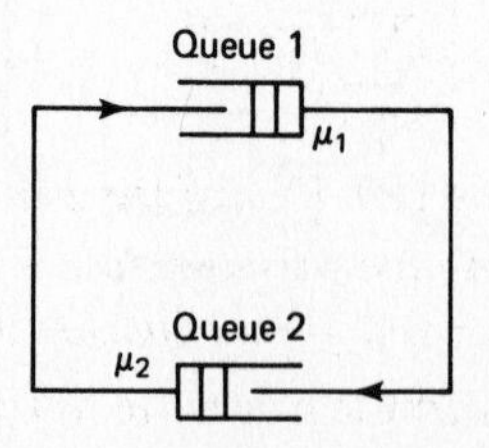

Figure 3.43 Closed queueing network for Problem 3.64.

APPENDIX A: REVIEW OF MARKOV CHAIN THEORY

The purpose of this appendix is to provide a brief summary of the results we need from discrete- and continuous-time Markov chain theory. We refer the reader to books on stochastic processes for detailed accounts.

3A.1 Discrete-Time Markov Chains

Consider a discrete-time stochastic process $\{X_n \mid n = 0, 1, 2, \ldots\}$ that takes values from the set of nonnegative integers, so the states that the process can be in are $i = 0, 1 \ldots$. The process is said to be a *Markov chain* if whenever it is in state i, there is a fixed probability P_{ij} that it will next be in state j regardless of the process history prior to arriving at i. That is, for all $n > 0$, $i_{n-1}, \ldots, i_0, i, j$,

$$\begin{aligned} P_{ij} &= P\{X_{n+1} = j \mid X_n = i,\ X_{n-1} = i_{n-1}, \ldots, X_0 = i_0\} \\ &= P\{X_{n+1} = j \mid X_n = i\} \end{aligned}$$

We refer to P_{ij} as the *transition probabilities*. They must satisfy

$$P_{ij} \geq 0, \ \sum_{j=0}^{\infty} P_{ij} = 1, \qquad i = 0, 1, \ldots$$

The corresponding transition probability matrix is denoted

$$P = \begin{bmatrix} P_{00} & P_{01} & P_{02} & \cdots \\ P_{10} & P_{11} & P_{12} & \cdots \\ \cdot & \cdot & \cdot & \cdot \\ P_{i0} & P_{i1} & P_{i2} & \cdots \\ \cdot & \cdot & \cdot & \cdot \end{bmatrix}$$

We will concentrate on the case where the number of states is infinite. There are analogous notions and results for the case where the number of states is finite.

Consider the n-step transition probabilities

$$P_{ij}^n = P\{X_{n+m} = j \mid X_m = i\}, \qquad n \geq 0, i, j \geq 0.$$

The *Chapman–Kolmogorov equations* provide a method for calculating P_{ij}^n. They are given by

$$P_{ij}^{n+m} = \sum_{k=0}^{\infty} P_{ik}^n P_{kj}^m, \qquad n, m \geq 0, i, j \geq 0$$

From these equations, we see that P_{ij}^n are the elements of the matrix P^n (the transition probability matrix P raised to the nth power).

We say that two states i and j *communicate* if for some n and n', we have $P_{ij}^n > 0$ and $P_{ji}^{n'} > 0$. If all states communicate, we say that the Markov chain is *irreducible*.

We say that a state i of a Markov chain is *periodic* if there exists some integer $m \geq 1$ such that $P_{ii}^m > 0$ and some integer $d > 1$ such that $P_{ii}^n > 0$ only if n is a multiple of d. A Markov chain is said to be *aperiodic* if none of its states is periodic. A probability distribution $\{p_j \mid j \geq 0\}$ is said to be a *stationary distribution* for the Markov chain if

$$p_j = \sum_{i=0}^{\infty} p_i P_{ij} \qquad j = 0, 1, \ldots \tag{3A.1}$$

We will restrict attention to irreducible and aperiodic Markov chains, since this is the only type we will encounter. For such a chain, it can be shown that the limit

$$p_j = \lim_{n \to \infty} P\{X_n = j \mid X_0 = i\}, \qquad i = 0, 1, \ldots$$

exists and is independent of the starting state $X_0 = i$. Furthermore, we have (with probability 1)

$$p_j = \lim_{k \to \infty} \frac{\text{Number of visits to state } j \text{ up to time } k}{k}$$

which leads to the interpretation that p_j is the proportion of time or the frequency with which the process visits j, a time average interpretation. Note again that this frequency does not depend on the starting state. The following result will be of primary interest:

Theorem. In an irreducible, aperiodic Markov chain, there are two possibilities for the scalars $p_j = \lim_{n \to \infty} P\{X_n = j \mid X_0 = i\}$:

1. $p_j = 0$ for all $j \geq 0$, in which case the chain has no stationary distribution.
2. $p_j > 0$ for all $j \geq 0$, in which case $\{p_j \mid j \geq 0\}$ is the unique stationary distribution of the chain [*i.e.*, it is the only probability distribution satisfying Eq. (3A.1)].

A typical example of case 1 is a queueing system where the arrival rate exceeds the service rate, and the number of customers in the system increases to ∞, so the steady-state probability p_j of having any finite number of customers j in the system is zero. Note that case 1 never arises when the number of states is finite. In particular, for every irreducible and aperiodic Markov chain with states $j = 0, 1, \ldots, m$, there exists a unique probability distribution $\{p_j \mid j = 0, 1, \ldots, m\}$ satisfying $p_j = \sum_{i=0}^{m} p_i P_{ij}$ and $p_j > 0$ for all j.

In case 2, there arises the issue of characterizing the stationary distribution $\{p_j \mid j \geq 0\}$. For queueing systems, the following equations are often useful. Multiplying the equation $\sum_{i=0}^{\infty} P_{ji} = 1$ by p_j and using Eq. (3A.1), we have

$$p_j \sum_{i=0}^{\infty} P_{ji} = \sum_{i=0}^{\infty} p_i P_{ij}, \qquad j = 0, 1, \ldots \tag{3A.2}$$

These equations are known as the *global balance equations*. Note that $p_i P_{ij}$ may be viewed as the long-term frequency of transitions from i to j. Thus the global balance equations state that at equilibrium, the frequency of transitions out of j [left side of Eq. (3A.2)] equals the frequency of transitions into j [right side of Eq. (3A.2)].

A typical approach for finding the stationary distribution of an irreducible, aperiodic Markov chain is to try to solve the global balance equations. If a distribution satisfying these equations is found, then by the preceding theorem, it must be the stationary distribution.

The global balance equations can be generalized to apply to an entire set of states. Consider a subset of states S. By adding Eq. (3A.2) over all $j \in S$, we obtain

$$\sum_{j \in S} p_j \sum_{i \notin S} P_{ji} = \sum_{i \notin S} p_i \sum_{j \in S} P_{ij} \qquad (3A.3)$$

An intuitive explanation of these equations is based on the fact that when the Markov chain is irreducible, the state (with probability 1) will return to the set S infinitely many times. Therefore, for each transition out of S there must be (with probability 1) a reverse transition into S at some later time. As a result, the frequency of transitions out of S equals the frequency of transitions into S. This is precisely the meaning of the global balance equations (3A.3).

3A.2 Detailed Balance Equations

As an application of the global balance equations, consider a Markov chain typical of queueing systems and, more generally, *birth–death* systems where two successive states can only differ by unity, that is, $P_{i,j} = 0$ if $|i - j| > 1$, cf. Fig. 3A.1. We assume that $P_{i,i+1} > 0$ and $P_{i+1,i} > 0$ for all i. This is a necessary and sufficient condition for the chain to be irreducible. Consider the sets of states

$$S = \{0, 1, \ldots, n\}$$

Application of Eq. (3A.3) yields

$$p_n P_{n,n+1} = p_{n+1} P_{n+1,n}, \qquad n = 0, 1, \ldots \qquad (3A.4)$$

(*i.e.*, in steady-state), the frequency of transitions from n to $n+1$ equals the frequency of transitions from $n+1$ to n. These equations can be very useful in computing the stationary distribution $\{p_j \mid j \geq 0\}$ (see Sections 3.3 and 3.4).

Equation (3A.4) is a special case of the equations

$$p_j P_{ji} = p_i P_{ij}, \qquad i, j \geq 0 \qquad (3A.5)$$

known as the *detailed balance equations*. These equations imply the global balance equations (3A.2) but need not hold in any given Markov chain. However, in many important special cases, they do hold and greatly simplify the calculation of the stationary

Figure 3A.1 Transition probability diagram for a birth–death process.

distribution. A common approach is to hypothesize the validity of the detailed balance equations and to try to solve them for the steady-state probabilities p_j, $j \geq 0$. There are two possibilities; either the system (3A.5) together with $\sum_j p_j = 1$ is inconsistent or else a distribution $\{p_j \mid j \geq 0\}$ satisfying Eq. (3A.5) will be found. In the latter case, this distribution will also satisfy the global balance equations (3A.2), so by the theorem given earlier, it is the unique stationary distribution.

3A.3 Partial Balance Equations

Some Markov chains have the property that their stationary distribution $\{p_j \mid j \geq 0\}$ satisfies a set of equations which is intermediate between the global and the detailed balance equations. For every node j, consider a partition $S_j^1, \ldots, S_j^k$ of the complementary set of nodes $\{i \mid i \geq 0, i \neq j\}$ and the equations

$$p_j \sum_{i \in S_j^m} P_{ji} = \sum_{i \in S_j^m} p_i P_{ij}, \qquad m = 1, 2, \ldots, k \tag{3A.6}$$

Equations of the form above are known as a set of *partial balance equations*. If a distribution $\{p_j \mid j \geq 0\}$ solves a set of partial balance equations, it will also solve the global balance equations, so it will be the unique stationary distribution of the chain. A technique that often proves useful is to guess the right set of partial balance equations satisfied by the stationary distribution and then proceed to solve them.

3A.4 Continuous-Time Markov Chains

A continuous-time Markov chain is a process $\{X(t) \mid t \geq 0\}$ taking values from the set of states $i = 0, 1, \ldots$ that has the property that each time it enters state i:

1. The time it spends in state i is exponentially distributed with parameter ν_i. We may view ν_i as the rate (in transitions/sec) at which the process makes a transition when at state i.
2. When the process leaves state i, it will enter state j with probability P_{ij}, where $\sum_j P_{ij} = 1$.

   We may view

$$q_{ij} = \nu_i P_{ij}$$

as the rate (in transitions/sec) at which the process makes a transition to j when at state i. Consequently, we call q_{ij} the *transition rate* from i to j.

We will be interested in chains for which the discrete-time Markov chain with transition probabilities P_{ij} (called the *embedded chain*) is irreducible. We also require a technical condition, namely that the number of transitions in any finite length of time is finite with probability 1; chains with this property are called *regular*. (Nonregular chains almost never arise in queueing systems of interest. For an example, see [Ros83], p. 142.)

Under the preceding conditions, it can be shown that the limit

$$p_j = \lim_{t \to \infty} P\{X(t) = j \mid X(0) = i\} \tag{3A.7}$$

exists and is independent of the initial state i. We refer to p_j as the steady-state occupancy probability of state j. It can be shown that if $T_j(t)$ is the time spent in state j up to time t, then, regardless of the initial state, we have with probability 1,

$$p_j = \lim_{t \to \infty} \frac{T_j(t)}{t} \tag{3A.8}$$

that is, p_j can be viewed as the long-term proportion of time the process spends in state j. It can be shown also that either the occupancy probabilities are all zero or else they are all positive and they sum to unity. Queueing systems where the arrival rate is larger than the service rate provide examples where all occupancy probabilities are zero.

The *global balance equations* for a continuous-time Markov chain take the form

$$p_j \sum_{i=0}^{\infty} q_{ji} = \sum_{i=0}^{\infty} p_i q_{ij}, \qquad j = 0, 1, \ldots \tag{3A.9}$$

It can be shown that if a probability distribution $\{p_j \mid j \geq 0\}$ satisfies these equations, then each p_j is the steady-state occupancy probability of state j.

To interpret the global balance equations, we note that since p_i is the proportion of time the process spends in state i, it follows that $p_i q_{ij}$ is the frequency of transitions from i to j (average number of transitions from i to j per unit time). It is seen therefore that the global balance equations (3A.9) express the natural fact that the frequency of transitions out of state j (the left-side term $p_j \sum_{i=1}^{\infty} q_{ji}$) is equal to the frequency of transitions into state j (the right-side term $\sum_{i=0}^{\infty} p_i q_{ij}$).

The continuous-time analog of the detailed balance equations for discrete-time chains is

$$p_j q_{ji} = p_i q_{ij}, \qquad i, j = 0, 1, \ldots$$

These equations hold in birth–death systems where $q_{ij} = 0$ for $|i - j| > 1$, but need not hold in other types of Markov chains. They express the fact that the frequencies of transitions from i to j and from j to i are equal. One can also write a set of partial balance equations and attempt to solve them for the distribution $\{p_j \mid j \geq 0\}$. If a solution can be found, it provides the stationary distribution of the continuous chain.

To understand the relationship between the global balance equations (3A.9) for continuous-time chains and the global balance equations (3A.2) for discrete-time chains, consider any $\delta > 0$, and the discrete-time Markov chain $\{X_n \mid n \geq 0\}$, where

$$X_n = X(n\delta), \qquad n = 0, 1, \ldots$$

The stationary distribution of $\{X_n\}$ is clearly $\{p_j \mid j \geq 0\}$, the occupancy distribution of the continuous chain [cf. Eq. (3A.7)]. The transition probabilities of $\{X_n \mid n \geq 0\}$ can be derived by using the properties of the exponential distribution and a derivation which is very similar to the one used in Section 3.3.1 for the Markov chain of the $M/M/1$ queue. We obtain

$$\overline{P}_{ij} = \delta q_{ij} + o(\delta), \qquad i \neq j$$

$$\overline{P}_{jj} = 1 - \delta \sum_{\substack{i=0 \\ i \neq j}}^{\infty} q_{ji} + o(\delta)$$

Using these expressions and Eq. (3A.1), which is equivalent to the global balance equations for the discrete chain, we obtain

$$p_j = \sum_{i=0}^{\infty} p_i \overline{P}_{ij} = p_j\big(1 - \delta \sum_{\substack{i=0 \\ i \neq j}}^{\infty} q_{ji} + o(\delta)\big) + \sum_{\substack{i=0 \\ i \neq j}} p_i\big(\delta q_{ij} + o(\delta)\big)$$

and dividing by δ and letting $\delta \to 0$, we obtain the global balance equations (3A.9) for the continuous chain.

3A.5 Drift and Stability

Suppose that we are given an irreducible, aperiodic, discrete-time Markov chain. In many situations one is particularly interested in whether the chain has a stationary distribution $\{p_j\}$. In this case, $p_j > 0$ for all j and the chain is "stable" in the sense that all states are visited infinitely often with probability 1. The notion of *drift*, defined as

$$D_i = E\{X_{n+1} - X_n \mid X_n = i\} = \sum_{k=-i}^{\infty} k P_{i(i+k)}, \qquad i = 0, 1, \ldots$$

is particularly useful in this respect. Roughly speaking, the sign of D_i indicates whether, starting at i, the state tends to increase ($D_i > 0$) or decrease ($D_i < 0$). Intuitively, the chain will be stable if the drift is negative for all large enough states. This is established in the following lemma:

Stability Lemma [Pak69]. Suppose that $D_i < \infty$ for all i, and that for some scalar $\delta > 0$ and integer $\bar{i} \geq 0$ we have

$$D_i \leq -\delta, \qquad \text{for all } i > \bar{i}$$

Then the Markov chain has a stationary distribution.

Proof: Let $\beta = \max_{i \leq \bar{i}} D_i$. We have for each state i

$$E\{X_n \mid X_0 = i\} - i = \sum_{k=1}^{n} E\{X_k - X_{k-1} \mid X_0 = i\}$$

$$= \sum_{k=1}^{n} \sum_{j=0}^{\infty} E\{X_k - X_{k-1} \mid X_{k-1} = j\} P\{X_{k-1} = j \mid X_0 = i\}$$

$$\leq \sum_{k=1}^{n} \left[\beta \sum_{j=0}^{\bar{i}} P\{X_{k-1} = j \mid X_0 = i\} \right.$$

$$\left. -\delta \left(1 - \sum_{j=0}^{\bar{i}} P\{X_{k-1} = j \mid X_0 = i\} \right) \right]$$

$$= (\beta + \delta) \sum_{k=1}^{n} \sum_{j=0}^{\bar{i}} P\{X_{k-1} = j \mid X_0 = i\} - n\delta$$

from which we obtain

$$0 \leq E\{X_n \mid X_0 = i\} \leq n \left[(\beta + \delta) \sum_{j=0}^{\bar{i}} \left(n^{-1} \sum_{k=1}^{n} P\{X_{k-1} = j \mid X_0 = i\} \right) - \delta \right] + i$$

Dividing by n and taking the limit as $n \to \infty$, we obtain

$$0 \leq (\beta + \delta) \sum_{j=0}^{\bar{i}} p_j - \delta$$

which implies that $p_j > 0$ for some $j \in \{0, \ldots, \bar{i}\}$. Since the chain is assumed irreducible and aperiodic, it follows from the theorem of Section 3A.1 that there exists a stationary distribution. **Q.E.D.**

We also give without proof a converse to the preceding lemma:

Instability Lemma [Kap79]. Suppose that there exist integers $\bar{i} \geq 0$ and k such that

$$D_i > 0, \qquad \text{for all } i > \bar{i}$$

and

$$P_{ij} = 0, \qquad \text{for all } i \text{ and } j \text{ such that } 0 \leq j \leq i - k$$

Then the Markov chain does not have a stationary distribution; that is, $p_j = 0$ for all j.

APPENDIX B: SUMMARY OF RESULTS

Notation

p_n = Steady-state probability of having n customers in the system

λ = Arrival rate (inverse of average interarrival time)

μ = Service rate (inverse of average service time)

N = Average number of customers in the system

N_Q = Average number of customers waiting in queue

T = Average customer time in the system

W = Average customer waiting time in queue (does not include service time)

$\overline{X}$ = Average service time

$\overline{X^2}$ = Second moment of service time

Little's Theorem

$$N = \lambda T$$

$$N_Q = \lambda W$$

Poisson distribution with parameter *m*

$$p_n = \frac{e^{-m} m^n}{n!}, \quad n = 0, 1, \ldots$$

$$\text{Mean} = \text{Variance} = m$$

Exponential distribution with parameter λ

$$P\{\tau \le s\} = 1 - e^{-\lambda s}, \qquad s \ge 0$$

$$\text{Density: } p(\tau) = \lambda e^{-\lambda \tau}$$

$$\text{Mean} = \frac{1}{\lambda}$$

$$\text{Variance} = \frac{1}{\lambda^2}$$

Summary of *M/M/1* system results

1. Utilization factor (proportion of time the server is busy)

$$\rho = \frac{\lambda}{\mu}$$

2. Probability of n customers in the system

$$p_n = \rho^n (1 - \rho), \qquad n = 0, 1, \ldots$$

3. Average number of customers in the system

$$N = \frac{\rho}{1 - \rho}$$

4. Average customer time in the system

$$T = \frac{\rho}{\lambda(1 - \rho)} = \frac{1}{\mu - \lambda}$$

5. Average number of customers in queue

$$N_Q = \frac{\rho^2}{1 - \rho}$$

6. Average waiting time in queue of a customer

$$W = \frac{\rho}{\mu - \lambda}$$

Summary of *M/M/m* system results

1. Ratio of arrival rate to maximal system service rate

$$\rho = \frac{\lambda}{m\mu}$$

2. Probability of n customers in the system

$$p_0 = \left[\sum_{k=0}^{m-1} \frac{(m\rho)^k}{k!} + \frac{(m\rho)^m}{m!(1-\rho)}\right]^{-1}, \qquad n = 0$$

$$p_n = \begin{cases} p_0 \dfrac{(m\rho)^n}{n!}, & n \leq m \\ p_0 \dfrac{m^m \rho^n}{m!}, & n > m \end{cases}$$

3. Probability that an arriving customer has to wait in queue (m customers or more in the system)

$$P_Q = \frac{p_0 (m\rho)^m}{m!(1-\rho)} \qquad \text{(Erlang C Formula)}$$

4. Average waiting time in queue of a customer

$$W = \frac{\rho P_Q}{\lambda(1-\rho)}$$

5. Average number of customers in queue

$$N_Q = \frac{\rho P_Q}{1-\rho}$$

6. Average customer time in the system

$$T = \frac{1}{\mu} + W$$

7. Average number of customers in the system

$$N = m\rho + \frac{\rho P_Q}{1-\rho}$$

Summary of *M/M/m/m* system results

1. Probability of m customers in the system

$$p_0 = \left[\sum_{n=0}^{m} \left(\frac{\lambda}{\mu}\right)^n \frac{1}{n!}\right]^{-1}$$

$$p_n = p_0 \left(\frac{\lambda}{\mu}\right)^n \frac{1}{n!}, \qquad n = 1, 2, \ldots, m$$

2. Probability that an arriving customer is lost

$$p_m = \frac{(\lambda/\mu)^m / m!}{\sum_{n=0}^{m} (\lambda/\mu)^n / n!} \qquad \text{(Erlang B Formula)}$$

Summary of *M/G/1* system results

1. Utilization factor

$$\rho = \frac{\lambda}{\mu}$$

2. Mean residual service time

$$R = \frac{\lambda \overline{X^2}}{2}$$

3. Pollaczek–Khinchin formula

$$W = \frac{R}{1-\rho} = \frac{\lambda \overline{X^2}}{2(1-\rho)}$$

$$T = \frac{1}{\mu} + W$$

$$N_Q = \frac{\lambda^2 \overline{X^2}}{2(1-\rho)}$$

$$N = \rho + \frac{\lambda^2 \overline{X^2}}{2(1-\rho)}$$

4. Pollaczek–Khinchin formula for $M/G/1$ queue with vacations

$$W = \frac{\lambda \overline{X^2}}{2(1-\rho)} + \frac{\overline{V^2}}{2\overline{V}}$$

$$T = \frac{1}{\mu} + W$$

where $\overline{V}$ and $\overline{V^2}$ are the first two moments of the vacation interval.

Summary of reservation/polling results

1. Average waiting time (m-user system, unlimited service)

$$W = \frac{\lambda \overline{X^2}}{2(1-\rho)} + \frac{(m-\rho)\overline{V}}{2(1-\rho)} + \frac{\sigma_V^2}{2\overline{V}} \qquad \text{(exhaustive)}$$

$$W = \frac{\lambda \overline{X^2}}{2(1-\rho)} + \frac{(m+\rho)\overline{V}}{2(1-\rho)} + \frac{\sigma_V^2}{2\overline{V}} \qquad \text{(partially gated)}$$

$$W = \frac{\lambda \overline{X^2}}{2(1-\rho)} + \frac{(m+2-\rho)\overline{V}}{2(1-\rho)} + \frac{\sigma_V^2}{2\overline{V}} \qquad \text{(gated)}$$

where $\rho = \lambda/\mu$, and $\overline{V}$ and σ_V^2 are the mean and variance of the reservation intervals, respectively, averaged over all users

$$\overline{V} = \frac{1}{m} \sum_{\ell=0}^{m-1} \overline{V}_\ell$$

$$\sigma_V^2 = \frac{1}{m} \sum_{\ell=0}^{m-1} \left(\overline{V_\ell^2} - \overline{V}_\ell^2 \right)$$

2. Average waiting time (m-user system, limited service)

$$W = \frac{\lambda \overline{X^2}}{2(1-\rho-\lambda\overline{V})} + \frac{(m+\rho)\overline{V}}{2(1-\rho-\lambda\overline{V})} + \frac{\sigma_V^2(1-\rho)}{2\overline{V}(1-\rho-\lambda\overline{V})} \quad \text{(partially gated)}$$

$$W = \frac{\lambda \overline{X^2}}{2(1-\rho-\lambda\overline{V})} + \frac{(m+2-\rho-2\lambda\overline{V})\overline{V}}{2(1-\rho-\lambda\overline{V})} + \frac{\sigma_V^2(1-\rho)}{2\overline{V}(1-\rho-\lambda\overline{V})} \quad \text{(gated)}$$

3. Average time in the system

$$T = \frac{1}{\mu} + W$$

Summary of priority queueing results

1. *Nonpreemptive priority*. Average waiting time in queue for class k customers

$$W_k = \frac{\sum_{i=1}^{n} \lambda_i \overline{X_i^2}}{2(1-\rho_1-\cdots-\rho_{k-1})(1-\rho_1-\cdots-\rho_k)}$$

2. *Nonpreemptive priority*. Average time in the system for class k customers

$$T_k = \frac{1}{\mu_k} + W_k$$

3. *Preemptive resume priority*. Average time in the system for class k customers

$$T_k = \frac{(1/\mu_k)(1-\rho_1-\cdots-\rho_k) + R_k}{(1-\rho_1-\cdots-\rho_{k-1})(1-\rho_1-\cdots-\rho_k)}$$

where

$$R_k = \frac{\sum_{i=1}^{k} \lambda_i \overline{X_i^2}}{2}$$

Heavy traffic approximation for the $G/G/1$ system

Average waiting time in queue satisfies

$$W \le \frac{\lambda(\sigma_a^2 + \sigma_b^2)}{2(1-\rho)}$$

where

σ_a^2 = Variance of the interarrival times

σ_b^2 = Variance of the service times

λ = Average interarrival time

ρ = Utilization factor λ/μ, where $1/\mu$ is the average service time

4

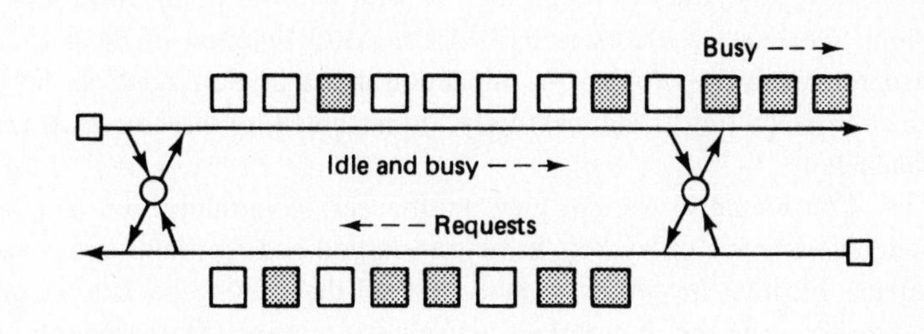

Multiaccess Communication

4.1 INTRODUCTION

The subnetworks considered thus far have consisted of nodes joined by point-to-point communication links. Each such link might consist physically of a pair of twisted wires, a coaxial cable, an optical fiber, a microwave radio link, and so on. The implicit assumption about point-to-point links, however, is that the received signal on each link depends only on the transmitted signal and noise on that link.

There are many widely used communication media, such as satellite systems, radio broadcast, multidrop telephone lines, and multitap bus systems, for which the received signal at one node depends on the transmitted signal at two or more other nodes. Typically, such a received signal is the sum of attenuated transmitted signals from a set of other nodes, corrupted by distortion, delay, and noise. Such media, called *multiaccess media*, form the basis for local area networks (LANs), metropolitan area networks (MANs), satellite networks, and radio networks.

The layering discussed in Chapters 1 and 2 is not quite appropriate for multiaccess media. One needs an additional sublayer, often called the *medium access control* (MAC) sublayer, beween the data link control (DLC) layer and the modem or physical layer. The purpose of this extra sublayer is to allocate the multiaccess medium among the various nodes. As we study this allocation issue, we shall see that the separation of functions between layers is not as clean as it is with point-to-point links. For example, feedback about transmission errors is part of the ARQ function of the DLC layer, but is often also central to the problem of allocation and thus flow control. Similarly, much of the function of routing is automatically implemented by the broadcast nature of multiaccess channels.

Conceptually, we can view multiaccess communication in queueing terms. Each node has a queue of packets to be transmitted and the multiaccess channel is a common server. Ideally, the server should view all the waiting packets as one combined queue to be served by the appropriate queueing discipline. Unfortunately, the server does not know which nodes contain packets; similarly, nodes are unaware of packets at other nodes. Thus, the interesting part of the problem is that knowledge about the state of the queue is distributed.

There are two extremes among the many strategies that have been developed for this generic problem. One is the "free-for-all" approach in which nodes normally send new packets immediately, hoping for no interference from other nodes. The interesting question here is when and how packets are retransmitted when collisions (*i.e.*, interference) occur. The other extreme is the "perfectly scheduled" approach in which there is some order (round robin, for example) in which nodes receive reserved intervals for channel use. The interesting questions here are: (1) what determines the scheduling order (it could be dynamic), (2) how long can a reserved interval last, and (3) how are nodes informed of their turns?

Sections 4.2 and 4.3 explore the free-for-all approach in a simple idealized environment that allows us to focus on strategies for retransmitting collided packets. Successively more sophisticated algorithms are developed that reduce delay, increase available throughput, and maintain stable operation. In later sections, these algorithms are adapted to take advantage of special channel characteristics so as to reduce delay and increase throughput even further. The more casual reader can omit Sections 4.2.3 and 4.3.

Section 4.4 explores carrier sense multiple access (CSMA). Here the free-for-all approach is modified; a packet transmission is not allowed to start if the channel is sensed to be busy. We shall find that this set of strategies is a relatively straightforward extension of the ideas in Sections 4.2 and 4.3. The value of these strategies is critically dependent on the ratio of propagation delay to packet transmission time, a parameter called β. If $\beta \ll 1$, CSMA can decrease delay and increase throughput significantly over the techniques of Sections 4.2 and 4.3. The casual reader can omit Sections 4.4.2 and 4.4.4.

Section 4.5 deals with scheduling, or reserving, the channel in response to the dynamic requirements of the individual nodes. We start with satellite channels in Section 4.5.1; here the interesting feature is dealing with $\beta \gg 1$. Next, Sections 4.5.2 to 4.5.4 treat the major approaches to LANs and MANs. These approaches can be viewed as reservation systems and differ in whether the reservations are scheduled in a free-for-all manner or in a round-robin manner. LANs are usually designed for the assumption that β is small, and Section 4.5.5 explores systems with higher speed or greater geographical coverage for which β is large.

Finally, Section 4.6 explores packet radio. Here the interesting issue is that each node interferes only with a limited subset of other nodes; thus, multiple nodes can transmit simultaneously without interference.

Before going into any of the topics above in detail, we briefly discuss some of the most widely used multiaccess media.

4.1.1 Satellite Channels

In a typical geosynchronous communication satellite system, many ground stations can transmit to a common satellite receiver, with the received messages being relayed to the ground stations (see Fig. 4.1). Such satellites often have separate antenna beams for different geographical areas, allowing independent reception and relaying between areas. Also, FDM (or TDM) can be used, permitting different earth stations within the region covered by a single antenna beam to be independently received.

It is thus possible to use a satellite channel as a collection of virtual point-to-point links, with two virtual links being separated either by antenna beam or multiplexing. The potential difficulty with this approach is the same as the difficulty with using FDM or

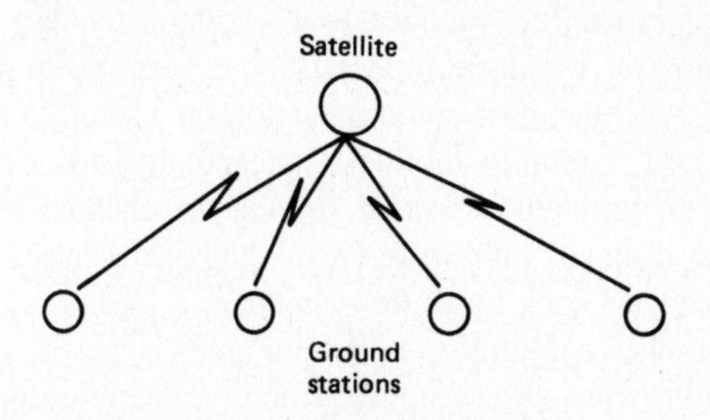

(a) Satellite multiaccess channel

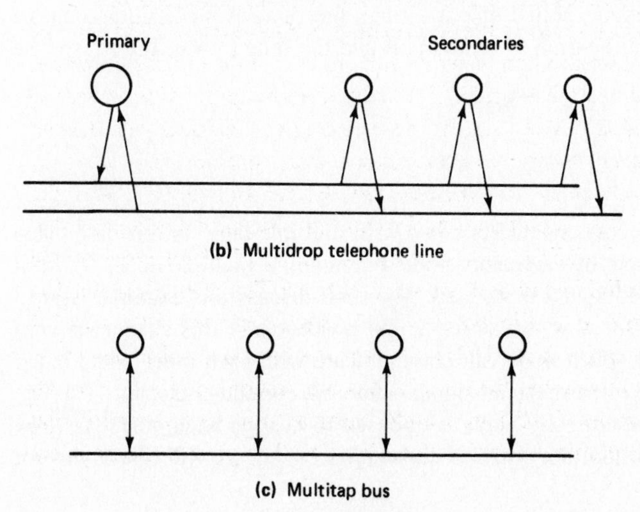

(b) Multidrop telephone line

(c) Multitap bus

Figure 4.1 Common multiaccess channels.

TDM for multiple sessions on a single point-to-point link, namely increased delay and underutilization of the medium.

In Section 4.5.1 we shall find that delay can be reduced and utilization increased by sharing the medium on a demand basis. This is more difficult than demand sharing (*i.e.*, statistical multiplexing) on a point-to-point link because the earth stations are not aware of the instantaneous traffic requirements of other earth stations. If several stations transmit at the same time (and in the same frequency band), their signals are garbled and incorrectly received.

4.1.2 Multidrop Telephone Lines

Another example of a multiaccess channel is a multidrop telephone line. Such lines connect one primary node with a number of secondary nodes; the signal transmitted by the primary node goes over one pair of wires and is received by all the secondary nodes. Similarly, there is a return pair of wires which carries the sum of the transmitted signals from all the secondary nodes to the primary node. Conceptually, this is like a satellite channel. The secondary nodes, or earth stations, share the path to the primary node, or satellite, whereas the primary node, or satellite, broadcasts to all the secondary nodes, or earth stations. Most communication on a multidrop phone line is intended to go from primary to secondary, or vice versa, whereas most communication on a satellite channel is relayed by the satellite from one earth station to another. Conceptually, this difference is not very important, since the major problem is that of sharing the channel from the secondary nodes to the primary, and it makes little difference whether the messages are removed at the primary node or broadcast back to all the secondary nodes.

The traditional mode of operation for multidrop telephone lines is for the primary node to poll (*i.e.*, request information from) each secondary node in some order. Each secondary node responds to its poll either by sending data back to the primary station or by indicating that it has no data to send. This strategy avoids interference between the secondary nodes, since nodes are polled one at a time, but there is a certain amount of inefficiency involved, both in the time to send a poll and the time to wait for a response from a node with no data.

4.1.3 Multitapped Bus

A third example of a multiaccess channel is a bus with multiple taps. In this case, each node can receive the signal sent by each other node, but again, if multiple nodes transmit at the same time, the received signal is garbled. We shall discuss this example later in the context of Ethernet, but for now, we observe that conceptually this channel is very similar to a satellite channel. Each node can communicate with each other node, but if nodes transmit simultaneously, the received signal cannot be correctly detected. The fact that nodes can hear each other directly here, as opposed to hearing each other via relay from the satellite, has some important practical consequences, but we will ignore this for now.

4.1.4 Packet Radio Networks

A fourth example of multiaccess communication is that of a packet radio network. Here each node is in reception range of some subset of other nodes. Thus, if only one node in the subset is transmitting, the given node can receive the transmission, whereas if multiple nodes are transmitting simultaneously in the same band, the reception will be garbled. Similarly, what one node transmits will be heard by a subset of the other nodes. In general, because of different noise conditions at the nodes, the subset of nodes from which a given node can receive is different from the subset to which the given node can transmit. The fact that each receiver hears a subset of transmitters rather than all transmitters makes packet radio far more complex than the other examples discussed. In the next four sections, we study multiaccess channels in which a receiver can hear all transmitters, and then in Section 4.6, we discuss packet radio networks in more detail.

4.2 SLOTTED MULTIACCESS AND THE ALOHA SYSTEM

Satellite channels, multidrop telephone lines, and multitap bus systems all share the feature of a set of nodes sharing a communication channel. If two or more nodes transmit simultaneously, the reception is garbled, and if none transmit, the channel is unused. The problem is somehow to coordinate the use of the channel so that exactly one node is transmitting for an appreciable fraction of the time. We start by looking at a highly idealized model. We shall see later that multiaccess channels can often be used in practice with much higher utilization than is possible in this idealized model, but we shall also see that these practical extensions can be understood more clearly in terms of our idealization.

4.2.1 Idealized Slotted Multiaccess Model

The idealized model to be developed allows us to focus on the problem of dealing with the contention that occurs when multiple nodes attempt to use the channel simultaneously. Conceptually, we view the system as in Fig. 4.1(a), with m transmitting nodes and one receiver.

It will be observed that aside from some questions of propagation delay, this same model can be applied with m nodes that can all hear each other (*i.e.*, the situation with a multitap bus). We first list the assumptions of the model and then discuss their implications.

1. *Slotted system.* Assume that all transmitted packets have the same length and that each packet requires one time unit (called a slot) for transmission. All transmitters are synchronized so that the reception of each packet starts at an integer time and ends before the next integer time.
2. *Poisson arrivals.* Assume that packets arrive for transmission at each of the m transmitting nodes according to independent Poisson processes. Let λ be the overall arrival rate to the system, and let λ/m be the arrival rate at each transmitting node.

3. *Collision or perfect reception.* Assume that if two or more nodes send a packet in a given time slot, then there is a *collision* and the receiver obtains no information about the contents or source of the transmitted packets. If just one node sends a packet in a given slot, the packet is correctly received.
4. *0,1,e Immediate feedback.* Assume that at the end of each slot, each node obtains feedback from the receiver specifying whether 0 packets, 1 packet, or more than one packet (e for error) were transmitted in that slot.
5. *Retransmission of collisions.* Assume that each packet involved in a collision must be retransmitted in some later slot, with further such retransmissions until the packet is successfully received. A node with a packet that must be retransmitted is said to be *backlogged.*

6a. *No buffering.* If one packet at a node is currently waiting for transmission or colliding with another packet during transmission, new arrivals at that node are discarded and never transmitted. An alternative to this assumption is the following.

6b. *Infinite set of nodes* ($m = \infty$). The system has an infinite set of nodes and each newly arriving packet arrives at a new node.

Discussion of assumptions. The slotted system assumption (1) has two effects. The first is to turn the system into a discrete-time system, thus simplifying analysis. The second is to preclude, for the moment, the possibility of carrier sensing or early collision detection. Carrier sensing is treated in Section 4.4 and early collision detection is treated in Section 4.5. Both allow much more efficient use of the multiaccess channel, but can be understood more clearly as an extension of the present model. Synchronizing the transmitters for slotted arrival at the receiver is not entirely trivial, but may be accomplished with relatively stable clocks, a small amount of feedback from the receiver, and some guard time between the end of a packet transmission and the beginning of the next slot.

The assumption of Poisson arrivals (2) is unrealistic for the case of multipacket messages. We discuss this issue in Section 4.5 in terms of nodes making reservations for use of the channel.

The assumption of collision or perfect reception (3) ignores the possibility of errors due to noise and also ignores the possibility of "capture" techniques, by which a receiver can sometimes capture one transmission in the presence of multiple transmissions.

The assumption of immediate feedback (4) is quite unrealistic, particularly in the case of satellite channels. It is made to simplify the analysis, and we shall see later that delayed feedback complicates multiaccess algorithms but causes no fundamental problems. We also discuss the effects of more limited types of feedback later.

The assumption that colliding packets must be retransmitted (5) is certainly reasonable in providing reliable communication. In light of this assumption, the no-buffering assumption (6a) appears rather peculiar, since new arrivals to backlogged nodes are thrown away with impunity, but packets once transmitted must be retransmitted until successful. In practice, one generally provides some buffering along with some form of flow control to ensure that not too many packets back up at a node. Our interest in this section, however, is in multiaccess channels with a large number of nodes, a relatively small arrival rate λ, and small required delay (*i.e.*, the conditions under which TDM does not suffice). Under these conditions, the fraction of backlogged nodes is typically small, and new arrivals at backlogged nodes are almost negligible. Thus, the delay for a system without buffering should be relatively close to that with buffering. Also, the delay for the unbuffered system provides a lower bound to the delay for a wide variety of systems with buffering and flow control.

The infinite-node assumption (6b) alternatively provides us with an upper bound to the delay that can be achieved with a finite number of nodes. In particular, given any multiaccess algorithm (*i.e.*, any rule that each node employs for selecting packet transmission times), each of a finite set of nodes can regard itself as a set of virtual nodes, one for each arriving packet. With the application of the given algorithm independently for each such packet, the situation is equivalent to that with an infinite set of nodes (*i.e.*, assumption 6b). In this approach, a node with several backlogged packets will sometimes send multiple packets in one slot, causing a sure collision. Thus, it appears that by avoiding such sure collisions and knowing the number of buffered packets at a node, a system with a finite number of nodes and buffering could achieve smaller delays than with $m = \infty$; in any case, however, the $m = \infty$ delay can always be achieved.

If the performance using assumption 6a is similar to that using 6b, then we are assured that we have a good approximation to the performance of a system with arbitrary assumptions about buffering. From a theoretical standpoint, assumption 6b captures the essence of multiaccess communication much better than 6a. We use 6a initially, however, because it is less abstract and it provides some important insights about the relationship between TDM and other algorithms.

4.2.2 Slotted Aloha

The Aloha network [Abr70] was developed around 1970 to provide radio communication between the central computer and various data terminals at the campuses of the University of Hawaii. This multiaccess approach will be described in Section 4.2.4, but first we discuss an improvement called slotted Aloha [Rob72]. The basic idea of this algorithm is that each unbacklogged node simply transmits a newly arriving packet in the first slot after the packet arrival, thus risking occasional collisions but achieving very small delay if collisions are rare. This approach should be contrasted with TDM, in which, with m nodes, an arriving packet would have to wait for an average of $m/2$ slots for its turn to transmit. Thus, slotted Aloha transmits packets almost immediately with occasional collisions, whereas TDM avoids collisions at the expense of large delays.

When a collision occurs in slotted Aloha, each node sending one of the colliding packets discovers the collision at the end of the slot and becomes backlogged. If each backlogged node were simply to retransmit in the next slot after being involved in a collision, then another collision would surely occur. Instead, such nodes wait for some random number of slots before retransmitting.

To gain some initial intuitive understanding of slotted Aloha, we start with an instructive but flawed analysis. With the infinite-node assumption (*i.e.*, 6b), the number of new arrivals transmitted in a slot is a Poisson random variable with parameter λ. If the retransmissions from backlogged nodes are sufficiently randomized, it is plausible to approximate the total number of retransmissions and new transmissions in a given slot as a Poisson random variable with some parameter $G > \lambda$. With this approximation, the probability of a successful transmission in a slot is Ge^{-G}. Finally, in equilibrium, the arrival rate, λ, to the system should be the same as the departure rate, Ge^{-G}. This relationship is illustrated in Fig. 4.2.

We see that the maximum possible departure rate (according to the argument above) occurs at $G = 1$ and is $1/e \approx 0.368$. We also note, somewhat suspiciously, that for any arrival rate less than $1/e$, there are two values of G for which the arrival rate equals the departure rate. The problem with this elementary approach is that it provides no insight into the dynamics of the system. As the number of backlogged packets changes, the parameter G will change; this leads to a feedback effect, generating further changes in the number of backlogged packets. To understand these dynamics, we will have to analyze the system somewhat more carefully. The simple picture below, however, correctly identifies the maximum throughput rate of slotted Aloha as $1/e$ and also shows that G, the mean number of attempted transmissions per slot, should be on the order of 1 to achieve a throughput close to $1/e$. If $G < 1$, too many idle slots are generated, and if $G > 1$, too many collisions are generated.

To construct a more precise model, assume that each backlogged node retransmits with some fixed probability q_r in each successive slot until a successful transmission occurs. In other words, the number of slots from a collision until a given node involved in the collision retransmits is a geometric random variable having value $i \geq 1$ with probability $q_r(1 - q_r)^{i-1}$. The original version of slotted Aloha employed a uniform distribution for retransmission, but this is more difficult to analyze and has no identifiable

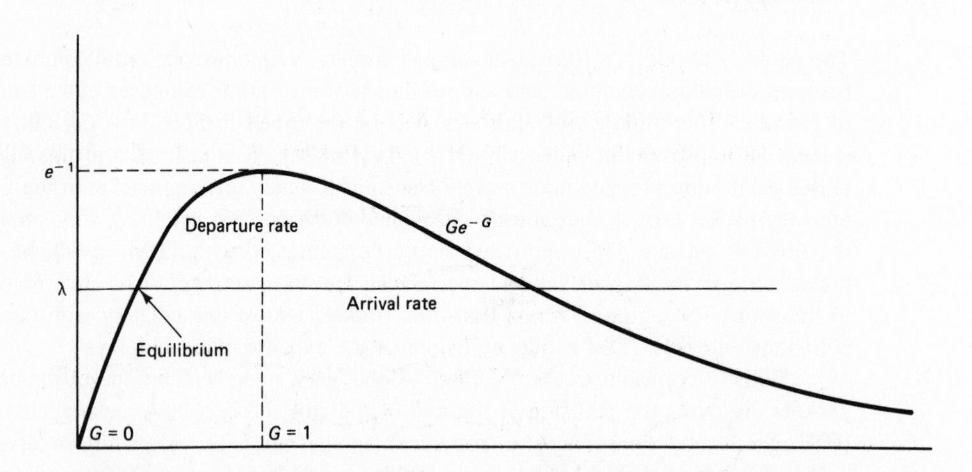

Figure 4.2 Departure rate as a function of attempted transmission rate G for slotted Aloha. Ignoring the dynamic behavior of G, departures (successful transmissions) occur at a rate Ge^{-G}, and arrivals occur at a rate λ, leading to a hypothesized equilibrium point as shown.

advantages over the geometric distribution. We will use the no-buffering assumption (6a) and switch later to the infinite node assumption (6b).

The behavior of slotted Aloha can now be described as a discrete-time Markov chain. Let n be the number of backlogged nodes at the beginning of a given slot. Each of these nodes will transmit a packet in the given slot, independently of each other, with probability q_r. Each of the $m-n$ other nodes will transmit a packet in the given slot if one (or more) such packets arrived during the previous slot. Since such arrivals are Poisson distributed with mean λ/m, the probability of no arrivals is $e^{-\lambda/m}$; thus, the probability that an unbacklogged node transmits a packet in the given slot is $q_a = 1 - e^{-\lambda/m}$. Let $Q_a(i,n)$ be the probability that i unbacklogged nodes transmit packets in a given slot, and let $Q_r(i,n)$ be the probability that i backlogged nodes transmit,

$$Q_a(i,n) = \binom{m-n}{i}(1-q_a)^{m-n-i}q_a^i \tag{4.1}$$

$$Q_r(i,n) = \binom{n}{i}(1-q_r)^{n-i}q_r^i \tag{4.2}$$

Note that from one slot to the next, the state (*i.e.*, the number of backlogged packets) increases by the number of new arrivals transmitted by unbacklogged nodes, less one if a packet is transmitted successfully. A packet is transmitted successfully, however, only if one new arrival and no backlogged packet, or no new arrival and one backlogged packet is transmitted. Thus, the state transition probability of going from state n to $n+i$ is given by

$$P_{n,n+i} = \begin{cases} Q_a(i,n), & 2 \le i \le (m-n) \\ Q_a(1,n)[1-Q_r(0,n)], & i = 1 \\ Q_a(1,n)Q_r(0,n) + Q_a(0,n)[1-Q_r(1,n)], & i = 0 \\ Q_a(0,n)Q_r(1,n), & i = -1 \end{cases} \tag{4.3}$$

Figure 4.3 illustrates this Markov chain. Note that the state can decrease by at most 1 in a single transition, and this makes it easy to calculate the steady-state probabilities

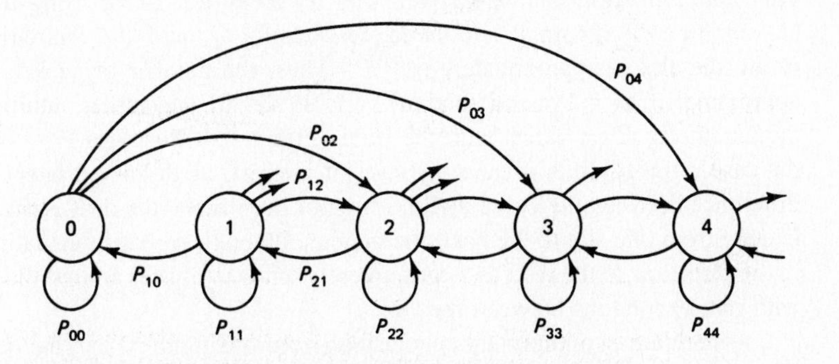

Figure 4.3 Markov chain for slotted Aloha. The state (*i.e.*, backlog) can decrease by at most one per transition, but can increase by an arbitrary amount.

iteratively, finding p_n for each successively larger n in terms of p_0 and finally, finding p_0 as a normalizing constant (see Problem 4.1). From this, the expected number of backlogged nodes can be found, and from Little's theorem, the average delay can be calculated.

Unfortunately, this system has some very strange properties for a large number of nodes, and the steady-state results above are very misleading. To get some intuitive understanding of this, note that we would like to choose the retransmission probability q_r to be moderately large, so as to avoid large delays after collisions. If the arrival rate is small and not many packets are involved in collisions, this works well and retransmissions are normally successful. On the other hand, if the system is afflicted with a run of bad luck and the number of backlogged packets n gets large enough to satisfy $q_r n >> 1$, then collisions occur in almost all successive slots and the system remains heavily backlogged for a long time.

To understand this phenomenon quantitatively, define the *drift* in state n (D_n) as the expected change in backlog over one slot time, starting in state n. Thus, D_n is the expected number of new arrivals accepted into the system [*i.e.*, $(m-n)q_a$] less the expected number of successful transmissions in the slot; the expected number of successful transmissions is just the probability of a successful transmission, defined as P_{succ}. Thus,

$$D_n = (m-n)q_a - P_{succ} \tag{4.4}$$

where

$$P_{succ} = Q_a(1,n)Q_r(0,n) + Q_a(0,n)Q_r(1,n) \tag{4.5}$$

Define the attempt rate $G(n)$ as the expected number of attempted transmissions in a slot when the system is in state n, that is,

$$G(n) = (m-n)q_a + nq_r$$

If q_a and q_r are small, P_{succ} is closely approximated as the following function of the attempt rate:

$$P_{succ} \approx G(n)e^{-G(n)} \tag{4.6}$$

This approximation can be derived directly from Eq. (4.5), using the approximation $(1-x)^y \approx e^{-xy}$ for small x in the expressions for Q_a and Q_r. Similarly, the probability of an idle slot is approximately $e^{-G(n)}$. Thus, the number of packets in a slot is well approximated as a Poisson random variable (as in the earlier intuitive analysis), but the parameter $G(n)$ varies with the state. Figure 4.4 illustrates Eqs. (4.4) and (4.6) for the case $q_r > q_a$ (this is the only case of interest, as discussed later). The drift is the difference between the curve and the straight line. Since the drift is the expected change in state from one slot to the next, the system, although perhaps fluctuating, tends to move in the direction of the drift and consequently tends to cluster around the two stable points with rare excursions between the two.

There are two important conclusions from this figure. First, the departure rate (*i.e.*, P_{succ}) is at most $1/e$ for large m. Second, the departure rate is almost zero for long periods whenever the system jumps to the undesirable stable point. Consider the effect of

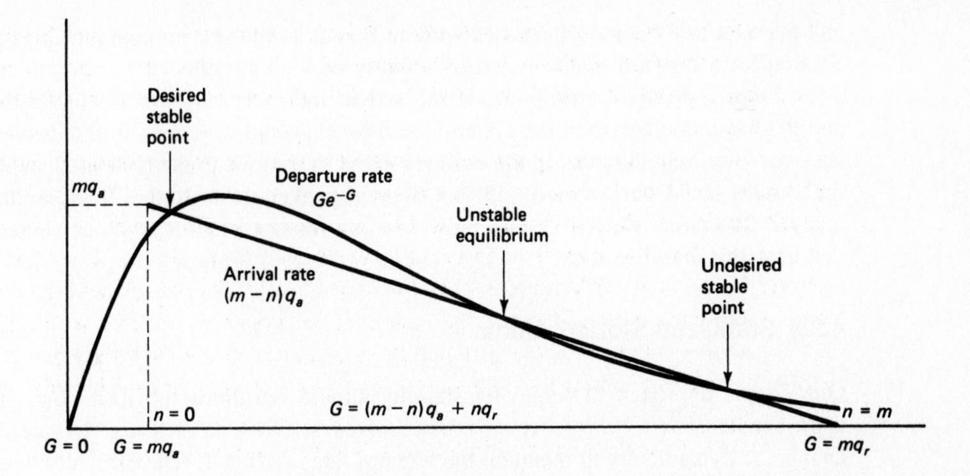

Figure 4.4 Instability of slotted Aloha. The horizontal axis corresponds to both the state n and the attempt rate G, which are related by the linear equation $G = (m-n)q_a + nq_r$ with $q_r > q_a$. For n to the left of the unstable point, D is negative and n drifts toward the desired stable point. For n to the right of the unstable point, D is positive and n drifts toward the undesired stable point.

changing q_r. As q_r is increased, the delay in retransmitting a collided packet decreases, but also the linear relationship between n and the attempt rate $G(n) = (m-n)q_a + nq_r$ changes [*i.e.*, $G(n)$ increases with n faster when q_r is increased]. If the horizontal scale for n is held fixed in Fig. 4.4, this change in attempt rate corresponds to a contraction of the horizontal $G(n)$ scale, and thus to a horizontal contraction of the curve Ge^{-G}. This means that the number of backlogged packets required to exceed the unstable equilibrium point decreases. Alternatively, if q_r is decreased, retransmission delay increases, but it becomes more difficult to exceed the unstable equilibrium point. If q_r is decreased enough (while still larger than q_a), the curve Ge^{-G} will expand enough in Fig. 4.4 that only one stable state will remain. At this stable point, and similarly when $q_r = q_a$, the backlog is an appreciable fraction of m, and this means both that an appreciable number of arriving packets are discarded and that the delay is excessively large.

The question of what values of q_r and arrival rate lead to stable behavior, particularly when q_r and arrival rate vary from node to node and infinite buffering is provided at each node, has received considerable theoretical attention (*e.g.*, [Tsy85]). These theoretical studies generally assume that nodes are considered backlogged immediately on arrival. Problem 4.8, however, illustrates that if q_r is small enough for stable operation, then the delay is considerably greater than that with TDM; thus these approaches are not of great practical importance.

If we replace the no-buffering assumption with the infinite-node assumption, the attempt rate $G(n)$ becomes $\lambda + nq_r$ and the straight line representing arrivals in Fig. 4.4 becomes horizontal. In this case, the undesirable stable point disappears, and once the state of the system passes the unstable equilibrium, it tends to increase without bound. In this case, the corresponding infinite-state Markov chain has no steady-state distribution and the expected backlog increases without bound as the system continues to run. (See Section 3A.5 for further discussion of stability of such systems.)

From a practical standpoint, if the arrival rate λ is very much smaller than $1/e$, and if q_r is moderate, then the system could be expected to remain in the desired stable state for very long periods. In the unlikely event of a move to the undesired stable point, the system could be restarted with backlogged packets lost. Rather than continuing to analyze this rather flawed system, however, we look at modifications of slotted Aloha that cure this stability issue.

4.2.3 Stabilized Slotted Aloha

One simple approach to achieving stability should be almost obvious now. P_{succ} is approximately $G(n)e^{-G(n)}$, which is maximized at $G(n) = 1$. Thus, it is desirable to change q_r dynamically to maintain the attempt rate $G(n)$ at 1. The difficulty here is that n is unknown to the nodes and can only be estimated from the feedback. There are many strategies for estimating n or the appropriate value of q_r (*e.g.*, [HaL82] and [Riv85]). All of them, in essence, increase q_r when an idle slot occurs and decrease q_r when a collision occurs.

Stability and maximum throughput. The notion of stability must be clarified somewhat before proceeding. Slotted Aloha was called unstable in the last subsection on the basis of the representation in Fig. 4.4. Given the no-buffering assumption, however, the system has a well-defined steady-state behavior for all arrival rates. With the infinite-node assumption, on the other hand, there is no steady-state distribution and the expected delay grows without bound as the system continues to run. With the no-buffering assumption, the system discards large numbers of arriving packets and has a very large but finite delay, whereas with the infinite-node assumption, no arrivals are discarded but the delay becomes infinite.

In the following, we shall use the infinite-node assumption (6b), and define a multiaccess system as stable for a given arrival rate if the expected delay per packet (either as a time average or an ensemble average in the limit of increasing time) is finite. Ordinary slotted Aloha is unstable, in this sense, for any arrival rate greater than zero. Note that if a system is stable, then for a sufficiently large but finite number of nodes m, the system (regarding each arrival as corresponding to a new virtual node) has a smaller expected delay than TDM, since the delay of TDM, for a fixed overall arrival rate, increases linearly with m.

The maximum stable throughput of a multiaccess system is defined as the least upper bound of arrival rates for which the system is stable. For example, the maximum stable throughput of ordinary slotted Aloha is zero. Our purpose with these definitions is to study multiaccess algorithms that do not require knowledge of the number of nodes m and that maintain small delay (for given λ) independent of m. Some discussion will be given later to modifications that make explicit use of the number of nodes.

Returning to the stabilization of slotted Aloha, note that when the estimate of backlog is perfect, and $G(n)$ is maintained at the optimal value of 1, then (according to the Poisson approximation) idles occur with probability $1/e \approx 0.368$, successes occur with probability $1/e$, and collisions occur with probability $1 - 2/e \approx 0.264$. Thus, the rule for changing q_r should allow fewer collisions than idles. The maximum stable throughput of such a system is at most $1/e$. To see this, note that when the backlog is large, the Poisson approximation becomes more accurate, the success rate is then limited to $1/e$, and thus the drift is positive for $\lambda > 1/e$. It is important to observe that this argument depends on all backlogged nodes using the same retransmission probability. We shall see in Section 4.3 that if nodes use both their own history of retransmissions and the feedback history in their decisions about transmitting, maximum stable throughputs considerably higher than $1/e$ are possible.

Pseudo-Bayesian algorithm. Rivest's pseudo-Bayesian algorithm [Riv85] is a particularly simple and effective way to stabilize Aloha. This algorithm is essentially the same as an earlier, independently derived algorithm by Mikhailov [Mik79], but Rivest's Bayesian interpretation simplifies understanding. The algorithm differs from slotted Aloha in that new arrivals are regarded as backlogged immediately on arrival. Rather than being transmitted with certainty in the next slot, they are transmitted with probability q_r in the same way as packets involved in previous collisions. Thus, if there are n backlogged packets (including new arrivals) at the beginning of a slot, the attempt rate is $G(n) = nq_r$, and the probability of a successful transmission is $nq_r(1 - q_r)^{n-1}$. For unstabilized Aloha, this modification would not make much sense, since q_r has to be relatively small and new arrivals would be unnecessarily delayed. For stabilized Aloha, however, q_r can be as large as 1 when the estimated backlog is negligible, so that new arrivals are held up only when the system is already estimated to be congested. In Problem 4.6 it is shown that this modification increases the probability of success if the backlog estimate is accurate.

The algorithm operates by maintaining an estimate $\hat{n}$ of the backlog n at the beginning of each slot. Each backlogged packet is then transmitted (independently) with probability $q_r(\hat{n}) = \min\{1, 1/\hat{n}\}$. The minimum operation limits q_r to at most 1, and subject to this limitation, tries to achieve an attempt rate $G = nq_r$ of 1. For each k, the estimated backlog at the beginning of slot $k + 1$ is updated from the estimated backlog and feedback for slot k according to the rule

$$\hat{n}_{k+1} = \begin{cases} \max\{\lambda, \hat{n}_k + \lambda - 1\}, & \text{for idle or success} \\ \hat{n}_k + \lambda + (e-2)^{-1}, & \text{for collision} \end{cases} \tag{4.7}$$

The addition of λ to the previous backlog accounts for new arrivals, and the max operation ensures that the estimate is never less than the contribution from new arrivals. On successful transmissions, 1 is subtracted from the previous backlog to account for the successful departure. Finally, subtracting 1 from the previous backlog on idle slots and adding $(e - 2)^{-1}$ on collisions has the effect of decreasing $\hat{n}$ when too many idles occur and increasing $\hat{n}$ when too many collisions occur. For large backlogs, if $\hat{n} = n$, each of the n backlogged packets is independently transmitted in a slot with probability $q_r = 1/n$. Thus $G(n)$ is 1, and, by the Poisson approximation, idles occur with probability $1/e$ and collisions with probability $(e-2)/e$, so that decreasing $\hat{n}$ by 1 on idles and increasing $\hat{n}$ by $(e-2)^{-1}$ on collisions maintains the balance between n and $\hat{n}$ on the average.

It is shown in Problem 4.9 that if the a priori probability distribution on n_k is Poisson with parameter $\hat{n}_k \geq 1$, then given an idle or successful transmission in slot k, the probability distribution of n_{k+1} is Poisson with parameter $\hat{n}_k + \lambda - 1$. Given a collision, the resulting distribution of n_{k+1} is not quite Poisson but is reasonably approximated as Poisson with parameter $\hat{n}_{k+1}$. For this reason, the algorithm is called pseudo-Bayesian.

Figure 4.5 provides some heuristic understanding of why the algorithm is stable for all $\lambda < 1/e$. The state of the system is characterized by n and $\hat{n}$, and the figure shows the expected drift in these variables from one slot to the next. If n and $\hat{n}$ are large and equal, the expected drift of each is $\lambda - 1/e$, which is negative. On the other hand, if the absolute value of $n - \hat{n}$ is large, the expected drift of n is positive, but the expected reduction in $|n - \hat{n}|$ is considerably larger. Thus, if the system starts at some arbitrary point in the $(n, \hat{n})$ plane, n might increase on the average for a number of slots, but eventually n and $\hat{n}$ will come closer together and then n will decrease on the average.

In applications, the arrival rate λ is typically unknown and slowly varying. Thus, the algorithm must either estimate λ from the time-average rate of successful transmissions or set its value within the algorithm to some fixed value. It has been shown by Mikhailov (see [Kel85]) and Tsitsiklis [Tsi87] that if the fixed value $1/e$ is used within the algorithm, stability is achieved for all actual $\lambda < 1/e$. Nothing has been proven about the behavior of the algorithm when a dynamic estimate of λ is used within the algorithm.

Approximate delay analysis. An exact analysis of expected delay for this algorithm, even with known λ, appears to be very difficult, but it is instructive to analyze an approximate model. Assume that λ is known and that the probability of successful transmission P_{succ} is $1/e$ whenever the backlog n is 2 or more, and that $P_{succ} = 1$ for

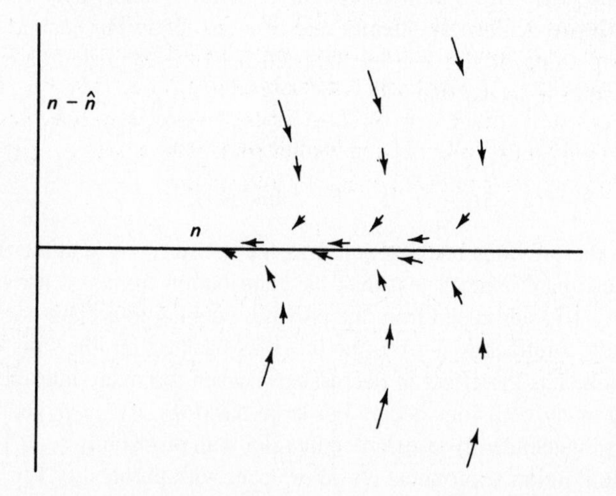

Figure 4.5 Drift of n and $n - \hat{n}$ for the pseudo-Bayesian stabilization algorithm. When the absolute value of $n - \hat{n}$ is large, it approaches 0 faster than n increases.

$n = 1$. This is a reasonable model for very small λ, since very few collisions occur and q_r is typically 1. It is also reasonable for large $\lambda < 1/e$, since typically n is large and $\hat{n} \approx n$.

Let W_i be the delay from the arrival of the ith packet until the beginning of the ith successful transmission. Note that if the system were first-come first-serve, W_i would be the queueing time of the ith arrival. By the same argument as in Problem 3.32, however, the average of W_i over all i is equal to the expected queueing delay W. Let n_i be the number of backlogged packets at the instant before i's arrival; n_i does not include any packet currently being successfully transmitted, but does include current unsuccessful transmissions. W_i can be expressed as

$$W_i = R_i + \sum_{j=1}^{n_i} t_j + y_i \tag{4.8}$$

R_i is the residual time to the beginning of the next slot, and t_1 (for $n_i > 0$) is the subsequent interval until the next successful transmission is completed. Similarly, t_j, $1 < j \leq n_i$, is the interval from the end of the $(j-1)^{\text{th}}$ subsequent success to the end of the jth subsequent success. After those n_i successes, y_i is the remaining interval until the beginning of the next successful transmission (*i.e.*, the ith transmission overall).

Observe that for each interval t_j, the backlog is at least two, counting the new i^{th} arrival and the n_i packets already backlogged. Thus, each slot is successful with probability $1/e$, and the expected value of each t_j is e. Next, observe that Little's formula relates the expected values of W_i and n_i (*i.e.*, W_i is the wait not counting the successful transmission, and n_i is the number in the system not counting successful transmissions in process). Finally, the expected value of R_i is $1/2$ (counting time in slots). Thus, taking the expected value of Eq. (4.8) and averaging over i, we get

$$W = 1/2 + \lambda e W + E\{y\} \tag{4.9}$$

Now consider the system at the first slot boundary at which both the $(i-1)^{\text{st}}$ departure has occurred and the ith arrival has occurred. If the backlog is 1 at that point (*i.e.*, only the ith packet is in the system), then y_i is 0. Alternatively, if $n > 1$, then $E\{y_i\} = e - 1$. Let p_n be the steady-state probability that the backlog is n at a slot boundary. Then, since a packet is always successfully transmitted if the state is 1 at the beginning of the slot, we see that p_1 is the fraction of slots in which the state is 1 and a packet is successfully transmitted. Since λ is the total fraction of slots with successful transmissions, p_1/λ is the fraction of packets transmitted from state 1 and $1 - p_1/\lambda$ is the fraction transmitted from higher-numbered states. It follows that

$$E\{y\} = \frac{(e-1)(\lambda - p_1)}{\lambda} \tag{4.10}$$

Finally, we must determine p_1. From the argument above, we see that $\lambda = p_1 + (1 - p_0 - p_1)/e$. Also, state 0 can be entered at a slot boundary only if no arrivals occurred in the previous slot and the previous state was 0 or 1. Thus, $p_0 = (p_0 + p_1)e^{-\lambda}$. Solving for p_1 gives

$$p_1 = \frac{(1 - \lambda e)(e^\lambda - 1)}{1 - (e-1)(e^\lambda - 1)} \tag{4.11}$$

Combining Eqs. (4.9) to (4.11) yields

$$W = \frac{e - 1/2}{1 - \lambda e} - \frac{(e^{\lambda} - 1)(e - 1)}{\lambda[1 - (e - 1)(e^{\lambda} - 1)]} \tag{4.12}$$

This value of W is quite close to delay values for the pseudo-Bayesian algorithm obtained through simulation. Figure 4.6 plots $E\{W\}$ for this approximate model and compares it with the queueing delay for TDM with 8 and 16 nodes. It is quite surprising that the delay is so small even for arrival rates relatively close to $1/e$. It appears that the assumption of immediate feedback is unnecessary for stabilization strategies of this type to be stable; the argument is that feedback delay will make the estimate $\hat{n}$ less accurate, but $n/\hat{n}$ will still stay close to 1 for large n.

Binary exponential backoff. In the packet radio networks to be discussed in Section 4.6, and in some other multiaccess situations, the assumption of 0,1,e feedback on all slots is unrealistic. In some systems, a node receives feedback only about whether or not its own packets were successfully transmitted; it receives no feedback about slots in which it does not transmit. Such limited feedback is sufficient for slotted Aloha but is insufficient for the backlog estimation of the pseudo-Bayesian strategy. Binary exponential backoff [MeB76] is a stabilization strategy used in Ethernet that employs only this more limited form of feedback.

This strategy is very simple; if a packet has been transmitted unsuccessfully i times, then the probability of transmission in successive slots is set at $q_r = 2^{-i}$ (or is uniformly distributed over the next 2^i slots after the ith failure). When a packet initially arrives in the system, it is transmitted immediately in the next slot.

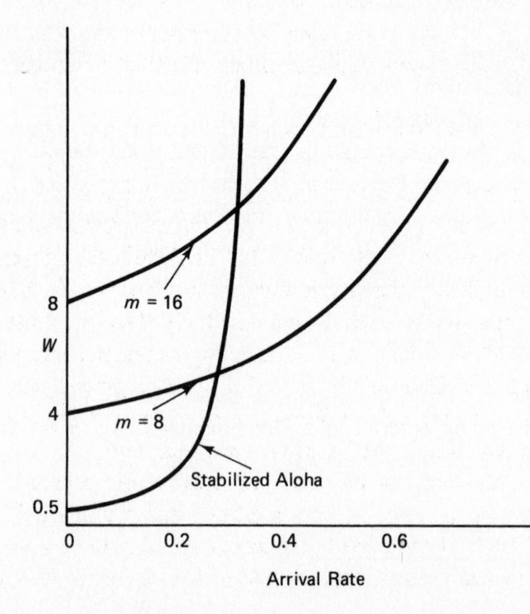

Figure 4.6 Comparison of expected waiting time W in slots, from arrival until beginning of successful transmission, for stabilized Aloha and for TDM with $m = 8$ and $m = 16$. For small arrival rates, the delay of stabilized Aloha is little more than waiting for the next slot, whereas as the arrival rate approaches $1/e$, the delay becomes unbounded.

Some rationale for this strategy can be seen by noting that when a packet first arrives (with this limited feedback), the node knows nothing of the backlog, so the immediate first transmission is reasonable. With successive collisions, any reasonable estimate of backlog would increase, motivating the decrease in the local q_r. To make matters worse, however, as q_r is reduced, the node gets less feedback per slot about the backlog, and thus, to play safe, it is reasonable to increase the backlog estimate by larger and larger amounts on each successive collision.

Unfortunately, in the limit as the number of nodes approaches infinity, this strategy is unstable for every arrival rate λ greater than 0 [Ald86]. It is unknown whether or not any strategy can achieve stability with this type of limited feedback. Problem 4.10, however, develops another strategy that can be used with this limited feedback and with a finite number of nodes with unlimited buffering to achieve finite expected delay for any λ less than 1. Unfortunately, the price of this high throughput is inordinately large delay.

4.2.4 Unslotted Aloha

Unslotted, or pure, Aloha [Abr70] was the precursor to slotted Aloha. In this strategy, each node, upon receiving a new packet, transmits it immediately rather than waiting for a slot boundary. Slots play no role in pure Aloha, so we temporarily omit the slotted system assumption. If a packet is involved in a collision, it is retransmitted after a random delay. Assume that if the transmission times for two packets overlap at all, the CRCs on those packets will fail and retransmission will be required. We assume that the receiver rebroadcasts the composite received signal (or that all nodes receive the composite signal), so that each node, after a given propagation delay, can determine whether or not its transmitted packets were correctly received. Thus, we have the same type of limited feedback discussed in the last subsection. Other types of feedback could be considered and Problem 4.11 develops some of the peculiar effects of other feedback assumptions.

Figure 4.7 shows that if one packet starts transmission at time t, and all packets have unit length, then any other transmission starting between $t-1$ and $t+1$ will cause a collision. For simplicity, assume an infinite number of nodes (*i.e.*, assumption 6b) and let n be the number of backlogged nodes at a given time. For our present purposes, a node is considered backlogged from the time it has determined that its previously transmitted packet was involved in a collision until the time that it attempts retransmission. Assume that the period until attempted retransmission τ is an exponentially distributed random variable with probability density $xe^{-x\tau}$, where x is an arbitrary parameter interpreted as a node's retransmission attempt rate. Thus, with an overall Poisson arrival rate of λ to the system, the initiation times of attempted transmissions is a time-varying Poisson process of rate $G(n) = \lambda + nx$ in which n is the backlog at a given time.

Consider the sequence of successive transmission attempts on the channel. For some given i, let τ_i be the duration of the interval between the initiations of the ith and $(i+1)^{\text{th}}$ transmission attempt. The ith attempt will be successful if both τ_i and τ_{i-1} exceed 1 (assuming unit length packets). Given the backlog in each intertransmission

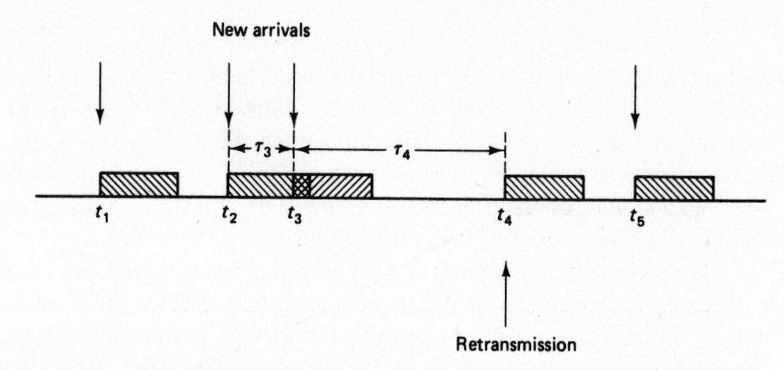

Figure 4.7 Unslotted Aloha. New arrivals are transmitted immediately and unsuccessful transmissions are repeated after a random delay. Packet transmission time is one unit, and two transmissions collide if their interdeparture interval is less than one unit.

interval, these intervals are independent. Thus, assuming a backlog of n for each interval, we have

$$P_{succ} = e^{-2G(n)} \tag{4.13}$$

Since attempted transmissions occur at rate $G(n)$, the throughput (*i.e.*, the expected number of successful transmissions per unit time) as a function of n is

$$\text{throughput}\,(n) = G(n)e^{-2G(n)} \tag{4.14}$$

Figure 4.8 illustrates this result. The situation is very similar to that of slotted Aloha, except that the throughput has a maximum of $1/(2e)$, achieved when $G(n) = 1/2$. The analysis above is approximate in the sense that Eq. (4.13) assumes that the backlog is the same in the intervals surrounding a given transmission attempt, whereas according to our definition of backlog, the backlog decreases by one whenever a backlogged packet initiates transmission and increases by one whenever a collided packet is detected. For small x (*i.e.*, large mean time before attempted retransmission), this effect is relatively small.

It can be seen from Fig. 4.8 that pure Aloha has the same type of stability problem as slotted Aloha. For the limited feedback assumed here, stability is quite difficult to achieve or analyze (as is the case with slotted Aloha). Very little is known about stabilization for pure Aloha, but if λ is very small and the mean retransmission time very large, the system can be expected to run for long periods without major backlog buildup.

One of the major advantages of pure Aloha is that it can be used with variable-length packets, whereas with slotted Aloha, long packets must be broken up to fit into slots and short packets must be padded out to fill up slots. This compensates for some of the inherent throughput loss of pure Aloha and gives it an advantage in simplicity. Some analysis, with simplifying assumptions, of the effect of variable-length packets appears in [Fer75] and [San80].

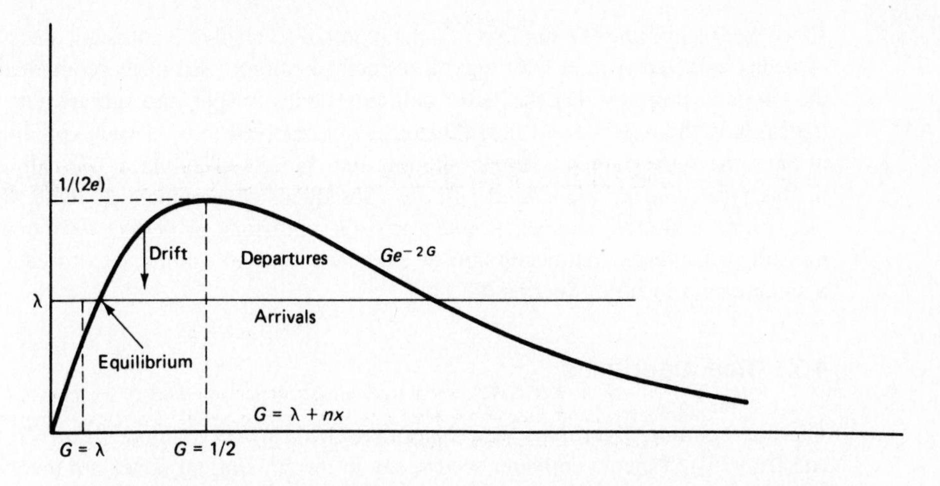

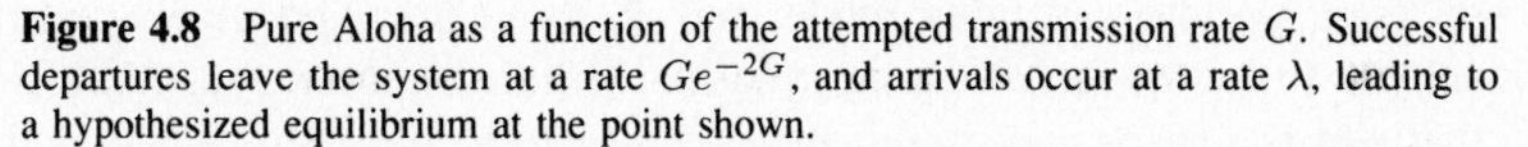
Figure 4.8 Pure Aloha as a function of the attempted transmission rate G. Successful departures leave the system at a rate Ge^{-2G}, and arrivals occur at a rate λ, leading to a hypothesized equilibrium at the point shown.

4.3 SPLITTING ALGORITHMS

We have seen that slotted Aloha requires some care for stabilization and is also essentially limited to throughputs of $1/e$. We now want to look at more sophisticated collision resolution techniques that both maintain stability without any complex estimation procedures and also increase the achievable throughput. To get an intuitive idea of how this can be done, note that with relatively small attempt rates, when a collision occurs, it is most likely between only two packets. Thus, if new arrivals could be inhibited from transmission until the collision was resolved, each of the colliding packets could be independently retransmitted in the next slot with probability $1/2$. This would lead to a successful transmission for one of the packets with probability $1/2$, and the other could then be transmitted in the following slot. Alternatively, with probability $1/2$, another collision or an idle slot occurs. In this case, each of the two packets would again be independently transmitted in the next slot with probability $1/2$, and so forth until a successful transmission occurred, which would be followed by the transmission of the remaining packet.

With the strategy above, the two packets require two slots with probability $1/2$, three slots with probability $1/4$, and i slots with probability $2^{-(i-1)}$. The expected number of slots for sending these two packets can thus be calculated to be three, yielding a throughput of $2/3$ for the period during which the collision is being resolved.

There are various ways in which the nodes involved in a collision could choose whether or not to transmit in successive slots. Each node could simply flip an unbiased coin for each choice. Alternatively (in a way to be described precisely later) each node could use the arrival time of its collided packet. Finally, assuming a finite set of nodes, each with a unique identifier represented by a string of bits, a node could use the successive bits of its identity to make the successive choices. This last alternative has the advantage of limiting the number of slots required to resolve a collision, since each pair of nodes must differ in at least one bit of their identifiers. All of these alternatives have the common property that the set of colliding nodes is split into subsets, one of which transmits in the next slot. If the collision is not resolved (*e.g.*, if each colliding node is in the same subset), then a further splitting into subsets takes place. We call algorithms of this type *splitting algorithms*. In the subsequent development of these algorithms, we assume a slotted channel, Poisson arrivals, collisions or perfect reception, $(0, 1, e)$ immediate feedback, retransmission of collisions, and an infinite set of nodes (*i.e.*, the assumptions 1 to 6b of Section 4.2.1).

4.3.1 Tree Algorithms

The first splitting algorithms were algorithms with a tree structure ([Cap77], [TsM78], and [Hay76]). When a collision occurs, say in the kth slot, all nodes not involved in the collision go into a waiting mode, and all those involved in the collision split into two subsets (*e.g.*, by each flipping a coin). The first subset transmits in slot $k+1$, and if that slot is idle or successful, the second subset transmits in slot $k+2$ (see Fig. 4.9). Alternatively, if another collision occurs in slot $k+1$, the first of the two subsets splits again, and the second subset waits for the resolution of that collision.

The rooted binary tree structure in Fig. 4.9 represents a particular pattern of idles, successes, and collisions resulting from such a sequence of splittings. S represents the

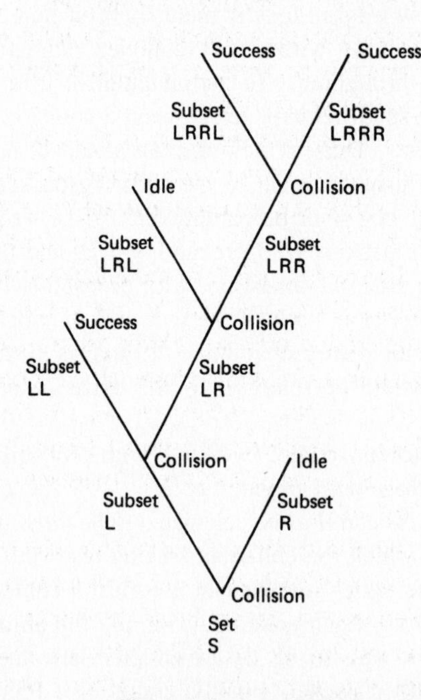

Slot	Xmit Set	Waiting Sets	Feedback
1	S	–	*e*
2	L	R	*e*
3	LL	LR, R	1
4	LR	R	*e*
5	LRL	LRR, R	0
6	LRR	R	*e*
7	LRRL	LRRR, R	1
8	LRRR	R	1
9	R	–	0

Figure 4.9 Tree algorithm. After a collision, all new arrivals wait and all nodes involved in the collision divide into subsets. Each successive collision in a subset causes that subset to again split into smaller subsets while other nodes wait.

set of packets in the original collision, and L (left) and R (right) represent the two subsets that S splits into. Similarly, LL and LR represent the two subsets that L splits into after L generates a collision. Each vertex in the tree corresponds to a subset (perhaps empty) of backlogged packets. Vertices whose subsets contain two or more packets have two upward branches corresponding to the splitting of the subset into two new subsets; vertices corresponding to subsets with 0 or 1 packet are leaf vertices of the tree.

The set of packets corresponding to the root vertex S is transmitted first, and after the transmission of the subset corresponding to any nonleaf vertex, the subset corresponding to the vertex on the left branch, and all of its descendant subsets, are transmitted before the subsets of the right branch. Given the immediate feedback we have assumed, it should be clear that each node, in principle, can construct this tree as the $0, 1, e$ feedback occurs; each node can keep track of its own subset in that tree, and thus each node can determine when to transmit its own backlogged packet.

The transmission order above corresponds to that of a stack. When a collision occurs, the subset involved is split, and each resulting subset is pushed on the stack (*i.e.*, each stack element is a subset of nodes); then the subset at the head of the stack (*i.e.*, the most recent subset pushed on the stack) is removed from the stack and transmitted. The list, from left to right, of waiting subsets in Fig. 4.9 corresponds to the stack elements starting at the head for the given slot. Note that a node with a backlogged packet can keep track of when to transmit by a counter determining the position of the packet's current subset on the stack. When the packet is involved in a collision, the counter is set to 0 or 1, corresponding to which subset the packet is placed in. When the counter is 0, the packet is transmitted, and if the counter is nonzero, it is incremented by 1 for each collision and decremented by 1 for each success or idle.

One problem with this tree algorithm is what to do with the new packet arrivals that come in while a collision is being resolved. A collision resolution period (CRP) is defined to be completed when a success or idle occurs and there are no remaining elements on the stack (*i.e.*, at the end of slot 9 in Fig. 4.9). At this time, a new CRP starts using the packets that arrived during the previous CRP. In the unlikely event that a great many slots are required in the previous CRP, there will be many new waiting arrivals, and these will collide and continue to collide until the subsets get small enough in this new CRP. The solution to this problem is as follows: At the end of a CRP, the set of nodes with new arrivals is immediately split into j subsets, where j is chosen so that the expected number of packets per subset is slightly greater than 1 (slightly greater because of the temporary high throughput available after a collision). These new subsets are then placed on the stack and the new CRP starts.

The tree algorithm is now essentially completely defined. Each node with a packet involved in the current CRP keeps track of its position in the stack as described above. All the nodes keep track of the number of elements on the stack and the number of slots since the end of the previous CRP. On the completion of that CRP, each node determines the expected number of waiting new arrivals, determines the new number j of subsets, and those nodes with waiting new arrivals randomly choose one of those j subsets and set their counter for the corresponding stack position.

The maximum throughput available with this algorithm, optimized over the choice of j as a function of expected number of waiting packets, is 0.43 packets per slot [Cap77]; we omit any analysis since we next show some simple ways of improving this throughput.

Improvements to the tree algorithm. First consider the situation in Fig. 4.10. Here, in slots 4 and 5, a collision is followed by an idle slot; this means that all the packets involved in the collision were assigned to the second subset. The tree algorithm would simply transmit this second subset, generating a guaranteed collision. An improvement results by omitting the transmission of this second subset, splitting it into two subsets, and transmitting the first subset. Similarly, if an idle again occurs, the second subset is again split before transmission, and so forth.

This improvement can be visualized in terms of a stack and implemented by manipulating counters just like the original tree algorithm. Each node must now keep track of an additional binary state variable that takes the value 1 if, for some $i \geq 1$, the last i slots contained a collision followed by $i - 1$ idles; otherwise, the state variable takes the value 0. If the feedback for the current slot is 0 and the state variable has the value 1, then the state variable maintains the value 1 and the subset on the top of the stack is split into two subsets that are pushed onto the stack in place of the original head element.

The maximum throughput with this improvement is 0.46 packets per slot [Mas80]. In practice, this improvement has a slight problem in that if an idle slot is incorrectly perceived by the receiver as a collision, the algorithm continues splitting indefinitely, never making further successful transmissions. Thus, in practice, after some number h of idle slots followed by splits, the algorithm should be modified simply to transmit the next subset on the stack without first splitting it; if the feedback is very reliable, h can be moderately large, whereas otherwise h should be small.

The next improvement in the tree algorithm not only improves throughput but also greatly simplifies the analysis. Consider what happens when a collision is followed by another collision in the tree algorithm (see slots 1 and 2 of Fig. 4.10). Let x be the number of packets in the first collision, and let x_L and x_R be the number of packets in the resultant subsets; thus, $x = x_L + x_R$. Assume that, a priori, before knowing that there is a collision, x is a Poisson random variable. Visualize splitting these x packets, by coin flip, say, into the two subsets with x_L and x_R packets, respectively, before knowing about the collision. Then a priori x_L and x_R are independent Poisson random variables each with half the mean value of x. Given the two collisions, then, x_L and x_R are independent Poisson random variables conditional on $x_L + x_R \geq 2$ and $x_L \geq 2$. The second condition implies the first, so the first can be omitted; this means that x_R, conditional on the feedback, is still Poisson with its original unconditional distribution. Problem 4.17 demonstrates this result in a more computational and less abstract way. Thus, rather than devoting a slot to this second subset, which has an undesirably small expected number of packets, it is better to regard the second subset as just another part of the waiting new arrivals that have never been involved in a collision.

When the idea above is incorporated into an actual algorithm, the first-come first-serve (FCFS) splitting algorithm, which is the subject of the next subsection, results. Before discussing this, we describe some variants of the tree algorithm.

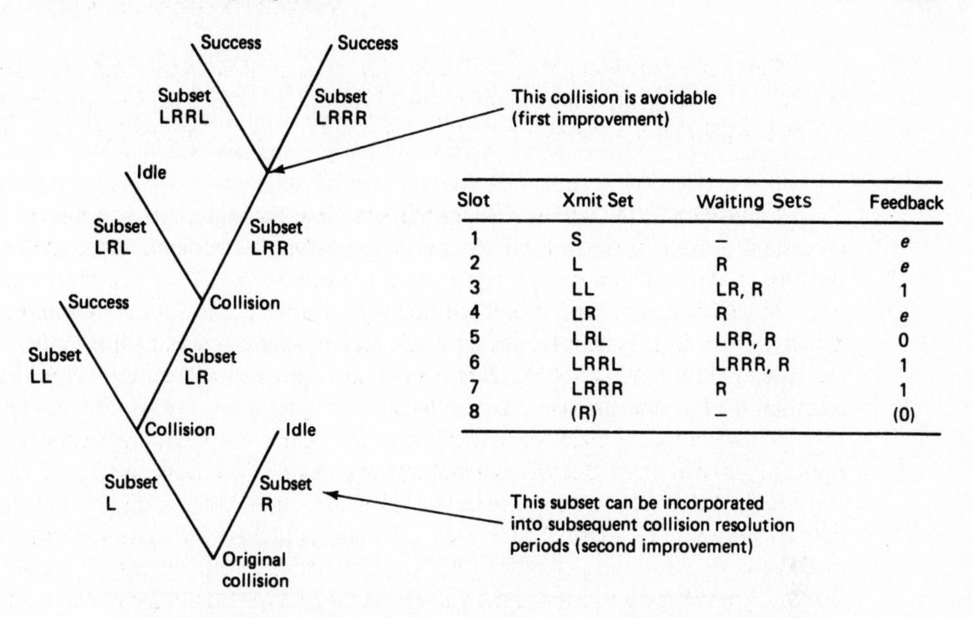

Slot	Xmit Set	Waiting Sets	Feedback
1	S	–	e
2	L	R	e
3	LL	LR, R	1
4	LR	R	e
5	LRL	LRR, R	0
6	LRRL	LRRR, R	1
7	LRRR	R	1
8	(R)	–	(0)

Figure 4.10 Improvements in the tree algorithm. Subset LRR can be split without first being transmitted since the feedback implies that it contains two or more packets. Also, subset R is better combined with new arrivals since the number of packets in it is Poisson with an undesirably low rate.

Variants of the tree algorithm. The tree algorithm as described above requires all nodes to monitor the channel feedback and to keep track of when each collision resolution period ends. This is a disadvantage if the receivers at nodes are turned off when there are no packets to send. One way to avoid this disadvantage while maintaining the other features of the tree algorithm is to have new arrivals simply join the subset of nodes at the head of the stack. Thus, only currently backlogged nodes need to monitor the channel feedback. Algorithms of this type are called *unblocked stack algorithms*, indicating that new arrivals are not blocked until the end of the current collision resolution period. In contrast, the tree algorithm is often called a blocked stack algorithm.

With the tree algorithm, new arrivals are split into a variable number of subsets at the beginning of each collision resolution period, and then subsets are split into two subsets after each collision. With an unblocked stack algorithm, on the other hand, new arrivals are constantly being added to the subset at the head of the stack, and thus, collisions involve a somewhat larger number of packets on the average. Because of the relatively large likelihood of three or more packets in a collision, higher maximum throughputs can be achieved by splitting collision sets into three subsets rather than two. The maximum throughput thus available for unblocked stack algorithms is 0.40 [MaF85].

4.3.2 First-Come First-Serve Splitting Algorithms

In the second improvement to the tree algorithm described above, the second subset involved in a collision is treated as if it had never been transmitted if the first subset contains two or more packets. Recall also that a set of colliding packets can be split into two subsets in a variety of ways (*e.g.*, by coin flipping, by node identities, or by arrival time). For our present purposes, the simplest choice is splitting by arrival time. By using this approach, each subset will consist of all packets that arrived in some given interval, and when a collision occurs, that interval will be split into two smaller intervals. By always transmitting the earlier arriving interval first, the algorithm will always transmit successful packets in the order of their arrival, leading to the name first-come first-serve (FCFS).

At each integer time k, the algorithm specifies the packets to be transmitted in slot k (*i.e.*, from k to $k+1$) to be the set of all packets that arrived in some earlier interval, say from $T(k)$ to $T(k)+\alpha(k)$. This interval is called the *allocation interval* for slot k (see Fig. 4.11). We can visualize the packets arriving after $T(k)+\alpha(k)$ as being in a

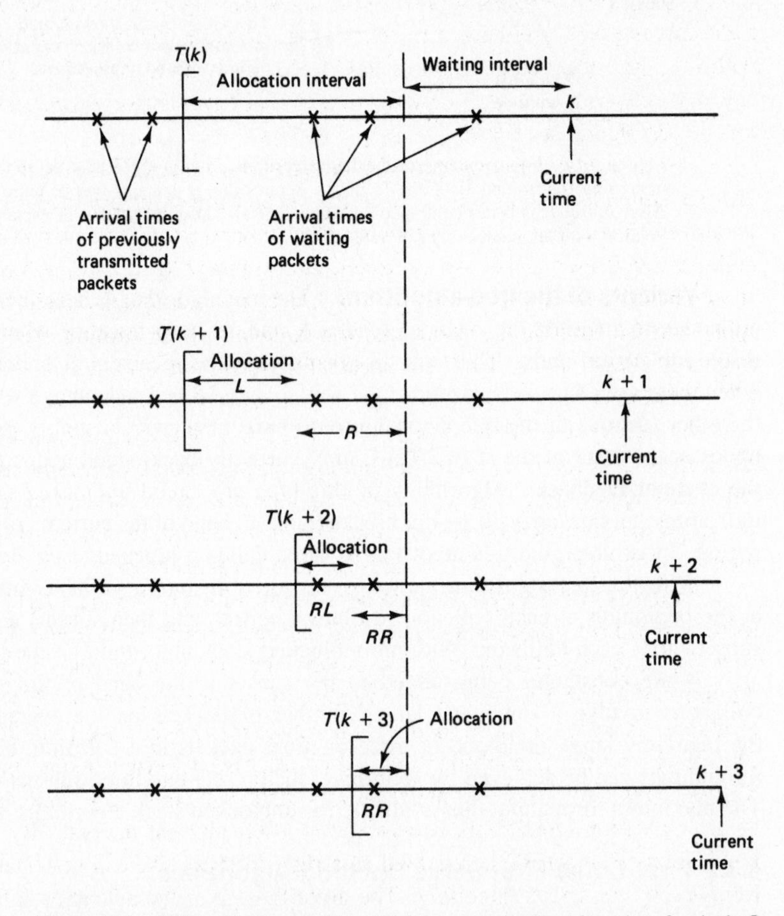

Figure 4.11 FCFS splitting algorithm. Packets are transmitted in order of arrival. On collisions, the allocation interval generating a collision is split into two subintervals, with the leftmost (earlier arrivals) transmitting first.

queue and those arriving between $T(k)$ and $T(k) + \alpha(k)$ as being in service. What is peculiar about this queue is that the number of packets in it is unknown, although the nodes all keep track of the allocation interval over which service (*i.e.*, transmission) is taking place.

The FCFS splitting algorithm is the set of rules by which the nodes calculate $T(k)$ and $\alpha(k)$ for each successive k on the basis of the feedback from the previous slot. These rules are simply the improved rules for the tree algorithm discussed previously, specialized to the case where sets of nodes are split in terms of packet arrival times.

Figure 4.11 illustrates these rules. When a collision occurs, as in slot k, the allocation interval is split into two equal subintervals and the leftmost (*i.e.*, longest waiting) subinterval L is the allocation interval in slot $k+1$. Thus, $T(k+1) = T(k)$ (*i.e.*, the left boundary is unchanged) and $\alpha(k+1) = \alpha(k)/2$. When an idle, as in slot $k+1$, follows a collision, the first improvement to the tree algorithm is employed. The previous rightmost interval R is known to contain two or more packets and is immediately split, with RL forming the allocation interval for slot $k+2$. Thus, $T(k+2) = T(k+1)+\alpha(k+1)$ and $\alpha(k+2) = \alpha(k+1)/2$. Finally, successful transmissions occur in slots $k+2$ and $k+3$, completing the CRP.

Next consider the example of Fig. 4.12. Here a collision in slot k is followed by another collision in slot $k+1$. Here the second improvement to the tree algorithm is employed. Since interval L contains two or more packets, the collision in slot k tells us nothing about interval R, and we would like to regard R as if it had never been part of an allocation interval. As shown for slot k + 2, this is simplicity itself. The interval L is split, with LL forming the next allocation interval and LR waiting; the algorithm simply forgets R. When LL and LR are successfully transmitted in slots $k+2$ and $k+3$, the CRP is complete.

In the tree algorithm, at the end of a CRP, all the waiting packets split into some number of subsets. Here, since the splitting is being done by allocation intervals in time, it is more appropriate simply to choose a new allocation interval of some given size, say α_0, to be discussed later. Note that this new interval, in slot $k+4$, includes the old interval R that previously lost its identity as a separate interval.

In terms of the tree of waiting subsets, the effect of having a right interval lose its identity whenever the corresponding left interval is split is to prune the tree so that it never has more than two leaves (correspondingly, the stack never remembers more than the top two elements). Whenever the allocation interval corresponds to the left subset of a split, there is a corresponding right subset that might have to be transmitted later. Conversely, when the allocation interval corresponds to a right subset, there are no more waiting intervals. Thus, the nodes in this algorithm need only remember the location of the allocation interval and whether it is a left or right interval. By convention, the initial interval of a CRP is regarded as a right interval. We can now state the algorithm followed by each node precisely. The algorithm gives the allocation interval [*i.e.*, $T(k)$ and $\alpha(k)$] and the status ($\sigma = L$ or R) for slot k in terms of the feedback, allocation interval, and status from slot $k-1$.

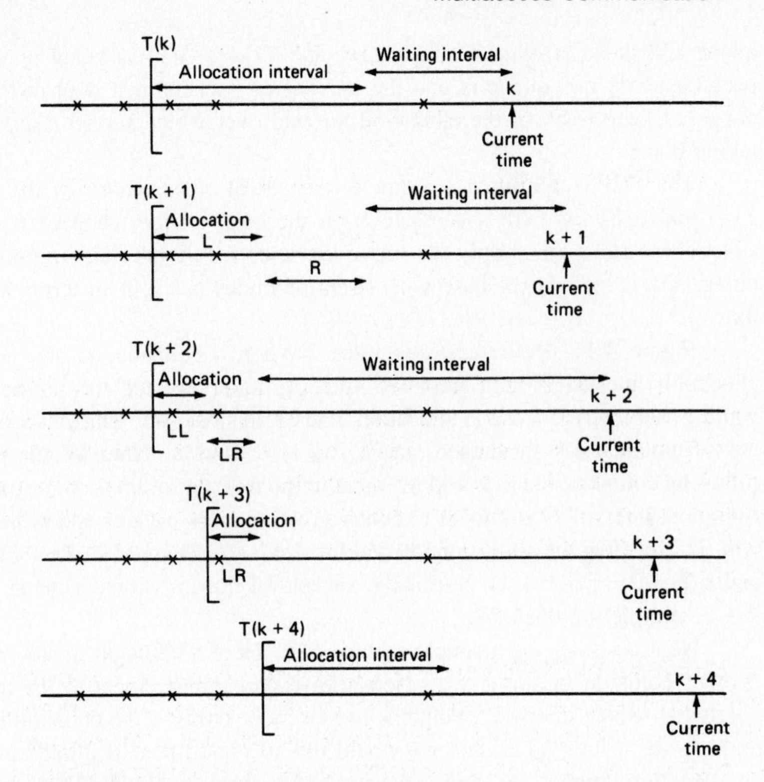

Figure 4.12 FCFS splitting algorithm. When a collision follows another collision, the interval on the right side of the second collision is returned to the waiting interval. The CRP is completed in slot $k+3$, and a new CRP is started with an allocation interval of fixed size.

If feedback $= e$, then

$$\begin{aligned} T(k) &= T(k-1) \\ \alpha(k) &= \frac{\alpha(k-1)}{2} \\ \sigma(k) &= L \end{aligned} \tag{4.15}$$

If feedback $= 1$ and $\sigma(k-1) = L$, then

$$\begin{aligned} T(k) &= T(k-1) + \alpha(k-1) \\ \alpha(k) &= \alpha(k-1) \\ \sigma(k) &= R \end{aligned} \tag{4.16}$$

If feedback $= 0$ and $\sigma(k-1) = L$, then

$$T(k) = T(k-1) + \alpha(k-1)$$

$$\alpha(k) = \frac{\alpha(k-1)}{2} \tag{4.17}$$
$$\sigma(k) = L$$

If feedback $= 0$ or 1 and $\sigma(k-1) = R$, then

$$T(k) = T(k-1) + \alpha(k-1)$$
$$\alpha(k) = \min[\alpha_0, k - T(k)] \tag{4.18}$$
$$\sigma(k) = R$$

The final statement, Eq. (4.18), is used at the end of a collision resolution or when no collisions are being resolved. The size of the new allocation interval in this case is some constant value α_0 which could be chosen either to minimize delay for a given arrival rate or to maximize the stable throughput. Naturally, when the queue becomes depleted, the allocation interval cannot include packets that have not yet arrived, so the interval size is limited to $k - T(k)$, as indicated by the min operation in Eq. (4.18).

Analysis of FCFS splitting algorithm. Figure 4.13 is helpful in visualizing the evolution of a collision resolution period; we shall see later that the diagram can be interpreted as a Markov chain. The node at the left side of the diagram corresponds to the initial slot of a CRP; the node is split in two as an artifice to visualize the beginning and end of a CRP. If an idle or success occurs, the CRP ends immediately and a new

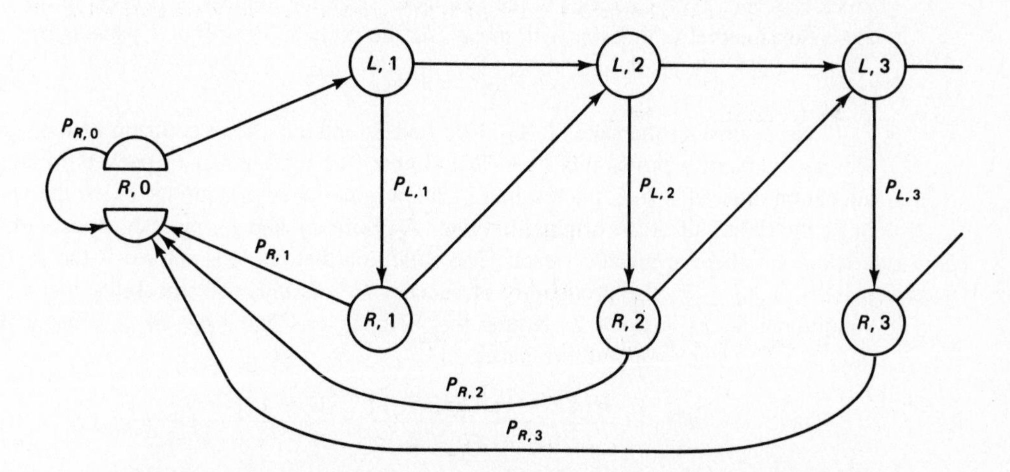

Figure 4.13 Markov chain for FCFS splitting algorithm. The top states are entered after splitting an interval and correspond to the transmission of the left side of that inteval. The lower states are entered after success on the left side and correspond to transmission of the right side. Transitions from top to bottom and from bottom back to $R, 0$ correspond to successes.

CRP starts on the next slot. Alternatively, if a collision occurs, a transition occurs to node $(L, 1)$; L indicates the status and the 1 indicates that one split in the allocation interval has occurred.

Each subsequent idle or collision from a left allocation interval generates one additional split with a smaller left allocation interval, corresponding to a transition in the diagram from (L, i) to $(L, i+1)$, where i is the number of times the original allocation interval has been split. A success from a left interval leads to a right interval with no additional split, corresponding to an (L, i) to (R, i) transition. A success from a right interval ends the CRP, with a transition back to $(R, 0)$, whereas a collision causes a new split, with a transition from (R, i) to $(L, i+1)$.

We now analyze a single CRP. Assume that the size of the initial allocation interval is α_0. Each splitting of the allocation interval decreases this by a factor of 2, so that nodes (L, i) and (R, i) in the diagram correspond to allocation intervals of size $2^{-i}\alpha_0$. Given our assumption of a Poisson arrival process of rate λ, the number of packets in the original allocation interval is a Poisson random variable with mean $\lambda\alpha_0$. Similarly, the a priori distributions on the numbers of packets in disjoint subintervals are independent and Poisson. Define G_i as the expected number of packets, a priori, in an interval that has been split i times,

$$G_i = 2^{-i}\lambda\alpha_0 \tag{4.19}$$

We next find the transition probabilities in Fig. 4.13 and show that they constitute a Markov chain (*i.e.*, that each transition is independent of the path used to reach the given node). Note that we are only interested (for now) in one period of collision resolution; we view the upper half of node $(R, 0)$ as the starting state and the lower half as the final state. We start from the left side and work to the right. $P_{R,0}$ is the probability of an idle or success (*i.e.*, 0 or 1 packet) in the first slot. Since the number of packets in the initial allocation interval is Poisson with mean G_0, the probability of 0 or 1 packets is

$$P_{R,0} = (1 + G_0)e^{-G_0} \tag{4.20}$$

Next consider the state $(L, 1)$. This state is entered after a collision in state $(R, 0)$, which occurs with probability $1 - P_{R,0}$. Let x_L be the number of packets in the new allocation interval L (*i.e.*, the left half of the original interval), and let x_R be the number in R, the right half of the original interval. A priori, x_L and x_R are independent Poisson random variables of mean G_1 each. The condition that $(L, 1)$ is entered is the condition that $x_L + x_R \geq 2$. The probability of success $P_{L,1}$ is thus the probability that $x_L = 1$ conditional on $x_L + x_R \geq 2$. Noting that both $x_L = 1$ and $x_L + x_R \geq 2$ occur if and only if $x_L = 1$ and $x_R \geq 1$, we have

$$P_{L,1} = \frac{P\{x_L = 1\}P\{x_R \geq 1\}}{P\{x_L + x_R \geq 2\}} = \frac{G_1 e^{-G_1}[1 - e^{-G_1}]}{1 - (1 + G_0)e^{-G_0}} \tag{4.21}$$

State $(R, 1)$ is entered if and only if the transition above takes place (*i.e.*, if $x_L = 1$ and $x_R \geq 1$). Thus, the probability of success $P_{R,1}$ in state $(R, 1)$ is

$$P_{R,1} = \frac{P\{x_R = 1\}}{P\{x_R \geq 1\}} = \frac{G_1 e^{-G_1}}{1 - e^{-G_1}} \tag{4.22}$$

We next show that Eqs. (4.21) and (4.22) generalize to $P_{L,i}$ and $P_{R,i}$ for all $i \geq 1$. That is,

$$P_{L,i} = \frac{G_i e^{-G_i}(1 - e^{-G_i})}{1 - (1 + G_{i-1})e^{-G_{i-1}}} \tag{4.23}$$

$$P_{R,i} = \frac{G_i e^{-G_i}}{1 - e^{-G_i}} \tag{4.24}$$

Consider state $(L, 2)$. This can be entered by a collision in state $(L, 1)$, an idle in $(L, 1)$, or a collision in $(R, 1)$. In the first case, interval L is split into LL and LR, and LL becomes the new allocation interval. In the second and third cases, R is split into RL and RR, with RL becoming the new allocation interval. For the first case, let x_{LL} and x_{LR} be the numbers of packets in LL and LR, respectively. A priori, these are independent Poisson random variables of mean G_2 each. The collision from $(L, 1)$ means that $x_L + x_R \geq 2$ and $x_L \geq 2$, which is equivalent to the single condition $x_L \geq 2$. The situation is thus the same as in finding $P_{L,1}$ except that the intervals are half as large, so $P_{L,2}$ in Eq. (4.23) is correct in this case.

Next, consider the second case, that of an idle in $(L, 1)$. This means that $x_L + x_R \geq 2$ and $x_L = 0$, which is equivalent to $x_R \geq 2$ and $x_L = 0$. $P_{L,2}$ in this case is the probability that RL, the left half of R, contains one packet given that R contains at least two; again Eq. (4.23) is correct. Finally, for the case of a collision in $(R, 1)$, we have $x_L + x_R \geq 2, x_L = 1$, and $x_R \geq 2$, or equivalently $x_L = 1, x_R \geq 2$; Eq. (4.23) is again correct for $(L, 2)$. Thus, the Markov condition is satisfied for $(L, 2)$.

Generally, no matter how $(L, 2)$ is entered, the given interval L or R preceding $(L, 2)$ is an interval of size $\alpha_0/2$ of a Poisson process, conditional on the given interval containing two or more packets. If a success occurs on the left half, the number of packets in the right half is Poisson, conditional on being one or more, yielding the expression for $P_{R,2}$ in Eq. (4.24). This argument repeats for $i = 3, 4, \ldots$ (or, more formally, induction can be applied). Thus, Fig. 4.13 is a Markov chain and Eqs. (4.20), (4.23), and (4.24) give the transition probabilities.

The analysis of this chain is particularly simple since no state can be entered more than once before the return to $(R, 0)$. The probabilities, $p(L, i)$ and $p(R, i)$, that (L, i) and (R, i), respectively, are entered before returning to $(R, 0)$ can be calculated iteratively from the initial state $(R, 0)$:

$$p(L, 1) = 1 - P_{R,0} \tag{4.25}$$

$$p(R, i) = P_{L,i}\, p(L, i); \qquad i \geq 1 \tag{4.26}$$

$$p(L, i + 1) = (1 - P_{L,i})\, p(L, i) + (1 - P_{R,i}) p(R, i); \qquad i \geq 1 \tag{4.27}$$

Let K be the number of slots in a CRP; thus, K is the number of states visited in the chain, including the initial state $(R, 0)$, before the return to $(R, 0)$,

$$E\{K\} = 1 + \sum_{i=1}^{\infty} [p(L, i) + p(R, i)] \tag{4.28}$$

We also must evaluate the change in $T(k)$ from one CRP to the next. For the assumed initial interval of size α_0, this change is at most α_0, but if left-hand intervals have collisions, the corresponding right-hand intervals are returned to the waiting interval, and the change is less than α_0. Let f be the fraction of α_0 returned in this way over a CRP, so that $\alpha_0(1-f)$ is the change in $T(k)$. The probability of a collision in state (L, i) is the probability that the left half-interval in state (L, i) contains at least two packets given that the right and left intervals together contain at least two; that is,

$$P\{e \mid (L,i)\} = \frac{1-(1+G_i)e^{-G_i}}{1-(1+G_{i-1})e^{-G_{i-1}}} \tag{4.29}$$

The fraction of the original interval returned on such a collision is 2^{-i}, so the expected value of f is

$$E\{f\} = \sum_{i=1}^{\infty} p(L,i)P\{e \mid (L,i)\}2^{-i} \tag{4.30}$$

Note that $E\{f\}$ and $E\{K\}$ are functions only of $G_i = \lambda\alpha_0 2^{-i}$, for $i \geq 1$, and hence are functions only of the product $\lambda\alpha_0$. For large i, $P_{L,i}$ tends to $1/2$, and thus $p(L,i)$ and $p(R,i)$ tend to zero with increasing i as 2^{-i}. Thus, $E\{f\}$ and $E\{K\}$ can be easily evaluated numerically as functions of $\lambda\alpha_0$.

Finally, define the drift D to be the expected change in the time backlog, $k - T(k)$, over a CRP (again assuming an initial allocation of α_0). This is the expected number of slots in a CRP less the expected change in $T(k)$; so

$$D = E\{K\} - \alpha_0(1 - E\{f\}) \tag{4.31}$$

The drift is negative if $E\{K\} < \alpha_0(1 - E\{f\})$, or equivalently, if

$$\lambda < \frac{\lambda\alpha_0(1 - E\{f\})}{E\{K\}} \tag{4.32}$$

The right side of Eq. (4.32), as a function of $\lambda\alpha_0$, has a numerically evaluated maximum of 0.4871 at $\lambda\alpha_0 = 1.266$. $\lambda\alpha_0$ is the expected number of packets in the original allocation interval; as expected, it is somewhat larger than 1 (which would maximize the initial probability of success) because of the increased probability of success immediately after a collision. If α_0 is chosen to be 2.6 (*i.e.*, 1.266/0.4871), then Eq. (4.32) is satisfied for all $\lambda < 0.4871$. Thus, the expected time backlog decreases (whenever it is initially larger than α_0), and we conclude* that the algorithm is stable for $\lambda < 0.4871$.

Expected delay is much harder to analyze than maximum stable throughput. Complex upper and lower bounds have been developed ([HuB85] and [TsM80]) and corre-

*For the mathematician, a more rigorous proof of stability is desirable. Define a busy period as a consecutive string of CRPs starting with a time backlog $k - T(k) < \alpha_0$ and continuing up to the beginning of the next CRP with $k - T(k) < \alpha_0$. The sequence of time backlogs at the beginnings of the CRPs in the busy period forms a random walk with the increments (except the first) having identical distributions with negative expectation for $\lambda < 0.4871$. Since $p(L,i)$ approaches 0 exponentially in i, the increments have a moment generating function, and from Wald's equality, the number N of CRPs in a busy period also has a moment generating function. Since the number of slots in a busy period is at most $N\alpha_0$, the number of slots also has a moment generating function, from which it follows that the expected delay per packet is finite.

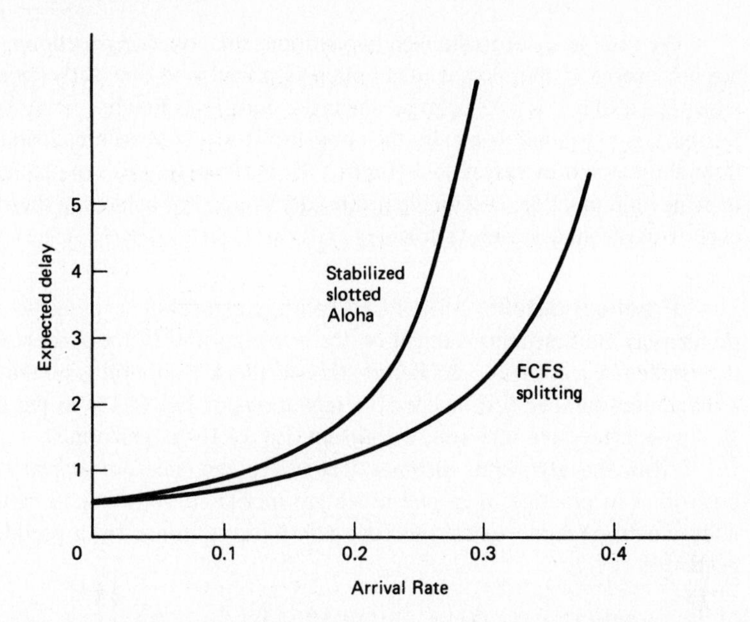

Figure 4.14 Comparison of expected delay for stabilized slotted Aloha and the FCFS splitting algorithm as a function of arrival rate. One becomes unbounded as the arrival rate approaches $1/e$, and the other as the arrival rate approaches 0.4871.

spond closely to simulation results. Figure 4.14 plots this delay and compares it with stabilized slotted Aloha.

Improvements in the FCFS splitting algorithm. Splitting intervals into equal-sized subintervals is slightly nonoptimal in achieving maximum stable throughput. When each interval is split into the optimally sized subintervals, the maximum stable throughput increases to 0.4878 ([MoH85] and [TsM80]). Another, even smaller, improvement of 3.6×10^{-7} results if in state (R, i) for large i, some of the waiting interval is appended to the right-side interval [VvP83]. While these improvements are not significant practically, they are of theoretical interest in determining optimality in terms of maximum stable throughput.

The maximum stable throughput, using assumptions 1 to 6b, is currently unknown. Considerable research has been devoted to finding upper bounds to throughput, and the tightest such bound is 0.587 [MiT81]. Thus, the maximum stable throughput achievable by any algorithm lies somewhere between 0.4878 and 0.587.

These questions of maximum throughput depend strongly on assumptions 1 to 6b. For any finite set of m nodes, we have seen that TDM can trivially achieve any throughput up to one packet per slot. This striking difference between finite and infinite m seems paradoxical until we recognize that with TDM, expected delay (for a given λ) increases linearly with m, whereas the algorithms that assume $m = \infty$ achieve a delay bounded independently of m.

We shall also see in the next two sections that much higher throughputs than 0.4878 are achievable if the slotted assumption is abandoned and early feedback is available when the channel is idle or experiencing a collision. Finally, rather surprisingly, if the feedback is expanded to specify the number of packets in each collision, maximum stable throughput again increases to 1 [Pip81]. Unfortunately, this type of feedback is difficult to achieve in practice, and no algorithms are known for achieving these high throughputs even if the feedback were available.

Practical details. The FCFS splitting algorithm is subject to the same deadlock problem as the first improvement on the tree algorithm if the feedback from an idle slot is mistaken as a collision. As before, this deadlock condition is eliminated by specifying a maximum number h of successive repetitions of Eq. (4.17) in the algorithm. On the $(h+1)$th successive idle after a collision, Eq. (4.16) is performed.

Also, the algorithm assumes that nodes can measure arrival times with infinite precision. In practice, if arrival times are measured with a finite number of bits, each node would generate extra bits, as needed for splitting, by a pseudo-random number generator.

Last-come first-serve (LCFS) splitting algorithm. The FCFS splitting algorithm requires all nodes to monitor the channel feedback at all times. A recent modification allows nodes to monitor the feedback only after receiving a packet to transmit ([Hum86] and [GeP85]). The idea is to send packets in approximately last-come first-serve (LCFS) order; thus, the most recently arrived packets need not know the length of the waiting set since they have first priority in transmission.

Figure 4.15 illustrates this variation. New arrivals are in a "prewaiting mode" until they receive enough feedback to detect the end of a CRP; they then join the waiting set. The end of a CRP can be detected by feedback equal to 1 in one slot, followed by either 0 or 1 in the next. Also, assuming the practical modification in which Eq. (4.17) can be repeated at most h successive times, feedback of h successive 0's followed by 1 or 0 also implies the end of a CRP.

After the waiting set has been extended on the right by the interval of time over which nodes can detect the end of the CRP, a new allocation set is chosen at the right end of the waiting set (thus, including part or all of the new interval). As shown in the figure, the waiting set, and consequently the allocation set, might consist of several disjoint intervals.

After joining the waiting set, backlogged nodes keep track of their distance from the right end of the waiting set. This distance includes only unresolved intervals and unallocated (left or right) intervals. Nodes in the allocation set similarly track their distance from the right end of the allocation set. When a collision occurs, the allocated set splits as before, but here the right half is allocated next. Upon a right-half collision, the corresponding left half is adjoined to the right end of the waiting set, increasing the distance of the right end from the previously waiting nodes. At the end of a CRP, the new arrival interval from which this event is detectable is first appended to the right end of the waiting set, and then the new allocation set is removed from the right end. Nodes

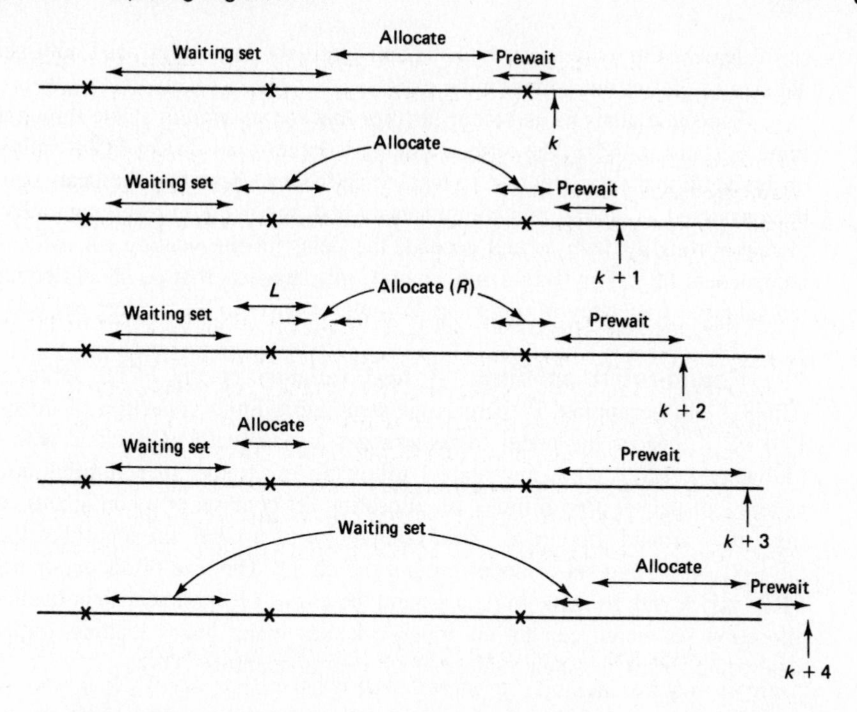

Figure 4.15 Last-come first-serve (LCFS) splitting algorithm. Arrivals wait for the beginning of a new CRP, but then are allocted in last-come first-serve fashion. Allocation sets and waiting sets can consist of more than one interval.

do not know where the left end of the waiting set is, and thus the allocation set always has size α_0 (not counting gaps), with perhaps dummy space at the left.

When the backlog is large, the LCFS splitting algorithm has the same drift as the FCFS splitting algorithm, and thus it is stable for the same range of λ. With increasing h [where h is the allowable number of repetitions of Eq. (4.17) in the algorithm], the upper limit of this range approaches 0.4871 as before. The expected delay is somewhat larger than that for FCFS, primarily because of the time packets spend in the prewaiting mode.

Delayed feedback. Assume that the feedback for the kth slot arrives sometime between the beginning of slot $k + j - 1$ and $k + j$ for some fixed $j > 1$. Visualize time-division multiplexing the slots on the channel between j different versions of the FCFS splitting algorithm, with the exception of maintaining a common waiting set for all j algorithms. Each node monitors the progress of each of the j current CRPs and tracks the extent of the waiting set. At the end of a CRP for one of the j versions, an allocation set of size α_0 (or less if the waiting set is smaller) is removed from the left end of the waiting set to start the next CRP on that version. Since the j versions experience different delays, packets are no longer served in first-come first-serve order.

When a right subset is returned to the waiting set due to a collision in a left subset for one of the j versions, it is appended to the left end of the waiting set. These returns can fragment the waiting set into disjoint intervals, but nodes need only keep track of the size of the set not counting the gaps.

The same analysis as before justifies that the maximum stable throughput for any finite j is still 0.4871. The expected delay is larger than that of FCFS splitting because of the additional time required to resolve collisions. Note that the delay can essentially be considered as having two components—first, the delay in resolving collisions, which increases roughly with j, and second, the delay in the waiting set, which is roughly independent of j. For large j and small λ, this suggests that α_0 should be reduced, thus reducing the frequency of collisions at some expense to the waiting set size.

Round-robin splitting. A final variation of the FCFS splitting algorithm [HlG81] can be applied if there is an identifiable finite collection of nodes numbered 1 to m. Consider the nodes to be arranged conceptually in a circle with node $i+1$ following i, $1 \leq i < m$, and node 1 following m. Rather than forming allocation sets in terms of packet arrival times, an allocation set consists of a contiguous set of nodes, say i to j around the circle. After completion of a CRP, the algorithm then allocates the next successive set of nodes around the circle. The size of allocation sets initiating CRPs varies with the time to pass around the circle, so that under light loading, an initial allocation set would contain all nodes, whereas under heavy loading, initial allocation sets would shrink to single nodes, which is equivalent to TDM.

4.4 CARRIER SENSING

In many multiaccess systems, such as local area networks, a node can hear whether other nodes are transmitting after a very small propagation and detection delay relative to a packet transmission time. The detection delay is the time required for a physical receiver to determine whether or not some other node is currently transmitting. This delay differs somewhat from the delay, first, in detecting the beginning of a new transmission, second, in synchronizing on the reception of a new transmission, and third, in detecting the end of an old transmission. We ignore these and other subtleties of the physical layer in what follows, and simply regard the medium as an intermittent synchronous multiaccess bit pipe on which idle periods can be distinguished (with delay) from packet transmission periods.

If nodes can detect idle periods quickly, it is reasonable to terminate idle periods quickly and to allow nodes to initiate packet transmissions after such idle detections. This type of strategy, called carrier sense multiple access (CSMA) [KlT75], does not necessarily imply the use of a carrier but simply the ability to detect idle periods quickly.

Let β be the propagation and detection delay (in packet transmission units) required for all sources to detect an idle channel after a transmission ends. Thus if τ is this time in seconds, C is the raw channel bit rate, and L is the expected number of bits in a data packet, then

$$\beta = \frac{\tau C}{L} \tag{4.33}$$

We shall see that the performance of CSMA degrades with increasing β and thus also degrades with increasing channel rate and with decreasing packet size.

Consider a slotted system in which, if nothing is being transmitted in a slot, the slot terminates after β time units and a new slot begins. This assumption of dividing idle periods into slots of length β is not realistic, but it provides a simple model with good insight. We have thus eliminated our previous assumption that time is slotted into equal-duration slots. We also eliminate our assumption that all data packets are of equal length, although we still assume a time normalization in which the expected packet transmission time is 1. In place of the instantaneous feedback assumption, we assume $0, 1, e$ feedback with a maximum delay β, as indicated above. For simplicity, we continue to assume an infinite set of nodes and Poisson arrivals of overall intensity λ. We first modify slotted Aloha for this new situation, then consider unslotted systems, and finally consider the FCFS splitting algorithm.

4.4.1 CSMA Slotted Aloha

The major difference between CSMA slotted Aloha and ordinary slotted Aloha is that idle slots in CSMA have a duration β. The other difference is that if a packet arrives at a node while a transmission is in progress, the packet is regarded as backlogged and begins transmission with probability q_r after each subsequent idle slot; packets arriving during an idle slot are transmitted in the next slot as usual. This technique was called nonpersistent CSMA in [KIT75] to distinguish it from two variations. In one variation, persistent CSMA, all arrivals during a busy slot simply postpone transmission to the end of that slot, thus causing a collision with relatively high probability. In the other, P-persistent CSMA, collided packets and new packets waiting for the end of a busy period use different probabilities for transmission. Aside from occasional comments, we will ignore these variations since they have no important advantages over nonpersistent CSMA.

To analyze CSMA Aloha, we can use a Markov chain again, using the number n of backlogged packets as the state and the ends of idle slots as the state transition times. Note that each busy slot (success or collision) must be followed by an idle slot, since nodes are allowed to start transmission only after detecting an idle slot. For simplicity, we assume that all data packets have unit length. The extension to arbitrary length packets is not difficult, however, and is treated in Problem 4.21. The time between successive state transitions is either β (in the case of an idle slot) or $1 + \beta$ (in the case of a busy slot followed by an idle). Rather than present the state transition equations, which are not particularly insightful, we simply modify the drift in Eq. (4.4) for this new model. At a transition into state n (*i.e.*, at the end of an idle slot), the probability of no transmissions in the following slot (and hence the probability of an idle slot) is $e^{-\lambda\beta}(1 - q_r)^n$. The first term is the probability of no arrivals in the previous idle slot, and the second is the probability of no transmissions by the backlogged nodes. Thus, the expected time between state transitions in state n is $\beta + [1 - e^{-\lambda\beta}(1 - q_r)^n]$. Similarly, the expected number of arrivals between state transitions is

$$E\{\text{arrivals}\} = \lambda[\beta + 1 - e^{-\lambda\beta}(1 - q_r)^n] \tag{4.34}$$

The expected number of departures between state transitions in state n is simply the probability of a successful transmission; assuming that $q_r < 1$, this is given by

$$P_{succ} = \left(\lambda\beta + \frac{q_r n}{1 - q_r}\right) e^{-\lambda\beta}(1 - q_r)^n \tag{4.35}$$

The drift in state n is defined as the expected number of arrivals less the expected number of departures between state transitions,

$$D_n = \lambda[\beta + 1 - e^{-\lambda\beta}(1 - q_r)^n] - \left(\lambda\beta + \frac{q_r n}{1 - q_r}\right) e^{-\lambda\beta}(1 - q_r)^n \tag{4.36}$$

For small q_r, we can make the approximation $(1 - q_r)^{n-1} \approx (1 - q_r)^n \approx e^{-q_r n}$, and D_n can be expressed as

$$D_n \approx \lambda\left(\beta + 1 - e^{-g(n)}\right) - g(n)e^{-g(n)} \tag{4.37}$$

where

$$g(n) = \lambda\beta + q_r n \tag{4.38}$$

is the expected number of attempted transmissions following a transition to state n. From Eq. (4.37), the drift in state n is negative if

$$\lambda < \frac{g(n)e^{-g(n)}}{\beta + 1 - e^{-g(n)}} \tag{4.39}$$

The numerator in Eq. (4.39) is the expected number of departures per state transition, and the denominator is the expected duration of a state transition period; thus, the ratio can be interpreted as departure rate (*i.e.*, expected departures per unit time) in

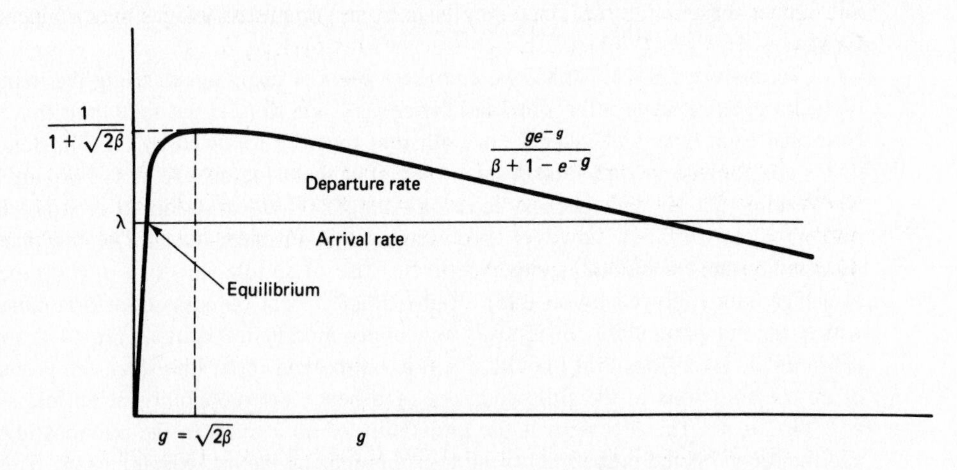

Figure 4.16 Departure rate, in packets per unit time, for CSMA slotted Aloha as a function of the attempted transmission rate g in packets per idle slot. If β, the duration of an idle slot as a fraction of a data slot, is small, the maximum departure rate is $1/(1 + \sqrt{2\beta})$.

state n. Figure 4.16 plots this ratio as a function of $g(n) = \lambda\beta + q_r n$. For small β, this function has a maximum of approximately $1/(1+\sqrt{2\beta})$ at $g(n) = \sqrt{2\beta}$. This can be seen by approximating $e^{g(n)}$ by $1 + g(n) + g^2(n)/2$ for small $g(n)$. To understand intuitively why the departure rate is maximized at $g = \sqrt{2\beta}$, note that for small β, very little time is wasted on a single idle slot, and significant time is wasted on a collision. The point $g = \sqrt{2\beta}$ is where idles occur so much more frequently than collisions that the same expected overall time is wasted on each.

Figure 4.16 also shows that CSMA Aloha has the same stability problem as ordinary slotted Aloha. For fixed q_r, $g(n)$ grows with the backlog n, and when n becomes too large, the departure rate is less than the arrival rate, leading to yet larger backlogs. From a practical standpoint, however, the stability problem is less serious for CSMA than for ordinary Aloha. Note that β/q_r is the expected idle time that a backlogged node must wait to attempt transmission, and for small β and modest λ, q_r can be quite small without causing appreciable delay. This means that the backlog must be very large before instability sets in, and one might choose simply to ignore the problem.

P-persistent CSMA, in which packets are transmitted after idle slots with probability p if they are new arrivals and transmitted with some much smaller probability q_r if they have had collisions, is a rudimentary way of obtaining a little extra protection against instability. The next section explores stabilization in a more fundamental way.

4.4.2 Pseudo-Bayesian Stabilization for CSMA Aloha

Consider all packets as backlogged immediately after entering the system. At the end of each idle slot, each backlogged packet is independently transmitted with probability q_r, which will vary with the estimated channel backlog $\hat{n}$. In state n, the expected number of packets transmitted at the end of an idle slot is $g(n) = nq_r$. Since we have seen that the packet departure rate (in packets per unit time) is maximized (for small β and q_r) by $g(n) = \sqrt{2\beta}$, we choose q_r, for a given estimated backlog $\hat{n}$, as

$$q_r(\hat{n}) = \min\left[\frac{\sqrt{2\beta}}{\hat{n}}, \sqrt{2\beta}\right] \tag{4.40}$$

The min operation prevents $q_r(\hat{n})$ from getting too large when $\hat{n}$ is small; we cannot expect $n/\hat{n}$ to approach 1 when the backlog is small, and it is desirable to prevent too many collisions in this case. The appropriate rule for updating the estimated backlog (again assuming unit length packets) is

$$\hat{n}_{k+1} = \begin{cases} \hat{n}_k[1 - q_r(\hat{n}_k)] + \lambda\beta & \text{for idle} \\ \hat{n}_k[1 - q_r(\hat{n}_k)] + \lambda(1+\beta) & \text{for success} \\ \hat{n}_k + 2 + \lambda(1+\beta) & \text{for collision} \end{cases} \tag{4.41}$$

This rule is motivated by the fact that if the a priori distribution of n_k is Poisson with mean $\hat{n}_k$, then, given an idle, the a posteriori distribution of n_k is Poisson with mean $\hat{n}_k[1 - q_r(\hat{n}_k)]$ (see Problem 4.20). Accounting for the Poisson arrivals in the idle slot of duration β, the resulting distribution of n_{k+1} is Poisson with mean $\hat{n}_{k+1}$ as shown above. Similarly, given a successful transmission, the a posteriori distribution on $n_k - 1$

(removing the successful packet) is Poisson with mean $\hat{n}_k[1 - q_r(\hat{n}_k)]$. Accounting for the Poisson arrivals in the successful slot and following idle slot, n_{k+1} is Poisson with mean $\hat{n}_{k+1}$ as shown. Finally, if a collision occurs, the a posteriori distribution of n_k is not quite Poisson, but is reasonably approximated as Poisson with mean $\hat{n}_k + 2$. Adding $\lambda(1 + \beta)$ for the new arrivals, we get the final expression in Eq. (4.41).

Note that when n_k and $\hat{n}_k$ are small, then q_r is large and new arrivals are scarcely delayed at all. When $\hat{n}_k \approx n_k$ and n_k is large, the departure rate is approximately $1/(1 + \sqrt{2\beta})$, so that for $\lambda < 1/(1 + \sqrt{2\beta})$, the departure rate exceeds the arrival rate, and the backlog decreases on the average. Finally, if $|n_k - \hat{n}_k|$ is large, the expected change in backlog can be positive, but the expected change in $|n_k - \hat{n}_k|$ is negative; Fig. 4.5 again provides a qualitative picture of the expected changes in n_k and $n_k - \hat{n}_k$.

We now give a crude analysis of delay for this strategy (and other similar stabilized strategies) by using the same type of analysis as in Section 4.2.3. Let W_i be the delay from the arrival of the ith packet until the beginning of the ith successful transmission. The average of W_i over all i is the expected queueing delay W. Let n_i be the number of backlogged packets at the instant before i's arrival, not counting any packet currently in successful transmission. Then

$$W_i = R_i + \sum_{j=1}^{n_i} t_j + y_i \tag{4.42}$$

where R_i is the residual time to the next state transition, t_j $(1 \le j \le n_i)$ is the sequence of subsequent intervals until each of the next n_i successful transmissions are completed, and y_i is the remaining interval until the ith successful transmission starts.

The backlog is at least 1 in all of the state transition intervals in the period on the right-hand side of Eq. (4.42), and we make the simplifying approximation that the number of attempted transmissions in each of these intervals is Poisson with parameter g. We later choose $g = \sqrt{2\beta}$, but for the moment it can be arbitrary. This approximation is somewhat different from that in Section 4.2.3, in which we assumed that a successful transmission always occurred with a backlog state of 1; the difference is motivated by Eq. (4.40), which keeps q_r small. The expected value for each t_j is given by

$$E\{t\} = e^{-g}(\beta + E\{t\}) + ge^{-g}(1 + \beta) + [1 - (1 + g)e^{-g}](1 + \beta + E\{t\}) \tag{4.43}$$

The first term corresponds to an idle transmission in the first state transition interval; this occurs with probability e^{-g}, uses time β, and requires subsequent time $E\{t\}$ to complete the successful transmission. The next two terms correspond similarly to a success and a collision, respectively. Solving for $E\{t\}$ gives

$$E\{t\} = \frac{\beta + 1 - e^{-g}}{ge^{-g}} \tag{4.44}$$

Note that this is the reciprocal of the expected number of departures per unit time in Eq. (4.39), as we would expect. $E\{t\}$ is thus approximately minimized by $g = \sqrt{2\beta}$. Averaging over i and using Little's result in Eq. (4.42), we get

$$W(1 - \lambda E\{t\}) = E\{R\} + E\{y\} \tag{4.45}$$

The expected residual time can be approximated by observing that the system spends a fraction $\lambda(1+\beta)$ of the time in successful state transition intervals. The expected residual time for arrivals in these intervals is $(1+\beta)/2$. The fraction of time spent in collision intervals is negligible (for small β) compared with that for success, and the residual time in idle intervals is negligible. Thus,

$$E\{R\} \approx \frac{\lambda(1+\beta)^2}{2} \tag{4.46}$$

Finally, $E\{y\}$ is just $E\{t\}$ less the successful transmission interval, so $E\{y\} = E\{t\} - (1+\beta)$. Substituting these expressions into Eq. (4.45) yields

$$W \approx \frac{\lambda(1+\beta)^2 + 2[E\{t\} - (1+\beta)]}{2[1 - \lambda E\{t\}]} \tag{4.47}$$

This expression is minimized over g by minimizing $E\{t\}$, and we have already seen that this minimum (for small β) is $1+\sqrt{2\beta}$, occurring at $g = \sqrt{2\beta}$. With this substitution, W is approximately

$$W \approx \frac{\lambda + 2\sqrt{2\beta}}{2[1 - \lambda(1+\sqrt{2\beta})]} \tag{4.48}$$

Note the similarity of this expression with the $M/D/1$ queueing delay given in Eq. (3.45). What we achieve by stabilizing CSMA Aloha is the ability to modify q_r with the backlog so as to maintain a departure rate close to $1/(1+\sqrt{2\beta})$ whenever a backlog exists.

4.4.3 CSMA Unslotted Aloha

In CSMA slotted Aloha, we assumed that all nodes were synchronized to start transmissions only at time multiples of β in idle periods. Here we remove that restriction and assume that when a packet arrives, its transmission starts immediately if the channel is sensed to be idle. If the channel is sensed to be busy, or if the transmission results in a collision, the packet is regarded as backlogged. Each backlogged packet repeatedly attempts to retransmit at randomly selected times separated by independent, exponentially distributed random delays τ, with probability density $xe^{-x\tau}$. If the channel is idle at one of these times, the packet is transmitted, and this continues until such a transmission is successful. We again assume a propagation and detection delay of β, so that if one transmission starts at time t, another node will not detect that the channel is busy until $t+\beta$, thus causing the possibility of collisions.

Consider an idle period that starts with a backlog of n. The time until the first transmission starts (with the $m = \infty$ assumption) is an exponentially distributed random variable with rate

$$G(n) = \lambda + nx \tag{4.49}$$

Note that $G(n)$ is the attempt rate in packets per unit time, whereas $g(n)$ in Section 4.4.2 was packets per idle slot. After the initiation of this first transmission, the backlog is either n (if a new arrival started transmission) or $n-1$ (if a backlogged packet started).

Thus the time from this first initiation until the next new arrival or backlogged node senses the channel is an exponentially distributed random variable of rate $G(n)$ or $G(n-1)$. A collision occurs if this next sensing is done within time β. Thus, the probability that this busy period is a collision is $1 - e^{-\beta G(n)}$ or $1 - e^{-\beta G(n-1)}$. This difference is small if βx is small, and we neglect it in what follows. Thus, we approximate the probability of a successful transmission following an idle period by $e^{-\beta G(n)}$.

The expected time from the beginning of one idle period until the next is $1/G(n)+(1+\beta)$; the first term is the expected time until the first transmission starts, and the second term $(1+\beta)$ is the time until the first transmission ends and the channel is detected as being idle again. If a collision occurs, there is a slight additional time, less than β, until the packets causing the collision are no longer detected; we neglect this contribution since it is negligible even with respect to β, which is already negligible. The departure rate during periods when the backlog is n is then given by

$$\text{departure rate}(n) = \frac{e^{-\beta G(n)}}{1/G(n) + (1+\beta)} \tag{4.50}$$

For small β, the maximum value of this departure rate is approximately $1/(1+2\sqrt{\beta})$, occurring when $G(n) \approx \beta^{-1/2}$. This maximum departure rate is slightly smaller than it is for the slotted case [see Eq. (4.39)]; the reason is the same as when CSMA is not being used—collisions are somewhat more likely for a given attempt rate in an unslotted system than a slotted system. For CSMA, with small β, however, this loss in departure rate is quite small. What is more, in a slotted system, β would have to be considerably larger than in an unslotted system to compensate for synchronization inaccuracies and worst-case propagation delays. Thus, unslotted Aloha appears to be the natural choice for CSMA.

CSMA unslotted Aloha has the same stability problems as all the Aloha systems, but it can be stabilized in the same way as CSMA slotted Aloha. The details are treated in Problem 4.22.

4.4.4 FCFS Splitting Algorithm for CSMA

We next investigate whether higher throughputs or smaller delays can be achieved by the use of splitting algorithms with CSMA. We shall see that relatively little can be gained, but it is interesting to understand why. We return to the assumption of idle slots of duration β and assume that $0, 1, e$ feedback is available. An idle slot occurs at the end of each success or collision to provide time for feedback. Thus, we regard successes and collisions as having a duration $1+\beta$; the algorithm is exercised at the end of each such elongated success or collision slot, and also at the end of each normal idle slot.

The same algorithm as in Eqs. (4.15) to (4.18) can be used, although the size α_0 of the initial interval in a CRP should be changed. Furthermore, as we shall see shortly, intervals with collisions should not be split into equal subintervals. Since collisions waste much more time than idle slots, the basic allocation interval α_0 should be small. This means in turn that collisions with more than two packets are negligible, and thus the analysis is simpler than before.

We first find the expected time and the expected number of successes in a CRP. Let $g = \lambda\alpha_0$ be the expected number of arrivals in an initial allocation interval of size α_0. With probability e^{-g}, an original allocation interval is empty, yielding a collision resolution time of β with no successes. With probability ge^{-g}, there is an initial success, yielding a collision resolution time $1+\beta$. Finally, with probability $(g^2/2)e^{-g}$, there is a collision, yielding a collision resolution time of $1+\beta+T$ for some T to be calculated later; since collisions with more than two packets are negligible, we assume two successes for each CRP with collisions. Thus,

$$E\{\text{time/CRP}\} \approx \beta e^{-g} + (1+\beta)ge^{-g} + (1+\beta+T)\frac{g^2}{2}e^{-g} \tag{4.51}$$

$$E\{\text{packets/CRP}\} \approx ge^{-g} + 2\frac{g^2}{2}e^{-g} \tag{4.52}$$

As before, the maximum stable throughput for a given g is

$$\lambda_{\max} = \frac{E\{\text{packets/CRP}\}}{E\{\text{time/CRP}\}} \approx \frac{g+g^2}{\beta + g(1+\beta) + (g^2/2)\,(1+\beta+T)} \tag{4.53}$$

We can now maximize the right-hand side of Eq. (4.53) over g (*i.e.*, over α_0). In the limit of small β, we get the asymptotic expressions

$$g \approx \sqrt{\frac{2\beta}{T-1}} \tag{4.54}$$

$$\lambda_{\max} \approx \frac{1}{1+\sqrt{2\beta(T-1)}} \tag{4.55}$$

Finally, we must calculate T, the time to resolve a collision after it has occurred. Let x be the fraction of an interval used in the first subinterval when the interval is split; we choose x optimally later. The first slot after the collision is detected is idle, successful, or a collision with probabilities $(1-x)^2, 2x(1-x)$, or x^2, respectively. The expected time required for each of these three cases is $\beta+T$, $2(1+\beta)$, and $1+\beta+T$. Thus,

$$T \approx (1-x)^2(\beta+T) + 4x(1-x)(1+\beta) + x^2(1+\beta+T) \tag{4.56}$$

from which T can be expressed as a function of x.

By setting the derivative dT/dx to 0, we find after a straightforward calculation that T is minimized by

$$x = \sqrt{\beta+\beta^2} - \beta \tag{4.57}$$

The resulting value of T, for small β, is $T \approx 2 + \sqrt{\beta}$. Substituting this in Eq. (4.55), we see that

$$\lambda_{\max} \approx \frac{1}{1+\sqrt{2\beta}} \tag{4.58}$$

For small β, then, the FCFS splitting algorithm has the same maximum throughput as slotted Aloha. This is not surprising, since without CSMA, the major advantage of the FCFS algorithm is its efficiency in resolving collisions, and with CSMA, collisions rarely occur. When collisions do occur, they are resolved in both strategies by retransmission with small probability. It is somewhat surprising at first that if we use the FCFS algorithm with equal subintervals (*i.e.*, $x = 1/2$), we find that we are limited to a throughput of $1/(1+\sqrt{3\beta})$. This degradation is due to a substantial increase in the number of collisions.

4.5 MULTIACCESS RESERVATIONS

We now look at a very simple way of increasing the throughput of multiaccess channels that has probably become apparent. If the packets of data being transmitted are long, why waste these long slot times either sending nothing or sending colliding packets? It would be far more efficient to send very short packets either in a contention mode or a TDM mode, and to use those short packets to reserve longer noncontending slots for the actual data. Thus, the slots wasted by idles or collisions are all short, leading to a higher overall efficiency. There are many different systems that operate in this way, and our major objective is not so much to explore the minor differences between them, but to see that they are all in essence the same.

To start, we explore a somewhat "canonic" reservation system. Assume that data packets require one time unit each for transmission and that reservation packets require $v \ll 1$ time units each for transmission. The format of a reservation packet is unimportant; it simply has to contain enough information to establish the reservation. For example, with the instantaneous feedback indicating idle, success, or collision that we have been assuming, the reservation packet does not have to contain any information beyond its mere existence. After a successful reservation packet is transmitted, either the next full time unit or some predetermined future time can be automatically allocated for transmission of the corresponding data packet. The reservation packets can use any strategy, including time-division multiplexing, slotted Aloha, or the splitting algorithm.

We can easily determine the maximum throughput S in data packets per time unit achievable in such a scheme. Let S_r be the maximum throughput, in successful reservation packets per reservation slot, of the algorithm used for the reservation packets (*i.e.*, $1/e$ for slotted Aloha, 0.478 for splitting, or 1 for TDM). Then, over a large number of reservations, the time required per reservation approaches v/S_r, and an additional one unit of time is required for each data packet. Thus, the total time per data packet approaches $1 + v/S_r$, and we see that

$$S = \frac{1}{1 + v/S_r} \tag{4.59}$$

This equation assumes that the reservation packet serves only to make the reservation and carries no data. As we shall see, in many systems, the reservation packet carries some of the data; thus for one time unit of data, it suffices to transmit the reservation packet of duration v followed by the rest of the data in time $1 - v$. In this case, the throughput becomes

$$S = \frac{1}{1 + v(1/S_R - 1)} \tag{4.60}$$

For example, with slotted Aloha, the throughput is $S = 1/[1 + v(e - 1)]$. It is apparent that if v is small, say on the order of 0.01, then the maximum throughput approaches 1 and is not highly dependent on the collision resolution algorithm or on whether the reservation packet carries part of the data. We shall see later that Ethernet local area networks [MeB76] can be modeled almost in this way, and v about 0.01 is typical. Thus, such networks can achieve very high throughputs without requiring much sophistication for collision resolution.

4.5.1 Satellite Reservation Systems

One of the simplest reservation systems, with particular application to satellite networks [JBH78], has a frame structure as shown in Fig. 4.17. A number of data slots are preceded by a reservation period composed of m reservation slots, one reservation slot per node. Let v be the duration of a reservation slot, where, as usual, time is normalized to the average duration of a data packet; thus, v is the ratio of the number of bits used to make a reservation (including guard space and overhead) to the expected number of bits in a data packet. The reservation period in each frame has an overall duration of $A = mv$. The minimum duration of a frame is set to exceed the round-trip delay 2β, so that the reservation slots at the beginning of one frame allocate the data slots for the next frame (see Fig. 4.17).

A frame is extended beyond the minimum length if need be to satisfy all the reservations. Note that the use of TDM for the reservation slots here makes a great deal more sense than it does for ordinary TDM applied to data slots. One reason is that the reservation slots are short, and therefore little time is wasted on a source with nothing to send. Another reason, for satellites, is that if collisions were allowed on the reservation slots, the channel propagation delay would make collision resolution rather

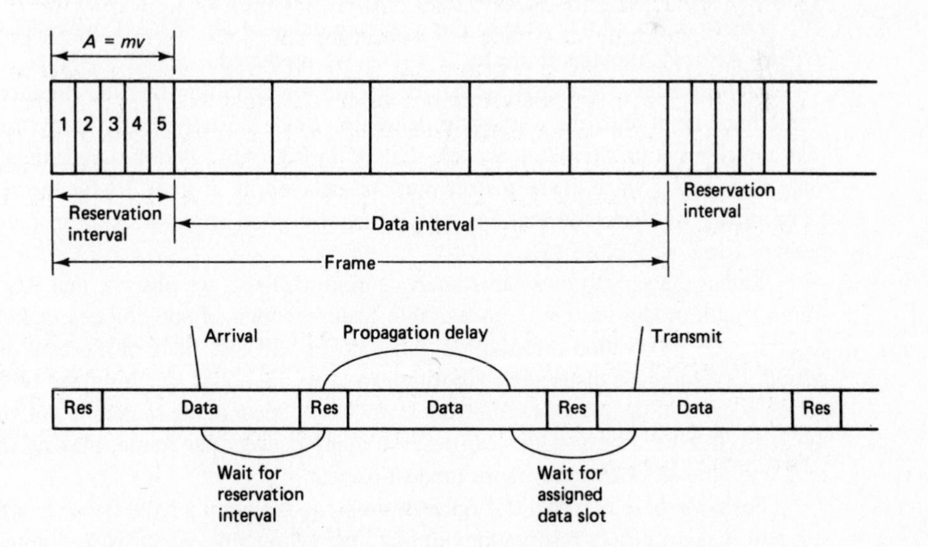

Figure 4.17 Satellite reservation system, using TDM to make reservations. Arrivals in one frame are transmitted in the second following frame.

slow. Equation (4.59) gives the maximum throughput of such a system as $1/(1+v)$ under the assumption that each reservation packet can make only one data packet reservation. If we assume, alternatively, that a reservation packet can make multiple reservations for its source, it is not hard to see that the throughput can approach 1 arbitrarily closely, since under heavy loading, the frames get very long and negligible time is used by the infrequent reservation periods.

With the assumption that each reservation packet can make multiple data packet reservations, this system is very similar to the single-user reservation system analyzed in Section 3.5.2. The major difference is that because of the propagation delay, the reservations requested in one reservation interval are for the data frame following the next reservation interval. Thus, if we assume Poisson packet arrivals, and if we neglect the minimum frame length requirement, the expected queueing delay for the ith packet becomes

$$E\{W_i\} = E\{R_i\} + \frac{E\{N_i\}}{\mu} + 2A \tag{4.61}$$

This expression is the same as Eq. (3.60), except that the last term is $2A$. Here $A = mv$ is the duration of the reservation interval, and it is multiplied by 2 since the packet has to wait for the next two reservation intervals before being transmitted. Using the same analysis as in Chapter 3, the expected queueing delay is

$$W = \frac{\lambda \overline{X^2}}{2(1-\lambda)} + \frac{A}{2} + \frac{2A}{1-\lambda} \tag{4.62}$$

This analysis allows the data packets to have a general length distribution with mean square $\overline{X^2}$ (which of course requires the reservations to contain length information), but we have normalized time to satisfy $\overline{X} = 1/\mu = 1$; thus, $\rho = \lambda$. Note that in the limit as v goes to 0, Eq. (4.62) goes to the queueing delay of an $M/G/1$ queue as we would expect. Also, W remains finite for $\lambda < 1$ as we predicted.

Unfortunately, this analysis has neglected the condition that the duration of each frame must be at least the round-trip delay 2β. Since 2β is typically many times larger than the reservation period A, we see that W in Eq. (4.62) is only larger than 2β for λ very close to 1. Since every packet must be delayed by at least 2β for the reservation to be made, we conclude that Eq. (4.62) is not a good approximation to delay except perhaps for λ very close to 1.

Rather than try to make this analysis more realistic, we observe that this variable-frame-length model has some undesirable features. First, if some nodes make errors in receiving the reservation information, those nodes will lose track of the next reservation period; developing a distributed algorithm to keep the nodes synchronized on the reservation periods in the presence of errors is not easy. Second, the system is not very fair in the sense that very busy nodes can reserve many packets per frame, making the frames long and almost locking out more modest users.

For both these reasons, it is quite desirable to maintain a fixed frame length. Nodes can still make multiple reservations in one reservation slot, which is desirable if only a few nodes are active. With a fixed frame, however, it is sometimes necessary to postpone packets with reservations from one frame to the next.

Note that all nodes are aware of all reservations after the delay of 2β suffered by reservation packets. Thus, conceptually, we have a common queue of packets with reservations, and the nodes can jointly exercise any desired queueing discipline, such as first-come first-serve, round robin, or some priority scheme. As long as a packet is sent in every data slot for which the queue of packets with reservations is not empty, and as long as the discipline is independent of packet length, the expected delay is independent of queueing discipline.

It is quite messy to find the expected delay for this system, but with the slight modification in Fig. 4.18, a good approximate analysis is very simple. Suppose that a fraction γ of the available bandwidth is set aside for making reservations, and that TDM is used within this bandwidth, giving each node one reservation packet in each round-trip delay period 2β. With v as the ratio of reservation packet length to data packet length, $\gamma = mv/2\beta$. An arriving packet waits $2\beta/2$ time units on the average until the beginning of its reservation packet, then $2\beta/m$ units for the reservation packet transmission, and then 2β time units until the reservation is received. Thus, after $2\beta(3/2 + 1/m)$ time units, on the average, a packet joins the common queue.

The arrival process of packets with reservations to the common queue is approximately Poisson (*i.e.*, the number of arrivals in different reservation slots are independent and have a Poisson distribution). Once a packet is in the common queue, it has a service time of $X/(1-\gamma)$, where X is the packet transmission time using the full bandwidth. The common queue is thus $M/G/1$ with $\rho = \lambda/(1-\gamma)$. The total queueing delay, adding the expected time to enter the common queue [*i.e.*, $2\beta(3/2+1/m)$] to the $M/G/1$ delay in the common queue, is then

$$W = 3\beta + \frac{2\beta}{m} + \frac{\lambda\overline{X^2}}{2(1-\gamma-\lambda)(1-\gamma)} \tag{4.63}$$

We see that this strategy essentially achieves perfect scheduling at the expense of the delay for making reservations. For small λ, this delay seems excessive, since a number of data slots typically are wasted in each frame. This leads to the clever idea of using the unscheduled slots in a frame in a contention mode [WiE80]. Reservations are made in the following reservation interval for packets sent in these unscheduled slots. The convention is that if the packet gets through, its reservation is canceled. This approach has the effect of achieving very small delay under light loading (because of no

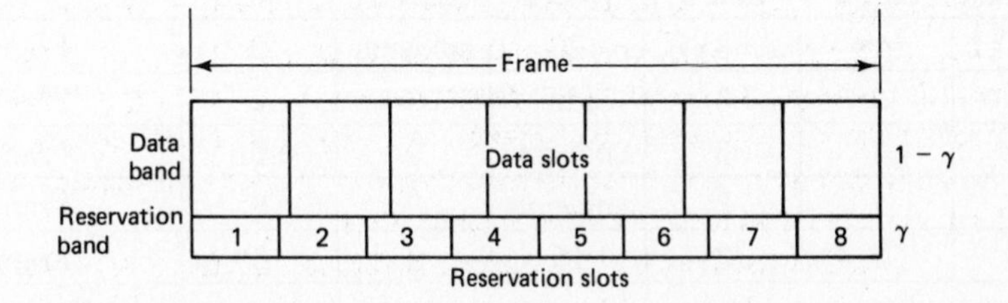

Figure 4.18 Reservation system with separate frequency band for reservations. Reservations are made by TDM in the reservation band.

round-trip delay for reservations) and very high efficiency [as given by Eq. (4.63)] under heavy loading.

All the strategies above use TDM for making reservations, and this makes it somewhat difficult to add new sources or delete old sources from the system. In addition, as m increases, γ increases and more of the bandwidth is used up by reservations. This suggests using the reservation slots in a contention resolution mode [JBH78]. It is difficult to analyze such a system, but it is clear that for large m and small λ, the delay could be reduced—we would use many fewer than m minislots in the reservation part of the frame.

Another somewhat simpler reservation idea is to allow the first packet of a message to make a reservation for the subsequent packets. In the context of Eq. (4.60), we view the first packet as the reservation packet and the entire message as the data packet. Thus Eq. (4.60) yields the throughput of this scheme if we take v as the inverse of the expected number of packets in a message. The appropriate length for a packet in this scheme is a trade-off between reservation inefficiency for long packets versus DLC inefficiency for short packets.

For satellite networks, these schemes are usually implemented with a frame structure (see Fig. 4.19), where a frame consists of a fixed number of packet slots [CRW73]. Enough packet slots are included in a frame so that the frame delay exceeds the round-trip delay.

When a source successfully transmits a packet in one of these slots, it can automatically reserve the corresponding slot in the next frame and each following frame until its message is completed. This can be done either by using a field in the packet header saying that another packet is to follow or, more simply and less efficiently, by automatically reserving that slot in each subsequent frame until that slot is first empty. After the end of the reservation period, the given slot in the frame is open for contention. Note that some care must be used in the contention algorithm. It does not suffice for all waiting sources simply to pounce on the first idle slot after a reservation period, since that would yield an unacceptably high collision probability.

Another variation on this scheme is to have each source "own" a particular slot within the frame [Bin75]. When a source is not using its own slot, other sources can capture it by contention, but when the source wants its own slot back, it simply transmits a packet in that slot, and if a collision occurs, the other sources are forbidden to use that

Slot 1	Slot 2	Slot 3	Slot 4	Slot 5	Slot 6	
15	idle	3	20	collision	2	Frame 1
15	7	3	idle	9	2	Frame 2
idle	7	3	collision	9	idle	Frame 3
18	7	3	collision	9	6	Frame 4
18	7	3	15	9	6	Frame 5

Figure 4.19 Reservation strategy with short packets and multiple packet messages. After a node captures a slot in a frame, it keeps that slot until finished.

slot on the next frame, letting the owner capture it. This variation is somewhat more fair than the previous scheme, but the delays are larger, both because of the less efficient contention resolution and because of the large number of slots in a frame if the number of users is large. Also, there is the usual problem of deleting and adding new sources to the system.

4.5.2 Local Area Networks: CSMA/CD and Ethernet

Section 4.5.1 treated satellite networks where round-trip propagation delay was an important consideration in each reservation scheme. For local area networks, at least with traditional technology, round-trip delay is a very small fraction of packet duration. Thus, instead of making a reservation for some slot far into the future, reservations can be made for the immediate future. Conceptually, this is not a big difference; it simply means that one must expect larger delays with satellite networks, since the feedback in any collision resolution strategy is always delayed by the round-trip propagation time. From a technological standpoint, however, the difference is more important since the physical properties of the media used for local area networks can simplify the implementation of reservation techniques.

Ethernet [MeB76] both illustrates this simplification and is a widely used technique for local area networks. A number of nodes are all connected onto a common cable so that when one node transmits a packet (and the others are silent), all the other nodes hear that packet. In addition, as in carrier sensing, a node can listen to the cable before transmitting (*i.e.*, conceptually, 0, 1, and idle can be distinguished on the bus). Finally, because of the physical properties of cable, it is possible for a node to listen to the cable while transmitting. Thus, if two nodes start to transmit almost simultaneously, they will shortly detect a collision in process and both cease transmitting. This technique is called CSMA/Collision Detection (CSMA/CD). On the other hand, if one node starts transmitting and no other node starts before the first node's signal has propagated throughout the cable, the first node is guaranteed to finish its packet without collision. Thus, we can view the first portion of a packet as making a reservation for the rest.

Slotted CSMA/CD. For analytic purposes, it is easiest to visualize Ethernet in terms of slots and minislots. The minislots are of duration β, which denotes the time required for a signal to propagate from one end of the cable to the other and to be detected. If the nodes are all synchronized into minislots of this duration, and if only one node transmits in a minislot, all the other nodes will detect the transmission and not use subsequent minislots until the entire packet is completed. If more than one node transmits in a minislot, each transmitting node will detect the condition by the end of the minislot and cease transmitting. Thus, the minislots are used in a contention mode, and when a successful transmission occurs in a minislot, it effectively reserves the channel for the completion of the packet.

CSMA/CD can be analyzed with a Markov chain in the same way as CSMA Aloha. We assume that each backlogged node transmits after each idle slot with probability q_r, and we assume at the outset that the number of nodes transmitting after an idle slot is Poisson with parameter $g(n) = \lambda\beta + q_r n$. Consider state transitions at the ends of idle slots; thus, if no transmissions occur, the next idle slot ends after time β. If one transmission occurs, the next idle slot ends after $1 + \beta$. We can assume variable-length packets here, but to correspond precisely to the model for idle slots, the packet durations should be multiples of the idle slot durations; we assume as before that the expected packet duration is 1. Finally, if a collision occurs, the next idle slot ends after 2β; in other words, nodes must hear an idle slot after the collision to know that it is safe to transmit.

The expected length of the interval between state transitions is then β, plus an additional 1 times the success probability, plus an additional β times the collision probability;

$$E\{\text{interval}\} = \beta + g(n)e^{-g(n)} + \beta[1 - (1 + g(n))e^{-g(n)}] \tag{4.64}$$

The expected number of arrivals between state transitions is λ times this interval, so the drift in state n is $\lambda E\{\text{interval}\} - P_{succ}$. The probability of success is simply $g(n)e^{-g(n)}$, so, as in Eq. (4.39), the drift in state n is negative if

$$\lambda < \frac{g(n)e^{-g(n)}}{\beta + g(n)e^{-g(n)} + \beta[1 - (1 + g(n))e^{-g(n)}]} \tag{4.65}$$

The right-hand side of Eq. (4.65) is interpreted as the departure rate in state n. This quantity is maximized over $g(n)$ at $g(n) = 0.77$ and the resulting value of the right-hand side is $1/(1 + 3.31\beta)$. Thus, if CSMA/CD is stabilized (this can be done, *e.g.*, by the pseudo-Bayesian technique), the maximum λ at which the system is stable is

$$\lambda < \frac{1}{1 + 3.31\beta} \tag{4.66}$$

The expected queueing delay for CSMA/CD, assuming the slotted model above and ideal stabilization, is calculated in the same way as for CSMA (see Problem 4.24). The result, for small β and mean-square packet duration $\overline{X^2}$, is

$$W \approx \frac{\lambda\overline{X^2} + \beta(4.62 + 2\lambda)}{2[1 - \lambda(1 + 3.31\beta)]} \tag{4.67}$$

The constant 3.31 in Eq. (4.66) is dependent on the detailed assumptions about the system. Different values are obtained by making different assumptions (see [Lam80], for example). If β is very small, as usual in Ethernet, this value is not very important. More to the point, however, is that the unslotted version of CSMA/CD makes considerably more sense than the slotted version, both because of the difficulty of synchronizing on short minislots and the advantages of capitalizing on shorter than maximum propagation delays when possible.

Unslotted CSMA/CD. Figure 4.20 illustrates why the analysis of an unslotted system is somewhat messy. Suppose that a node at one end of the cable starts to transmit and then, almost β time units later, a node at the other end starts. This second node ceases its transmission almost immediately upon hearing the first node, but nonetheless causes errors in the first packet and forces the first node to stop transmission another β time units later. Finally, another β time units go by before the other end of the line is

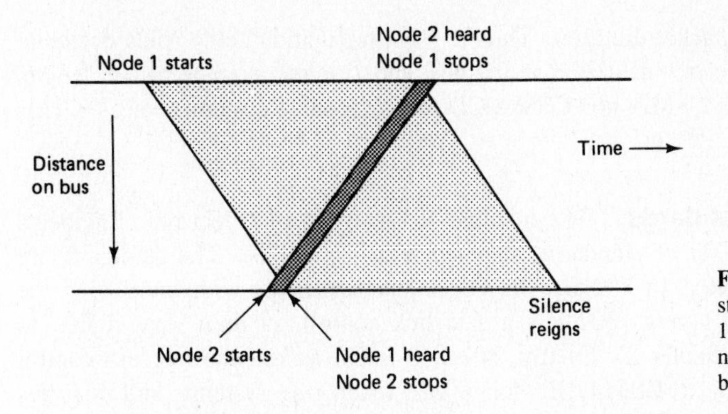

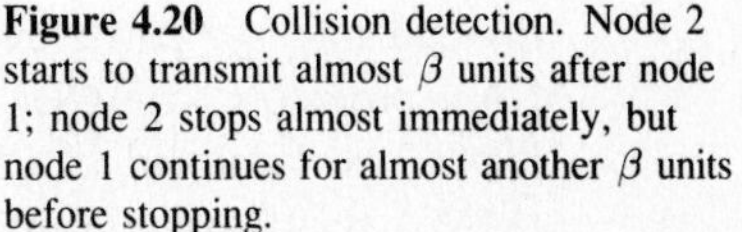
Figure 4.20 Collision detection. Node 2 starts to transmit almost β units after node 1; node 2 stops almost immediately, but node 1 continues for almost another β units before stopping.

quiet. Another complication is that nodes closer together on the cable detect collisions faster than those more spread apart. As a result, the maximum throughput achievable with Ethernet depends on the arrangement of nodes on the cable and is very complex to calculate exactly.

To get a conservative bound on maximum throughput, however, we can find bounds on all the relevant parameters from the end of one transmission (either successful or aborted) to the end of the next. Assume that each node initiates transmissions according to an independent Poisson process whenever it senses the channel idle, and assume that G is the overall Poisson intensity. All nodes sense the beginning of an idle period at most β after the end of a transmission, and the expected time to the beginning of the next transmission is at most an additional $1/G$. This next packet will collide with some later starting packet with probability at most $1 - e^{-\beta G}$ and the colliding packets will cease transmission after at most 2β. On the other hand, the packet will be successful with probability at least $e^{-\beta G}$ and will occupy 1 time unit. The departure rate S for a given G is the success probability divided by the expected time of a success or collision; so

$$S > \frac{e^{-\beta G}}{\beta + 1/G + 2\beta(1 - e^{-\beta G}) + e^{-\beta G}} \tag{4.68}$$

Optimizing the right-hand side of Eq. (4.68) over G, we find that the maximum occurs at $\beta G = (\sqrt{13} - 1)/6 = 0.43$; the corresponding maximum value is

$$S > \frac{1}{1 + 6.2\beta} \tag{4.69}$$

This analysis is very conservative, but if β is small, throughputs very close to 1 can be achieved and the difference between Eqs. (4.66) and (4.69) is not large. Note that maximum stable throughput approaches 1 with decreasing β as a constant times β for CSMA/CD, whereas the approach is as a constant times $\sqrt{\beta}$ for CSMA. The reason for this difference is that collisions are not very costly with CSMA/CD, and thus much higher attempt rates can be used. For the same reason, persistent CSMA (where new arrivals during a data slot are transmitted immediately at the end of the data slot) works reasonably for CSMA/CD but quite poorly for CSMA.

CSMA/CD (and CSMA) becomes increasingly inefficient with increasing bus length, with increasing data rate, and with decreasing data packet size. To see this, recall that

β is in units of the data packet duration. Thus if τ is propagation delay (plus detection time) in seconds, C is the raw data rate on the bus, and L is the average packet length, then $\beta = \tau C/L$. Neither CSMA nor CSMA/CD are reasonable system choices if β is more than a few tenths.

The IEEE 802 standards. The Institute of Electrical and Electronic Engineers (IEEE) has developed a set of standards, denoted 802, for LANs. The standards are divided into six parts, 802.1 to 802.6. The 802.1 standard deals with interfacing the LAN protocols to higher layers. 802.2 is a data link control standard very similar to HDLC as discussed in Chapter 2. Finally, 802.3 to 802.6 are medium access control (MAC) standards referring to CSMA/CD, token bus, token ring systems, and dual bus systems, respectively. The 802.3 standard is essentially the same as Ethernet, using unslotted persistent CSMA/CD with binary exponential backoff. We discuss the 802.5, 802.4, and 802.6 standards briefly in the next three subsections.

4.5.3 Local Area Networks: Token Rings

Token ring networks ([FaN69] and [FaL72]) constitute another popular approach to local area networks. In such networks, the nodes are arranged logically in a ring with each node transmitting to the next node around the ring (see Fig. 4.21). Normally, each node simply relays the received bit stream from the previous node on to the next. It does this with at least a one bit delay, allowing the node to read and regenerate the incoming binary digit before sending it on to the next node. Naturally, when a node transmits its own packet to the next node, it must discard what is being received. For the system to work correctly, we must ensure that what is being received and discarded is a packet that has already reached its destination. Conceptually, we visualize a "token" which exists in the net and which is passed from node to node. Whatever node has the token is allowed to transmit a packet, and when the packet is finished, the token is passed on to the next node. Nodes with nothing to send are obligated to pass the token on rather than saving it.

When we look at the properties that a token must have, we see that they are essentially the properties of the flags we studied for DLC, and that the same flag could be used as a token and to indicate the end of a packet. That is, whenever the node that is currently transmitting a packet finishes the transmission, it could place the token or flag, for example 01111110, at the end of the packet as usual. When the next node reads this token, it simply passes the token on if it has no packet to send, but if it does have a packet to send, it inverts the last token bit, turning the token into 01111111. This modified token, 01111111, is usually called a *busy token*, and the original, 01111110, is called a *free* (or *idle*) *token*. The node then follows this busy token with its own packet. Bit stuffing by inserting a 0 after 011111 is used within the data packets to avoid having either type of token appear in the data. Thus, every node can split the received stream into packets by recognizing the free and busy tokens, and the free token constitutes the passing of permission to send from one node to the next. Token rings in practice generally have longer tokens with extra information in them; there is usually more than a single bit of delay in a node also.

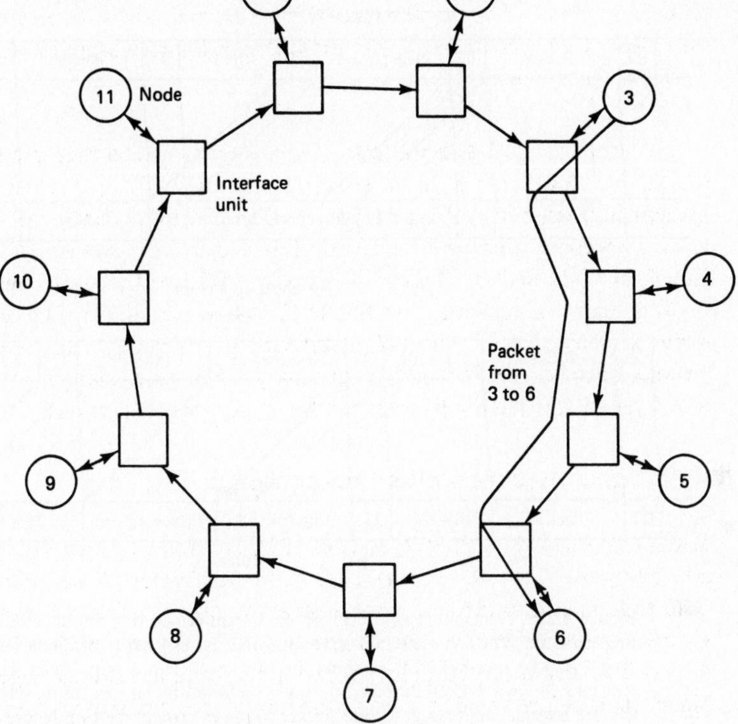

Figure 4.21 Ring network. Travel of data around the ring is unidirectional. Each node either relays the received bit stream to the next node with a one bit delay or transmits its own packet, discarding the incoming bit stream.

Let us look closely at how packets travel around the ring. Suppose, for example, that at time 0, node 1 receives a free token, inverts the last bit to form a busy token, and then starts to transmit a packet [see Fig. 4.22(a)]. Each subsequent node around the ring simply delays this bit stream by one bit per node and relays it on to the next node. The intended recipient of the packet both reads the packet into the node and relays it around the ring.

After a round-trip delay, the bit stream gets back to the originator, node 1 for our example. A round-trip delay (often called the ring latency) is defined as the propagation delay of the ring plus mk bits, where m is the number of nodes and k is the number of bit delays in a node. Assuming that the packet length is longer than the round-trip delay (in bits), the first part of the incoming packet is automatically removed by node 1, since node 1 is still transmitting a subsequent portion of the packet. When node 1 completes sending the packet, it appends a free token and then sends idle fill while the remainder of the just transmitted packet is returning to node 1 on the ring. After the last bit of the packet has returned, node 1 starts to relay what is coming in with a k bit

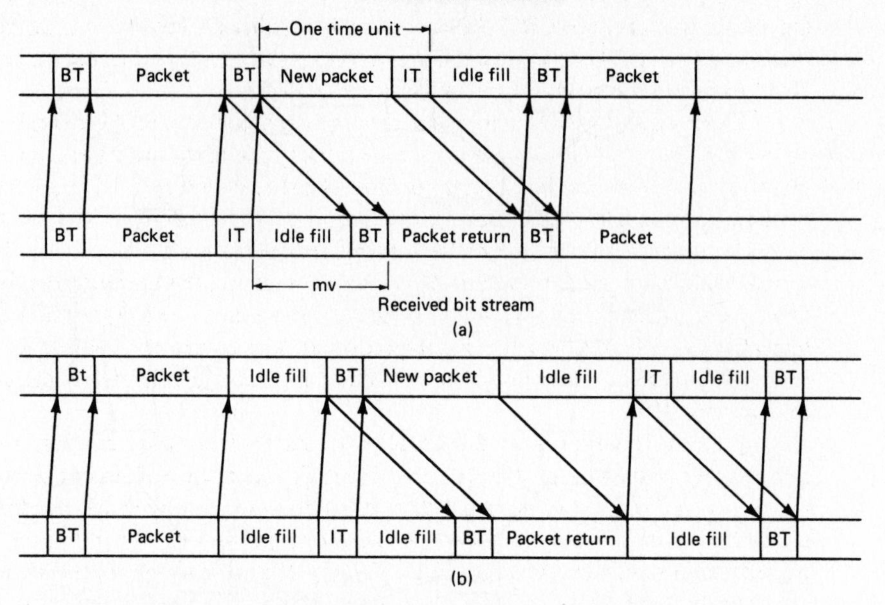

Figure 4.22 Transmitted and received bit stream at one ring interface unit of a token ring network. The interface unit transmits what is received with a one bit delay until seeing an idle token (IT). It converts the idle token into a busy token (BT) and then transmits its own packet. In part (a), this is followed by an idle token, whereas in part (b), the interface unit waits for the packet to return around the ring before transmitting the idle token. In each case, idle fill is transmitted after the idle token until the token (either busy or idle) returns around the ring, and then the unit reverts to relaying what is received with a one bit delay.

delay. If some other node has a packet to send, the first thing relayed through node 1 is a busy token followed by that packet; if no other node has a packet to send, the free token is relayed through node 1 and continues to circulate until some node has a packet to send.

Since all nodes follow this same strategy, when the idle token arrives at node 1 in the received bit stream, it must be followed by idle fill. This idle fill persists until the node sending that idle fill relays the busy token sent by node 1 (see Fig. 4.22). Thus, busy tokens are always followed by packets and idle tokens are always followed by enough idle fill to make up the round-trip delay on the ring.

Note that the round-trip delay must be at least as large as the token length; otherwise, a node, on completing a packet transmission and sending a free token, will discard the first part of the token as it returns to the node through the ring. One way to look at this is that the storage around the ring (*i.e.*, the propagation length in bits plus number of nodes) must be sufficient to store the token. Since the node transmitting a packet also reads it before removing it from the network, it is possible, either by checking bit by bit or by checking a CRC, for the transmitting node to verify that the packet was correctly received. It is also possible for the receiving node to set a bit in a given position at the end of the packet if the CRC checks. Actually, neither of these techniques is foolproof, since an error could occur between the receiving node and its ring interface or an error could occur on the given bit set by the receiving node.

There are many variations in the detailed operation of a ring net. Each node could be restricted to sending a single packet each time it acquires the token, or a node could empty its queue of waiting packets before releasing the token to the next node. Also, priorities can be introduced fairly easily in a ring net by having a field for priorities in a fixed position after the free or busy token; any node could alter the bits in this field to indicate a high-priority packet. Other nodes, with lower-priority traffic, would relay free tokens rather than using them, so as to allow nodes with high-priority packets to transmit first. Obviously, when a node transmits its high-priority packets, it would then reduce the value in the priority field. Another approach to priorities is discussed in the section on FDDI.

Another variation is in the handling of ARQ. If a node transmits a free token immediately at the end of a packet transmission, that node will no longer have control of the channel if the packet is not delivered error-free. An alternative is for a node to send idle fill after completing a packet transmission until verifying whether or not the packet was correctly received [see Fig. 4.22(b)]. If correct reception occurs, a free token is transmitted; otherwise, a busy token is sent followed by a retransmission. As seen in the figure, each busy or idle token is preceded by a round-trip delay of idle fill. Idle tokens also have idle fill following them. This alternative makes it somewhat easier to see what is happening on a ring (since at most one packet is active at a time), but it lowers efficiency and increases delay, particularly for a large ring.

Yet another variation is in the physical layout of the ring. If the cable making up the ring is put into a star configuration, as shown in Fig. 4.23, a number of benefits accrue. First, a disadvantage of a ring net is that each node requires an active interface in which each bit is read and retransmitted. If an interface malfunctions, the entire ring fails. By physically connecting each interface at a common location, it is easier to find a failed interface and bypass it at the central site. If the interface is also located at the central site (which is not usually done), the propagation delay around the ring is materially reduced.

The most important variation in a ring network is the treatment of free and busy token failures. If a free token is destroyed by noise, or if multiple free tokens are created, or if a busy token is created and circulates indefinitely, the system fails. One obvious solution to this problem is to give a special node responsibility for recreating a lost free token or destroying spurious tokens; this is rather complex because of the possibility of the special node failing or leaving the network.

IEEE 802.5 token ring standard. The more common solution to token failure, which is used in the IEEE 802.5 standard, is for each node to recognize the loss of a token or existence of multiple tokens after a time-out. If a node has a packet to transmit after a time-out occurs, it simply transmits a busy token followed by the packet followed by a free token, simultaneously purging the ring of all other tokens. If successful, the ring is again functional; if unsuccessful, due to two or more nodes trying to correct the situation at the same time, the colliding nodes try again after random delays.

The IEEE 802.5 standard also uses the star configuration and postpones releasing the token until the current packet is acknowledged. This standard uses a 24-bit token in place of the 8-bit token and contains elaborate procedures to recover from many possible malfunctions. The standard has been implemented in VLSI chips to implement a token ring running at either 4 or 16 megabits per second; the complexity of these chips, given the simplicity of the token ring concept, is truly astonishing.

Expected delay for token rings. Let us analyze the expected delay on a token ring. Assume first that each node, upon receiving a free token, empties its queue before passing the free token to the next node. Assume that there are m nodes, each with independent Poisson input streams of rate λ/m. Let v be the average propagation delay from one node to the next plus the relaying delay (usually one or a few bits) at a node. View the system conceptually from a central site, observing the free token passing

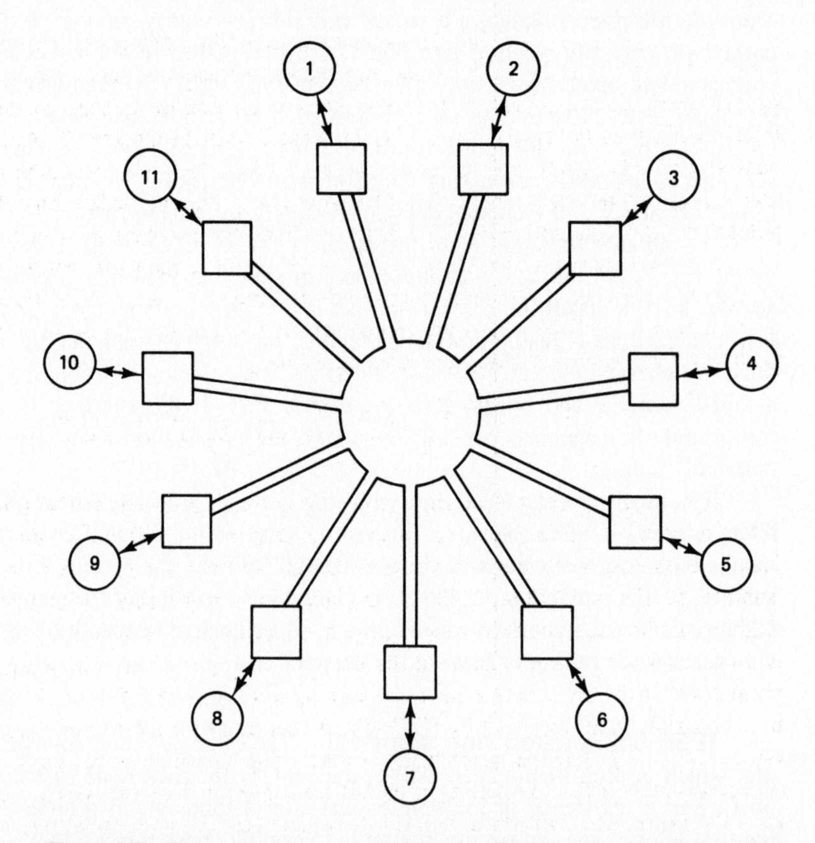

Figure 4.23 Ring network in a star configuration. Nodes can be bypassed or added from the central site.

around the ring. Let $\overline{X} = 1$ be the mean transmission time for a packet (including its busy token). Thus, $\rho = \lambda$. The mean queueing delay W is that of an exhaustive, multiuser system; the delay is given by Eq. (3.69) as

$$W = \frac{\lambda \overline{X^2}}{2(1-\lambda)} + \frac{(m-\lambda)v}{2(1-\lambda)} \tag{4.70}$$

Note that the first term is the usual $M/G/1$ queueing delay. The second term gives us the delay under no loading, $mv/2$. Observe that mv is the propagation delay around the entire ring, β, plus m times the one or few bits delay within each interface to a node. If β is small relative to a packet transmission time and if λ is relatively large, W is very close to the $M/G/1$ delay.

Next we look at the situation in which a node can transmit at most one packet with each free token. The system is then the partially gated, limited service system of Section 3.5.2. The delay is given by Eq. (3.76) as

$$W = \frac{\lambda \overline{X^2} + (m+\lambda)v}{2(1-\lambda-\lambda v)} \tag{4.71}$$

In this case, we note that the maximum stable throughput has been reduced somewhat to $1/(1+v)$. In the analysis above, we have included the token length as part of the packet overhead and included in v only the propagation delay plus bit delay at a node interface. This is necessary to use the queueing results of Chapter 3. According to that analysis, v is the delay inserted at a node even when the node has nothing to send, and that delay does not include the entire token but only the one or few bits of transit delay within the node. Equation (4.71) can be modified for the case in which a transmitting node waits for the packet to return (as in the IEEE 802.5 standard) before passing on the free token. In essence, this adds one round-trip delay (*i.e.*, mv) to the transmission time of each packet, and the maximum throughput is reduced to $1/(1+mv+v)$. As shown in Problem 4.27, the resulting expected delay is

$$W = \frac{\lambda(\overline{X^2} + 2mv + m^2v^2) + [m + \lambda(1+mv)]v}{2[1-\lambda(1+mv+v)]} \tag{4.72}$$

In comparing the token ring with CSMA/CD, note that if the propagation and detection delay is small relative to the packet transmission time, both systems have maximum throughputs very close to 1 packet per unit time. The token ring avoids stability problems, whereas CSMA/CD avoids the complexities of lost tokens and has slightly smaller delay under very light loads. Both are well-established technologies, and one chooses between them on the basis of cost, perceived reliability, and personal preference. If the propagation delay is large ($\beta > 1$), CSMA/CD loses its advantage over pure collision resolution and the IEEE 802.5 version of the token ring degrades similarly (since $mv > \beta$). On the other hand, a token ring in which nodes pass the free token on immediately after completing a packet transmission does not suffer this degradation [*i.e.*, the maximum throughput is $1/(1+v)$ and v can be very much less than β if m is large].

FDDI. FDDI, which stands for "fiber distributed data interface," is a 100-Mbps token ring using fiber optics as the transmission medium [Ros86]. Because of the high speed and relative insensitivity to physical size, FDDI can be used both as a backbone net for slower local area networks or as a metropolitan area network. There is an enhanced version of FDDI called FDDI-II that provides a mix of packet-switched and circuit-switched data. We discuss only FDDI itself here, however, since that is where the interesting new ideas are contained.

There is a rather strange character code used in FDDI in which all characters are five bits in length; 16 of the possible five bit characters represent four bits of data each, and the other characters are either special communication characters or forbidden. This character code was chosen to improve the dc balance on the physical fiber and to simplify synchronization. The data rate on the fiber is 125 Mbps in terms of these five bit characters, and thus 100 Mbps in terms of the actual data bits. The idle character is one of these special characters. It is repeated about 16 times at the beginning of each frame to allow for clock slippage between adjacent nodes.

All frames, including the token, are delimited by a start field and an end field, each of which consists of two of the special communication characters (see Fig. 4.24). The frame control field (FC) distinguishes the token from data frames and includes supplementary control information such as the length of the source address and destination address fields, which can be 16 or 48 data bits long. The CRC is the standard 32-bit CRC used in the IEEE 802 standards and as an option in HDLC; it checks on the frame control, addresses, and data. Finally, the frame status field (FS) at the end of a frame provides several flags for the destination to indicate that it has received the frame correctly. Note that when a node reads the token (i.e., what we have previously called the free token) on the ring, it can simply change the frame control field and then add the rest of its packet, thus maintaining a small delay within the node. Thus, what we called the busy token previously is just the beginning of the frame header.

There are two particularly interesting features about FDDI. The first is that nodes send the token immediately after sending data, thus corresponding to Fig. 4.22(a) rather than (b) (recall that IEEE 802.5 waits for the frame to return before releasing the token). As will be seen in the next section, this is what allows FDDI to maintain high throughput efficiency in the presence of much higher data rates and much larger distances than is possible for the 802.5 ring.

The second interesting feature about FDDI is its handling of priorities. In particular, high-priority traffic receives guaranteed throughput and guaranteed delay, thus making FDDI suitable for digitized voice, real-time control, and other applications requiring guaranteed service. The essence of the scheme is that each node times the interval between successive token arrivals at that node. High-priority traffic from the node can be sent whenever the token arrives, but low-priority traffic can be sent only if the intertoken

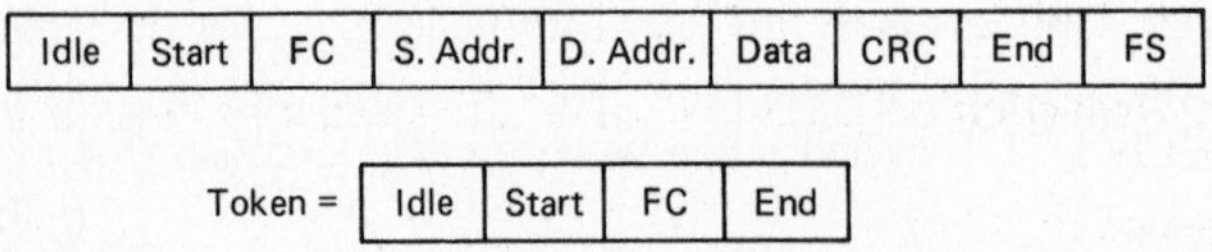

Idle	Start	FC	S. Addr.	D. Addr.	Data	CRC	End	FS

Token =

Idle	Start	FC	End

Figure 4.24 Frame format for FDDI.

interval is sufficiently small. What this means is that the low-priority traffic is decreased or even blocked when congestion starts to build up.

To provide guaranteed service to high-priority traffic, it should be clear that the network must impose constraints on that traffic. These constraints take the form of a limitation on how much high-priority traffic each node can send per received token. In particular, let the m nodes be numbered $0, 1, \ldots, m-1$ in order around the ring. Let α_i be the allocated time for node i to send its high-priority traffic, including delay to reach the next node. Thus, if a token reaches node i at time t and node i sends its allocated amount of traffic, the token reaches node $i+1$ at time $t+\alpha_i$.

When the ring is initialized, there is a parameter τ called *target token rotation time.* This parameter is used by the nodes in deciding when to send low-priority traffic, and as we shall see, τ is an upper bound on the time-average intertoken arrival time. The allocated transmission times $\alpha_0 \ldots, \alpha_{m-1}$ are allocated in such a way that $\alpha_0 + \alpha_1 + \ldots + \alpha_{m-1} \leq \tau$. To describe and analyze the algorithm precisely, let $t_0, t_1, \ldots, t_{m-1}$ be the times at which the token reaches nodes 0 to $m-1$ for some given cycle. Similarly, let $t_m, \ldots, t_{2m-1}$ be the times at which the token reaches nodes 0 to $m-1$ in the next cycle, and so forth. Thus $t_i, i \geq 0$ is the time at which the token reaches node (i mod m) in cycle $\lfloor i/m \rfloor$ (where we denote the given cycle as cycle 0 and where $\lfloor x \rfloor$ denotes the integer part of x). Finally, let t_{-m} to t_{-1} be the times at which the token reaches nodes 0 to m_{-1} in the cycle previous to the given one. At time $t_i, i \geq 0$ the node (i mod m), having just received the token, measures $t_i - t_{i-m}$, which is the elapsed time since its previous receipt of the token. If $t_i - t_{i-m} < \tau$, the node is allowed to send low-priority traffic for $\tau - (t_i - t_{i-m})$ seconds. If $t_i - t_{i-m} \geq \tau$, no low-priority traffic is allowed. In both cases, the allocated high-priority traffic is allowed. The time at which the token reaches the next node is then upper bounded by

$$t_{i+1} \leq t_{i-m} + \tau + \alpha_i, \qquad \text{for } t_i - t_{i-m} < \tau;\ i \geq 0$$

$$t_{i+1} \leq t_i + \alpha_i, \qquad \text{for } t_i - t_{i-m} \geq \tau;\ i \geq 0$$

where $\alpha_i = \alpha_{i \bmod m}$ is the allocated transmission plus propagation time for node (i mod m). Combining these conditions, we have

$$t_{i+1} \leq \max(t_i, t_{i-m} + \tau) + \alpha_i;\quad i \geq 0 \tag{4.73}$$

Note that equality holds if node (i mod m) sends as much data as allowed.

We first look at a simple special case of this inequality where $\alpha_i = 0$ for all i. Let τ' be the target token rotation time and t'_i be t_i for this special case. Thus

$$t'_{i+1} \leq \max(t'_i, t'_{i-m} + \tau');\ i \geq 0 \tag{4.74}$$

Since t'_i must be non-decreasing in i, we have $t'_{i-m} \leq t'_i$. for all $i \geq 0$. Substituting this in Eq. (4.74) yields $t'_{i+1} \leq t'_i + \tau'$. Similarly, for $1 \leq j \leq m+1$, we have

$$t'_{i+j} \leq \max(t'_{i+j-1}, t'_i + \tau') \leq t'_i + \tau'$$

where the last step follows by induction on j. For $j = m+1$, this equation reduces to

$$t'_{i+m+1} \leq t'_i + \tau' \qquad \text{for all } i \geq 0 \tag{4.75}$$

Iterating this over multiples of $m+1$, we obtain

$$t'_i \le t'_{i \bmod (m+1)} + \lfloor i/(m+1) \rfloor \tau'; \qquad \text{all } i \ge 0 \tag{4.76}$$

The appearance of $m+1$ in Eq. (4.76) is somewhat surprising. To see what is happening, assume that $t'_i = 0$ for $0 \le i \le m$, and assume that all nodes are heavily loaded. Node 0 then transmits for time τ' starting at t'_m. This prevents all other nodes from transmitting for a complete cycle of the token. Since $t'_{2m} = \tau'$ and $t'_m = 0$, node 0 is also prevented from transmitting at t'_{2m} (note that the token travels around the ring in zero time since we are ignoring propagation delay here). Node 1 is then allowed to transmit for time τ' in this new cycle, and then no other node is allowed to transmit until node 2 transmits in the next cycle. This helps explain why events occur spaced $m+1$ token passings apart.

Now consider the elapsed time for the token to make $m+1$ cycles around the ring starting from some node j. From Eq. (4.76) we have

$$t'_{j+(m+1)m} - t'_j \le m\tau' \tag{4.77}$$

Note that $t'_{j+(m+1)m} - t'_j$ is the sum of the $m+1$ cycle rotation times measured by node j at times $t'_{j+m}, t'_{j+2m}, \ldots, t'_{j+(m+1)m}$. Since each cycle rotation requires at most τ' seconds, we see that $(m+1)\tau' - (t'_{j+(m+1)m} - t'_j)$ is equal to the total time offered to j at the above $m+1$ token receipt times. Combining this with Eq. (4.77), we see that over the $m+1$ token receipt times above, node j is offered an aggregate of at least τ' seconds for its own transmission; also, the corresponding $m+1$ cycles occupy at most time $m\tau'$. It is also seen from this that the average token rotation time is at most $[m/(m+1)]\tau'$ rather than the target time τ'. Using the same argument, it can be seen that over a sequence of any number n of cycles, the total amount of time allocated to node j is at least as great as that offered to any other node during its $n-1$ token receptions within the given n cycles at node j. Thus each node receives a fair share of the resources of the ring. (This type of fairness is known as max-min fairness and is discussed in detail in Chapter 6.)

Although there is a certain elegance to this analysis, one should observe that in this special case of no high-priority traffic, the same kind of results could be obtained much more simply by allowing each node to transmit for τ' seconds on each receipt of the token. Fortunately, however, the results above make it quite easy to understand the general case of Eq. (4.73), in which node i is allocated α_i units of high-priority traffic. If we subtract $\sum_{j=0}^{i} \alpha_j$ from each side of Eq. (4.73) for $i \ge m$, we get

$$t_{i+1} - \sum_{j=0}^{i} \alpha_j \le \max\left(t_i - \sum_{j=0}^{i-1} \alpha_j, \qquad t_{i-m} - \sum_{j=0}^{i-m-1} \alpha_j + \tau - \sum_{j=i-m}^{i-1} \alpha_j \right) \tag{4.78}$$

Since $\alpha_j = \alpha_{j \bmod m}$, we see that $\sum_{j=i-m}^{i-1} \alpha_j$ is independent of i and is equal to the total allocated traffic, $T = \sum_{j=0}^{m-1} \alpha_j$. We now define

$$t_i' = t_i - \sum_{j=0}^{i-1} \alpha_j \text{ for all } i \geq 0; \qquad \tau' = \tau - T \tag{4.79}$$

Substituting this in Eq. (4.78), we obtain Eq. (4.74) for $i \geq m$,

$$t_{i+1}' \leq \max(t_i', t_{i-m}' + \tau')$$

To obtain initial conditions for this, assume that $t_0 = 0$ and that $t_i \leq 0$ for $i < 0$. Iterating Eq. (4.73) from $i = 0$ to $m - 1$, and using $t_i \leq t_{i+1}$, we get

$$0 \leq t_i \leq \tau + \sum_{j=0}^{i-1} \alpha_j; \qquad 1 \leq i \leq m$$

From Eq. (4.79), then, we have $0 \leq t_i' \leq \tau$ for $1 \leq i \leq m$. Using τ as an upper bound to t_i' for $0 \leq i \leq m$, induction over i can be applied to Eq. (4.74) to yield

$$t_i' \leq \tau + \left\lfloor \frac{i}{m+1} \right\rfloor \tau'$$

$$t_i \leq \sum_{j=0}^{i-1} \alpha_j + (\tau + \tau) - T \left\lfloor \frac{i}{m+1} \right\rfloor \tag{4.80}$$

We can rewrite $\sum_{j=0}^{i-1} \alpha_j$ as $\sum_{j=0}^{i-1 \bmod m} \alpha_j + \lfloor (i-1)/m \rfloor T$, so Eq. (4.80) becomes

$$t_i \leq \sum_{j=0}^{(i-1) \bmod m} \alpha_j + \tau \left(1 + \left\lfloor \frac{i}{m+1} \right\rfloor\right) + T \left(\left\lfloor \frac{i-1}{m} \right\rfloor - \left\lfloor \frac{i}{m+1} \right\rfloor \right) \tag{4.81}$$

We now focus on the time mk at which node 0 receives its kth token

$$t_{mk} \leq \tau \left(1 + \left\lfloor \frac{mk}{m+1} \right\rfloor\right) + T \left(k - \left\lfloor \frac{mk}{m+1} \right\rfloor \right) \tag{4.82}$$

We see that $t_m - t_0 \leq \tau + T$ (which we already knew), and this bound can be met with equality if all the nodes are idle up to time t_0 and then all become busy. Since node 0 could be any node and t_0 the time of any token receipt at that node, this says in general that the intertoken interval at a node can be as large as $\tau + T$ but no larger. We also see that $\limsup_{k \to \infty} (t_{mk}/k) \leq \tau m/(m+1) + T/(m+1)$. Thus the time-average round-trip token time is at most τ, and is somewhat smaller than τ if $T < \tau$.

Finally, node 0 is allowed to send at least $k\tau - (t_{mk} - t_0)$ seconds of low-priority traffic at the k token receipt times ending with t_{mk}. From Eq. (4.82), the average such traffic per token receipt is at least $(\tau - T)/(m+1)$. Thus each node receives the same amount of guaranteed low-priority traffic and Eq. (4.82) indicates how much is guaranteed within any given number of token receipts.

There is a trade-off between throughput efficiency and delay for FDDI. Assuming that the high-priority traffic is stream-type traffic within the allocated rate, we have seen that the delay is bounded by $\tau + T$, which is more loosely bounded by 2τ. On the other hand, each time the token travels around the ring, there is a period mv, corresponding to the propagation time plus delay within the nodes, during which no data are sent. Thus the throughput, as a fraction of the 100 Mbps, is at most $1 - mv/\tau$, and actually slightly smaller than this because the actual token rotation time is slightly faster than τ. As an example, if $\tau = 10$ msec and the ring circumference is 100 miles, then throughputs of about 94% are possible. Thus this type of system maintains high throughput and low guaranteed delay for large metropolitan area networks.

The maximum packet size for FDDI is set at 36,000 bits (because of physical synchronization requirements). Since the low-priority traffic is sometimes allowed in rather small pieces, however, a user with smaller packets will get better service than one with large packets.

Slotted rings and register insertion rings. Assuming that traffic on a ring is uniformly distributed between different source–destination pairs, a packet need be transmitted on only half of a ring's links on the average. Since in the token ring, a packet travels on every one of the links, we see that half the system's transmission capability is potentially wasted. It is therefore conceivable that a different control strategy could achieve twice the throughput.

Slotted rings and register insertion rings allow this higher potential throughput, at least in principle. A slotted ring is best viewed as a conveyor belt of packet slots; the ring is extended by shift registers within the nodes to provide the desired number of slots. When a node has a packet to send, it looks for an empty slot in the conveyor belt and places the packet in that slot, marking the slot as full. When the destination node sees the packet, it removes it and marks the slot as empty again.

One disadvantage of a slotted ring is that all packets must have equal length (as contrasted with token rings and CSMA/CD). Another disadvantage is significant delay (due to the conveyor belt length) even under light load. This can be compensated for by making the packets very short, but the added DLC overhead caused by short packets loses much of the potential throughput gain. Finally, to accomplish ARQ, it is common to leave packets in their slots until they return to the sending node; this automatically throws away the potential doubling of throughput.

The register insertion ring provides true store-and-forward buffering of ring traffic within the nodes. Each node has a buffer for transit traffic and a buffer for new arrivals (see Fig. 4.25). When a new arrival is being transmitted, incoming ring traffic that must be forwarded is saved in the transit buffer. When the transit traffic is being transmitted, the buffer gradually empties as either idle fill or packets destined for the given node arrive at the ring input. New arrivals are inhibited from transmission whenever the transit buffer does not have enough space to store the input from the ring while the new packet is being inserted on the ring.

The register insertion ring is capable of higher throughputs and has only a slightly greater delay for light loading than the token ring. Its greatest disadvantage is that it loses the fair allocation and guaranteed access provided by the token ring's round-robin packet service. The token ring is much more popular for applications, probably because maximum throughput is not the dominant consideration for most local area networks.

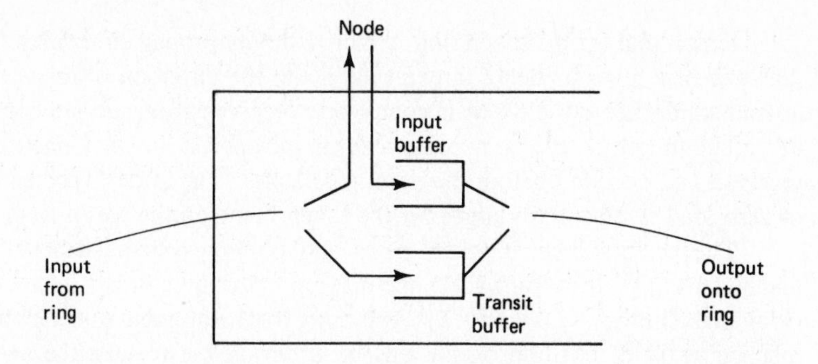

Figure 4.25 Interface to a register insertion ring. The output onto the ring comes from either the input buffer or the transit buffer. The input from the ring goes to the transit buffer or directly to the node, if addressed there.

4.5.4 Local Area Networks: Token Buses and Polling

The general idea of the token ring is used in a wide variety of communication situations. The idea is that there are m nodes with ordered identities and the nodes are offered service one at a time in round-robin order. The differences between these systems lie in the question of how one node knows when the previous node has finished service (or refused the offer of service). In other words, what mechanism performs the role of the token, and what is the delay, v, in passing this virtual token from one node to the next?

As discussed briefly in Section 4.1.2, polling is a common example of such a system. The polls sent by the central node to each of the secondary nodes act as tokens. The token-passing delay in a polling system is quite large, involving first a communication from a polled node back to the central node, and then a new polling request from the central node to the next secondary node.

Hub polling is a way of avoiding the double delays above. The central station polls (*i.e.*, passes the token to) the first secondary node; each secondary node, after using the channel, passes the token on to the next secondary node. If the nodes are ordered in terms of distance on the multidrop telephone line or bus, then, of course, token passing delay is reduced even further.

A token bus can be implemented on the same type of physical bus as a CSMA/CD system. The nodes are ordered in a round-robin fashion, and when a node finishes sending its packet or packets, it sends a token to the next node, giving it permission to send next. A node that has nothing to send simply sends the token to the next node. Conceptually, it can be seen that a token bus is essentially the same as a hub polling system. Polling is the more common terminology with a central node, and token bus is more common for a fully distributed system. Equations (4.70) and (4.71) give the delay associated with these systems under the assumptions of sending all queued packets per poll and one packet per poll, respectively. Both equations assume equal average traffic for all nodes. The parameter v in these equations can be interpreted in general as the delay, in an empty system, from token arrival at one node to the next, averaged over the nodes.

Recall that for the token ring, v included only propagation delay from node to node plus one or a few bits delay through the node; the initiation of token transmission from a node starts before the token is completely received from the previous node. Here, the token length (which can be considerable) is included in v, since one token must be fully received and decoded before the next node starts. Thus, the expected delay in the limit of zero load is inherently larger for the token bus than the token ring.

The performance of token buses and polling systems depends critically on the parameter v [*i.e.*, from Eq. (4.71), the maximum throughput is $1/(1+v)$ and the expected delay in the limit $\lambda \to 0$ is $mv/2$]. Assuming for the moment that the nodes are numbered independently of position on the bus, it is reasonable to take the average propagation delay as $\beta/3$ (see Problem 4.29), where $\beta = \tau C/L$ is the normalized propagation time from one end of the bus to the other (*i.e.*, the time measured as a fraction of the average packet length). Here τ is propagation time in seconds, C is the channel bit rate, and L is the expected packet length. Similarly, the normalized token transmission time is k/L, where k is the length of the token. Thus,

$$v = \frac{\tau C}{3L} + \frac{k}{L} + \delta \tag{4.83}$$

where δ is the normalized time for the receiver to detect a token. In our previous discussions, we included δ as part of the propagation delay β, but here we count this time separately.

The quantity τC is the number of bits that can travel along the bus at one time. If $\tau C/3$ is small relative to k, improving the performance of the algorithm depends on decreasing the length of the token. Conversely, if $\tau C/3$ is large, k is relatively unimportant and improvements depend on reducing the effects of propagation delay. One obvious way to reduce the effect of propagation delay is to number the nodes sequentially from one end of the bus to the other. If this is done, the sum of the propagation delays over all nodes is 2β; that is, there is a cumulative propagation delay of β moving down the bus to poll all nodes, and then another β to return. Thus, the average value of v is

$$v = \frac{2\tau C}{mL} + \frac{k}{L} + \delta \tag{4.84}$$

This is a major reduction in propagation delay, but it makes it somewhat more difficult to add new nodes to the bus. Note that the propagation delays are much smaller than the average as reservation opportunities move down the bus, but then there is a long reservation interval of duration β to return from the end of the bus to the beginning. We recall from Section 3.5.2 that Eqs. (4.70) and (4.71) are valid using the average value of v.

IEEE 802.4 token bus standard. The IEEE 802.4 standard corresponds essentially to the system we have just described. We will briefly describe some of its features. To allow new nodes to enter the round-robin token structure, each node already in the structure periodically sends a special control packet inviting waiting nodes to join. All waiting nodes respond, and if more than one, a splitting algorithm is used to select one. The new node enters the round robin after the inviting node, and the new node subsequently addresses the token to the node formerly following the inviting node. An old node can drop out of the round robin simply by sending a control packet to its predecessor directing the predecessor to send subsequent tokens to the successor of the node that is dropping out. Finally, failure recovery is essentially accomplished by all nodes dropping out, then one starting by contention, and finally the starting node adding new nodes by the procedure above.

Implicit tokens: CSMA/CA. Consider how to reduce the token size. One common approach is to replace the token with an implicit token represented by the channel becoming idle. Two of the better known acronyms for this are BRAM [CFL79] and MSAP [KIS80]. In these schemes, when a node completes a packet transmission, it simply goes idle. The next node in sequence, upon detecting the idle channel, starts transmission if it has a packet or otherwise remains idle. Successive nodes in the sequence wait for successively longer times, after hearing an idle, before starting transmission, thus giving each of the earlier stations an opportunity to transmit if it has packets. These schemes are often called CSMA/Collision Avoidance (CSMA/CA) schemes.

To see how long a node must hear an idle channel before starting to transmit, consider the worst case, in which the node that finishes a packet is at one end of the bus, the next node is at the opposite end, and the second node is at the first end again. Then the second node will detect the idle channel almost immediately at the end of the transmission, but the first node will not detect the event until $\beta + \delta$ units later, and the second node will not know whether the first node is going to transmit until an additional delay of $\beta + \delta$. Thus, the second node must wait for $2(\beta + \delta)$ before starting to transmit. By the same argument, we see that each successive node must wait an additional increment of $(\beta + \delta)$. Thus, this scheme replaces an explicit token of duration k/L with an implicit token of duration $(\beta + \delta) = (\tau C/L + \delta)$. The scheme is promising, therefore, in situations where τC and δ are small.

If the nodes are ordered on the bus, and the delays are known and built into the algorithm, the durations of the implicit tokens are greatly reduced, as in Eq. (4.70), but the complexity is greatly increased. There are many variations on this scheme dealing with the problems of maintaining synchronization after long idle periods and recovering from errors. (See [FiT84] for an excellent critical summary.)

4.5.5 High-Speed Local Area Networks

Increasing requirements for data communications, as well as the availability of high-data-rate communication media, such as optical fiber, coaxial cable, and CATV systems, motivate the use of higher- and higher-speed local area networks. For our purposes, we define a high-speed local area network as one in which β exceeds 1. Recall that β is the ratio of propagation delay to average packet transmission time, so $\beta > 1$ means that a transmitter will have finished sending a packet before a distant receiver starts to hear it. Since $\beta = \tau C/L$, we see that increasing propagation delay τ, increasing data rate C, and decreasing expected packet length L all contribute to making the net high speed. There has also been great interest in extending local area techniques to wider area coverages, such as metropolitan areas; the larger areas here can cause a network to become high speed even at modest data rates.

We have already seen the effect of β on the performance of local area networks. CSMA/CD degrades rapidly with increasing β and becomes pointless for $\beta > 1$. Similarly, for token rings such as IEEE 802.5 that release the token only after an ack is received, we see from Eq. (4.72) that the maximum throughput degrades as $1/(1+\beta)$ as β gets large. [Note that mv in Eq. (4.72) is essentially equal to β when β is large.] For token rings such as FDDI that release the token after completing transmission, we see from Eq. (4.71) that the maximum throughput degrades as $1/(1+\beta/m)$. Thus in essence it is the propagation delay between successive nodes that is important rather than the propagation delay around the ring.

As a practical matter, Eq. (4.71) must be used with considerable caution. It assumes that all nodes on the ring have Poisson packet arrivals at the same rate. For local area networks, however, there are often only a very small number of nodes actively using the net in any given interval of a few minutes. In such situations, m should be taken as the number of active nodes rather than the total number of nodes, thus giving rise to much greater throughput degradation. Fortunately, FDDI, with its token rotation timer, avoids this problem by allowing each node to transmit more data per token when the number of active nodes is small. FDDI was designed as a high-speed local area network, but it was described in the last section since it is a high-performance type of token ring.

The token bus, with implicit or explicit tokens, also degrades as $1/(1+\beta/3)$ if the nodes are numbered arbitrarily. If the nodes are numbered sequentially with respect to physical bus location, then, as seen by Eq. (4.84), the maximum throughput is degraded substantially only when β is a substantial fraction of the number of nodes m. If an ack is awaited after each packet, however, this throughput advantage is lost.

One of the most attractive possibilities for high-speed local networks lies in the use of buses that propagate signals in only one direction. Optical fibers have this property naturally, and cable technology is well developed for unidirectional transmission.

A unidirectional bus is particularly easy to use if the data frames are restricted to be all of the same size. In this case, the bus can be considered to be slotted. Empty slots are generated at the head of the bus and there is a "busy bit" at the front of the slot that has the value 0 if the slot is empty and 1 if the slot is full. Each node in turn, then, can read this busy bit; if the busy bit is 0 and the node has a frame to send, the node changes the busy bit to 1 and sends the frame. Nodes farther down the bus receive the busy bit as 1 and read the frame but do not modify it. Note that the fixed-length slot is necessary here, since otherwise a node could still be transmitting a longer frame when a frame arrives from upstream on the bus.

The slotted unidirectional bus structure above is almost trivial from a logical point of view, but has some very powerful characteristics. First it has the same advantage as Ethernet in that there is almost no delay under light loading. Here, of course, there is a half-slot delay on the average, but if the slots are short, this is negligible. Second, there is the advantage of ideal efficiency. As long as nodes are backlogged, every slot will be utilized, and aside from frame headers and trailers, the maximum data rate is equal to the bus rate.

There are also a few disadvantages to this trivial scheme. The first is that data can flow in only one direction, and the second is an inherent unfairness in that nodes close to the head of the bus get priority over those farther down the bus. The following dual bus architecture solves both these problems.

Distributed queue dual bus (IEEE 802.6). The distributed queue dual bus architecture (DQDB) (also known by its older name QPSX) is being standardized as the IEEE 802.6 metropolitan area network ([NBH88] and [HCM90]). Each node is connected to two unidirectional 150 Mbps buses (optical fibers) going in opposite directions (Fig. 4.26). Thus a node uses the right-moving bus to send frames to nodes on its right and uses the left-moving bus for nodes on its left (thus solving one of the problems with the trivial single bus structure above). The frames have a fixed length of 53 bytes and fit into slots generated at the head ends of the buses. The frame length was chosen for compatibility with ATM (see Section 2.10).

The dual bus architecture attempts to achieve fairness by adding a request bit to the overhead. If a node has a frame to send on the right bus, it sets the request bit in a frame on the left bus. Similarly, if it has a frame for the left bus, it sets the request bit in a frame on the right bus. In what follows, we focus on data traffic in a single direction, denoted downstream, and refer to the request bits as moving upstream (relative to the data direction of interest). DQDB allows for a set of different priorities for different classes of traffic, and each slot contains a one bit request field for each priority. For simplicity of exposition, we ignore these priorities, thus assuming that each slot contains only a single request bit.

Each request bit seen by a node on the upstream bus serves as an indication of a waiting frame at a more downstream node. These request bits are not removed as they pass upstream, and thus all upstream nodes see each such request bit. Each node views its own frames, plus the downstream frames indicated by request bits, as forming a virtual queue. This virtual queue is served in first-come first-serve order. This means that if a frame from the given node is at the front of the queue, that frame replaces the first idle slot to arrive on the downstream bus. Alternatively, suppose that a request from downstream is at the front of the queue. In this case, when the next idle slot arrives in the downstream direction, the request is removed but the slot is left idle, thus allowing a downstream node to transmit a frame.

To prevent any node from hogging the bus, each node is allowed to enter only one frame at a time on its virtual queue. Other frames at that node are forced to wait in a supplementary queue. As soon as the frame in the virtual queue is transmitted, a frame from the supplementary queue (or a new frame) can be entered into the virtual queue.

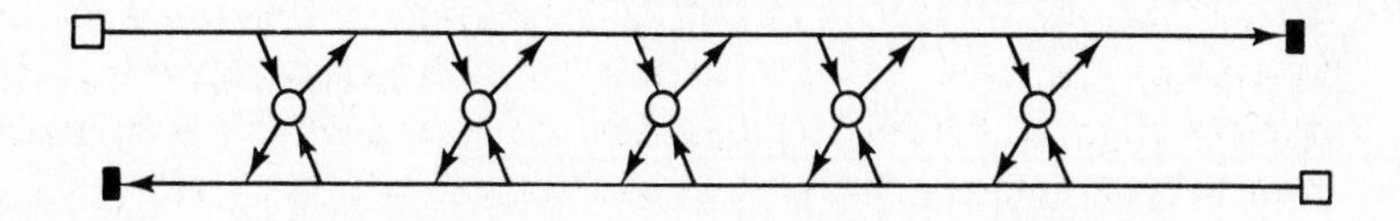

Figure 4.26 Bus structure for IEEE 802.6 dual bus.

A request bit is sent upstream at each instant when a frame enters the virtual queue. Since the virtual queue consists of at most one actual frame plus an arbitrary number of requests before and after that frame, the virtual queue can be implemented simply as two counters giving the number of requests before and behind the frame, respectively.

The description above of sending request bits upstream was oversimplified. The problem is that each slot contains only one request bit (ignoring priorities), and thus if some downstream node has already set the request bit in a slot, a node farther upstream will have to wait to send its own request. If there is a long string of requests coming from downstream nodes and a long string of idle slots coming from upstream nodes, it is possible that the waiting frame will be sent downstream before the request bit is sent upstream. In this case, the request bit continues to be queued waiting to be sent upstream. If a new frame enters the virtual queue waiting to be sent downstream, a new request bit is generated and queued behind the old request bit, and such a queue can continue to grow.

To explain the reason for a queue of request bits at a node, note that the request bits do not indicate which node has a frame to send. Thus we could visualize a node as sending its own request bit and queueing request bits from downstream; this would correspond to the position of the frame in the virtual queue. Another rationale is that there is a certain cleanness in ensuring that one and only one request bit is sent upstream for each frame that is transmitted. This makes it possible for each node to determine the loading on the bus by summing the busy bits traveling downstream (the number of frames sent by upstream nodes) to the request bits traveling upstream (the number of frames from downstream nodes for which request bits have been received).

We now give two examples of the operation of this system which show that the system as explained is quite unfair in terms of giving much higher rates to some nodes than others under certain conditions. We then explain a modification (see [HCM90]), which is now part of the proposed standard. These examples demonstrate the peculiar behavior that results from large propagation delays on the bus. Since DQDB is designed as a metropolitan area network with a bit rate of 150 Mbps, large propagation delays relative to the slot time are expected. For example a 30 km bus has about 50 slots in simultaneous flight.

As the first example, suppose that initially only one node is sending data, and then another node farther downstream becomes active. Initially, the upstream node is filling all the slots with data, which it is allowed to do since it sees no request bits on the bus in the upstream direction. Figure 4.27 illustrates the situation shortly after the downstream node becomes active and sends a request bit which is propagating on the

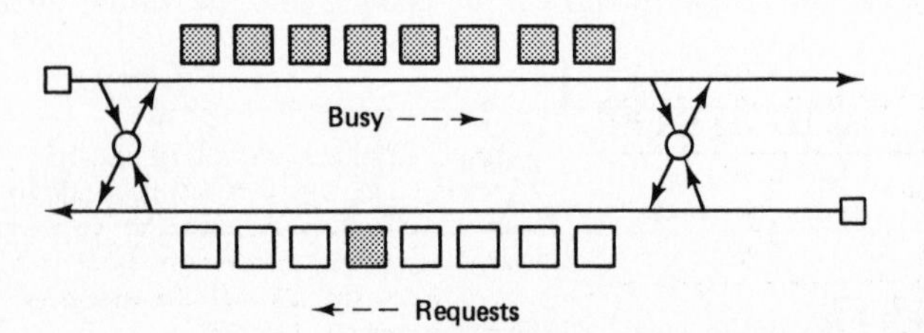

Figure 4.27 The upstream node on the left has a large file of data and is filling each downstream slot. The downstream node then receives a large file to send. The first frame is in the virtual queue and the request bit is traveling upstream.

upstream bus. The downstream node cannot send its first frame until an idle slot appears on the downstream bus, and cannot send another request bit upstream until the first frame is sent.

When the request bit reaches the busy upstream node, that node allows a slot to remain idle, and that slot propagates along the downstream bus (see Fig. 4.28). When the idle slot reaches the busy downstream node, a frame is transmitted, a new frame enters the virtual queue, and a new request bit starts to propagate upstream. It is evident that the downstream node sends only one frame out of each round-trip propagation delay. Thus the downstream node transmits one slot out of each $2n$, where n is the number of slots on the bus between the upstream and downstream node. With a separation of 30 km between the nodes, the downstream node receives less than one slot out of a hundred.

The next example is somewhat more complex and shows that a downstream node is also capable of hogging the bus if it starts sending a long file before any upstream node has anything to send. Figure 4.29 shows the situation before the upstream node becomes busy. The upstream bus is full of slots containing request bits.

After the upstream node becomes active, it puts a frame in its virtual queue. Assuming that this queue is initially empty, the frame is transmitted in the next slot and the arriving request bit is queued. When the next frame enters the virtual queue, it must wait for the queued request before being transmitted, and during this time two new requests join the virtual queue. Figure 4.30 illustrates the situation before the first busy slot from the busy upstream node arrives at the busy downstream node.

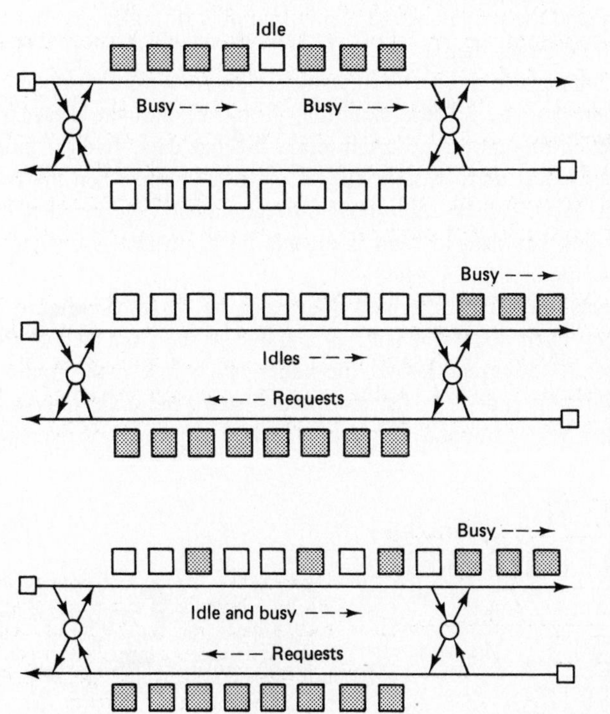

Figure 4.28 Continuation of Fig. 4.27 in which the upstream node allows an idle slot to propagate downstream after receiving a request from the downstream node.

Figure 4.29 Illustration of situation in which a downstream node is sending a frame in every slot just before an upstream node becomes active. Note that the upstream bus is full of slots carrying request bits.

Figure 4.30 Continuation of Fig. 4.29 showing the scenario as the first busy slot arrives at the downstream node. Note that the first busy slot between the two nodes is followed by one idle slot, the second by two idle slots, and so forth.

Each time a busy slot from the busy upstream node passes the busy downstream node, the downstream node cannot enter a new frame into its virtual queue and thus cannot place a request bit in the corresponding upstream slot. Thus the sequence of requests and no requests on the upstream bus can be viewed as an extension of the idle and busy slots, respectively, on the downstream bus, circling around at the busy downstream node (see Fig. 4.31). It turns out that in steady state, the downstream node obtains approximately $\sqrt{2n}$ slots for each slot obtained by the upstream node (see [HCM90]).

There are several observations that can be made from these examples. First, in the case of two users with large files, the first user to start transmission receives an inordinate share of the slots, and this unfairness will persist for arbitrarily long files. Second, in both examples, each request sent upstream by the downstream node eventually returns downstream in the form of an idle slot, and each such idle slot gives rise to a further request going upstream. The total number p of these requests and idle slots between the two nodes (including the requests queued at the upstream node) remains fixed for the duration of the file transfers. We call p the number of reservations captured by the downstream node. In the first example above, $p = 1$, and in the second, $p = 2n$. For fair operation, it would be desirable for p to be approximately n.

Note that something very peculiar has happened; only one frame is allowed into the virtual queue at a time (thus making it look like the system is providing some sort of round-robin service to the nodes), but in fact a large number p of reservations can be pipelined through the upstream nodes for service at the given node. In fact, because of the large propagation delay and the prevalence of a small number of nodes sending large files, it is essential for p to be larger than 1.

Consider now what happens in the second example if the downstream node completes its file transfer. There will be $2n$ idle slots that propagate past the downstream node and never get used, even though the upstream node still has data. Even if fairness were achieved between the nodes, there would still be n wasted slots when the downstream node becomes idle. What can be said about throughput, however, is that if the farthest downstream node that has data to send is always busy, no slot is ever wasted (*i.e.*, that node will fill any idle slot that it sees).

Fortunately, the fairness problem described above can be cured. One cure is to change the rule followed by each node when it has a frame at the front of the virtual queue and sees an idle slot. As described above, the node always inserts the frame into the idle slot under these circumstances. In the modified system, the node only inserts its frame into the idle slot a given fraction $1 - f$ of the time. In terms of example 1,

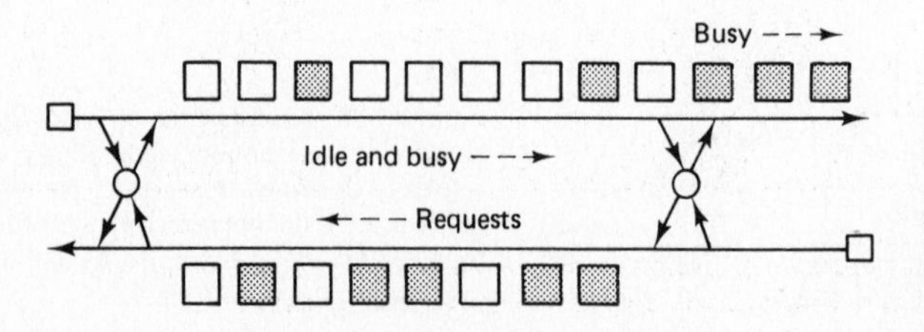

Figure 4.31 Continuation of Fig. 4.29 showing the scenario as the first nonrequest slot arrives at the upstream node. Note that successive nonrequest (busy) slots are followed by increasing numbers of request (idle) slots.

this modification forces the upstream node to insert occasional extra idle slots, which allows the downstream node to increase its number of reservations gradually. In terms of example 2, the downstream node, when inserting extra idles, is also inhibited from sending requests upstream; this reduces its number of reservations. Thus, as shown in [HCM90], the number of reservations approaches the fair value. This modification is now a part of the 802.6 draft standard.

The problem with the modification above is that it wastes part of the throughput of the system. For the case of two nodes with large files, for example, the fraction of the throughput that is wasted can be shown to be $f/(f+2)$, where f is the fraction of time that a node leaves an idle slot empty when it has a frame at the front of the virtual queue. One can choose f arbitrarily small, but this slows the rate at which the system converges to fair operation (see [HCM90]).

Both DQDB and FDDI have considerable promise as high-speed networks. FDDI has the advantage of simple guarantees on rate and delay, whereas DQDB has the advantage of compatibility with ATM. Neither use optical fiber in an ideal way, however; since each node must read all the data, the speed of these networks is limited by the electronic processing at the nodes.

Expressnet. There are other approaches to high-speed local area networks using unidirectional buses but allowing variable-length frames. The general idea in these approaches is to combine the notion of an implicit token (*i.e.*, silence on the bus) with collision detection. To see how this is done, consider Fig. 4.32, and for now ignore the problem of how packets on the bus get to their destinations. Assume that a node on the left end of the bus has just finished transmission. Each subsequent node that has a packet to send starts to transmit carrier as soon as it detects the channel to be idle.

If δ is the time required for a node to detect idle and to start transmitting carrier, we see from the figure that if two nodes have packets to send at the completion

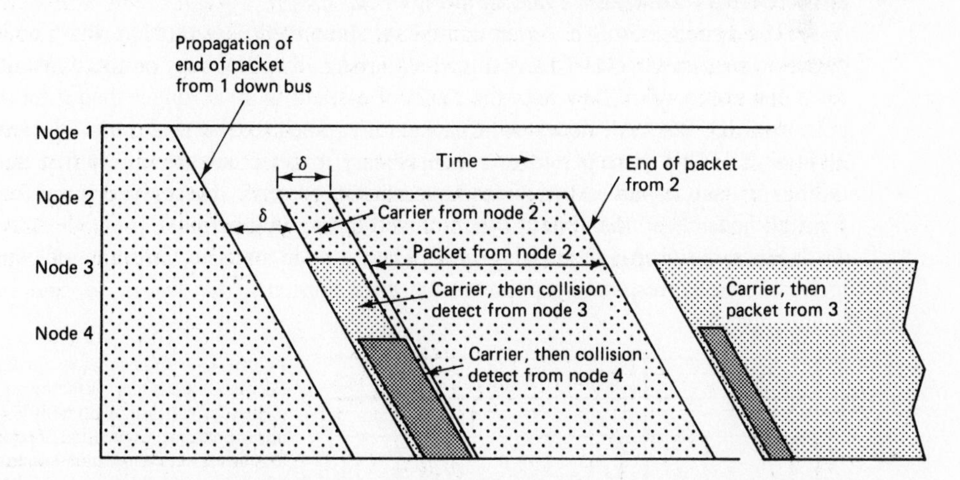

Figure 4.32 Implicit tokens on a unidirectional bus. Each node with traffic transmits carrier on hearing silence, then defers to upstream nodes on hearing carrier.

of a given packet, the second node will start to send carrier at a delay δ after the end of the previous packet passes that node. The carrier sent by the first node will reach the second node at essentially the same time as the second node starts to send carrier. Since the second node can detect whether or not the first node is sending carrier after another delay of δ, the second node (and all further nodes) can terminate the transmission of carrier after this delay δ. Thus, if each node starts to transmit its packet after sending carrier for δ, the node is assured that no earlier node can be transmitting, and that all subsequent nodes will cease transmitting before any of the packet information arrives.

We see that the implicit token is simply the termination of transmission on the line. The time v for this token to travel from one node to the next is simply the propagation time from one node to the next. We view the length of the packet as containing δ units of time for detection at the end and δ units of time for carrier at the beginning. In a sense, this is the ideal system with TDM reservations. The time for making a reservation consists only of propagation delay (essentially the implicit token has zero length), and the packets follow immediately after the reservations.

The discussion above ignored the questions of how to receive the transmitted packets and how to return to the beginning of the bus and restart the process after all nodes have their turns. We first illustrate how these problems are solved in the Expressnet system [TBF83], and then briefly describe several alternative approaches.

Expressnet uses a unidirectional bus with two folds as shown in Fig. 4.33. Packets are transmitted on the first portion of the bus, as explained above, and nodes read the packets from the third portion of the bus. When a silence interval longer than δ is detected on the third portion of the bus, it is clear that all nodes have had a chance to transmit and it is time to start a new cycle. Thus, all nodes with packets to send again start to send carrier. Because of the double fold in the bus, however, nodes on the left end of the bus hear the idle signal before those on the right, and the carrier signals will again overlap for the next cycle as in Fig. 4.32.

The system is still not quite complete, since there is a problem if no node has a packet to send in a cycle. This is solved by having all nodes, busy or not, transmit carrier for a duration δ when they hear the onset of a silent interval longer than δ on the third portion of the bus. All nodes with packets must then extend their burst of carrier to a duration 2δ. This extra period of δ is necessary to detect silence on the first portion of the bus if none of the earlier nodes have packets to send. These extra bursts of carrier from all nodes keep the system synchronized during idle periods.

Since the propagation delay from when a node starts to transmit until the signal arrives at the corresponding point on the third portion of the bus is 2β, and since the

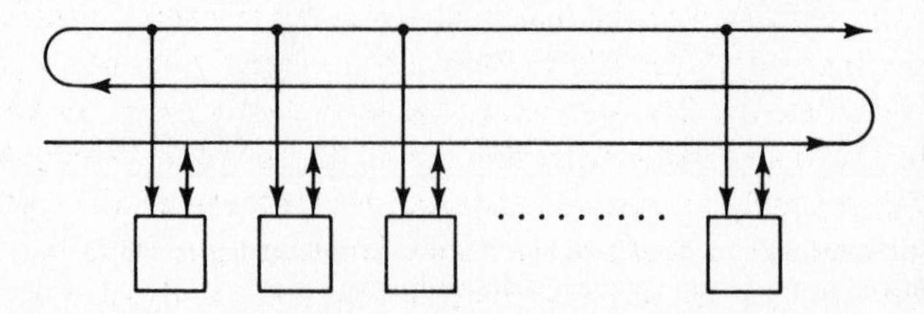

Figure 4.33 Bus structure for Expressnet. Each node transmits on the lower portion of the bus and listens on both lower and upper portions. Contention is resolved on the lower portion with the leftmost node taking priority. Reception and detection of the end of a cycle takes place from the upper portion.

transmission of the burst of carrier and its detection take time 2δ, we see that the average reservation interval is

$$v = 2\frac{(\beta + \delta)}{m} \tag{4.85}$$

Equations (4.70) and (4.71) again give the expected queueing delay for multiple packets per reservation and a single packet per reservation, respectively.

There are many possible variations on how to use this unidirectional bus. One could replace the double-folded structure with a single fold, reading packets on the return trip on the bus, and using a special node at the left end of the bus to start new cycles. One could also use two buses, one in each direction. A node could then send traffic to nodes on its right on the right-going bus and traffic to nodes on its left on the left-going bus as in DQDB. Special nodes on each end of the two buses would then interchange information for starting new cycles in each direction. Other possibilities include using a separate control wire, or a separate frequency band, to take the place of the implicit reservation tokens. [FiT84] contrasts a large set of these possible approaches.

Homenets. The previous approaches to higher-speed local networks were based on the use of a unidirectional bus. CATV systems, on the other hand, typically have a tree structure with transmissions from the root of the tree being broadcast throughout the tree. To use such networks for data, there is usually a separate frequency band for communication from the leaves of the tree in toward the root. When packets are sent inward toward the root from different leaves, there is the usual problem of collisions, but the use of implicit reservation tokens based on bus position no longer works.

Homenets [MaN85] provide an interesting approach to coping with the problem of large β on such a net. The idea is to break up the CATV net into subnets called homenets. Each homenet forms its own subtree in the overall network tree, as shown in Fig. 4.34, and each homenet has its own frequency bands, one for propagation inward toward the root and the other for propagation outward. This strategy cures the problem of large β in two ways. First, the propagation delay within a homenet is greatly reduced from its value in the entire net, and second, by using a restricted bandwidth, the data rate within the homenet is reduced. With this twofold decrease in β, it becomes reasonable to use CSMA/CD within the homenet.

This leaves us with two unanswered questions. First, since the nodes transmit their packets inward toward the root, how do other outlying nodes in the same homenet hear collisions? Second, how are the packets received by nodes outside the homenet? The solution to the first question is for the node at the root of a given homenet to receive the signal on the homenet's incoming frequency band and both to forward it toward the root of the entire net and convert it to the homenet's outgoing frequency band and broadcast it back out to the homenet. This allows the nodes to detect collisions within the propagation delay of the homenet. The solution to the second question is for all of the incoming frequency bands to be forwarded to the root of the entire net, which then converts these bands to the outgoing frequency bands and rebroadcasts all the signals back through the entire network. The root node for each homenet then filters out the signal from the overall root on its own outgoing frequency (since that signal is just a

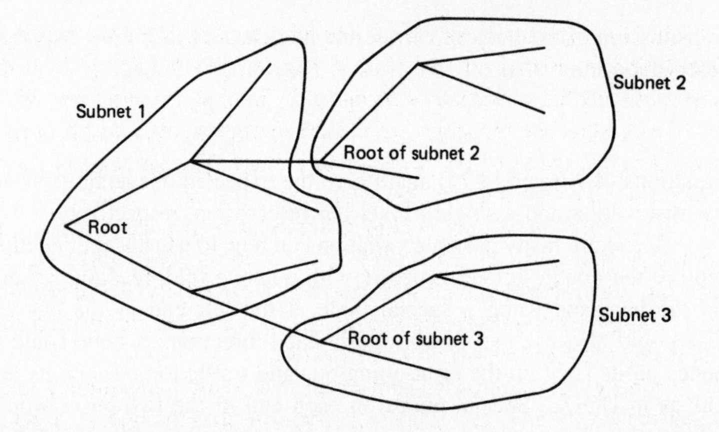

Figure 4.34 A CATV network divided into subnets in the Homenet strategy. Each subnet has one incoming and one outgoing frequency band and uses CSMA/CD to resolve collisions with a subnet.

delayed replica of what it has already transmitted outward) and forwards the other bands over its own homenet.

One additional advantage of this strategy is that the nodes need only transmit and receive within these restricted frequency bands. When a session is set up between two nodes, the receiver at each node must know of the outgoing frequency band for the sender's homenet, and then it simply receives the desired packets out of the signal in that band.

4.5.6 Generalized Polling and Splitting Algorithms

Section 4.5.5 treated multiaccess systems in the presence of large propagation delays. Here, we look at the opposite case in which $\beta \ll 1$. One example of this arises with very short buses, such as in the back plane of a multimicroprocessor system. Another example occurs for polling on multidrop telephone lines; here the data rates are small, leading to packet transmission times much greater than the propagation delay.

In these situations, we see from Eqs. (4.83) and (4.84) that the reservation time per node v is linearly increasing with the detection delay δ and with the (implicit or explicit) token delay. From the queueing delay formulas, Eqs. (4.70) and (4.71), the queueing delay is $mv/2$ in the limit of light load; furthermore, if mv is large relative to the packet transmission time, delay increases at least linearly with mv at all stable loads. Thus, we want to reduce mv, the reservation overhead per cycle.

For simplicity, we now model the multiaccess channel, at least for the purpose of reservations, as a bit synchronous binary "or" channel. That is, at each bit time, the output of the channel is 1 if the input for one or more nodes is 1; otherwise, the output is 0. This is a reasonable model if a separate control wire is used for reservations; it is also reasonable if nodes use short bursts of carrier to request reservations (*i.e.*, the existence of carrier can be detected, but not the number of nodes sending carrier).

This model is very similar to our original slotted multiaccess model in Section 4.2. One difference is that the slot time is reduced to a single bit time; the other difference is that the feedback, instead of being 0,1,e, is now 0 or positive. We regard the slots as reservation slots and recognize (as in the examples above), that packets can be transmitted at higher rates than one bit per reservation slot. The question of interest is: How many reservation slots are required to make a reservation?

The simplest strategy within this model is to make reservations by TDM within the reservation slots. Under heavy loading, most nodes will have a packet to send at each turn, and almost each reservation slot successfully establishes a reservation. In the light loading limit, an arriving packet has to wait $m/2$ reservation slots on the average; this is the situation we would like to avoid.

A better strategy, under light loading, is to use a logarithmic search for a node with a packet (see [NiS74] and [Hay76]). Suppose that the nodes are numbered 0 to $m-1$ and let $n_1, \ldots, n_k$ be the binary representation of any given node number where

$$k = \lceil \log_2 m \rceil$$

The algorithm proceeds in successive collision resolution periods (CRP) to find the lowest-numbered node that has a packet to send. In the first slot of a CRP, all active nodes (*i.e.*, all nodes with waiting packets) send a 1 and all other nodes send 0 (*i.e.*, nothing). If the channel output, say y_0, is 0, the CRP is over and a new CRP starts on the next slot to make reservations for any packets that may have arrived in the interim. If $y_0 = 1$ on the other hand, one or more nodes have packets and the logarithmic search starts.

Assuming that $y_0 = 1$, all active nodes with $n_1 = 0$ send a 1 in the next slot. If y_1, the output in this slot, is 1, it means that the lowest-numbered node's binary representation starts with 0; otherwise, it starts with 1. In the former case, all active nodes with $n_1 = 1$ become inactive and wait for the end of the CRP to become active again. In the latter case, all active nodes have $n_1 = 1$ and all remain active.

The general rule on the $(i+1)^{\text{st}}$ slot of the CRP, $1 \leq i \leq k$ (assuming that $y_0 = 1$), is that all active nodes with $n_i = 0$ send a 1; at the end of the slot, if the channel output y_i is 1, all nodes with $n_i = 1$ become inactive. Figure 4.35 shows an example of this strategy. It should be clear that at the end of the $(k+1)^{\text{st}}$ slot, only the lowest-numbered node is active and the binary representation of that node's number is the bitwise complement of $y_1, \ldots, y_k$.

To maintain fairness and to serve the nodes in round-robin order, the nodes should be renumbered after each reservation is made. If the reservation is for node n, say, each node subtracts $n+1$ modulo m from its current number. Thus, a node's number is one less than its round-robin distance from the last node that transmitted; obviously, this rule could be modified by priorities in any desired way.

Assuming that a packet is sent immediately after a reservation is made and that the next CRP starts after that transmission, it is easy to find the expected delay of this algorithm. Each CRP that starts when the system is busy lasts for $k+1$ reservation slots. Thus, we regard the packet transmission time as simply being extended by these $k+1$ slots. If the system is empty at the beginning of a CRP, the CRP lasts for one reservation

Nodes								Output	
000	001	010	011	100	101	110	111	y	
0	0	1	1	0	0	1	0	1	Slot 1
0	0	1	1	0	0	0	0	1	Slot 2
0	0	0	0	0	0	0	0	0	Slot 3
0	0	1	0	0	0	0	0	1	Slot 4

Figure 4.35 Logarithmic search for the lowest-numbered active node. Nodes 010, 011, and 110 have packets. At the end of slot 2, node 110 becomes inactive; at the end of slot 4, node 011 becomes inactive. Outputs from slot 2 to 4, complemented, are 010.

slot and this can be regarded as the server going on vacation for one reservation slot. Equation (3.55) gives the queueing delay for this system.

In contrasting this logarithmic search reservation strategy with TDM reservations, we see that logarithmic search reduces delay from $m/2$ to $k+1$ reservation slots per packet for light loads but increases delay from 1 to $k+1$ reservation slots per packet at heavy loads. The obvious question is: How do we combine these strategies to have the best of both worlds? The answer to this question comes from viewing this problem as a source coding problem, much like the framing issue in Chapter 2. The TDM strategy here corresponds to the use of a unary code there to encode frame lengths. The logarithmic search here corresponds to the use of ordinary binary representation. We noticed the advantage there of going to a combined unary–binary representation. In the current context, this means that each CRP should test only a limited number of nodes, say the 2^j lowest-numbered nodes, at a time. If the output is 1, the lowest-numbered node in that set is resolved as above in j additional slots. If the output is 0, 2^j is subtracted from each node's number and the next CRP starts. Problem 4.30 finds the expected number of reservation slots per packet and the optimal choice of j under several different assumptions.

Note the similarity of these algorithms to the splitting algorithms of Section 4.3. The ideas are the same, and the only difference is that "success" cannot be distinguished from collision here; thus, even though only one node in an active subset contains a packet, the subset must be split until only one node remains.

4.6 PACKET RADIO NETWORKS

Section 4.1.4 briefly described packet radio networks as multiaccess networks in which not all nodes could hear the transmissions of all other nodes. This feature is characteristic both of line-of-sight radio communication in the UHF band (300 to 3,000 MHz) and also of non-line-of-sight communication at HF (3 to 30 MHz). Our interest here is in the effect of partial connectivity on multiaccess techniques rather than the physical characteristics of the radio broadcast medium.

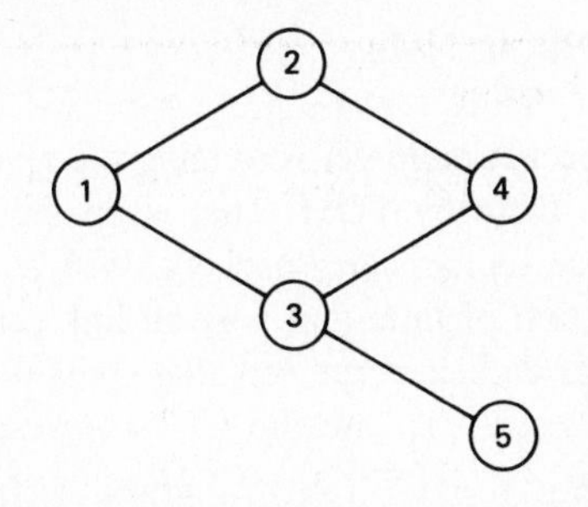

Figure 4.36 Packet radio network; each edge indicates that two nodes at the ends of the edge can hear each other's transmission.

The topology of a radio network can be described by a graph as in Fig. 4.36. The graph, $G = (N, L)$, contains a set of nodes N and a set of links L. Each link in L corresponds to an ordered pair of nodes, say (i, j), and indicates that transmissions from i can be heard at j. In some situations, node j might be able to hear i, but i is unable to hear j. In such a case $(i, j) \in L$ but $(j, i) \notin L$. This asymmetry does not occur in Fig. 4.36, where each edge denotes two links, one in each direction.

Our assumption about communication in this multiaccess medium is that if node i transmits a packet, that packet will be correctly received by node j if and only if

1. There is a link from i to j [*i.e.*, $(i, j) \in L$], and
2. No other node k for which $(k, j) \in L$ is transmitting while i is transmitting, and
3. j itself is not transmitting while i is transmitting.

Thus, for Fig. 4.36, we see that if nodes 1 and 3 are transmitting simultaneously, node 2 will correctly receive the packet from 1 and node 4 will correctly receive the packet from 3. On the other hand, if nodes 2 and 3 are transmitting simultaneously, nodes 1 and 4 will each see a collision, but node 5 will correctly receive the packet from 3.

It can be seen from this example that having a large number of links in a graph is not necessarily desirable. A large number of links increases the number of pairs of nodes that can communicate directly, but also increases the likelihood of collisions. This trade-off is explored further in Section 4.6.3.

One interesting question that can now be posed is how much traffic can be carried in such a network. Define a *collision-free set* as a set of links that can carry packets simultaneously with no collisions at the receiving ends of the links. For example, [(1,2), (3,4)] and [(2,1), (5,3)] are both collision-free sets; also, the empty set and each set consisting of a single link are collision-free sets. It is convenient to order the links in some arbitrary order and represent each collision-free set as a vector of 0's and 1's called a *collision-free vector* (CFV). The ℓth component of a CFV is 1 if and only if the ℓth link is in the corresponding collision-free set. For example, some CFVs are listed below for the graph of Fig. 4.36.

(1,2)	(2,1)	(1,3)	(3,1)	(2,4)	(4,2)	(3,4)	(4,3)	(3,5)	(5,3)
1	0	0	0	0	0	1	0	0	0
1	0	0	0	0	0	0	0	1	0
0	1	0	0	0	0	0	1	0	0
0	1	0	0	0	0	0	0	0	1

4.6.1 TDM for Packet Radio Nets

One simple way to use a packet radio net is to choose a given collection of collision-free sets and to cycle between them by TDM. That is, in the ith slot of a TDM cycle, all links in the ith collision-free set can carry packets. With such a TDM strategy, there are no collisions, and the fraction of time that a given link can carry packets is simply the fraction of the collection of collision-free sets that contain that link.

More compactly, if $x_1, \ldots, x_J$ are the CFVs corresponding to a collection of J collision-free sets, the vector $f = \left(\sum_j x_j\right)/J$ gives the fraction of time that each link can be used. As a slight generalization, the TDM frame could be extended and each collision-free set could be used some arbitrary number of times in the frame. If α_j is the fraction of frame slots using the jth collision-free set, then

$$f = \sum_j \alpha_j x_j \tag{4.86}$$

gives the fractional utilization of each link. A vector of the form $\sum_j \alpha_j x_j$, in which $\sum_j \alpha_j = 1$ and $\alpha_j \geq 0$ for $1 \leq j \leq J$, is called a convex combination of the vectors $x_1, \ldots, x_J$. What we have just seen is that any convex combination of CFVs can be approached arbitrarily closely as a fractional link utilization vector through the use of TDM.

Suppose that instead of using TDM as above, we use some sort of collision resolution approach in the network. At any given time, the vector of links that are transmitting packets successfully is a CFV. Averaging this vector of successful link transmissions over time, we get a vector whose ℓ^{th} component is the fraction of time that the ℓth link is carrying packets successfully. This is also a convex combination of CFVs. Thus, we see that any link utilization that is achievable with collision resolution is also achievable by TDM.

One difficulty with TDM is that delays are longer than necessary for a lightly loaded network. This is not as serious as when all nodes are connected to a common receiver, since if all nodes have only a small number of incoming links, many links can transmit simultaneously and the waiting for a TDM slot is reduced.

A more serious problem with the TDM approach is that the nodes in a packet radio network are usually mobile, and thus the topology of the network is constantly changing. This means that the collision-free sets keep changing, requiring frequent updates of the TDM schedule. This is a difficult problem since even for a static network, the problem of determining whether a potential vector of link utilizations is a convex combination of CFVs falls into a class of difficult problems known as *NP complete* [Ari84]. See [PaS82] for an introduction to the theory of NP complete problems; for our purposes, this simply indicates that the worst-case computational effort to solve the problem increases very rapidly with the number of links in the network. The essential reason for this difficulty is that the number of different collision-free sets typically increases exponentially with the number of links in the network.

Frequency-division multiplexing (FDM) can also be used for packet radio networks in a way very similar to TDM. All links in a collision-free set can use the same frequency band simultaneously, so in principle the links can carry the same amount of traffic as in

TDM. This approach is used in cellular radio networks for mobile voice communication. Here the area covered by the network is divided into a large number of local areas called cells, with each cell having a set of frequency bands for use within the cell. The set of frequency bands used by one cell can be reused by other cells that are sufficiently separated from one another to avoid interference. This provides a simple and practical way to choose a collection of collision-free sets. This same cellular separation principle could be used for TDM.

The discussion so far has ignored the question of how to route packets from source to destination. With the use of a given TDM or FDM structure, each link in the net has a given rate at which it can send packets, and the resource-sharing interaction between links has been removed. Thus, the problem of routing is essentially the same as in conventional networks with dedicated links between nodes; this problem is treated in Chapter 5. When collision resolution strategies are used, however, we shall see that routing is considerably more complicated than in the conventional network case.

4.6.2 Collision Resolution for Packet Radio Nets

Collision resolution is quite a bit trickier for packet radio nets than for the single-receiver systems studied before. The first complication is obtaining feedback information. For the example of Fig. 4.36, suppose that links (2,4) and (3,5) contain packets in a given slot. Then node 4 perceives a collision and node 5 correctly receives a packet. If nodes 5 and 4 send feedback information, node 3 will experience a feedback collision. A second problem is that if a node perceives a collision, it does not know if any of the packets were addressed to it. For both reasons, we cannot assume the perfect 0,1,e feedback that we assumed previously. It follows that the splitting algorithms of Section 4.3 cannot be used and the stabilization techniques of Section 4.2.3 require substantial revisions.

Fortunately, slotted and unslotted Aloha are still applicable, and to a certain extent, some of the ideas of carrier sensing and reservation can still be used. We start by analyzing how slotted Aloha can work in this environment. When an unbacklogged node receives a packet to transmit (either a new packet entering the network, or a packet in transit that has to be forwarded to another node), it sends the packet in the next slot. If no acknowledgment (ack) of correct reception arrives within some time-out period, the node becomes backlogged and the packet is retransmitted after a random delay. Finally, a backlogged node becomes unbacklogged when all of its packets have been transmitted and acked successfully.

There are a number of ways in which acks can be returned to the transmitting node. The simplest is that if node i sends a packet to j that must be forwarded on to some other node k, then if i hears j's transmission to k, that serves as an ack of the (i, j) transmission. This technique is somewhat defective in two ways. First, some other technique is required to ack packets whose final destination is j. Second, suppose that j successfully relays the packet to k, but i fails to hear the transmission because of a collision. This causes an unnecessary retransmission from i to j, and also requires some way for j to ack, since j has already forwarded the packet to k. Another approach, which can be used in conjunction with the implicit acks above, is for each node to include explicit acks for the last few packets it has received in each outgoing packet. This approach requires a node to send a dummy packet carrying ack information if the node has no data to send for some period. A third approach, which seems somewhat inferior to the approach above, is to provide time at the end of each slot for explicit acks of packets received within the slot.

Let us now analyze what happens in slotted Aloha for a very heavily loaded network. In particular, assume that all nodes are backlogged all the time and have packets to send on all outgoing links at all times. We can assume that the nodes have infinite buffers to store the backlogged packets, but for the time being, we are not interested in the question of delay. This assumption of constant backlogging is very different from our assumptions in Section 4.2, but the reasons for this will be discussed later. For all nodes i and j, let q_{ij} be the probability that node i transmits a packet to node j in any given slot, and let Q_i be the probability that node i transmits to any node. Thus,

$$Q_i = \sum_j q_{ij} \tag{4.87}$$

To simplify notation, we simply assume that q_{ij} is zero if (i, j) is not in the set of links L. Let p_{ij} be the probability that a transmission on (i, j) is successful. Under our assumption of heavy loading, each node transmits or not in a slot independently of all other nodes. Since p_{ij} is the probability that none of the other nodes in range of j, including j itself, is transmitting, we have

$$p_{ij} = (1 - Q_j) \prod_{\substack{k:(k,j)\in L \\ k\neq i}} (1 - Q_k) \tag{4.88}$$

Finally, the rate f_{ij} of successful packet transmissions per slot (*i.e.*, the throughput) on link (i, j) is

$$f_{ij} = q_{ij}\, p_{ij} \tag{4.89}$$

Equations (4.87) to (4.89) give us the link throughputs in terms of the attempt rates q_{ij} under the heavy-loading assumption. The question of greater interest, however, is to find the attempt rates q_{ij} that will yield a desired set of throughputs (if that set of throughputs is feasible).

This problem can be solved through an iterative approach. To simplify notation, let q denote a vector whose components are the attempt rates q_{ij}, let p and f be vectors whose components are p_{ij} and f_{ij}, respectively, and let Q be a vector with components Q_i. Given a desired throughput vector f, we start with an initial q^0 which is a vector of 0's. We then use Eqs. (4.87) and (4.88) to find Q^0 and p^0 (Q^0 is thus a vector of 0's and p^0 a vector of 1's). Equation (4.89) is then used to obtain the next iteration for q; that is, the components of q^1 are given by

$$q_{ij}^1 = \frac{f_{ij}}{p_{ij}^0} \tag{4.90}$$

For each successive iteration, Q^n is found from Eq. (4.87) using q^n, and p^n is found from Eq. (4.88) using Q^n; then q^{n+1} is found from Eq. (4.89) using p^n. Note that $q^1 \geq q^0$ (*i.e.*, each component of q^1 is greater than or equal to the corresponding component of q^0). Thus, $Q^1 \geq Q^0$ and $p^1 \leq p^0$. From Eq. (4.89) it can then be seen that $q^2 \geq q^1$. Continuing this argument, it is seen that as long as none of the components of Q exceed 1, q is nondecreasing with successive iterations and p is nonincreasing. It follows that either some component of Q must exceed 1 at some iteration or else q approaches a limit, say q^*, and in this limit Eqs. (4.87) to (4.89) are simultaneously satisfied with the resulting Q^* and p^*.

We now want to show that if (4.87) to (4.89) have any other solution, say q', Q', p' (subject, of course, to $q' \geq 0, Q' \leq 1$), then $q' \geq q^*, Q' \geq Q^*$, and $p' \leq p^*$. To see this, we simply observe that $q^0 \leq q'$, $Q^0 \leq Q'$, and $p^0 \geq p'$. From Eq. (4.89), then, $q^1 \leq q'$. Continuing this argument over successive iterations, $q^n \leq q', Q^n \leq Q'$, and $p^n \geq p'$ for all n, so the result holds in the limit. This argument also shows that if some component of Q^n exceeds 1 for some n, then Eqs. (4.87) to (4.89) have no solution (*i.e.*, that the given f is infeasible).

Next, assume we know the input rates to the network and know the routes over which the sessions will flow, so that in principle we can determine the steady-state rates f'_{ij} at which the links must handle traffic. We would like to choose the throughputs of each link under heavy load to exceed these steady-state rates so that the backlogs do not build up indefinitely. One approach then is to find the largest number $\beta > 1$ for which $f = \beta f'$ is feasible under the heavy-load assumption. Given this largest f, and the corresponding attempt rates q, we can then empty out the backlog as it develops.

There is one difficulty here, and that is that if some nodes are backlogged and others are not, the unbacklogged nodes no longer choose their transmission times independently. Thus, it is conceivable in bizarre cases that some backlogged nodes fare more poorly when other nodes are unbacklogged than they do when all nodes are backlogged. Problem 4.32 gives an example of this phenomenon. One way to avoid this difficulty is for new packets at a node to join the backlog immediately rather than being able to transmit in the next slot. This, of course, increases delay under light-loading conditions. The other approach is to live dangerously and hope for the best. To a certain extent, one has to do this anyway with packet radio, since with a changing topology, one cannot maintain carefully controlled attempt rates.

Our reason for focusing on the heavily loaded case is that the number of links entering each node is usually small for a packet radio net, and thus the attempt rates can be moderately high even under the heavy-loading assumption. For the single-receiver case, on the other hand, the number of nodes tends to be much larger, and thus the attempt rates appropriate for heavy loading tend to create large delays. The other reason is that stabilization is a much harder problem here than in the single-receiver case; a node cannot help itself too much by adjusting its own attempt rates, since other nodes might be causing congestion but not experiencing any congestion themselves (see Problem 4.32).

4.6.3 Transmission Radii for Packet Radio

In the previous subsections, we viewed the set of links in a packet radio net as given. It can be seen, however, that if a node increases its transmitter power, its transmission will be heard by a larger set of nodes. The following qualitative argument shows that it is desirable to keep the power level relatively small so that each node has a moderately small set of incoming and outgoing links. Assume for simplicity that we have a symmetric net in which each node has exactly n incoming links and n outgoing links. Suppose further that each link has an identical traffic-carrying requirement. It is not hard to convince oneself that Eqs. (4.87) to (4.89) are satisfied by an identical attempt rate q on each link. Each Q_i is then nq, each p_{ij} is given by $(1 - nq)^n$, and finally,

$$f = q(1 - nq)^n \tag{4.91}$$

It is easy to verify that f is maximized by choosing $q = 1/[n(n+1)]$, and the resulting value of f is approximately $1/(en^2)$. Each node then sends packets successfully at a rate of $1/en$. If there are m nodes in the network and the average number of links on the path from source to destination is J, the rate at which the network can deliver packets is $m/(Jen)$ packets per slot.

Now let us look at what happens when the transmission radius R over which a node can be heard varies. The number of nodes within radius R of a given node will vary roughly as R^2; so the rate at which an individual node sends packets successfully will decrease as $1/R^2$. On the other hand, as R increases, the routing will presumably be changed to send the packets as far as possible toward the destination on each link of the path. Thus, we expect the number of links on a path to decrease as $1/R$. Thus, if J is proportional to $1/R$ and n is proportional to R^2, the rate at which the network can deliver packets is proportional to $1/R$, leading us to believe that R should be kept very small.

The very crude analysis above leaves out two important factors. First, when R and n are large, a packet can move almost a distance R toward its destination on each link of a well-chosen path, so that J is essentially proportional to $1/R$ in that region. When R gets small, however, the paths become very circuitous and thus, J increases with decreasing R much faster than $1/R$. Second, when R is too small, the network is not very well connected, and some links might have to carry very large amounts of traffic. This leads us to the conclusion that R should be small, but not too small, so that n is considerably larger than 1. Takagi and Kleinrock [TaK85] have done a much more careful analysis (although still with some questionable assumptions) and have concluded that the radius should be set so that n is on the order of 8.

4.6.4 Carrier Sensing and Busy Tones

We saw in Section 4.4 that carrier sensing yielded a considerable improvement over slotted Aloha in the situation where all nodes could hear all other nodes and the propagation delay is small. For line-of-sight radio, the propagation delay is typically small relative to packet transmission times, so it is reasonable to explore how well carrier sensing will work here. Unfortunately, if node i is transmitting to node j, and node k wants to transmit to j, there is no assurance that k can hear i. There might be an obstruction between i and k, or they might simply be out of range of each other. Thus, carrier sensing will serve to prevent some collisions from occurring, but cannot prevent others.

To make matters worse, with carrier sensing, there is no uniform slotting structure, and thus, carrier sensing loses some of the advantage that slotted Aloha has over pure Aloha. Finally, radio transmission is subject to fading and variable noise, so that the existence of another transmitting node, even within range, is hard to detect in a short time. For these reasons, carrier sensing is not very effective for packet radio.

A busy tone ([ToK75] and [SiS81]) is one approach to improving the performance of carrier sensing in a packet radio network. Whenever any node detects a packet being transmitted, it starts to send a signal, called a busy tone, in a separate frequency band. Thus, when node i starts to send a packet to node j, node j (along with all other nodes that can hear i) will start to send a busy tone. All the nodes that can hear j will thus avoid transmitting; thus, assuming reciprocity (*i.e.*, the nodes that can hear j are the same as the nodes that j can hear), it follows that j will experience no collision.

A problem with the use of busy tones is that when node i starts to send a packet, *all* the nodes in range of i will start to send busy tones, and thus every node within range of any node in range of i will be inhibited from transmitting. Using the very crude type of analysis in the last subsection, and assuming a transmission radius of R, we see that when node i starts to transmit, most of the nodes within radius $2R$ of i will be inhibited. This number will typically be about four times the number of nodes within radius R of the receiving node, which is the set of nodes that should be inhibited. Thus, from a throughput standpoint, this is not a very promising approach.

Another variation on the busy tone approach is for a node to send a busy tone only after it receives the address part of the packet and recognizes itself as the intended recipient. Aside from the complexity, this greatly increases β, the time over which another node could start to transmit before hearing the busy tone.

It can be seen that packet radio is an area in which many more questions than answers exist, both in terms of desirable structure and in terms of analysis. Questions of modulation and detection of packets make the situation even more complex. In military applications, it is often desirable to use spread-spectrum techniques for sending packets. One of the consequences of this is that if two packets are being received at once, the receiver can often lock on to one, with the other acting only as wideband noise. If a different spread-spectrum code is used for each receiver, the situation is even better, since the receiver can look for only its own sequence and thus reject simultaneous packets sent to other receivers. Unfortunately, attenuation is often quite severe in line-of-sight communication, so that unwanted packets can arrive at a node with much higher power levels than the desired packets, and still cause a collision.

SUMMARY

The central problem of multiaccess communication is that of sharing a communication channel between a multiplicity of nodes where each node has sporadic service requirements. This problem arises in local area networks, metropolitan area networks, satellite networks, and various types of radio networks.

Collision resolution is one approach to such sharing. Inherently, collision resolution algorithms can achieve small delay with a large number of lightly loaded nodes, but stability is a major concern. The joint issues of stability, throughput, and delay are studied most cleanly with the infinite node assumption. This assumption lets one study collision resolution without the added complication of individual queues at each node. Under this assumption, we found that throughputs up to $1/e$ packets per slot were possible with stabilized slotted Aloha, and throughputs up to 0.487 packets per slot were possible with splitting algorithms.

Reservations provide the other major approach to multiaccess sharing. The channel can be reserved by a prearranged fixed allocation (*e.g.*, TDM or FDM) or can be reserved dynamically. Dynamic reservations further divide into the use of collision resolution and the use of TDM (or round-robin ordering) to make the reservations for channel use. CSMA/CD (*i.e.*, Ethernet) is a popular example of the use of collision resolution to make (implicit) reservations. Token rings, token buses, and their elaborations are examples of the use of round-robin ordering to make reservations.

There are an amazing variety of ways to use the special characteristics of particular multiaccess media to make reservations in round-robin order. Some of these variations require a time proportional to β (the propagation delay) to make a reservation, and some require a time proportional to β/m, where m is the number of nodes. The latter variations are particularly suitable for high-speed systems with $\beta > 1$.

Packet radio systems lie at an intermediate point between pure multiaccess systems, where all nodes share the same medium (and thus no explicit routing is required), and point-to-point networks, where routing but no multiaccess sharing is required. Radio networks are still in a formative and fragmentary stage of research.

NOTES, SOURCES, AND SUGGESTED READING

Section 4.2. The Aloha network was first described in [Abr70] and the slotted improvement in [Rob72]. The problem of stability was discussed in [Met73], [LaK75], and [CaH75]. Binary exponential backoff was developed in [MeB76]. Modern approaches to stability are treated in [HaL82] and [Riv85].

Section 4.3. The first tree algorithms are due to [Cap77], [TsM78], and [Hay76]. [Mas80] provided improvements and simple analysis techniques. The FCFS splitting algorithm is due to [Gal78] and, independently, [TsM80]. Upper bounds on maximum throughput (with assumptions 1 to 6b of Section 4.2.1) are in [Pip81] and [MiT81]. The March 1985 issue of the *IEEE Transactions on Information Theory* is a special issue on random-access communications; the articles provide an excellent snapshot of the status of work related to splitting algorithms.

Section 4.4. The classic works on CSMA are [KlT75] and [Tob74].

Section 4.5. The literature on local area networks and satellite networks is somewhat overwhelming. [Sta85] provides a wealth of practical details. Good source references in the satellite area are [JBH78], [CRW73], [Bin75], and [WiE80]. For local area networks, [MeB76] is the source work on Ethernet, and [FaN69] and [FaL72] are source works on ring nets. [CPR78] and [KuR82] are good overview articles and [Ros86] provides a readable introduction to FDDI. [FiT84] does an excellent job of comparing and contrasting the many approaches to implicit and explicit tokens and polling on buses. Finally, DQDB is treated in [NBH88] and [HCM90].

Section 4.6. [KGB78] provides a good overview of packet radio. Busy tones are described in [ToK75] and [SiS81]. Transmission radii are discussed in [TaK85].

PROBLEMS

4.1 **(a)** Verify that the steady-state probabilities p_n for the Markov chain in Fig. 4.3 are given by the solution to the equations

$$p_n = \sum_{i=0}^{n+1} p_i P_{in}$$

$$\sum_{n=0}^{m} p_n = 1$$

(b) For $n < m$, use part (a) to express p_{n+1} in terms of $p_0, p_1, \ldots, p_n$.

(c) Express p_1 in terms of p_0 and then p_2 in terms of p_0.

(d) For $m = 2$, solve for p_0 in terms of the transition probabilities.

4.2 **(a)** Show that P_{succ} in Eq. (4.5) can be expressed as

$$P_{succ} = \left[\frac{(m-n)q_a}{1-q_a} + \frac{nq_r}{1-q_r}\right](1-q_a)^{m-n}(1-q_r)^n$$

(b) Use the approximation $(1-x)^y \approx e^{-xy}$ for small x to show that for small q_a and q_r,

$$P_{succ} \approx G(n)e^{-G(n)}$$

where $G(n) = (m-n)q_a + nq_r$.

(c) Note that $(1-x)^y = e^{y\ln(1-x)}$. Expand $\ln(1-x)$ in a power series and show that

$$\frac{(1-x)^y}{e^{-xy}} = \exp\left(-\frac{x^2y}{2} - \frac{x^3y}{3}\cdots\right)$$

Show that this ratio is close to 1 if $x \ll 1$ and $x^2y \ll 1$.

4.3 **(a)** Redraw Fig. 4.4 for the case in which $q_r = 1/m$ and $q_a = 1/me$.

(b) Find the departure rate (*i.e.*, P_{succ}) in the fully backlogged case $n = m$.

(c) Note that there is no unstable equilibrium or undesired stable point in this case and show (graphically) that this holds true for any value of q_a.

(d) Solve numerically (using $q_a = 1/me$) for the value of G at which the stable point occurs.

(e) Find n/m at the stable point. Note that this is the fraction of the arriving packets that are not accepted by the system at this typical point.

4.4 Consider the idealized slotted multiaccess model of Section 4.2.1 with the no-buffering assumption. Let n_k be the number of backlogged nodes at the beginning of the kth slot and let $\overline{n}$ be the expected value of n_k over all k. Note that $\overline{n}$ will depend on the particular way in which collisions are resolved, but we regard $\overline{n}$ as given here (see [HIG81]).

(a) Find the expected number of *accepted* arrivals per slot, $\overline{N}_a$, as a function of $\overline{n}, m$, and q_a, where m is the number of nodes and q_a is the arrival rate per node.

(b) Find the expected departure rate per slot, $\overline{P}_{succ}$, as a function of $\overline{n}, m$, and q_a. *Hint:* How is $\overline{N}_a$ related to $\overline{P}_{succ}$? Recall that both are averages over time.

(c) Find the expected number of packets in the system, $\overline{N}_{sys}$, immediately after the beginning of a slot (the number in the system is the backlog plus the accepted new arrivals).

(d) Find the expected delay T of an accepted packet from its arrival at the beginning of a slot until the completion of its successful transmission at the end of a slot. *Hint:* Use Little's theorem; it may be useful to redraw the diagram used to prove Little's theorem.

(e) Suppose that the strategy for resolving collisions is now modified and the expected backlog $\overline{n}$ is reduced to $\overline{n}' < \overline{n}$. Show that $\overline{N}_a$ increases, $\overline{P}_{succ}$ increases, $\overline{N}_{sys}$ decreases, and T decreases. Note that this means that improving the system with respect to one of these parameters improves it with respect to all.

4.5 Assume for simplicity that each transmitted packet in a slotted Aloha system is successful with some fixed probability p. New packets are assumed to arrive at the beginning of a slot and are transmitted immediately. If a packet is unsuccessful, it is retransmitted with probability q_r in each successive slot until successfully received.

(a) Find the expected delay T from the arrival of a packet until the completion of its successful transmission. *Hint:* Given that a packet has not been transmitted successfully before, what is the probability that it is both transmitted and successful in the ith slot ($i > 1$) after arrival?

(b) Suppose that the number of nodes m is large, and that q_a and q_r are small. Show that in state n, the probability p that a given packet transmission is successful is approximately $p = e^{-G(n)}$, where $G(n) = (m-n)q_a + nq_r$.

(c) Now consider the stable equilibrium state n^* of the system where $G = G(n^*)$; $Ge^{-G} = (m-n^*)q_a$. Substitute (b) into your expression for T for (a), using $n = n^*$, and show that

$$T = 1 + \frac{n^*}{q_a(m-n^*)}$$

(Note that if n^* is assumed to be equal to $\overline{n}$ in Problem 4.4, this is the same as the value of T found there.)

(d) Solve numerically for T in the case where $q_a m = 0.3$ and $q_r m = 1$; show that $n^* \approx m/8$, corresponding to $1/8$ loss of incoming traffic, and $T \approx m/2$, giving roughly the same delay as TDM.

4.6 **(a)** Consider P_{succ} as given exactly in Eq. (4.5). For given $q_a < 1/m$, $n > 1$, show that the value of q_r that maximizes P_{succ} satisfies

$$\frac{1}{1-q_r} - \frac{q_a(m-n)}{1-q_a} - \frac{q_r n}{1-q_r} = 0$$

(b) Consider the value of q_r that satisfies the equation above as a function of q_a, say $q_r(q_a)$. Show that $q_r(q_a) > q_a$ (assume that $q_a < 1/m$).

(c) Take the total derivative of P_{succ} with respect to q_a, using $q_r(q_a)$ for q_r, and show that this derivative is negative. *Hint*: Recall that $\partial P_{succ}/\partial q_r$ is 0 at $q_r(q_a)$ and compare $\partial P_{succ}/\partial q_a$ with $\partial P_{succ}/\partial q_r$.

(d) Show that if q_r is chosen to maximize P_{succ} and $q_r < 1$, then P_{succ} is greater if new arrivals are treated immediately as backlogged than if new arrivals are immediately transmitted. *Hint:* In the backlog case, a previously unbacklogged node transmits with probability $q_a q_r < q_a$.

4.7 Consider a slotted Aloha system with "perfect capture." That is, if more than one packet is transmitted in a slot, the receiver "locks onto" one of the transmissions and receives it correctly; feedback immediately informs each transmitting node about which node was successful and the unsuccessful packets are retransmitted later.

(a) Give a convincing argument why expected delay is minimized if all waiting packets attempt transmission in each slot.

(b) Find the expected system delay assuming Poisson arrivals with overall rate λ. *Hint:* Review Example 3.16.

(c) Now assume that the feedback is delayed and that if a packet is unsuccessful in the slot, it is retransmitted on the kth subsequent slot rather than the first subsequent slot. Find the new expected delay as a function of k. *Hint:* Consider the system as k subsystems, the ith subsystem handling arrivals in slots j such that $j \bmod k = i$.

4.8 Consider a slotted system in which all nodes have infinitely large buffers and all new arrivals (at Poisson rate λ/m per node) are allowed into the system, but are considered as backlogged immediately rather than transmitted in the next slot. While a node contains one or more packets, it independently transmits one packet in each slot, with probability q_r. Assume that any given transmission is successful with probability p.

(a) Show that the expected time from the beginning of a backlogged slot until the completion of the first success at a given node is $1/pq_r$. Show that the second moment of this time is $(2 - pq_r)/(pq_r)^2$.

(b) Note that the assumption of a constant success probability allows each node to be considered independently. Assume that λ/m is the Poisson arrival rate at a node, and use the service-time results of part (a) to show that the expected delay is

$$T = \frac{1}{q_r p(1-\rho)} + \frac{1-2\rho}{2(1-\rho)}$$

$$\rho = \frac{\lambda}{mpq_r}$$

(c) Assume that $p = 1$ (this yields a smaller T than any other value of p, and corresponds to very light loading). Find T for $q_r = 1/m$; observe that this is roughly twice the delay for TDM if m is large.

4.9 Assume that the number of packets n in a slotted Aloha system at a given time is a Poisson random variable with mean $\hat{n} \geq 1$. Suppose that each packet is independently transmitted in the next slot with probability $1/\hat{n}$.

(a) Find the probability that the slot is idle.

(b) Show that the a posteriori probability that there were n packets in the system, given an idle slot, is Poisson with mean $\hat{n} - 1$.

(c) Find the probability that the slot is successful.

(d) Show that the a posteriori probability that there were $n+1$ packets in the system, given a success, is $e^{-(\hat{n}-1)}(\hat{n}-1)^n/n!$ (*i.e.*, the number of remaining packets is Poisson with mean $\hat{n} - 1$).

4.10 Consider a slotted Aloha system satisfying assumptions 1 to 6a of Section 4.2.1 *except* that each of the nodes has a limitless buffer to store all arriving packets until transmission *and* that nodes receive immediate feedback only about whether or not their own packets were successfully transmitted. Each node is in one of two different modes. In mode 1, a node transmits (with probability 1) in each slot, repeating unsuccessful packets, until its buffer of waiting packets is exhausted; at that point the node goes into mode 2. In mode 2, a node transmits a "dummy packet" in each slot with probability q_r until a successful transmission occurs, at which point it enters mode 1 (dummy packets are used only to simplify the mathematics). Assume that the system starts with all nodes in mode 2. Each node has Poisson arrivals of rate λ/m.

(a) Explain why at most one node at a time can be in mode 1.

(b) Given that a node is in mode 1, find its probability, p_1, of successful transmission. Find the mean time $\overline{X}$ between successful transmissions and the second moment $\overline{X^2}$ of this time. *Hint:* Review the ARQ example in Section 3.5.1, with $N = 1$.

(c) Given that all nodes are in mode 2, find the probability p_2 that some dummy packet is successfully transmitted in a given slot. Find the mean time $\overline{v}$ until the completion of such a successful transmission and its second moment $\overline{v^2}$.

(d) Regard the intervals of time when all nodes are in mode 2 as reservation intervals. Show that the mean time a packet must wait in queue before first attempting transmission is

$$W = \frac{R + E\{S\}\overline{v}}{1-\rho}, \qquad \rho = \lambda\overline{X}$$

where R is the mean residual time until completion of a service in mode 1 or completion of a reservation interval, and S is the number of whole reservation intervals until the node at which the packet arrived is in mode 1.

(e) Show that

$$W = \frac{\lambda(2-p_1)}{2p_1^2(1-\rho)} + \frac{2-p_2}{2p_2} + \frac{m-1}{p_2(1-\rho)}$$

Show that W is finite if $q_r < (1-\lambda)^{1/(m-1)}$.

4.11 Consider the somewhat unrealistic feedback assumption for unslotted Aloha in which all nodes are informed, precisely τ time units after the beginning of each transmission whether or not that transmission was successful. Thus, in the event of a collision, each node knows how many packets were involved in the collision, and each node involved in the collision knows how many other nodes started transmission before itself. Assume that each transmission lasts one time unit and assume that $m = \infty$. Consider a retransmission strategy in which the first node involved in a collision waits one time unit after receiving feedback on its collision and then transmits its packet. Successive nodes in the collision retransmit in order spaced one time unit apart. All new arrivals to the system while these retransmissions are taking place wait until the retransmissions are finished. At the completion of the retransmissions, each backlogged node chooses a time to start its transmission uniformly distributed over the next time unit. All new arrivals after the end of the retransmissions above start transmission immediately.

(a) Approximate the system above as a reservation system with reservation intervals of duration $1 + \tau$ (note that this is an approximation in the sense that successful transmissions will sometimes occur in the reservation intervals, but the approximation becomes more accurate as the loading becomes higher). Find the expected packet delay for this approximation (assume Poisson arrivals at rate λ).

(b) Show that the delay above remains finite for all $\lambda < 1$.

4.12 This problem illustrates that the maximum throughput of unslotted Aloha can be increased up to e^{-1} at an enormous cost in delay. Consider a finite but large set m of nodes with unlimited buffering at each node. Each node waits until it has accumulated k packets and then transmits them one after the other in the next k time units. Those packets involved in collisions, plus new packets, are then retransmitted a random time later, again k at a time. Assume that the starting time of transmissions from all nodes collectively is a Poisson process with parameter G (*i.e.*, ignore stability issues).

(a) Show that the probability of success on the jth of the k packets in a sequence is $e^{-G(k+1)}$. *Hint:* Consider the intervals between the initiation of the given sequence and the preceding sequence and subsequent sequence.

(b) Show that the throughput is $kGe^{-G(k+1)}$, and find the maximum throughput by optimizing over G.

4.13 **(a)** Consider a CRP that results in the feedback pattern $e, 0, e, e, 1, 1, 0$ when using the tree algorithm as illustrated in Fig. 4.9. Redraw this figure for this feedback pattern.

(b) Which collision or collisions would have been avoided if the first improvement to the tree algorithm had been used?

(c) What would the feedback pattern have been for the CRP if both improvements to the tree algorithms had been used?

4.14 Consider the tree algorithm in Fig. 4.9. Given that k collisions occur in a CRP, determine the number of slots required for the CRP. Check your answer with the particular example of Fig. 4.9. *Hint 1:* Note that each collision corresponds to a nonleaf node of the rooted tree. Consider "building" any given tree from the root up, successively replacing leaf nodes by internal nodes with two upward-going edges. *Hint 2:* For another approach, consider what happens in the stack for each collision, idle, or success.

4.15 Consider the tree algorithm in Fig. 4.9. Assume that after each collision, each packet involved in the collision flips an unbiased coin to determine whether to go into the left or right subset.

(a) Given a collision of k packets, find the probability that i packets go into the left subset.

(b) Let A_k be the expected number of slots required in a CRP involving k packets. Note that $A_0 = A_1 = 1$. Show that for $k \geq 2$,

$$A_k = 1 + \sum_{i=0}^{k} \binom{k}{i} 2^{-k}(A_i + A_{k-i})$$

(c) Simplify your answer in part (b) to the form

$$A_k = c_{kk} + \sum_{i=0}^{k-1} c_{ik} A_i$$

and find the coefficients c_{ik}. Evaluate A_2 and A_3 numerically. For more results on A_k for large k, and the use of A_k in evaluating maximum throughput, see [Mas80].

4.16 **(a)** Consider the first improvement to the tree algorithm as shown in Fig. 4.10. Assume that each packet involved in a collision flips an unbiased coin to join either the left or right subset. Let B_k be the expected number of slots required in a CRP involving k packets; note that $B_0 = B_1 = 1$. Show that for $k \geq 2$,

$$B_k = 1 + \sum_{i=1}^{k} \binom{k}{i} 2^{-k}(B_i + B_{k-i}) + 2^{-k} B_k$$

(b) Simplify your answer to the form

$$B_k = C'_{kk} + \sum_{i=1}^{k-1} C'_{ik} B_i$$

and evaluate the constants C'_{ik}. Evaluate B_2 and B_3 numerically (see [Mas80]).

4.17 Let X_L and X_R be independent Poisson distributed random variables, each with mean G. Use the definition of conditional probabilities to evaluate the following:
(a) $P\{X_L = 0 \mid X_L + X_R \geq 2\}$
(b) $P\{X_L = 1 \mid X_L + X_R \geq 2\}$
(c) $P\{X_L \geq 2 \mid X_L + X_R \geq 2\}$
(d) $P\{X_R = 1 \mid X_L = 1, X_L + X_R \geq 2\}$
(e) $P\{X_R = i \mid X_L = 0, X_L + X_R \geq 2\} \quad (i \geq 2)$
(f) $P\{X_R = i \mid X_L \geq 2, X_L + X_R \geq 2\}$

4.18 Suppose that at time k, the FCFS splitting algorithm allocates a new interval from $T(k)$ to $T(k) + \alpha_0$. Suppose that this interval contains a set of packets with arrival times $T(k) + 0.1\alpha_0, T(k) + 0.6\alpha_0, T(k) + 0.7\alpha_0$, and $T(k) + 0.8\alpha_0$.
(a) Find the allocation intervals for each of the subsequent times until the CRP is completed.
(b) Indicate which of the rules of Eqs. (4.15) to (4.18) are used in determining each of these allocation intervals.
(c) Indicate the path through the Markov chain in Fig. 4.13 for this sequence of events.

4.19 **(a)** Show that $\overline{n}$, the expected number of packets successfully transmitted in a CRP of the FCFS splitting algorithm, is given by

$$\overline{n} = 1 - e^{-G_0} + \sum_{i=1}^{\infty} p(R, i)$$

(Assume that the initial allocation interval is α_0, with $G_0 = \alpha_0 \lambda$.)
(b) Show that

$$\overline{n} = \lambda \alpha_0 (1 - E\{f\})$$

where $E\{f\}$ is the expected fraction of α_0 returned to the waiting interval in a CRP. (This provides an alternative way to calculate $E\{f\}$.)

4.20 Show, for Eq. (4.41), that if n_k is a Poisson random variable with mean $\hat{n}_k$, then the a posteriori distribution of n_k, given an idle, is Poisson with mean $\hat{n}_k[1 - q_r(\hat{n}_k)]$. Show that the a posteriori distribution of $n_k - 1$, given a success, is Poisson with mean $\hat{n}_k[1 - q_r(\hat{n}_k)]$.

4.21 *Slotted CSMA with Variable Packet Lengths.* Assume that the time to transmit a packet is a random variable X; for consistency with the slotted assumption, assume that X is discrete, taking values that are integer multiples of β. Assume that all transmissions are independent and identically distributed (IID) with mean $\overline{X} = 1$.
(a) Let Y be the longer of two IID transmissions X_1 and X_2 [*i.e.*, $Y = \max(X_1, X_2)$]. Show that the expected value of Y satisfies $\overline{Y} \leq 2\overline{X}$. Show that if X takes the value of β with large probability and $k\beta$ (for a large k) with small probability, this bound is close to equality.
(b) Show that the expected time between state transitions, given a collision of two packets, is at most $2 + \beta$.

(c) Let $g(n) = \lambda\beta + q_r n$ be the expected number of attempted transmissions following a state transition to state n, and assume that this number of attempts is Poisson with mean $g(n)$. Show that the expected time between state transitions in state n is at most

$$\beta e^{-g(n)} + (1+\beta)g(n)e^{-g(n)} + (1+\beta/2)g^2(n)e^{-g(n)}$$

Ignore collisions of more than two packets as negligible.

(d) Find a lower bound to the expected number of departures per unit time in state n [see Eq. (4.39)].

(e) Show that this lower bound is maximized (for small β) by

$$g(n) \approx \sqrt{\beta}$$

with a corresponding throughput of approximately $1 - 2\sqrt{\beta}$.

4.22 *Pseudo-Bayesian Stabilization for Unslotted CSMA.* Assume that at the end of a transmission, the number n of backlogged packets in the system is a Poisson random variable with mean $\hat{n}$. Assume that in the idle period until the next transmission starts, each backlogged packet attempts transmission at rate x and each new arrival starts transmission immediately. Thus, given n, the time τ until the next transmission starts has probability density $p(\tau \mid n) = (\lambda + xn)e^{-(\lambda+xn)\tau}$.

(a) Find the unconditional probability density $p(\tau)$.

(b) Find the a posteriori probability $P\{n, b \mid \tau\}$ that there were n backlogged packets and one of them started transmission first, given that the transmission starts at τ.

(c) Find the a posteriori probability $P\{n, a \mid \tau\}$ that there were n backlogged packets and a new arrival started transmission first, given that this transmission starts at τ.

(d) Let n' be the number of backlogged packets immediately after the next transmission starts (not counting the packet being transmitted); that is, $n' = n - 1$ if a backlogged packet starts and $n' = n$ if a new arrival starts. Show that, given τ, n' is Poisson with mean $\hat{n}' = \hat{n}e^{-\tau x}$.

This means that the pseudo-Bayesian rule for updating estimated backlog (assuming unit time transmission) is to estimate the backlog at the end of the $(k+1)^{\text{st}}$ transmission in terms of the estimate at the end of the kth transmission and the idle period τ_k between the transmissions by

$$\hat{n}_{k+1} = \begin{cases} \hat{n}_k e^{-\tau_k x_k} + \lambda(1+\beta); & \text{success} \\ \hat{n}_k e^{-\tau_k x_k} + 2 + \lambda(1+\beta); & \text{collision} \end{cases}$$

$$x_k = \beta^{-\frac{1}{2}} \min\left(\frac{1}{\hat{n}_k}, 1\right)$$

4.23 Give an intuitive explanation of why the maximum throughput, for small β, is approximately the same for CSMA slotted Aloha and FCFS splitting with CSMA. Show that the optimal expected number of packets transmitted after a state transition in Aloha is the same as that at the beginning of a CRP for FCFS splitting. Note that after a collision, the expected numbers are slightly different in the two systems, but the difference is unimportant since collisions are rare.

4.24 *Delay of Ideal Slotted CSMA/CD.* Assume that for all positive backlogs, the number of packets attempting transmission in an interval is Poisson with mean g.

(a) Start with Eq. (4.42), which is valid for CSMA/CD as well as CSMA. Show that for CSMA/CD,

$$E\{t\} = \frac{\beta e^{-g} + (1+\beta)ge^{-g} + 2\beta[1-(1+g)e^{-g}]}{ge^{-g}}$$

Show that $E\{t\}$ is minimized over $g > 0$ by $g = 0.77$, and

$$\min_g E\{t\} = 1 + 3.31\beta$$

(b) Show that for this g and mean packet length 1,

$$W = \frac{\overline{R} + \overline{y}}{1 - \lambda(1 + 3.31\beta)}$$

(c) Evaluate $\overline{R}$ and $\overline{y}$ to verify Eq. (4.67) for small β.

(d) Discuss the assumption of a Poisson-distributed number of packets attempting transmission in each interval, particularly for a backlog of 1.

4.25 Show that for unslotted CSMA/CD, the maximum interval of time over which a transmitting node can hear a collision is 2β. (Note in Fig. 4.20 that the time when a collision event starts at one node until it ends at another node can be as large as 3β.)

4.26 Consider an unslotted CSMA/CD system in which the propagation delay is negligible compared to the time β required for a node to detect that the channel is idle or busy. Assume that each packet requires one time unit for transmission. Assume that β time units after either a successful transmission or a collision ends, all backlogged nodes attempt transmission after a random delay and that the composite process of initiation times is Poisson of rate G (up to time β after the first initiation). For simplicity, assume that each collision lasts for β time units.

(a) Find the probability that the first transmission initiation after a given idle detection is successful.

(b) Find the expected time from one idle detection to the next.

(c) Find the throughput (for the given assumptions).

(d) Optimize the throughput numerically over G.

4.27 Modify Eq. (4.71) for the case in which a node, after transmitting a packet, waits for the packet to return before transmitting the free token. *Hint:* View the added delay as an increase in the packet transmission time.

4..28 Show by example that a node in a register insertion ring might have to wait an arbitrarily long time to transmit a packet once its transit buffer is full.

4.29 Suppose that two nodes are randomly placed on a bus; that is, each is placed independently, and the position of each is chosen from a uniform distribution over the length of the bus. Assuming that the length of the bus is 1 unit, show that the expected distance between the nodes is $1/3$.

4.30 Assume, for the analysis of generalized polling in Section 4.5.6, that each node has a packet to send with probability q.

(a) Find the probability $P\{i\}$ that node i, $i \geq 0$, is the lowest-numbered node with a packet to send.

(b) Assume that the CRP tests only the first 2^j lowest-numbered nodes at a time for some j. Represent the lowest-numbered node with a packet as $i = k2^j + r$, where k is an integer, and $0 \le r < 2^j$. Show that, given i, the number of reservation slots needed to find i is $k + 1 + j$.

(c) Assume that the total number of nodes is infinite and approximate k above by $i2^{-j}$. Find the expected number of reservation slots to find the lowest-numbered node i containing a packet.

(d) Find the integer value of j that minimizes your answer in part (c). *Hint:* Find the smallest value of j for which the expected number of reservation slots is less than the expected number for $j + 1$.

4.31 Consider the simple packet radio network shown in Fig. 4.37 and let f_ℓ be the throughput on link ℓ ($\ell = 1, 2, 3$); the links are used only in the direction shown.

(a) Note that at most one link can be in any collision-free set. Using generalized TDM [as in Eq. (4.86)], show that any set of throughputs satisfying $f_1 + f_2 + f_3 \le 1, f_\ell \ge 0$ ($\ell = 1, 2, 3$) can be achieved.

(b) Suppose that $f_1 = f_2 = f$ and $f_3 = 2f$ for some f. Use Eqs. (4.88) and (4.89) to relate f_1, f_2, f_3 to the attempt rates q_1, q_2, q_3 for slotted Aloha. Show that $q_1 = q_2$ and $q_3 = 2f$.

(c) Find the maximum achievable value of f for part (b).

4.32 Consider the packet radio network shown in Fig. 4.38 in which the links are used only in the direction shown. Links 5 and 6 carry no traffic but serve to cause collisions for link 7 when packets are transmitted on links 3 and 4, respectively.

(a) Show that throughputs $f_1 = f_2 = f_3 = f_4 = 1/3, f_7 = 4/9$ are feasible for the assumption of heavy loading and independent attempts on each link. Find the corresponding attempt rates and success rates from Eqs. (4.88) and (4.89).

(b) Now assume that links 3 and 4 are used to forward incoming traffic and always transmit in the slot following a successful incoming packet. Show that with $f_1 = f_2 = f_3 = f_4 = 1/3$, f_7 is restricted to be no more than 1/3.

(c) Assume finally that packet attempts on links 1 and 2 are in alternating order and links 3 and 4 operate as in part (b). Show that $f_1 = f_2 = f_3 = f_4 = 1/2$ is feasible, but that f_7 must then be 0.

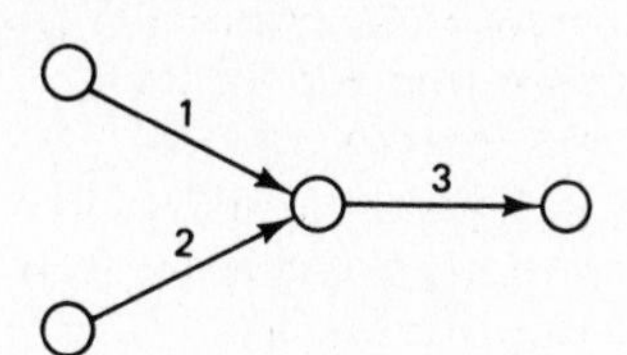

Figure 4.37

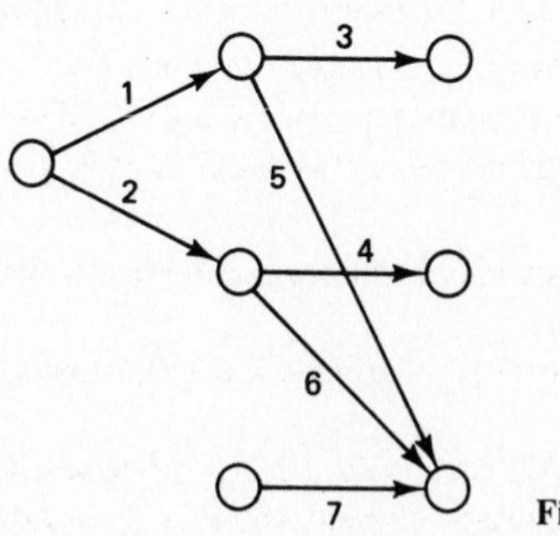

Figure 4.38

4.33 Suppose that an FDDI ring has two users. The target token rotation time is 3 msec and $\alpha_1 = \alpha_2 = 1$ msec. Assume that neither node has any traffic to send up to time 0, and then both nodes have an arbitrarily large supply of both high- and low-priority traffic. Node 0 is the first node to send traffic on the ring starting at time 0. Find the sequence of times t_i at which each node captures the token; ignore propagation delays. Explain any discrepancy between your solution and the upper bound of Eq. (4.81).

4.34 **(a)** Consider an FDDI ring with m nodes, with target token rotation time τ and high-priority allocations $\alpha_0, \ldots, \alpha_{m-1}$. Assume that every node has an arbitrarily large backlog of both high-priority and low-priority traffic. In the limit of long-term operation, find the fraction of the transmitted traffic that goes to each node for each priority. Ignore propagation and processing delay and use the fact that the bound in Eq. (4.81) is met with equality given the assumed initial conditions and given arbitarily large backlogs. Assume that $T = \alpha_0 + \alpha_1 + \cdots + \alpha_{m-1}$ is strictly less than τ.

(b) Now assume that each node k can fill only a fraction α_k/τ of the ring with high-priority traffic and find the fractions in part (a) again.

4.35 Give the rules that the two counters implementing the virtual queue in DQDB must follow. Assume there is also a binary variable F that is one when a frame is in the virtual queue. Assume that when $F = 0$, the first counter is kept at value 0.

5

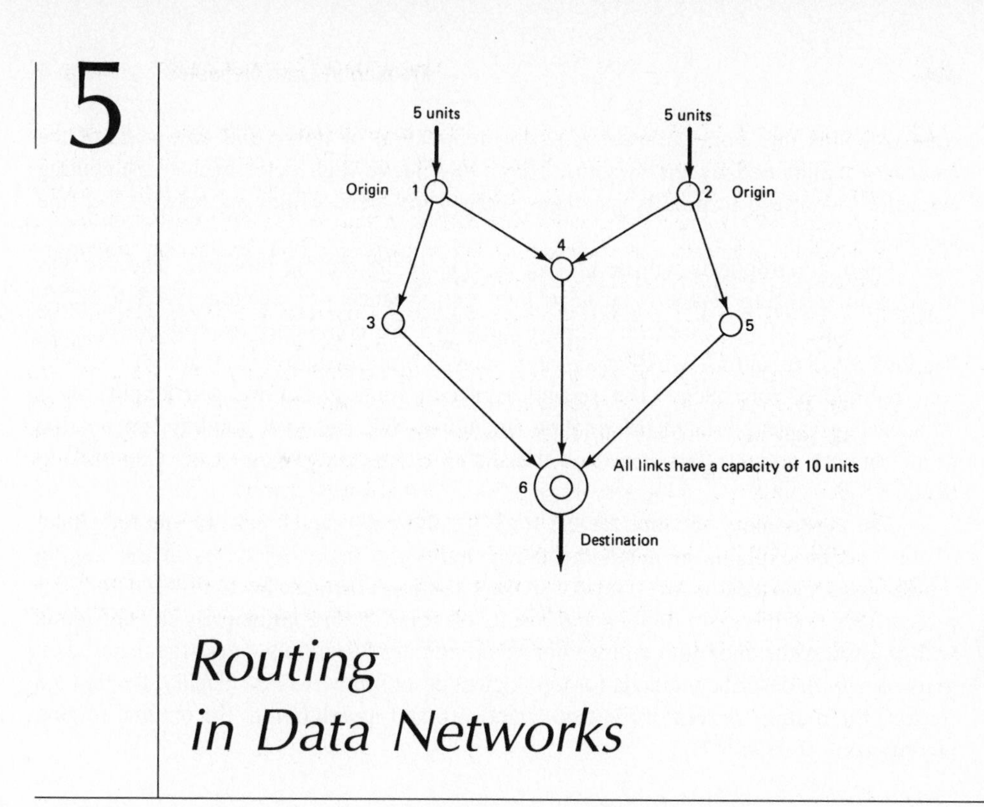

Routing in Data Networks

5.1 INTRODUCTION

We have frequently referred to the routing algorithm as the network layer protocol that guides packets through the communication subnet to their correct destination. The times at which routing decisions are made depend on whether the network uses datagrams or virtual circuits. In a datagram network, two successive packets of the same user pair may travel along different routes, and a routing decision is necessary for each individual packet (see Fig. 5.1). In a virtual circuit network, a routing decision is made when each virtual circuit is set up. The routing algorithm is used to choose the communication path for the virtual circuit. All packets of the virtual circuit subsequently use this path up to the time that the virtual circuit is either terminated or rerouted for some reason (see Fig. 5.2).

Routing in a network typically involves a rather complex collection of algorithms that work more or less independently and yet support each other by exchanging services or information. The complexity is due to a number of reasons. First, routing requires coordination between all the nodes of the subnet rather than just a pair of modules as, for example, in data link and transport layer protocols. Second, the routing system must cope with link and node failures, requiring redirection of traffic and an update of the databases maintained by the system. Third, to achieve high performance, the routing algorithm may need to modify its routes when some areas within the network become congested.

The main emphasis will be on two aspects of the routing problem. The first has to do with selecting routes to achieve high performance. In Sections 5.2.3 to 5.2.5, we discuss algorithms based on shortest paths that are commonly used in practice. In Sections 5.5, 5.6, and 5.7 we describe sophisticated routing algorithms that try to achieve near optimal performance. The second aspect of routing that we will emphasize is broadcasting routing-related information (including link and node failures and repairs) to all network nodes. This issue and the subtleties associated with it are examined in Section 5.3.

The introductory sections set the stage for the main development. The remainder of this section explains in nonmathematical terms the main objectives in the routing problem and provides an overview of current routing practice. Sections 5.2.1 to 5.2.3 present some of the main notions and results of graph theory, principally in connection with shortest paths and minimum weight spanning trees. Section 5.4 uses the material on graph theory to describe methods for topological design of networks. Finally, Section 5.8 reviews the routing system of the Codex network and its relation to the optimal routing algorithm of Section 5.7.

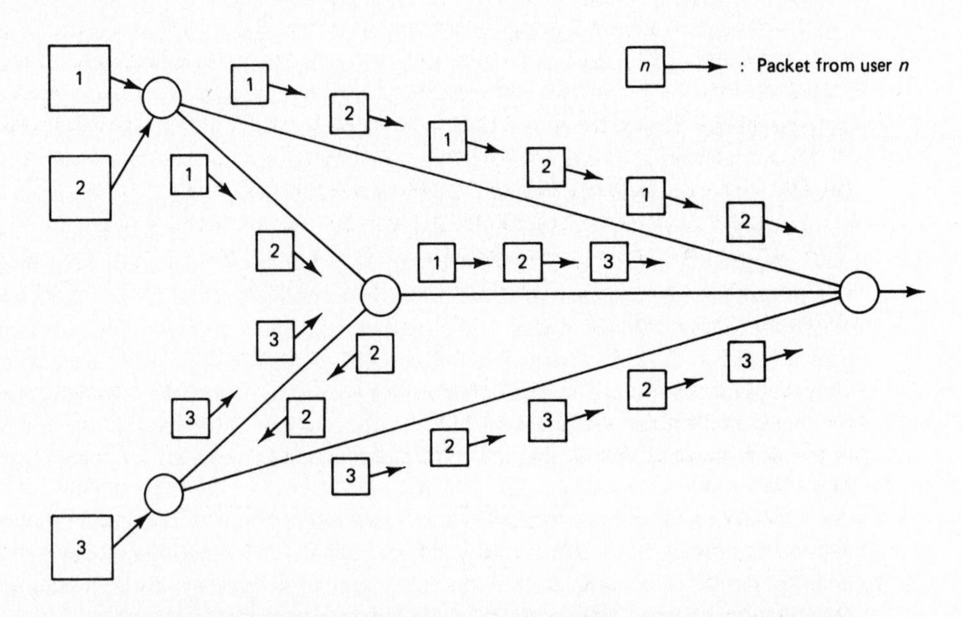

Figure 5.1 Routing in a datagram network. Two packets of the same user pair can travel along different routes. A routing decision is required for each individual packet.

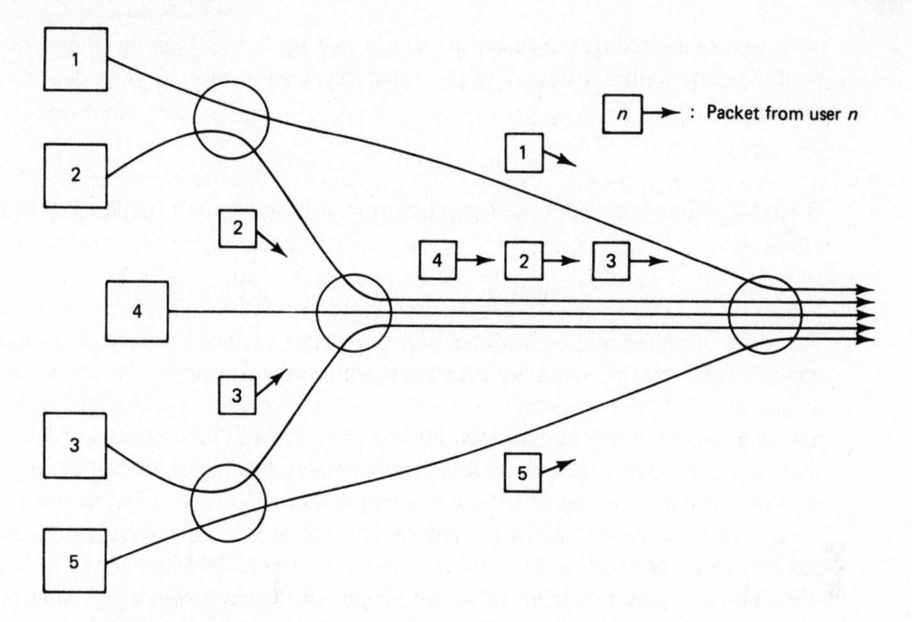

Figure 5.2 Routing in a virtual circuit network. All packets of each virtual circuit use the same path. A routing decision is required only when a virtual circuit is set up.

5.1.1 Main Issues in Routing

The two main functions performed by a routing algorithm are the selection of routes for various origin–destination pairs and the delivery of messages to their correct destination once the routes are selected. The second function is conceptually straightforward using a variety of protocols and data structures (known as routing tables), some of which will be described in the context of practical networks in Section 5.1.2. The focus will be on the first function (selection of routes) and how it affects network performance.

There are two main performance measures that are substantially affected by the routing algorithm—*throughput* (quantity of service) and *average packet delay* (quality of service). Routing interacts with flow control in determining these performance measures

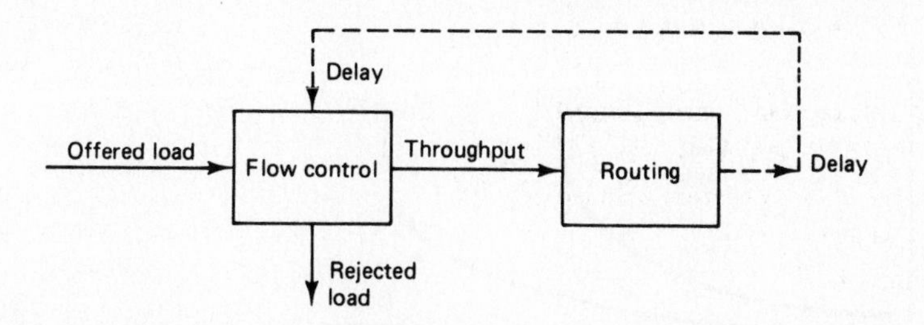

Figure 5.3 Interaction of routing and flow control. As good routing keeps delay low, flow control allows more traffic into the network.

by means of a feedback mechanism shown in Fig. 5.3. When the traffic load offered by the external sites to the subnet is relatively low, it will be fully accepted into the network, that is,

$$\text{throughput} = \text{offered load}$$

When the offered load is excessive, a portion will be rejected by the flow control algorithm and

$$\text{throughput} = \text{offered load} - \text{rejected load}$$

The traffic accepted into the network will experience an average delay per packet that will depend on the routes chosen by the routing algorithm. However, throughput will also be greatly affected (if only indirectly) by the routing algorithm because typical flow control schemes operate on the basis of striking a balance between throughput and delay (*i.e.*, they start rejecting offered load when delay starts getting excessive). Therefore, *as the routing algorithm is more successful in keeping delay low, the flow control algorithm allows more traffic into the network.* While the precise balance between delay and throughput will be determined by flow control, the effect of good routing under high offered load conditions is to realize a more favorable delay-throughput curve along which flow control operates, as shown in Fig. 5.4.

The following examples illustrate the discussion above:

Example 5.1

In the network of Fig. 5.5, all links have capacity 10 units. (The units by which link capacity and traffic load is measured is immaterial and is left unspecified.) There is a single destination (node 6) and two origins (nodes 1 and 2). The offered load from each of nodes 1 and 2 to node 6 is 5 units. Here, the offered load is light and can easily be accommodated with small delay by routing along the leftmost and rightmost paths, $1 \to 3 \to 6$ and $2 \to 5 \to 6$, respectively. If instead, however, the routes $1 \to 4 \to 6$ and $2 \to 4 \to 6$ are used, the flow on link (4,6) will equal capacity, resulting in very large delays.

Example 5.2

For the same network, assume that the offered loads at nodes 1 and 2 are 5 and 15 units, respectively (see Fig. 5.6). If routing from node 2 to the destination is done along a single path, then at least 5 units of offered load will have to be rejected since all path capacities

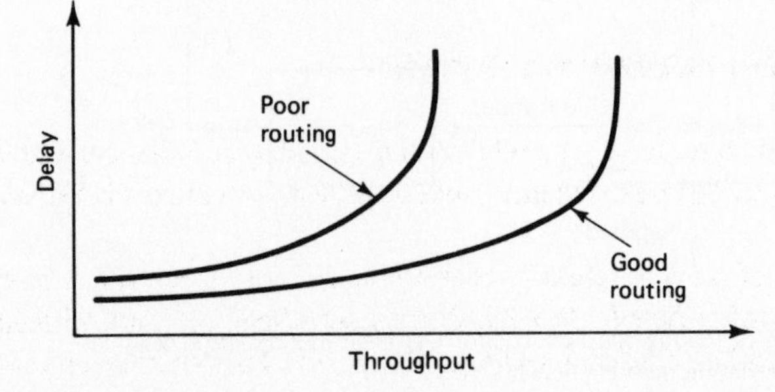

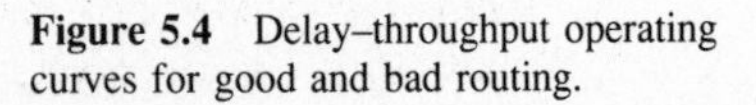
Figure 5.4 Delay–throughput operating curves for good and bad routing.

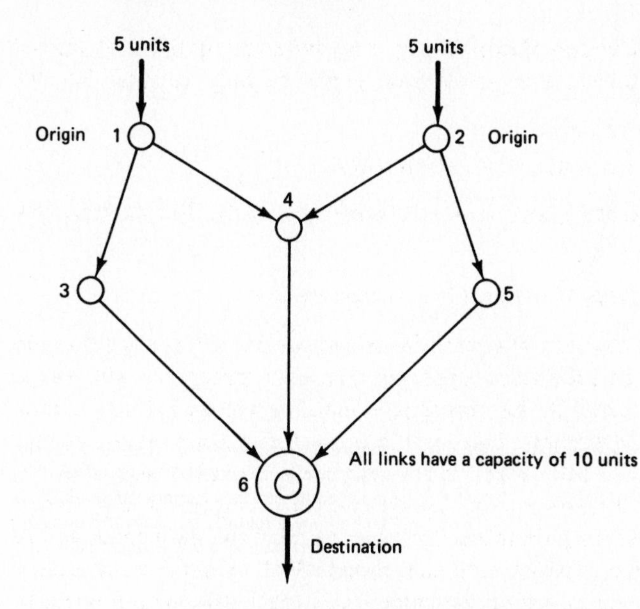

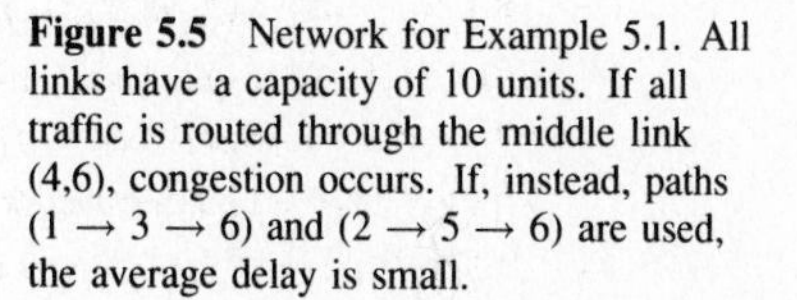

Figure 5.5 Network for Example 5.1. All links have a capacity of 10 units. If all traffic is routed through the middle link (4,6), congestion occurs. If, instead, paths ($1 \rightarrow 3 \rightarrow 6$) and ($2 \rightarrow 5 \rightarrow 6$) are used, the average delay is small.

equal 10. Thus, the total throughput can be no more than 15 units. On the other hand, suppose that the traffic originating at node 2 is evenly split between the two paths $2 \rightarrow 4 \rightarrow 6$ and $2 \rightarrow 5 \rightarrow 6$, while the traffic originating at node 1 is routed along $1 \rightarrow 3 \rightarrow 6$. Then, the traffic arrival rate on each link will not exceed 75% of capacity, the delay per packet will be reasonably small, and (given a good flow control scheme) no portion of the offered load will be rejected. Arguing similarly, it is seen that when the offered loads at nodes 1 and 2 are both large, the maximum total throughput that this network can accommodate is between 10 and 30 units, depending on the routing scheme. This example also illustrates that to achieve high throughput, the traffic of some origin–destination pairs may have to be divided among more than one route.

In conclusion, *the effect of good routing is to increase throughput for the same value of average delay per packet under high offered load conditions and to decrease average delay per packet under low and moderate offered load conditions.* Furthermore, it is evident that the routing algorithm should be operated so as to keep average delay per packet as low as possible for any given level of offered load. While this is easier said than done, it provides a clear-cut objective that can be expressed mathematically and dealt with analytically.

5.1.2 Wide Area Network Routing: An Overview

The purpose of this section is to survey current routing practice in wide area networks, to introduce classifications of different schemes, and to provide a context for the analysis presented later.

There are a number of ways to classify routing algorithms. One way is to divide them into *centralized* and *distributed.* In centralized algorithms, all route choices are made at a central node, while in distributed algorithms, the computation of routes is

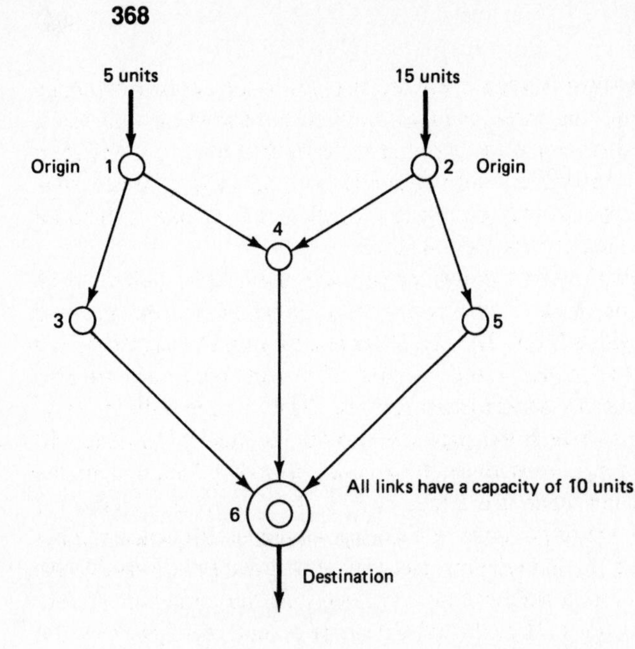

Figure 5.6 Network for Example 5.2. All links have a capacity of 10 units. The input traffic can be accommodated with multiple-path routing, but at least 5 units of traffic must be rejected if a single-path routing is used.

shared among the network nodes with information exchanged between them as necessary. Note that this classification relates mostly to the implementation of an algorithm, and that a centralized and a distributed routing algorithm may be equivalent at some level of mathematical abstraction.

Another classification of routing algorithms relates to whether they change routes in response to the traffic input patterns. In *static* routing algorithms, the path used by the sessions of each origin–destination pair is fixed regardless of traffic conditions. It can only change in response to a link or node failure. This type of algorithm cannot achieve a high throughput under a broad variety of traffic input patterns. It is recommended for either very simple networks or for networks where efficiency is not essential. Most major packet networks use some form of *adaptive* routing where the paths used to route new traffic between origins and destinations change occasionally in response to congestion. The idea here is that congestion can build up in some part of the network due to changes in the statistics of the input traffic load. Then, the routing algorithm should try to change its routes and guide traffic around the point of congestion.

There are many routing algorithms in use with different levels of sophistication and efficiency. This variety is partly due to historical reasons and partly due to the diversity of needs in different networks. In this section we provide a nonmathematical description of some routing techniques that are commonly used in practice. We illustrate these techniques in terms of the routing algorithms of three wide area networks (ARPANET, TYMNET, and SNA). The routing algorithm of another wide area network, the Codex network, will be described in Section 5.8, because this algorithm is better understood after studying optimal routing in Sections 5.5 to 5.7.

Flooding and broadcasting During operation of a data network, it is often necessary to broadcast some information, that is, to send this information from an origin node to all other nodes. An example is when there are changes in the network topology due to link failures and repairs, and these changes must be transmitted to the entire network. Broadcasting could also be used as a primitive form of routing packets from a single transmitter to a single receiver or, more generally, to a subset of receivers; this use is generally rather inefficient, but may be sensible because it is simple or because the receivers' locations within the network are unknown.

A widely used broadcasting method, known as *flooding*, operates as follows. The origin node sends its information in the form of a packet to its neighbors (the nodes to which it is directly connected with a link). The neighbors relay it to their neighbors, and so on, until the packet reaches all nodes in the network. Two additional rules are also observed, which limit the number of packet transmissions. First, a node will not relay the packet back to the node from which the packet was obtained. Second, a node will transmit the packet to its neighbors at most once; this can be ensured by including on the packet the ID number of the origin node and a sequence number, which is incremented with each new packet issued by the origin node. By storing the highest sequence number received for each origin node, and by not relaying packets with sequence numbers that are less than or equal to the one stored, a node can avoid transmitting the same packet more than once on each of its incident links. Note that with these rules, links need not preserve the order of packet transmissions; the sequence numbers can be used to recognize the correct order. Figure 5.7(a) gives an example of flooding and illustrates that the total number of packet transmissions per packet broadcast lies between L and $2L$, where L is the number of bidirectional links of the network. We note also that one can implement a flooding-like algorithm without using sequence numbers. This possibility is described in Section 5.3, where flooding and the topology broadcast problem are discussed in more detail.

Another broadcasting method, based on the use of a *spanning tree*, is illustrated in Fig. 5.7(b). A spanning tree is a connected subgraph of the network that includes all nodes and has no cycles (see a more detailed discussion in the next section). Broadcasting on a spanning tree is more communication-efficient than flooding. It requires a total of only $N - 1$ packet transmissions per packet broadcast, where N is the number of nodes.

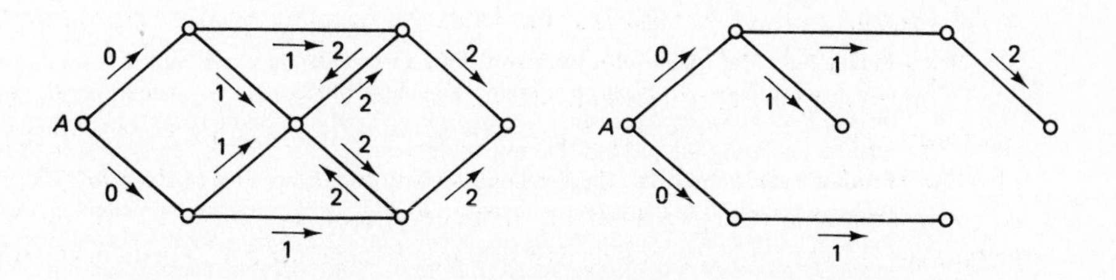

Figure 5.7 Packet broadcasting from node A to all other nodes by using flooding [as in (a)] or a spanning tree [as in (b)]. Arrows indicate packet transmissions at the times shown. Each packet transmission time is assumed to be one unit. Flooding requires at least as many packet transmissions as the spanning tree method and usually many more. In this example, the time required for the broadcast packet to reach all nodes is the same for the two methods. In general, however, depending on the choice of spanning tree, the time required with flooding may be less than with the spanning tree method.

The price for this saving is the need to maintain and update the spanning tree in the face of topological changes.

We note two more uses of spanning trees in broadcasting and routing. Given a spanning tree rooted at the origin node of the broadcast, it is possible to implement flooding as well as routing without the use of sequence numbers. The method is illustrated in Fig. 5.8. Note that flooding can be used to construct a spanning tree rooted at a node, as shown in Fig. 5.8. Also given a spanning tree, one can perform selective broadcasting, that is, packet transmission from a single origin to a limited set of destinations. For this it is necessary that each node knows which of its incident spanning tree links leads to any given destination (see Fig. 5.9). The spanning tree can also be used for routing packets with a single destination and this leads to an important method for bridged local area networks; see the next subsection.

Shortest path routing. Many practical routing algorithms are based on the notion of a *shortest path* between two nodes. Here, each communication link is assigned a positive number called its *length.* A link can have a different length in each direction. Each path (*i.e.*, a sequence of links) between two nodes has a length equal to the sum of the lengths of its links. (See Section 5.2 for more details.) A shortest path routing algorithm routes each packet along a minimum length (or shortest) path between the origin and destination nodes of the packet. The simplest possibility is for each link to have unit length, in which case a shortest path is simply a path with minimum number of links (also called a *min-hop path*). More generally, the length of a link may depend on its transmission capacity and its projected traffic load. The idea here is that a shortest

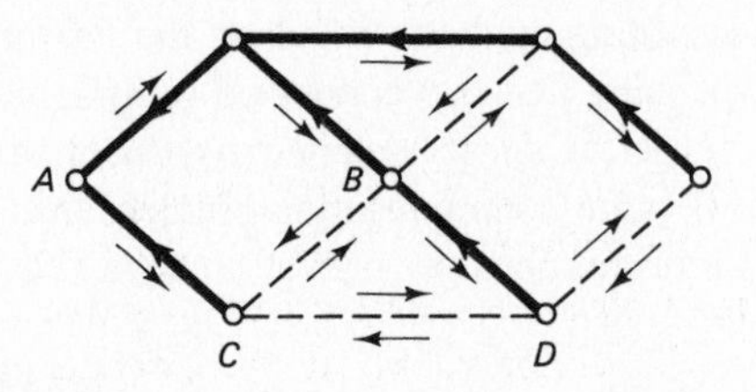

Figure 5.8 Illustration of a method for broadcasting using a tree rooted at the broadcast's origin node A. Each node knows its unique predecessor (or parent) on the tree, but need not know its successors. The tree can be used for routing packets to A from the other nodes using the paths of the tree in the reverse direction. It can also be used for flooding packets from A. The flooding rule for a node other than A is the following: a packet received from the parent is broadcast to all neighbors except the parent; all other packets are ignored. Thus in the figure, node D broadcasts the packet received from its parent B but ignores the packet received from C. Since only the packets transmitted on the spanning tree in the direction pointing away from the root are relayed further, there is no indefinite circulation of copies, and therefore, no need for sequence numbers.

Flooding can also be used to construct the tree rooted at A. Node A starts the process by sending a packet to all its neighbors; the neighbors must send the packet to their neighbors, and so on. All nodes should mark the transmitter of the first packet they receive as their parent on the tree. The nodes should relay the packet to their neighbors only once (after they first receive the packet from their parent); all subsequent packet receptions should be ignored.

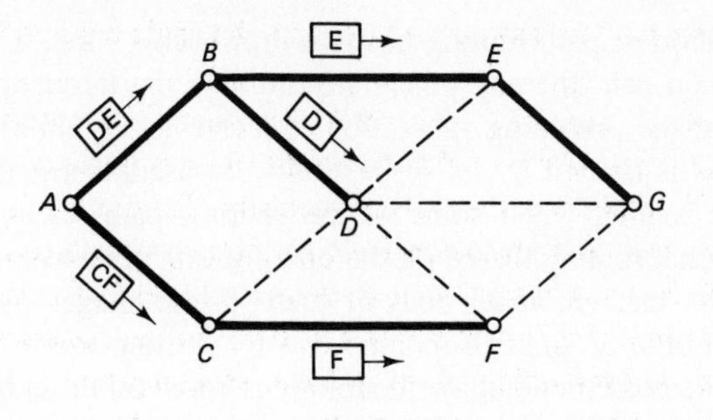

Figure 5.9 Illustration of selective broadcasting using a spanning tree. We assume that each node knows which of the incident spanning tree links lead to any given destination. In this example, node A wishes to broadcast a packet to nodes C, D, E, and F. Node A sends a copy of the packet to each of its incident links that lead to intended destinations, and each copy carries in its header the ID numbers of its intended destinations. Upon reception of a copy, a node strips the ID number from the header (if it is an intended destination) and repeats the process. This eliminates some unnecessary transmissions, for example the transmission on link EG in the figure.

path should contain relatively few and uncongested links, and therefore be desirable for routing.

A more sophisticated alternative is to allow the length of each link to change over time and to depend on the prevailing congestion level of the link. Then a shortest path may adapt to temporary overloads and route packets around points of congestion. This idea is simple but also contains some hidden pitfalls, because by making link lengths dependent on congestion, we introduce a feedback effect between the routing algorithm and the traffic pattern within the network. It will be seen in Section 5.2.5 that this can result in undesirable oscillations. However, we will ignore this possibility for the time being.

An important distributed algorithm for calculating shortest paths to a given destination, known as the *Bellman–Ford method*, has the form

$$D_i := \min_j [d_{ij} + D_j]$$

where D_i is the estimated shortest distance of node i to the destination and d_{ij} is the length of link (i, j). Each node i executes periodically this iteration with the minimum taken over all of its neighbors j. Thus $d_{ij} + D_j$ may be viewed as the estimate of shortest distance from node i to the destination subject to the constraint of going through j, and $\min_j [d_{ij} + D_j]$ may be viewed as the estimate of shortest distance from i to the destination going through the "best" neighbor.

We discuss the Bellman–Ford algorithm in great detail in the next section, where we show that it terminates in a finite number of steps with the correct shortest distances under reasonable assumptions. In practice, the Bellman–Ford iteration can be implemented as an iterative process, that is, as a sequence of communications of the current value of D_j of nodes j to all their neighbors, followed by execution of the shortest distance estimate updates $D_i := \min_j [d_{ij} + D_j]$. A remarkable fact is that this process is very flexible with respect to the choice of initial estimates D_j and the ordering of communications and updates; it works correctly, finding the shortest distances in a finite number of steps, for an essentially arbitrary choice of initial conditions and for an arbitrary order of communications and updates. This allows an asynchronous, real-time distributed implementation of the Bellman-Ford method, which can tolerate changes of the link lengths as the algorithm executes (see Section 5.2.4).

Optimal routing. Shortest path routing has two drawbacks. First, it uses only one path per origin–destination pair, thereby potentially limiting the throughput of the network; see the examples of the preceding subsection. Second, its capability to adapt to changing traffic conditions is limited by its susceptibility to oscillations; this is due to the abrupt traffic shifts resulting when some of the shortest paths change due to changes in link lengths. Optimal routing, based on the optimization of an average delay-like measure of performance, can eliminate both of these disadvantages by splitting any origin–destination pair traffic at strategic points, and by shifting traffic gradually between alternative paths. The corresponding methodology is based on the sophisticated mathematical theory of optimal multicommodity flows, and is discussed in detail in Sections 5.4 to 5.7. Its application to the Codex network is described in Section 5.8.

Hot potato (deflection) routing schemes. In networks where storage space at each node is limited, it may be important to modify the routing algorithm so as to minimize buffer overflow and the attendant loss of packets. The idea here is for nodes to get rid of their stored packets as quickly as possible, transmitting them on whatever link happens to be idle—not necessarily one that brings them closer to their destination.

To provide an example of such a scheme, let us assume that all links of the communication network can be used simultaneously and in both directions, and that each packet carries a destination address and requires unit transmission time on every link. Assume also that packets are transmitted in slots of unit time duration, and that slots are synchronized so that their start and end are simultaneous at all links. In a typical routing scheme, each node, upon reception of a packet destined for a different node, uses table lookup to determine the next link on which the packet should be transmitted; we refer to this link as the *assigned link of the packet*. It is possible that more than one packet with the same assigned link is received by a node during a slot; then at most one of these packets can be transmitted by the node in the subsequent slot, and the remaining packets must be stored by the node in a queue. The storage requirement can be eliminated by modifying the routing scheme so that all these packets are transmitted in the next slot; one of them is transmitted on its assigned link, and the others are transmitted on some other links chosen, possibly at random, from the set of links that are not assigned to any packet received in the previous slot. It can be seen that with this modification, at any node with d incident links, there can be at most d packets received in any one slot, and out of these packets, the ones that are transient (are not destined for the node) will be transmitted along some link (not necessarily their assigned one) in the next slot. Therefore, assuming that at most $d - k$ new packets are generated at a node in a slot where k transient packets are to be transmitted, there will be no queueing, and the storage space at the node need not exceed $2d$ packets. A scheme of this type is used in the Connection Machine, a massively parallel computing system [Hil85]. A variation is obtained when storage space for more than $2d$ packets is provided, and the transmission of packets on links other than the ones assigned to them is allowed only when the available storage space falls below a certain threshold.

The type of method just described was suggested in [Bar64] under the name *hot potato routing*. It has also been known more recently as *deflection routing*. Its drawback is that successive packets of the same origin–destination pair may be received out of order, and some packets may travel on long routes to their destinations; indeed, one should take precautions to ensure that a packet cannot travel on a cycle indefinitely. For some recent analyses of deflection routing, see [BrC91a], [BrC91b], [HaC90], [Haj91], [GrH89], [Max87], [Syz90], [Var90].

Cut-through routing. We have been implicitly assuming thus far that a packet must be fully received at a node before that node starts relaying it to another node. There is, however, an incentive to split a long message into several smaller packets in order to reduce the message's delay on multiple link paths, taking advantage of pipelining (see the discussion in Section 2.5.5). The idea of message splitting, when carried to its extreme, leads to a transmission method called *cut-through routing*, whereby a node can start relaying any portion of a packet to another node without waiting to receive the packet in its entirety. In the absence of additional traffic on the path, the delay of the packet is equal to the transmission time on the slowest link of the path plus the propagation delay, regardless of the number of links on the path. Thus, delay can be reduced by as much as a factor of n on a path with n links.

Note that the nature of cut-through routing is such that error detection and retransmission cannot be done on a link-by-link basis; it must be done on an end-to-end basis. Another issue is that pieces of the same packet may simultaneously be traveling on different links while other pieces are stored at different nodes. To keep the pieces of the packet together, one imposes the requirement that once a packet starts getting transmitted on a link, the link is reserved until all bits of the packet are transmitted. The reservation is made when the link transmits the packet's first bit, and it is released when the link transmits the packet's last bit. Thus, at any time a packet is reserving a portion of its path consisting of several links as shown in Fig. 5.10. Portions of the packet may be stored simultaneously at several nodes either because two or more links on the path have unequal transmission capacities, or because the packet's header had to wait in queue for a link to become available. To avoid the need for complex error recovery procedures, it is generally advisable to ensure that when the packet's first bit is transmitted by a link,

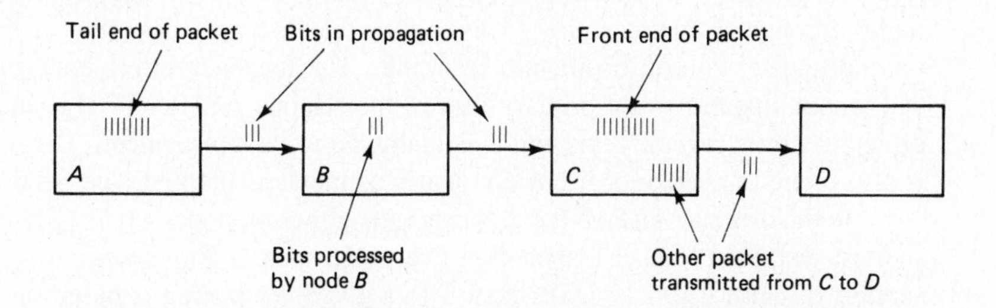

Figure 5.10 Illustration of cut-through routing. Here a packet originates at node A and is destined for node D, after going through nodes B and C. The header of the packet traveled to node B, found the link BC free, and continued to node C without waiting for the rest of the packet to arrive at B. Upon arrival to C, the header found that the link CD was in use, so it was forced to wait at node C. The header reserves links as it travels through them. A link reservation is relinquished only after the last bit of the packet goes through the link, so a packet stays together even though portions of it may be residing in several different nodes.

there is enough buffer space to store the entire packet at the other end of the link. Note that cut-through routing encounters a serious problem if a slow link is followed by a fast link along a packet's path. Then the fast link must effectively be slowed down to the speed of the slow link, and the associated idle fill issues may be hard to resolve.

A variation of cut-through routing, motivated by storage concerns, drops a packet once its first bit encounters a busy link along its path. Thus a packet is either successful in crossing its path without any queueing delay, or else it is discarded and must be retransmitted in its entirety, similar to circuit switching. Indeed, with some thought, it can be seen that this version of cut-through routing is equivalent to a form of circuit switching; the circuit's path is set up by the first bit of the packet, and once the circuit is established, its duration is equal to the packet's transmission time from origin to destination.

ARPANET: An example of datagram routing. The routing algorithm of the ARPANET, first implemented in 1969, has played an important historical role in the development of routing techniques. This was an ambitious, distributed, adaptive algorithm that stimulated considerable research on routing and distributed computation in general. On the other hand, the algorithm had some fundamental flaws that were finally corrected in 1979 when it was replaced by a new version. By that time, however, the original algorithm had been adopted by several other networks.

Shortest path routing is used in both ARPANET algorithms. The length of each link is defined by some measure of traffic congestion on the link, and is updated periodically. Thus, the ARPANET algorithms are adaptive and, since the ARPANET uses datagrams, two successive packets of the same session may follow different routes. This has two undesirable effects. First, packets can arrive at their destination out of order and must, therefore, be put back in order upon arrival. (This can also happen because of the data link protocol of the ARPANET, which uses eight logical channels; see the discussion in Chapter 2.) Second, the ARPANET algorithms are prone to oscillations. This phenomenon is explained in some detail in Section 5.2.5, but basically the idea is that selecting routes through one area of the network increases the lengths of the corresponding links. As a result, at the next routing update the algorithm tends to select routes through different areas. This makes the first area desirable at the subsequent routing update with an oscillation resulting. The feedback effect between link lengths and routing updates was a primary flaw of the original ARPANET algorithm that caused difficulties over several years and eventually led to its replacement. The latest algorithm is also prone to oscillations, but not nearly as much as the first (see Section 5.2.5).

In the original ARPANET algorithm, neighboring nodes exchanged their estimated shortest distances to each destination every 625 msec. The algorithm for updating the shortest distance estimate D_i of node i to a given destination is based on the Bellman–Ford method

$$d_i := \min_j [d_{ij} + D_j]$$

outlined previously and discussed further in Sections 5.2.3 and 5.2.4. Each link length d_{ij} was made dependent on the number of packets waiting in the queue of link (i, j) at the time of the update. Thus, link lengths were changing very rapidly, reflecting statistical traffic fluctuations as well as the effect of routing updates. To stabilize oscillations, a large positive constant was added to the link lengths. Unfortunately, this reduced the sensitivity of the algorithm to traffic congestion.

In the second version of the ARPANET algorithm [MRR80], known as *Shortest Path First* (or SPF for short), the length of each link is calculated by keeping track of the delay of each packet in crossing the link. Each link length is updated periodically every 10 sec, and is taken to be the average packet delay on the link during the preceding 10-sec period. The delay of a packet on a link is defined as the time between the packet's arrival at the link's start node and the time the packet is correctly delivered to the link's end node (propagation delay is also included). Each node monitors the lengths of its outgoing links and broadcasts these lengths throughout the network at least once every 60 sec by using a flooding algorithm (see Section 5.3.1). Upon reception of a new link length, each node recalculates a shortest path from itself to each destination, using an incremental form of Dijkstra's algorithm (see Section 5.2.3). The algorithm is asynchronous in nature in that link length update messages are neither transmitted nor received simultaneously at all nodes. It turns out that this tends to improve the stability properties of the algorithm (see Section 5.2.5). Despite this fact, with increased traffic load, oscillations became serious enough to warrant a revision of the method for calculating link lengths. In the current method, implemented in 1987, the range of possible link lengths and the rate at which these lengths can change have been drastically reduced, with a substantial improvement of the algorithm's dynamic behavior ([KhZ89] and [ZVK89]).

Note that a node is not concerned with calculating routing paths that originate at other nodes even though the information available at each node (network topology plus lengths of all links) is sufficient to compute a shortest path from every origin to every destination. In fact, the routing tables of an ARPANET node contain only the first outgoing link of the shortest path from the node to each destination (see Fig. 5.11). When the node receives a new packet, it checks the packet's destination, consults its routing table, and places the packet in the corresponding outgoing link queue. If all nodes visited by a packet have calculated their routing tables on the basis of the same link length information, it is easily seen that the packet will travel along a shortest path. It is actually possible that a packet will occasionally travel in a loop because of changes in routing tables as the packet is in transit. However, looping of this type is rare, and when it happens, it is unlikely to persist.

The ARPANET routing tables can also be used conveniently for broadcasting a packet to a selected set of destinations. Multiple copies of the packet are transmitted along shortest paths to the destinations, with each copy guided by a header that includes a subset of the destinations. However, care is taken to avoid unnecessary proliferation of the copies. In particular, let D be the set of destinations and for every link l incident to the packet's origin, let D_l be the subset of destinations for which l is the first link on the shortest path from the origin (the origin knows D_l by looking at its routing tables). The origin starts the broadcast by transmitting a copy of the packet on each link l for which D_l is nonempty and includes the destination set D_l in the copy's header. Each node s that subsequently receives a copy of the packet repeats this process with D replaced by the set of remaining destinations (the set of destinations in the copy's header other than s; if s is the only destination on the copy, the node does not relay it further, as shown

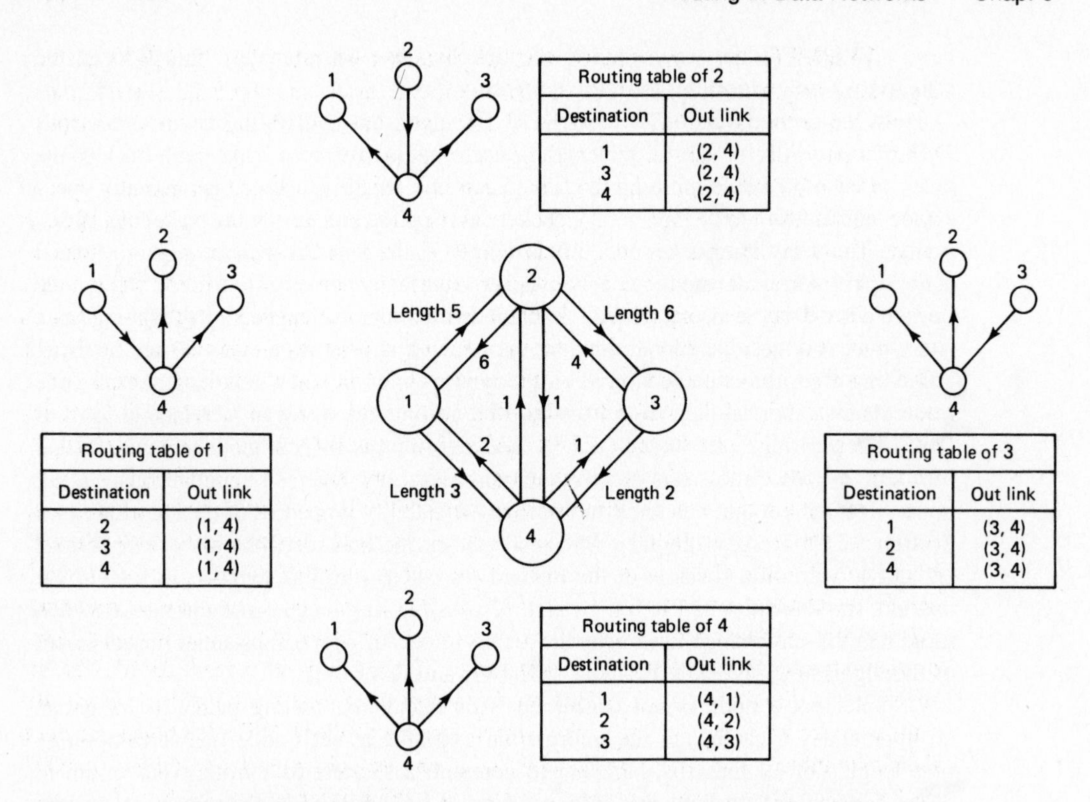

Routing table of 2	
Destination	Out link
1	(2, 4)
3	(2, 4)
4	(2, 4)

Routing table of 1	
Destination	Out link
2	(1, 4)
3	(1, 4)
4	(1, 4)

Routing table of 3	
Destination	Out link
1	(3, 4)
2	(3, 4)
4	(3, 4)

Routing table of 4	
Destination	Out link
1	(4, 1)
2	(4, 2)
3	(4, 3)

Figure 5.11 Routing tables maintained by the nodes in the ARPANET algorithm. Each node calculates a shortest path from itself to each destination, and enters the first link on that path in its routing table.

in Fig. 5.12). It can be seen with a little thought that if the routing tables of the nodes are consistent in the sense that they are based on the same link length information, the packet will be broadcast along the shortest path tree to all its destinations and with a minimal number of transmissions.

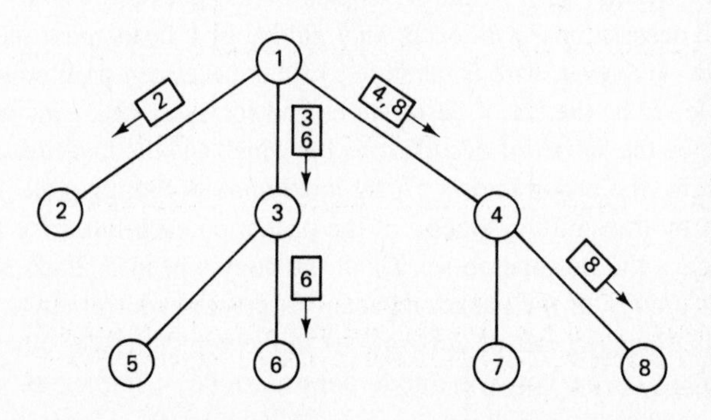

Figure 5.12 Broadcasting a packet to a selected set of destinations using the ARPANET routing tables. In this example, the broadcast is from node 1 to nodes 2, 3, 4, 6, and 8. The figure shows the links on which a copy of the packet is transmitted, the header of each copy, and the shortest path tree from node 1 to all destinations (each node actually maintains only its outgoing links on the tree). The number of transmissions is five. If a separate copy were to be transmitted to each destination, the number of transmissions would be seven.

TYMNET: An example of virtual circuit routing. The TYMNET routing algorithm, originally implemented in 1971, is based on the shortest path method, and is adaptive like the ARPANET algorithms. However, the implementation of the adaptive shortest path idea is very different in the two networks.

In the TYMNET, the routing algorithm is centralized and is operated at a special node called the supervisor. TYMNET uses virtual circuits, so a routing decision is needed only at the time a virtual circuit is set up. A virtual circuit request received by the supervisor is assigned to a shortest path connecting the origin and destination nodes of the virtual circuit. The supervisor maintains a list of current link lengths and does all the shortest path calculations. The length of each link depends on the load carried by the link as well as other factors, such as the type of the link and the type of virtual circuit being routed [Tym81]. While the algorithm can be classified as adaptive, it does not have the potential for oscillations of the ARPANET algorithms. This is due primarily to the use of virtual circuits rather than datagrams, as explained in Section 5.2.5.

Once the supervisor has decided upon the path to be used by a virtual circuit, it informs the nodes on the path and the necessary information is entered in each node's routing tables. These tables are structured as shown in Fig. 5.13. Basically, a virtual circuit is assigned a channel (or port) number on each link it crosses. The routing table at each node matches each virtual circuit's channel number on the incoming link with a channel number on the outgoing link. In the original version of TYMNET (now called TYMNET I), the supervisor maintains an image of the routing tables of all nodes and explicitly reads and writes the tables in the nodes. In the current version (TYMNET II), the nodes maintain their own tables. The supervisor establishes a new virtual circuit by sending a "needle" packet to the origin node. The needle contains routing information and threads its way through the network, building the circuit as it goes, with the user data following behind it.

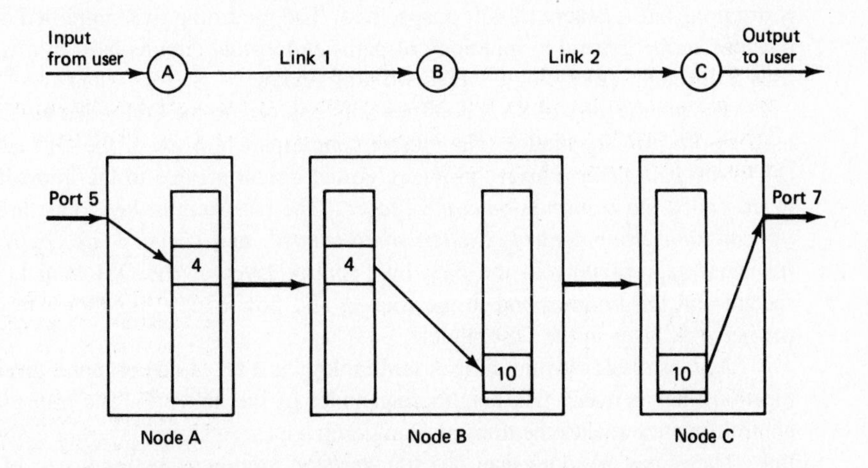

Figure 5.13 Structure of TYMNET routing tables. For the virtual circuit on the path ABC, the routing table at node A maps input port 5 onto channel 4 on link 1. At node B, incoming channel 4 is mapped into outgoing channel 10 on link 2. At node C, incoming channel 10 is mapped into output port 7.

The routing information includes the node sequence of the routing path, and some flags indicating circuit class. When a needle enters a TYMNET II node, its contents are checked. If the circuit terminates at this node, it is assigned to the port connected with the appropriate external site. Otherwise, the next node on the route is checked. If it is unknown (because of a recent link failure or some other error), the circuit is zapped back to its origin. Otherwise, the routing tables are updated and the needle is passed to the next node.

The TYMNET algorithm is well thought out and has performed well over the years. Several of its ideas were implemented in the Codex network, which is a similarly structured network (see Section 5.8). The Codex algorithm, however, is distributed and more sophisticated. A potential concern with the TYMNET algorithm is its vulnerability to failure of the supervisor node. In TYMNET, this is handled by using backup supervisor nodes. A distributed alternative, used in the Codex network, is to broadcast all link lengths to all nodes (as in the ARPANET), and to let each node choose the routing path for each circuit originating at that node. This path may be established by updating the routing tables of the nodes on the path, just as in TYMNET. Another possibility is to let the originating node include the routing path on each packet. This makes the packets longer but may save substantially in processing time at the nodes, when special switching hardware are used. For high-speed networks where the packet volume handled by the nodes is very large, it is essential to use such special hardware and to trade communication efficiency for reduced computation.

Routing in SNA. Routing in IBM's SNA is somewhat unusual in that the method for choosing routes is left partially to the user. SNA provides instead a framework within which a number of routing algorithm choices can be made. Specifically, for every origin–destination pair, SNA provides several paths along which virtual circuits can be built. The rule for assignment of virtual circuits to paths is subject to certain restrictions but is otherwise left unspecified. The preceding oversimplified description is couched in the general terminology of paths and virtual circuits used in this book. SNA uses a different terminology, which we now review.

The architecture of SNA does not fully conform to the OSI seven-layer architecture used as our primary model. The closest counterpart in SNA of the OSI network layer, called the *path control layer*, provides virtual circuit service to the immediately higher layer, called the *transmission control layer*. The path control layer has three functions: *transmission group control*, *explicit route control*, and *virtual route control*. The first fits most appropriately in the data link control layer of the OSI model, whereas the second and third correspond to the routing and flow control functions, respectively, of the network layer in the OSI model.

A transmission group in SNA terminology is a set of one or more physical communication lines between two neighboring nodes of the network. The transmission group control function makes the transmission group appear to higher layers as a single physical link. There may be more than one transmission group connecting a pair of nodes. The protocol places incoming packets on a queue, sends them out on the next available physical link, and sequences them if they arrive out of order. Resequencing of packets is done at every node, unlike the ARPANET, which resequences only at the destination node.

An explicit route in SNA terminology is what we have been referring to as a path within a subnet. Thus, an explicit route is simply a sequence of transmission groups providing a physical path between an origin and a destination node. There are several explicit routes provided (up to eight in SNA 4.2) for each origin–destination pair, which can be modified only under supervisory control. The method for choosing these explicit routes is left to the network manager. Each node stores the next node for each explicit route in its routing tables, and each packet carries an explicit route number. Therefore, when a packet arrives at a node, the next node on the packet's path (if any) is determined by a simple table lookup.

A virtual route in SNA terminology is essentially what we have been calling a virtual circuit. A user pair conversation (or session) uses only one virtual route. However, a virtual route can carry multiple sessions. Each virtual route belongs to a priority (or service) class. Each transmission group can serve up to three service classes, each with up to eight virtual routes, for a maximum of 24 virtual routes. When a user requests a session through a subnet entry node, a service class is specified on the basis of which the entry node will attempt to establish the session over one of its virtual routes. If all the communication lines of a transmission group fail, every session using that transmission group must be rerouted. If no alternative virtual route can be found, the session is aborted. When new transmission groups are established, all nodes are notified via special control packets, so each node has a copy of the network topology and knows which explicit routes are available and which are not.

Routing in circuit switching networks. Routing in circuit switching networks is conceptually not much different than routing in virtual circuit networks. In both cases a path is established at the time the circuit is set up, and all subsequent communication follows that path. In both cases the circuit will be blocked when the number of existing circuits on some link of the path reaches a certain threshold, although the choice of threshold may be based on different considerations in each case.

Despite these similarities, in practice, telephone networks have used quite different routing methods than virtual circuit networks. A common routing method in circuit switching networks, known as *hierarchical routing*, has been to use a fixed first-choice route, and to provide for one or more alternative routes, to be tried in a hierarchical order, for the case where the first-choice route is blocked. However, the differences in routing philosophy between circuit switching and virtual circuit networks have narrowed recently. In particular, there have been implementations of adaptive routing in telephone networks, whereby the path used by a call is chosen from several alternative paths on the basis of a measure of congestion (*e.g.*, the number of circuits these paths carry). We refer to [Ash90], [KeC90], [Kri90], [ReC90], and [WaO90] for detailed descriptions and discussion of such implementations.

5.1.3 Interconnected Network Routing: An Overview

As networks started to proliferate, it became necessary to interconnect them using various interface devices. In the case of wide-area networks, the interfaces are called *gateways*, and usually perform fairly sophisticated network layer tasks, including routing. Gateways typically operate at the internet sublayer of the network layer. A number of engineering and economic factors, to be discussed shortly, motivated the interconnection of local area networks (LANs) at the MAC sublayer. The devices used for interconnection, known as *bridges*, perform a primitive form of routing. LANs can also be connected with each other or with wide-area networks using more sophisticated devices called *routers*. These provide fairly advanced routing functions at the network layer, possibly including adaptive and multiple-path routing. In this section we survey various routing schemes for interconnected networks, with a special emphasis on bridged LANs.

Conceptually, one may view an interconnected network of subnetworks as a single large network, where the interfaces (gateways, bridges, or routers) are additional nodes, each connected to two or more of the constituent subnetworks (see Fig. 5.14). Routing based on this conceptual model is called *nonhierarchical* and is similar to routing in a wide area network.

It is also possible to adopt a coarser-grain view of an interconnected network of subnetworks, whereby each subnetwork is considered to be a single node connected to interface nodes. Routing based on this model is called *hierarchical*, and consists of two levels. At the lower level, there is a separate routing algorithm for each subnetwork that handles local subnetwork traffic and also directs internetwork traffic to the subnetwork's interfaces with other subnetworks. At the higher level, there is a routing algorithm that determines the sequence of subnetworks and interface nodes that each internetwork packet must follow (see Fig. 5.15). Basically, the low-level algorithms handle "local" packets, and the high-level algorithm delivers "long-distance" packets to the appropriate low-level algorithm. An example of a hierarchical network is the Internet discussed in Section 2.8.4. Hierarchical routing is generally recommended for large internetworks; for example, telephone networks are typically organized hierarchically. We contrast the two approaches to internetworking by means of an example.

Example 5.3 Hierarchical versus Nonhierarchical Routing

Consider a country with several cities each having its own data network, and suppose that the city networks are connected into a national network using gateways. Each packet must have an internetwork destination address, consisting of a city address (or city ID number) and a local address (an ID number for the destination node within the destination city). Assume for concreteness that we want to use an ARPANET-like algorithm, where each

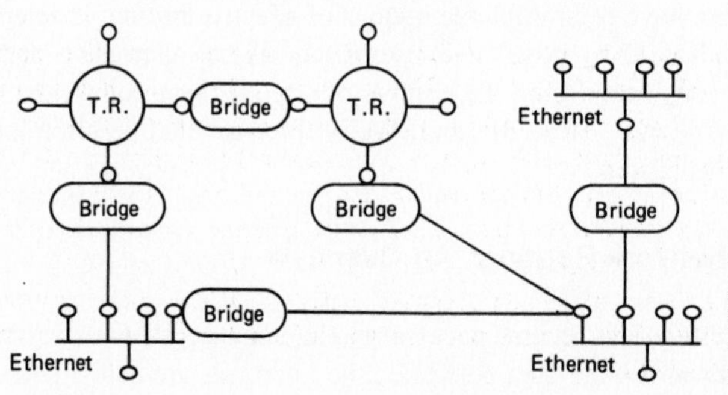

Figure 5.14 Conceptual view of an interconnected network of subnetworks. In a nonhierarchical approach to routing the internetwork is viewed as a single large network.

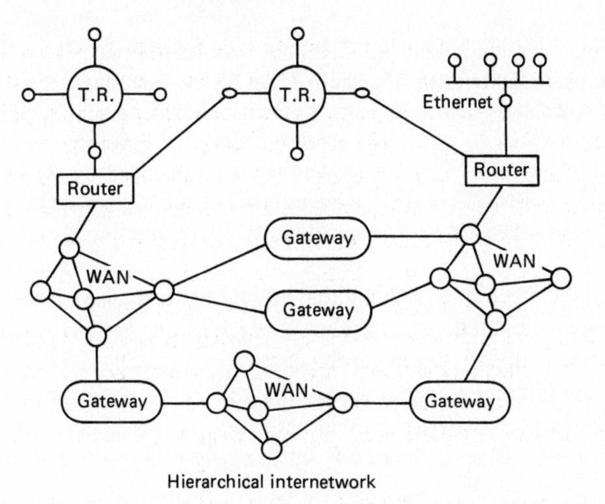

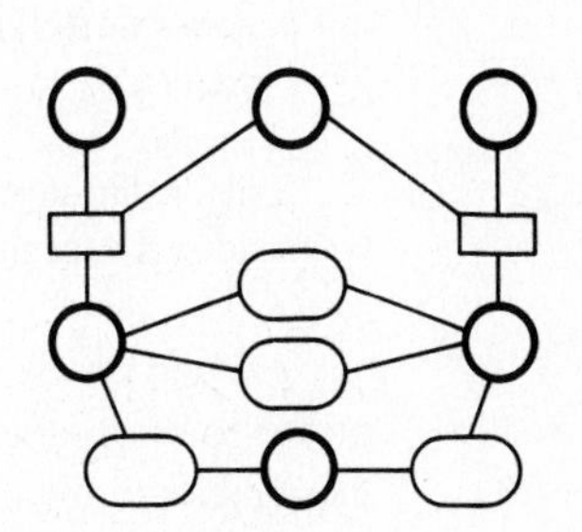

Figure 5.15 Conceptual view of hierarchical routing in an interconnected network of subnetworks. There is a lower-level algorithm that handles routing within each subnetwork and a higher-level algorithm that handles routing between subnetworks.

node has a unique entry in its routing tables for every destination (the next outgoing link on its shortest path to the destination).

In a nonhierachical routing approach, there is no distinction between the city networks; each node of each city must have a separate entry in its routing tables for every node in every other city. Thus the size of the routing tables is equal to the total number of nodes in the national network. Furthermore, a shortest path calculation to update the routing tables must be performed for the entire national network. Thus we see the main drawbacks of the nonhierarchical approach: the size of the routing tables and the calculations to update them may become excessive.

In a hierarchical approach, there is a lower-level (local) ARPANET-like routing algorithm within each city, and a higher-level ARPANET-like routing algorithm between cities. The local routing algorithm of a city operates only at the nodes of that city. Every node of a city has a routing table entry for every node of the same city and a routing table entry for every other city. The higher-level algorithm provides for routing between cities and operates only at the gateways. Each gateway has a routing table entry for every destination city. The latter entry is the next city network to which an internetwork packet must be routed on its way to its destination.

To see how this algorithm works, consider a packet generated at a node of some city A and destined for a node of a different city. The origin node checks the packet's address, consults its routing table entry for the packet's destination city, and sends the packet on the corresponding outgoing link. Each node of city A subsequently visited by the packet does the same, until the packet arrives at an appropriate gateway. At this point, the higher-level routing algorithm takes over and determines (based on the gateway's routing table) the next city network within which the packet will travel. The packet is then guided by the latter network's local routing algorithm, and the process is repeated until the packet arrives at the destination city's network. Then the packet is routed to its destination node by the local routing algorithm of the destination city.

The routing tables of a city network algorithm may be generated or updated using an adaptive shortest-path algorithm such as the ARPANET SPF algorithm. To do this, the nodes of the city network must have a destination gateway for every other city (the gateway connected to the city network to which packets destined for the other city will be guided by the local routing algorithm). This information must be provided by the higher-level algorithm that operates at the gateways. The higher-level algorithm may itself be static or adaptive, adjusting its intercity routes to temporary congestion patterns.

It can be seen from the preceding example that in the hierarchical approach, the size of the routing tables and the calculations to update them are considerably reduced over the nonhierarchical approach. In fact, the hierarchical approach may be recommended for a single large network (no internetworking) to reduce the routing algorithm's overhead and storage. The disadvantage of the hierarchical approach is that it may not result in routing that is optimal for the entire internetwork with respect to a shortest path or some other criterion.

Note that hierarchical networks may be organized in more than two levels. For example, consider an international network interconnecting several national networks. In a three-level hierarchy, there could be a routing algorithm between countries at the highest level, a routing algorithm between cities of each country at the middle level, and a routing algorithm within each city at the lowest level.

Bridged local area networks. As we discussed in Chapter 4, the performance of a local area network tends to degrade as its number of users increases. In the case of Ethernet-based LANs the performance also degrades as the ratio of propagation delay to transmission capacity increases. A remedy is to connect several LANs so that a large geographical area with many users can be covered without a severe performance penalty.

Routing for interconnected LANs has a special character because LANs operate in a "promiscuous mode," whereby all nodes, including the interface nodes connected to a LAN, hear all the traffic broadcast within the LAN. Furthermore, in a LAN internetwork, interface nodes must be fast and inexpensive, consistent with the characteristics of the LANs connected. This has motivated the use of simple, fast, and relatively inexpensive bridges. A bridge listens to all the packets transmitted within two or more LANs through connections known as *ports*. It also selectively relays some packets from one LAN to another. In this way, packets can travel from one user to another, going through a sequence of LANs and bridges.

An important characteristic of bridges is that they connect LANs at the MAC sublayer. The resulting internetwork is, by necessity, nonhierarchical; it may be viewed as a single network, referred to as the *extended LAN*. To alleviate the disadvantages of nonhierarchical routing mentioned earlier, bridges use some special routing techniques, the two most prominent of which are known as *spanning tree routing* and *source routing*. Both of these schemes assume that each station in the extended LAN has a unique ID number known through a directory that can be accessed by all stations. A station's ID number need not provide information regarding the LAN to which the station is connected. Also, the location of a station is not assumed known, thus allowing stations to be turned on and off and to be moved from one LAN to another at will. To appreciate the difficulties of this, think of a telephone company trying to operate a network where customers can change their residence without notifying the company, while maintaining their phone numbers. We now describe the two major routing appoaches.

Spanning tree routing in bridged local area networks. In the spanning tree method, the key data structure is a spanning tree, which is used by a bridge to determine whether and to which LAN it should relay a packet (see Fig. 5.16). The spanning tree consists of a subset of ports that connect all LANs into a network with no cycles. An important property of a spanning tree is that each of its links separates the tree in two parts; that is, each node of the tree lies on exactly one "side" of the link. From this it is seen that there is a unique path from every LAN to every other LAN on the spanning tree. This path is used for communication of all packets between the LANs. The ports lying on the spanning tree are called *designated* or *active* ports. The spanning tree may need to be modified due to bridge failures, but we postpone a discussion of this issue for later, assuming for the time being that the spanning tree is fixed.

To understand how routing works, note that each bridge can communicate with any given station through a unique active port (see Fig. 5.16). Therefore, to implement spanning tree routing, it is sufficient to maintain for each active port a list of the stations with which the port communicates; this is known as the *forwarding database* (FDB) of the port. Assuming that the FDBs at each bridge are complete (they contain all stations), routing would be implemented as follows: If a packet is heard on the LAN corresponding to an active port A of a bridge, then if the destination ID of the packet is included in the FDB of an active bridge port B different from A, the packet is transmitted on the LAN connected to B; otherwise, the packet is discarded by the bridge (see Fig. 5.17).

Unfortunately, the routing method just described does not quite work because the FDBs of the active ports are not complete at all times. The reason is that stations may be turned on and off or change location, and also the spanning tree may change.

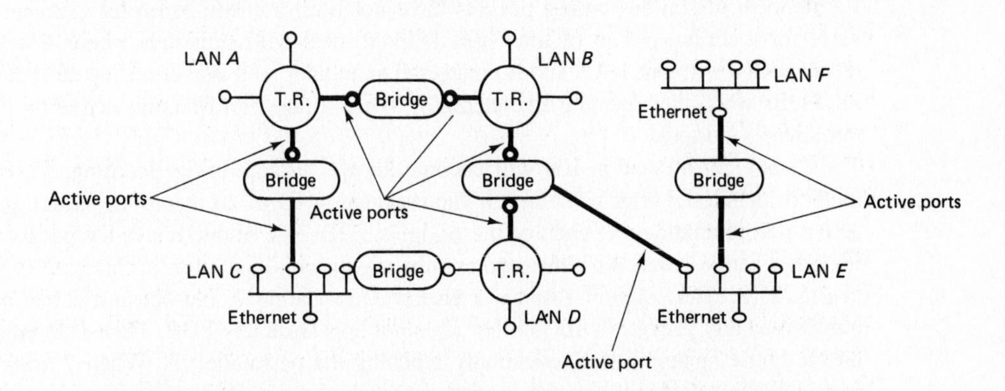

Figure 5.16 Illustration of spanning tree in a network of LANs interconnected with bridges. The spanning tree consists of a set of bridge ports (called active), which connect all LANs into a network without cycles. Only active ports are used for routing packets and thus there is a unique route connecting a pair of LANs. For example, a packet of a station connected to LAN C will go to a station in LAN E through LANs A and B.

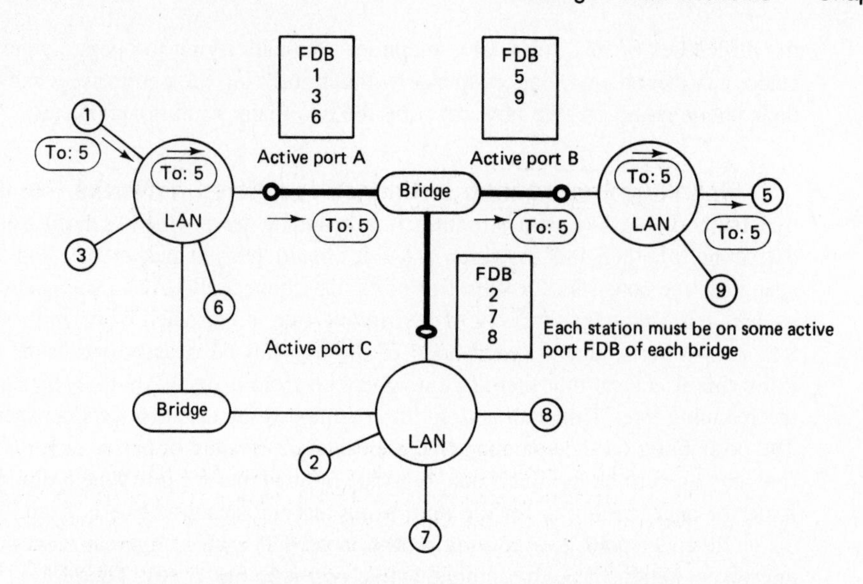

Figure 5.17 Using forwarding databases (FDBs) to effect spanning tree routing. Each active port maintains an FDB, that is, a list of the stations with which it communicates. An active port, say A, checks the destination of each packet transmitted within its associated LAN. If the destination appears in the FDB of another active port of the same bridge, say B, the bridge broadcasts the packet on the LAN connected with port B.

For this reason, the FDBs are constantly being updated through a process known as *bridge learning*. Each active port adds to its FDB the source ID numbers of the packets transmitted on the corresponding LAN. Also each active port deletes from its FDB the ID numbers of stations whose packets have not been transmitted on the corresponding LAN for a certain period of time; this helps to deal with situations where a station is disconnected with one LAN and is connected to another, and also cleans up automatically the FDBs when the spanning tree changes. When a port first becomes active its FDB is initialized to empty.

Since not all station IDs will become known through bridge learning, the routing method includes a search feature. If the destination ID of an incoming packet at some active port is unknown at each of the bridge's FDBs, the bridge relays the packet on all the other active ports. An illustration of the algorithm's operation in this case is shown in Fig. 5.18. Here station 1 sends a packet P to station 2, but because 2 has not yet transmitted any packet, its ID number does not appear on any FDB. Then P is broadcast on the entire spanning tree, eventually reaching the destination 2. When 2 answers P with a packet R of its own, this packet will follow the unique spanning tree route to 1 (because 1's ID number was entered in all of the appropriate FDBs of active ports of the spanning tree when P was broadcast through the spanning tree). As a result, 2's ID number will be entered in the FDBs of the active ports along the route that connects 2 with 1, and a subsequent packet from 1 to 2 will follow this route.

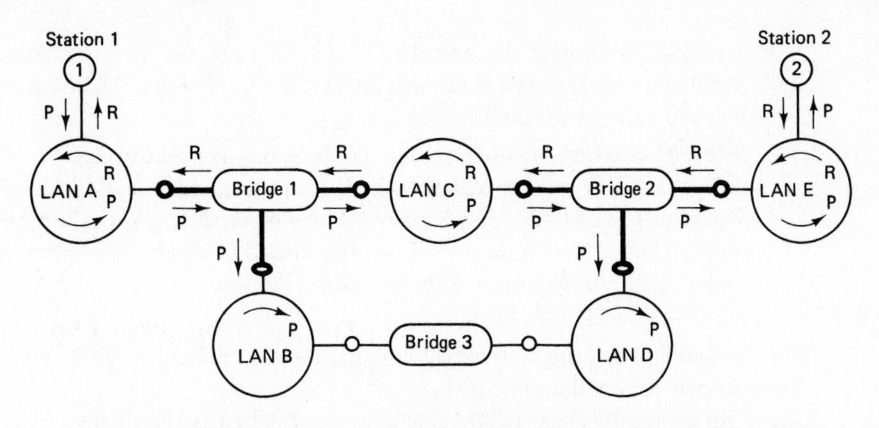

Figure 5.18 Illustration of bridge learning and FDB updating. Station 1 sends a packet P to station 2, but the ID of station 2 does not appear in the FDB of any active port of bridge 1 (because, for example, station 2 has not transmitted any packet for a long time). Packet P is broadcast into LANs B and C by bridge 1, and then it is broadcast into LANs D and E by bridge 2. In the process the ID of node 1 is entered in the left-side FDBs of bridges 1 and 2 (if not already there). Station 2 replies with a packet R, which goes directly to its destination station 1 through bridges 2 and 1, because the ID number of station 1 appears in the left-side FDBs of these bridges.

The method by which the spanning tree is updated in the event of a bridge or LAN failure or repair remains to be discussed. There are several algorithms for constructing spanning trees, some of which are discussed in Section 5.2. What is needed here, however, is a distributed algorithm that can work in the face of multiple and unpredictable failures and repairs. A relatively simple possibility is to use a spanning tree of shortest paths from every LAN and bridge to a particular destination (or root). We will see in Section 5.2 that the Bellman–Ford algorithm is particularly well suited for this purpose. There is still, however, an extra difficulty, namely how to get all bridges to agree on the root and what to do when the root itself fails. Such issues involve subtle difficulties, which are unfortunately generic in distributed algorithms dealing with component failures. The difficulty can be resolved by numbering the bridges and by designating as root of the tree the bridge with the smallest ID number. Each bridge keeps track of the smallest bridge ID number that it knows of. A bridge must periodically initiate a spanning tree construction algorithm if it views itself as the root based on its own ID number and the ID numbers of other bridges it is aware of. Also, each bridge ignores messages from bridges other than its current root. To cope with the situation where the root fails, bridges may broadcast periodically special "hello" messages to other bridges. These messages may also carry additional information relating to the spanning tree construction and updating algorithm. (See [Bac88] for a description of the corresponding IEEE 802.1 MAC Bridge Draft Standard.)

Source routing in bridged local area networks. Source routing is an alternative to the spanning tree method for bridged LANs. The idea is to discover a route between two stations A and B wishing to communicate and then to include the route on the header of each packet exchanged between A and B. There are two phases here:

1. Locating B through the extended LAN. This can be done by broadcasting an exploratory packet from A through the extended LAN (either flooding or a spanning tree can be used for this).
2. Selecting a route out of the many possible that connect A and B. One way of doing this is to send a packet from B to A along each possible loop-free route. The sequence of LANs and bridges is recorded on the packet as it travels along its route. Thus A gets to know all possible routes to B and can select one of these routes using, for example, a shortest path criterion.

Note the similarity of source routing with virtual circuit routing; in both cases, all packets of a user pair follow the same path.

The main advantage of the spanning tree method is that it does not require any cooperation of the user/stations in the routing algorithm. By contrast, in source routing, the origin station must select a route along which to establish a virtual circuit. Thus the software of the stations must provide for this capability. This is unnecessary in spanning tree routing, where the routing process is transparent to the stations.

The main advantage of source routing is that with each packet having its route stamped on it, there is no need for routing table lookup at the bridges. This simplifies packet processing and may result in bridges with higher packet throughput. Another advantage of source routing is that it may use the available network transmission capacity more efficiently. While spanning tree routing uses a single possibly nonshortest path for each origin–destination pair, with source routing, shortest path and multiple-path routing can be used. Figure 5.19 shows an example where spanning tree–based routing is inefficient.

We finally mention that both spanning tree and source routing suffer from a common disadvantage. They must often broadcast packets throughout the network either to update FDBs or to establish new connections between user pairs. These broadcasts are needed

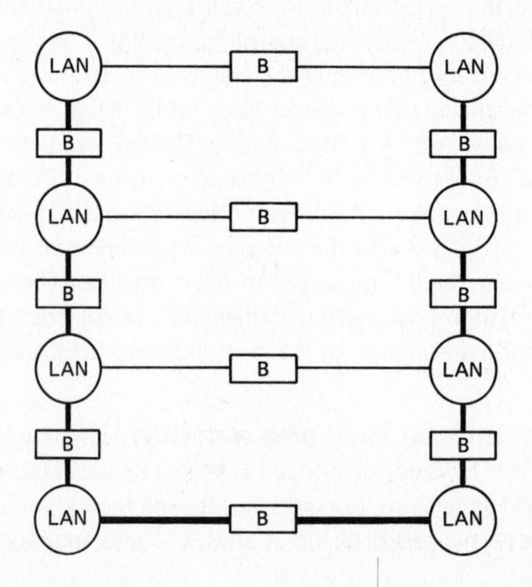

Figure 5.19 Example of inefficient spanning tree routing. The routing paths are good for origin LANs near the bottom of the figure but poor for origin and destination LANs near the top of the figure.

to allow dynamic relocation of the stations while keeping the bridge hardware simple. They may, however, become a source of congestion.

5.2 NETWORK ALGORITHMS AND SHORTEST PATH ROUTING

Routing methods involve the use of a number of simple graph-theoretic problems such as the shortest path problem, discussed in the preceding section. In this section we consider shortest path routing in detail together with related problems of interest. We start by introducing some basic graph-theoretic notions.

5.2.1 Undirected Graphs

We define a *graph*, $G = (\mathcal{N}, \mathcal{A})$, to be a finite nonempty set $\mathcal{N}$ of *nodes* and a collection $\mathcal{A}$ of pairs of distinct nodes from $\mathcal{N}$. Each pair of nodes in $\mathcal{A}$ is called an *arc*. Several examples of graphs are given in Fig. 5.20. The major purpose of the formal definition $G = (\mathcal{N}, \mathcal{A})$ is to stress that the location of the nodes and the shapes of the arcs in a pictorial representation are totally irrelevant. If n_1 and n_2 are nodes in a graph and (n_1, n_2) is an arc, this arc is said to be *incident* on n_1 and n_2. Some authors define graphs so as to allow arcs to have both ends incident on the same node, but we have intentionally disallowed such loops. We have also disallowed multiple arcs between the same pair of nodes.

A *walk* in a graph G is a sequence of nodes $(n_1, n_2, \ldots, n_\ell)$ of nodes such that each of the pairs (n_1, n_2), (n_2, n_3), ..., $(n_{\ell-1}, n_\ell)$ are arcs of G. A walk with no repeated nodes is a *path*. A walk $(n_1, ..., n_\ell)$ with $n_1 = n_\ell$, $\ell > 3$, and no repeated nodes other than $n_1 = n_\ell$ is called a *cycle*. These definitions are illustrated in Fig. 5.21.

We say that a graph is *connected* if for each node i there is a path $(i = n_1, n_2, \ldots, n_\ell = j)$ to each other node j. Note that the graphs in Fig. 5.20(a) and (c) are connected, whereas the graph in Fig. 5.20(b) is not connected. We spot an absence of connectivity in a graph by seeing two sets of nodes with no arcs between them. The following lemma captures this insight. The proof is almost immediate and is left as an exercise.

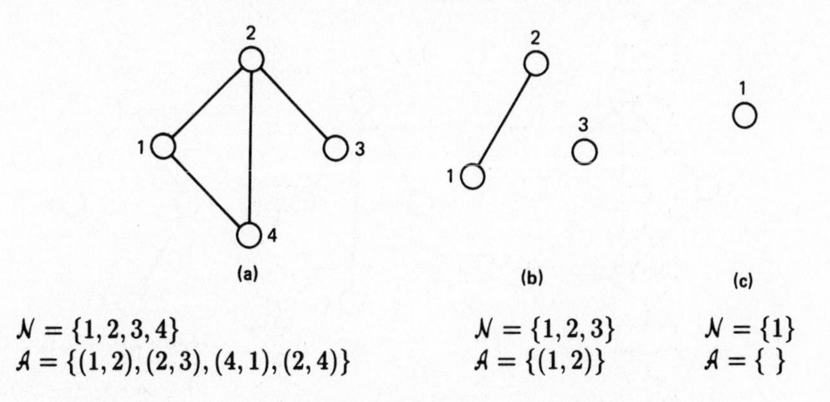

Figure 5.20 Examples of graphs with a set of nodes $\mathcal{N}$ and a set of arcs $\mathcal{A}$.

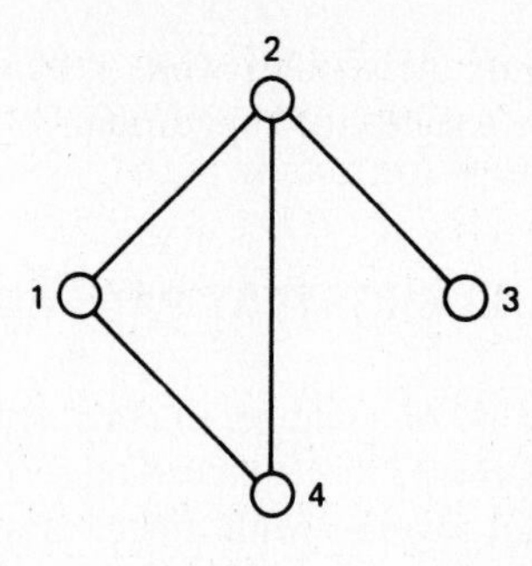

Figure 5.21 Illustration of walks, paths, and cycles. The sequences (1,4,2,3), (1,4,2,1), (1,4,2,1,4,1), (2,3,2), and (2) are all walks. The sequences (1,4,2,3) and (2) are each paths; and (1,4,2,1) is a cycle. Note that (2) and (2,3,2) are not considered cycles.

Lemma. Let G be a connected graph and let S be any nonempty strict subset of the set of nodes N. Then at least one arc (i, j) exists with $i \in S$ and $j \notin S$.

We say that $G' = (\mathcal{N}', \mathcal{A}')$ is a *subgraph* of $G = (\mathcal{N}, \mathcal{A})$ if G' is a graph, $\mathcal{N}' \subset \mathcal{N}$, and $\mathcal{A}' \subset \mathcal{A}$. For example, the last three graphs in Fig. 5.22 are subgraphs of the first.

A *tree* is a connected graph that contains no cycles. A *spanning tree* of a graph G is a subgraph of G that is a tree and that includes all the nodes of G. For example, the subgraphs in Fig. 5.22(b) and (c) are spanning trees of the graph in Fig. 5.22(a). The following simple algorithm constructs a spanning tree of an arbitrary connected graph $G = (\mathcal{N}, \mathcal{A})$:

1. Let n be an arbitrary node in $\mathcal{N}$. Let $\mathcal{N}' = \{n\}$, and let $\mathcal{A}'$ be the empty set $\emptyset$.
2. If $\mathcal{N}' = \mathcal{N}$, then stop [$G' = (\mathcal{N}', \mathcal{A}')$ is a spanning tree]; else go to step 3.
3. Let $(i, j) \in \mathcal{A}$ be an arc with $i \in \mathcal{N}'$, $j \in \mathcal{N} - \mathcal{N}'$. Update $\mathcal{N}'$ and $\mathcal{A}'$ by

$$\mathcal{N}' := \mathcal{N}' \cup \{j\}$$
$$\mathcal{A}' := \mathcal{A}' \cup \{(i, j)\}$$

Go to step 2.

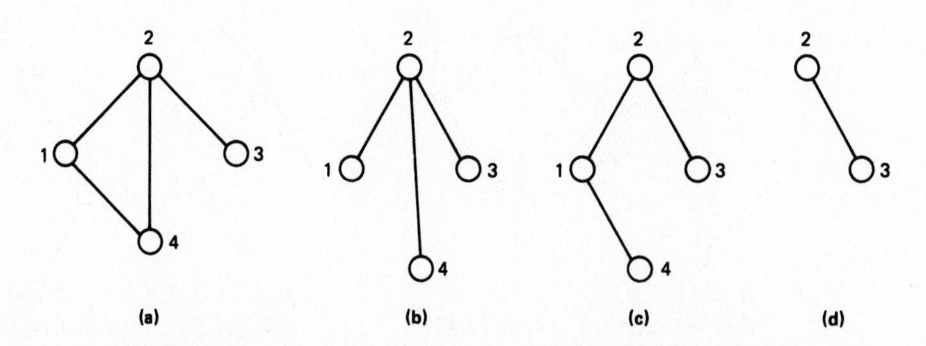

Figure 5.22 A graph (a) and three subgraphs (b), (c), and (d). Subgraphs (b) and (c) are spanning trees.

To see why the algorithm works, note that step 3 is entered only when $\mathcal{N}'$ is a proper subset of $\mathcal{N}$, so that the earlier lemma guarantees the existence of the arc (i, j). We use induction on successive executions of step 3 to see that $G' = (\mathcal{N}', \mathcal{A}')$ is always a tree. Initially, $G' = (\{n\}, \emptyset)$ is trivially a tree, so assume that $G' = (\mathcal{N}', \mathcal{A}')$ is a tree before the update of step 3. This ensures that there is a path between each pair of nodes in $\mathcal{N}'$ using arcs of $\mathcal{A}'$. After adding node j and arc (i, j), each node has a path to j simply by adding j to the path to i, and similarly j has a path to each other node. Finally, node j cannot be in any cycles since (i, j) is the only arc of G' incident to j, and there are no cycles not including j by the inductive hypothesis. Figure 5.23 shows G' after each execution of step 3 for one possible choice of arcs.

Observe that the algorithm starts with a subgraph of one node and zero arcs and adds one node and one arc on each execution of step 3. This means that the spanning tree, G', resulting from the algorithm always has N nodes, where N is the number of nodes in G, and $N - 1$ arcs. Since G' is a subgraph of G, the number of arcs, A, in G must satisfy $A \geq N - 1$; this is true for every connected graph G. Next, assume that $A = N - 1$. This means that the algorithm uses all arcs of G in the spanning tree G', so that $G = G'$, and G must be a tree itself. Finally, if $A \geq N$, then G contains at least one arc (i, j) not in the spanning tree G' generated by the algorithm. Letting $(j, \ldots, i)$ be the path from j to i in G', it is seen that $(i, j, \ldots, i)$ is a cycle in G and G cannot be a tree. The following proposition summarizes these results:

Proposition. Let G be a connected graph with N nodes and A arcs. Then:

1. G contains a spanning tree.
2. $A \geq N - 1$.
3. G is a tree if and only if $A = N - 1$.

Figure 5.24 shows why connectedness is a necessary assumption for the last part of the proposition.

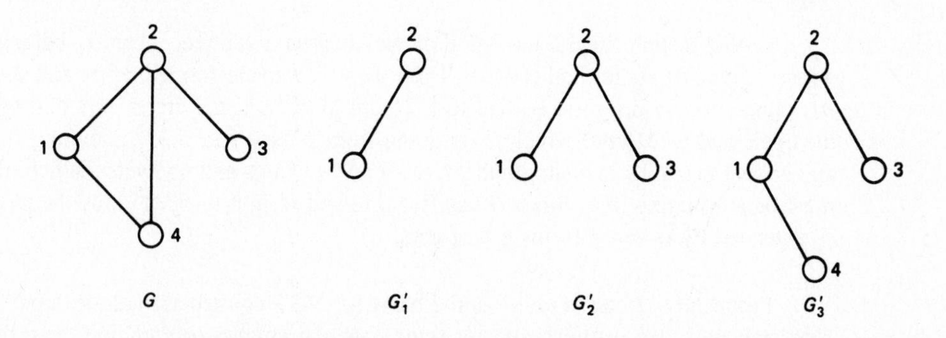

Figure 5.23 Algorithm for constructing a spanning tree of a given graph G.

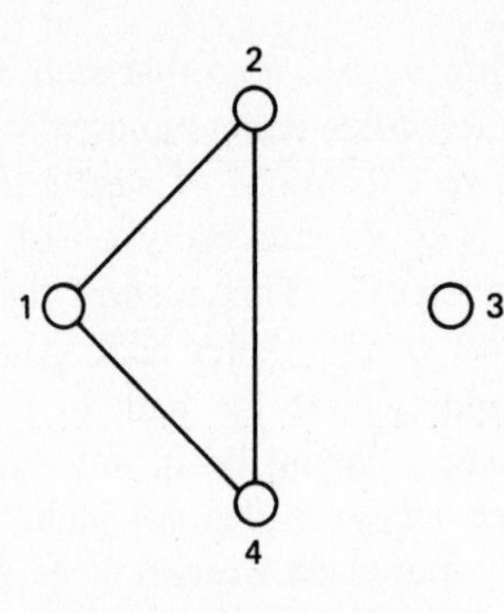

Figure 5.24 Graph with $A = N - 1$, which is both disconnected and contains a cycle.

5.2.2 Minimum Weight Spanning Trees

It is possible to construct a spanning tree rooted at some node by sending a packet from the node to all its neighbors; the neighbors must send the packet to their neighbors, and so on, with all nodes marking the transmitter of the first packet they receive as their "father" on the tree, as described in Fig. 5.8. However, such a tree has no special properties. If the communication cost or delay of different links should be taken into account in constructing a spanning tree, we may consider using a minimum weight spanning tree, which we now define.

Consider a connected graph $G = (\mathcal{N}, \mathcal{A})$ with N nodes and A arcs, and a weight w_{ij} for each arc $(i, j) \in \mathcal{A}$. A *minimum weight spanning tree* (MST for short) is a spanning tree with minimum sum of arc weights. An arc weight represents the communication cost of a message along the arc in either direction, and the total spanning tree weight represents the cost of broadcasting a message to all nodes along the spanning tree.

Any subtree (*i.e.*, a subgraph that is a tree) of an MST will be called a *fragment*. Note that a node by itself is considered a fragment. An arc having one node in a fragment and the other node not in this fragment is called an *outgoing arc* from the fragment. The following proposition is of central importance for MST algorithms.

Proposition 1. Given a fragment F, let $\alpha = (i, j)$ be a minimum weight outgoing arc from F, where the node j is not in F. Then F, extended by arc α and node j, is a fragment.

Proof: Denote by M the MST of which F is a subtree. If arc α belongs to M, we are done, so assume otherwise. Then there is a cycle formed by α and the arcs of M. Since node j does not belong to F, there must be some arc $\beta \neq \alpha$ that belongs to the cycle and to M, and which is outgoing from F (see Fig. 5.25). Deleting β from M and adding α results in a subgraph M' with $(N - 1)$ arcs and no cycles which, therefore, must be a spanning tree. Since α has less or equal weight to β, M' must be an MST, so F extended by α and j forms a fragment. **Q.E.D.**

Proposition 1 can be used as the basis for MST construction algorithms. The idea is to start with one or more disjoint fragments and enlarge or combine them by adding minimum weight outgoing arcs. One method (the Prim–Dijkstra algorithm) starts with an

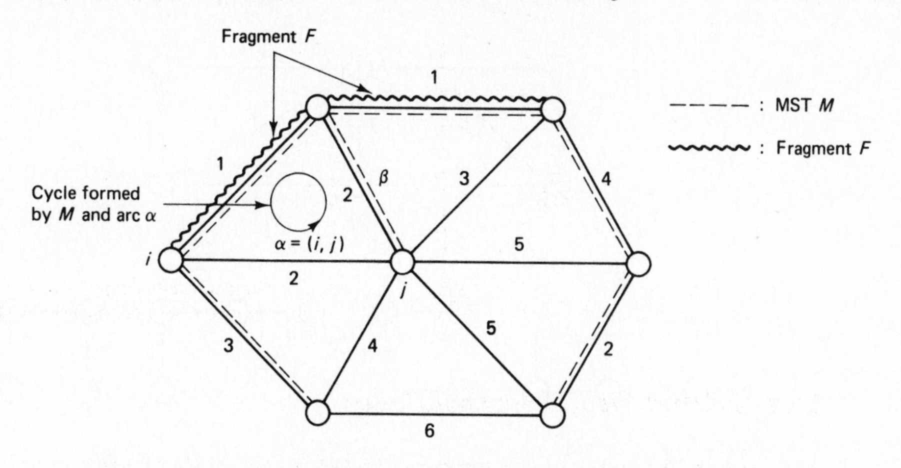

Figure 5.25 Proof of Proposition 1. The numbers next to the arcs are the arc weights. F is a fragment which is a subtree of an MST M. Let α be a minimum weight outgoing arc from F not belonging to M. Let $\beta \neq \alpha$ be an arc that is outgoing from F and simultaneously belongs to M and to the cycle formed by α and M. Deleting β from M and adding α in its place forms another MST M'. When F is extended by α, it forms a fragment.

arbitrarily selected single node as a fragment and enlarges the fragment by successively adding a minimum weight outgoing arc. Another method (Kruskal's algorithm) starts with each node being a single node fragment; it then successively combines two of the fragments by using the arc that has minimum weight over all arcs that when added to the current set of fragments do not form a cycle (see Fig. 5.26). Both of these algorithms terminate in $N - 1$ iterations.

Kruskal's algorithm proceeds by building up simultaneously several fragments that eventually join into an MST; however, only one arc at a time is added to the current set of fragments. Proposition 1 suggests the possibility of adding a minimum weight outgoing arc simultaneously to each fragment in a distributed algorithmic manner. This is possible if there is a unique MST, as we now discuss.

A distributed algorithm for constructing the MST in a graph with a unique MST is as follows. Start with a set of fragments (these can be the nodes by themselves, for example). Each fragment determines its minimum weight outgoing arc, adds it to itself, and informs the fragment that lies at the other end of this arc. It can be seen that as long as the arc along which two fragments join is indeed a minimum weight outgoing arc for some fragment, the algorithm maintains a set of fragments of the MST at all times, and no cycle is ever formed. Furthermore, new arcs will be added until there are no further outgoing arcs and there is only one fragment (by necessity the MST). Therefore, the algorithm cannot stop short of finding the MST. Indeed, for the algorithm to work correctly, it is not necessary that the procedure for arc addition be synchronized for all fragments. What is needed, however, is a scheme for the nodes and arcs of a fragment to determine the minimum weight outgoing arc. There are a number of possibilities along these lines, but it is of interest to construct schemes that accomplish this with low

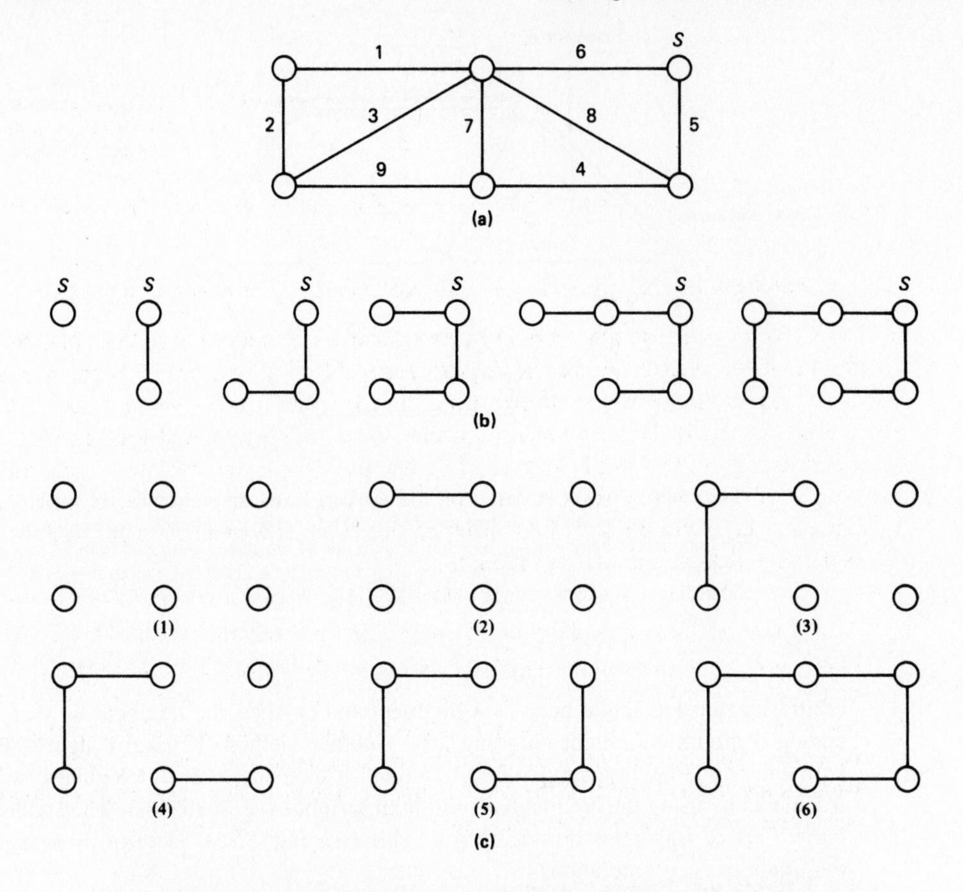

Figure 5.26 Minimum weight spanning tree construction. (a) Graph with arc weights as indicated. (b) Successive iterations of the Prim–Dijkstra algorithm. The starting fragment consists of the single node marked S. The fragment is successively extended by adding a minimum weight outgoing arc. (c) Successive iterations of the Kruskal algorithm. The algorithm starts with all nodes as fragments. At each iteration, we add the arc that has minimum weight out of all arcs that are outgoing from one of the fragments.

communication overhead. This subject is addressed in [GHS83], to which we refer for further details. Reference [Hum83] considers the case where the arc weights are different in each direction. A distributed MST algorithm of the type just described has been used in the PARIS network [CGK90] (see Section 6.4 for a description of this network).

To see what can go wrong in the case of nonunique MSTs, consider the triangular network of Fig. 5.27 where all arc lengths are unity. If we start with the three nodes as fragments and allow each fragment to add to itself an arbitrary, minimum weight outgoing arc, there is the possibility that the arcs (1,2), (2,3), and (3,1) will be added simultaneously by nodes 1, 2, and 3, respectively, with a cycle resulting.

The following proposition points the way on how to handle the case of nonunique MSTs.

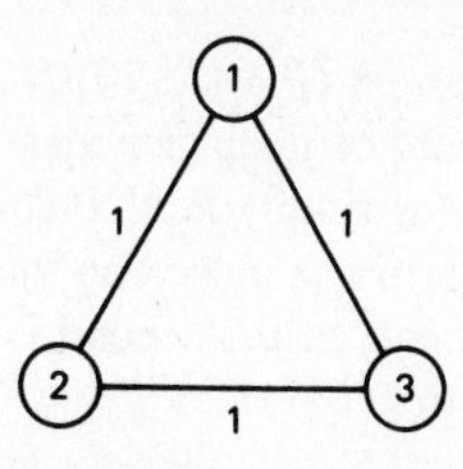

Figure 5.27 Counterexample for distributed MST construction with nondistinct arc weights. Nodes 1, 2, and 3 can add simultaneously to their fragments arcs (1,2), (2,3), and (3,1), respectively, with a cycle resulting.

Proposition 2. If all arc weights are distinct, there exists a unique MST.

Proof: Suppose that there exist two distinct MSTs denoted M and M'. Let α be the minimum weight arc that belongs to either M or M' but not to both. Assume for concreteness that α belongs to M. Consider the graph formed by the union of M' and α (see Fig. 5.28). It must contain a cycle (since $\alpha \notin M'$), and at least one arc of this cycle, call it β, does not belong to M (otherwise M would contain a cycle). Since the weight of α is strictly smaller than that of β, it follows that deleting β from M' and adding α in its place results in a spanning tree of strictly smaller weight than M'. This is a contradiction since M' is optimal. **Q.E.D.**

Consider now an MST problem where the arc weights are not distinct. The ties between arcs with the same weight can be broken by using the identities of their nodes; that is, if arcs (i, j) and (k, l) with $i < j$ and $k < l$ have equal weight, prefer arc (i, j) if $i < k$ or if $i = k$ and $j < l$, and prefer arc (k, l) otherwise. Thus, by implementing, if necessary, the scheme just described, one can assume without loss of generality that arc weights are distinct and that the MST is unique, thereby allowing the use of distributed MST algorithms that join fragments along minimum weight outgoing arcs.

5.2.3 Shortest Path Algorithms

In what follows we shall be interested in traffic flowing over the arcs of a network. This makes it necessary to distinguish the direction of the flow, and thus to give a reference direction to the arcs themselves.

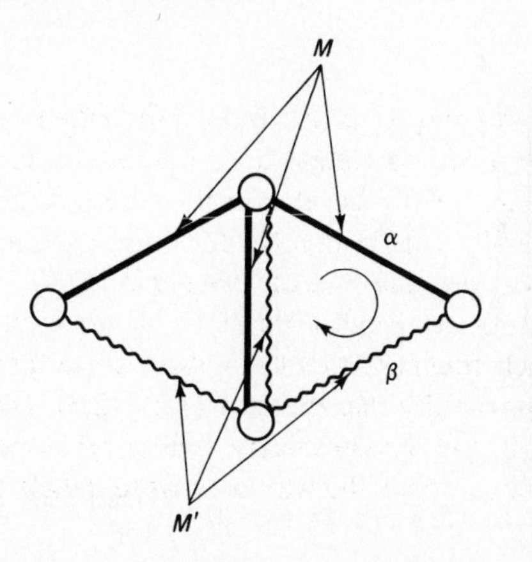

Figure 5.28 Proof of MST uniqueness when all arc weights are distinct. Let M and M' be two MSTs and assume that α is the minimum weight arc that belongs to M and M' but not both. Suppose that $\alpha \in M$ and let β be an arc of the cycle of $M' \cup \{\alpha\}$ which does not belong to M. By deleting β from M' and adding α in its place, we obtain a spanning tree because the corresponding subgraph has $N - 1$ arcs and is seen to be connected. The weight of this spanning tree is smaller than the one of M', because the weight of α is smaller than the weight of β—a contradiction.

A *directed graph* or *digraph* $G = (\mathcal{N}, \mathcal{A})$ is a finite nonempty set $\mathcal{N}$ of nodes and a collection $\mathcal{A}$ of *ordered* pairs of distinct nodes from $\mathcal{N}$; each ordered pair of nodes in $\mathcal{A}$ is called a *directed arc* (or simply arc). Pictorially, a digraph is represented in the same way as a graph, but an arrow is placed on the representation of the arc, going from the first node of the ordered pair to the second (see Fig. 5.29). Note in Fig. 5.29 that (2,4) and (4,2) are different arcs.

Given any digraph $G = (\mathcal{N}, \mathcal{A})$, there is an associated (undirected) graph $G' = (\mathcal{N}', \mathcal{A}')$, where $\mathcal{N}' = \mathcal{N}$ and $(i, j) \in \mathcal{A}'$ if either $(i, j) \in \mathcal{A}$, or $(j, i) \in \mathcal{A}$, or both. We say that $(n_1, n_2, ..., n_\ell)$ is a walk, path, or cycle in a digraph if it is a walk, path, or cycle in the associated graph. In addition, $(n_1, n_2, ..., n_\ell)$ is a *directed walk* in a digraph G if (n_i, n_{i+1}) is a directed arc in G for $1 \leq i \leq \ell - 1$. A *directed path* is a directed walk with no repeated nodes, and a *directed cycle* is a directed walk $(n_1, \ldots, n_\ell)$, for $\ell > 2$ with $n_1 = n_\ell$ and no repeated nodes. Note that (n_1, n_2, n_1) is a directed cycle if (n_1, n_2) and (n_2, n_1) are both directed arcs, whereas (n_1, n_2, n_1) cannot be an undirected cycle if (n_1, n_2) is an undirected arc.

A digraph is *strongly connected* if for each pair of nodes i and j there is a directed path $(i = n_1, n_2, \ldots, n_\ell = j)$ from i to j. A digraph is *connected* if the associated graph is connected. The first graph in Fig. 5.30 is connected but not strongly connected since there is no directed path from 3 to 2. The second graph is strongly connected.

Consider a directed graph $G = (\mathcal{N}, \mathcal{A})$ with number of nodes N and number of arcs A, in which each arc (i, j) is assigned some real number d_{ij} as the length or distance of the arc. Given any directed path $p = (i, j, k, \ldots, \ell, m)$, the *length* of p is defined as $d_{ij} + d_{jk} + \cdots + d_{\ell m}$. The length of a directed walk or cycle is defined analogously. Given any two nodes i, m of the graph, the shortest path problem is to find a minimum length (*i.e.*, shortest) directed path from i to m.

The shortest path problem appears in a surprisingly large number of contexts. If d_{ij} is the cost of using a given link (i, j) in a data network, then the shortest path from i to m is the minimum cost route over which to send data. Thus, if the cost of a link equals the average packet delay to cross the link, the minimum cost route is also a

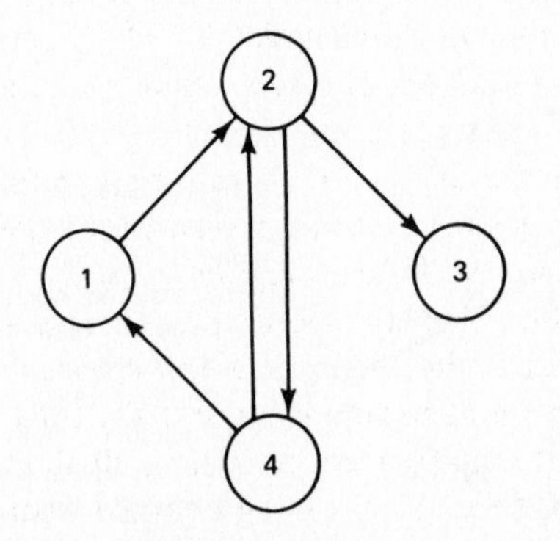

$\mathcal{N} = \{1, 2, 3, 4\}$

$\mathcal{A} = \{(1,2), (2,3), (2,4), (4,2), (4,1)\}$

Figure 5.29 Representation of a directed graph.

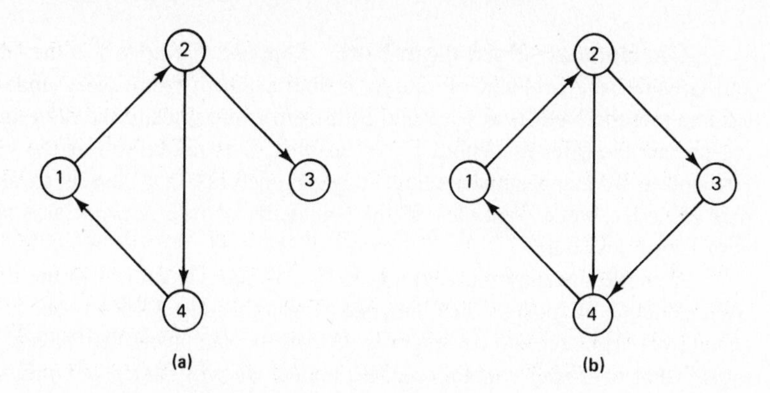

Figure 5.30 Both digraphs (a) and (b) are connected; (b) is strongly connected, but (a) is not strongly connected since there is no directed path from 3 to 2.

minimum delay route. Unfortunately, in a data network, the average link delay depends on the traffic load carried by the link, which in turn depends on the routes selected by the routing algorithm. Because of this feedback effect, the minimum average delay routing problem is more complex than just solving a shortest path problem; however, the shortest path problem is still an integral part of all the formulations of the routing problem to be considered. As another example, if p_{ij} is the probability that a given arc (i, j) in a network is usable, and each arc is usable independently of all the other arcs, then finding the shortest path between i and m with arc distances $(-\ln p_{ij})$ is equivalent to finding the most reliable path from i to m.

Another application of shortest paths is in the PERT networks used by organizations to monitor the progress of large projects. The nodes of the network correspond to subtasks, and an arc from subtask i to j indicates that the completion of task j is dependent on the completion of i. If t_{ij} is the time required to complete j after i is completed, the distance for (i, j) is taken as $d_{ij} = -t_{ij}$. The shortest path from project start to finish is then the most time-consuming path required for completion of the project, and the shortest path indicates the critical subtasks for which delays would delay the entire project. Yet another example is that of discrete dynamic programming problems, which can be viewed as shortest path problems [Ber87]. Finally, many more complex graph-theoretic problems require the solution of shortest path problems as subproblems.

In the following, we develop three standard algorithms for the shortest path problem: the Bellman–Ford algorithm, the Dijkstra algorithm, and the Floyd–Warshall algorithm. The first two algorithms find the shortest paths from all nodes to a given destination node, and the third algorithm finds the shortest paths from all nodes to all other nodes. Note that the problem of finding shortest paths from a given origin node to all other nodes is equivalent to the problem of finding all shortest paths to a given destination node; one version of the problem can be obtained from the other simply by reversing the direction of each arc while keeping its length unchanged. In this subsection we concentrate on centralized shortest path algorithms. Distributed algorithms are considered in Section 5.2.4.

The Bellman–Ford algorithm. Suppose that node 1 is the "destination" node and consider the problem of finding a shortest path from every node to node 1. We assume that there exists at least one path from every node to the destination. To simplify the presentation, let us denote $d_{ij} = \infty$ if (i, j) is not an arc of the graph. Using this convention we can assume without loss of generality that there is an arc between every pair of nodes, since walks and paths consisting of true network arcs are the only ones with length less than ∞.

A shortest walk from a given node i to node 1, subject to the constraint that the walk contains at most h arcs and goes through node 1 only once, is referred to as a *shortest* ($\leq h$) *walk* and its length is denoted by D_i^h. Note that such a walk may not be a path, that is, it may contain repeated nodes; we will later give conditions under which this is not possible. By convention, we take

$$D_1^h = 0, \qquad \text{for all } h$$

We will prove shortly that D_i^h can be generated by the iteration

$$D_i^{h+1} = \min_j \left[d_{ij} + D_j^h\right], \qquad \text{for all } i \neq 1 \tag{5.1}$$

starting from the initial conditions

$$D_i^0 = \infty, \qquad \text{for all } i \neq 1 \tag{5.2}$$

This is the Bellman–Ford algorithm, illustrated in Fig. 5.31. Thus, we claim that the Bellman–Ford algorithm first finds the one-arc shortest walk lengths, then finds the two-arc shortest walk lengths, and so forth. Once we show this, we will argue that the shortest walk lengths are equal to the shortest path lengths, under the additional assumption that all cycles not containing node 1 have nonnegative length. We say that the algorithm terminates after h iterations if

$$D_i^h = D_i^{h-1}, \qquad \text{for all } i$$

The following proposition provides the main results.

Proposition. Consider the Bellman–Ford algorithm (5.1) with the initial conditions $D_i^0 = \infty$ for all $i \neq 1$. Then:

(a) The scalars D_i^h generated by the algorithm are equal to the shortest ($\leq h$) walk lengths from node i to node 1.

(b) The algorithm terminates after a finite number of iterations if and only if all cycles not containing node 1 have nonnegative length. Furthermore, if the algorithm terminates, it does so after at most $h \leq N$ iterations, and at termination, D_i^h is the shortest path length from i to 1.

Proof: (a) We argue by induction. From Eqs. (5.1) and (5.2) we have

$$D_i^1 = d_{i1}, \qquad \text{for all } i \neq 1$$

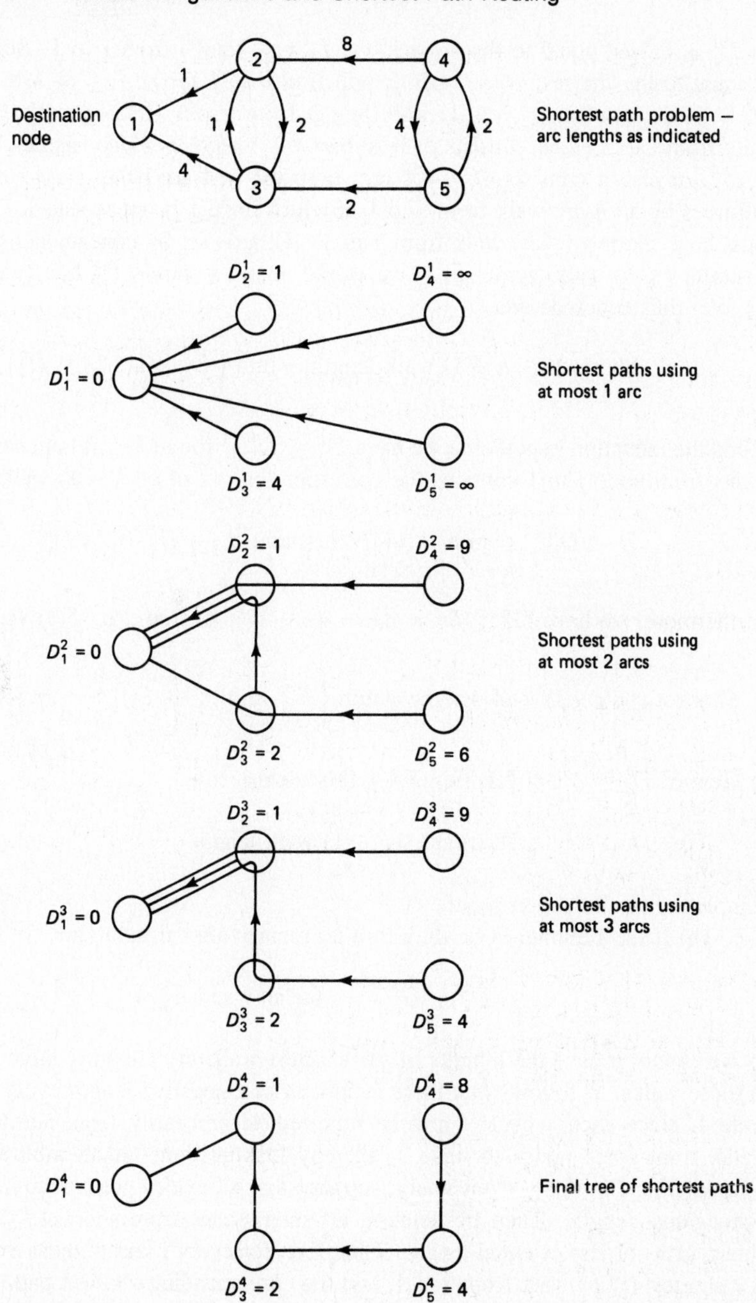

Figure 5.31 Successive iterations of the Bellman–Ford method. In this example, the shortest ($\leq h$) walks are paths because all arc lengths are positive and therefore all cycles have positive length. The shortest paths are found after $N - 1$ iterations, which is equal to 4 in this example.

so D_i^1 is indeed equal to the shortest (≤ 1) walk length from i to 1. Suppose that D_i^k is equal to the shortest ($\leq k$) walk length from i to 1 for all $k \leq h$. We will show that D_i^{h+1} is the shortest ($\leq h+1$) walk length from i to 1. Indeed, a shortest ($\leq h+1$) walk from i to 1 either consists of less than $h+1$ arcs, in which case its length is equal to D_i^h, or else it consists of $h+1$ arcs with the first arc being (i,j) for some $j \neq 1$, followed by an h-arc walk from j to 1 in which node 1 is not repeated. The latter walk must be a shortest ($\leq h$) walk from j to 1. [Otherwise, by concatenating arc (i,j) and a shorter ($\leq h$) walk from j to 1, we would obtain a shorter ($\leq h+1$) walk from i to 1.] We thus conclude that

$$\text{Shortest } (\leq h+1) \text{ walk length} = \min\left\{D_i^h, \min_{j \neq 1}\left[d_{ij} + D_j^h\right]\right\} \tag{5.3}$$

Using the induction hypothesis, we have $D_j^k \leq D_j^{k-1}$ for all $k \leq h$ [since the set of ($\leq k$) walks from node j to 1 contains the corresponding set of ($\leq k-1$) walks]. Therefore,

$$D_i^{h+1} = \min_j\left[d_{ij} + D_j^h\right] \leq \min_j\left[d_{ij} + D_j^{h-1}\right] = D_i^h \tag{5.4}$$

Furthermore, we have $D_i^h \leq D_i^1 = d_{i1} = d_{i1} + D_1^h$, so from Eq. (5.3) we obtain

$$\text{Shortest } (\leq h+1) \text{ walk length} = \min\left\{D_i^h, \min_j\left[d_{ij} + D_j^h\right]\right\} = \min\left\{D_i^h, D_i^{h+1}\right\}$$

In view of $D_i^{h+1} \leq D_i^h$ [cf. Eq. (5.4)], this yields

$$\text{Shortest } (\leq h+1) \text{ walk length} = D_i^{h+1}$$

completing the induction proof.

(b) If the Bellman-Ford algorithm terminates after h iterations, we must have

$$D_i^k = D_i^h, \qquad \text{for all } i \text{ and } k \geq h \tag{5.5}$$

so we cannot reduce the lengths of the shortest walks by allowing more and more arcs in these walks. It follows that there cannot exist a negative-length cycle not containing node 1, since such a cycle could be repeated an arbitrarily large number of times in walks from some nodes to node 1, thereby making their length arbitrarily small and contradicting Eq. (5.5). Conversely, suppose that all cycles not containing node 1 have nonnegative length. Then by deleting all such cycles from shortest ($\leq h$) walks, we obtain paths of less or equal length. Therefore, for every i and h, there exists a path that is a shortest ($\leq h$) walk from i to 1, and the corresponding shortest path length is equal to D_i^h. Since paths have no cycles, they can contain at most $N-1$ arcs. It follows that

$$D_i^N = D_i^{N-1}, \qquad \text{for all } i$$

implying that the algorithm terminates after at most N iterations. **Q.E.D.**

Note that the preceding proposition is valid even if there is no path from some nodes i to node 1 in the original network. Upon termination, we will simply have for those nodes $D_i^h = \infty$.

To estimate the computation required to find the shortest path lengths, we note that in the worst case, the algorithm must be iterated N times, each iteration must be done for $N - 1$ nodes, and for each node the minimization must be taken over no more than $N - 1$ alternatives. Thus, the amount of computation grows at worst like N^3, which is written as $O(N^3)$. Generally, the notation $O\big(p(N)\big)$, where $p(N)$ is a polynomial in N, is used to indicate a number depending on N that is smaller than $cp(N)$ for all N, where c is some constant independent of N. Actually, a more careful accounting shows that the amount of computation is $O(mA)$, where A is the number of arcs and m is the number of iterations required for termination (m is also the maximum number of arcs contained in a shortest path).

The example in Fig. 5.32 shows the effect of negative-length cycles not involving node 1, and illustrates that one can test for existence of such cycles simply by comparing D_i^N with D_i^{N-1} for each i. As implied by part (b) of the preceding proposition, there exists such a negative-length cycle if and only if $D_i^N < D_i^{N-1}$ for some i.

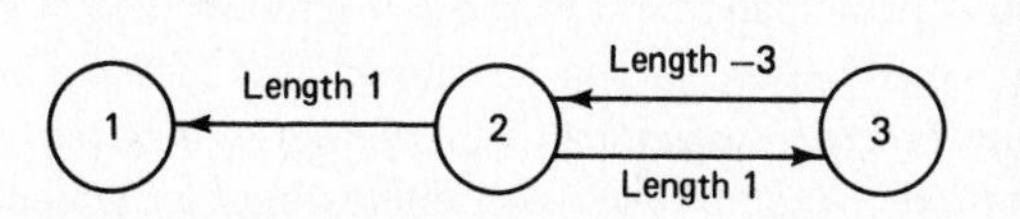

Figure 5.32 Graph with a negative cycle. The shortest path length from 2 to 1 is 1. The Bellman–Ford algorithm gives $D_2^2 = 1$ and $D_2^3 = -1$, indicating the existence of a negative-length cycle.

Bellman's equation and shortest path construction. Assume that all cycles not containing node 1 have nonnegative length and denote by D_i the shortest path length from node i to 1. Then upon termination of the Bellman–Ford algorithm, we obtain

$$D_i = \min_j \left[d_{ij} + D_j\right], \qquad \text{for all } i \neq 1 \tag{5.6a}$$

$$D_1 = 0 \tag{5.6b}$$

This is called *Bellman's equation* and expresses that the shortest path length from node i to 1 is the sum of the length of the arc to the node following i on the shortest path plus the shortest path length from that node to node 1. From this equation it is easy to find the shortest paths (as opposed to the shortest path lengths) if all cycles not including node 1 have a positive length (as opposed to zero length). To do this, select, for each $i \neq 1$, one arc (i, j) that attains the minimum in the equation $D_i = \min_j[d_{ij} + D_j]$, and consider the subgraph consisting of these $N - 1$ arcs (see Fig. 5.33). To find the shortest path from any node i, start at i and follow the corresponding arcs of the subgraph until node 1 is reached. Note that the same node cannot be reached twice before reaching node 1, since a cycle would be formed that (on the basis of the equation $D_i = \min_j[d_{ij} + D_j]$) would have zero length [let $(i_1, i_2, \ldots, i_k, i_1)$ be the cycle and add the equations

$$D_{i_1} = d_{i_1 i_2} + D_{i_2}$$

$$\cdots$$

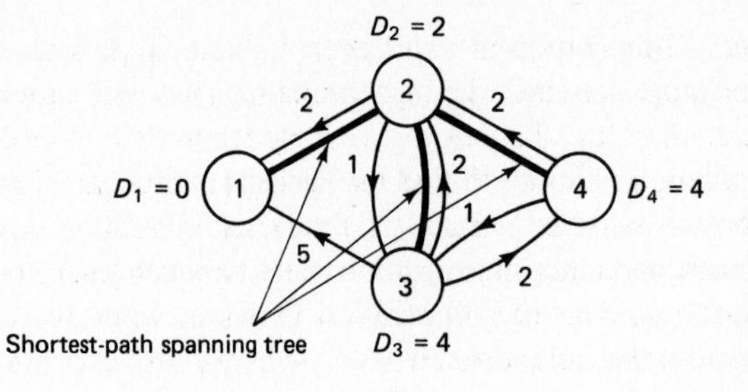

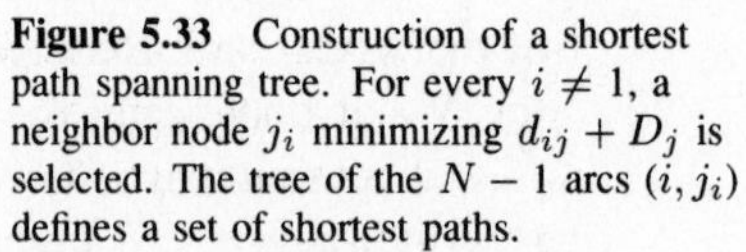
Figure 5.33 Construction of a shortest path spanning tree. For every $i \neq 1$, a neighbor node j_i minimizing $d_{ij} + D_j$ is selected. The tree of the $N - 1$ arcs (i, j_i) defines a set of shortest paths.

$$D_{i_{k-1}} = d_{i_{k-1}i_k} + D_{i_k}$$

$$D_{i_k} = d_{i_k i_1} + D_{i_1}$$

obtaining $d_{i_1 i_2} + \cdots + d_{i_k i_1} = 0$]. Since the subgraph connects every node to node 1 and has $N - 1$ arcs, it must be a spanning tree. We call this subgraph the *shortest path spanning tree* and note that it has the special structure of having a root (node 1), with every arc of the tree directed toward the root. Problem 5.7 shows how such a spanning tree can be obtained when there are cycles of zero length. Problem 5.4 explores the difference between a shortest path spanning tree and a minimum weight spanning tree.

Using the preceding construction, it can be shown that *if there are no zero (or negative)-length cycles, then Bellman's equation (5.6) (viewed as a system of N equations with N unknowns) has a unique solution.* This is an interesting fact which will be useful in Section 5.2.4 when we consider the Bellman–Ford algorithm starting from initial conditions other than ∞ [cf. Eq. (5.2)]. For a proof, we suppose that $\tilde{D}_i, i = 1, \ldots, N$, are another solution of Bellman's equation (5.6) with $\tilde{D}_1 = 0$, and we show that $\tilde{D}_i$ are equal to the shortest path lengths D_i. Let us repeat the path construction of the preceding paragraph with $\tilde{D}_i$ replacing D_i. Then $\tilde{D}_i$ is the length of the corresponding path from node i to node 1, showing that $\tilde{D}_i \geq D_i$. To show the reverse inequality, consider the Bellman–Ford algorithm with two different initial conditions. The first initial condition is $D_i^0 = \infty$, for $i \neq 1$, and $D_1^0 = 0$, in which case the true shortest path lengths D_i are obtained after at most $N - 1$ iterations, as shown earlier. The second initial condition is $D_i^0 = \tilde{D}_i$, for all i, in which case $\tilde{D}_i$ is obtained after every iteration (since the $\tilde{D}_i$ solve Bellman's equation). Since the second initial condition is, for every i, less than or equal to the first, it is seen from the Bellman–Ford iteration $D_i^{h+1} = \min_j [d_{ij} + D_j^h]$ that $\tilde{D}_i \leq D_i$, for all i. Therefore, $\tilde{D}_i = D_i$, and the only solution of Bellman's equation is the set of the true shortest path lengths D_i. It is also possible to show that if there are zero-length cycles not involving node 1, then Bellman's equation has a nonunique solution [although the Bellman-Ford algorithm still terminates with the correct shortest path lengths from the initial conditions $D_i^0 = \infty$ for all $i \neq 1$; see Problem 5.8].

It turns out that the Bellman–Ford algorithm works correctly even if the initial conditions D_i^0 for $i \neq 1$ are arbitrary numbers, and the iterations are done in parallel for different nodes in virtually any order. This fact will be shown in the next subsection and accounts in part for the popularity of the Bellman–Ford algorithm for distributed applications.

Dijkstra's algorithm. This algorithm requires that all arc lengths are nonnegative (fortunately, the case for most data network applications). Its worst-case computational requirements are considerably less than those of the Bellman–Ford algorithm. The general idea is to find the shortest paths in order of increasing path length. The shortest of the shortest paths to node 1 must be the single-arc path from the closest neighbor of node 1, since any multiple-arc path cannot be shorter than the first arc length because of the nonnegative-length assumption. The next shortest of the shortest paths must either be the single-arc path from the next closest neighbor of 1 or the shortest two-arc path through the previously chosen node, and so on. To formalize this procedure into an algorithm, we view each node i as being labeled with an estimate D_i of the shortest path length to node 1. When the estimate becomes certain, we regard the node as being *permanently labeled* and keep track of this with a set P of permanently labeled nodes. The node added to P at each step will be the closest to node 1 out of those that are not yet in P. Figure 5.34 illustrates the main idea. The detailed algorithm is as follows: Initially, $P = \{1\}$, $D_1 = 0$, and $D_j = d_{j1}$ for $j \neq 1$.

Step 1: (Find the next closest node.) Find $i \notin P$ such that

$$D_i = \min_{j \notin P} D_j$$

Set $P := P \cup \{i\}$. If P contains all nodes, then stop; the algorithm is complete.

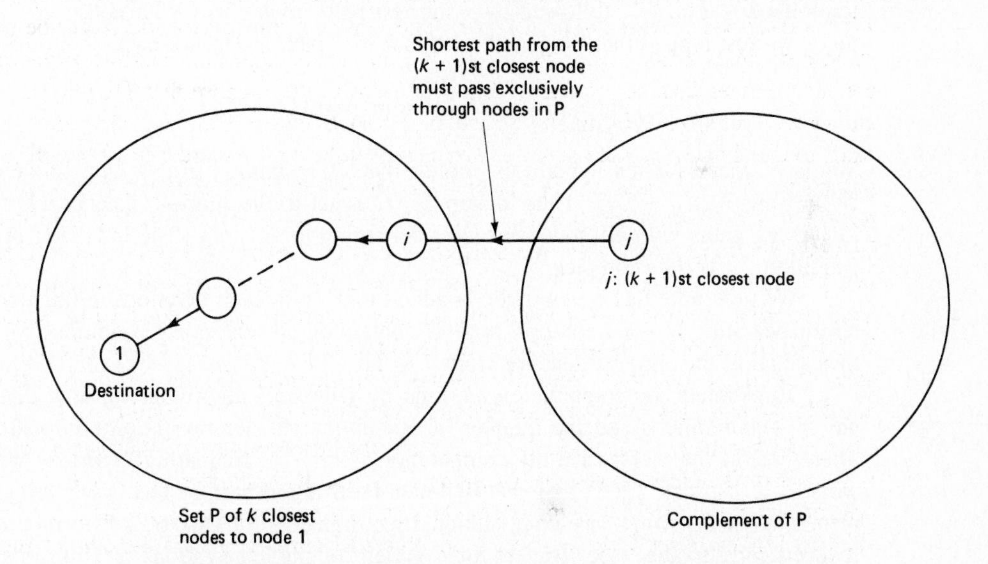

Figure 5.34 Basic idea of Dijkstra's algorithm. At the kth step we have the set P of the k closest nodes to node 1 as well as the shortest distance D_i from each node i in P to node 1. Of all paths connecting some node not in P with node 1, there is a shortest one that passes exclusively through nodes in P (since $d_{ij} \geq 0$). Therefore, the $(k+1)$st closest node and the corresponding shortest distance are obtained by minimizing over $j \notin P$ the quantity $\min_{i \in P}\{d_{ji} + D_i\}$. This calculation can be organized efficiently as discussed in the text, resulting in an $O(N^2)$ computational complexity.

Step 2: (Updating of labels.) For all $j \notin P$ set

$$D_j := \min\,[D_j, d_{ji} + D_i]$$

Go to step 1.

To see why the algorithm works, we must interpret the estimates D_j. We claim that at the beginning of each step 1:

(a) $D_i \leq D_j$ for all $i \in P$ and $j \notin P$.

(b) D_j is, for each node j, the shortest distance from j to 1 using paths with all nodes except possibly j belonging to the set P.

Indeed, condition (a) is satisfied initially, and since $d_{ji} \geq 0$ and $D_i = \min_{j \notin P} D_j$, it is preserved by the formula $D_j := \min[D_j, d_{ji} + D_i]$ for all $j \notin P$, in step 2. We show condition (b) by induction. It holds initially. Suppose that it holds at the beginning of some step 1, let i be the node added to P at that step, and let D_k be the label of each node k at the beginning of that step. Then condition (b) holds for $j = i$ by the induction hypothesis. It is also seen to hold for all $j \in P$, in view of condition (a) and the induction hypothesis. Finally, for a node $j \notin P \cup \{i\}$, consider a path from j to 1 which is shortest among those with all nodes except j belonging to the set $P \cup \{i\}$, and let D'_j be the corresponding shortest distance. Such a path must consist of an arc (j, k) for some $k \in P \cup \{i\}$, followed by a shortest path from k to 1 with nodes in $P \cup \{i\}$. Since we just argued that the length of this k to 1 path is D_k, we have

$$D'_j = \min_{k \in P \cup \{i\}} [d_{jk} + D_k] = \min\left[\min_{k \in P} [d_{jk} + D_k], d_{ji} + D_i\right]$$

Similarly, the induction hypothesis implies that $D_j = \min_{k \in P}[d_{jk} + D_k]$, so we obtain $D'_j = \min[D_j, d_{ji} + D_i]$. Thus in step 2, D_j is set to the shortest distance D'_j from j to 1 using paths with all nodes except j belonging to $P \cup \{i\}$. The induction proof of condition (b) is thus complete.

We now note that a new node is added to P with each iteration, so the algorithm terminates after $N - 1$ iterations, with P containing all nodes. By condition (b), D_j is then equal to the shortest distance from j to 1.

To estimate the computation required by Dijkstra's algorithm, we note that there are $N - 1$ iterations and the number of operations per iteration is proportional to N. Therefore, in the worst case the computation is $O(N^2)$, comparing favorably with the worst-case estimate $O(N^3)$ of the Bellman–Ford algorithm; in fact, with proper implementation the worst-case computational requirements for Dijkstra's algorithm can be reduced considerably (see [Ber91] for a variety of implementations of Dijkstra's algorithm). However, there are many problems where $A \ll N^2$, and the Bellman–Ford algorithm terminates in very few iterations (say $m \ll N$), in which case its required computation $O(mA)$ can be less than the $O(N^2)$ requirement of the straightforward implementation of Dijkstra's algorithm. Generally, for nondistributed applications, efficiently implemented variants of the Bellman-Ford and Dijkstra algorithms appear to be competitive (see [Ber91] for a more detailed discussion and associated codes).

The Dijkstra and Bellman–Ford algorithms are contrasted on an example problem in Fig. 5.35.

The Floyd–Warshall algorithm. This algorithm, unlike the previous two, finds the shortest paths between all pairs of nodes together. Like the Bellman–Ford algorithm, the arc distances can be positive or negative, but again there can be no negative-length cycles. All three algorithms iterate to find the final solution, but each iterates on something different. The Bellman–Ford algorithm iterates on the number of arcs in a path, the Dijkstra algorithm iterates on the length of the path, and finally, the Floyd–Warshall algorithm iterates on the set of nodes that are allowed as intermediate nodes on the paths. The Floyd–Warshall algorithm starts like both other algorithms with single arc distances

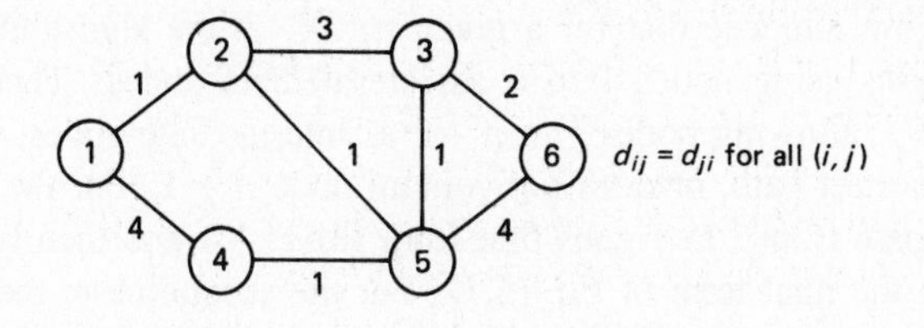

Bellman-Ford

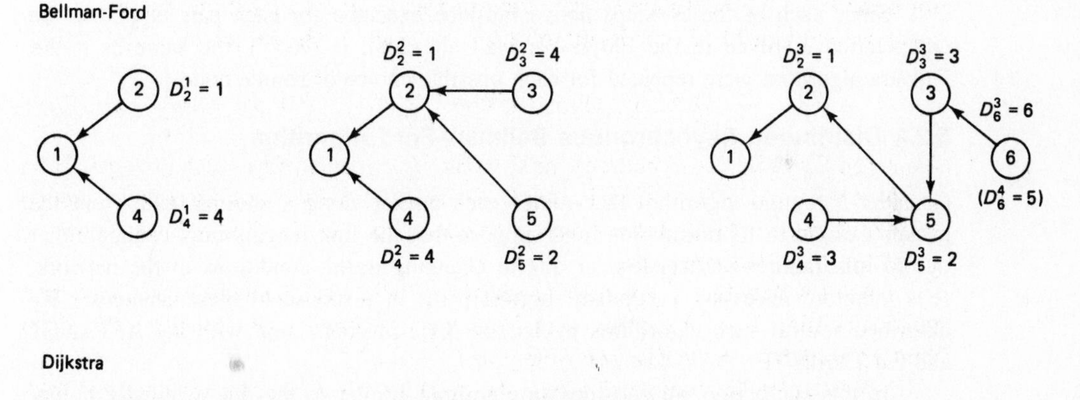

Dijkstra

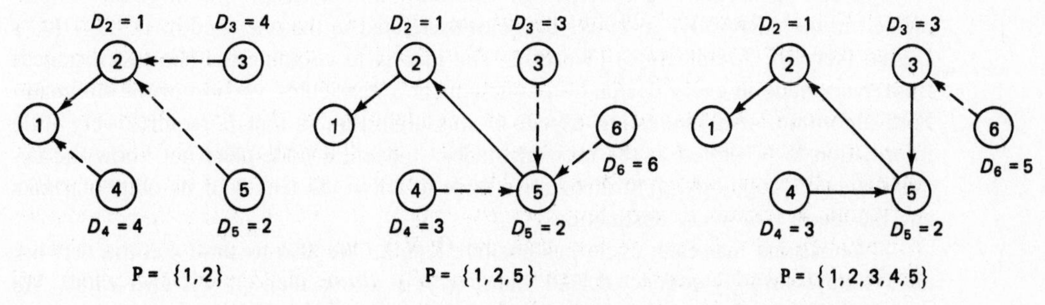

Figure 5.35 Example using the Bellman–Ford and Dijkstra algorithms.

(*i.e.*, no intermediate nodes) as starting estimates of shortest path lengths. It then calculates shortest paths under the constraint that only node 1 can be used as an intermediate node, and then with the constraint that only nodes 1 and 2 can be used, and so forth.

To state the algorithm more precisely, let D_{ij}^n be the shortest path length from node i to j with the constraint that only nodes $1, 2, \ldots, n$ can be used as intermediate nodes on paths. The algorithm then is as follows: Initially,

$$D_{ij}^0 = d_{ij}, \qquad \text{for all } i,\ j, \qquad i \neq j$$

For $n = 0, 1, \ldots, N-1$,

$$D_{ij}^{n+1} = \min\left[D_{ij}^n, D_{i(n+1)}^n + D_{(n+1)j}^n\right], \qquad \text{for all } i \neq j \tag{5.7}$$

To see why this works, we use induction. For $n = 0$, the initialization clearly gives the shortest path lengths subject to the constraint of no intermediate nodes on paths. Now, suppose that for a given n, D_{ij}^n in the algorithm above gives the shortest path lengths using nodes 1 to n as intermediate nodes. Then the shortest path length from i to j, allowing nodes 1 to $n+1$ as intermediate nodes, either contains node $n+1$ on the shortest path, or does not contain node $n+1$. For the first case, the constrained shortest path from i to j goes first from i to $n+1$ and then from $n+1$ to j, giving the length in the final term of Eq. (5.7). For the second case, the constrained shortest path is the same as the one using nodes 1 to n as intermediate nodes, yielding the length of the first term in the minimization of Eq. (5.7).

Since each of the N steps above must be executed for each pair of nodes, the computation involved in the Floyd–Warshall algorithm is $O(N^3)$, the same as if the Dijkstra algorithm were repeated for each possible choice of source node.

5.2.4 Distributed Asynchronous Bellman–Ford Algorithm

Consider a routing algorithm that routes each packet along a shortest path from the packet's origin to its destination, and suppose that the link lengths may change either due to link failures and repairs, or due to changing traffic conditions in the network. It is therefore necessary to update shortest paths in response to these changes. We described several such algorithms in Section 5.1.2 in connection with the ARPANET and the TYMNET.

In this subsection we consider an algorithm similar to the one originally implemented in the ARPANET in 1969. It is closely related to the one used in DNA (DEC's Digital Network Architecture) [Wec80]. The idea is to compute the shortest distances from every node to every destination by means of a distributed version of the Bellman–Ford algorithm. An interesting aspect of this algorithm is that it requires very little information to be stored at the network nodes. Indeed, a node need not know the detailed network topology. It suffices for a node to know the length of its outgoing links and the identity of every destination.

We assume that each cycle has positive length. We also assume that the network always stays strongly connected, and that if (i, j) is a link, then (j, i) is also a link. We envision a practical situation where the lengths d_{ij} can change with time. In the analysis, however, it is assumed that the lengths d_{ij} are fixed while the initial conditions for the algorithm are allowed to be essentially arbitrary. These assumptions provide an adequate model for a situation where the link lengths stay fixed after some time t_0 following a number of changes that occurred before t_0.

We focus on the shortest distance D_i from each node i to a generic destination node taken for concreteness to be node 1. (In reality, a separate version of the algorithm must be executed for each destination node.) Under our assumptions, these distances are the unique solution of Bellman's equation,

$$D_i = \min_{j \in N(i)} [d_{ij} + D_j], \qquad i \neq 1$$

$$D_1 = 0$$

where $N(i)$ denotes the set of current neighbors of node i (*i.e.*, the nodes connected with i via an up link). This was shown in Section 5.2.3; see the discussion following Eq. (5.6).

The Bellman–Ford algorithm is given by

$$D_i^{h+1} = \min_{j \in N(i)} \left[d_{ij} + D_j^h\right], \qquad i \neq 1 \tag{5.8}$$

$$D_1^{h+1} = 0 \tag{5.9}$$

In Section 5.2.3, we showed convergence to the correct shortest distances for the initial conditions

$$D_i^0 = \infty, \qquad i \neq 1 \tag{5.10}$$

$$D_1^0 = 0 \tag{5.11}$$

The algorithm is well suited for distributed computation since the Bellman–Ford iteration (5.8) can be executed at each node i in parallel with every other node. One possibility is for all nodes i to execute the iteration simultaneously, exchange the results of the computation with their neighbors, and execute the iteration again with the index h incremented by one. When the infinite initial conditions of Eqs. (5.10) and (5.11) are used, the algorithm will terminate after at most N iterations (where N is the number of nodes), with each node i knowing both the shortest distance, D_i, and the outgoing link on the shortest path to node 1.

Unfortunately, implementing the algorithm in such a synchronous manner is not as easy as it appears. There is a twofold difficulty here. First, a mechanism is needed for getting all nodes to agree to start the algorithm. Second, a mechanism is needed to abort the algorithm and start a new version if a link status or length changes as the algorithm is running. Although it is possible to cope successfully with these difficulties [Seg81], the resulting algorithm is far more complex than the pure Bellman–Ford method given by Eqs. (5.8) and (5.9).

A simpler alternative is to use an asynchronous version of the Bellman–Ford algorithm that does not insist on maintaining synchronism between nodes, and on starting with the infinite initial conditions of Eqs. (5.10) and (5.11). This eliminates the need for either an algorithm initiation or an algorithm restart protocol. The algorithm simply operates indefinitely by executing from time to time at each node $i \neq 1$ the iteration

$$D_i := \min_{j \in N(i)} [d_{ij} + D_j] \tag{5.12}$$

using the last estimates D_j received from the neighbors $j \in N(i)$, and the latest status and lengths of the outgoing links from node i. The algorithm also requires that each node i transmit from time to time its latest estimate D_i to all its neighbors. However, there is no need for either the iterations or the message transmissions to be synchronized at all nodes. Furthermore, no assumptions are made on the initial values D_j, $j \in N(i)$ available at each node i. The only requirement is that a node i will eventually execute the Bellman–Ford iteration (5.12) and will eventually transmit the result to the neighbors. Thus, a totally asynchronous mode of operation is envisioned.

It turns out that the algorithm is still valid when executed asynchronously as described above. It will be shown that if a number of link length changes occur up to some time t_0, and no other changes occur subsequently, then within finite time from t_0, the asynchronous algorithm finds the correct shortest distance of every node i. The shortest distance estimates available at time t_0 can be arbitrary numbers, so it is not necessary to reinitialize the algorithm after each link status or link length change.

The original 1969 ARPANET algorithm was based on the Bellman–Ford iteration (5.12), and was implemented asynchronously much like the scheme described above. Neighboring nodes exchanged their current shortest distance estimates D_j needed in the iteration every 625 msec, but this exchange was not synchronized across the network. Furthermore, the algorithm was not restarted following a link length change or a link failure. The major difference between the ARPANET algorithm and the one analyzed in the present subsection is that the link lengths d_{ij} were changing very frequently in the ARPANET algorithm, and the eventual steady state assumed in our analysis was seldom reached.

We now state formally the distributed, asynchronous Bellman–Ford algorithm and proceed to establish its validity. At each time t, a node $i \neq 1$ has available:

$D_j^i(t) =$ Estimate of the shortest distance of each neighbor node $j \in N(i)$ which was last communicated to node i

$D_i(t) =$ Estimate of the shortest distance of node i which was last computed at node i according to the Bellman–Ford iteration

The distance estimates for the destination node 1 are defined to be zero, so

$$D_1(t) = 0, \qquad \text{for all } t \geq t_0$$

$$D_1^i(t) = 0, \qquad \text{for all } t \geq t_0, \text{ and } i \text{ with } 1 \in N(i)$$

Each node i also has available the link lengths d_{ij}, for all $j \in N(i)$, which are assumed constant after the initial time t_0. We assume that the distance estimates do not change except at some times $t_0, t_1, t_2, \ldots$, with $t_{m+1} > t_m$, for all m, and $t_m \to \infty$ as $m \to \infty$, when at each processor $i \neq 1$, one of three events happens:

1. Node i updates $D_i(t)$ according to

$$D_i(t) := \min_{j \in N(i)} \left[d_{ij} + D_j^i(t) \right]$$

and leaves the estimates $D_j^i(t)$, $j \in N(i)$, unchanged.

2. Node i receives from one or more neighbors $j \in N(i)$ the value of D_j which was computed at node j at some earlier time, updates the estimate D_j^i, and leaves all other estimates unchanged.
3. Node i is idle, in which case all estimates available at i are left unchanged.

Let T^i be the set of times for which an update by node i as in case 1 occurs, and T_j^i the set of times when a message is received at i from node j, as in case 2. We assume the following:

Assumption 1. Nodes never stop updating their own estimates and receiving messages from all their neighbors [*i.e.*, T^i and T_j^i have an infinite number of elements for all $i \neq 1$ and $j \in N(i)$].

Assumption 2. Old distance information is eventually purged from the system [*i.e.*, given any time $\bar{t} \geq t_0$, there exists a time $\tilde{t} > \bar{t}$ such that estimates D_j computed at a node j prior to time $\bar{t}$ are not received at any neighbor node i after time $\tilde{t}$].

The following proposition shows that the estimates $D_i(t)$ converge to the correct shortest distances within finite time. The proof (which can be skipped without loss of continuity) is interesting in that it serves as a model for proofs of validity of several other asynchronous distributed algorithms. (See the convergence proof of the algorithm of Section 5.3.3 and references [Ber82a], [Ber83], and [BeT89].)

Proposition. Suppose that each cycle has positive length and let the initial conditions $D_i(t_0)$, $D_j^i(t_0)$ be arbitrary numbers for $i = 2, \ldots, N$ and $j = 2, \ldots, N$. Then there is a time t_m such that

$$D_i(t) = D_i, \qquad \text{for all } t \geq t_m, \; i = 1, \ldots, N$$

where D_i is the correct shortest distance from node i to the destination node 1.

Proof: The idea of the proof is to define for every node i two sequences $\{\underline{D}_i^k\}$ and $\{\overline{D}_i^k\}$ with

$$\underline{D}_i^k \leq \underline{D}_i^{k+1} \leq D_i \leq \overline{D}_i^{k+1} \leq \overline{D}_i^k \tag{5.13}$$

and

$$\underline{D}_i^k = D_i = \overline{D}_i^k, \qquad \text{for all } k \text{ sufficiently large} \tag{5.14}$$

These sequences are obtained from the Bellman–Ford algorithm by starting at two different initial conditions. It is then shown that for every k and i, the estimates $D_i(t)$ satisfy

$$\underline{D}_i^k \leq D_i(t) \leq \overline{D}_i^k, \qquad \text{for all } t \text{ sufficiently large} \tag{5.15}$$

A key role in the proof is played by the monotonicity property of the Bellman–Ford iteration. This property states that if for some scalars $\overline{D}_j$ and $\tilde{D}_j$,

$$\overline{D}_j \geq \tilde{D}_j, \qquad \text{for all } j \in N(i)$$

then the direction of the inequality is preserved by the iteration, that is,

$$\min_{j \in N(i)} [d_{ij} + \overline{D}_j] \geq \min_{j \in N(i)} [d_{ij} + \tilde{D}_j]$$

A consequence of this property is that if D_i^k are sequences generated by the Bellman–Ford iteration (5.8) and (5.9) starting from some initial condition $D_i^0, i = 1, \ldots, N$, and we have $D_i^1 \geq D_i^0$ for each i, then $D_i^{k+1} \geq D_i^k$, for each i and k. Similarly, if $D_i^1 \leq D_i^0$, for each i, then $D_i^{k+1} \leq D_i^k$, for each i and k.

Consider the Bellman–Ford algorithm given by Eqs. (5.8) and (5.9). Let $\overline{D}_i^k, i = 1, \ldots, N$ be the kth iterate of this algorithm when the initial condition is

$$\overline{D}_i^0 = \infty, \qquad i \neq 1 \tag{5.16a}$$

$$\overline{D}_1^0 = 0 \tag{5.16b}$$

and let $\underline{D}_i^k, i = 1, \ldots, N$ be the kth iterate when the initial condition is

$$\underline{D}_i^0 = D_i - \delta, \qquad i \neq 1 \tag{5.17a}$$

$$\underline{D}_1^0 = 0 \tag{5.17b}$$

where δ is a positive scalar large enough so that $\underline{D}_i^0$ is smaller than all initial node estimates $D_i(t_0)$, $D_j^i(t_0)$, $j \in N(i)$ and all estimates D_i that were communicated by node i before t_0 and will be received by neighbors of i after t_0.

Lemma. The sequences $\{\underline{D}_i^k\}$ and $\{\overline{D}_i^k\}$ defined above satisfy Eqs. (5.13) and (5.14).

Proof: Relation (5.13) is shown by induction using the monotonicity property of the Bellman–Ford iteration, and the choice of initial conditions above. To show Eq. (5.14), first note that from the convergence analysis of the Bellman–Ford algorithm of the preceding section, for all i,

$$\overline{D}_i^k = D_i, \qquad k \geq N - 1 \tag{5.18}$$

so only $\underline{D}_i^k = D_i$, for sufficiently large k, remains to be established. To this end, we first note that by an induction argument it follows that $\underline{D}_i^k$ is, for every k, the length of a walk from i to some other node j involving no more than k links, plus $\underline{D}_j^0$. Let $L(i,k)$ and $n(i,k)$ be the length and the number of links of this walk, respectively. We can decompose this walk into a path from i to j involving no more than $N-1$ links plus a number of cycles each of which must have positive length by our assumptions. Therefore, if $\lim_{k \to \infty} n(i,k) = \infty$ for any i, we would have $\lim_{k \to \infty} L(i,k) = \infty$, which is impossible since $D_i \geq \underline{D}_i^k = L(i,k) + \underline{D}_j^0$ and D_i is finite (the network is assumed strongly connected). Therefore, $n(i,k)$ is bounded with respect to k and it follows that the number of all possible values of $\underline{D}_i^k$, for $i = 1, \ldots, N$ and $k = 0, 1, \ldots$, is finite. Since $\underline{D}_i^k$ is monotonically nondecreasing in k for all i, it follows that for some h

$$\underline{D}_i^{h+1} = \underline{D}_i^h, \qquad \text{for all } i \tag{5.19}$$

Therefore, the scalars $\underline{D}_i^h$ satisfy Bellman's equation, which as shown in Section 5.2.3, has as its unique solution the shortest distances D_i. It follows that

$$\underline{D}_i^k = D_i, \qquad \text{for all } i, \quad k \geq h \tag{5.20}$$

Equations (5.18) and (5.20) show Eq. (5.14). **Q.E.D.**

We now complete the proof of the proposition by showing by induction that for every k there exists a time $t(k)$ such that for all $t \geq t(k)$,

$$\underline{D}_i^k \leq D_i(t) \leq \overline{D}_i^k, \qquad i = 1, 2, \ldots, N \tag{5.21}$$

$$\underline{D}_j^k \leq D_j^i(t) \leq \overline{D}_j^k, \qquad j \in N(i), \quad i = 1, 2, \ldots, N \tag{5.22}$$

and for all $t \in T_j^i$, $t \geq t(k)$,

$$\underline{D}_j^k \leq D_j\left[\tau_j^i(t)\right] \leq \overline{D}_j^k, \qquad j \in N(i), \quad i = 1, 2, \ldots, N \tag{5.23}$$

where $\tau_j^i(t)$ is the largest time at which the estimate $D_j^i(t)$ available at node i at time t was computed using the iteration $D_j := \min_{k \in N(j)}[d_{jk} + D_k]$ at node j. [Formally, $\tau_j^i(t)$ is defined as the largest time in T^j that is less than t, and is such that $D_j[\tau_j^i(t)] = D_j^i(t)$.]

Indeed, the induction hypothesis is true for $k = 0$ [for $t(0) = t_0$] because δ was chosen large enough in the initial conditions (5.17). Assuming that the induction hypothesis is true for a given k, it will be shown that there exists a time $t(k+1)$ with the required properties. Indeed, from relation (5.22) and the monotonicity of the Bellman–Ford iteration, we have that for every $t \in T^i, t \geq t(k)$ [*i.e.*, a time t for which $D_i(t)$ is updated via the Bellman–Ford iteration],

$$\underline{D}_i^{k+1} \leq D_i(t) \leq \overline{D}_i^{k+1}$$

Therefore, if $t'(k)$ is the smallest time $t \in T^i$, with $t \geq t(k)$, then

$$\underline{D}_i^{k+1} \leq D_i(t) \leq \overline{D}_i^{k+1}, \qquad \text{for all } t \geq t'(k), \quad i = 1, 2, \ldots, N \tag{5.24}$$

Since $\tau_j^i(t) \to \infty$ as $t \to \infty$ (by Assumptions 1 and 2) we can choose a time $t(k+1) \geq t'(k)$ so that $\tau_j^i(t) \geq t'(k)$, for all i, $j \in N(i)$, and $t \geq t(k+1)$. Then, in view of Eq. (5.24), for all $t \geq t(k+1)$,

$$\underline{D}_j^{k+1} \leq D_j^i(t) \leq \overline{D}_j^{k+1}, \qquad j \in N(i), \quad i = 1, 2, \ldots, N \tag{5.25}$$

and for all $t \in T_j^i$, $t \geq t(k+1)$,

$$\underline{D}_j^{k+1} \leq D_j\left[\tau_j^i(t)\right] \leq \overline{D}_j^{k+1}, \qquad j \in N(i), \quad i = 1, 2, \ldots, N \tag{5.26}$$

This completes the induction proof. **Q.E.D.**

The preceding analysis assumes that there is a path from every node to node 1. If not, the analysis applies to the portion of the network that is connected with node 1. For a node i that belongs to a different network component than node 1 and has at least one neighbor (so that it can execute the algorithm), it can be seen that $D_i(t) \to \infty$ as $t \to \infty$. This property can be used by a node to identify the destinations to which it is not connected.

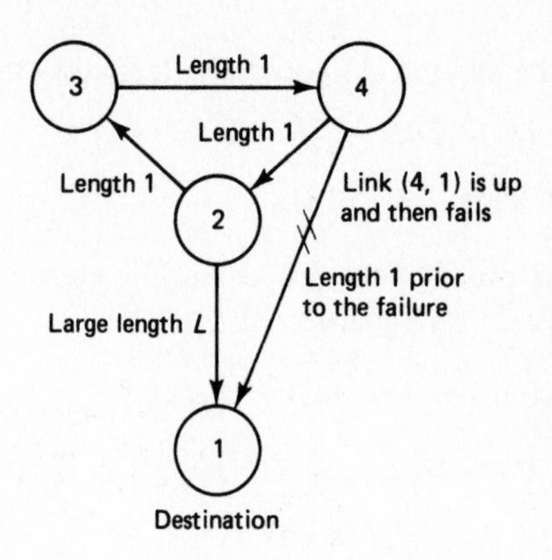

Figure 5.36 Example where the number of iterations of the (synchronous) Bellman–Ford algorithm is excessive. Suppose that the initial conditions are the shortest distances from nodes 2, 3, and 4 to node 1 before link (4,1) fails ($D_2 = 3$, $D_3 = 2$, $D_4 = 1$). Then, after link (4,1) fails, nearly L iterations will be required before node 2 realizes that its shortest path to node 1 is the direct link (2,1). This is an example of the so-called "bad news phenonenon," whereby the algorithm reacts slowly to a sudden increase in one or more link lengths.

We close this section with a discussion of two weaknesses of the asynchronous Bellman–Ford method. The first is that in the worst case, the algorithm may require an excessive number of iterations to terminate (see the example of Fig. 5.36). This is not due to the asynchronous nature of the algorithm, but rather to the arbitrary choice of initial conditions [an indication is provided by the argument preceding Eq. (5.19)]. A heuristic remedy is outlined in Problem 5.6; for other possibilities, see [JaM82], [Gar87], and [Hum91]. The second weakness, demonstrated in the example of Fig. 5.37, is that in the worst case, the algorithm requires an excessive number of message transmissions. The example of Fig. 5.37 requires an unlikely sequence of events. An analysis given in [TsS90] shows that under fairly reasonable assumptions, the average number of messages required by the algorithm is bounded by a polynomial in N.

5.2.5 Stability of Adaptive Shortest Path Routing Algorithms

We discussed in earlier sections the possibility of using link lengths that reflect the traffic conditions on the links in the recent past. The idea is to assign a large length to a congested link so that the shortest path algorithm will tend to exclude it from a routing path. This sounds attractive at first, but on second thought one gets alerted to the possibility of oscillations. We will see that this possibility is particularly dangerous in datagram networks.

Stability issues in datagram networks. For a simple example of oscillation, consider a datagram network, and suppose that there are two paths along which an origin can send traffic to a destination. Routing along a path during some time period will increase its length, so the other path will tend to be chosen for routing in the next time period, resulting in an oscillation between the two paths. It turns out that a far worse type of oscillation is possible, as illustrated in the following example.

Example 5.4

Consider the 16-node network shown in Fig. 5.38, where node 16 is the only destination. Let the traffic input (in data units/sec) at each node $i = 1, \dots, 7, 9, \dots, 15$ be one unit and

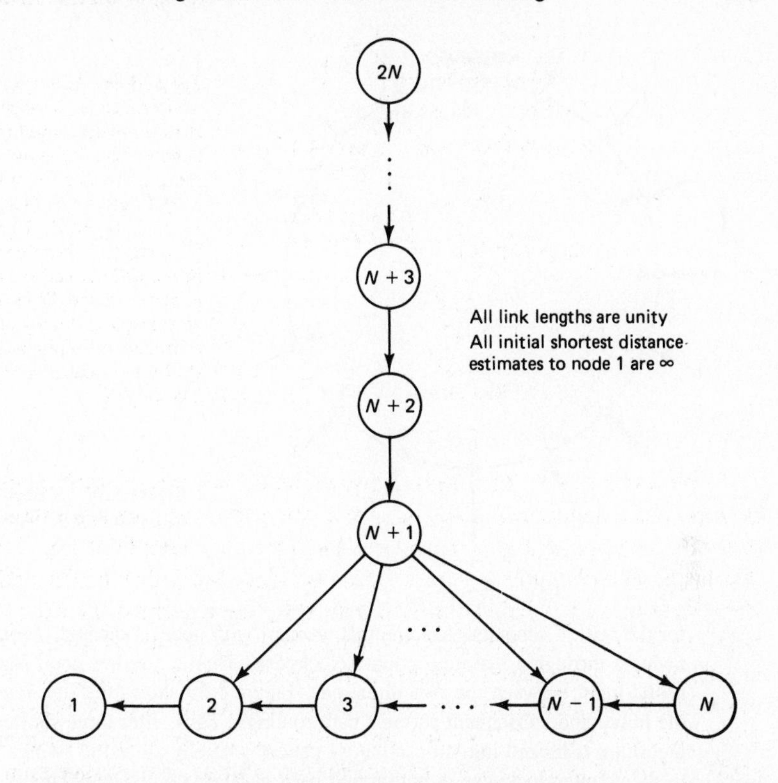

Figure 5.37 Example where the asynchronous version of the Bellman–Ford algorithm requires many more message transmissions than the synchronous version. The initial estimates of shortest distance of all nodes is ∞. Consider the following sequence of events, where all messages are received with zero delay:

1. Node 2 updates its shortest distance and communicates the result to 3. Node 3 updates and communicates the result to 4.... Node $N-1$ updates and communicates the result to N. Node N updates and communicates the result to $N+1$. (These are $N-1$ messages.)
2. Node $N+1$ updates and communicates the result to $N+2$.... Node $2N-1$ updates and communicates the result to $2N$. Node $2N$ updates. (These are $N-1$ messages.)
3. For $i = N-1, N-2, \ldots, 2$, in that order, node i communicates its update to $N+1$ and sequence 2 above is repeated. [These are $N(N-2)$ messages.]

The total number of messages is N^2-2. If the algorithm were executed synchronously, with a message sent by a node to its neighbors only when its shortest distance estimate changes, the number of messages needed would be $3(N-1)-1$. The difficulty in the asynchronous version is that a lot of unimportant information arrives at node $N+1$ early and triggers a lot of unnecessary messages starting from $N+1$ and proceeding all the way to $2N$.

let the traffic input of node 8 be $\epsilon > 0$, where ϵ is very small. Assume that the length of link (i, j) is

$$d_{ij} = F_{ij}$$

where F_{ij} is the arrival rate at the link counting input and relayed traffic. Suppose that all nodes compute their shortest path to the destination every T seconds using as link lengths the arrival rates F_{ij} during the preceding T seconds, and route all their traffic along that path

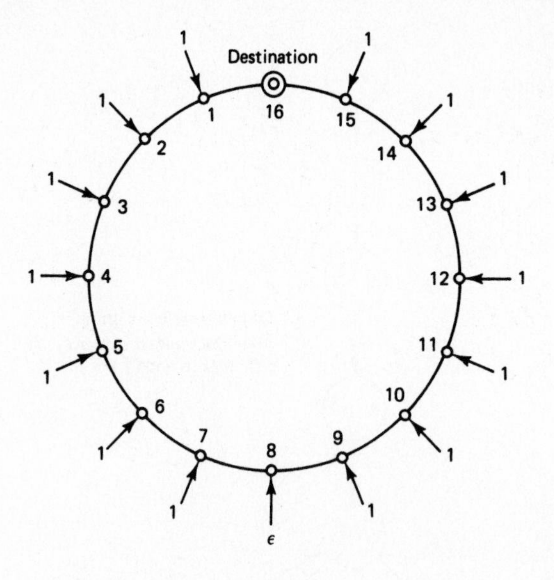

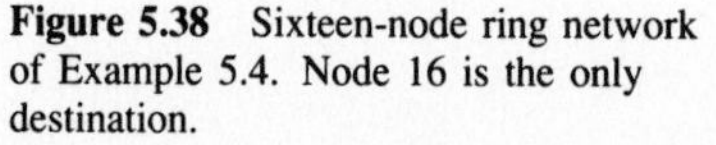
Figure 5.38 Sixteen-node ring network of Example 5.4. Node 16 is the only destination.

for the next T seconds. Assume that we start with nodes 1 through 7 routing clockwise and nodes 8 through 15 routing counterclockwise. This is a rather good routing, balancing the traffic input between the two directions. Figure 5.39 shows the link rates corresponding to the initial and subsequent shortest path routings. Thus, after three shortest path updates, the algorithm is locked into an oscillatory pattern whereby all traffic swings from the clockwise to the counterclockwise direction and back at alternate updates. This is certainly the worst type of routing performance that could occur.

The difficulty in Example 5.4 is due to the fact that link arrival rates depend on routing, which in turn depends on arrival rates via the shortest path calculation, with a feedback effect resulting. This is similar to the stability issue in feedback control theory, and can be analyzed using a related methodology [Ber82b]. Actually, it can be shown that the type of instability illustrated above will occur generically if the length d_{ij} of link (i, j) increases continuously and monotonically with the link arrival rate F_{ij}, and $d_{ij} = 0$ when $F_{ij} = 0$. It is possible to damp the oscillations by adding a positive constant to the link length so that $d_{ij} = \alpha > 0$ when $F_{ij} = 0$. The scalar α (link length at zero load) is known as a *bias* factor.

Figure 5.39 Oscillations in a ring network for link lengths d_{ij} equal to the link arrival rates F_{ij}. Each node sends one unit of input traffic to the destination except for the middle node 8, which sends $\epsilon > 0$, where ϵ is very small. The numbers next to the links are the link rates in each of the two directions. As an example of the shortest path calculations, at the first iteration the middle node 8 computes the length of the clockwise path as 28 $(= 0 + 1 + 2 + \cdots + 7)$, and the length of the counterclockwise path as $28 + 8\epsilon$ $(= \epsilon + 1 + \epsilon + 2 + \epsilon + \cdots + 7 + \epsilon)$, and switches its traffic to the shortest (clockwise) path at the second routing. The corresponding numbers for node 9 are 28 and $28 + 7\epsilon$, so node 9 also switches its traffic to the clockwise path. All other nodes find that the path used at the first routing is shortest, and therefore they do not switch their traffic to the other path.

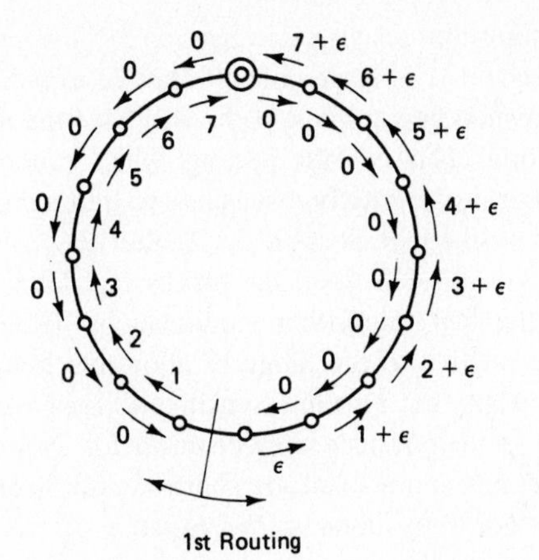

1st Routing

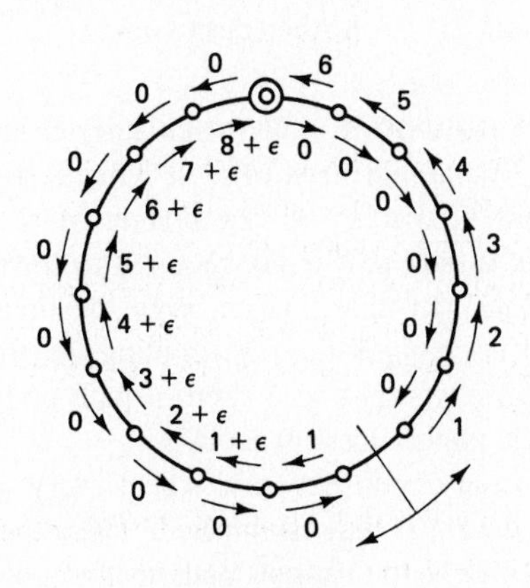

2nd Routing

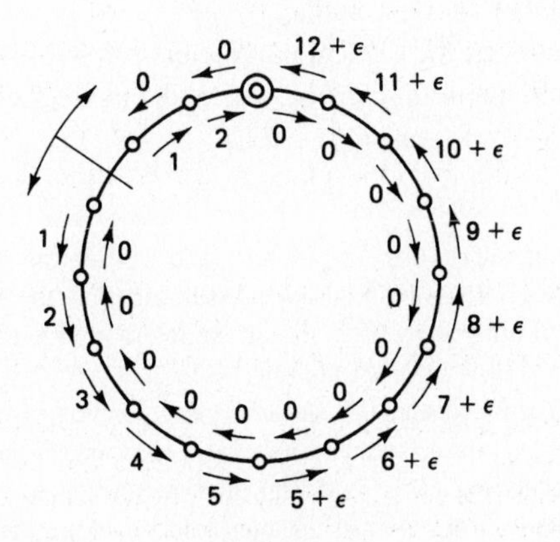

3rd Routing

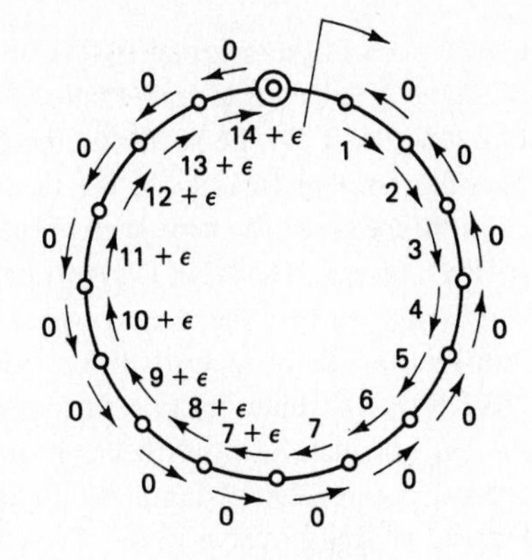

4th Routing

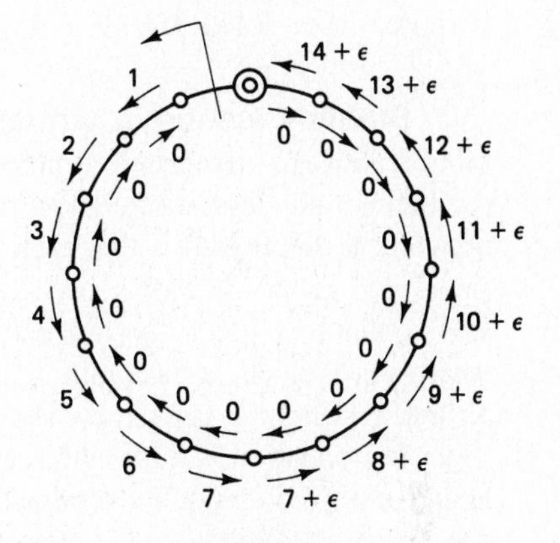

5th Routing

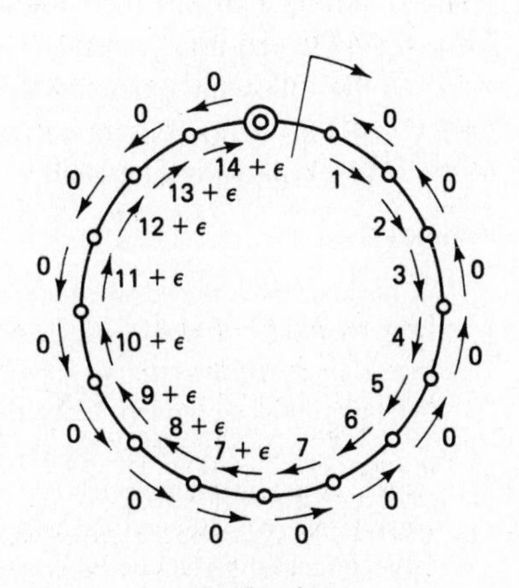

6th Routing

Oscillations in the original ARPANET algorithm discussed in Section 5.1.2 were damped by using a substantial bias factor. Indeed, if α is large enough relative to the range of link lengths, it can be seen that the corresponding shortest paths will have the minimum possible number of links to the destination. This is static routing, which cannot exhibit any oscillation of the type seen earlier, but is also totally insensitive to traffic congestion. In the current ARPANET algorithm described in Section 5.1.2, the bias α in effect equals the sum of the average packet transmission time, the processing delay, and the propagation delay along a link. With the 1987 algorithm modifications ([KhZ89] and [ZVK89]), the bias became fairly large relative to the range of allowable link lengths. As a result the algorithm became quite stable but also less sensitive to congestion. The second possibility to damp oscillations is to introduce a mechanism for averaging the lengths of links over a time period spanning more than one shortest path update. This tends to improve the stability of the algorithm, albeit at the expense of reducing its speed of response to congestion. It turns out that asynchronous shortest path updating by the network nodes results in a form of length averaging that is beneficial for stability purposes. (See [MRR78], [Ber79b], and [Ber82b] for further discussion.)

Stability issues in virtual circuit networks. The oscillatory behavior exhibited above is associated principally with datagram networks. It will be shown that oscillations are less severe in virtual circuit networks. A key feature of a datagram network in this regard is that each packet of a user pair is not required to travel on the same path as the preceding packet. Therefore, the time that an origin–destination pair will continue to use a shortest path after it is changed due to a routing update is very small. As a result, a datagram network reacts very fast to a shortest path update, with all traffic switching to the new shortest paths almost instantaneously.

The situation is quite different in a virtual circuit network, where every session is assigned a fixed communication path at the time it is first established. There the average duration of a virtual circuit is often large relative to the shortest path updating period. As a result, the network reaction to a shortest path update is much more gradual since old sessions continue to use their established communication paths and only new sessions are assigned to the most recently calculated shortest paths.

As the following example demonstrates, the critical parameters for stability are the "speed" of the virtual circuit arrival and departure processes, and the frequency with which shortest paths are updated.

Example 5.5

Consider the simple two-link network with one origin and one destination shown in Fig. 5.40(a). Suppose that the arrival rate is r bits/sec. Assuming that the two links have equal capacity C, an "optimal" routing algorithm should somehow divide the input r equally between the two links, thereby allowing a throughput up to $2C$.

Consider a typical adaptive algorithm based on shortest paths for this example network. The algorithm divides the time axis into T-second intervals. It measures the average arrival rate (bits/second) on both links during each T-second interval, and directs all traffic (datagrams or virtual circuits) generated during every T-second interval along the link that had the smallest rate during the preceding T-second interval.

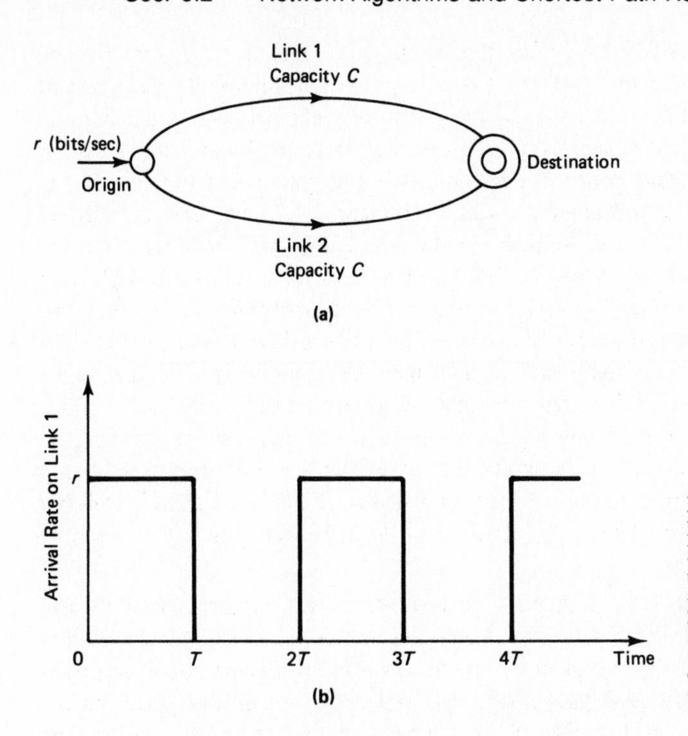

Figure 5.40 (a) Two-link network of Example 5.5. (b) Arrival rate on link 1 in Example 5.5 using the shortest path rule in the datagram case. Essentially, only one path is used for routing at any one time if the shortest path update period is much larger than the time required to empty the queue of waiting packets at the time of an update.

If the network uses datagrams, then, assuming that T is much larger than the time required to empty the queue of waiting packets at the time of an update, each link will essentially carry either no traffic or all the input traffic r at alternate time intervals, as shown in Fig. 5.40(b).

Next consider the case where the network uses virtual circuits, which are generated according to a Poisson process at a rate λ per second. Each virtual circuit uses the link on which it was assigned by the routing algorithm for its entire duration, assumed exponentially distributed with mean $1/\mu$ seconds. Therefore, according to the $M/M/\infty$ queueing results (cf. Section 3.4.2), the number of active virtual circuits is Poisson distributed with mean λ/μ. If γ is the average communication rate (bits/sec) of a virtual circuit, we must have $r = (\lambda/\mu)\gamma$, or

$$\gamma = \frac{r\mu}{\lambda} \tag{5.27}$$

Suppose that the shortest path updating interval T is small relative to the average duration of the virtual circuits $1/\mu$. Then, approximately a fraction μT of the virtual circuits that were carried on each link at the beginning of a T-second interval will be terminated at the end of the interval. There will be λT virtual circuits on the average added on the link that carried the least traffic during the preceding interval. This amounts to an added arrival rate of $\gamma\lambda T$ bits/sec or, using the fact that $\gamma = r\mu/\lambda$ [cf. Eq. (5.27)], $r\mu T$ bits/sec. Therefore, the average rates x_1^k and x_2^k (bits/sec) on the two links at the kth interval will evolve approximately according to

$$x_i^{k+1} = \begin{cases} (1-\mu T)x_i^k + r\mu T, & \text{if } i \text{ is shortest } (i.e., x_i^k = \min\{x_1^k, x_2^k\}) \\ (1-\mu T)x_i^k, & \text{otherwise} \end{cases} \tag{5.28}$$

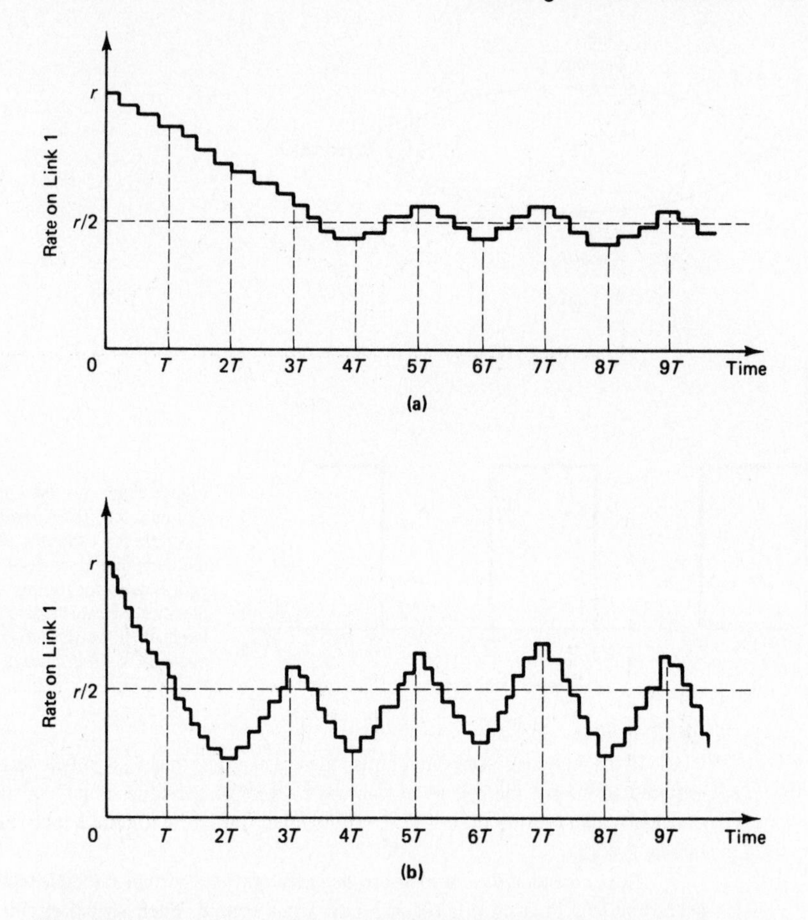

Figure 5.41 Arrival rate on link 1 in Example 5.5 when virtual circuits are used. (a) Virtual circuits last a long time, and the shortest paths are updated frequently (μT is small). (b) Virtual circuits last a short time, and the shortest paths are updated frequently (μT is moderate).

Figures 5.41 and 5.42 provide examples of the evolution of the average rate on link 1. From Eq. (5.28) it can be seen that as $k \to \infty$, the average rates x_1 and x_2 will tend to oscillate around $r/2$ with a magnitude of oscillation roughly equal to $r\mu T$. Therefore, if the average duration of a virtual circuit is large relative to the shortest path updating interval ($\mu T \ll 1$), the routing algorithm performs almost optimally, keeping traffic divided in nearly equal proportions among the two links. Conversely, if the product μT is large, the analysis above indicates considerable oscillation of the link rates. Figures 5.41 and 5.42 demonstrate the relation between μ, T, and the magnitude of oscillation.

Example 5.5 illustrates behavior that has been demonstrated analytically for general virtual circuit networks [GaB83]. It is also shown in [GaB83] that the average duration of a virtual circuit is a critical parameter for the performance of adaptive shortest path

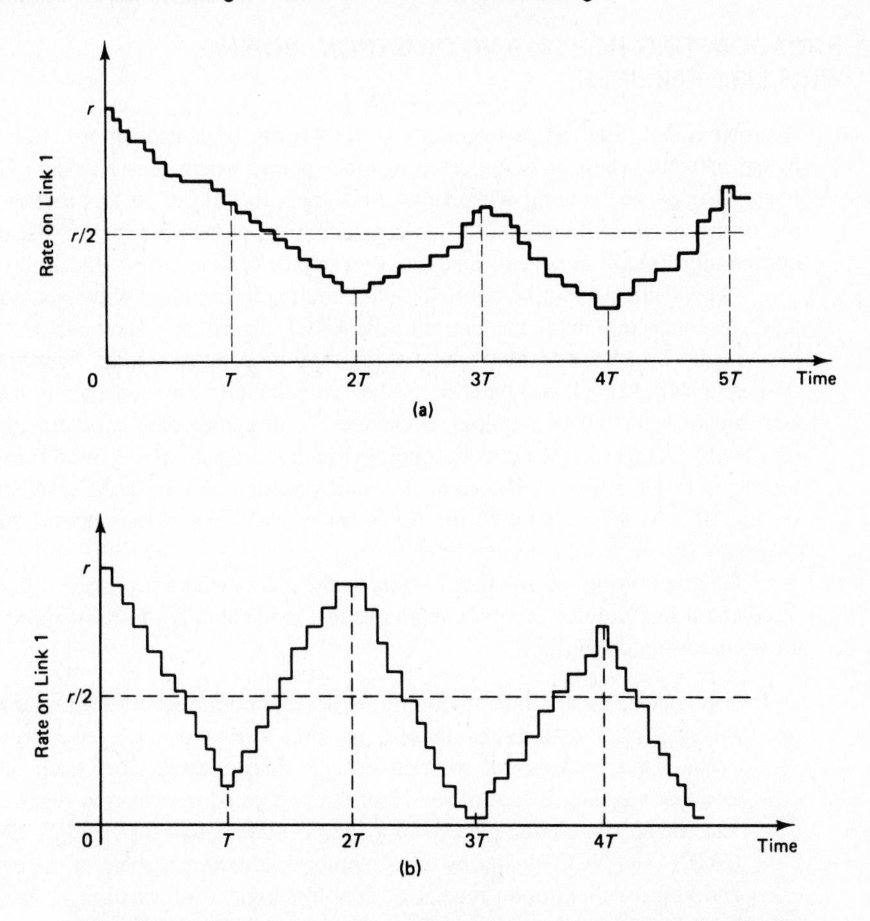

Figure 5.42 Arrival rate on link 1 in Example 5.5 when virtual circuits are used. (a) Virtual circuits last a long time, and the shortest paths are updated infrequently (μT is moderate). (b) Virtual circuits last a short time, and shortest paths are updated infrequently (μT is large). In the limit as $\mu \to \infty$, the datagram case is obtained.

routing. If it is very large, the rate of convergence of the algorithm will be slow, essentially because virtual circuits that are misplaced on congested paths persist for a long time [cf. Figs. 5.41(a) and 5.42(a)]. If it is very small, the shortest path update interval must be accordingly small for the oscillation around optimality to be small [cf. Fig. 5.41(b)]. Unfortunately, there is a practical limit on how small the update interval can be, because frequent updates require more overhead and because sufficient time between updates is needed to measure accurately the current link lengths. The problem with a large duration of virtual circuits would not arise if virtual circuits could be rerouted during their lifetime. In the Codex network described in Section 5.8, such rerouting is allowed, thereby resulting in an algorithm that is more efficient than the one described in this section.

5.3 BROADCASTING ROUTING INFORMATION: COPING WITH LINK FAILURES

A problem that often arises in routing is the transfer of control information from points in the network where it is collected to other points where it is needed. This problem is surprisingly challenging when links are subject to failure. In this section we explain the nature of the difficulties and we consider ways to address them. We also take the opportunity to look at several instructive examples of distributed algorithms.

One practical example of broadcasting routing information was discussed in Section 5.1.2 in connection with the current ARPANET algorithm. Here all link lengths are periodically broadcast to all nodes, which then proceed to update their routing tables through a shortest path computation. Other cases include situations where it is desired to alert higher layers about topological changes, or to update data structures that might be affected by changes in the network topology, such as a spanning tree used for broadcasting messages to all nodes. [The term *network topology* will be used extensively in this section. It is synonymous with the list of nodes and links of the network together with the status (up or down) of each link.]

Getting routing information reliably to the places where it is needed in the presence of potential link failures involves subtleties that are generally not fully appreciated. Here are some of the difficulties:

1. Topological update information must be communicated over links that are themselves subject to failure. Indeed, in some networks, one must provide for the event where portions of the network get disconnected from each other. As an example consider a *centralized* algorithm where all information regarding topological changes is communicated to a special node called the network control center (NCC). The NCC maintains in its memory the routing tables of the entire system, and updates them upon receipt of new topological information by using some algorithm that is of no concern here. It then proceeds to communicate to the nodes affected all information needed for local routing operations. Aside from the generic issue of collecting and disseminating information over failure-prone links, a special problem here is that the NCC may fail or may become disconnected from a portion of the network. In some cases it is possible to provide redundancy and ensure that such difficulties will occur very rarely. In other cases, however, resolution of these difficulties may be neither simple nor foolproof, so that it may be advisable to abandon the centralized approach altogether and adopt a distributed algorithm.
2. One has to deal with multiple topological changes (such as a link going down and up again within a short time), so there is a problem of distinguishing between old and new update information. As an example, consider a flooding method for broadcasting topological change information (cf. Section 5.1.2). Here a node monitors the status of all its outgoing links, and upon detecting a change, sends a packet to all its neighbors reporting that change. The neighbors send this packet to their neighbors, and so on. Some measures are needed to prevent update packets from circulating indefinitely in the network, but this question is discussed

later. Flooding works when there is a single topological change, but can fail in its pure form when there are multiple changes, as shown in the example of Fig. 5.43. Methods for overcoming this difficulty are discussed later in this section, where it is shown that apparently simple solutions sometimes involve subtle failure modes.

3. Topological update information will be processed by some algorithm that computes new routing paths (*e.g.*, a shortest path algorithm). It is possible, however, that new topological update information arrives as this algorithm is running. Then, either the algorithm must be capable of coping with changes in the problem data as it executes, or else it must be aborted and a new version must be started upon receipt of new data. Achieving either of these may be nontrivial, particularly when the algorithm is distributed.

4. The repair of a single link can cause two parts of the network which were disconnected to reconnect. Each part may have out-of-date topology information about the other part. The algorithm must ensure that eventually the two parts agree and adopt the correct network topology.

The difficulties discussed above arise also in the context of broadcasting information related to the congestion status of each link. An important difference, however, is that incorrect congestion information typically has less serious consequences than does incorrect topological information; the former can result at worst in the choice of inferior routes, while the latter can result in the choice of nonexistent routes. Thus, one can afford to allow for a small margin of error in an algorithm that broadcasts routing information other than topological updates, if that leads to substantial algorithmic simplification or reduction in communication overhead. Note, however, that in some schemes, such as the ARPANET flooding algorithm to be discussed shortly, topological update information is embedded within congestion information, and the same algorithm is used to broadcast both.

As we approach the subject of broadcasting topological update information, we must first recognize that it is impossible for every node to know the correct network topology at all times. Therefore, the best that we can expect from an algorithm is that it

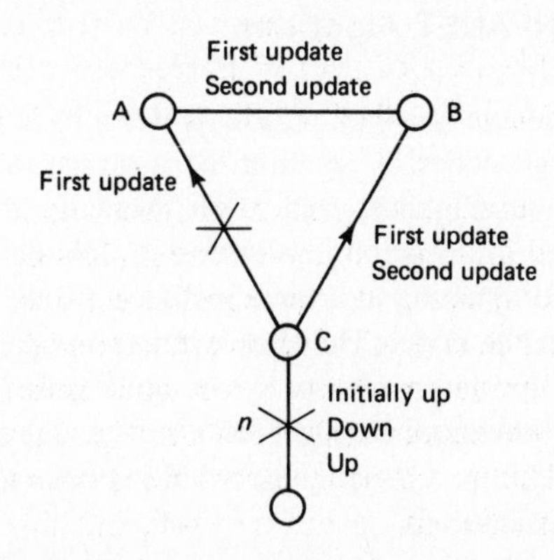

Figure 5.43 Example where the flooding algorithm in its pure form fails. Link n is initially up, then it goes down, then up again. Suppose that the two updates travel on the path CBA faster than the first update travels on the link CA. Suppose also that link CA fails after the first update, but before the second update travels on it. Then, the last message received by A asserts that link n is down, while the link is actually up. The difficulty here is that outdated information is mistakenly taken to be new.

can cope successfully with any finite number of topological changes within finite time. By this we mean that if a finite number of changes occur up to some time and no other changes occur subsequently, all nodes within each connected portion of the network should know the correct status of each link in that portion within finite time.

In the discussion of validity of various schemes in the sense described above, we will assume the following:

1. Network links preserve the order and correctness of transmissions. Furthermore, nodes maintain the integrity of messages that are stored in their memory.
2. A link failure is detected by both end nodes of the link, although not necessarily simultaneously. By this we mean that any link declared down by one end node will also eventually be declared down by the other, before the first end node declares it up again.
3. There is a data link protocol for recognizing, at the ends of a failed link, when the link is again operational. If one of the end nodes declares the link to be up, then within finite time, either the opposite end node also declares the link up or the first declares it down again.
4. A node can crash, in which case each of its incident links is, within finite time, declared down by the node at the opposite end of the link.

The preceding assumptions provide a basis for discussion and analysis of specific schemes, but are not always satisfied in practice. For example, on rare occasions, Assumption 1 is violated because damaged data frames may pass the error detection test of Data Link Control, or because a packet may be altered inside a node's memory due to hardware malfunction. Thus, in addition to analysis of the normal case where these assumptions are satisfied, one must weigh the consequences of the exceptional case where they are not. Keep in mind here that topology updating is a low-level algorithm on which other algorithms rely for correct operation.

Note that it is not assumed that the network will always remain connected. In fact, the topology update algorithm together with some protocol for bringing up links should be capable of starting up the network from a state where all links are down.

5.3.1 Flooding—The ARPANET Algorithm

Flooding was described previously as an algorithm whereby a node broadcasts a topological update message to all nodes by sending the message to its neighbors, which in turn send the message to their neighbors, and so on. Actually, the update messages may include other routing-related information in addition to link status. For example, in the ARPANET, each message originating at a node includes a time average of packet delay on each outgoing link from the node. The time average is taken over 10-sec intervals. The time interval between update broadcasts by the node varies in the ARPANET from 10 to 60 sec, depending on whether there has been a substantial change in any single-link average delay or not. In addition, a message is broadcast once a status change in one of the node's outgoing links is detected.

A serious difficulty with certain forms of flooding is that they may require a large number of message transmissions. The example of Fig. 5.44 demonstrates the difficulty and shows that it is necessary to store enough information in update messages and network nodes to ensure that each message is transmitted by each node only a finite number of times (and preferably only once). In the ARPANET, update messages from each node are marked by a *sequence number*. When a node j receives a message that originated at some node i, it checks to see if its sequence number is greater than the sequence number of the message last received from i. If so, the message together with its sequence number is stored in memory, and its contents are transmitted to all neighbors of j except the one from which the message was received. Otherwise, the message is discarded. Assuming that the sequence number field is large enough, one can be sure that wraparound of the sequence numbers will never occur under normal circumstances. (With a 48-bit field and one update per millisecond, it would take more than 5000 years for wraparound to occur.)

The use of sequence numbers guarantees that each topological update message will be transmitted at most once by each node to its neighbors. Also, it resolves the difficulty of distinguishing between new and old information, which was discussed earlier and illustrated in the example of Fig. 5.43. Nonetheless, there are subtle difficulties with sequence numbers relating to exceptional situations, such as when portions of the network become disconnected or when equipment malfunctions. For example, when a complete or partial network reinitialization takes place (perhaps following a crash of one or more nodes), it may be necessary to reset some sequence numbers to zero, since a node may not remember the sequence number it was using in the past. Consider a time period when two portions of the network are disconnected. During this period, each portion cannot learn about topological changes occurring in the other portion. If sequence numbers are

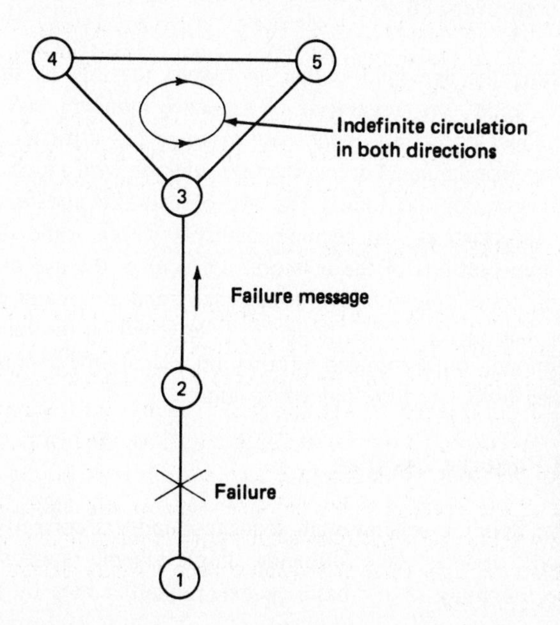

Figure 5.44 Example of a form of flooding where the transmission of messages never terminates. The rule is that a node that receives a message relays it to all of its neighbor nodes except the one from which the message was received. A failure of link (1, 2) is communicated to node 3, which triggers an indefinite circulation of the failure message along the loop (3,4,5) in both directions. This form of flooding works only if the network has no cycles.

reset in one of the two portions while they are disconnected, there may be no way for the two portions to figure out the correct topology after they connect again by relying on the sequence number scheme alone. Another concern is that a sequence number could be altered either inside a node's memory due to a malfunction, or due to an undetected error in transmission. Suppose, for example, that an update message originating at node i reaches another node j, has its sequence number changed accidentally to a high number, and is flooded throughout the network. Then the erroneous sequence number will take over, and in the absence of a correcting mechanism, node i will not be listened to until its own (correct) sequence number catches up with the wrong one. Similarly, the sequence number of a node that is accidentally set to a high number inside the node itself can wrap around, at which time all subsequent update messages from the node will be ignored by the network.

The ARPANET resolves the problems associated with sequence numbers by using two devices:

1. Each update message includes an *age field*, which indicates the amount of time that the message has been circulating inside the network. Each node visited by a message keeps track of the message's arrival time and increments its age field (taking into account the transmission and propagation time) before transmitting it to its neighbors. (The age field is incremented slightly differently in the ARPANET; see [Per83].) Thus, a node can calculate the age of all messages in its memory at any time. A message whose age exceeds a prespecified limit is not transmitted further. Otherwise, the message is transmitted to all neighbor nodes except the one from which the message was received.
2. Each node is required to transmit update messages *periodically* in addition to the messages transmitted when a link status change is detected. (There is at least one of these from every node every 60 sec.)

The rule used regarding the age field is that an "aged" message is superseded by a message that has not "aged" yet, regardless of sequence numbers. (A message that has not "aged" yet is superseded only if the new message has a higher sequence number.) This rule guarantees that damaged or incorrect information with a high sequence number will not be relied upon for too long. The use of periodic update messages guarantees that up-to-date information will become available within some fixed time following reconnection of two portions of the network. Of course, the use of periodic updates implies a substantial communication overhead penalty and is a major drawback of the ARPANET scheme. The effect of this, however, is lessened by the fact that the update packets include additional routing-related information, namely the average packet delay on the node's outgoing links since the preceding update.

5.3.2 Flooding without Periodic Updates

It is possible to operate the flooding scheme with sequence numbers correctly without using an age field or periodic updates. The following simple scheme (suggested by P. Humblet) improves on the reliability of the basic sequence number idea by providing a mechanism for coping with node crashes and some transmission errors. It requires, however, that the sequence number field be so large that wraparound never occurs. The idea is to modify the flooding rules so that the difficulties with reconnecting disconnected network components are dealt with adequately. We describe briefly the necessary modifications, leaving some of the details for the reader to work out. We concentrate on the case of broadcasting topological information, but a similar scheme can be constructed to broadcast other routing information.

As before, we assume that when the status of an outgoing link from a node changes, that node broadcasts an update message containing the status of all its outgoing links to all its current neighbors. The message is stamped by a sequence number which is either zero or is larger by one than the last sequence number used by the node in the past. There is a restriction here, namely that a zero sequence number is allowed only when the node is recovering from a crash (defined as a situation where all of the node's incident links are down and the node is in the process of bringing one or more of these links up). As before, there is a separate sequence number associated with each origin node.

The first flooding rule modification has to do with bringing up links that have been down. When this happens, the end nodes of the link, in addition to broadcasting a regular update message on all of their outgoing links, should exchange their current views of the network topology. By this we mean that they should send to each other all the update messages stored in their memory that originated at other nodes together with the corresponding sequence numbers. These messages are then propagated through the network using the modified flooding rules described below. This modification guarantees that, upon reconnection, the latest update information will reach nodes in disconnected network portions. It also copes with a situation where, after a crash, a node does not remember the sequence number it used in the past. Such a node must use a zero sequence number following the crash, and then, through the topology exchange with its neighbors that takes place when its incident links are brought up, become aware of the highest sequence number present in the other nodes' memories. The node can then increment that highest sequence number and flood the network with a new update message.

For the scheme to work correctly we modify the flooding rule regarding the circumstances under which an update message is discarded. To this end we order, for each node i, the topological update messages originated at i. For two such messages A and B, we say that $A > B$ if A has a greater sequence number than B, or if A and B have identical sequence numbers and the content of A is greater than the content of B according to some lexicographic rule (*e.g.*, if the content of A interpreted as a binary number is greater than the similarly interpreted content of B). Any two messages A and B originating at the same node can thus be compared in the sense that $A > B$, or $B > A$, or $A = B$, the last case occurring only if the sequence numbers and contents of A and B are identical.

Suppose that node j receives an update message A that has originated at node i, and let B be the message originated at node i and currently stored in node j's memory. The message is discarded if $A < B$ or $A = B$. If $A > B$, the flooding rule is now as follows:

1. If $j \neq i$, node j stores A in its memory in place of B and sends a copy of A on all its incident links except the one on which A was received.

2. If $j = i$ (*i.e.*, node i receives a message it issued some time in the past), node i sends to all its neighbors a new update message carrying the current status of all its outgoing links together with a sequence number that is 1 plus the sequence number of A.

Figure 5.45 shows by example why it is necessary to compare the contents of A and B in the case of equal sequence numbers.

The algorithm of this section suffers from a difficulty that seems generic to all event-driven algorithms that do not employ periodic updates: it is vulnerable to memory and transmission errors. Suppose, for example, that a message's sequence number is altered to a very high number either during a transmission or while residing inside a node's memory. Then the message's originating node may not be listened to until its sequence number catches up with the high numbers that are unintentionally stored in some nodes' memories. It turns out that this particular difficulty can be corrected by modifying the flooding rule discussed earlier. Specifically, if node j receives A and has B with $A < B$ in its memory, then message A is discarded as in the earlier rule, but in addition message B is sent back to the neighbor from which A was received. (The neighbor will then propagate B further.) Suppose now that a node issues an update message carrying a sequence number which is less than the sequence number stored in some other node's memory. Then, because of the modification just described, if the two nodes are connected by a path of up links, the higher sequence number (call it k) will be propagated back to the originator node, which can then issue a new update message with sequence number $k + 1$ according to rule 2 above. Unfortunately, there is still a problem associated with a possible sequence number wraparound (*e.g.*, when k above is the highest number possible within the field of possible sequence numbers). There is also still the problem of update messages themselves being corrupted by memory or

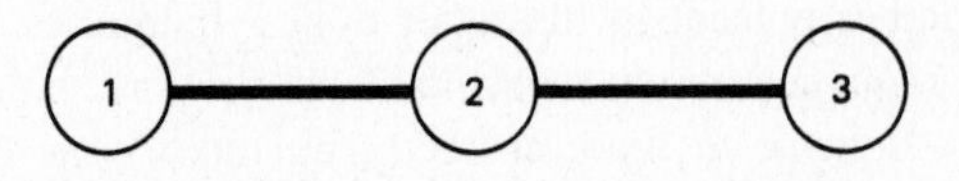

Figure 5.45 Example showing why it is necessary to compare the contents of A and B in the case of equal sequence numbers in the flooding algorithm without periodic updates of Section 5.3.2. Here, there are three nodes, 1, 2, and 3, connected in tandem with two (undirected) links (1,2) and (2,3). Suppose, initially, that both links are up, and the messages stored in all the nodes' memories carry the correct link status information and have a sequence number of zero. Consider a scenario whereby link (2,3) goes down, then link (1,2) goes down, and then link (2,3) comes up while node 2 resets its sequence number to zero. Then, nodes 2 and 3 exchange their (conflicting) view of the status of the directed links (1,2) and (2,1), but if the earlier flooding rules are used, both nodes discard each other's update message since it carries a sequence number zero which is equal to the one stored in their respective memories. By contrast, if the steps of the modified flooding algorithm described above are traced, it is seen that (depending on the lexicographic rule used) either the (correct) view of node 2 regarding link (2,1) will prevail right away, or else node 2 will issue a new update message with sequence number 1 and its view will again prevail. Note that node 3 may end up with an incorrect view of the status of link (1,2). This happens because nodes 1 and 3 are not connected with a path of up links. For the same reason, however, the incorrect information stored by node 3 regarding the outgoing link of node 1 is immaterial.

transmission errors (rather than just their sequence numbers being corrupted). It seems that there is no clean way to address these difficulties other than ad hoc error recovery schemes using an age field and periodic updates (see Problem 5.13).

5.3.3 Broadcast without Sequence Numbers

In this subsection we discuss a flooding-like algorithm called the *Shortest Path Topology Algorithm* (SPTA), which, unlike the previous schemes, does not use sequence numbers and avoids the attendant reset and wraparound problems.

To understand the main idea of the algorithm, consider a generic problem where a special node, called the *center*, has some information that it wishes to broadcast to all other nodes. We model this information as a variable V taking values in some set. As a special case, V could represent the status of the incident links of the center, as in the topology broadcast problem. We assume that after some initial time, several changes of V and several changes of the network topology occur, but there is some time after which there are no changes of either V or the network topology. We wish to have an algorithm by which all nodes get to know the correct value of V after a finite time from the last change.

Consider a scheme whereby each node i stores in memory a value V_i, which is an estimate of the correct value of V. Each time V_i changes, its value is transmitted to all neighbors of i along each of the currently operating outgoing links of i. Also, when a link (i, j) becomes operational after being down, the end nodes i and j exchange their current values V_i and V_j. Each node i stores the last value received by each of its neighbors j, denoted as V_j^i.

We now give a general rule for changing the estimate V_i of each node i. Suppose that we have an algorithm running in the network that maintains a tree of directed paths from all nodes to the center (*i.e.*, a tree rooted at the center; see Fig. 5.46). What we mean here is that this algorithm constructs such a tree within a finite time following the last change in the network topology. Every node i except for the center has a unique successor $s(i)$ in such a tree and we assume that this successor eventually becomes known to i. The rule for changing V_i then is for node i to make it equal to the value $V_{s(i)}^i$ last transmitted by the successor $s(i)$ within a finite time after either the current successor transmits a new value or the successor itself changes. It is evident under these assumptions that each node i will have the correct value V_i within a finite time following the last change in V or in the network topology.

It should be clear that there are many candidate algorithms that maintain a tree rooted at the center in the face of topological changes. One possibility is the asynchronous Bellman–Ford algorithm described in the preceding section, in which case the tree rooted at the center is a shortest path tree. The following SPTA algorithm constructs a separate shortest path tree for each node/center more economically than the asynchronous Bellman–Ford algorithm. The algorithm is given for the special case where the broadcast information is the network topology; in this case the tree construction algorithm and the topological change broadcast procedure are integrated into a single algorithm. The SPTA algorithm can be generalized, however, to broadcast any kind of information.

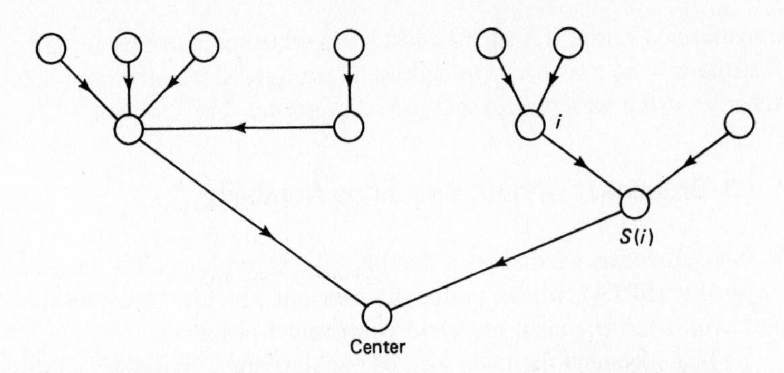

Figure 5.46 Information broadcast over a tree rooted at the center that initially holds the value V. Here there is a unique directed path from every processor to the center and a unique successor $s(i)$ of every node i on the tree. The idea of the broadcast algorithm is to maintain such a tree in the presence of topological changes and to require each node i to adopt (eventually) the latest received message from its successor $s(i)$ (i.e., $V_i = V^i_{s(i)}$). This guarantees that the correct information will eventually be propagated and adopted along the tree even after a finite number of topological changes that may require the restructuring of the tree several times.

Throughout this subsection we will make the same assumptions as with flooding about links preserving the order and correctness of transmissions, and nodes being able to detect incident link status changes.

The data structures that each node i maintains in the SPTA are (see Fig. 5.47):

1. The *main topology table* T_i where node i stores the believed status of every link. This is the "official" table used by the routing algorithm at the node. The main topology tables of different nodes may have different entries at different times. The goal of the SPTA is to get these tables to agree within finite time from the last topological change.
2. The *port topology tables* T^i_j. There is one such table for each neighbor node j. Node i stores the status of every link in the network, as communicated latest by neighbor node j, in T^i_j.

The algorithm consists of five simple rules:

Communication Rules

1. When a link status entry of a node's main topology table changes, the new entry is transmitted on each of the node's operating incident links.
2. When a failed link comes up, each of its end nodes transmits its entire main topology table to the opposite end node over that link. Upon reception of this table the opposite end node enters the new link status in its main and port topology tables.

Topology Table Update Rules

3. When an incident link of a node fails, the failed status is entered into the node's main and port topology tables.

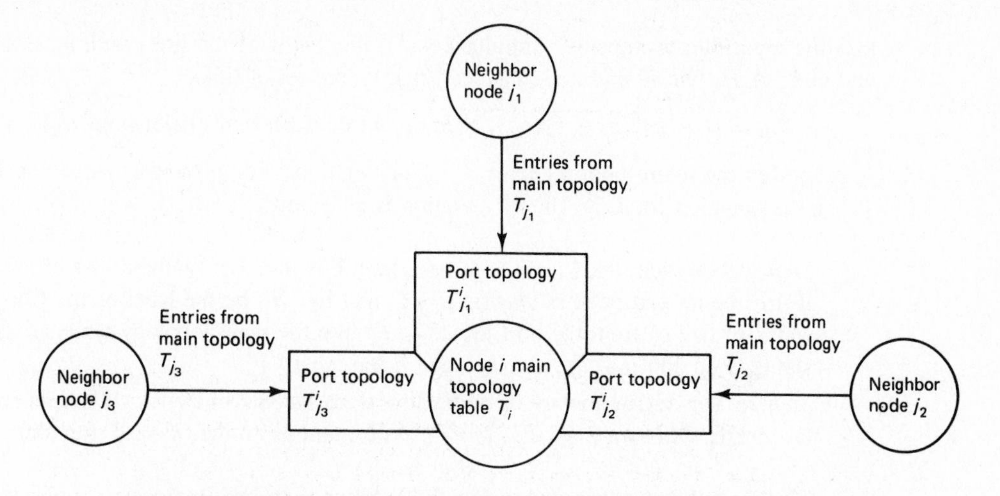

Figure 5.47 Data structures of SPTA at node i are the main topology table T_i and the port topology tables $T^i_{j_1}$, $T^i_{j_2}$, and $T^i_{j_3}$. Changes in the main topology tables of the neighbor nodes j_1, j_2, and j_3 are transferred with some delay to the corresponding port topology tables of node i, except if the change corresponds to a link that is incident to i. Such a change is entered directly in its main and port topology tables. Also, when an incident failed link comes up, node i transmits its entire main topology table over that link. Finally, when an entry of a port topology changes, the entire main topology table is recalculated using the main topology update algorithm.

4. When a node receives a link status message from a neighbor, it enters the message into the port topology table associated with that neighbor.
5. When an entry of a main topology T_i changes due to a status change of an incident link, or an entry of a port topology T^i_j changes due to a communication from neighbor node j, node i updates its main topology table by using the following algorithm. (It is assumed that the algorithm is restarted if an incident link changes status or a new message is received from a neighbor node while the algorithm executes.) The idea in the algorithm is for node i to believe, for each link ℓ, the status communicated by a neighbor that lies on a shortest path from node i to link ℓ. A shortest path is calculated on the basis of a length of unity for each up link and a length of infinity for each down link.

Main Topology Update Algorithm at Node i. The algorithm proceeds in iterations that are similar to those of Dijkstra's shortest path algorithm when all links have a unit length (Section 5.2.3). At the start of the kth iteration, there is a set of nodes P_k. Each node in P_k carries a label which is just the ID number of some neighbor node of i. The nodes in P_k are those that can be reached from i via a path of k links or fewer that are up according to node i's main topology table at the start of the kth iteration. The label of a node $m \in P_k$ is the ID number of the neighbor of i that lies on a path from i to m that has a minimum number of (up) links. In particular, P_1 consists of the nodes connected with i via an up link, and each node in P_1 is labeled with its own ID number. During the kth iteration, P_k is either augmented by some new nodes to form P_{k+1} or else the algorithm terminates. Simultaneously, the status of the links with at least one end node in P_k but no end node in P_{k-1}, that is, the set of links

$$L_k = \{(m,n) \mid m \notin P_{k-1}, n \notin P_{k-1}, \ m \text{ or } n \text{ (or both) belongs to } P_k\}$$

is entered in the main topology table T_i. (For notational completeness, we define $P_0 = \{i\}$ in the equation for L_1.) The kth iteration is as follows:

Step 1: For each link $(m, n) \in L_k$, do the following: Let (without loss of generality) m be an end node belonging to P_k, and let "j" be the label of m. Copy the status of (m, n) from the port topology T_j^i into the main topology table T_i. If this status is up and $n \notin P_k$, give to node n the label "j".

Step 2: Let M_k be the set of nodes that were labeled in step 1. If M_k is empty, terminate. Otherwise, set $P_{k+1} = P_k \cup M_k$, and go to the $(k+1)^{\text{st}}$ iteration.

The algorithm is illustrated in Fig. 5.48. Since each link is processed only once in step 1 of the algorithm, it is clear that the computational requirements are proportional to the number of links. It is straightforward to verify the following properties of the main topology update algorithm:

1. The set P_k, for $k \geq 1$, is the set of nodes m with the property that in the topology T_i finally obtained, there is a path of k or fewer up links connecting m with i.
2. The set L_k, $k \geq 1$, is the set of links ℓ with the property that in the topology T_i finally obtained, all paths of up links connecting one of the end nodes of ℓ with node i has no fewer than k links, and there is exactly one such path with k links.
3. The final entry of T_i for a link $\ell \in L_k$ is the entry from the port topology of a neighbor that is on a path of k up links from i to one of the end nodes of ℓ. If there are two such neighbors with conflicting entries, the algorithm arbitrarily chooses the entry of one of the two.
4. When the main topology update algorithm executes at node i in response to failure of an incident link (i, j), the information in the port topology T_j^i is in effect disregarded. This is due to the fact that node j is not included in the set P_1, and therefore no node is subsequently given the label "j" during execution of the algorithm.

The operation of the SPTA for the case of a single-link status change is shown in Fig. 5.49. The algorithm works roughly like flooding in this case, but there is no need to number the update messages to bound the number of message transmissions. Figure 5.50 illustrates the operation of the SPTA for the case of multiple-link status changes, and illustrates how the SPTA handles the problem of distinguishing old from new information (cf. Fig. 5.43). A key idea here is that when a link (i, j) fails, all information that came over that link is effectively disregarded, as discussed above.

We now provide a proof of correctness of the SPTA. At any given time the correct topology table (as seen by an omniscient observer) is denoted by T^*. We say that at that time *link ℓ is connected with node i* if there is a path connecting i with one of the end nodes of ℓ, and consisting of links that are up according to T^*. We say that at that

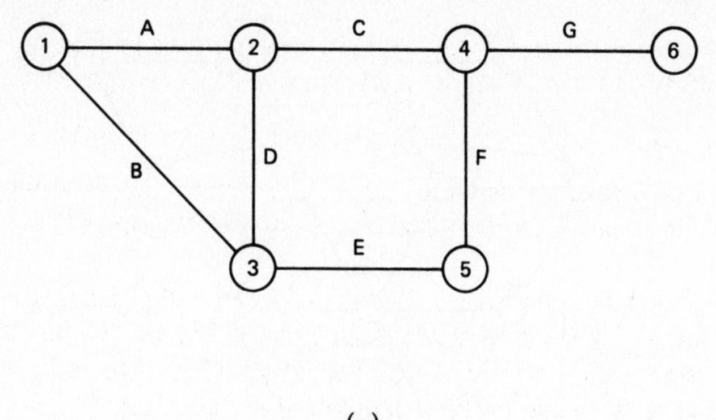

(a)

$$P_1 = \{2,3\},\quad P_2 = \{2,3,5\},\quad P_3 = \{2,3,5,4\},\quad P_4 = \{2,3,5,4,6\}$$
$$L_1 = \{D,C,E\},\quad L_2 = \{F\},\quad L_3 = \{G\}$$

(b)

Link	Port Topology T_2^1	Port Topology T_3^1	Distance	Believes Neighbor	Final Main Topology of Node 1
A	—	—	0	—	U
B	—	—	0	—	U
C	D	U	1	2	D
D	U	U	1	2,3	U
E	U	U	1	3	U
F	U	U	2	3	U
G	D	U	3	3	U

(c)

Figure 5.48 Illustration of the main topology update algorithm for node 1 in the network (a). The node sets P_k and the link sets L_k are given in (b). The contents of the port topologies T_2^1 and T_3^1 at node 1 are given in the table in (c) (where D means down, U means up). There is conflicting information on the status of links C and G. Node 1 believes node 2 on the status of link C since node 2 is closer to this link. Node 1 believes node 3 on the status of link G since node 3 is closer to this link (given that link C was declared down by node 1).

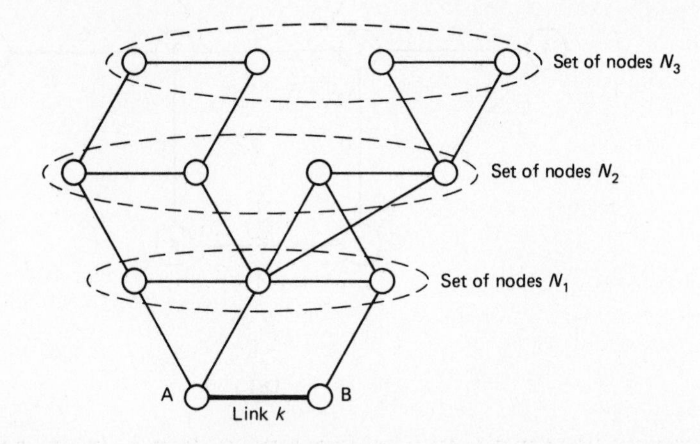

Figure 5.49 Operation of the SPTA for a single-link status change. The status change information propagates to nodes in the set N_1 (one link away from A or B or both), then to nodes in N_2 (two links away from A or B or both), then to N_3, and so on. Specifically, when link k changes status, the change is entered in the topology tables of A and B and is broadcast to the nodes in N_1. It is entered in the port topology tables T_A^i and/or T_B^i of nodes i in N_1. Since the distance to link k through nodes $j \neq A, B$ is greater than zero, the new status of link k will eventually be entered in the main topology tables of each node in N_1. It will then be broadcast to nodes in N_2 and entered in the corresponding port topology tables, and so on. The number of required message transmissions does not exceed $2L$, where L is the number of undirected network links.

time, node j is an *active neighbor* of node i if link (i, j) is up according to T*. We say that at that time, *node i knows the correct topology* if the main topology table T_i agrees with T^* on the status of all links connected with i. We assume that at some initial time t_s, each node knows the correct topology (*e.g.*, at network reset time when all links are down), and all active port topologies T_j^i agree on all entries with the corresponding main topologies T_j. After that time, several links change status. However, there is a time t_0 after which no link changes status and the end nodes of each link are aware of the correct link status. We will show the following:

Proposition. The SPTA works correctly in the sense that under the preceding assumptions, there is a time $t_f \geq t_0$ such that each node knows the correct topology for all $t \geq t_f$.

Figure 5.50 Operation of the SPTA for the topology change scenario of Fig. 5.43. The example demonstrates how the SPTA copes with a situation where pure flooding fails. (a) Example network with four links and four nodes. Initially, all links are up. Three changes occur: (1) link (3,4) goes down, (2) link (3,4) goes up, and (3) link (1,3) goes down. (b) Assumed sequence of events, message transmissions, and receptions of the SPTA. All topology table updating is assumed to take place instantaneously following an event. (c) Main and port topology table contents at each node following each event. U and D mean up and down, respectively. For example, a table entry UUUD means that links (1,2), (2,3), and (1,3) are up, and link (3,4) is down. Tables that change after an event are shown in bold. Note that the last event changes not only the entry for link (1,3) in T_1, but also the entry for link (3,4). Also, after the last event, the port topology table T_1^3 is incorrect, but this is inconsequential since the connecting link (1,3) is down.

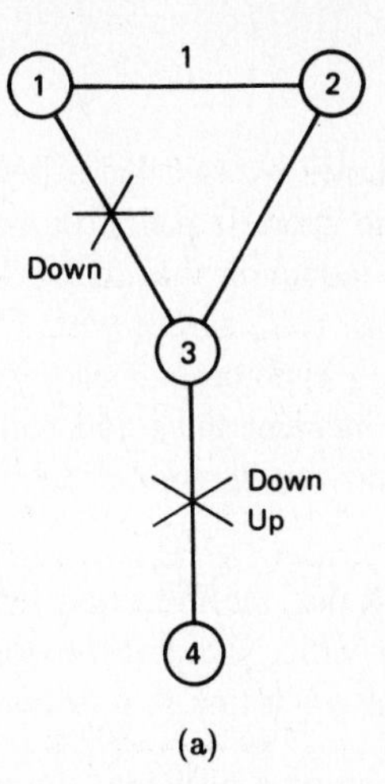

(a)

Event Number	1	2	3
Event	Link (3,4) goes down. Message is transmitted from 3 to 1 and 2. Message is received at 2.	Message is transmitted from 2 to 1. Message is received at 1.	Link (3,4) goes up. Message is transmitted from 3 to 1 and 2. Message is received at 2.
Event Number	**4**	**5**	**6**
Event	Message is transmitted from 2 to 1. Message is received at 1.	Message from 3 to 1 sent during Event 1 is received.	Link (1,3) goes down. Messages from 1 and 3 are sent to 2 and 4. Messages received at 2 and 4.

(b)

Event Number		1	2	3	4	5	6
	T_1	UUUU	UUUU	UUUU	UUUU	**UUUD**	**UUDU**
Tables at Node 1	T_1^2	UUUU	**UUUD**	UUUD	**UUUU**	UUUU	**UUDU**
	T_1^3	UUUU	UUUU	UUUU	UUUU	**UUUD**	**UUDD**
	T_2	**UUUD**	UUUD	**UUUU**	UUUU	UUUU	**UUDU**
Tables at Node 2	T_2^1	UUUU	UUUU	UUUU	UUUU	UUUU	**UUDU**
	T_2^3	**UUUD**	UUUD	**UUUU**	UUUU	UUUU	**UUDU**
	T_3	**UUUD**	UUUD	**UUUU**	UUUU	UUUU	**UUDU**
Tables at Node 3	T_3^1	**UUUD**	UUUD	**UUUU**	UUUU	UUUU	**UUDU**
	T_1^2	**UUUD**	UUUD	**UUUU**	UUUU	UUUU	**UUDU**
	T_1^4	**UUUD**	UUUD	**UUUU**	UUUU	UUUU	**UUDU**
Tables at Node 4	T_4	**UUUD**	UUUD	**UUUU**	UUUU	UUUU	**UUDU**
	T_4^3	**UUUD**	UUUD	**UUUU**	UUUU	UUUU	**UUDU**

(c)

Proof: In what follows we say that a link ℓ is at distance n away from i if in the graph defined by T^*, the shortest path from i to the closest end node of ℓ is n links long. We will show by induction that for each integer $n \geq 0$, there is a time $t_n \geq t_0$ after which, for each node i, T_i agrees with T^* for all links that are at a distance n or less from i. The induction hypothesis is clearly true for $n = 0$, since each node i knows the correct status of its incident links and records them in its main topology table T_i. We first establish the following lemma.

Lemma. Assume that the induction hypothesis is true for time t_n. Then, there is a time $t'_{n+1} \geq t_n$ after which the port topology table T^i_j for each active neighbor j of node i agrees with T^* for each link at a distance n or less from j.

Proof: Consider waiting a sufficient amount of time after t_n for all messages which were sent from j to i before t_n to arrive. By rules 1, 2, and 4 of the algorithm, T^i_j agrees with T_j for all links which are not incident to i. Therefore, by the induction hypothesis, T^i_j agrees with T^* for each link at a distance of n or less from j. This proves the lemma.

To complete the proof of the proposition, one must show that there is a time $t_{n+1} \geq t'_{n+1}$ such that for all $t \geq t_{n+1}$ and nodes i, T_i agrees with T^* for each link ℓ which is at a distance $n + 1$ from i. Consider the first time that link ℓ is processed by the topology update algorithm after all port topologies T^i_j and T^* agree on all links at distance n or less from j as per the preceding lemma. Then link ℓ will belong to the set L_{n+1}. Also, the closest end node(s) of ℓ to node i will belong to the set P_{n+1} and will have a label which is one of the active neighbors of i (say j) that is at distance n from ℓ. By the lemma, the port topology T^i_j agrees with T^* on all links at distance n from j. Therefore, the entry of T^i_j for link ℓ, which will be copied into T_i when link ℓ is processed, will agree with the corresponding entry of T^*. Because the conclusion of the lemma holds for all t after t'_{n+1}, the entry for link ℓ in T_i will also be correct in all subsequent times at which link ℓ will be processed by the main topology update algorithm. **Q.E.D.**

Note that the lemma above establishes that the active port topology tables also eventually agree with the corresponding main topology tables. This shows that the initial conditions for which validity of the algorithm was shown reestablish themselves following a sufficiently long period for which no topological changes occur.

The main advantage of the SPTA over the ARPANET flooding algorithm is that it does not require the use of sequence numbers with the attendant reset difficulties, and the use of an age field, the proper size of which is network dependent and changes as the network expands. Furthermore, the SPTA is entirely event driven and does not require the substantial overhead associated with the regular periodic updates of the ARPANET algorithm.

Consider now the behavior of SPTA when we cannot rely on the assumption that errors in transmission or in nodes' memories are always detected and corrected. Like all event-driven algorithms, SPTA has no ability to correct such errors. Without periodic retransmission, an undetected error can persist for an arbitrary length of time. Thus, in situations where extraordinary reliability is needed, a node should periodically retransmit its main table to each of its neighbors. This would destroy the event-driven property of SPTA, but would maintain its other desirable characteristics and allow recovery from undetected database errors.

As mentioned earlier, the SPTA can be generalized to broadcast information other than link status throughout the network. For example, in the context of an adaptive routing algorithm, one might be interested in broadcasting average delay or traffic arrival rate information for each link. What is needed for the SPTA is to allow for additional, perhaps nonbinary, information entry in the main and port topology tables in addition to the up–down status of links. When this additional information depends on the link direction, it is necessary to make a minor modification to the SPTA so that there are separate entries in the topology tables for each link direction. This matter is considered in Problem 5.14.

5.4 FLOW MODELS, OPTIMAL ROUTING, AND TOPOLOGICAL DESIGN

To evaluate the performance of a routing algorithm, we need to quantify the notion of traffic congestion. In this section we formulate performance models based on the traffic arrival rates at the network links. These models, called *flow models* because of their relation to network flow optimization models, are used to formulate problems of optimal routing that will be the subject of Sections 5.5 to 5.7. Flow models are also used in our discussion of the design of various parts of the network topology in Sections 5.4.1 to 5.4.3.

Traffic congestion in a data network can be quantified in terms of the statistics of the arrival processes of the network queues. These statistics determine the distributions of queue length and packet waiting time at each link. It is evident that desirable routing is associated with a small mean and variance of packet delay at each queue. Unfortunately, it is generally difficult to express this objective in a single figure of merit suitable for optimization. A principal reason is that, as seen in Chapter 3, there is usually no accurate analytical expression for the means or variances of the queue lengths in a data network.

A convenient but somewhat imperfect alternative is to measure congestion at a link in terms of the *average* traffic carried by the link. More precisely, we assume that the statistics of the arrival process at each link (i, j) change due only to routing updates, and that we measure congestion on (i, j) via the traffic arrival rate F_{ij}. We call F_{ij} the *flow* of link (i, j), and we express it in data units/sec where the data units can be bits, packets, messages, and so on. Sometimes, it is meaningful to express flow in units that are assumed directly proportional to data units/sec, such as virtual circuits traversing the link.

Implicit in flow models is the assumption that the statistics of the traffic entering the network do not change over time. This is a reasonable hypothesis when these statistics change very slowly relative to the average time required to empty the queues in the network. A typical network where such conditions hold is one accommodating a large number of users for each origin–destination pair, with the traffic rate of each of these users being small relative to the total rate (see Section 3.4.2).

An expression of the form

$$\sum_{(i,j)} D_{ij}(F_{ij}) \tag{5.29}$$

where each function D_{ij} is monotonically increasing, is often appropriate as a cost function for optimization. A frequently used formula is

$$D_{ij}(F_{ij}) = \frac{F_{ij}}{C_{ij} - F_{ij}} + d_{ij}F_{ij} \tag{5.30}$$

where C_{ij} is the transmission capacity of link (i, j) measured in the same units as F_{ij}, and d_{ij} is the processing and propagation delay. With this formula, the cost function (5.29) becomes the average number of packets in the system based on the hypothesis that each queue behaves as an $M/M/1$ queue of packets—a consequence of the Kleinrock independence approximation and Jackson's Theorem discussed in Sections 3.6 and 3.8. Although this hypothesis is typically violated in real networks, the cost function of Eqs. (5.29) and (5.30) represents a useful measure of performance in practice, principally because it expresses qualitatively that congestion sets in when a flow F_{ij} approaches the corresponding link capacity C_{ij}. Another cost function with similar qualitative properties is given by

$$\max_{(i,j)} \left\{ \frac{F_{ij}}{C_{ij}} \right\} \tag{5.31}$$

(*i.e.*, maximum link utilization). A computational study [Vas79] has shown that it typically makes little difference whether the cost function of Eqs. (5.29) and (5.30) or that of Eq. (5.31) is used for routing optimization. This indicates that one should employ the cost function that is easiest to optimize. In what follows we concentrate on cost functions of the form $\sum_{(i,j)} D_{ij}(F_{ij})$. The analysis and computational methods of Sections 5.5 to 5.7 extend to more general cost functions (see Problem 5.27).

We now formulate a problem of optimal routing. For each pair $w = (i, j)$ of distinct nodes i and j [also referred to as an origin–destination (or OD) pair], the input traffic arrival process is assumed stationary with rate r_w.* Thus r_w (measured in data units/sec) is the arrival rate of traffic entering the network at node i and destined for node j. The routing objective is to divide each r_w among the many paths from origin to destination in a way that the resulting total link flow pattern minimizes the cost function (5.29). More precisely, denote

W = Set of all OD pairs

P_w = Set of all directed paths connecting the origin and destination nodes of OD pair w (in a variation of the problem, P_w is a given subset of the set of all directed paths connecting origin and destination of w; the optimality conditions and algorithms to be given later apply nearly verbatim in this case)

*Sometimes it is useful to adopt a broader view of an OD pair and consider it simply as a class of users that shares the same set of paths. This allows modeling of multiple priority classes of users. In this context we would allow several OD pairs to have the same origin and destination nodes. The subsequent analysis and algorithms extend to this broader context almost verbatim. To simplify the following exposition, however, we assume that there is at most one OD pair associated with each pair of nodes.

x_p = Flow (data units/sec) of path p

Then the collection of all path flows $\{x_p \mid w \in W, p \in P_w\}$ must satisfy the constraints

$$\sum_{p \in P_w} x_p = r_w, \qquad \text{for all } w \in W,$$

$$x_p \geq 0, \qquad \text{for all } p \in P_w, w \in W$$

as shown in Fig. 5.51. The total flow F_{ij} of link (i, j) is the sum of all path flows traversing the link

$$F_{ij} = \sum_{\substack{\text{all paths } p \\ \text{containing } (i,j)}} x_p$$

Consider a cost function of the form

$$\sum_{(i,j)} D_{ij}(F_{ij}) \tag{5.32}$$

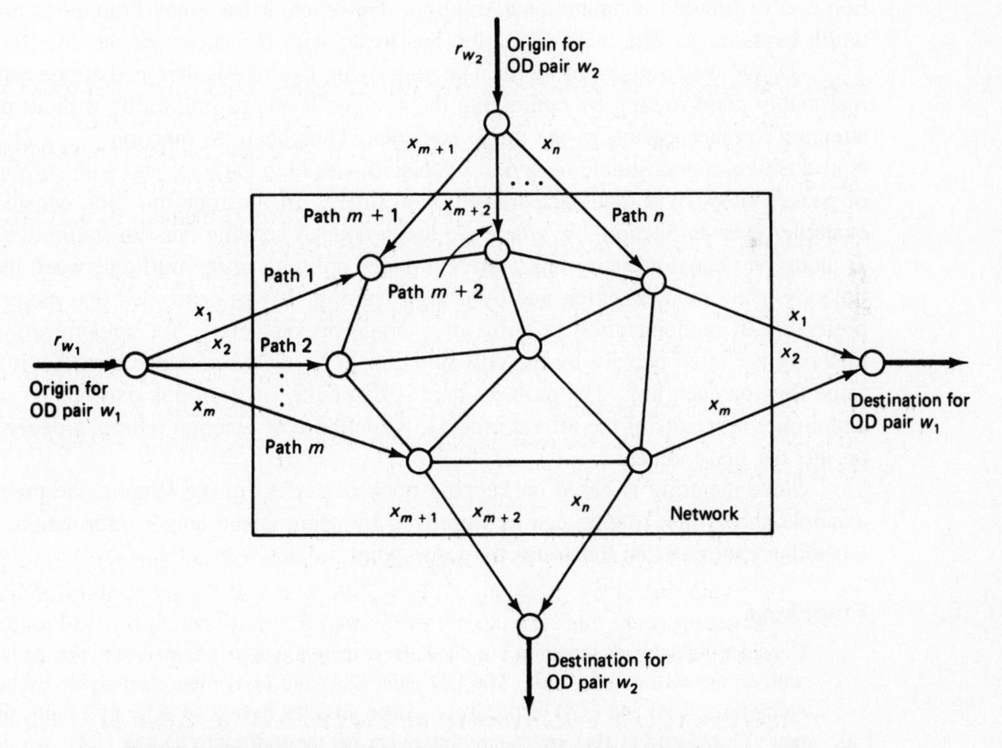

Figure 5.51 Schematic representation of a network with two OD pairs w_1 and w_2. The paths of the OD pairs are $P_{w_1} = \{1, 2, \ldots, m\}$ and $P_{w_2} = \{m + 1, m + 2, \ldots, n\}$. The traffic inputs r_{w_1} and r_{w_2} are to be divided into the path flows $x_1, \ldots, x_m$ and $x_{m+1}, \ldots, x_n$, respectively.

and the problem of finding the set of path flows $\{x_p\}$ that minimize this cost function subject to the constraints above.

By expressing the total flows F_{ij} in terms of the path flows in the cost function (5.32), the problem can be written as

$$\text{minimize} \sum_{(i,j)} D_{ij}\left[\sum_{\substack{\text{all paths } p \\ \text{containing } (i,j)}} x_p\right]$$

$$\text{subject to} \sum_{p \in P_w} x_p = r_w, \qquad \text{for all } w \in W \tag{5.33}$$

$$x_p \geq 0, \quad \text{for all } p \in P_w, w \in W$$

Thus, the problem is formulated in terms of the unknown path flows $\{x_p \mid p \in P_w, w \in W\}$. This is the main optimal routing problem that will be considered. A characterization of its optimal solution is given in Section 5.5, and interestingly, it is expressed in terms of shortest paths. Algorithms for its solution are given in Sections 5.6 and 5.7.

The optimal routing problem just formulated is amenable to analytical investigation and distributed computational solution. However, it has some limitations that are worth explaining. The main limitation has to do with the choice of the cost function $\sum_{(i,j)} D_{ij}(F_{ij})$ as a figure of merit. The underlying hypothesis here is that one achieves reasonably good routing by optimizing the average levels of link traffic without paying attention to other aspects of the traffic statistics. Thus, the cost function $\sum_{(i,j)} D_{ij}(F_{ij})$ is insensitive to undesirable behavior associated with high variance and with correlations of packet interarrival times and transmission times. To illustrate this fact, consider the example given in Section 3.6, where we had a node A sending Poisson traffic to a node B along two equal-capacity links. We compared splitting of the traffic between the two links by using randomization and by using metering. It was concluded that metering is preferable to randomization in terms of average packet delay. Yet randomization and metering are rated equally by the cost function $\sum_{(i,j)} D_{ij}(F_{ij})$, since they result in the same flow on each link. The problem here is that delay on each link depends on second and higher moments of the arrival process, while the cost function reflects a dependence on just the first moment.

Since metering is based on keeping track of current queue lengths, the preceding example shows that routing can be improved by using queue length information. Here is another example that illustrates the same point.

Example 5.6

Consider the network shown in Fig. 5.52. Here there are three origin–destination pairs, each with an arrival rate of 1 unit. The OD pairs (2,4) and (3,4) route their traffic exclusively along links (2,4) and (3,4) respectively. Since all links have a capacity of 2 units, the two paths (1,2,4) and (1,3,4) are equally attractive for the traffic of OD pair (1,4). An optimal routing algorithm based on time average link rates would alternate sending packets of OD pair (1,4) along the two paths, thereby resulting in a total rate of 1.5 units on each of the bottleneck links (2,4) and (3,4). This would work reasonably well if the traffic of the OD

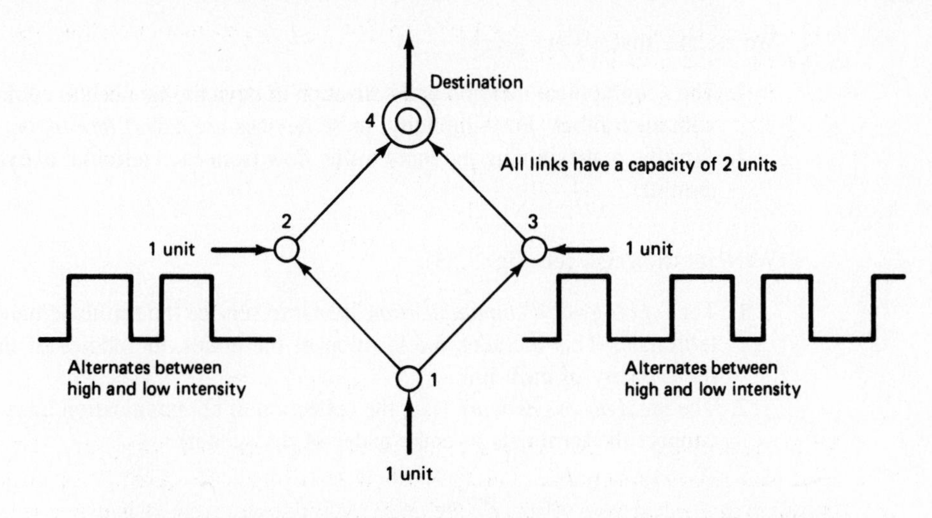

Figure 5.52 Network of Example 5.6 where routing based on queue length information is beneficial. The traffic originating at nodes 2 and 3 alternates in intensity between 2 units and 0 units over large time intervals. A good form of routing gets information on the queue sizes at links (2,4) and (3,4) and routes the traffic originating at node 1 on the path of least queue size.

pairs (2,4) and (3,4) had a Poisson character. Consider, however, the situation where instead, this traffic alternates in intensity between 2 units and 0 units over large time intervals. Then the queues in link (2,4) and (3,4) would build up (not necessarily simultaneously) over some large time intervals and empty out during others. Suppose now that a more dynamic form of routing is adopted, whereby node 1 is kept informed about queue sizes at links (2,4) and (3,4), and routes packets on the path of least queue size. Queueing delay would then be greatly reduced.

Unfortunately, it is impractical to keep nodes informed of all queue lengths in a large network. Even if the overhead for doing so were not prohibitive, the delays involved in transferring the queue length information to the nodes could make this information largely obsolete. It is not known at present how to implement effective and practical routing based on queue length information, so we will not consider the subject further. For an interesting but untested alternative, see Problem 5.38.

The remainder of this section considers the use of network algorithms, flow models, and the optimal routing problem in the context of topological design of a network. The reader may skip this material without loss of continuity.

5.4.1 An Overview of Topological Design Problems

In this subsection we discuss algorithms for designing the topology of a data network. Basically, we are given a set of traffic demands, and we want to put together a network that will service these demands at minimum cost while meeting some performance requirements. A common, broadly stated formulation of the problem is as follows:

We assume that we are given:

1. The geographical location of a collection of devices that need to communicate with each other. For simplicity, these devices are called *terminals*.
2. A traffic matrix giving the input traffic flow from each terminal to every other terminal.

We want to design (cf. Fig. 5.53):

1. The *topology of a communication subnet* to service the traffic demands of the terminals. This includes the location of the nodes, the choice of links, and the capacity of each link.
2. The *local access network* (*i.e.*, the collection of communication lines that will connect the terminals to entry nodes of the subnet).

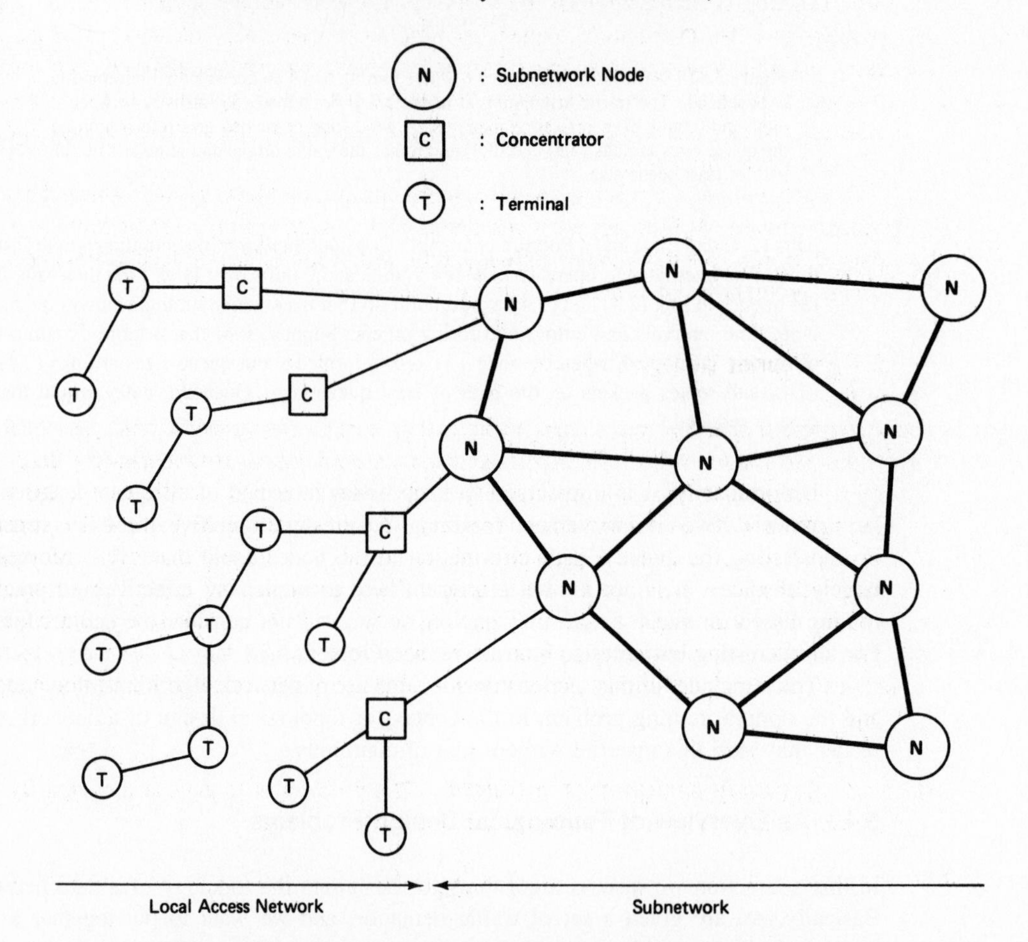

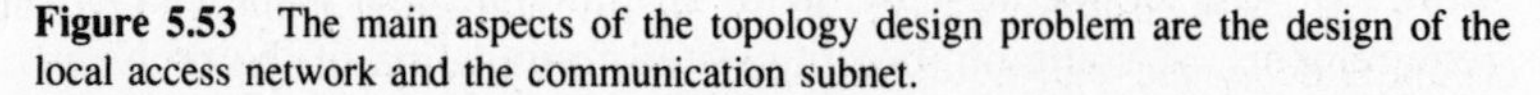

Figure 5.53 The main aspects of the topology design problem are the design of the local access network and the communication subnet.

The design objectives are:

1. Keep the average delay per packet or message below a given level (for the given nominal traffic demands and assuming some type of routing algorithm)
2. Satisfy some reliability constraint to guarantee the integrity of network service in the face of a number of link and node failures
3. Minimize a combination of capital investment and operating costs while meeting objectives 1 and 2

It is evident that this is, in general, a very broad problem that may be difficult to formulate precisely, let alone solve. However, in many cases, the situation simplifies somewhat. For example, part of the problem may already be solved because portions of the local access network and/or the subnet might already be in place. There may be also a natural decomposition of the problem—for example, the subnet topology design problem may be naturally decoupled from the local access network design problem. In fact, this will be assumed in the presentation. Still, one seldom ends up with a clean problem that can be solved exactly in reasonable time. One typically has to be satisfied with an approximate solution obtained through heuristic methods that combine theory, trial and error, and common sense.

In Section 5.4.2 we consider the subnet design problem, assuming that the local access network has been designed and therefore, the matrix of input traffic flow for every pair of subnet nodes is known. Subsequently, in Section 5.4.3, the local access network design problem is considered.

5.4.2 Subnet Design Problem

Given the location of the subnet nodes and an input traffic flow for each pair of these nodes, we want to select the capacity and flow of each link so as to meet some delay and reliability constraints while minimizing costs. Except in very simple cases, this turns out to be a difficult combinatorial problem. To illustrate the difficulties, we consider a simplified version of the problem, where we want to choose the link capacities so as to minimize a linear cost. We impose a delay constraint but neglect any reliability constraints. Note that by assigning zero capacity to a link we effectively eliminate the link. Therefore, the capacity assignment problem includes as a special case the problem of choosing the pairs of subnet nodes that will be directly connected with a communication line.

Capacity assignment problem. The problem is to choose the capacity C_{ij} of each link (i,j) so as to minimize the linear cost

$$\sum_{(i,j)} p_{ij} C_{ij} \tag{5.34}$$

where p_{ij} is a known positive price per unit capacity, subject to the constraint that the average delay per packet should not exceed a given constant T.

The flow on each link (i, j) is denoted F_{ij} and is expressed in the same units as capacity. We adopt the $M/M/1$ model based on the Kleinrock independence approximation, so we can express the average delay constraint as

$$\frac{1}{\gamma}\sum_{(i,j)} \frac{F_{ij}}{C_{ij} - F_{ij}} \leq T \tag{5.35}$$

where γ is the total arrival rate into the network. We assume that there is a given input flow for each origin–destination pair and γ is the sum of these. The link flows F_{ij} depend on the known input flows and the scheme used for routing. We will first assume that routing and therefore also the flows F_{ij} are known, and later see what happens when this assumption is relaxed.

When the flows F_{ij} are known, the problem is to minimize the linear cost $\sum_{(i,j)} p_{ij} C_{ij}$ over the capacities C_{ij} subject to the constraint (5.35), and it is intuitively clear that the constraint will be satisfied as an equality at the optimum. We introduce a Lagrange multiplier β and form the Lagrangian function

$$L = \sum_{(i,j)} \left(p_{ij} C_{ij} + \frac{\beta}{\gamma} \frac{F_{ij}}{C_{ij} - F_{ij}} \right)$$

In accordance with the Lagrange multiplier technique, we set the partial derivatives $\partial L / \partial C_{ij}$ to zero:

$$\frac{\partial L}{\partial C_{ij}} = p_{ij} - \frac{\beta F_{ij}}{\gamma (C_{ij} - F_{ij})^2} = 0$$

Solving for C_{ij} gives

$$C_{ij} = F_{ij} + \sqrt{\frac{\beta F_{ij}}{\gamma p_{ij}}} \tag{5.36}$$

and substituting in the constraint equation, we obtain

$$T = \frac{1}{\gamma}\sum_{(i,j)} \frac{F_{ij}}{C_{ij} - F_{ij}} = \sum_{(i,j)} \sqrt{\frac{p_{ij} F_{ij}}{\beta \gamma}}$$

From the last equation,

$$\sqrt{\beta} = \frac{1}{T} \sum_{(i,j)} \sqrt{\frac{p_{ij} F_{ij}}{\gamma}} \tag{5.37}$$

which when substituted in Eq. (5.36) yields the optimal solution:*

$$C_{ij} = F_{ij} + \frac{1}{T}\sqrt{\frac{F_{ij}}{\gamma p_{ij}}} \sum_{(m,n)} \sqrt{\frac{p_{mn}F_{mn}}{\gamma}}$$

The solution, after rearranging, can also be written as

$$C_{ij} = F_{ij}\left(1 + \frac{1}{\gamma T}\frac{\sum_{(m,n)}\sqrt{p_{mn}F_{mn}}}{\sqrt{p_{ij}F_{ij}}}\right) \tag{5.38}$$

Finally, by substitution in the cost function $\sum_{(i,j)} p_{ij}C_{ij}$, the optimal cost is expressed as

$$\text{optimal cost} = \sum_{(i,j)} p_{ij}F_{ij} + \frac{1}{\gamma T}\left(\sum_{(i,j)}\sqrt{p_{ij}F_{ij}}\right)^2 \tag{5.39}$$

Consider now the problem of optimizing the network cost with respect to both the capacities C_{ij} and the flows F_{ij} (equivalently, the routing). This problem can be solved by minimizing the cost (5.39) with respect to F_{ij} and then obtaining the optimal capacities from the closed-form expression (5.38). Unfortunately, it turns out that the cost (5.39) has many local minima and is very difficult to minimize. Furthermore, in these local minima, there is a tendency for many of the flows F_{ij} and corresponding capacities C_{ij} to be zero. The resulting networks tend to have low connectivity (few links with large capacity) and may violate reliability constraints. The nature of this phenomenon can be understood by considering the simple network of Fig. 5.54 involving two nodes and n links. Suppose that we want to choose the capacities $C_1, \ldots, C_n$ so as to minimize the sum $C_1 + \cdots + C_n$ while dividing the input rate λ among the n links so as to meet an average delay constraint. From our comparison of statistical and time-division multiplexing (Sections 3.1 and 3.3), we know that delay increases if a transmission line is divided into smaller lines, each serving a fraction of the total traffic. It is therefore evident that the optimal solution is the minimal connectivity topology whereby all links are eliminated, except one which has just enough capacity to meet the delay constraint.

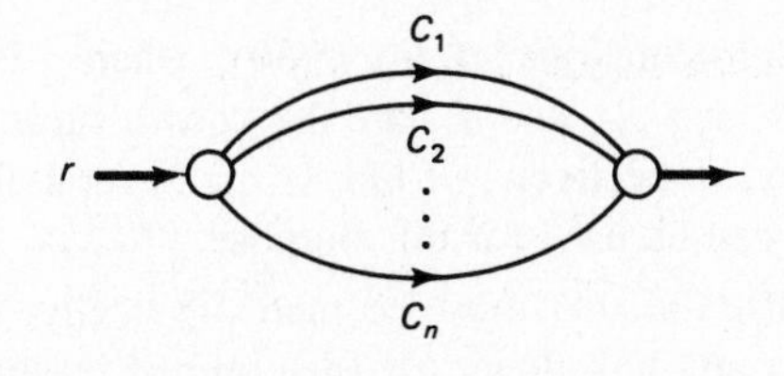

Figure 5.54 Minimizing the total capacity $C_1 + \cdots + C_n$ while meeting an average delay constraint results in a minimum connectivity network (all links are eliminated except one).

*It is rigorous to argue at this point that the capacities C_{ij} of Eq. (5.38) are optimal. The advanced reader can verify this by first showing that the average delay expression of Eq. (5.35) is convex and differentiable over the set of all C_{ij} with $C_{ij} > F_{ij}$, and by arguing that the equation $\partial L/\partial C_{ij} = 0$ is a sufficient condition for optimality that is satisfied by the positive Lagrange multiplier β of Eq. (5.37) and the capacities C_{ij} of Eq. (5.38) (see [Las70], p. 84, [Roc70], p. 283, or [BeT89], p. 662).

It makes no difference which lines are eliminated, so this problem has n local minima. Each of these happens to be also a global minimum, but this is a coincidence due to the extreme simplicity of the example.

The conclusion from the preceding discussion is that the simultaneous optimization of link flows and capacities is a hard combinatorial problem. Furthermore, even if this problem is solved, the network topology obtained will tend to concentrate capacity in a few large links, and possibly violate reliability constraints. The preceding formulation is also unrealistic because, in practice, capacity costs are not linear. Furthermore, the capacity of a leased communication line usually must be selected from a finite number of choices. In addition, the capacity of a line is typically equal in both directions, whereas this is not necessarily true for the optimal capacities of Eq. (5.38). These practical constraints enhance further the combinatorial character of the problem and make an exact solution virtually impossible. As a result, the only real option for addressing the problem subject to these constraints is to use the heuristic schemes that are described next.

Heuristic methods for capacity assignment. Typically, these methods start with a network topology and then successively perturb this topology by changing one or more link capacities at a time. Thus, these methods search "locally" around an existing topology for another topology that satisfies the constraints and has lower cost. Note that the term "topology" here means the set of capacities of all potential links of the network. In particular, a link with zero capacity represents a link that in effect does not exist. Usually, the following are assumed:

1. The nodes of the network and the input traffic flow for each pair of nodes are known.
2. A routing model has been adopted that determines the flows F_{ij} of all links (i, j) given all link capacities C_{ij}. The most common possibility here is to assume that link flows minimize a cost function $\sum_{(i,j)} D_{ij}(F_{ij})$ as in Eq. (5.33). In particular, F_{ij} can be determined by minimizing the average packet delay,

$$D = \frac{1}{\gamma} \sum_{(i,j)} \left(\frac{F_{ij}}{C_{ij} - F_{ij}} + d_{ij} F_{ij} \right) \tag{5.40}$$

   based on $M/M/1$ approximations [cf. Eq. (5.30)], where γ is the known total input flow into the network, and C_{ij} and d_{ij} are the known capacity and the processing and propagation delay, respectively, of link (i, j). Several algorithms described in Sections 5.6 and 5.7 can be used for this purpose.
3. There is a delay constraint that must be met. Typically, it is required that the chosen capacities C_{ij} and link flows F_{ij} (determined by the routing algorithm as in 2 above) result in a delay D given by the $M/M/1$ formula (5.40) (or a similar formula) that is below a certain threshold.
4. There is a reliability constraint that must be met. As an example, it is common to require that the network be 2-connected (*i.e.*, after a single node failure, all other nodes remain connected). More generally, it may be required that after $k-1$ nodes

fail, all other nodes remain connected—such a network is said to be k-connected. We postpone a discussion of how to evaluate the reliability of a network for later.

5. There is a cost criterion according to which different topologies are ranked.

The objective is to find a topology that meets the delay and reliability constraints as per 3 and 4 above, and has as small a cost as possible as per 5 above. We describe a prototype iterative heuristic method for addressing the problem. At the start of each iteration, there is available a *current best topology* and a *trial topology*. The former topology satisfies the delay and reliability constraints and is the best one in terms of cost that has been found so far, while the latter topology is the one that will be evaluated in the current iteration. We assume that these topologies are chosen initially by some ad hoc method—for example, by establishing an ample number of links to meet the delay and reliability constraints. The steps of the iteration are as follows:

Step 1: (Assign Flows). Calculate the link flows F_{ij} for the trial topology by means of some routing algorithm as per assumption 2.

Step 2: (Check Delay). Evaluate the average delay per packet D [as given, for example, by the $M/M/1$ formula (5.40)] for the trial topology. If

$$D \leq T$$

where T is a given threshold, go to step 3; else go to step 5.

Step 3: (Check Reliability). Test whether the trial topology meets the reliability constraints. (Methods for carrying out this step will be discussed shortly.) If the constraints are not met, go to step 5; else go to step 4.

Step 4: (Check Cost Improvement). If the cost of the trial topology is less than the cost of the current best topology, replace the current best topology with the trial topology.

Step 5: (Generate a New Trial Topology). Use some heuristic to change one or more capacities of the current best topology, thereby obtaining a trial topology that has not been considered before. Go to step 1.

Note that a trial topology is adopted as the current best topology in step 4 only if it satisfies the delay and reliability constraints in steps 2 and 3 and improves the cost (step 4). The algorithm terminates when no new trial topology can be generated or when substantial further improvement is deemed unlikely. Naturally, there is no guarantee that the final solution is optimal. One possibility to attempt further improvement is to repeat the algorithm with a different starting topology. Another possibility is to modify step 4 so that on occasion a trial topology is accepted as a replacement of the current best topology even if its cost is greater than the best found so far. A popular scheme of this type, known as *simulated annealing*, can be used to find a globally optimal solution under mild assumptions ([KGV83], [Haj88], and [Tsi89]). The amount of computing time required, however, may be excessive. For a method that uses this approach see [FeA89].

There are a number of heuristic rules proposed in the literature for generating new trial topologies (step 5). One possibility is simply to lower the capacity of a link that seems underutilized (low F_{ij}/C_{ij} and F_{ji}/C_{ji}) or to eliminate the link altogether. Another possibility is to increase the capacity of some overutilized link when the delay constraint is not met. A combination of these possibilities is the *branch exchange heuristic*, whereby one link is deleted and another link is added. A useful way to choose the links to be deleted and added is the so-called *saturated cut method.* The idea here is to identify a partition (or cut) of the nodes into two sets N_1 and N_2 such that the links joining N_1 and N_2 are highly utilized. It then seems plausible that adding a link between a node in N_1 and a node in N_2 will help reduce the high level of utilization across the cut. The method works as follows (see Fig. 5.55):

1. Prepare a list of all undirected links (i, j), sorted in order of decreasing utilization as measured by $\max\{F_{ij}/C_{ij}, F_{ji}/C_{ji}\}$.
2. Find a link k such that (a) if all links above k on the list are eliminated, the network remains connected, and (b) if k is eliminated together with all links above it on the list, the network separates in two disconnected components N_1 and N_2.
3. Remove the most underutilized link in the network, and replace it with a new link that connects a node in the component N_1 and a node in the component N_2.

There are a number of variations of the scheme described above. For example, in selecting the link to remove, one may take into account the cost of capacity for the link. Needless to say, prior knowledge about the network and the cost structure for link capacity can be effectively incorporated into the heuristics. In addition, the design process can be enhanced by using an interactive program that can be appropriately guided by the designer. We now turn our attention to methods for evaluating reliability of a network.

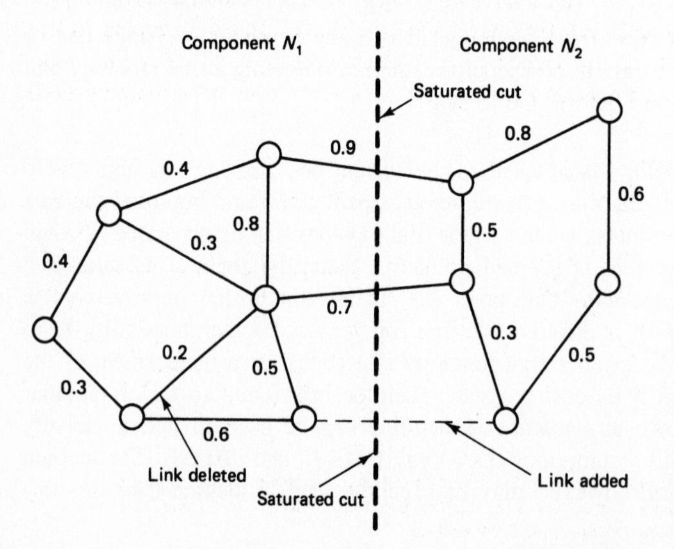

Figure 5.55 Illustration of the cut saturation method for performing a branch exchange heuristic. The number next to each arc is the link utilization. Links of high utilization are temporarily removed until the network separates into two components, N_1 and N_2. A link is then added that connects a node in N_1 with a node in N_2 while a link of low utilization is deleted.

Network reliability issues. We have already mentioned that a common reliability requirement for a network is that it must be k-connected. We now define k-connectivity more formally.

We say that *two nodes i and j* in an undirected graph are *k-connected* if there is a path connecting i and j in every subgraph obtained by deleting $(k - 1)$ nodes other than i and j together with their adjacent arcs from the graph. We say that *the graph is k-connected* if every pair of nodes is k-connected. These definitions are illustrated in Fig. 5.56. Another notion of interest is that of *arc connectivity*, which is defined as the minimum number of arcs that must be deleted before the graph becomes disconnected. A lower bound k on arc connectivity can be established by checking for k-(node) connectivity in an expanded graph where each arc (i, j) is replaced by two arcs (i, n) and (n, j), where n is a new node. This is evident by noting that failure of node n in the expanded graph can be associated with failure of the arc (i, j) in the original graph.

It is possible to calculate the number of paths connecting two nodes i and j which are node-disjoint in the sense that any two of them share no nodes other than i and j. This can be done by setting up and solving a max-flow problem, a classical combinatorial problem, which is treated in detail in many sources (*e.g.*, [PaS82] and [Ber91]). We do not discuss here the max-flow problem, since it is somewhat tangential to the purposes of this section. The reader who is already familiar with this problem can see its connection with the problem of k-connectivity of two nodes from Fig. 5.57. Using the theory of the max-flow problem, it can be shown that i and j are k-connected if and only if either i and j are connected with an arc or there are at least k node-disjoint paths connecting i and j.

One way to check k-connectivity of a graph is to check k-connectivity of every pair of nodes. There are, however, more efficient methods. One particular method, due to Kleitman [Kle69], operates as follows:

Choose an arbitrary node n_0 and check k-connectivity between that node and every other node. Delete n_0 and its adjacent arcs from the graph, choose another node n_1, and check $(k - 1)$-connectivity between that node and every other node. Continue in this manner until either node n_{k-1} is checked to be 1-connected to every remaining node, or $(k - i)$-connectivity of some node $n_i, i = 0, 1, \ldots, k - 1$, to every remaining node

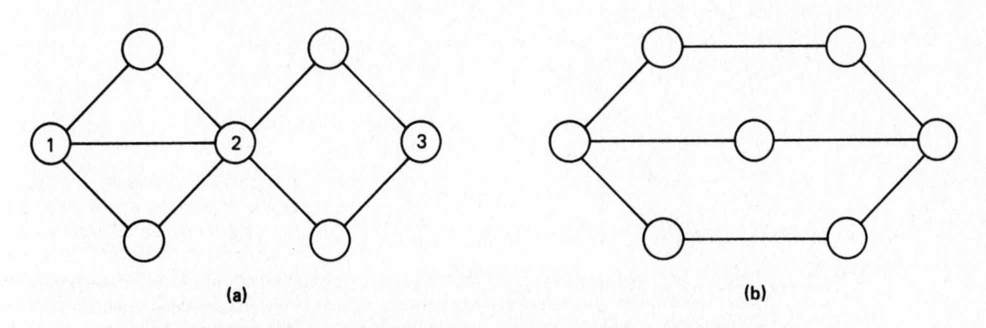

Figure 5.56 Illustration of the definition of k-connectivity. In graph (a), nodes 1 and 2 are 6-connected, nodes 2 and 3 are 2-connected, and nodes 1 and 3 are 1-connected. Graph (a) is 1-connected. Graph (b) is 2-connected.

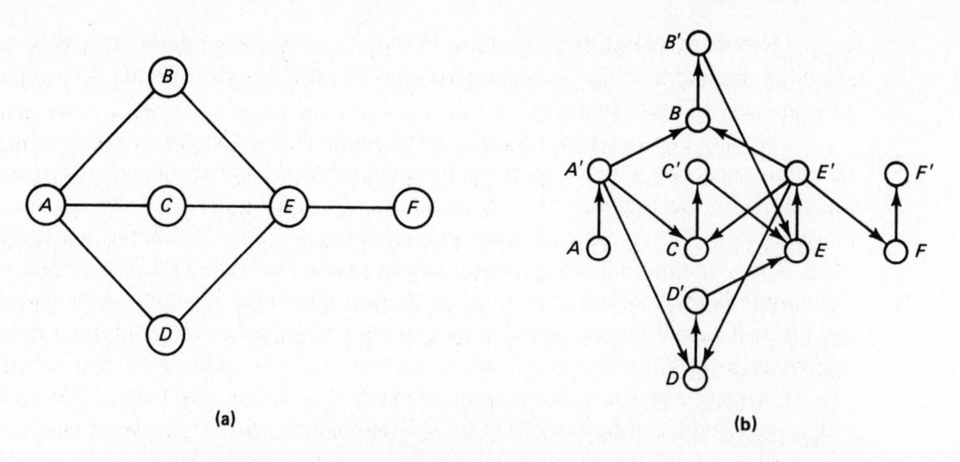

Figure 5.57 Illustration of test of k-connectivity for those familiar with the max-flow problem. To find the number of node-disjoint paths from node A to node F in graph (a), do the following: (1) Replace each node i with two nodes i and i' and a directed arc (i, i'). (2) Replace each arc (i, j) of graph (a) with two directed arcs (i', j) and (j', i). (3) Delete all arcs incoming to A and outgoing from F', thereby obtaining graph (b). (4) Assign infinite capacity to arcs (A, A') and (F, F'), and unit capacity to all other arcs. Then the maximum flow that can be sent from node A to node F' in graph (b) equals the number of node-disjoint paths connecting nodes A and F in graph (a).

cannot be verified. In the latter case, it is evident that the graph is not k-connected: If node n_i is not $(k - i)$-connected to some other remaining node, $(k - i - 1)$ nodes can be found such that their deletion along with nodes $n_0, \dots, n_{i-1}$ disconnects the graph. The process is demonstrated in Fig. 5.58.

To establish the validity of Kleitman's algorithm, it must be shown that when nodes $n_0, \dots, n_{k-1}$ can be found as described above, the graph is k-connected. The argument is by contradiction. If the graph is not k-connected, there must exist nodes i and j, and

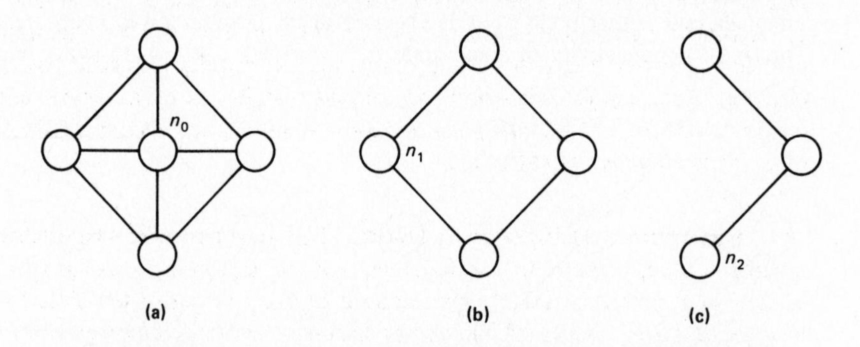

Figure 5.58 Verifying 3-connectivity of graph (a) by Kleitman's algorithm. An arbitrary node n_0 is chosen, and 3-connectivity between n_0 and every other node is verified. Then node n_0 and its adjacent arcs are deleted, thereby obtaining graph (b). An arbitrary node n_1 of (b) is chosen, and 2-connectivity between n_1 and every other node in (b) is verified. Then node n_1 and its adjacent arcs are deleted, thereby obtaining graph (c). Since graph (c) is connected, it follows that the original graph (a) is 3-connected.

a set of $(k-1)$ nodes $\mathcal{F}_0$ ($i \notin \mathcal{F}_0, j \notin \mathcal{F}_0$), such that deletion of $\mathcal{F}_0$ leaves no path connecting i and j. Then n_0 must belong to $\mathcal{F}_0$ because if $n_0 \notin \mathcal{F}_0$, then since n_0 is k-connected to every other node, it follows that there are paths P_i and P_j connecting n_0, i, and j even after the set $\mathcal{F}_0$ is deleted. Therefore, a path from i to j can be constructed by first using P_i to go from i to n_0, and then using P_j to go from n_0 to j. Next, consider the sets $\mathcal{F}_1 = \mathcal{F}_0 - \{n_0\}$, $\mathcal{F}_2 = \mathcal{F}_1 - \{n_1\}, \ldots$. Similarly, n_1 must belong to $\mathcal{F}_1$, n_2 must belong to $\mathcal{F}_2$, and so on. After this argument is used $(k-1)$ times, it is seen that $\mathcal{F}_0 = \{n_0, n_1, \ldots, n_{k-2}\}$. Therefore, after $\mathcal{F}_0$ is deleted from the graph, node n_{k-1} is still intact, and by definition of n_{k-1}, so are the paths from n_{k-1} to i and j. These paths can be used to connect i and j, contradicting the hypothesis that deletion of $\mathcal{F}_0$ leaves i and j disconnected.

Kleitman's algorithm requires a total of $\sum_{i=1}^{k}(N-i) = kN - k(k+1)/2$ connectivity tests of node pairs, where N is the number of nodes. There is another algorithm which requires roughly N connectivity tests; see [Eve75] for details.

The preceding formulation of the reliability problem addresses the worst case where the most unfavorable combination of k node failures occurs. An alternative formulation is to assign failure probabilities to all network elements (nodes and links) and evaluate the probability that a given pair of nodes becomes disconnected. To get an appreciation of the difficulties involved in this approach, suppose that there are n failure-prone elements in the network. Then the number of all distinct combinations of surviving elements is 2^n. Let s_k denote the kth combination and p_k denote the probability of its occurrence. The probability that the network remains connected is

$$P_C = \sum_{s_k \in C} p_k \tag{5.41}$$

where C is the set of combinations of surviving elements for which the network remains connected. Thus, evaluation of P_C involves some form of implicit or explicit enumeration of the set C. This is in general a very time-consuming procedure. However, shortcuts and approximations have been found making the evaluation of the survival probability P_C feasible for some realistic networks ([BaP83] and [Ben86]). An alternative approach evaluates a lower bound for P_C by ordering the possible combinations of surviving elements according to their likelihood, and then adding p_k over a subset of most likely combinations from C ([LaL86] and [LiS84]). A similar approach using the complement of C gives an upper bound to P_C.

Spanning tree topology design. For some networks where reliability is not a serious issue, a spanning tree topology may be appropriate. Using such a topology is consistent with the idea of concentrating capacity on just a few links to reduce the average delay per packet (cf. the earlier discussion on capacity assignment).

It is possible to design a spanning tree topology by assigning a weight w_{ij} to each possible link (i, j) and by using the minimum weight spanning tree (MST) algorithms of Section 5.2.2. A potential difficulty arises when there is a constraint on the amount of traffic that can be carried by any one link. This gives rise to the *constrained MST problem*, where a matrix of input traffic from every node to every other node is given and an MST is to be designed subject to the constraint that the flow on each link will not exceed a given upper bound. Because of this constraint, the problem has a combinatorial character and is usually addressed using heuristic methods.

One possibility is to modify the Kruskal or Prim–Dijkstra algorithms, described in Section 5.2.2, so that at each iteration, when a new link is added to the current fragments, a check is made to see if the flow constraints on the links of the fragments are satisfied. If not, the link is not added and another link is considered. This type of heuristic can also be combined with some version of the branch exchange heuristic discussed earlier in connection with the general subnet design problem.

A special case of the constrained MST problem can be addressed using the *Essau–Williams algorithm* [EsW66]. In this case, there is a central node denoted 0, and N other nodes. All traffic must go through the central node, so it can be assumed that there is input traffic only from the noncentral nodes to the central node and the reverse. Problems of this type arise also in the context of the local access design problem, with the central node playing the role of a traffic concentrator.

The Essau–Williams algorithm is, in effect, a branch exchange heuristic. One starts with the spanning tree where the central node is directly connected with each of the N other nodes. (It is assumed that this is a feasible solution.) At each successive iteration, a link $(i, 0)$ connecting some node i with the central node 0 is deleted from the current spanning tree, and a link (i, j) is added. These links are chosen so that:

1. No cycle is formed.
2. The capacity constraints of all the links of the new spanning tree are satisfied.
3. The saving $w_{i0} - w_{ij}$ in link weight obtained by exchanging $(i, 0)$ with (i, j) is positive and is maximized over all nodes i and j for which 1 and 2 are satisfied.

The algorithm terminates when there are no nodes i and j for which requirements 1 to 3 are satisfied when $(i, 0)$ is exchanged with (i, j). The operation of the algorithm is illustrated in Fig. 5.59.

5.4.3 Local Access Network Design Problem

Here we assume that a communication subnet is available, and we want to design a network that connects a collection of terminals with known demands to the subnet. This problem is often addressed in the context of a hierarchical strategy, whereby groups of terminals are connected, perhaps through local area networks, to various types of concentrators, which are in turn connected to higher levels of concentrators, and so on. It is difficult to recommend a global design strategy without knowledge of the given practical situation. However, there are a few subproblems that arise frequently. We will discuss one such problem, known as the *concentrator location problem.*

In this problem, there are n sources located at known geographical points. For example, a source may be a gateway of a local area network that connects several terminals within a customer's facility; or it could be a host computer through which several time-sharing terminals access some network. The nature of the sources is not

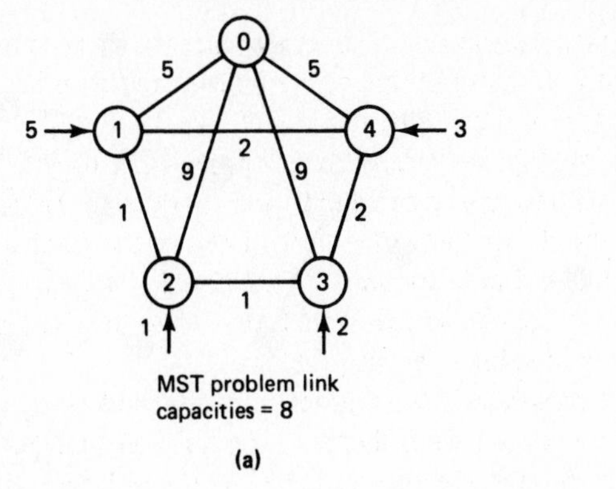

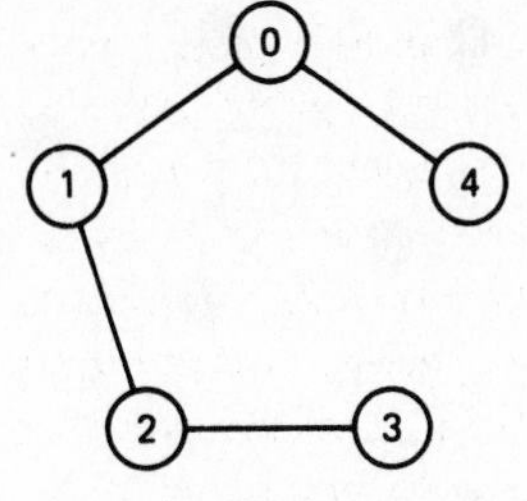

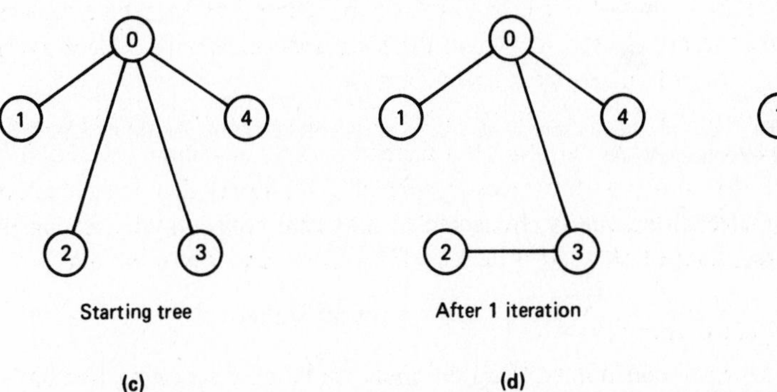

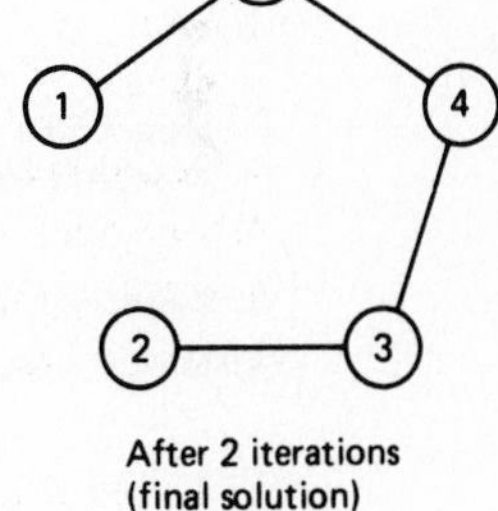

Figure 5.59 Illustration of the Essau–Williams, constrained MST heuristic algorithm. The problem data are given in (a). Node 0 is the central node. The link weights are shown next to the links. The input flows from nodes 1, 2, 3, and 4 to the central node are shown next to the arrows. The flow on each link is constrained to be no more than 8. The algorithm terminates after two iterations, with the tree (e) having a total weight of 13. Termination occurs because when link (1,0) or (4,0) is removed and a link that is not adjacent to node 0 is added, some link capacity constraint (link flow $\leq$ 8) will be violated. The optimal tree, shown in (b), has a total weight of 12.

material to the discussion. The problem is to connect the n sources to m concentrators that are themselves connected to a communication subnet (see Fig. 5.60). We denote by a_{ij} the cost of connecting source i to concentrator j. Consider the variables x_{ij} where

$$x_{ij} = \begin{cases} 1, & \text{if source } i \text{ is connected to concentrator } j \\ 0, & \text{otherwise} \end{cases}$$

Then the total cost of a source assignment specified by a set of variables $\{x_{ij}\}$ is

$$\text{cost} = \sum_{i=1}^{n} \sum_{j=1}^{m} a_{ij} x_{ij} \,. \tag{5.42}$$

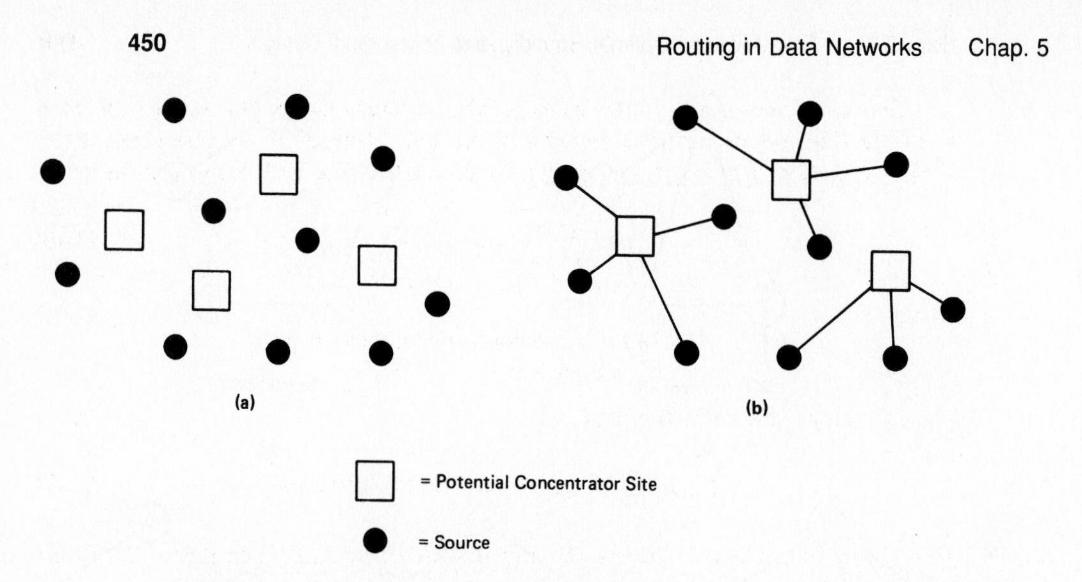

Figure 5.60 (a) Collection of sources and potential concentrator locations. (b) Possible assignment of sources to concentrators.

We assume that each source is connected to only one concentrator, so that there is the constraint

$$\sum_{j=1}^{m} x_{ij} = 1, \qquad \text{for all sources } i \tag{5.43}$$

Also, there is a maximum number of sources K_j that can be handled by concentrator j. This is expressed by the constraint

$$\sum_{i=1}^{n} x_{ij} \leq K_j, \qquad \text{for all concentrators } j \tag{5.44}$$

The problem is to select the variables x_{ij} so as to minimize the cost (5.42) subject to the constraints (5.43) and (5.44).

Even though the variables x_{ij} are constrained to be 0 or 1, the problem above is not really a difficult combinatorial problem. If the integer constraint

$$x_{ij} = 0 \text{ or } 1$$

is replaced by the noninteger constraint

$$0 \leq x_{ij} \leq 1 \tag{5.45}$$

the problem becomes a linear transportation problem that can be solved by very efficient algorithms such as the simplex method. It can be shown that the solution obtained from the simplex method will be integer (*i.e.*, x_{ij} will be either 0 or 1; see [Dan63] and [Ber91]). Therefore, solving the problem without taking into account the integer constraints automatically obtains an integer optimal solution. There are methods other than simplex that can be used for solving the problem ([Ber85], [Ber91], [BeT89], [FoF62], and [Roc84]).

Next, consider a more complicated problem whereby the location of the concentrators is not fixed but instead is subject to optimization. There are m potential concentrator sites, and there is a cost b_j for locating a concentrator at site j. The cost then becomes

$$\text{cost} = \sum_{i=1}^{n}\sum_{j=1}^{m} a_{ij}x_{ij} + \sum_{j=1}^{m} b_j y_j \tag{5.46}$$

where

$$y_j = \begin{cases} 1, & \text{if a concentrator is located at site } j \\ 0, & \text{otherwise} \end{cases}$$

The constraint (5.44) now becomes

$$\sum_{i=1}^{n} x_{ij} \le K_j y_j, \qquad \text{for all concentrators } j \tag{5.47}$$

Thus, the problem is to minimize the cost (5.46) subject to the constraints (5.43) and (5.47) and the requirement that x_{ij} and y_j are either 0 or 1.

This problem has been treated extensively in the operations research literature, where it is known as the warehouse location problem. It is considered to be a difficult combinatorial problem that can usually be solved only approximately. A number of exact and heuristic methods have been proposed for its solution. We refer the reader to the literature for further details ([AkK77], [AlM76], [CFN77], [KeH63], [Khu72], and [RaW83]).

5.5 CHARACTERIZATION OF OPTIMAL ROUTING

We now return to the optimal routing problem formulated at the beginning of Section 5.4. The main objective in this section is to show that optimal routing directs traffic exclusively along paths which are shortest with respect to some link lengths that depend on the flows carried by the links. This is an interesting characterization which motivates algorithms discussed in Sections 5.6 and 5.7.

Recall the form of the cost function

$$\sum_{(i,j)} D_{ij}(F_{ij}) \tag{5.48}$$

Here F_{ij} is the total flow (in data units/sec) carried by link (i,j) and given by

$$F_{ij} = \sum_{\substack{\text{all paths } p \\ \text{containing } (i,j)}} x_p \tag{5.49}$$

where x_p is the flow (in data units/sec) of path p. For every OD pair w, there are the constraints

$$\sum_{p \in P_w} x_p = r_w \tag{5.50}$$

$$x_p \ge 0, \qquad \text{for all } p \in P_w \tag{5.51}$$

where r_w is the given traffic input of the OD pair w (in data units/sec) and P_w is the set of directed paths of w. In terms of the unknown path flow vector $x = \{x_p \mid p \in P_w, w \in W\}$, the problem is written as

$$\text{minimize} \sum_{(i,j)} D_{ij} \left[\sum_{\substack{\text{all paths } p \\ \text{containing } (i,j)}} x_p \right]$$

$$\text{subject to} \sum_{p \in P_w} x_p = r_w, \qquad \text{for all } w \in W \tag{5.52}$$

$$x_p \geq 0, \qquad \text{for all } p \in P_w, w \in W$$

In what follows we will characterize an optimal routing in terms of the first derivatives D'_{ij} of the functions D_{ij}. We assume that each D_{ij} is a differentiable function of F_{ij} and is defined in an interval $[0, C_{ij})$, where C_{ij} is either a positive number (typically representing the capacity of the link) or else infinity. Let x be the vector of path flows x_p. Denote by $D(x)$ the cost function of the problem of Eq. (5.52),

$$D(x) = \sum_{(i,j)} D_{ij} \left[\sum_{\substack{\text{all paths } p \\ \text{containing } (i,j)}} x_p \right]$$

and by $\partial D(x)/\partial x_p$ the partial derivative of D with respect to x_p. Then

$$\frac{\partial D(x)}{\partial x_p} = \sum_{\substack{\text{all links } (i,j) \\ \text{on path } p}} D'_{ij}$$

where the first derivatives D'_{ij} are evaluated at the total flows corresponding to x. It is seen that $\partial D/\partial x_p$ *is the length of path p when the length of each link* (i, j) *is taken to be the first derivative* D'_{ij} *evaluated at* x. Consequently, in what follows $\partial D/\partial x_p$ is called the *first derivative length of path p*.

Let $x^* = \{x_p^*\}$ be an optimal path flow vector. Then if $x_p^* > 0$ for some path p of an OD pair w, we must be able to shift a small amount $\delta > 0$ from path p to any other path p' of the same OD pair without improving the cost; otherwise, the optimality of x^* would be violated. To first order, the change in cost from this shift is

$$\delta \frac{\partial D(x^*)}{\partial x_{p'}} - \delta \frac{\partial D(x^*)}{\partial x_p}$$

and since this change must be nonnegative, we obtain

$$x_p^* > 0 \quad \Rightarrow \quad \frac{\partial D(x^*)}{\partial x_{p'}} \geq \frac{\partial D(x^*)}{\partial x_p}, \qquad \text{for all } p' \in P_w \tag{5.53}$$

In words, *optimal path flow is positive only on paths with a minimum first derivative length.* Furthermore, at an optimum, the paths along which the input flow r_w of OD pair w is split must have *equal* length (and less or equal length to that of all other paths of w).

The condition (5.53) is a necessary condition for optimality of x^*. It can also be shown to be sufficient for optimality if the functions D_{ij} are convex; for example, when the second derivatives D''_{ij} exist and are positive in $[0, C_{ij})$, the domain of definition of D_{ij} (see Problem 5.24).

Example 5.7

Consider the two-link network shown in Fig. 5.61, where nodes 1 and 2 are the only origin and destination, respectively. The given input r is to be divided into the two path flows x_1 and x_2 so as to minimize a cost function based on the $M/M/1$ approximation

$$D(x) = D_1(x_1) + D_2(x_2)$$

where for $i = 1, 2$,

$$D_i(x_i) = \frac{x_i}{C_i - x_i}$$

and C_i is the capacity of link i. For the problem to make sense, we must assume that r is less than the maximum throughput $C_1 + C_2$ that the network can sustain.

At the optimum, the shortest path condition (5.53) must be satisfied as well as the constraints

$$x_1^* + x_2^* = r, \quad x_1^* \geq 0, \quad x_2^* \geq 0$$

Assume that $C_1 \geq C_2$. Then, from elementary reasoning, the optimal flow x_1 cannot be less than x_2 (it makes no sense to send more traffic on the slower link). The only possibilities are:

1. $x_1^* = r$ and $x_2^* = 0$. According to the shortest path condition (5.53), we must have

$$\frac{dD_1(r)}{dx_1} \leq \frac{dD_2(0)}{dx_2}$$

Equivalently [since the derivative of $x/(C - x)$ is $C/(C - x)^2$],

$$\frac{C_1}{(C_1 - r)^2} \leq \frac{1}{C_2}$$

or

$$r \leq C_1 - \sqrt{C_1 C_2} \tag{5.54}$$

2. $x_1^* > 0$ and $x_2^* > 0$. In this case, the shortest path condition (5.53) implies that the lengths of paths 1 and 2 are equal, that is,

$$\frac{dD_1(x_1^*)}{dx_1} = \frac{dD_2(x_2^*)}{dx_2}$$

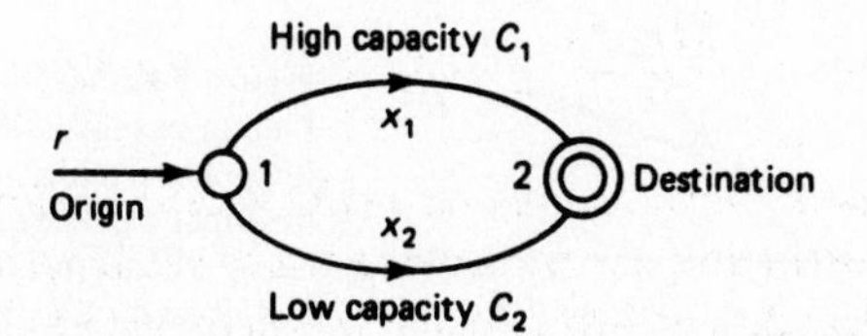

Figure 5.61 Example routing problem involving a single OD pair and two paths.

or, equivalently,

$$\frac{C_1}{(C_1 - x_1^*)^2} = \frac{C_2}{(C_2 - x_2^*)^2} \tag{5.55}$$

This equation together with the constraint $x_1^* + x_2^* = r$ determine the values of x_1^* and x_2^*. A straightforward calculation shows that the solution is

$$x_1^* = \frac{\sqrt{C_1}\left[r - (C_2 - \sqrt{C_1 C_2})\right]}{\sqrt{C_1} + \sqrt{C_2}}$$

$$x_2^* = \frac{\sqrt{C_2}\left[r - (C_1 - \sqrt{C_1 C_2})\right]}{\sqrt{C_1} + \sqrt{C_2}}$$

The optimal solution is shown in Fig. 5.62 for r in the range $[0, C_1 + C_2)$ of possible inputs. It can be seen that for

$$r \le C_1 - \sqrt{C_1 C_2}$$

the faster link 1 is used exclusively. When r exceeds the threshold value $C_1 - \sqrt{C_1 C_2}$ for which the first derivative lengths of the two links are equalized and relation (5.54) holds as an equation, the slower link 2 is also utilized. As r increases, the flows on both links increase while the equality of the first derivative lengths, as in Eq. (5.55), is maintained. This behavior is typical of optimal routing when the cost function is based on the $M/M/1$ approximation; *for low input traffic, each OD pair tends to use only one path for routing (the fastest in terms of packet transmission time), and as traffic input increases, additional paths are used to avoid overloading the fastest path.* More generally, this type of behavior is associated with link cost functions D_{ij} with the property that the derivative $D'_{ij}(0)$ at zero flow depends only on the link capacity and decreases as the link capacity increases (see Problem 5.36).

We were able to solve the preceding example analytically only because of its extreme simplicity. Unfortunately, in more complex problems we typically have to resort to a computational solution—either centralized or distributed. The following sections deal with some of the possibilities.

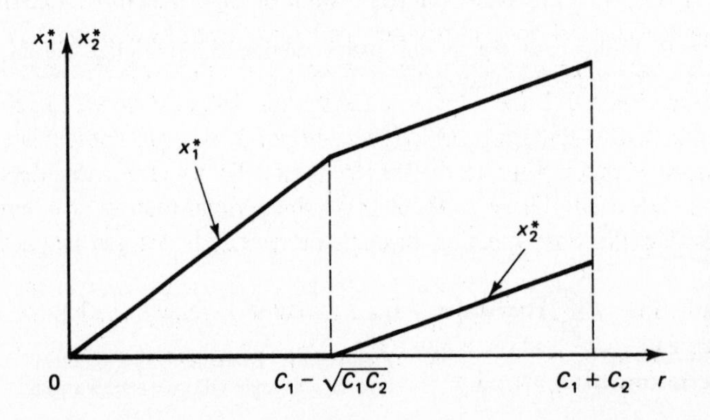

Figure 5.62 Optimal path flows for the routing example. When the traffic input is low, only the high-capacity link is used. As the input increases beyond the threshold $C_1 - \sqrt{C_1 C_2}$, some traffic is routed on the low-capacity link.

Finally, note that while the optimal routing problem was formulated in terms of a fixed set of input rates, the optimal solution $\{x_p^*\}$ can be implemented naturally even when the input rates are time varying. This can be done by working with the *fractions* of flow

$$\xi_p = \frac{x_p^*}{r_w}, \qquad \text{for all } p \in P_w$$

and by requiring each OD pair w to divide traffic (packets or virtual circuits) among the available paths according to these fractions. In virtual circuit networks, where the origin nodes are typically aware of the detailed description of the paths used for routing, it is easy to implement the routing process in terms of the fractions ξ_p. The origins can estimate the data rate of virtual circuits and can route new virtual circuits so as to bring about a close match between actual and desired fractions of traffic on each path. In some datagram networks or in networks where the detailed path descriptions are not known at the origin nodes, it may be necessary to maintain in a routing table at each node i a routing variable $\phi_{ik}(j)$ for each link (i, k) and destination j. This routing variable is defined as the fraction of all flow arriving at node i, destined for node j, and routed along link (i, k). Mathematically,

$$\phi_{ik}(j) = \frac{f_{ik}(j)}{\sum_m f_{im}(j)}, \qquad \text{for all } (i, k) \text{ and } j \tag{5.56}$$

where $f_{ik}(j)$ is the flow that travels on link (i, k) and is destined for node j. Given an optimal solution of the routing problem in terms of the path flow variables $\{x_p^*\}$, it is possible to determine the corresponding link flow variables $f_{ik}(j)$ and, by Eq. (5.56), the corresponding optimal routing variables $\phi_{ik}(j)$. A direct formulation and distributed computational solution of the optimal routing problem in terms of the variables $\phi_{ik}(j)$ is given in references [Gal77], [Ber79a], [Gaf79], and [BGG84].

5.6 FEASIBLE DIRECTION METHODS FOR OPTIMAL ROUTING

In Section 5.5 it was shown that optimal routing results only if flow travels along minimum first derivative length (MFDL) paths for each OD pair. Equivalently, a set of path flows is strictly suboptimal only if there is a positive amount of flow that travels on a non-MFDL path. This suggests that suboptimal routing can be improved by shifting flow to an MFDL path from other paths for each OD pair. The adaptive shortest path method for datagram networks of Section 5.2.5 does that in a sense, but shifts *all* flow of each OD pair to the shortest path, with oscillatory behavior resulting. It is more appropriate to shift only *part* of the flow of other paths to the shortest path. This section considers methods based on this idea. Generally, these methods solve the optimal routing problem computationally by decreasing the cost function through incremental changes in path flows.

Given a feasible path flow vector $x = \{x_p\}$ (*i.e.*, a vector x satisfying the constraints of the problem), consider changing x along a direction $\Delta x = \{\Delta x_p\}$. There are two requirements imposed on the direction Δx:

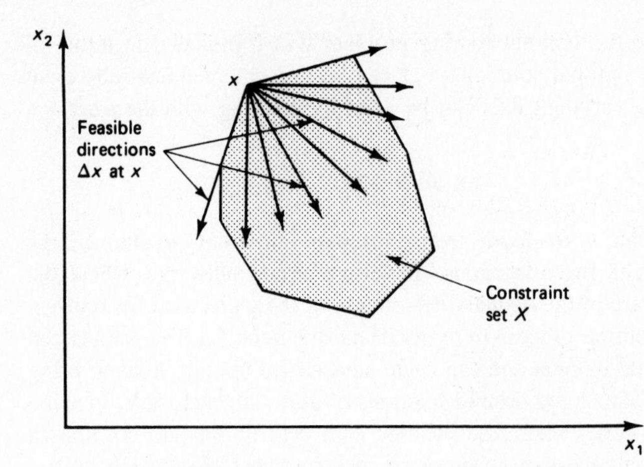

Figure 5.63 Feasible directions Δx at a feasible point x. Δx is a feasible direction if changing x by a small amount along the direction Δx still maintains feasibility.

1. The first requirement is that Δx should be a *feasible direction* in the sense that small changes along Δx maintain the feasibility of the path flow vector x (see Fig. 5.63). Mathematically, it is required that for some $\overline{\alpha} > 0$ and all $\alpha \in [0, \overline{\alpha}]$, the vector $x + \alpha\Delta x$ is feasible or equivalently,

$$\sum_{p \in P_w} \Delta x_p = 0, \quad \text{for all } w \in W \tag{5.57}$$

$$\Delta x_p \geq 0, \quad \text{for all } p \in P_w,\ w \in W \text{ for which } x_p = 0 \tag{5.58}$$

Equation (5.57) follows from the feasibility requirement

$$\sum_{p \in P_w} (x_p + \alpha\Delta x_p) = r_w$$

and the fact that x is feasible, which implies that

$$\sum_{p \in P_w} x_p = r_w$$

It simply expresses that to maintain feasibility, all increases of flow along some paths must be compensated by corresponding decreases along other paths of the same OD pair. One way to obtain feasible directions is to select another feasible vector $\overline{x}$ and take

$$\Delta x = \overline{x} - x$$

In fact, a little thought reveals that all feasible directions can be obtained in this way up to scalar multiplication.

2. The second requirement is that Δx should be a *descent direction* in the sense that the cost function can be decreased by making small movements along the direction Δx starting from x (see Fig. 5.64). Since the gradient vector $\nabla D(x)$ is normal to the equal cost surfaces of the cost function D, it is clear from Fig. 5.64 that the

descent condition translates to the condition that the inner product of $\nabla D(x)$ and Δx is negative, that is,

$$\sum_{w \in W} \sum_{p \in P_w} \frac{\partial D(x)}{\partial x_p} \Delta x_p < 0 \tag{5.59}$$

For a mathematical verification, note that the inner product in Eq. (5.59) is equal to the first derivative of the function $G(\alpha) = D(x + \alpha \Delta x)$ at $\alpha = 0$, so Eq. (5.59) is equivalent to $G(\alpha)$ being negative for sufficiently small positive α. Note that the partial derivative $\partial D(x)/\partial x_p$ is given by

$$\frac{\partial D(x)}{\partial x_p} = \sum_{\substack{\text{all links } (i,j) \\ \text{on path } p}} D'_{ij}(F_{ij}),$$

and can be viewed as the first derivative length of path p [the length of link (i, j) is $D'_{ij}(F_{ij})$, cf. Section 5.5]. One way to satisfy the descent condition of Eq. (5.59), which is in fact commonly used in algorithms, is to require that Δx satisfies the conservation of flow condition $\sum_{p \in P_w} \Delta x_p = 0$ [cf. Eq. (5.57)], and that

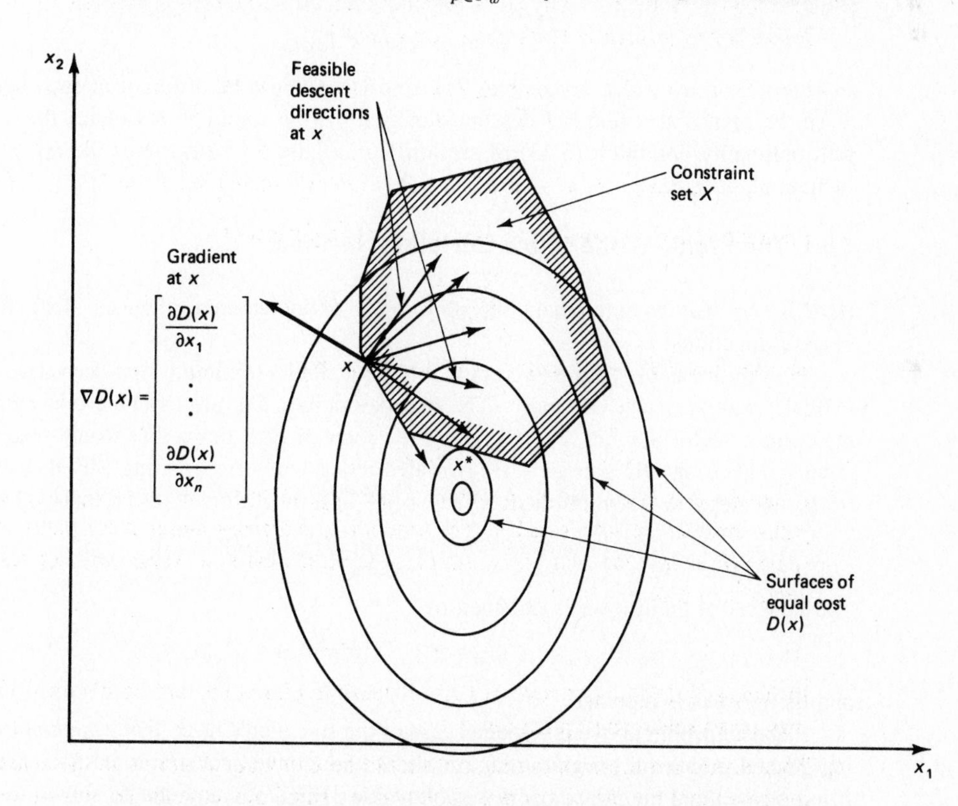

Figure 5.64 Descent directions at a feasible point x form an angle greater than 90° with the gradient $\nabla D(x)$. At the optimal point x^*, there are no feasible descent directions.

$$\begin{aligned}
\Delta x_p \le 0, \quad & \text{for all paths } p \text{ that are nonshortest} \\
& \text{in the sense } \partial D(x)/\partial x_p > \partial D(x)/\partial x_{\overline{p}} \text{ for some} \\
& \text{path } \overline{p} \text{ of the same OD pair} \\
\Delta x_p < 0, \quad & \text{for at least one nonshortest path } p.
\end{aligned} \tag{5.60}$$

In words, the conditions above together with the condition $\sum_{p \in P_w} \Delta x_p = 0$ state that some positive flow is shifted from nonshortest paths to the shortest paths [with respect to the lengths $D'_{ij}(F_{ij})$], and no flow is shifted from shortest paths to nonshortest ones. Since $\partial D(x)/\partial x_p$ is minimal for those (shortest) paths $\overline{p}$ for which $\Delta x_{\overline{p}} > 0$, it is seen that these conditions imply the descent condition (5.59).

We thus obtain a broad class of iterative algorithms for solving the optimal routing problem. Their basic iteration is given by

$$x := x + \alpha \Delta x$$

where Δx is a feasible descent direction [*i.e.*, satisfies conditions (5.57) to (5.59) or (5.57), (5.58), and (5.60)], and α is a positive stepsize chosen so that the cost function is decreased, that is,

$$D(x + \alpha \Delta x) < D(x)$$

and the vector $x + \alpha \Delta x$ is feasible. The stepsize α could be different in each iteration. It can be seen that a feasible descent direction can be found if x violates the shortest path optimality condition (5.53) of section 5.5. Figure 5.65 illustrates the operation of such an algorithm.

5.6.1 The Frank–Wolfe (Flow Deviation) Method

Here is one way to implement the philosophy of incremental changes along feasible descent directions:

Given a feasible path flow vector $x = \{x_p\}$, find a minimum first derivative length (MFDL) path for each OD pair. (The first derivatives D'_{ij} are evaluated, of course, at the current vector x.) Let $\overline{x} = \{\overline{x}_p\}$ be the vector of path flows that would result if all input r_w for each OD pair $w \in W$ is routed along the corresponding MFDL path. Let α^* be the stepsize that minimizes $D[x + \alpha(\overline{x} - x)]$ over all $\alpha \in [0, 1]$, that is,

$$D\big[x + \alpha^*(\overline{x} - x)\big] = \min_{\alpha \in [0,1]} D\big[x + \alpha(\overline{x} - x)\big] \tag{5.61}$$

The new set of path flows is obtained by

$$x_p := x_p + \alpha^*(\overline{x}_p - x_p), \qquad \text{for all } p \in P_w, w \in W \tag{5.62}$$

and the process is repeated.

The algorithm above is a special case of the so-called *Frank–Wolfe method* for solving general, nonlinear programming problems with convex constraint sets (see [Zan69]). It has been called the *flow deviation method* (see [FGK73]), and can be shown to reduce the value of the cost function to its minimum in the limit (see Problem 5.31) although its convergence rate near the optimum tends to be very slow.

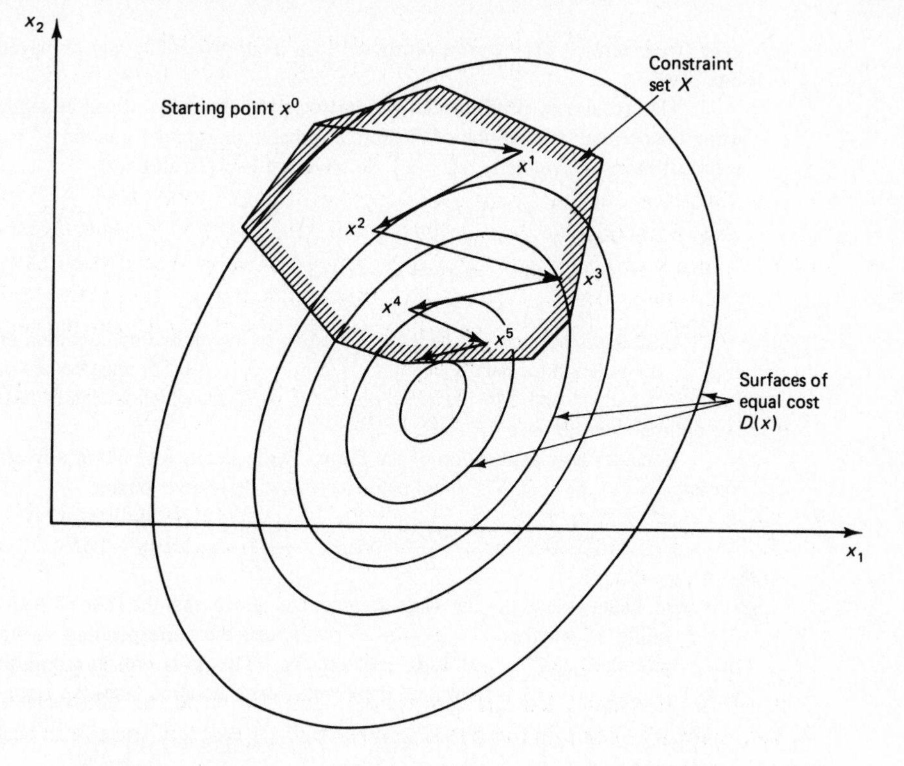

Figure 5.65 Sample path of an iterative descent method based on feasible descent directions. At each iteration, a feasible point of lower cost is obtained.

Note that the feasible direction used in the Frank–Wolfe iteration (5.62) is of the form (5.60); that is, a proportion α^* of the flow of all the nonshortest paths (those paths for which $x_p = 0$) is shifted to the shortest path (the one for which $\overline{x}_p = r_w$) for each OD pair w. The characteristic property here is that flow is shifted from the nonshortest paths in *equal* proportions. This distinguishes the Frank–Wolfe method from the gradient projection methods discussed in the next section. The latter methods also shift flow from the nonshortest paths to the shortest paths; however, they do so in generally unequal proportions.

Here is a simple example illustrating the Frank–Wolfe method:

Example 5.8

Consider the three-link network with one origin and one destination, shown in Fig. 5.66. There are three paths with corresponding flows x_1, x_2, and x_3, which must satisfy the constraints

$$x_1 + x_2 + x_3 = 1, \quad x_1 \geq 0, \quad x_2 \geq 0, \quad x_3 \geq 0$$

The cost function is

$$D(x) = \frac{1}{2}\left(x_1^2 + x_2^2 + 0.1x_3^2\right) + 0.55x_3$$

(The linear term $0.55x_3$ can be attributed to a large processing and propagation delay on link 3.)

This is an easy problem that can be solved analytically. It can be argued (or shown using the optimality condition of Section 5.5) that at an optimal solution $x^* = (x_1^*, x_2^*, x_3^*)$, we must have by symmetry $x_1^* = x_2^*$, so there are two possibilities:

(a) $x_3^* = 0$ and $x_1^* = x_2^* = 1/2$.
(b) $x_3^* = \beta > 0$ and $x_1^* = x_2^* = (1-\beta)/2$.

Case (b) is not possible because according to the optimality condition of Section 5.5, if $x_3^* > 0$, the length of path 3 [$= \partial D(x^*)/\partial x_3 = 0.1\beta + 0.55$] must be less or equal to the lengths of paths 1 and 2 [$= \partial D(x^*)/\partial x_1 = (1-\beta)/2$], which is clearly false. Therefore, the optimal solution is $x^* = (1/2, 1/2, 0)$.

Consider now application of the Frank–Wolfe iteration (5.68) at a feasible path flow vector $x = (x_1, x_2, x_3)$. The three paths have first derivative lengths

$$\frac{\partial D(x)}{\partial x_1} = x_1, \qquad \frac{\partial D(x)}{\partial x_2} = x_2, \qquad \frac{\partial D(x)}{\partial x_3} = 0.1x_3 + 0.55$$

It is seen, using essentially the same argument as above, that the shortest path is either 1 or 2, depending on whether $x_1 \le x_2$ or $x_2 < x_1$, and the corresponding shortest-path flows are $\overline{x} = (1,0,0)$ and $\overline{x} = (0,1,0)$, respectively. (The tie is broken arbitrarily in favor of path 1 in case $x_1 = x_2$.) Therefore, the Frank–Wolfe iteration takes the form

$$\begin{bmatrix} x_1 \\ x_2 \\ x_3 \end{bmatrix} := \begin{bmatrix} x_1 \\ x_2 \\ x_3 \end{bmatrix} + \alpha^* \begin{bmatrix} 1 - x_1 \\ -x_2 \\ -x_3 \end{bmatrix} = \begin{bmatrix} x_1 + \alpha^*(x_2 + x_3) \\ (1-\alpha^*)x_2 \\ (1-\alpha^*)x_3 \end{bmatrix}, \qquad \text{if } x_1 \le x_2$$

$$\begin{bmatrix} x_1 \\ x_2 \\ x_3 \end{bmatrix} := \begin{bmatrix} x_1 \\ x_2 \\ x_3 \end{bmatrix} + \alpha^* \begin{bmatrix} -x_1 \\ 1 - x_2 \\ -x_3 \end{bmatrix} = \begin{bmatrix} (1-\alpha^*)x_1 \\ x_2 + \alpha^*(x_1 + x_3) \\ (1-\alpha^*)x_3 \end{bmatrix}, \qquad \text{if } x_2 < x_1$$

Thus, at each iteration, a proportion α^* of the flows of the nonshortest paths is shifted to the shortest path.

The stepsize α^* is obtained by line minimization over [0,1] [cf. Eq. (5.67)]. Because $D(x)$ is quadratic, this minimization can be done analytically. We have

$$D\big[x + \alpha(\overline{x} - x)\big] = \frac{1}{2}\Big(\big[x_1 + \alpha(\overline{x}_1 - x_1)\big]^2 + \big[x_2 + \alpha(\overline{x}_2 - x_2)\big]^2$$
$$+0.1\big[x_3 + \alpha(\overline{x}_3 - x_3)\big]^2\Big) + 0.55\big[x_3 + \alpha(\overline{x}_3 - x_3)\big]$$

Differentiating with respect to α and setting the derivative to zero we obtain the unconstrained minimum $\overline{\alpha}$ of $D\big[x + \alpha(\overline{x} - x)\big]$ over α. This is given by

$$\overline{\alpha} = -\frac{x_1(\overline{x}_1 - x_1) + x_2(\overline{x}_2 - x_2) + (0.1x_3 + 0.55)(\overline{x}_3 - x_3)}{(\overline{x}_1 - x_1)^2 + (\overline{x}_2 - x_2)^2 + 0.1(\overline{x}_3 - x_3)^2}$$

where $\overline{x} = (\overline{x}_1, \overline{x}_2, \overline{x}_3)$ is the current shortest path flow vector [(1,0,0) or (0,1,0), depending on whether $x_1 \leq x_2$ or $x_2 < x_1$]. Since $(\overline{x} - x)$ is a descent direction, we have $\overline{\alpha} \geq 0$. Therefore, the stepsize α^*, which is the constrained minimum over [0,1] [cf. Eq. (5.67)], is given by

$$\alpha^* = \min[1, \overline{\alpha}]$$

(In this example, it can be verified that $\overline{\alpha} < 1$, implying that $\alpha^* = \overline{a}$.)

Figures 5.66 and 5.67 show successive iterates of the method. It can be seen that the rate of convergence becomes slow near the optimal solution. The reason is that as x^* is approached, the directions of search tend to become orthogonal to the direction leading from the current iterate x^k to x^*. In fact, it can be seen from Fig. 5.66 that the ratio of successive cost errors

$$\frac{D(x^{k+1}) - D(x^*)}{D(x^k) - D(x^*)}$$

while always less than 1, actually converges to 1 as $k \to \infty$. This is known as *sublinear* convergence rate [Ber82d], and is typical of the Frank–Wolfe method ([CaC68] and [Dun79]).

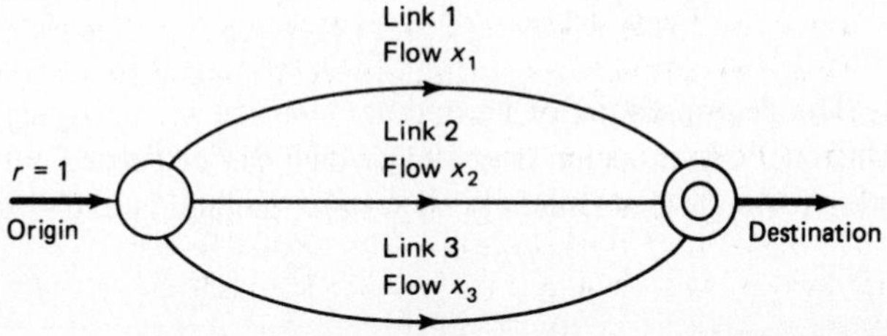

Iteration # k	0	10	20	40	80	160	320
x_1	0.4	0.4593	0.4702	0.4795	0.4866	0.4917	0.4950
x_2	0.3	0.4345	0.4562	0.4717	0.4823	0.4893	0.4938
x_3	0.3	0.1061	0.0735	0.0490	0.0310	0.0189	0.0110
Cost	0.2945	0.2588	0.2553	0.2532	0.2518	0.2510	0.2506
Error Ratio $\frac{D(x^{k+1}) - D(x^*)}{D(x^k) - D(x^*)}$	0.7164	0.9231	0.9576	0.9774	0.9882	0.9939	0.9969

Figure 5.66 Example problem and successive iterates of the Frank–Wolfe method. The cost function is

$$D(x) = \frac{1}{2}\left(x_1^2 + x_2^2 + x_3^2\right) + 0.55x_3$$

The optimal solution is $x^* = \left(\frac{1}{2}, \frac{1}{2}, 0\right)$. As the optimal solution is approached, the method becomes slower. In the limit, the ratio of successive errors $D(x^{k+1}) - D(x^*)/D(x^k) - D(x^*)$ tends to unity.

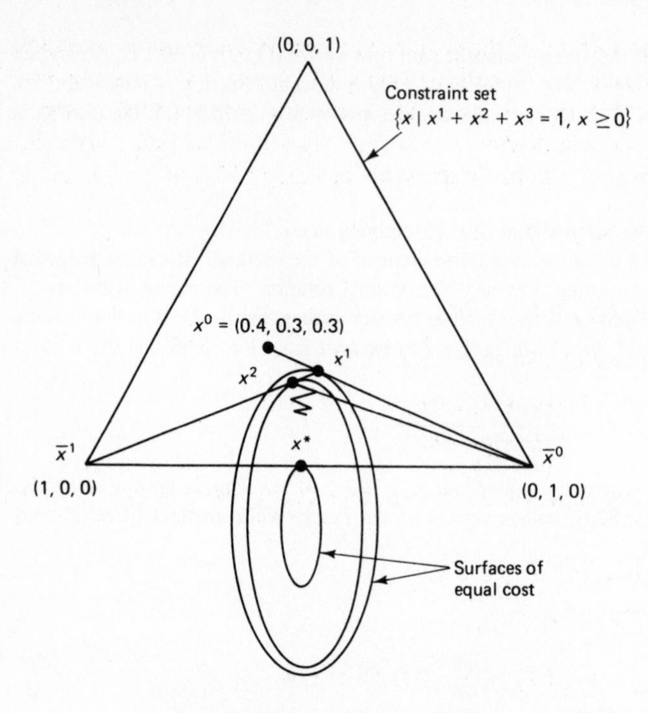

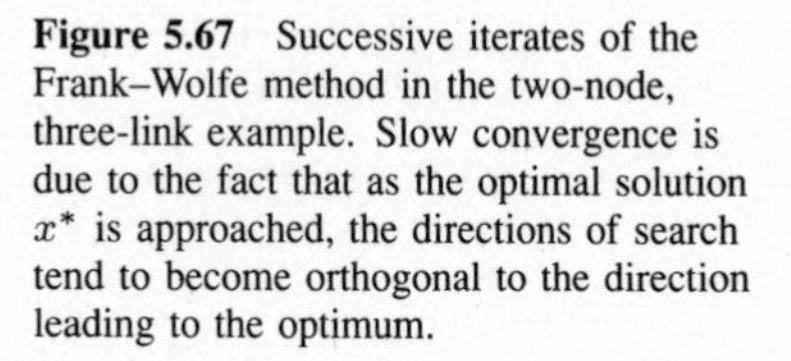
Figure 5.67 Successive iterates of the Frank–Wolfe method in the two-node, three-link example. Slow convergence is due to the fact that as the optimal solution x^* is approached, the directions of search tend to become orthogonal to the direction leading to the optimum.

The determination of an optimal stepsize α^* satisfying Eq. (5.61) requires a one-dimensional minimization over [0,1] which can be carried out through any one of several existing methods (see [Lue84]). A simpler method is to choose the stepsize α^* by means of

$$\alpha^* = \min\left[1, -\frac{\sum_{(i,j)}(\overline{F}_{ij} - F_{ij})D'_{ij}}{\sum_{(i,j)}(\overline{F}_{ij} - F_{ij})^2 D''_{ij}}\right] \tag{5.63}$$

Here $\{F_{ij}\}$ and $\{\overline{F}_{ij}\}$ are the sets of total link flows corresponding to $\{x_p\}$ and $\{\overline{x}_p\}$, respectively [*i.e.*, F_{ij} (or $\overline{F}_{ij}$) is obtained by adding all flows x_p (or $\overline{x}_p$) of paths p traversing the link (i, j)]. The first and second derivatives, D'_{ij} and D''_{ij}, respectively, are evaluated at F_{ij}. The formula for α^* in Eq. (5.63) is obtained by making a second-order Taylor approximation $\tilde{G}(\alpha)$ of $G(\alpha) = D\big[x + \alpha(\overline{x} - x)\big]$ around $\alpha = 0$:

$$\tilde{G}(\alpha) = \sum_{(i,j)} \left\{ D_{ij}(F_{ij}) + \alpha D'_{ij}(F_{ij})(\overline{F}_{ij} - F_{ij}) + \frac{\alpha^2}{2} D''_{ij}(F_{ij})(\overline{F}_{ij} - F_{ij})^2 \right\}$$

and by minimizing $\tilde{G}(\alpha)$ with respect to α over the interval [0,1] as in the earlier example.

It can be shown that the algorithm with the choice of Eq. (5.63) for the stepsize converges to the optimal set of total link flows provided that the starting set of total link flows is sufficiently close to the optimal. For the type of cost functions used in routing problems [*e.g.*, the $M/M/1$ delay formula; cf. Eq. (5.30) in Section 5.4], it appears that the simple stepsize choice of Eq. (5.63) typically leads to convergence even when the starting total link flows are far from optimal. This stepsize choice is also well suited for distributed implementation. However, even with this rule, the method appears inferior for distributed application to the projection method of the next section primarily because it is essential for the network nodes to synchronize their calculations in order to obtain the stepsize.

The Frank–Wolfe method has an insightful geometrical interpretation. Suppose that we have a feasible path flow vector $x = \{x_p\}$, and we try to find a path flow variation $\Delta x = \{\Delta x_p\}$ which is feasible in the sense of Eqs. (5.57) and (5.58), that is,

$$\sum_{p \in P_w} \Delta x_p = 0, \quad \text{for all } w \in W$$
$$x_p + \Delta x_p \geq 0, \quad \text{for all } p \in P_w, w \in W \tag{5.64}$$

and along which the initial rate of change [cf. Eq. (5.59)]

$$\sum_{w \in W} \sum_{p \in P_w} \frac{\partial D(x)}{\partial x_p} \Delta x_p$$

of the cost function D is most negative. Such a variation Δx is obtained by solving the optimization problem

$$\begin{aligned} \text{minimize} \quad & \sum_{w \in W} \sum_{p \in P_w} \frac{\partial D(x)}{\partial x_p} \Delta x_p \\ \text{subject to} \quad & \sum_{p \in P_w} \Delta x_p = 0, \quad \text{for all } w \in W \\ & x_p + \Delta x_p \geq 0, \quad \text{for all } p \in P_w, w \in W \end{aligned} \tag{5.64}$$

and it is seen that $\Delta x = \overline{x} - x$ is an optimal solution, where $\overline{x}$ is the shortest path flow vector generated by the Frank–Wolfe method. The process of finding Δx can be visualized as in Fig. 5.68. It can be seen that $\overline{x}$ is a vertex of the polyhedron of feasible path flow vectors that lies farthest out along the negative gradient direction $-\nabla D(x)$, thereby minimizing the inner product of $\nabla D(x)$ and Δx subject to the constraints (5.64).

Successive iterations of the Frank–Wolfe method are shown in Fig. 5.69. At each iteration, a vertex of the feasible set (*i.e.*, the shortest path flow $\overline{x}$) is obtained by solving a shortest path problem; then the next path flow vector is obtained by a search along the line joining the current path flow vector with the vertex. The search ensures that the new path flow vector has lower cost than the preceding one.

We finally mention an important situation where the Frank–Wolfe method has an advantage over other methods. Suppose that one is not interested in obtaining optimal path flows, but that the only quantities of interest are the optimal total link flows F_{ij} or just the value of optimal cost. (This arises in topological design studies; see Section 5.4.2.) Then the Frank–Wolfe method can be implemented in a way that only the current total link flows together with the current shortest paths for all OD pairs are maintained in memory at each iteration. This can be done by computing the total link flows corresponding to the shortest path flow vector $\overline{x}$ and executing the stepsize minimization indicated in Eq. (5.61) in the space of total link flows. The amount of storage required in this implementation is relatively small, thereby allowing the solution of very large network problems. If, however, one is interested in optimal path flows as well as total

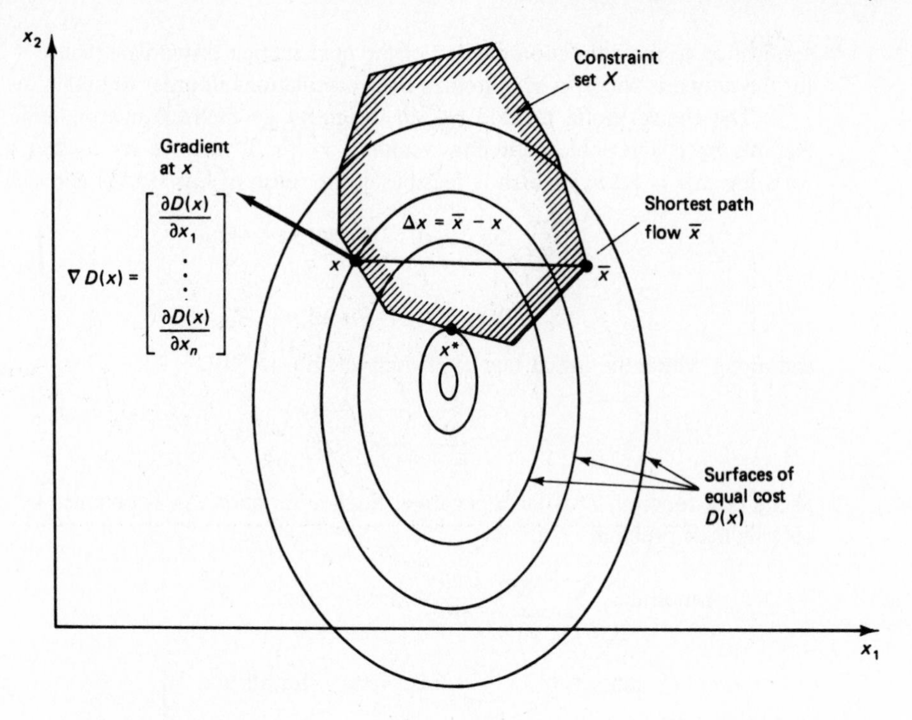

Figure 5.68 Finding the feasible descent direction $\Delta x = \bar{x} - x$ at a point x in the Frank–Wolfe method. $\bar{x}$ is the shortest path flow (or extreme point of the feasible set X) that lies farthest out along the negative gradient direction $-\nabla D(x)$.

link flows, the storage requirements of the Frank–Wolfe method are about the same as those of other methods, including the projection methods discussed in the next section.

5.7 PROJECTION METHODS FOR OPTIMAL ROUTING

We now consider a class of feasible direction algorithms for optimal routing that are faster than the Frank–Wolfe method and lend themselves more readily to distributed implementation. These methods are also based on shortest paths and determine a minimum first derivative length (MFDL) path for every OD pair at each iteration. An increment of flow change is calculated for each path on the basis of the relative magnitudes of the path lengths and, sometimes, the second derivatives of the cost function. If the increment is so large that the path flow becomes negative, the path flow is simply set to zero (*i.e.*, it is "projected" back onto the positive orthant). These routing algorithms may be viewed as constrained versions of common, unconstrained optimization methods, such as steepest descent and Newton's method, which are given in nonlinear programming texts (*e.g.*, [Zan69], [Pol71], [Ber82d], and [Lue84]). We first present these methods briefly in a general, nonlinear optimization setting, and subsequently specialize them to the routing problem.

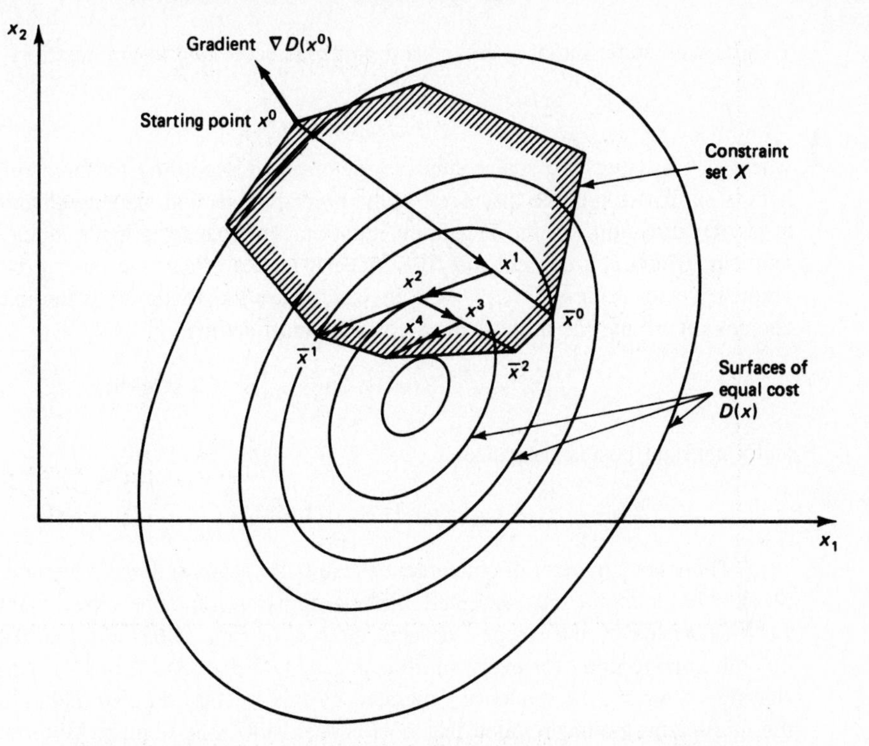

Figure 5.69 Successive iterations of the Frank–Wolfe method. At each iteration, a vertex of the feasible set (a shortest path flow) is found. The next path flow vector is obtained by a search on the line segment connecting the current path flow vector with the vertex.

5.7.1 Unconstrained Nonlinear Optimization

Let f be a twice differentiable function of the n-dimensional vector $x = (x_1, \ldots, x_n)$, with a gradient and Hessian matrix at any x denoted $\nabla f(x)$ and $\nabla^2 f(x)$:

$$\nabla f(x) = \begin{bmatrix} \dfrac{\partial f(x)}{\partial x_1} \\ \vdots \\ \dfrac{\partial f(x)}{\partial x_n} \end{bmatrix}, \qquad \nabla^2 f(x) = \begin{bmatrix} \dfrac{\partial^2 f(x)}{(\partial x_1)^2} & \cdots & \dfrac{\partial^2 f(x)}{\partial x_1 \partial x_n} \\ \vdots & & \vdots \\ \dfrac{\partial^2 f(x)}{\partial x_n \partial x_1} & \cdots & \dfrac{\partial^2 f(x)}{(\partial x_n)^2} \end{bmatrix}$$

We assume that for all x, $\nabla^2 f(x)$ is a positive semidefinite matrix that depends continuously on x.* The *method of steepest descent* for finding an unconstrained minimum of

*A symmetric $n \times n$ matrix A with elements A_{ij} is said to be *positive semidefinite* if the quadratic form

$$\sum_{i=1}^{n} \sum_{j=1}^{n} A_{ij} z_i z_j$$

is nonnegative for all vectors $z = (z_1, \ldots, z_n)$. The matrix A is said to be *positive definite* if this quadratic form is positive for all vectors $z \neq 0$.

f starts with some initial guess x^0 and proceeds according to the iteration

$$x^{k+1} = x^k - \alpha^k \nabla f(x^k), \qquad k = 0, 1, \ldots \tag{5.65}$$

where α^k is a positive scalar stepsize, determined according to some rule. The idea here is similar to the one discussed in the preceding section, namely changing x^k along a descent direction. In the preceding iteration, the change is made along the negative gradient, which is a descent direction since it makes a negative inner product with the gradient vector [unless $\nabla f(x^k) = 0$, in which case the vector x^k is optimal]. Common choices for α^k are the minimizing stepsize determined by

$$f\left[x^k - \alpha^k \nabla f(x^k)\right] = \min_{\alpha > 0} f\left[x^k - \alpha \nabla f(x^k)\right] \tag{5.66}$$

and a constant positive stepsize $\overline{\alpha}$

$$\alpha^k \equiv \overline{\alpha}, \qquad \text{for all } k \tag{5.67}$$

There are a number of convergence results for steepest descent methods. For example, if f has a unique unconstrained minimizing point, it may be shown that the sequence $\{x^k\}$ generated by the steepest descent method of Eqs. (5.65) and (5.66) converges to this minimizing point for every starting x^0 (see [Ber82d] and [Lue84]). Also, given any starting vector x^0, the sequence generated by this steepest descent method converges to the minimizing point provided that $\overline{\alpha}$ is chosen sufficiently small. Unfortunately, however, the speed of convergence of $\{x^k\}$ can be quite slow. It can be shown ([Lue84], p. 218) that for the case of the line minimization rule of Eq. (5.66), if f is a positive definite quadratic function, there holds

$$\frac{f(x^{k+1}) - f^*}{f(x^k) - f^*} \leq \left(\frac{M - m}{M + m}\right)^2 \tag{5.68}$$

where $f^* = \min_x f(x)$, and M, m are the largest and smallest eigenvalues of $\nabla^2 f(x)$, respectively. Furthermore, there exist starting points x^0 such that Eq. (5.68) holds with equality for every k. So if the ratio M/m is large (this corresponds to the equal-cost surfaces of f being very elongated ellipses), the rate of convergence is slow. Similar results can be shown for the steepest descent method with the constant stepsize (5.67), and these results also hold in a qualitatively similar form for functions f with an everywhere continuous and positive definite Hessian matrix.

The rate of convergence of the steepest descent method can be improved by premultiplying the gradient by a suitable positive definite scaling matrix B^k, thereby obtaining the iteration

$$x^{k+1} = x^k - \alpha^k B^k \nabla f(x^k), \qquad k = 0, 1, \ldots \tag{5.69}$$

This method is also based on the idea of changing x^k along a descent direction. [The direction of change $-B^k \nabla f(x^k)$ makes a negative inner product with $\nabla f(x^k)$, since B^k is a positive definite matrix.] From the point of view of rate of convergence, the best method is obtained with the choice

$$B^k = \left[\nabla^2 f(x^k)\right]^{-1} \tag{5.70}$$

[assuming that $\nabla^2 f(x^k)$ is invertible]. This is *Newton's method*, which can be shown to have a very fast (superlinear) rate of convergence near the minimizing point when the stepsize α^k is taken as unity. Unfortunately, this excellent convergence rate is achieved at the expense of the potentially substantial overhead associated with the inversion of $\nabla^2 f(x^k)$. It is often useful to consider other choices of B^k which approximate the "optimal" choice $[\nabla^2 f(x^k)]^{-1}$ but do not require as much overhead. A simple choice that often works well is to take B^k as a diagonal approximation to the inverse Hessian, that is,

$$B^k = \begin{bmatrix} \left[\frac{\partial^2 f(x^k)}{(\partial x_1)^2}\right]^{-1} & 0 & \cdots & 0 \\ 0 & \left[\frac{\partial^2 f(x^k)}{(\partial x_2)^2}\right]^{-1} & \cdots & 0 \\ \vdots & \vdots & \ddots & \vdots \\ 0 & 0 & \cdots & \left[\frac{\partial^2 f(x^k)}{(\partial x_n)^2}\right]^{-1} \end{bmatrix} \tag{5.71}$$

With this choice, the scaled gradient iteration (5.69) can be written in the simple form

$$x_i^{k+1} = x_i^k - \alpha^k \left[\frac{\partial^2 f(x^k)}{(\partial x_i)^2}\right]^{-1} \frac{\partial f(x^k)}{\partial x_i}, \qquad i = 1, \ldots, n \tag{5.72}$$

5.7.2 Nonlinear Optimization over the Positive Orthant

Next consider the problem of minimizing the function f subject to the nonnegativity constraints $x_i \geq 0$, for $i = 1, \ldots, n$, that is, the problem

$$\begin{aligned} &\text{minimize } f(x) \\ &\text{subject to } x \geq 0 \end{aligned} \tag{5.73}$$

A straightforward analog of the steepest descent method known as the *gradient projection method* ([Gol64] and [LeP65]), is given by

$$x^{k+1} = \left[x^k - \alpha^k \nabla f(x^k)\right]^+, \qquad k = 0, 1, \ldots \tag{5.74}$$

where for any vector z, we denote by $[z]^+$ the projection of z onto the positive orthant

$$[z]^+ = \begin{bmatrix} \max\{0, z_1\} \\ \max\{0, z_2\} \\ \vdots \\ \max\{0, z_n\} \end{bmatrix} \tag{5.75}$$

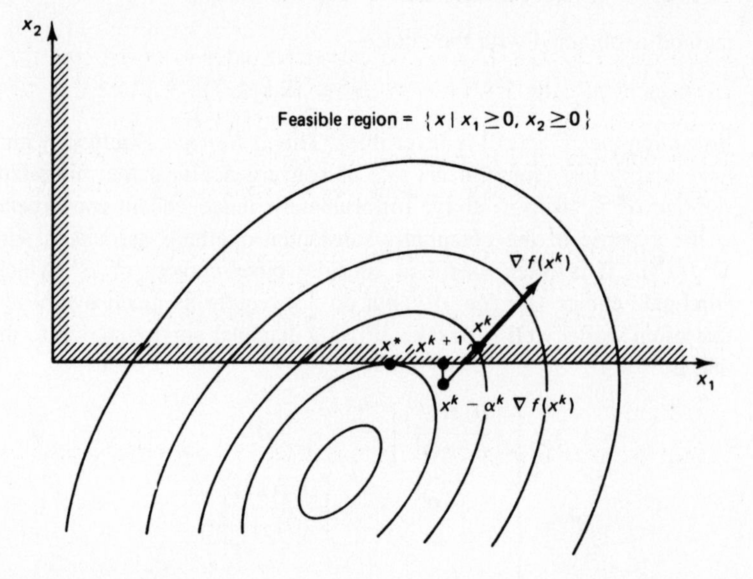

Figure 5.70 Operation of the gradient projection method. A step is made in the direction of the negative gradient, and the result is orthogonally projected on the positive orthant. It can be shown that the step $(x^{k+1} - x^k)$ is a descent direction [makes a negative inner product with $\nabla f(x^k)$], and if the stepsize α is sufficiently small, the iteration improves the cost [*i.e.*, $f(x^{k+1}) < f(x^k)$].

An illustration of how this method operates is given in Fig. 5.70. It can be shown that the convergence results mentioned earlier in connection with the unconstrained steepest descent method of Eq. (5.65) also hold true for its constrained analog of Eq. (5.74) (see Problem 5.32 and [Ber76]). The same is true for the method

$$x^{k+1} = \left[x^k - \alpha^k B^k \nabla f(x^k)\right]^+$$

where B^k is a *diagonal*, positive definite scaling matrix (see Problem 5.33). When B^k is given by the diagonal inverse Hessian approximation of Eq. (5.71), the following iteration is obtained:

$$x_i^{k+1} = \max\left\{0, x_i^k - \alpha^k \left[\frac{\partial^2 f(x^k)}{(\partial x_i)^2}\right]^{-1} \frac{\partial f(x^k)}{\partial x_i}\right\}, \qquad i = 1, \ldots, n \tag{5.76}$$

5.7.3 Application to Optimal Routing

Consider now the optimal routing problem

$$\begin{aligned} &\text{minimize } D(x) \overset{\Delta}{=} \sum_{(i,j)} D_{ij}(F_{ij}) \\ &\text{subject to } \sum_{p \in P_w} x_p = r_w, \quad x_p \geq 0, \qquad \text{for all } p \in P_w, w \in W \end{aligned} \tag{5.77}$$

where each total link flow F_{ij} is expressed in terms of the path flow vector $x = \{x_p\}$ as the sum of path flows traversing the link (i,j). Assume that the second derivatives of D_{ij}, denoted by $D''_{ij}(F_{ij})$, are positive for all F_{ij}. Let $x^k = \{x_p^k\}$ be the path flow vector obtained after k iterations, and let $\{F_{ij}^k\}$ be the corresponding set of total link flows. For each OD pair w, let $\overline{p}_w$ be an MFDL path [with respect to the current link lengths $D'_{ij}(F_{ij}^k)$].

The optimal routing problem (5.77) can be converted (for the purpose of the next iteration) to a problem involving only positivity constraints by expressing the flows of the MFDL paths $\overline{p}_w$ in terms of the other path flows, while eliminating the equality constraints

$$\sum_{p \in P_w} x_p = r_w$$

in the process. For each w, $x_{\overline{p}_w}$ is substituted in the cost function $D(x)$ using the equation

$$x_{\overline{p}_w} = r_w - \sum_{\substack{p \in P_w \\ p \neq \overline{p}_w}} x_p \tag{5.78}$$

thereby obtaining a problem of the form

$$\begin{aligned} &\text{minimize } \tilde{D}(\tilde{x}) \\ &\text{subject to} \quad x_p \geq 0, \qquad \text{for all } w \in W,\ p \in P_w,\ p \neq \overline{p}_w \end{aligned} \tag{5.79}$$

where $\tilde{x}$ is the vector of all path flows which are *not* MFDL paths.

We now calculate the derivatives that will be needed to apply the scaled projection iteration (5.76) to the problem of Eq. (5.79). Using Eq. (5.78) and the definition of $\tilde{D}(\tilde{x})$, we obtain

$$\frac{\partial \tilde{D}(\tilde{x}^k)}{\partial x_p} = \frac{\partial D(x^k)}{\partial x_p} - \frac{\partial D(x^k)}{\partial x_{\overline{p}_w}}, \qquad \text{for all } p \in P_w,\ p \neq \overline{p}_w \tag{5.80}$$

for all $w \in W$. In Section 5.5 it was shown that $\partial D(x)/\partial x_p$ is the first derivative length of path p, that is,

$$\frac{\partial D(x^k)}{\partial x_p} = \sum_{\substack{\text{all links} \\ (i,j) \text{ on path } p}} D'_{ij}(F_{ij}^k), \tag{5.81}$$

Regarding second derivatives, a straightforward differentiation of the first derivative expressions (5.80) and (5.81) shows that

$$\frac{\partial^2 \tilde{D}(\tilde{x}^k)}{(\partial x_p)^2} = \sum_{(i,j) \in L_p} D''_{ij}(F_{ij}^k), \qquad \text{for all } w \in W,\ p \in P_w,\ p \neq \overline{p}_w \tag{5.82}$$

where, for each p,

L_p = Set of links belonging to either p, or the corresponding MFDL path $\bar{p}_w$, but not both

Expressions for both the first and second derivatives of the "reduced" cost $\tilde{D}(\tilde{x})$, are now available and thus the scaled projection method of Eq. (5.76) can be applied. The iteration takes the form [cf. Eqs. (5.76), (5.80), and (5.82)]

$$x_p^{k+1} = \max\{0, x_p^k - \alpha^k H_p^{-1}(d_p - d_{\overline{p}_w})\}, \qquad \text{for all } w \in W,\ p \in P_w,\ p \neq \overline{p}_w \tag{5.83}$$

where d_p and $d_{\overline{p}_w}$ are the first derivative lengths of the paths p and $\overline{p}_w$ given by [cf. Eq. (5.81)]

$$d_p = \sum_{\substack{\text{all links } (i,j) \\ \text{on path } p}} D'_{ij}(F_{ij}^k), \qquad d_{\overline{p}_w} = \sum_{\substack{\text{all links } (i,j) \\ \text{on path } \overline{p}_w}} D'_{ij}(F_{ij}^k), \tag{5.84}$$

and H_p is the "second derivative length"

$$H_p = \sum_{(i,j)\in L_p} D''_{ij}(F_{ij}^k) \tag{5.85}$$

given by Eq. (5.82). The stepsize α^k is some positive scalar which may be chosen by a variety of methods. For example, α^k can be constant or can be chosen by some form of line minimization. Stepsize selection will be discussed later. We refer to the iteration of Eqs. (5.83) to (5.85) as the *projection algorithm*.

The following observations can be made regarding the projection algorithm:

1. Since $d_p \geq d_{\overline{p}_w}$ for all $p \neq \overline{p}_w$, all the nonshortest path flows that are positive will be reduced with the corresponding increment of flow being shifted to the MFDL path $\overline{p}_w$. If the stepsize α^k is large enough, all flow from the nonshortest paths will be shifted to the shortest path. Therefore, the projection algorithm may also be viewed as a generalization of the adaptive routing method based on shortest paths of Section 5.2.5, with α^k, H_p, and $(d_p - d_{\overline{p}_w})$ determining the amounts of flow shifted to the shortest path.
2. Those nonshortest path flows x_p, $p \neq \overline{p}_w$ that are zero will stay at zero. Therefore, the path flow iteration of Eq. (5.83) should only be carried out for paths that carry positive flow.
3. Only paths that carried positive flow at the starting flow pattern or were MFDL paths at some previous iteration can carry positive flow at the beginning of an iteration. This is important since it tends to keep the number of flow carrying paths small, with a corresponding reduction in the amount of calculation and bookkeeping needed at each iteration.

There are several ways to choose the stepsize α^k. One possibility is to keep α^k constant ($\alpha^k \equiv \alpha$, for all k). With this choice it can be shown that given any starting set of path flows, there exists $\overline{\alpha} > 0$ such that if $\alpha \in (0, \overline{\alpha}]$, a sequence generated by the projection algorithm converges to the optimal cost of the problem (see Problem 5.32). A crucial question has to do with the magnitude of the constant stepsize. It is known from nonlinear programming experience and analysis that a stepsize equal to 1 usually works well with Newton's method as well as diagonal approximations to Newton's method [cf. Eq. (5.71)] that employ scaling based on second derivatives ([Lue84] and [Ber82d]).

Experience has verified that choosing α^k in the path flow iteration (5.83) close to 1 typically works quite well regardless of the values of the input flows r_w [BGV79]. Even better performance is usually obtained if the iteration is carried out *one OD pair (or one origin) at a time*; that is, first carry out the iteration with $\alpha^k = 1$ for a single OD pair (or origin), adjust the corresponding total link flows to account for the effected change in the path flows of this OD pair (or origin), and continue with the next OD pair until all OD pairs are taken up cyclically. The rationale for this is that dropping the off-diagonal terms of the Hessian matrix [cf. Eqs. (5.69) and (5.71)], in effect, neglects the interaction between the flows of different OD pairs. In other words, the path flow iteration (5.83) is based to some extent on the premise that each OD pair will adjust its own path flows while the other OD pairs will keep theirs unchanged. Carrying out this iteration one OD pair at a time reduces the potentially detrimental effect of the neglected off-diagonal terms and increases the likelihood that the unity stepsize is appropriate and effective.

Using a constant stepsize is well suited for distributed implementation. For centralized computation, it is possible to choose α^k by a simple form of line search. For example, start with a unity stepsize, evaluate the corresponding cost, and if no reduction is obtained over $D(x^k)$, successively reduce the stepsize until a cost reduction $D(x^{k+1}) < D(x^k)$ is obtained. (It is noted for the theoretically minded that this scheme cannot be shown to converge to the optimal solution. However, typically, it works well in practice. Similar schemes with better theoretical convergence properties are described in [Ber76] and [Ber82c].) Still another possibility for stepsize selection is discussed in Problem 5.32.

The projection algorithm typically yields rapid convergence to a neighborhood of an optimal solution. Once it comes near a solution (how "near" depends on the problem), it tends to slow down. Its progress is often satisfactory near a solution and usually far better than that of the Frank–Wolfe method. Figure 5.71 provides a comparison of the projection method and the Frank–Wolfe method for the example of the preceding section

To obtain faster convergence near an optimal solution, it is necessary to modify the projection algorithm so that the off-diagonal terms of the Hessian matrix are taken into account. Surprisingly, it is possible to implement sophisticated methods of this type (see [BeG83]), but we will not go into this further. Suffice to say that these methods are based on a more accurate approximation of a constrained version of Newton's method (using the conjugate gradient method) near an optimal solution. However, when far from a solution, their speed of convergence is usually only slightly superior to that of the projection algorithm. So, if one is only interested in getting fast near an optimal solution in few iterations, but the subsequent rate of progress is of little importance (as is often the case in practical routing problems), the projection algorithm is typically satisfactory.

We finally note that the projection algorithm is well suited for distributed implementation. The most straightforward possibility is for all nodes i to broadcast to all other nodes the current total flows F_{ij}^k of their outgoing links (i, j), using, for example, a flooding algorithm or the SPTA of Section 5.3.3. Each node then computes the MFDL

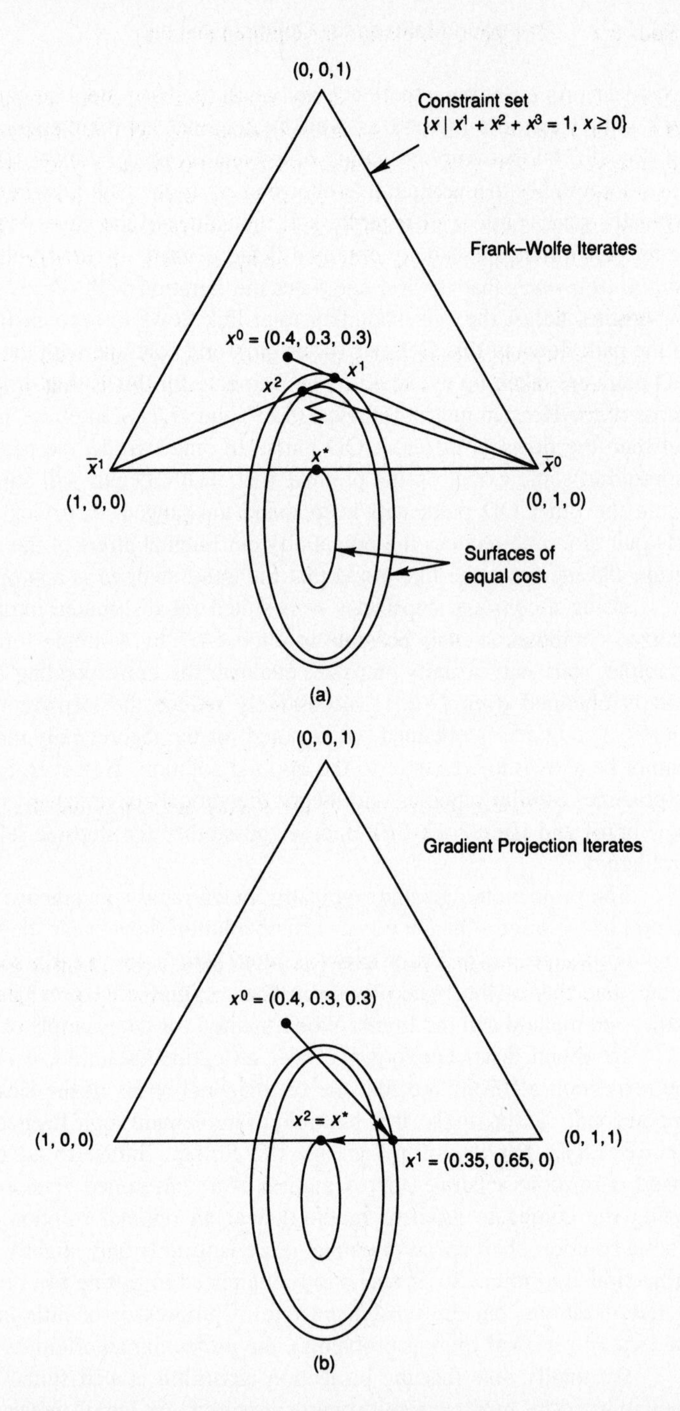

Figure 5.71 Iterates of (a) the Frank–Wolfe and (b) the gradient projection method for the example of the preceding section. The gradient projection method converges much faster. It sets the flow x_3 to the correct value 0 in one iteration and (because the Hessian matrix here is diagonal and the cost is quadratic) requires one more iteration to converge.

paths of OD pairs for which it is the origin and executes the path flow iteration (5.83) for some fixed stepsize. This corresponds to an "all OD pairs at once" mode of implementation. The method can also be implemented in an asynchronous, distributed format, whereby the computation and information reception are not synchronized at each node. The validity of the method under these conditions is shown in [TsB86] and [BeT89].

We close this section with an example illustrating the projection algorithm:

Example 5.9

Consider the network shown in Fig. 5.72. There are only two OD pairs (1,5) and (2,5) with corresponding inputs $r_1 = 4$ and $r_2 = 8$. Consider the following two paths for each OD pair:

Paths of OD pair $(1,5)$

$$p_1(1) = \{1,4,5\}$$

$$p_2(1) = \{1,3,4,5\}$$

Paths of OD pair $(2,5)$

$$p_1(2) = \{2,4,5\}$$

$$p_2(2) = \{2,3,4,5\}$$

Consider the instance of the routing problem where the link cost functions are all identical and are given by

$$D_{ij}(F_{ij}) = \frac{1}{2}(F_{ij})^2, \qquad \text{for all } (i,j)$$

Consider an initial path flow pattern whereby each OD pair input is routed through the middle link (3,4). This results in a flow distribution given in Tables 5.1 and 5.2.

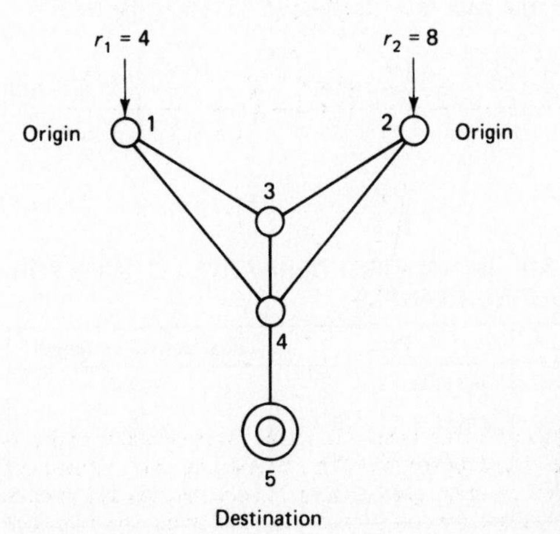

Figure 5.72 The network, for the example, where node 5 is the only destination and there are two origins, nodes 1 and 2.

TABLE 5.1 INITIAL PATH FLOWS FOR THE ROUTING EXAMPLE

OD pair	Path	Path flow
(1,5)	$p_1(1) = \{1, 4, 5\}$	0
	$p_2(1) = \{1, 3, 4, 5\}$	4
(2,5)	$p_1(2) = \{2, 4, 5\}$	0
	$p_2(2) = \{2, 3, 4, 5\}$	8

TABLE 5.2 INITIAL TOTAL LINK FLOWS FOR THE ROUTING EXAMPLE

Link	Total link flow
(1,3)	4
(1,4)	0
(2,3)	8
(2,4)	0
(3,4)	12
(4,5)	12
Others	0

The first derivative length of each link is given by

$$D'_{ij}(F_{ij}) = F_{ij}$$

so the total link flows given in Table 5.2 are also the link lengths for the current iteration. The corresponding first derivative lengths of paths are given in Table 5.3. Therefore, the shortest paths for the current iteration are $p_1(1)$ and $p_1(2)$ for OD pairs (1,5) and (2,5), respectively.

The form of the projection algorithm of Eqs. (5.83) to (5.85) is illustrated for the first OD pair. Here, for the nonshortest path $p = p_2(1)$ and the shortest path $\overline{p} = p_1(1)$ we have $d_p = 28$ and $d_{\overline{p}} = 12$. Also, $H_p = 3$ [each link has a second derivative length $D''_{ij} = 1$, and there are three links that belong to either $p_1(1)$ or $p_2(1)$, but not to both; cf. Eqs. (5.82) and (5.85). Therefore, the path flow iteration (5.83) takes the form

$$x_p := \max\left\{0, 4 - \frac{\alpha^k}{3}(28 - 12)\right\} = \max\left\{0, 4 - \frac{16\alpha^k}{3}\right\} = 4 - \min\left\{4, \frac{16\alpha^k}{3}\right\}$$

and

$$x_{\overline{p}} := r_1 - x_p$$

TABLE 5.3 INITIAL FIRST DERIVATIVE LENGTHS FOR THE ROUTING EXAMPLE

OD pair	Path	First derivative length
(1,5)	$p_1(1) = \{1, 4, 5\}$	12
	$p_2(1) = \{1, 3, 4, 5\}$	28
(2,5)	$p_1(2) = \{2, 4, 5\}$	12
	$p_2(2) = \{2, 3, 4, 5\}$	32

More generally, let $x_1(1), x_2(1), x_1(2)$, and $x_2(2)$ denote the flows along the paths $p_1(1)$, $p_2(1)$, $p_1(2)$, and $p_2(2)$, respectively, at the beginning of an iteration. The corresponding path lengths are

$$d_{p_1(1)} = x_1(1) + r_1 + r_2$$
$$d_{p_2(1)} = 2x_2(1) + x_2(2) + r_1 + r_2$$
$$d_{p_1(2)} = x_1(2) + r_1 + r_2$$
$$d_{p_2(2)} = 2x_2(2) + x_2(1) + r_1 + r_2$$

The second derivative length H_p of Eq. (5.85) equals 3. The projection algorithm (5.83) to (5.85) takes the following form. For $i = 1, 2$,

$$x_1(i) := \begin{cases} x_1(i) - \min\left[x_1(i), \frac{\alpha^k}{3}[d_{p_1(i)} - d_{p_2(i)}]\right], & \text{if } d_{p_1(i)} > d_{p_2(i)} \\ x_1(i) + \min\left[x_2(i), \frac{\alpha^k}{3}[d_{p_2(i)} - d_{p_1(i)}]\right], & \text{otherwise} \end{cases}$$

$$x_2(i) := r_i - x_1(i)$$

Notice that the presence of link (4,5) does not affect the form of the iteration, and indeed this should be so since the total flow of link (4,5) is always equal to $r_1 + r_2$ independent of the routing.

Table 5.4 gives sequences of successive cost function values obtained by the algorithm for different stepsizes and the "all OD pairs at once" and "one OD pair at a time" modes of implementation. The difference between these two implementation modes is that in the all-at-once mode, the OD pairs are processed simultaneously during an iteration using the link and path flows obtained at the end of the preceding iteration. In the one-at-a-time mode the OD pairs are processed sequentially, and following the iteration of one OD pair, the link flows are adjusted to reflect the results of the iteration before carrying out an iteration for the other OD pair. The stepsize is chosen to be constant at one of three possible values ($\alpha^k \equiv 0.5$, $\alpha^k \equiv 1$, and $\alpha^k \equiv 1.8$). It can be seen that for a unity stepsize, the convergence to a neighborhood of a solution is very fast in both the one-at-a-time and the all-at-once modes. As the stepsize is increased, the danger of divergence increases, with divergence typically occurring first for the all-at-once mode. This can be seen from the table, where for $\alpha^k \equiv 1.8$, the algorithm converges (slowly) in the one-at-a-time mode but diverges in the all-at-once mode.

TABLE 5.4 SEQUENCE OF COSTS GENERATED BY THE GRADIENT PROJECTION METHOD FOR THE ROUTING EXAMPLE AND FOR A VARIETY OF STEPSIZES

Iteration	All-at-once mode			One-at-a-time mode		
k	$\alpha^k \equiv 0.5$	$\alpha^k \equiv 1.0$	$\alpha^k \equiv 1.8$	$\alpha^k \equiv 0.5$	$\alpha^k \equiv 1.0$	$\alpha^k \equiv 1.8$
0	184.00	184.00	184.00	184.00	184.00	184.00
1	110.88	104.00	112.00	114.44	101.33	112.00
2	102.39	101.33	118.72	103.54	101.00	109.15
3	101.28	101.03	112.00	101.63		101.79
4	101.09	101.00	118.72	101.29		101.56
5	101.03		112.00	101.08		101.33
6	101.01		118.72	101.03		101.18
7	101.00		112.00	101.01		101.12
8			118.72	101.00		101.09
9			112.00			101.06
10			118.72			101.03

5.8 ROUTING IN THE CODEX NETWORK

In this section the routing system of a network marketed by Codex, Inc., is discussed. References [HuS86] and [HSS86] describe the system in detail. The Codex network uses virtual circuits internally for user traffic and datagrams for system traffic (accounting, control, routing information, etc.). It was decided to use datagrams for system traffic because the alternative, namely establishing a virtual circuit for each pair of nodes, was deemed too expensive in terms of network resources. Note that for each user session, there are two separate virtual circuits carrying traffic in opposite directions, but not necessarily over the same set of links. Routing decisions are done separately in each direction.

There are two algorithms for route selection. The first, used for datagram routing of internal system traffic, is a simple shortest path algorithm of the type discussed in Section 5.2. The shortest path calculations are done as part of the second algorithm, which is used for selecting routes for new virtual circuits and for rerouting old ones. We will focus on the second algorithm, which is more important and far more sophisticated than the first.

The virtual circuit routing algorithm has much in common with the gradient projection method for optimal routing of the preceding section. There is a cost function D_{ij} for each link (i, j) that depends on the link flow, but also on additional parameters, such as the link capacity, the processing and propagation delay, and a priority factor for the virtual circuits currently on the link. Each node monitors the parameters of its adjacent links and broadcasts them periodically to all other nodes. If all virtual circuits have the same priority, the link cost function has the form

$$D_{ij}(F_{ij}) = \frac{F_{ij}}{C_{ij} - F_{ij}} + d_{ij}F_{ij}$$

where F_{ij} is the total data rate on the link, d_{ij} is the processing and propagation delay, and C_{ij} is the link capacity. This formula is based on the $M/M/1$ delay approximation; see the discussion of Sections 3.6 and 5.4. When there are more than one, say M, priority levels for virtual circuits, the link cost function has the form

$$D_{ij}(F_{ij}^1, \ldots, F_{ij}^M) = \left(\sum_{k=1}^{M} p_k F_{ij}^k\right)\left(\frac{1}{C_{ij} - \sum_{k=1}^{M} F_{ij}^k} + d_{ij}\right)$$

where F_{ij}^k is the total link flow for virtual circuits of priority k, and p_k is a positive weighting factor. The form of this expression is motivated by the preceding formula, but otherwise has no meaningful interpretation based on a queueing analysis.

Routing of a new virtual circuit is done by assignment on a path that is shortest for the corresponding OD pair. This path is calculated at the destination node of the virtual circuit on the basis of the network topology, and the flow and other parameters last received for each network link. The choice of link length is motivated by the fact that the communication rate of a virtual circuit may not be negligible compared with the total flow of the links on its path. The length used is the link cost difference between when the virtual circuit is routed through the link and when it is not. Thus, if the new virtual circuit has an estimated data rate Δ and priority class k, the length of link (i, j) is

$$D_{ij}\left(F_{ij}^1, \ldots, F_{ij}^k + \Delta, \ldots, F_{ij}^M\right) - D_{ij}(F_{ij}^1, \ldots, F_{ij}^M) \quad (5.86)$$

As a result, routing of a new virtual circuit on a shortest path with respect to the link length (5.86) results in a minimum increase in total cost. When all virtual circuits have the same priority, the link length (5.86) becomes

$$D_{ij}(F_{ij} + \Delta) - D_{ij}(F_{ij}) \quad (5.87)$$

If Δ is very small, the link lengths (5.87) are nearly equal to $\Delta D'_{ij}(F_{ij})$. Otherwise (by the mean value theorem of calculus), they are proportional to the corresponding derivatives of D_{ij} at some intermediate point between the current flow and the flow resulting when the virtual circuit is routed through (i, j). Thus, the link lengths (5.87), divided by Δ, are approximations to the first derivatives of link costs that are used in the gradient projection method. The link lengths (5.86) admit a similar interpretation (see Problem 5.27).

Rerouting of old virtual circuits is done in a similar manner. If there is a link failure, virtual circuits crossing the link are rerouted as if they are new. Rerouting with the intent of alleviating congestion is done by all nodes gradually once new link flow information is received. Each node scans the virtual circuits terminating at itself. It selects one of the virtual circuits as a candidate for rerouting and calculates the length of each link as the difference of link cost with and without that virtual circuit on the link. The virtual circuit is then rerouted on the shortest path if it does not already lie on the shortest path. An important parameter here is the number of virtual circuits chosen for rerouting between successive link flow broadcasts. This corresponds to the stepsize parameter in the gradient projection method, which determines how much flow is shifted on a shortest path at each iteration. The problem, of course, is that too many virtual circuits may be rerouted simultaneously by several nodes acting without coordination, thereby resulting in oscillatory behavior similar to that discussed in Section 5.2.5. The Codex network uses a heuristic rule whereby only a fraction of the existing virtual circuits are (pseudorandomly) considered for rerouting between successive link flow broadcasts.

SUMMARY

Routing is a sophisticated data network function that requires coordination between the network nodes through distributed protocols. It affects the average packet delay and the network throughput.

Our main focus was on methods for route selection. The most common approach, shortest path routing, can be implemented in a variety of ways, as exemplified by the ARPANET and TYMNET algorithms. Depending on its implementation, shortest path routing may cause low throughput, poor response to traffic congestion, and oscillatory behavior. These drawbacks are more evident in datagram than in virtual circuit networks. A more sophisticated alternative is optimal routing based on flow models. Several algorithms were given for computation of an optimal routing, both centralized and distributed.

As the statistics of the input arrival processes change more rapidly, the appropriateness of the type of optimal routing we focused on diminishes. In such cases it is difficult to recommend routing methods that are simultaneously efficient and practical.

Another interesting aspect of the routing problem relates to the dissemination of routing-related information over failure-prone links. We described several alternative algorithms based on flooding ideas.

Finally, routing must be taken into account when designing a network's topology, since the routing method determines how effectively the link capacities are utilized. We described exact and heuristic methods for addressing the difficulties of topological design.

NOTES, SOURCES, AND SUGGESTED READING

Section 5.1. Surveys of routing, including descriptions of some practical routing algorithms, can be found in [Eph86] and [ScS80]. Routing in the ARPANET is described in [McW77], [MRR78], and [MRR80]. The TYMNET system is described in [Tym81]. Routing in SNA is described in [Ahu79] and [Atk80]. A route selection method for SNA networks is described in [GaH83]. Descriptions of other routing systems may be found in [Wec80] and [SpM81]. Topology broadcast algorithms are discussed in [HuS88]. Extensive discussions of interconnected local area networks can be found in a special issue of *IEEE Network* [IEE88].

Section 5.2. There are many sources for the material on graphs, spanning trees, and shortest paths (*e.g.*, [Ber91] and [PaS82]). Particularly fast shortest path algorithms are described in [Pap74], [DGK79], and [GaP88].

An analysis of some asynchronous min-hop algorithms was first given in [Taj77]. The material on asynchronous shortest path algorithms given here is taken from [Ber82a]. For an extensive treatment of asynchronous distributed algorithms, see [BeT89].

For more on stability issues of shortest path routing algorithms, see [MRR78], [Ber79b], [Ber82b], and [GaB83].

Section 5.3. The difficulties with the ARPANET flooding algorithm and some remedies are described in [Ros81] and [Per83]. The SPTA appeared in the thesis [Spi85]; see also [SpG89]. For related work, see [EpB81], [Fin79], [GaB81], [Seg83], and [SoH86].

Section 5.4. There is considerable literature on routing based on queue state information. These works consider data networks but also relate to routing and control problems in other contexts. See [EVW80], [FoS78], [HaO84], [MoS82], [Ros86], [SaH87], [SaO82], [Sar82], [StK85], and [Yum81]. A survey of the application of control and optimization methods in communication network problems is given in [EpV89].

Surveys on topological design of data networks are provided in [BoF77], [GeK77], [KeB83], [McS77], and [MoS86]. The special issue [IEE89] describes several recent works on the subject. For further material on network reliability, see [BaP83], [Bal79], [Ben86], [LaL86], [LiS84], [PrB84], and the references quoted therein. For some alternative approaches to the problem of capacity assignment and routing, see [Gav85], [NgH87], and [LeY89].

Section 5.5. The optimal routing problem considered is a special case of a multicommodity network flow problem. The problem arises also in the context of transportation networks in a form that is very similar to the one discussed here; see [AaM81], [Daf71], [Daf80], [DeK81], [FlN74], and [LaH84].

Section 5.6. The Frank–Wolfe method [FrW56] has been proposed in the context of the routing problem in [FGK73]. Problems 5.31 and 5.32 give a self-contained presentation of the convergence properties of this method as well as those of the gradient projection method.

Section 5.7. The gradient projection method was applied to multicommodity flow problems based on a path flow formulation in [Ber80]. Extensions were described in [BeG82], [BeG83], and [GaB84a]. A public domain FORTRAN implementation of the gradient projection method is given in [BGT84]. Distributed asynchronous versions of the algorithm are analyzed in [TsB86] and [Tsa89]. The optimal routing algorithm of [Gal77], which adjusts link flow fractions (see the end of Section 5.5), bears a relationship to the gradient projection method (see [BGG84]). Extensions of the algorithm are given in [Ber79a], [Gaf79], and [BGG84]. Computational and simulation results relating to this algorithm and its extensions may be found in [BGV79] and [ThC86].

Section 5.8. The comments of P. Humblet, one of the principal designers of the Codex network, were very helpful in preparing this section. Additional material can be found in [HuS86] and [HSS86].

PROBLEMS

5.1 Find a minimum weight spanning tree of the graph in Fig. 5.73 using the Prim–Dijkstra and the Kruskal algorithms.

5.2 Find the shortest path tree from every node to node 1 for the graph of Fig. 5.74 using the Bellman–Ford and Dijkstra algorithms.

5.3 The number shown next to each link of the network in Fig. 5.75 is the probability of the link failing during the lifetime of a virtual circuit from node A to node B. It is assumed that links fail independently of each other. Find the most reliable path from A to B, that

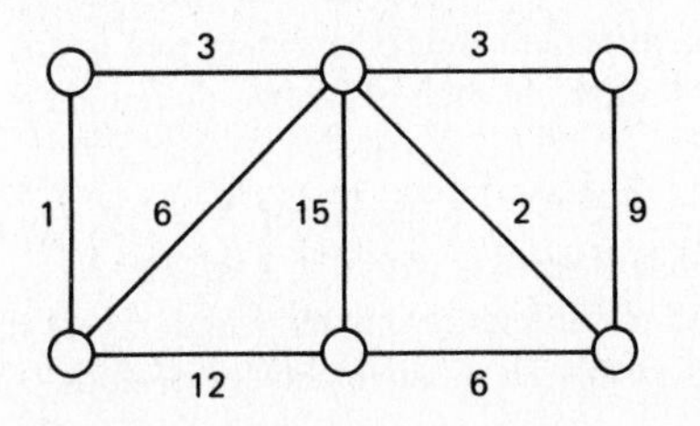

Figure 5.73 Graph for Problem 5.1.

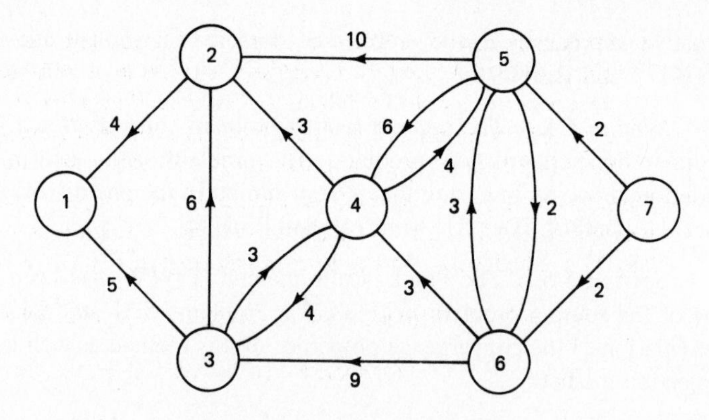

Figure 5.74 Graph for Problem 5.2.

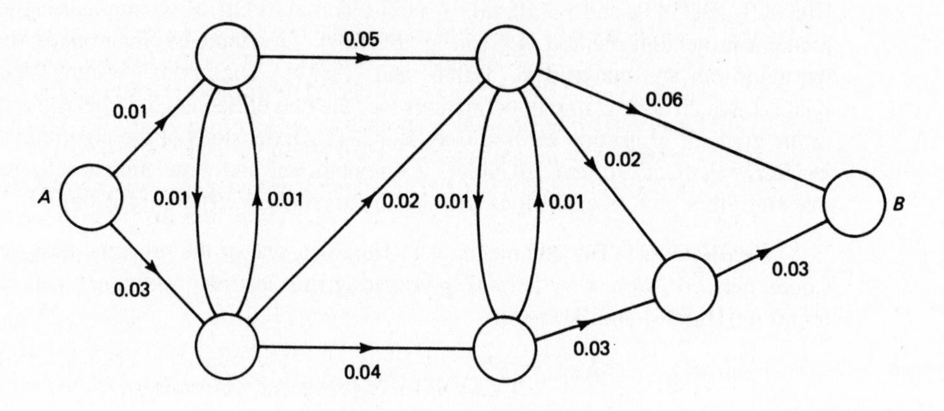

Figure 5.75 Graph for Problem 5.3.

is, the path for which the probability that all its links stay intact during the virtual circuit's lifetime is maximal. What is this probability?

5.4 A shortest path spanning tree (Section 5.2.3) differs from a minimum weight spanning tree in that the arc lengths refer to directed arcs, while in the minimum weight spanning tree problem, the arcs weights refer to undirected arcs or, equivalently, the arc weights are assumed equal in both directions. However, even if all arcs have equal length in both directions, a minimum weight spanning tree (for arc weights equal to the corresponding lengths) need not be a shortest path spanning tree. To see this, consider a network with three nodes A, B, and C, and three bidirectional arcs AB, BC, and CA. Choose a weight for each arc so that the minimum weight spanning tree is not a shortest path tree with root node C. (The length of an arc in each of its two directions is equal to the arc weight.)

5.5 Consider Example 5.4.

(a) Repeat the example with $N = 6$ instead of $N = 16$. (Nodes $1, 2, 4$, and 5 send 1 unit to node 6, while node 3 sends ϵ with $0 < \epsilon \ll 1$.)

(b) Repeat part (a) with the difference that the length d_{ij} of link (i, j) is $\alpha + F_{ij}$ for $\alpha = 1$ (instead of F_{ij}). Consider all possible choices of initial routing.

(c) What is the minimum value of α for which the shortest paths of all nodes except node 3 eventually stay constant regardless of the choice of initial routing?

(d) Repeat part (a) with the difference that at each iteration after the first, the length of each link is the average of the link arrival rates corresponding to the current and the preceding routings (instead of the arrival rates at just the current routing).

5.6 Consider the shortest path problem assuming that all cycles not containing node 1 have nonnegative length. Let $\tilde{D}_j$ be the shortest path distances corresponding to a set of link lengths $\tilde{d}_{ij}$, and let d_{ij} be a new set of link lengths. For $k = 1, 2, \ldots$, define

$$N_1 = \left\{ i \neq 1 \mid \text{for some arc } (i,j) \text{ with } d_{ij} > \tilde{d}_{ij} \text{ we have } \tilde{D}_i = \tilde{d}_{ij} + \tilde{D}_j \right\}$$

$$N_{k+1} = \left\{ i \neq 1 \mid \text{for some arc } (i,j) \text{ with } j \in N_k \text{ we have } \tilde{D}_i = \tilde{d}_{ij} + \tilde{D}_j \right\}$$

(a) Show that the Bellman–Ford algorithm starting from the initial conditions $D_1^0 = 0$, $D_i^0 = \tilde{D}_i$ for $i \notin \cup_k N_k \cup \{1\}$, and $D_i^0 = \infty$ for $i \in \cup_k N_k$ terminates after at most N iterations.

(b) Based on part (a), suggest a heuristic method for alleviating the "bad news phenomenon" of Fig. 5.36, which arises in the asynchronous Bellman–Ford algorithm when some link lengths increase. *Hint*: Whenever the length of a link on the current shortest path increases, the head node of the link should propagate an estimated distance of ∞ along the shortest path tree in the direction away from the destination.

5.7 Consider the Bellman–Ford algorithm assuming that all cycles not containing node 1 have nonnegative length. For each node $i \neq 1$, let (i, j_i) be an arc such that j_i attains the minimum in the equation

$$D_i^{h_i} = \min_j \left[d_{ij} + D_j^{h_i - 1} \right]$$

where h_i is the largest h such that $D_i^h \neq D_i^{h-1}$. Consider the subgraph consisting of the arcs (i, j_i), for $i \neq 1$, and show that it is a spanning tree consisting of shortest paths. *Hint*: Show that $h_i > h_{j_i}$.

5.8 Consider the shortest path problem. Show that if there is a cycle of zero length not containing node 1, but no cycles of negative length not containing node 1, Bellman's equation has more than one solution.

5.9 Suppose that we have a directed graph with no directed cycles. We are given a length d_{ij} for each directed arc (i, j) and we want to compute a shortest path to node 1 from all other nodes, assuming there exists at least one such path. Show that nodes $2, 3, \ldots, N$ can be renumbered so that there is an arc from i to j only if $i > j$. Show that once the nodes are renumbered, Bellman's equation can be solved with $O(N^2)$ operations at worst.

5.10 Consider the shortest path problem from every node to node 1 and Dijkstra's algorithm.

(a) Assume that a positive lower bound is known for all arc lengths. Show how to modify the algorithm to allow more than one node to enter the set of permanently labeled nodes at each iteration.

(b) Suppose that we have already calculated the shortest path from every node to node 1, and assume that a single arc length *increases*. Modify Dijkstra's algorithm to recalculate as efficiently as you can the shortest paths.

5.11 *A Generic Single-Destination Multiple-Origin Shortest Path Algorithm.* Consider the shortest path problem from all nodes to node 1. The following algorithm maintains a list of nodes V and a vector $D = (D_1, D_2, \ldots, D_N)$, where each D_j is either a real number or ∞.

Initially,

$$V = \{1\}$$

$$D_1 = 0, \qquad D_i = \infty, \qquad \text{for all } i \neq 1$$

The algorithm proceeds in iterations and terminates when V is empty. The typical iteration (assuming that V is nonempty) is as follows:

Remove a node j from V. For each incoming arc $(i,j) \in \mathcal{A}$, with $i \neq 1$, if $D_i > d_{ij} + D_j$, set

$$D_i = d_{ij} + D_j$$

and add i to V if it does not already belong to V.

(a) Show that at the end of each iteration: (1) if $D_j < \infty$, then D_j is the length of some path starting at j and ending at 1, and (2) if $j \notin V$, then either $D_j = \infty$ or else

$$D_i \leq d_{ij} + D_j, \qquad \text{for all } i \text{ such that } (i,j) \in \mathcal{A}$$

(b) If the algorithm terminates, then upon termination, for all $i \neq 1$ such that $D_i < \infty$, D_i is the shortest distance from i to 1 and

$$D_i = \min_{(i,j)\in\mathcal{A}} [d_{ij} + D_j]$$

Furthermore, $D_i = \infty$ for all i such that there is no path from i to 1.

(c) If the algorithm does not terminate, there exist paths of arbitrarily small length from at least one node i to node 1.

(d) Show that if at each iteration the node j removed from V satisfies $D_j = \min_{i\in V} D_i$, the algorithm is equivalent to Dijkstra's algorithm.

5.12 *A Generic Single-Destination Single-Origin Shortest Path Algorithm.* Consider the problem of finding a shortest path from a given node t to node 1. Suppose that for all i we have an underestimate u_i of the shortest distance from t to i. The following algorithm maintains a list of nodes V and a vector $D = (D_1, D_2, \ldots, D_N)$, where each D_i is either a real number or ∞. Initially,

$$V = \{1\}$$

$$D_1 = 0, \qquad D_i = \infty, \qquad \text{for all } i \neq 1$$

The algorithm proceeds in iterations and terminates when V is empty. The typical iteration (assuming that V is nonempty) is as follows:

Remove a node j from V. For each incoming arc $(i,j) \in \mathcal{A}$, with $i \neq 1$, if $\min\{D_i, D_t - u_i\} > d_{ij} + D_j$, set

$$D_i = d_{ij} + D_j$$

and add i to V if it does not already belong to V.

(a) If the algorithm terminates, then upon termination, either $D_t < \infty$, in which case D_t is the shortest distance from t to 1, or else there is no path from t to 1.

(b) If the algorithm does not terminate, there exist paths of arbitrarily small length from at least one node i to node 1.

5.13 Consider the second flooding algorithm of Section 5.3.2. Suppose that there is a known upper bound on the time an update message requires to reach every node connected with the originating node. Devise a scheme based on an age field to supplement the algorithm so that it works correctly even if the sequence numbers can wrap around due to a memory or communication error. *Note*: Packets carrying an age field should be used only in exceptional circumstances after the originating node detects an error.

5.14 Modify the SPTA so that it can be used to broadcast one-directional link information other than status throughout the network (cf. the remarks at the end of Section 5.3.3). The link information is collected by the start node of the link. Justify the modified algorithm. *Hint*: Add to the existing algorithm additional main and port tables to hold the directional link information. Formulate update rules for these tables using the node labels provided by the main topology update algorithm of Section 5.3.3.

5.15 **(a)** Give an example where the following algorithm for broadcasting topological update information fails. Initially, all nodes know the correct status of all links.

Update rules:

1. When an adjacent link changes status, the node sends the new status in a message on all operating adjacent links.
2. When a node receives a message about a nonadjacent link which differs from its view of the status of the link, it changes the status of the link in its topology table. It also sends the new status in a message on all operating adjacent links *except* the one on which it received the message.
3. When a node receives a message about an adjacent link which differs from its view of the status of the link, it sends the correct status on the link on which it received the message.

Hint: Failure occurs with simple network topologies, such as the one shown in Fig. 5.43.

(b) Give an example of failure when the word "except" in rule 2 is changed to "including."

5.16 Describe what, if anything, can go wrong in the topology broadcast algorithms of Section 5.3 if the assumption that messages are received across links in the order transmitted is relaxed. Consider the ARPANET flooding algorithm, the two algorithms of Section 5.3.2, and the SPTA of Section 5.3.3.

5.17 *Distributed Computation of the Number of Nodes in a Network.* Consider a strongly connected communication network with N nodes and A (bidirectional) links. Each node knows its identity and the set of its immediate neighbors but not the network topology. Node 1 wishes to determine the number of nodes in the network. As a first step, it initiates an algorithm for finding a *directed, rooted spanning tree with node 1 as the root*. By this we mean a tree each link (i, k) of which is directed and oriented toward node 1 along the unique path on the tree leading from i to 1 (see the example tree shown in Fig. 5.76).

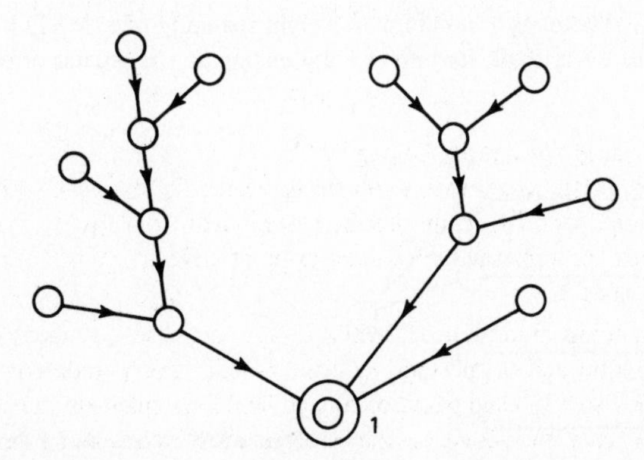

Figure 5.76 Example tree for Problem 5.17.

(a) Devise a distributed algorithm involving exchange of messages between nodes that constructs such a tree. The algorithm is initiated by node 1 and should involve no more than $O(A)$ message transmissions. Communication along any link is assumed error-free. At the end of the algorithm, the end nodes of each link should know whether the link is part of the tree, and, if so, they should know its direction.

(b) Supplement the algorithm derived in part (a) by another algorithm involving no more than $O(N)$ message transmissions by means of which node 1 gets to know N.

(c) Assuming that each message transmission takes an equal amount of time T, derive an upper bound for the time needed to complete the algorithms in parts (a) and (b).

5.18 Consider the minimum weight spanning tree problem. Show that the following procedure implements the Prim–Dijkstra algorithm and that it requires $O(N^2)$ arithmetic operations. Let node 1 be the starting node, and initially set $P = \{1\}$, $T = \emptyset$, $D_1 = 0$, and $D_j = w_{1j}$, $a_j = 1$ for $j \neq 1$.

Step 1. Find $i \notin P$ such that

$$D_i = \min_{j \notin P} D_j$$

and set $T := T \cup \{(j, a_i)\}$, $P := P \cup \{i\}$. If P contains all nodes, stop.

Step 2. For all $j \notin P$, if $w_{ij} < D_j$, set $D_j := w_{ij}$, $a_j := i$. Go to step 1.

Note: Observe the similarity with Dijkstra's algorithm for shortest paths.

5.19 Use Kleitman's algorithm to prove or disprove that the network shown in Fig. 5.77 is 3-connected. What is the maximum k for which the network is k-connected?

5.20 Suppose you had an algorithm that would find the maximum k for which two nodes in a graph are k-connected (*e.g.*, the max-flow algorithm test illustrated in Fig. 5.57). How would you modify Kleitman's algorithm to find the maximum k for which the graph is k-connected? Apply your modified algorithm to the graph of Fig. 5.77.

5.21 Use the Essau–Williams heuristic algorithm to find a constrained MST for the network shown in Fig. 5.78. Node 0 is the central node. The link weights are shown next to the links. The input flows from each node to the concentrator are shown next to the arrows. All link capacities are 10.

5.22 *MST Algorithm [MaP88].* We are given an undirected graph $G = (N, A)$ with a set of nodes $N = \{1, 2, \ldots, n\}$, and a weight a_{ij} for each arc (i, j) $(a_{ij} = a_{ji})$. We refer to the maximum over all the weights of the arcs contained in a given walk as the *critical weight* of the walk.

(a) Show that an arc (i, j) belongs to a minimum weight spanning tree (MST) if and only if the critical weight of every walk starting at i and ending at j is greater or equal to a_{ij}.

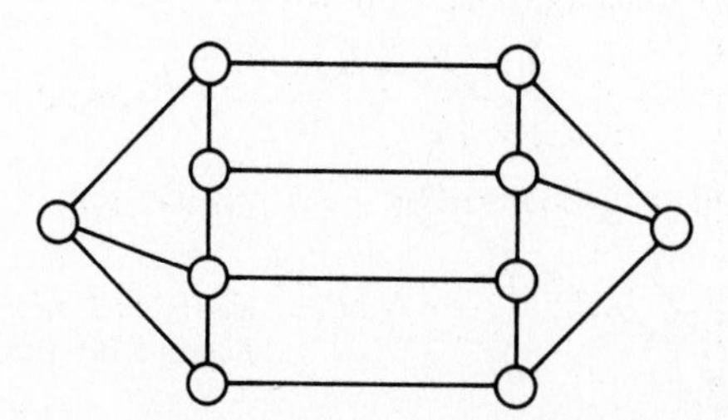

Figure 5.77 Network for Problem 5.19.

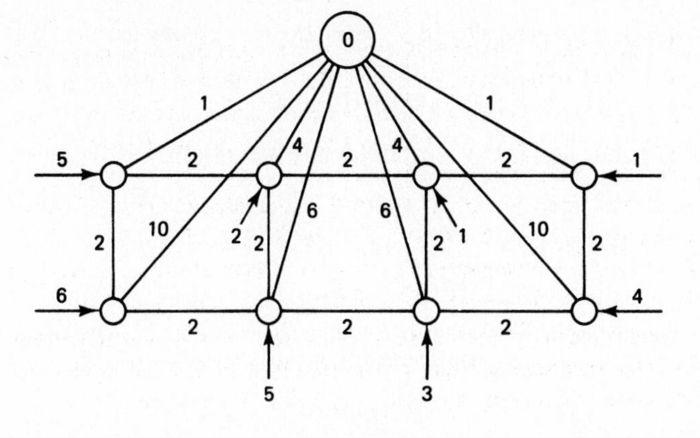

Figure 5.78 Network for Problem 5.21.

(b) Consider the following iterative algorithm:

$$x_{ij}^0 = \begin{cases} a_{ij}, & \text{if } (i,j) \in A, \\ \infty, & \text{otherwise,} \end{cases}$$

$$x_{ij}^{k+1} = \begin{cases} \min\left\{x_{ij}^k,\ \max\left\{x_{i(k+1)}^k, x_{(k+1)j}^k\right\}\right\}, & \text{if } j \neq i, \\ \infty, & \text{otherwise.} \end{cases}$$

Show that x_{ij}^k is the minimum over all critical weights of walks that start at i, end at j, and use only nodes 1 to k as intermediate nodes.

5.23 *Suboptimal Selective Broadcasting.* We are given a connected undirected graph $G = (N, A)$ with a nonnegative weight a_{ij} for each arc $(i, j) \in A$. Consider a tree in G which spans a given subset of nodes $\tilde{N}$ and has minimum weight over all such trees. Let W^* be the weight of this tree. Consider the graph $I(G)$, which has node set $\tilde{N}$ and is complete (has an arc connecting every pair of its nodes). Let the weight for each arc (i, j) of $I(G)$ be equal to the shortest distance in the graph G from the node $i \in \tilde{N}$ to the node $j \in \tilde{N}$. [Here, G is viewed as a directed graph with bidirectional arcs and length a_{ij} for each arc $(i, j) \in A$ in each direction.] Let T be a minimum weight spanning tree of $I(G)$.

(a) Show that the weight of T is no greater than $2W^*$. *Hint*: A *tour* in a graph is a cycle with no repeated nodes that passes through all the nodes. Consider a minimum weight tour in $I(G)$. Show that the weight of this tour is no less than the weight of T and no more than $2W^*$.

(b) Provide an example where the weight of T lies strictly between W^* and $2W^*$.

(c) Develop an algorithm for selective broadcasting from one node of $\tilde{N}$ to all other nodes of $\tilde{N}$ using T. Show that this algorithm comes within a factor of 2 of being optimal based on the given weights of the arcs of G.

5.24 Show that the necessary optimality condition of Section 5.5,

$$x_p^* > 0 \quad \Rightarrow \quad \frac{\partial D(x^*)}{\partial x_{p'}} \geq \frac{\partial D(x^*)}{\partial x_p}, \qquad \text{for all } p' \in P_w$$

implies that for all feasible path flow vectors $x = \{x_p\}$, we have

$$\sum_{p \in P_w} \left(x_p - x_p^*\right) \frac{\partial D(x^*)}{\partial x_p} \geq 0, \qquad \text{for all OD pairs } w$$

[The latter condition is sufficient for optimality of x^* when $D(x)$ is a convex function, that is,

$$D\big(\alpha x + (1-\alpha)x'\big) \le \alpha D(x) + (1-\alpha)D(x')$$

for all feasible x and x' and all α with $0 \le \alpha \le 1$. To see this, note that if $D(x)$ is convex, we have

$$D(x) \ge D(x^*) + \sum_{w\in W}\sum_{p\in P_w}(x_p - x_p^*)\frac{\partial D(x^*)}{\partial x_p}$$

for all x ([Lue84], p. 178), so when the shortest path condition above holds, x^* is guaranteed to be optimal.] *Hint*: Let $D_w^* = \min_{p\in P_w} \partial D(x^*)/\partial x_p$. We have for every feasible x,

$$0 = \sum_{p\in P_w}(x_p - x_p^*)D_w^* \le \sum_{\{p\in P_w | x_p > x_p^*\}}(x_p - x_p^*)\frac{\partial D(x^*)}{\partial x_p} + \sum_{\{p\in P_w | x_p < x_p^*\}}(x_p - x_p^*)D_w^*$$

5.25 Consider the network shown in Fig. 5.79 involving one OD pair with a given input rate r and three links with capacities C_1, C_2, and C_3, respectively. Assume that $C_1 = C_3 = C$, and $C_2 > C$, and solve the optimal routing problem with cost function based on the $M/M/1$ delay approximation and r between 0 and $2C + C_2$.

5.26 Consider the problem of finding the optimal routing of one unit of input flow in a single-OD-pair, three-link network for the cost function

$$D(x) = \frac{1}{2}\left[(x_1^2) + 2(x_2^2) + (x_3^2)\right] + 0.7x_3$$

Mathematically, the problem is

$$\text{minimize } D(x)$$
$$\text{subject to } x_1 + x_2 + x_3 = 1$$
$$x_1, x_2, x_3 \ge 0$$

(a) Show that $x_1^* = 2/3$, $x_2^* = 1/3$, and $x_3^* = 0$ is the optimal solution.

(b) Carry out several iterations of the Frank–Wolfe method and the projection method starting from the point $(1/3, 1/3, 1/3)$. Do enough iterations to demonstrate a clear trend in rate of convergence. (Use a computer for this if you wish.) Plot the successive iterates on the simplex of feasible flows.

5.27 *Extensions of the Gradient Projection Method*

(a) Show how the optimality condition of Section 5.5 and the gradient projection method can be extended to handle the situation where the cost function is a general twice differentiable function of the path flow vector x.

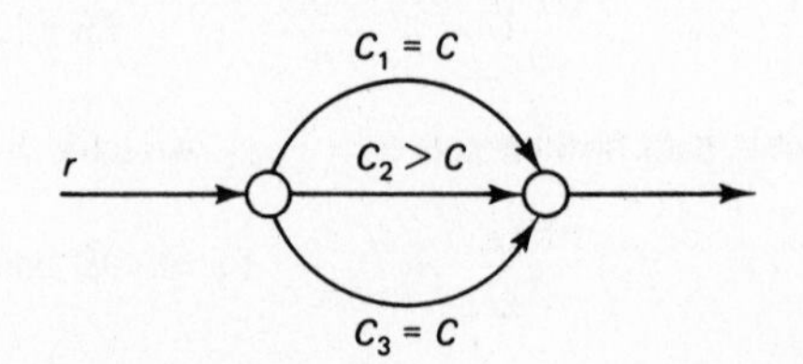

Figure 5.79 Network for Problem 5.25.

(b) Discuss the special case where the cost of link (i,j) is $D_{ij}(\tilde{F}_{ij}, F_{ij})$ where

$$\tilde{F}_{ij} = \sum_{k=1}^{M} p_k F_{ij}^k, \qquad F_{ij} = \sum_{k=1}^{M} F_{ij}^k$$

p_k is a positive scalar weighting factor for priority class k, and F_{ij}^k is the total flow of priority class k crossing link (i,j) (cf. the Codex algorithm cost function).

5.28 *Optimal Broadcast Routing along Spanning Trees.* Consider the optimal routing problem of Section 5.4. Suppose that in addition to the regular OD pair input rates r_w that are to be routed along directed paths, there is additional traffic R to be broadcast from a single node, say 1, to all other nodes along one or more directed spanning trees rooted at node 1. The spanning trees to be used for routing are to be selected from a given collection T, and the portion of R to be broadcast along each is subject to optimization.

(a) Characterize the path flow rates and the portions of R broadcast along spanning trees in T that minimize a total link cost, such as the one given by Eq. (5.29) in Section 5.4.

(b) Extend the gradient projection method to solve this problem.

(c) Extend your analysis to the case where there are several root nodes and there is traffic to be broadcast from each of these nodes to all other nodes.

Hint: Consider the first derivative total weights of spanning trees in addition to the first derivative lengths of paths.

5.29 *Optimal Routing with VCs Using the Same Links in Both Directions.* Consider the optimal routing problem of Section 5.4 with the additional restriction that the traffic originating at node i and destined for node i' must use the same path (in the reverse direction) as the traffic originating at i' and destined for i. In particular, assume that if w and w' are the OD pairs (i,i') and (i',i), respectively, there is a proportionality constant $c_{ww'}$ such that if $p \in P_w$ is a path and $p' \in P_{w'}$ is the reverse path, and x_p and $x_{p'}$ are the corresponding path flows, then $x_p = c_{ww'} x_{p'}$. Derive the conditions characterizing an optimal routing, and appropriately extend the gradient projection method.

5.30 *Use of the $M/M/1$-Based Cost Function with the Gradient Projection Method.* When the cost function

$$\sum_{(i,j)} D_{ij}(F_{ij}) = \sum_{(i,j)} \frac{F_{ij}}{C_{ij} - F_{ij}}$$

is minimized using the gradient projection method, there is potential difficulty due to the fact that some link flow F_{ij} may exceed the capacity C_{ij}. One way to get around this is to replace D_{ij} with a function $\tilde{D}_{ij}$ that is identical with D_{ij} for F_{ij} in the interval $[0, \rho C_{ij}]$, where ρ is some number less than one (say, $\rho = 0.99$), and is quadratic for $F_{ij} > \rho C_{ij}$. The quadratic function is chosen so that D_{ij} and $\tilde{D}_{ij}$ have equal values and first two derivatives at the point $F_{ij} = \rho C_{ij}$. Derive the formula for $\tilde{D}_{ij}$. Show that if the link flows F_{ij}^* that minimize

$$\sum_{(i,j)} D_{ij}(F_{ij})$$

result in link utilizations not exceeding ρ [*i.e.*, $F_{ij}^* \le \rho C_{ij}$ for all (i,j)], then F_{ij}^* also minimize

$$\sum_{(i,j)} \tilde{D}_{ij}(F_{ij})$$

5.31 *Convergence of the Frank–Wolfe Algorithm.* The purpose of this problem and the next one is to guide the advanced reader through convergence proofs of the Frank–Wolfe and the gradient projection methods. Consider the problem

$$\text{minimize } f(x)$$

$$\text{subject to } x \in X$$

where f is a differentiable function of the n-dimensional vector x and X is a closed, bounded, convex set. Assume that the gradient of f satisfies

$$|\nabla f(x) - \nabla f(y)| \leq L|x - y|, \quad \text{for all } x, y \in X$$

where L is a positive constant and $|z|$ denotes the Euclidean norm of a vector z, that is,

$$|z| = \sqrt{\sum_{i=1}^{n} (z_i)^2}$$

(This assumption can be shown to hold if $\nabla^2 f$ exists and is continuous over X.)

(a) Show that for all vectors x, Δx, and scalars α, such that $x \in X$, $x + \Delta x \in X$, $\alpha \in [0, 1]$,

$$f(x + \alpha\Delta x) \leq f(x) + \alpha\nabla f(x)^T \Delta x + \frac{\alpha^2 L}{2}|\Delta x|^2$$

where the superscript T denotes transpose. *Hint*: Use the Taylor formula

$$f(x + y) = f(x) + \nabla f(x)^T y + \int_0^1 \left[\nabla f(x + ty) - \nabla f(x)\right]^T y \, dt$$

(b) Use part (a) to show that if Δx is a descent direction at x [*i.e.*, $\nabla f(x)^T \Delta x < 0$], then

$$\min_{\alpha \in [0,1]} f(x + \alpha\Delta x) \leq f(x) + \delta$$

where

$$\delta = \begin{cases} \frac{1}{2}\nabla f(x)^T \Delta x, & \text{if} \quad \nabla f(x)^T \Delta x + L|\Delta x|^2 < 0 \\ -\frac{|\nabla f(x)^T \Delta x|^2}{2LR^2}, & \text{otherwise} \end{cases}$$

where R is the diameter of X, that is,

$$R = \max_{x,\, y \in X} |x - y|$$

Hint: Minimize over $\alpha \in [0, 1]$ in both sides of the inequality of part (a).

(c) Consider the Frank–Wolfe method

$$x^{k+1} = x^k + \alpha^k \Delta x^k$$

where $x^0 \in X$ is the starting vector, Δx^k solves the problem

$$\text{minimize } \nabla f(x^k)^T \Delta x$$

$$\text{subject to } x^k + \Delta x \in X$$

and α^k is a minimizing stepsize

$$f(x^k + \alpha^k \Delta x^k) = \min_{\alpha \in [0,1]} f(x^k + \alpha \Delta x^k)$$

Show that every limit x^* of the sequence $\{x^k\}$ satisfies the optimality condition

$$\nabla f(x^*)^T (x - x^*) \geq 0, \qquad \text{for all } x \in X$$

Hint: Argue that if $\{x^k\}$ has a limit point and δ^k corresponds to x^k as in part (b), then $\delta^k \to 0$, and therefore also $\nabla f(x^k)^T \Delta x^k \to 0$. Taking the limit in the relation $\nabla f(x^k)^T \Delta x \leq \nabla f(x^k)^T (x - x^k)$ for all $x \in X$, argue that a limit point x^* must satisfy $0 \leq \nabla f(x^*)^T (x - x^*)$ for all $x \in X$.

5.32 *Convergence of the Gradient Projection Method.* Consider the minimization problem and the assumptions of Problem 5.31. Consider also the iteration

$$x^{k+1} = x^k + \alpha^k (\bar{x}^k - x^k)$$

where $\bar{x}^k$ is the projection of $x^k - s\nabla f(x^k)$ on X and s is some fixed positive scalar, that is, $\bar{x}^k$ solves

$$\text{minimize } \left| x - x^k + s\nabla f(x^k) \right|^2$$

$$\text{subject to } x \in X$$

(a) Assume that α^k is a minimizing stepsize

$$f\left[x^k + \alpha^k (\bar{x}^k - x^k)\right] = \min_{\alpha \in [0,1]} f\left[x^k + \alpha (\bar{x}^k - x^k)\right]$$

and show that every limit point x^* of $\{x^k\}$ satisfies the optimality condition

$$\nabla f(x^*)^T (x - x^*) \geq 0, \qquad \text{for all } x \in X$$

Hint: Follow the hints of Problem 5.31. Use the following necessary condition satisfied by the projection

$$\left[x^k - s\nabla f(x^k) - \bar{x}^k\right]^T (x - \bar{x}^k) \leq 0, \qquad \text{for all } x \in X$$

to show that $s\nabla f(x^k)^T (\bar{x}^k - x^k) \leq -|\bar{x}^k - x^k|^2$.

(b) Assume that $\alpha^k = 1$ for all k. Show that if $s < 2/L$, the conclusion of part (a) holds.

5.33 *Gradient Projection Method with a Diagonal Scaling Matrix.* Consider the problem

$$\text{minimize } f(x)$$

$$\text{subject to } x \geq 0$$

of Section 5.7. Show that the iteration

$$x_i^{k+1} = \max\left\{0, x_i^k - \alpha^k b_i \frac{\partial f(x^k)}{\partial x_i}\right\}, \qquad i = 1, \ldots, n$$

with b_i some positive scalars, is equivalent to the gradient projection iteration

$$y_i^{k+1} = \max\left\{0, y_i^k - \alpha^k \frac{\partial h(y^k)}{\partial y_i}\right\}, \qquad i = 1, \ldots, n$$

where the variables x_i and y_i are related by $x_i = \sqrt{b_i} y_i$, and $h(y) = f(Ty)$, and T is the diagonal matrix with the ith diagonal element equal to $\sqrt{b_i}$.

5.34 This problem illustrates the relation between optimal routing and adaptive shortest path routing for virtual citcuits; see also Section 5.2.5 and [GaB83]. Consider a virtual circuit (VC) routing problem involving a single origin–destination pair and two links as shown in the figure. Assume that a VC arrives at each of the times $0, \tau, 2\tau, \ldots,$ and departs exactly H time units later. Each VC is assigned to one of the two links and remains assigned on that link for its entire duration. Let $N_i(t)$, $i = 1, 2$, be the number of VCs on link i at time t. We assume that initially the system is empty, and that an arrival and/or departure at time t is not counted in $N_i(t)$, so, for example, $N_i(0) = 0$. The routing algorithm is of the adaptive shortest path type and works as follows: For $kT \le t < (k+1)T$, a VC that arrives at time t is routed on link 1 if $\gamma_1 N_1(kT) \le \gamma_2 N_2(kT)$ and is routed on link 2 otherwise. Here T, γ_1, and γ_2 are given positive scalars.

Assume that $T \le H$ and that τ divides evenly H. Show that:

(a) $N_1(t) + N_2(t) = H/\tau$ for all $t > H$.

(b) For all $t > H$ we have

$$\frac{|N_1(t) - N_1^*|}{N_1^*} \le \frac{\gamma_1 + \gamma_2}{\gamma_2} \frac{T}{H}, \qquad \frac{|N_2(t) - N_2^*|}{N_2^*} \le \frac{\gamma_1 + \gamma_2}{\gamma_1} \frac{T}{H}$$

where (N_1^*, N_2^*) solve the optimal routing problem

$$\text{minimize} \quad \gamma_1 N_1^2 + \gamma_2 N_2^2$$

$$\text{subject to} \quad N_1 + N_2 = H/\tau, \quad N_1 \ge 0, \ N_2 \ge 0$$

Hint: Show that for $t > H$ we have

$$\gamma_1 N_1(t) \le \gamma_2 N_2(t) \qquad \Longleftrightarrow \qquad N_1(t) \le N_1^*$$

5.35 *Virtual Circuit Routing by Flooding.* In some networks where it may be hard to keep track of the topology as it changes (*e.g.*, packet radio networks), it is possible to use flooding to set up virtual circuits. The main idea is that a node m, which wants to set up a virtual circuit to node n, should flood an "exploratory" packet through the network. A node that rebroadcasts this packet stamps its ID number on it so that when the destination node n receives an exploratory packet, it knows the route along which it came. The destination node can then choose one of the potentially many routes carried by the copies of the exploratory packet received and proceed to set up the virtual circuit. Describe one or more flooding protocols that will make this process workable. Address the issues of indefinite or excessive message circulation in the network, appropriate numbering of exploratory packets, unpredictable link failures and delays, potential confusion between the two end nodes of virtual circuits, and so on.

5.36 Consider the optimal routing problem of Section 5.5. Assume that each link cost is chosen to be the same function $D(F)$ of the link flow F, where the first link cost derivative at zero flow $D'(0)$ is positive.

(a) For sufficiently small values of origin–destination pair input rates r_w show that optimal routings use only minimum-hop paths from origins to destinations.

(b) Construct an example showing that optimal routings do not necessarily have the minimum-hop property for larger values of r_w.

5.37 Consider the Frank–Wolfe method with the stepsize of Eq. (5.63) in Section 5.6. Describe an implementation as a distributed synchronous algorithm involving communication from origins to links and the reverse. At each iteration of this algorithm, each origin should calculate the stepsize of Eq. (5.63) and update accordingly the path flows that originate at itself.

5.38 *Optimal Dynamic Routing Based on Window Ratio Strategies.* Consider the optimal routing problem of Section 5.4 in a datagram network. Suppose that we want to implement a set of path flows $\{x_p\}$ calculated on the basis of a nominal set of traffic inputs $\{r_w\}$. The usual solution, discussed in Section 5.5, is to route the traffic of each OD pair w in a way that matches closely the actual fractions of packets routed on the paths $p \in P_w$ with the desired fractions x_p/r_w. With this type of implementation each OD pair w takes into account changes in its own input traffic r_w, but makes no effort to adapt to changes in the input traffic of other OD pairs or to queue lengths inside the network. This problem considers the following more dynamic alternative.

Suppose that each OD pair w calculates the average number of packets N_p traveling on each path $p \in P_w$ by means of Little's theorem

$$N_p = x_p T_p, \qquad \text{for all } p \in P_w, w \in W$$

where T_p is the estimated average round-trip packet delay on path p (time between introduction of the packet into the network and return of an end-to-end acknowledgment for the packet). The origin of each OD pair w monitors the *actual* number of routed but unacknowledged packets $\tilde{N}_p$ on path p and routes packets in a way that roughly equalizes the ratios $\tilde{N}_p/N_p$ for all paths $p \in P_w$. Note that this scheme is sensitive to changes in delay on the paths of an OD pair due to statistical fluctuations of the input traffic and/or the routing policies of other OD pairs. The difficulty with this scheme is that the delays T_p are unknown when the numbers N_p are calculated, and therefore T_p must be estimated somehow. The simplest possibility is to use the measured value of average delay during a preceding time period. You are asked to complete the argument in the following analysis of this scheme for a simple special case.

Suppose that there is a single OD pair and only two paths with flows and path delays denoted x_1, x_2 and $T_1(x)$, $T_2(x)$, respectively, where $x = (x_1, x_2)$. The form of the path delay functions $T_1(x)$, $T_2(x)$ is unknown. We assume that $x_1 + x_2 = r$ for some constant r. Suppose that we measure x_1, x_2, $T_1(x)$, $T_2(x)$, then calculate new nominal path flows $\overline{x}_1$, $\overline{x}_2$ using some unspecified algorithm, and then implement them according to the ratio scheme described above.

(a) Argue that the actual path flow vector $\tilde{x} = (\tilde{x}_1, \tilde{x}_2)$ is determined from $\tilde{x}_1 + \tilde{x}_2 = r$ and the relation

$$\frac{\tilde{x}_1 T_1(\tilde{x})}{\overline{x}_1 T_1(x)} = \frac{\tilde{x}_2 T_2(\tilde{x})}{\overline{x}_2 T_2(x)}$$

(b) Assume that T_1 and T_2 are monotonically increasing in the sense that if $z = (z_1, z_2)$ and $z' = (z_1', z_2')$ are such that $z_1 > z_1'$, $z_2 < z_2'$ and $z_1 + z_2 = z_1' + z_2'$, then $T_1(z) > T_1(z')$ and $T_2(z) < T_2(z')$. Show that the path flows $\tilde{x}_1$, $\tilde{x}_2$ of part (a) lie in the interval between x_1 and $\overline{x}_1$ and x_2 and $\overline{x}_2$, respectively.

(c) Consider a convex cost function of the form $\sum_{(i,j)} D_{ij}(F_{ij})$. Under the assumption in part (b) show that if $\overline{x}$ gives a lower cost than x, the same is true for $\tilde{x}$.

5.39 *Routing in Networks with Frequently Changing Topology [GaB81], [BeT89].* In some situations (*e.g.*, mobile packet radio networks) topological changes are so frequent that topology broadcast algorithms are somewhat impractical. The algorithms of this problem are designed to cope with situations of this type. Consider a connected undirected graph with a special node referred to as the *destination*. Consider the collection C of directed acyclic graphs obtained by assigning a unique direction to each of the undirected links. A graph G in C is said to be *rooted at the destination* if for every node there is a directed path in G starting at the node and ending at the destination. Show that the following two distributed asynchronous algorithms will yield a graph in C that is rooted at the destination starting from any other graph in C.

Algorithm A. A node other than the destination with no outgoing links reverses the direction of all its adjacent links. (This is done repeatedly until every node other than the destination has an outgoing link.)

Algorithm B. Every node i other than the destination keeps a list of its neighboring nodes j that have reversed the direction of the corresponding links (i, j). At each iteration, each node i that has no outgoing link reverses the directions of the links (i, j), for all j that do not appear on its list, and empties the list. If no such j exists (*i.e.*, the list is full), node i reverses the directions of all incoming links and empties the list. Initially, all lists are empty.

Hint: For algorithm A, assign to each node a distinct number. With each set of node numbers, associate the graph in C where each link is directed from a higher to a lower number. Consider an algorithm where each node reverses link directions by changing its number. Use a similar idea for algorithm B.

5.40 *Verifying the Termination of the Distributed Bellman–Ford Algorithm [DiS80], [BeT89].* The objective of the following algorithm (based on the Bellman–Ford method) is to compute in a distributed way a shortest path from a single origin (node 1) to all other nodes *and* to notify node 1 that the computation has terminated. Assume that all links (i, j) are bidirectional, have nonnegative length d_{ij}, maintain the order of messages sent on them, and operate with no errors or failures. Each node i maintains an estimate D_i of the shortest distance from node 1 to itself. Initially, $D_i = \infty$ for all nodes $i \neq 1$, and $D_1 = 0$. Node 1 initiates the computation by sending the estimate $D_1 + d_{1j}$ to all neighbor nodes j. The algorithmic rules for each node $j \neq 1$ are as follows:

1. When node j receives an estimate $D_i + d_{ij}$ from some other node i, after some unspecified finite delay (but before processing a subsequently received estimate), it does the following:

(a) If $D_j \leq D_i + d_{ij}$, node j sends an ACK to node i.

(b) If $D_j > D_i + d_{ij}$, node j sets $D_j = D_i + d_{ij}$, marks node i as its best current predecessor on a shortest path, sends an ACK to its previous best predecessor (if any), and sends the estimate $D_j + d_{jk}$ to each neighbor k.

2. Node j sends an ACK to its best current predecessor once it receives an ACK for each of the latest estimates sent to its neighbors.

We assume that each ACK is uniquely associated with a previously sent estimate, and that node 1 responds with an ACK to any length estimate it receives.

(a) Show that eventually node 1 will receive an ACK from each of its neighbors, and at that time D_i will be the correct shortest distance for each node i.

(b) What are the advantages and disadvantages of this algorithm compared with the distributed, asynchronous Bellman–Ford algorithm of Section 5.2.4?

6

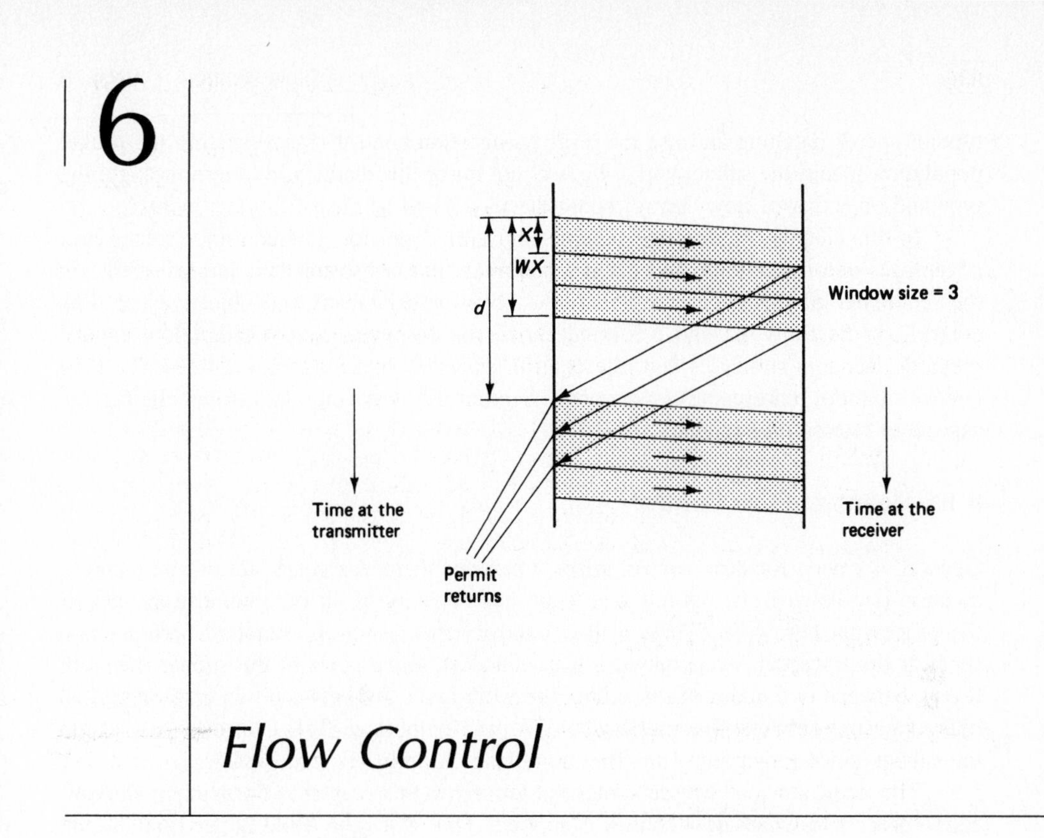

Flow Control

6.1 INTRODUCTION

In most networks, there are circumstances in which the externally offered load is larger than can be handled even with optimal routing. Then, if no measures are taken to restrict the entrance of traffic into the network, queue sizes at bottleneck links will grow and packet delays will increase, possibly violating maximum delay specifications. Furthermore, as queue sizes grow indefinitely, the buffer space at some nodes may be exhausted. When this happens, some of the packets arriving at these nodes will have to be discarded and later retransmitted, thereby wasting communication resources. As a result, a phenomenon similar to a highway traffic jam may occur whereby, as the offered load increases, the actual network throughput decreases while packet delay becomes excessive. It is thus necessary at times to prevent some of the offered traffic from entering the network to avoid this type of congestion. This is one of the main functions of flow control.

Flow control is also sometimes necessary between two users for speed matching, that is, for ensuring that a fast transmitter does not overwhelm a slow receiver with more packets than the latter can handle. Some authors reserve the term "flow control" for this type of speed matching and use the term "congestion control" for regulating the packet population within the subnetwork. We will not make this distinction in terminology; the type and objective of flow control being discussed will be clear from the context.

In this chapter we describe schemes currently used for flow control, explain their advantages and limitations, and discuss potential approaches for their improvement. In the remainder of this section we identify the principal means and objectives of flow control. In Sections 6.2 and 6.3 we describe the currently most popular flow control methods; window strategies and rate control schemes. In Section 6.4 we describe flow control in some representative networks. Section 6.5 is devoted to various algorithmic aspects of rate control schemes.

6.1.1 Means of Flow Control

Generally, a need for flow control arises whenever there is a constraint on the communication rate between two points due to limited capacity of the communication lines or the processing hardware. Thus, a flow control scheme may be required between two users at the transport layer, between a user and an entry point of the subnet (network layer), between two nodes of the subnet (network layer), or between two gateways of an interconnected network (internet layer). We will emphasize flow control issues within the subnet, since flow control in other contexts is in most respects similar.

The term "session" is used somewhat loosely in this chapter to mean any communication process to which flow control is applied. Thus a session could be a virtual circuit, a group of virtual circuits (such as all virtual circuits using the same path), or the entire packet flow originating at one node and destined for another node. Often, flow control is applied independently to individual sessions, but there is a strong interaction between its effects on different sessions because the sessions share the network's resources.

Note that different sessions may have radically different service requirements. For example, sessions transferring files may tolerate considerable delay but may require strict error control, while voice and video sessions may have strict minimum data rate and maximum end-to-end delay requirements, but may tolerate occasional packet losses.

There are many approaches to flow control, including the following:

1. *Call blocking*. Here a session is simply blocked from entering the network (its access request is denied). Such control is needed, for example, when the session requires a minimum guaranteed data rate that the network cannot provide due to limited uncommitted transmission capacity. A typical situation is that of the voice telephone network and, more generally, circuit switched networks, all of which use flow control of this type. However, a call blocking option is also necessary in integrated voice, video, and data networks, at least with respect to those sessions requiring guaranteed rate. In a more general view of call blocking one may admit a session only after a negotiation of some "service contract," for example, an agreement on some service parameters for the session's input traffic (maximum rate, minimum rate, maximum burst size, priority level, etc.)

2. *Packet discarding.* When a node with no available buffer space receives a packet, it has no alternative but to discard the packet. More generally, however, packets may be discarded while buffer space is still available if they belong to sessions that are using more than their fair share of some resource, are likely to cause congestion for higher-priority sessions, are likely to be discarded eventually along their path, and so on. (Note that if a packet has to be discarded anyway, it might as well be discarded as early as possible to avoid wasting additional network resources unnecessarily.) When some of a session's packets are discarded, the session may need to take some corrective action, depending on its service requirements. For sessions where all packets carry essential information (*e.g.*, file transfer sessions), discarded packets must be retransmitted by the source after a suitable timeout; such sessions require an acknowledgment mechanism to keep track of the packets that failed to reach their destination. On the other hand, for sessions such as voice or video, where delayed information is useless, there is nothing to be done about discarded packets. In such cases, packets may be assigned different levels of priority, and the network may undertake the obligation never to discard the highest-priority packets—these are the packets that are sufficient to support the minimum acceptable quality of service for the session. The data rate of the highest-priority packets (the minimum guaranteed rate) may then be negotiated between the network and the source when the session is established. This rate may also be adjusted in real time, depending on the congestion level in the network.

3. *Packet blocking.* When a packet is discarded at some node, the network resources that were used to get the packet to that node are wasted. It is thus preferable to restrict a session's packets from entering the network if after entering they are to be discarded. If the packets carry essential information, they must wait in a queue outside the network; otherwise, they are discarded at the source. In the latter case, however, the flow control scheme must honor any agreement on a minimum guaranteed rate that the session may have negotiated when it was first established.

4. *Packet scheduling.* In addition to discarding packets, a subnetwork node can exercise flow control by selectively expediting or delaying the transmission of the packets of various sessions. For example, a node may enforce a priority service discipline for transmitting packets of several different priorities on a given outgoing link. As another example, a node may use a (possibly weighted) round-robin scheduling strategy to ensure that various sessions can access transmission lines in a way that is consistent both with some fairness criterion and also with the minimum data rate required by the sessions. Finally, a node may receive information regarding congestion farther along the paths used by some sessions, in which case it may appropriately delay the transmission of the packets of those sessions.

In subsequent sections we discuss specific strategies for throttling sources and for restricting traffic access to the network.

6.1.2 Main Objectives of Flow Control

We look now at the main principles that guide the design of flow control algorithms. Our focus is on two objectives. First, *strike a good compromise between throttling sessions (subject to minimum data rate requirements) and keeping average delay and buffer overflow at a reasonable level.* Second, *maintain fairness between sessions in providing the requisite quality of service.*

Limiting delay and buffer overflow. We mentioned earlier that for important classes of sessions, such as voice and video, packets that are excessively delayed are useless. For such sessions, a limited delay is essential and should be one of the chief concerns of the flow control algorithm; for example, such sessions may be given high priority.

For other sessions, a small average delay per packet is desirable but it may not be crucial. For these sessions, network level flow control does not necessarily reduce delay; it simply shifts delay from the network layer to higher layers. That is, by restricting entrance to the subnet, flow control keeps packets waiting outside the subnet rather than in queues inside it. In this way, however, flow control avoids wasting subnet resources in packet retransmissions and helps prevent a disastrous traffic jam inside the subnet. Retransmissions can occur in two ways: first, the buildup of queues causes buffer overflow to occur and packets to be discarded; second, slow acknowledgments can cause the source node to retransmit some packets because it thinks mistakenly that they have been lost. Retransmissions waste network resources, effectively reduce network throughput, and cause congestion to spread. The following example (from [GeK80]) illustrates how buffer overflow and attendant retransmissions can cause congestion.

Example 6.1

Consider the five-node network shown in Fig. 6.1(a). There are two sessions, one from top to bottom with a Poisson input rate of 0.8, and the other from left to right with a Poisson input rate f. Assume that the central node has a large but finite buffer pool that is shared on a first-come first-serve basis by the two sessions. If the buffer pool is full, an incoming packet is rejected and then retransmitted by the sending node. For small f, the buffer rarely fills up and the total throughput of the system is $0.8 + f$. When f is close to unity (which is the capacity of the rightmost link), the buffer of the central node is almost always full, while the top and left nodes are busy most of the time retransmitting packets. Since the left node is transmitting 10 times faster than the top node, it has a 10-fold greater chance of capturing a buffer space at the central node, so the left-to-right throughput will be roughly 10 times larger than the top-to-bottom throughput. The left-to-right throughput will be roughly unity (the capacity of the rightmost link), so that the total throughput will be roughly 1.1. This is illustrated in more detail in Fig. 6.1(b), where it can be seen that the total throughput decreases toward 1.1 as the offered load f increases.

This example also illustrates how with buffer overflow, some sessions can capture almost all the buffers and nearly prevent other sessions from using the network. To avoid this, it is sometimes helpful to implement a *buffer management scheme*. In such a scheme, packets are divided in different classes based, for example, on origin, destination,

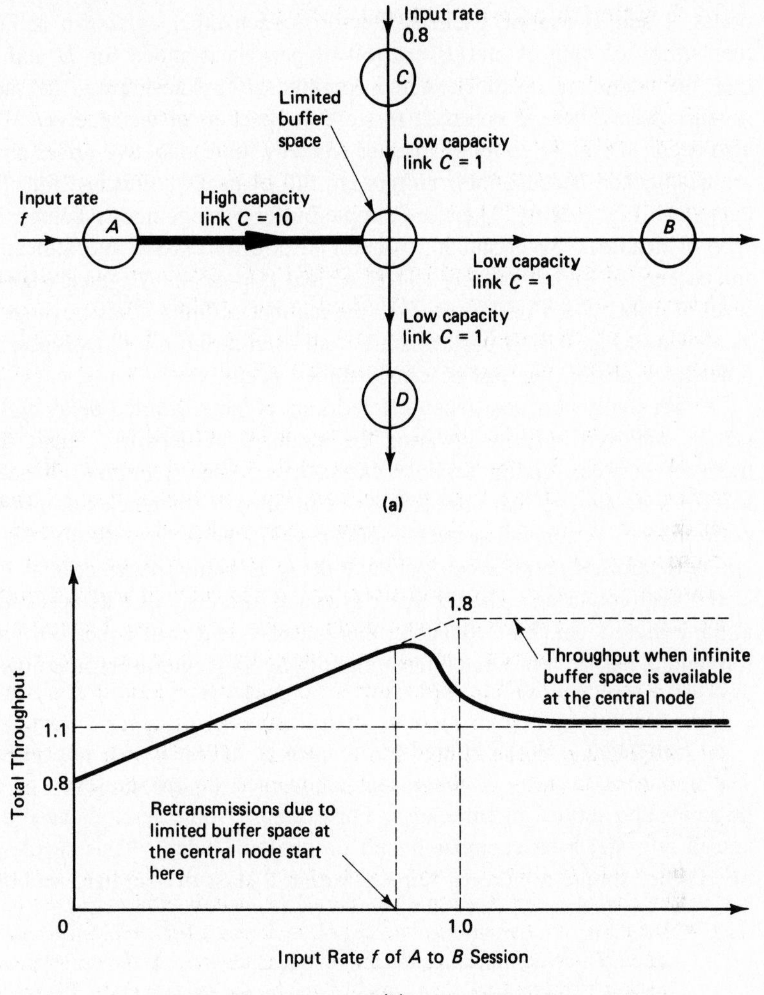

Figure 6.1 Example demonstrating throughput degradation due to retransmissions caused by buffer overflow. (a) For f approaching unity, the central node buffer is almost always full, thereby causing retransmissions. Because the A-to-B session uses a line 10 times faster than the C-to-D session, it has a 10-fold greater chance of capturing a free buffer and getting a packet accepted at the central node. As a result, the throughput of the A-to-B session approaches unity, while the throughput of the C-to-D session approaches 0.1. (b) Total throughput as a function of the input rate of the A-to-B session.

or number of hops traveled so far; at each node, separate buffer space is reserved for different classes, while some buffer space is shared by all classes.

Proper buffer management can also help avoid deadlocks due to buffer overflow. Such a deadlock can occur when two or more nodes are unable to move packets due to unavailable space at all potential receivers. The simplest example of this is two nodes A and B routing packets directly to each other as shown in Fig. 6.2(a). If all the buffers of both A and B are full of packets destined for B and A, respectively, then the nodes are deadlocked into continuously retransmitting the same packets with no success, as there is no space to store the packets at the receiver. This problem can also occur in a more complex manner whereby more than two nodes arranged in a cycle are deadlocked because their buffers are full of packets destined for other nodes in the cycle [see Fig. 6.2(b)]. There are simple buffer management schemes that preclude this type of deadlock by organizing packets in priority classes and allocating extra buffers for packets of higher priority ([RaH76] and [Gop85]). A typical choice is to assign a level of priority to a packet equal to the number of links it has traversed in the network, as shown in Fig. 6.3. If packets are not allowed to loop, it is then possible to show that a deadlock of the type just described cannot occur.

We finally note that when offered load is large, limited delay and buffer overflow can be achieved only by lowering the input to the network. Thus, there is a natural trade-off between giving sessions free access to the network and keeping delay at a level low enough so that retransmissions or other inefficiencies do not degrade network performance. A somewhat oversimplified guideline is that, ideally, flow control should not be exercised at all when network delay is below some critical level, and, under heavy load conditions, should reject as much offered traffic as necessary to keep delay at the critical level. Unfortunately, this is easier said than done, since neither delay nor throughput can be represented meaningfully by single numbers in a flow control context.

Fairness. When offered traffic must be cut back, it is important to do so fairly. The notion of fairness is complicated, however, by the presence of different session priorities and service requirements. For example, some sessions need a minimum guaranteed rate and a strict upper bound on network delay. Thus, while it is appropriate to consider simple notions of fairness within a class of "similar" sessions, the notion of

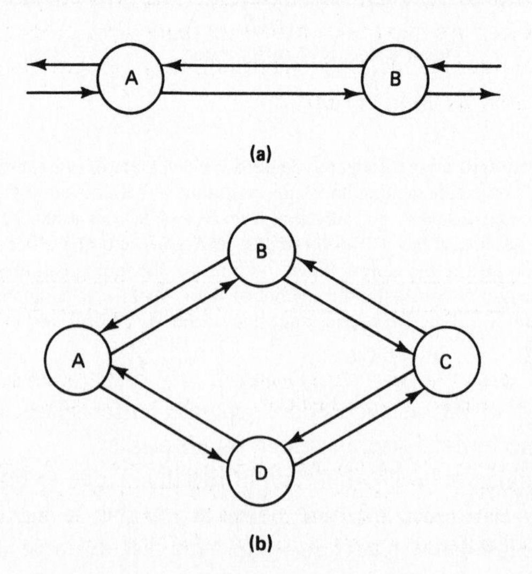

Figure 6.2 Deadlock due to buffer overflow. In (a), all buffers of A and B are full of packets destined for B and A, respectively. As a result, no packet can be accepted at either node. In (b), all buffers of A, B, C, and D are full of packets destined for C, D, A, and B, respectively.

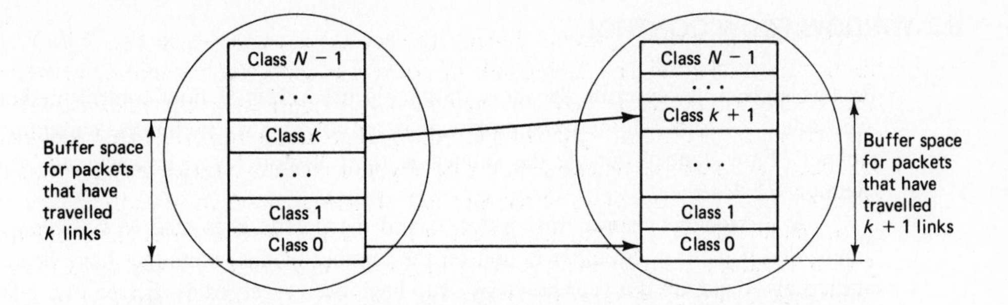

Figure 6.3 Organization of node memory in buffer classes to avoid deadlock due to buffer overflow. A packet that has traveled k links is accepted at a node only if there is an available buffer of class k or lower, where k ranges from 0 to $N-1$ (where N is the number of nodes). Assuming that packets that travel more than $N-1$ links are discarded as having traveled in a loop, no deadlock occurs. The proof consists of showing by induction (starting with $k = N-1$) that at each node the buffers of class k cannot fill up permanently.

fairness between classes is complex and involves the requirements of those classes. Even within a class, there are several notions of fairness that one may wish to adopt. The example of Fig. 6.4 illustrates some of the contradictions inherent in choosing a fairness criterion. There are $n+1$ sessions each offering 1 unit/sec of traffic along a sequence of n links with capacity of 1 unit/sec. One session's traffic goes over all n links, while the rest of the traffic goes over only one link. A maximum throughput of n units/sec can be achieved by accepting all the traffic of the single-link sessions while shutting off the n-link session. However, if our objective is to give equal rate to all sessions, the resulting throughput is only $(n+1)/2$ units/sec. Alternatively, if our objective is to give equal resources to all sessions, the single link sessions should get a rate of $n/(n+1)$ units/sec, while the n-link session should get a rate of $1/(n+1)$ units/sec.

Generally, a compromise is required between equal access to a network resource and throttling those sessions that are most responsible for using up that resource. Achieving the proper balance, however, is a design decision that may be hard to quantify; often such decisions must be reached by trial and error.

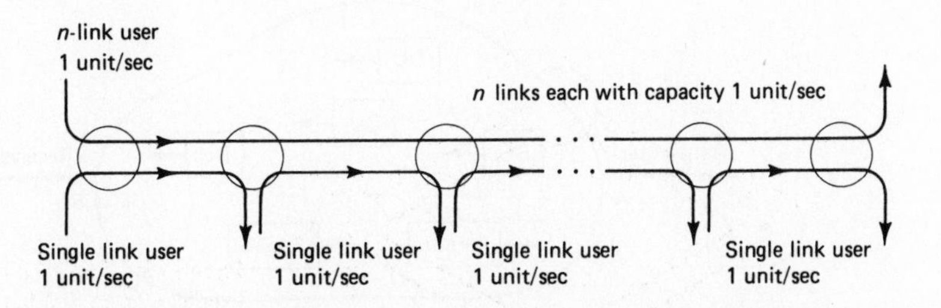

Figure 6.4 Example showing that maximizing total throughput may be incompatible with giving equal throughput to all sessions. A maximum throughput of n units/sec can be achieved if the n-link session is shut off completely. Giving equal rates of $1/2$ unit/sec to all sessions achieves a throughput of only $(n+1)/2$ units/sec.

6.2 WINDOW FLOW CONTROL

In this section we describe the most frequently used class of flow control methods. In Sections 6.2.1 to 6.2.3, the main emphasis is on flow control within the communication subnet. Flow control outside the subnet, at the transport layer, is discussed briefly in Section 6.2.4.

A session between a transmitter A and a receiver B is said to be *window flow controlled* if there is an upper bound on the number of data units that have been transmitted by A and are not yet known by A to have been received by B (see Fig. 6.5). The upper bound (a positive integer) is called the *window size* or, simply, the *window*. The transmitter and receiver can be, for example, two nodes of the communication subnet, a user's machine and the entry node of the communication subnet, or the users' machines at the opposite ends of a session. Finally, the data units in a window can be messages, packets, or bytes, for example.

The receiver B notifies the transmitter A that it has disposed of a data unit by sending a special message to A, which is called a *permit* (other names in the literature are *acknowledgment, allocate message*, etc.). Upon receiving a permit, A is free to send one more data unit to B. Thus, a permit may be viewed as a form of passport that a data unit must obtain before entering the logical communication channel between A and B. The number of permits in use should not exceed the window size.

Permits are either contained in special control packets, or are piggybacked on regular data packets. They can be implemented in a number of ways; see the practical examples of Section 6.4 and the following discussion. Note also that a window flow control scheme for a given session may be combined with an error control scheme for the session, where the permits also play the role of acknowledgments; see Section 2.8.2 and the descriptions of the ARPANET and the Codex network in Section 6.4.

The general idea in the window strategy is that the input rate of the transmitter is reduced when permits return slowly. Therefore, if there is congestion along the communication path of the session, the attendant large delays of the permits cause a natural slowdown of the transmitter's data rate. However, the window strategy has an additional dimension, whereby the receiver may intentionally delay permits to restrict

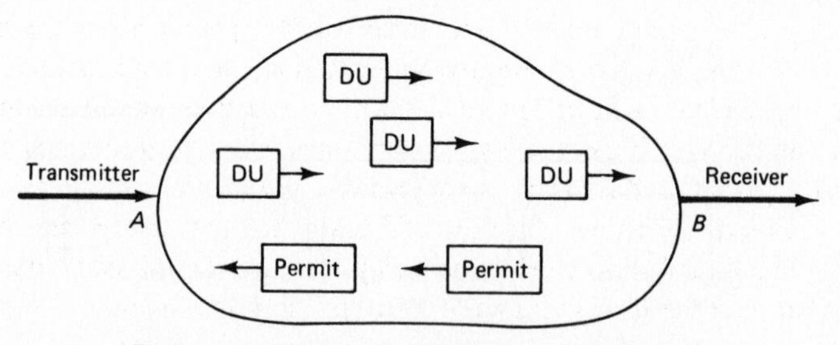

Figure 6.5 Window flow control between a transmitter and a receiver consists of an upper bound W_{AB} on the number of data units and permits in transit inside the network.

the transmission rate of the session. For example, the receiver may do so to avoid buffer overflow.

In the subsequent discussion, we consider two strategies, *end-to-end* and *node-by-node* windowing. The first strategy refers to flow control between the entry and exit subnet nodes of a session, while the second strategy refers to flow control between every pair of successive nodes along a virtual circuit's path.

6.2.1 End-to-End Windows

In the most common version of end-to-end window flow control, the window size is αW, where α and W are some positive numbers. Each time a new batch of α data units is received at the destination node, a permit is sent back to the source allocating a new batch of α data units. In a variation of this scheme, the destination node will send a new α-data unit permit upon reception of just the first of an α-data unit batch. (See the SNA pacing scheme description in Section 6.3.) To simplify the following exposition, we henceforth assume that $\alpha = 1$, but our conclusions are valid regardless of the value of α. Also, for concreteness, we talk in terms of packets, but the window maintained may consist of other data units such as bytes.

Usually, a numbering scheme for packets and permits is used so that permits can be associated with packets previously transmitted and loss of permits can be detected. One possibility is to use a sliding window protocol similar to those used for data link control, whereby a packet contains a sequence number and a request number. The latter number can serve as one or more permits for flow control purposes (see also the discussion in Section 2.8.2). For example, suppose that node A receives a packet from node B with request number k. Then A knows that B has disposed of all packets sent by A and numbered less than k, and therefore A is free to send those packets up to number $k+W-1$ that it has not sent yet, where W is the window size. In such a scheme, both the sequence number and the request number are represented modulo m, where $m \geq W+1$. One can show that if packet ordering is preserved between transmitter and receiver, this representation of numbers is adequate; the proof is similar to the corresponding proof for the goback n ARQ system. In some networks the end-to-end window scheme is combined with an end-to-end retransmission protocol, and a packet is retransmitted if following a suitable timeout, the corresponding permit has not returned to the source.

To simplify the subsequent presentation, the particular manner in which permits are implemented will be ignored. It will be assumed that the source node simply counts the number x of packets it has already transmitted but for which it has not yet received back a permit, and transmits new packets only as long as $x < W$.

Figure 6.6 shows the flow of packets for the case where the round-trip delay d, including round-trip propagation delay, packet transmission time, and permit delay is smaller than the time required to transmit the full window of W packets, that is,

$$d \leq WX$$

where X is the transmission time of a single packet. Then the source is capable of transmitting at the full speed of $1/X$ packets/sec, and flow control is not active. (To simplify the following exposition, assume that all packets have equal transmission time and equal round-trip delay.)

The case where flow control is active is shown in Fig. 6.7. Here

$$d > WX$$

and the round-trip delay d is so large that the full allocation of W packets can be transmitted before the first permit returns. Assuming that the source always has a packet waiting in queue, the rate of transmission is W/d packets/sec.

If the results of Figs. 6.6 and 6.7 are combined, it is seen that the maximum rate of transmission corresponding to a round-trip delay d is given by

$$r = \min\left\{\frac{1}{X}, \frac{W}{d}\right\} \tag{6.1}$$

Figure 6.8 illustrates the flow control mechanism; the source transmission rate is reduced in response to congestion and the attendant large delay. Furthermore, assuming that W is relatively small, the window scheme reacts fast to congestion—within at most W packets' transmission time. This fast reaction, coupled with low overhead, is the major advantage of window strategies over other (nonwindow) schemes.

Limitations of end-to-end windows. One drawback of end-to-end windows is that they cannot guarantee a minimum communication rate for a session. Thus windows are inadequate for sessions that require a minimum guaranteed rate, such as voice and video sessions. Other drawbacks of the end-to-end window strategy have to do with window sizes. There is a basic trade-off here: One would like to make window sizes small to limit the number of packets in the subnet, thus avoiding large delays and congestion, and one would also like to make window sizes large to allow full-speed transmission and maximal throughput under light-to-moderate traffic conditions. Determining the proper window size and adjusting that size in response to congestion is not easy. This delay–throughput trade-off is particularly acute for high-speed networks, where because of high propagation delay relative to packet transmission time, window sizes should be large to allow high data rates [cf. Eq. (6.1) and Fig. 6.8]. One should remember, however, that the existence of a high-speed network does not necessarily make

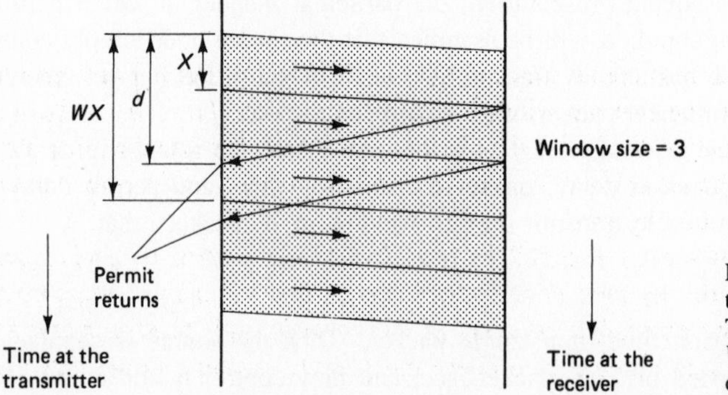

Figure 6.6 Example of full-speed transmission with a window size $W = 3$. The round-trip delay d is less than the time WX required to transmit the full window of W packets.

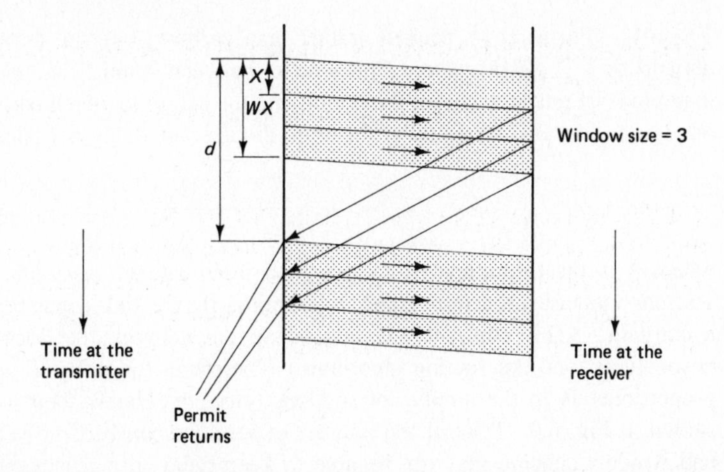

Figure 6.7 Example of delayed transmission with a window size $W = 3$. The round-trip delay d is more than the time WX required to transmit the full window of W packets. As a result, window flow control becomes active, restricting the input rate to W/d.

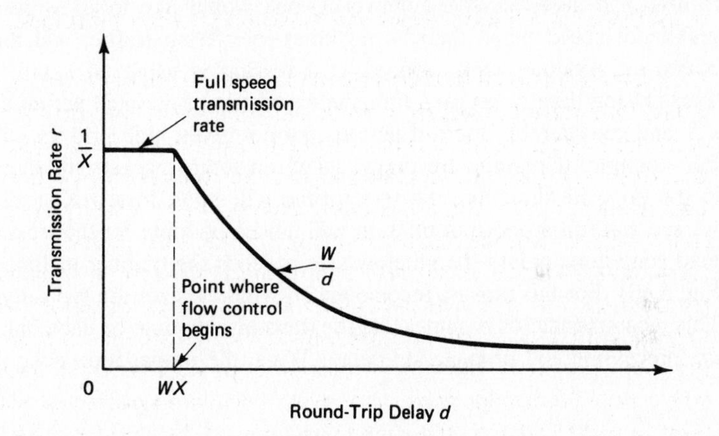

Figure 6.8 Transmission rate versus round-trip delay in a window flow control system. This oversimplified relationship assumes that all packets require equal transmission time at the source and have equal round-trip packet/permit delay.

it desirable for individual sessions to transmit at the full network speed. For high-speed networks, it is generally true that the window size of the session should be proportional to the round-trip propagation delay for the session. However, the window size should also be proportional to the session's maximum *desired* rate rather than the maximum rate allowed by the network; cf. Eq. (6.1).

End-to-end windows may also fail to provide adequte control of packet delay. To understand the relation between window size, delay, and throughput, suppose that there are n actively flow controlled sessions in the network with fixed window sizes $W_1, \ldots, W_n$. Then the total number of packets and permits traveling in the network is

$\sum_{i=1}^{n} W_i$. Focusing on packets (rather than permits), we see that their number in the network is $\sum_{i=1}^{n} \beta_i W_i$, where β_i is a factor between 0 and 1 that depends on the relative magnitude of return delay for the permit and the extent to which permits are piggybacked on regular packets. By Little's Theorem, the average delay per packet is given by

$$T = \frac{\sum_{i=1}^{n} \beta_i W_i}{\lambda}$$

where λ is the throughput (total accepted input rate of sessions). As the number of sessions increases, the throughput λ is limited by the link capacities and will approach a constant. (This constant will depend on the network, the location of sources and destinations, and the routing algorithm.) Therefore, the delay T will roughly increase proportionately to the number of sessions (more accurately their total window size) as shown in Fig. 6.9. Thus, if the number of sessions can become very large, the end-to-end window scheme may not be able to keep delay at a reasonable level and prevent congestion.

One may consider using small window sizes as a remedy to the problem of very large delays under high load conditions. Unfortunately, in many cases (particularly in low- and moderate-speed networks) one would like to allow sessions to transmit at maximum speed when there is no other interfering traffic, and this imposes a lower bound on window sizes. Indeed, if a session is using an n-link path with a packet transmission time X on each link, the round-trip packet and permit delay will be at least nX and considerably more if permits are not given high priority on the return channel. For example, if permits are piggybacked on return packets traveling on the same path in the opposite direction, the return time will be at least nX also. So from Fig. 6.8, we see that full-speed transmission will not be possible for that session even under light load conditions unless the window size exceeds the number of links n on the path (see Fig. 6.10). For this reason, recommended window sizes are typically between n and $3n$. This recommendation assumes that the transmission time on each link is much larger than the processing and propagation delay. When the propagation delay is much larger than

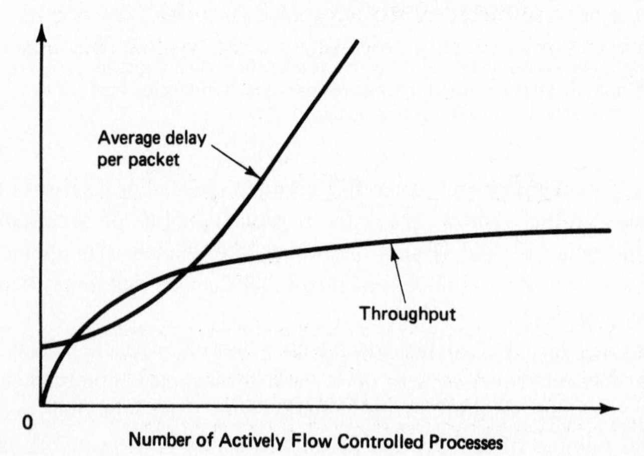

Figure 6.9 Average delay per packet and throughput as a function of the number of actively window flow controlled sessions in the network. When the network is heavily loaded, the average delay per packet increases approximately linearly with the number of active sessions, while the total throughput stays approximately constant. (This assumes that there are no retransmissions due to buffer overflow and/or large permit delays. In the presence of retransmissions, throughput may decrease as the number of active sessions increases.)

the transmission time, as in satellite links and some high-speed networks, the appropriate window size might be much larger. This is illustrated in the following example.

Example 6.2

Consider a transmission line with capacity of 1 gigabit per second (10^9 bits/sec) connecting two nodes which are 50 miles apart. The round-trip propagation delay can be roughly estimated as 1 millisecond. Assuming a packet size of 1000 bits, it is seen that a window size of at least 1000 packets is needed to sustain full-speed transmission; it is necessary to have 1000 packets and permits simultaneously propagating along the transmission line to "keep the pipeline full," with permits returning fast enough to allow unimpeded transmission of new packets. By extrapolation it is seen that Atlantic coast to Pacific coast end-to-end transmission over a distance of, say, 3000 miles requires a window size of at least 60,000 packets! Fortunately, for most sessions, such unimpeded transmission is neither required nor desirable.

Example 6.2 shows that windows should be used with care when high-speed transmission over a large distance is involved; they require excessive memory and they respond to congestion relatively slowly when the round-trip delay time is very long relative to the packet transmission time. For networks with relatively small propagation delays, end-to-end window flow control may be workable, particularly if there is no requirement for a minimum guaranteed rate. However, to achieve a good trade-off between delay and throughput, dynamic window adjustment is necessary. Under light load conditions, windows should be large and allow unimpeded transmission, while under heavy load conditions, windows should shrink somewhat, thereby not allowing delay to become excessive. This is not easy to do systematically, but some possibilities are examined in Section 6.2.5.

End-to-end windows can also perform poorly with respect to fairness. It was argued earlier that when propagation delay is relatively small, the proper window size of a session should be proportional to the number of links on its path. This means that at a heavily loaded link, long-path sessions can have many more packets awaiting transmission than short-path sessions, thereby obtaining a proportionately larger throughput. A typical situation is illustrated in Fig. 6.11. Here the windows of all sessions accumulate at the heavily loaded link. If packets are transmitted in the order of their arrival, the rate of transmission obtained by each session is roughly proportional to its window size, and this favors the long-path sessions.

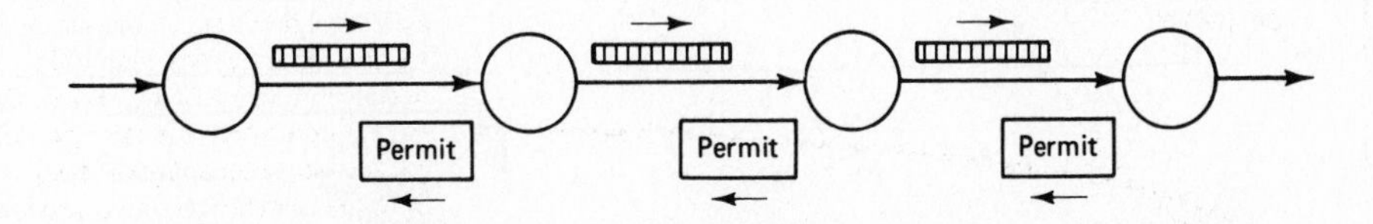

Figure 6.10 The window size must be at least equal to the number of links on the path to achieve full-speed transmission. (Assuming equal transmission time on each link, a packet should be transmitted at each link along the path simultaneously to achieve nonstop transmission.) If permit delay is comparable to the forward delay, the window size should be doubled. If the propagation delay is not negligible, an even larger window size is needed.

The fairness properties of end-to-end windows can be improved if flow-controlled sessions of the same priority class are served via a weighted round-robin scheme at each transmission queue. Such a scheme should take into account the priorities as well as the minimum guaranteed rate of different sessions. Using a round-robin scheme is conceptually straightforward when each session is a virtual circuit, but not in a datagram network, where it may not be possible to associate packets with particular flow-controlled sessions.

6.2.2 Node-by-Node Windows for Virtual Circuits

In this strategy, there is a separate window for every virtual circuit and pair of adjacent nodes along the path of the virtual circuit. Much of the discussion on end-to-end windows applies to this scheme as well. Since the path along which flow control is exercised is effectively one link long, the size of a window measured in packets is typically two or three for moderate-speed terrestrial links. For high-speed networks, the required window size might be much larger, thereby making the node-by-node strategy less attractive. For this reason, the following discussion assumes that a window size of about two is a reasonable choice.

Let us focus on a pair of successive nodes along a virtual circuit's path; we refer to them as the transmitter and the receiver. The main idea in the node-by-node scheme is that the receiver can avoid the accumulation of a large number of packets into its memory by slowing down the rate at which it returns permits to the transmitter. In the most common strategy, the receiver maintains a W-packet buffer for each virtual circuit and returns a permit to the transmitter as soon as it releases a packet from its W-packet buffer. A packet is considered to be released from the W-packet buffer once it is either delivered to a user outside the subnet or is entered in the data link control (DLC) unit leading to the subsequent node on the virtual circuit's path.

Consider now the interaction of the windows along three successive nodes ($i-1$, i, and $i+1$) on a virtual circuit's path. Suppose that the W-packet buffer of node i is full. Then node i will send a permit to node $i-1$ once it delivers an extra packet to the DLC of the $(i, i+1)$ link, which in turn will occur once a permit sent by node $(i+1)$ is received at node i. Thus, there is coupling of successive windows along the path of a virtual circuit. In particular, suppose that congestion develops at some link. Then the W-packet window

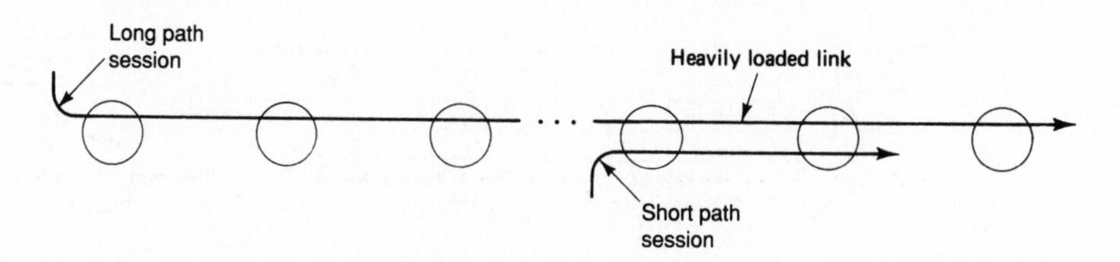

Figure 6.11 End-to-end windows discriminate in favor of long-path sessions. It is necessary to give a large window to a long-path session to achieve full-speed transmission. Therefore, a long-path session will typically have more packets waiting at a heavily loaded link than will a short-path session, and will receive proportionally larger service (assuming that packets are transmitted on a first-come first-serve basis).

at the start node of the congested link will fill up for each virtual circuit crossing the link. As a result, the W-packet windows of nodes lying upstream of the congested link will progressively fill up, including the windows of the origin nodes of the virtual circuits crossing the congested link. At that time, these virtual circuits will be actively flow controlled. The phenomenon whereby windows progressively fill up from the point of congestion toward the virtual circuit origins is known as *backpressure* and is illustrated in Fig. 6.12.

One attractive aspect of node-by-node windows can be seen from Fig. 6.12. In the worst case, where congestion develops on the last link (say, the nth) of a virtual circuit's path, the total number of packets inside the network for the virtual circuit will be approximately nW. If the virtual circuit were flow controlled via an end-to-end window, the total number of packets inside the network would be roughly comparable. (This assumes a window size of $W = 2$ in the node-by-node case, and of $W \simeq 2n$ in the end-to-end case based on the rule of thumb of using a window size that is twice the number of links of the path between transmitter and receiver.) The important point, however, is that these packets will be uniformly distributed along the virtual circuit's path in the node-by-node case, but will be concentrated at the congested link in the end-to-end case. Because of this the amount of memory required at each node to prevent buffer overflow may be much smaller for node-by-node windows than for end-to-end windows.

Distributing the packets of a virtual circuit uniformly along its path also alleviates the fairness problem, whereby large window sessions monopolize a congested link at the expense of small window sessions (cf. Fig. 6.11). This is particularly true when the window sizes of all virtual circuits are roughly equal as, for example, when the circuits involve only low-speed terrestrial links. A fairness problem, however, may still arise when satellite links (or other links with relatively large propagation delay) are involved. For such links, it is necessary to choose a large window size to achieve unimpeded transmission when traffic is light because of the large propagation delay. The difficulty arises at a node serving both virtual circuits with large window size that come over a satellite link and virtual circuits with small window size that come over a terrestrial link (see Fig. 6.13). If these circuits leave the node along the same transmission line, a fairness problem may develop when this line gets heavily loaded. A reasonable way to address this difficulty is to schedule transmissions of packets from different virtual circuits on a weighted round-robin basis (with the weights accounting for different priority classes).

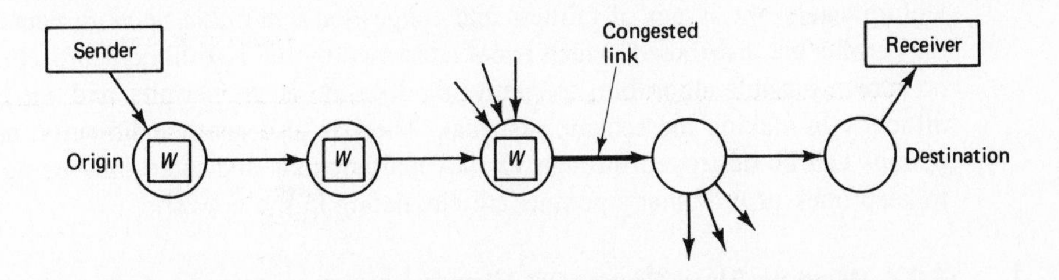

Figure 6.12 Backpressure effect in node-by-node flow control. Each node along a virtual circuit's path can store no more than W packets for that virtual circuit. The window storage space of each successive node lying upstream of the congested link fills up. Eventually, the window of the origin node fills, at which time transmission stops.

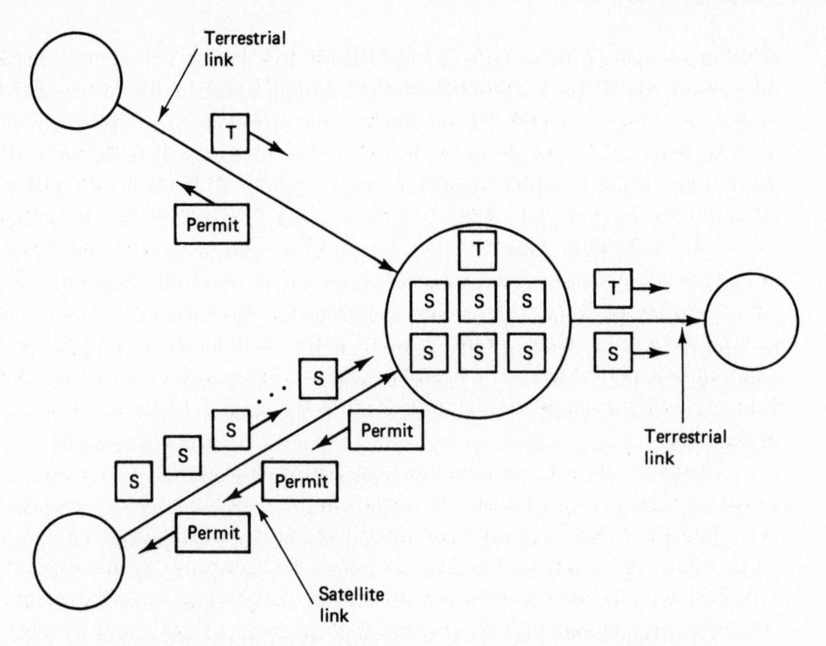

Figure 6.13 Potential fairness problem at a node serving virtual circuits that come over a satellite link (large window) and virtual circuits that come over a terrestrial link (small window). If virtual circuits are served on a first-come first-serve basis, the virtual circuits with large windows will receive better service on a subsequent transmission line. This problem can be alleviated by serving virtual circuits via a round-robin scheme.

6.2.3 The Isarithmic Method

The isarithmic method may be viewed as a version of window flow control whereby there is a single global window for the entire network. The idea here is to limit the total number of packets in the network by having a fixed number of permits circulating in the subnet. A packet enters the subnet after capturing one of these permits. It then releases the permit at its destination node. The total number of packets in the network is thus limited by the number of permits. This has the desirable effect of placing an upper bound on average packet delay that is independent of the number of sessions in the network. Unfortunately, the issues of fairness and congestion within the network depend on how the permits are distributed, which is not addressed by the isarithmic approach. There are no known sensible algorithms to control the location of the permits, and this is the main difficulty in making the scheme practical. There is also another difficulty, namely that permits can be destroyed through various malfunctions and there may be no easy way to keep track of how many permits are circulating in the network.

6.2.4 Window Flow Control at Higher Layers

Much of what has been described so far about window strategies is applicable to higher layer flow control of a session, possibly communicating across several interconnected subnetworks. Figure 6.14 illustrates a typical situation involving a single subnetwork. A user sends data out of machine A to an entry node NA of the subnet, which is forwarded to the exit node NB and is then delivered to machine B. There is (network layer) flow control between the entry and exit nodes NA and NB (either end-to-end or node-by-node involving a sequence of nodes). There is also (network layer) window flow control between machine A and entry node NA, which keeps machine A from swamping node NA with more data than the latter can handle. Similarly, there is window flow control between exit node NB and machine B, which keeps NB from overwhelming B with too much data. Putting the pieces together we see that there is a network layer flow control system extending from machine A to machine B, which operates much like the node-by-node window flow control system of Section 6.2.2. In essence, we have a three-link path where the subnet between nodes NA and NB is viewed conceptually as the middle link.

It would appear that the system just described would be sufficient for flow control purposes, and in many cases it is. For example, it is the only one provided in the TYMNET and the Codex network described later on a virtual circuit basis. There may be a need, however, for additional flow control at the transport layer, whereby the input traffic rate of a user at machine A is controlled directly by the receiver at machine B. One reason is that the network layer window flow control from machine A to node NA may apply collectively to multiple user-pair sessions. These sessions could, for a variety of reasons, be multiplexed into a single traffic stream but have different individual flow control needs. For example, in SNA there is a transport layer flow control algorithm, which is the same as the one used in the network layer except that it is applied separately for each session rather than to a group of sessions. (See the description given in Section 6.4).

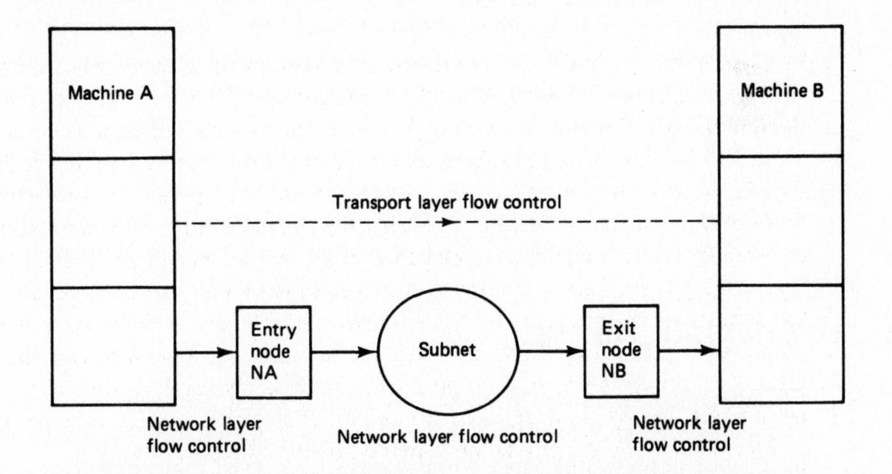

Figure 6.14 User-to-user flow control. A user sends data from machine A to a user in machine B via the subnet using the entry and exit nodes NA and NB. There is a conceptual node-by-node network layer flow control system along the connection A–NA–NB–B. There may also be direct user-to-user flow control at the transport layer or the internet sublayer, particularly if several users with different flow control needs are collectively flow controlled within the subnetwork.

In the case where several networks are interconnected with gateways, it may be useful to have windows for the gateway-to-gateway traffic of sessions that span several networks. If we take a higher-level view of the interconnected network where the gateways correspond to nodes and the networks correspond to links (Section 5.1.3), this amounts to using node-by-node windows for flow control in the internet sublayer. The gateway windows serve the purpose of distributing the total window of the internetwork sessions, thereby alleviating a potential congestion problem at a few gateways. However, the gateway-to-gateway windows may apply to multiple transport layer sessions, thereby necessitating transport layer flow control for each individual session.

6.2.5 Dynamic Window Size Adjustment

We mentioned earlier that it is necessary to adjust end-to-end windows dynamically, decreasing their size when congestion sets in. The most common way of doing this is through feedback from the point of congestion to the appropriate packet sources.

There are several ways by which this feedback can be obtained. One possibility is for nodes that sense congestion to send a special packet, sometimes called a *choke packet*, to the relevant sources. Sources that receive such a packet must reduce their windows. The sources can then attempt to increase their windows gradually following a suitable timeout. The method by which this is done is usually ad hoc in practice, and is arrived at by trial and error or simulation. The circumstances that will trigger the generation of choke packets may vary: for example, buffer space shortage or excessive queue length.

It is also possible to adjust window sizes by keeping track of permit delay or packet retransmissions. If permits are greatly delayed or if several retransmissions occur at a given source within a short period of time, this is likely to mean that packets are being excessively delayed or are getting lost due to buffer overflow. The source then reduces the relevant window sizes, and subsequently attempts to increase them gradually following a timeout.

Still another way to obtain feedback is to collect congestion information on regular packets as they traverse their route from origin to destination. This congestion information can be used by the destination to adjust the window size by withholding the return of some permits. A scheme of this type is used in SNA and is described in Section 6.4.

6.3 RATE CONTROL SCHEMES

We mentioned earlier that window flow control is not very well suited for high-speed sessions in high-speed wide area networks because the propagation delays are relatively large, thus necessitating large window sizes. An even more important reason is that windows do not regulate end-to-end packet delays well and do not guarantee a minimum data rate. Voice, video, and an increasing variety of data sessions require upper bounds on delay and lower bounds on rate. High-speed wide area networks increasingly carry such traffic, and many lower-speed networks also carry such traffic, making windows inappropriate.

An alternative form of flow control is based on giving each session a guaranteed data rate, which is commensurate to its needs. This rate should lie within certain limits that depend on the session type. For example, for a voice session, the rate should lie between the minimum needed for language intelligibility and a maximum beyond which the quality of voice cannot be further improved.

The main considerations in setting input session rates are:

1. *Delay–throughput trade-off.* Increasing throughput by setting the rates too high runs the risk of buffer overflow and excessive delay.
2. *Fairness.* If session rates must be reduced to accommodate some new sessions, the rate reduction must be done fairly, while obeying the minimum rate requirement of each session.

We will discuss various rate adjustment schemes focusing on these considerations in Section 6.5.

Given an algorithm that generates desired rates for various sessions, the question of implementing these rates arises. A strict implementation of a session rate of r packets/sec would be to admit 1 packet each $1/r$ seconds. This, however, amounts to a form of time-division multiplexing and tends to introduce large delays when the offered load of the sessions is bursty. A more appropriate implementation is to admit as many as W packets ($W > 1$) every W/r seconds. This allows a burst of as many as W packets into the network without delay, and is better suited for a dynamically changing load. There are several variations of this scheme. The following possibility is patterned after window flow control.

An allocation of W packets (a window) is given to each session, and a count x of the unused portion of this allocation is kept at the session origin. Packets from the session are admitted into the network as long as $x > 0$. Each time a packet is admitted, the count is decremented by 1, and W/r seconds later (r is the rate assigned to the session), the count is incremented by 1 as shown in Fig. 6.15. This scheme, called *time window flow control*, is very similar to window flow control with window size W except that the count is incremented W/r seconds after admitting a packet instead of after a round-trip delay when the corresponding permit returns.

A related method that regulates the burstiness of the transmitted traffic somewhat better is the so-called *leaky bucket scheme*. Here the count is incremented periodically, every $1/r$ seconds, up to a maximum of W packets. Another way to view this scheme is to imagine that for each session, there is a queue of packets without a permit and a bucket of permits at the session's source. The packet at the head of the packet queue obtains a permit once one is available in the permit bucket and then joins the set of packets with permits waiting to be transmitted (see Fig. 6.16). Permits are generated at the desired input rate r of the session (one permit each $1/r$ seconds) as long as the number in the permit bucket does not exceed a certain threshold W. The leaky bucket scheme is used in PARIS, an experimental high-speed network developed by IBM ([CiG88]; see Section 6.4). A variation, implemented in the ARPANET, is to allocate $W > 1$ permits initially to a session and subsequently restore the count back to W every W/r seconds, whether or not the session used any part of the allocation.

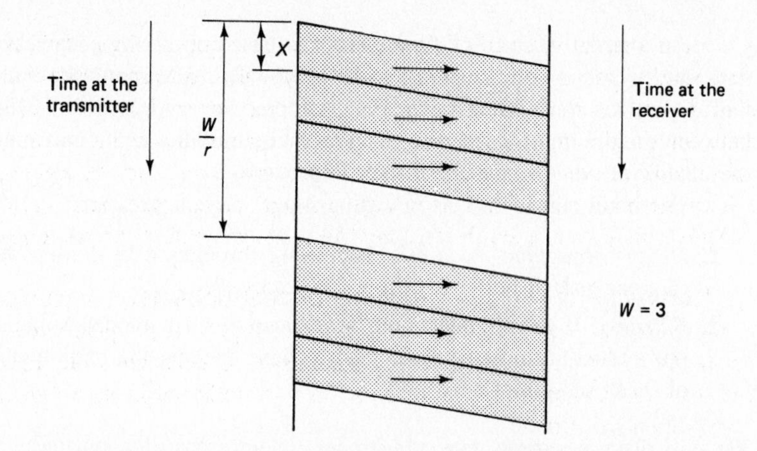

Figure 6.15 Time window flow control with $W = 3$. The count of packet allocation is decremented when a packet is transmitted and incremented W/r seconds later.

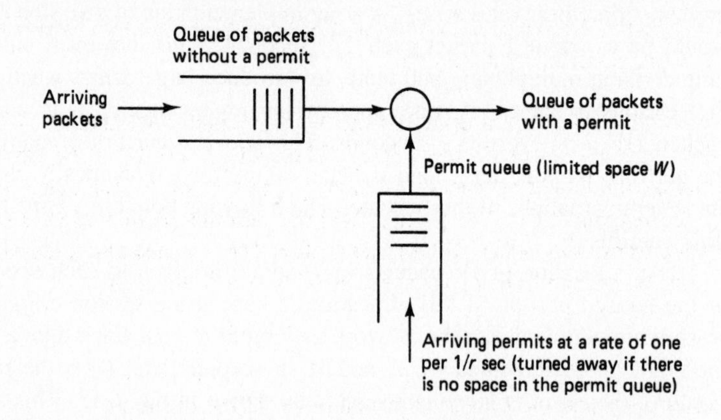

Figure 6.16 Leaky bucket scheme. To join the transmission queue, a packet must get a permit from the permit queue. A new permit is generated every $1/r$ seconds, where r is the desired input rate, as long as the number of permits does not exceed a given threshold.

The leaky bucket scheme does not necessarily preclude buffer overflow and does not guarantee an upper bound on packet delay in the absence of additional mechanisms to choose and implement the session rates inside the network; for some insight into the nature of such mechanisms, see Problem 6.19. However, with proper choice of the bucket parameters, buffer overflow and maximum packet delay can be reduced. In particular, the bucket size W is an important parameter for the performance of the leaky bucket scheme. If W is small, bursty traffic is delayed waiting for permits to become available ($W = 1$ resembles time-division multiplexing). If W is too large, long bursts of packets will be allowed into the network; these bursts may accumulate at a congested node downstream and cause buffer overflow.

The preceding discussion leads to the idea of dynamic adjustment of the bucket size. In particular, a congested node may send a special control message to the corresponding sources instructing them to shrink their bucket sizes. This is similar to the choke packet discussed in connection with dynamic adjustment of window sizes. The dynamic adjustment of bucket sizes may be combined with the dynamic adjustment of permit rates and merged into a single algorithm. Note, however, that in high-speed networks, the effectiveness of the feedback control messages from the congested nodes may be diminished because of relatively large propagation delays. Therefore, some predictive mechanism may be needed to issue control messages before congestion sets in.

Queueing analysis of the leaky bucket scheme. To provide some insight into the behavior of the leaky bucket scheme of Fig. 6.16, we give a queueing analysis. One should be very careful in drawing quantitative conclusions from this analysis, since some data sessions are much more bursty than is reflected in the following assumption of Poisson packet arrivals.

Let us assume the following:

1. Packets arrive according to a Poisson process with rate λ.
2. A permit arrives every $1/r$ seconds, but if the permit pool contains W permits, the arriving permit is discarded.

We view this system as a discrete-time Markov chain with states $0, 1, \ldots$. (For a quite similar formulation, see Problem 3.13.) The states $i = 0, 1, \ldots, W$ correspond to $W - i$ permits available and no packets without permits waiting. The states $i = W + 1, W + 2, \ldots$, correspond to $i - W$ packets without permits waiting and no permits available. The state transitions occur at the times $0, 1/r, 2/r, \ldots$, just after a permit arrival. Let us consider the probabilities of k packet arrivals in $1/r$ seconds,

$$a_k = \frac{e^{-\lambda/r}(\lambda/r)^k}{k!}$$

It can be seen that the transition probabilities of the chain are

$$P_{0i} = \begin{cases} a_{i+1}, & \text{if } i \geq 1 \\ a_0 + a_1, & \text{if } i = 0 \end{cases}$$

and for $j \geq 1$,

$$P_{ji} = \begin{cases} a_{i-j+1}, & \text{if } j \leq i + 1 \\ 0, & \text{otherwise} \end{cases}$$

(see Fig. 6.17). The global balance equations yield

$$p_0 = a_0 p_1 + (a_0 + a_1) p_0$$

$$p_i = \sum_{j=0}^{i+1} a_{i-j+1} p_j, \qquad i \geq 1$$

These equations can be solved recursively. In particular, we have

$$p_1 = a_2 p_0 + a_1 p_1 + a_0 p_2$$

so by using the equation $p_1 = (1 - a_0 - a_1)p_0/a_0$, we obtain

$$p_2 = \frac{p_0}{a_0}\left(\frac{(1 - a_0 - a_1)(1 - a_1)}{a_0} - a_2\right)$$

Similarly, we may use the global balance equation for p_2, and the computed expressions for p_1 and p_2, to express p_3 in terms of p_0, and so on.

The steady-state probabilities can now be obtained by noting that

$$p_0 = \frac{r - \lambda}{r a_0}$$

To see this, note that the permit generation rate averaged over all states is $(1 - p_0 a_0)r$, while the packet arrival rate is λ. Equating the two rates, we obtain $p_0 = (r - \lambda)/(r a_0)$. The system is stable, that is, the packet queue stays bounded, if $\lambda < r$. The average delay for a packet to obtain a permit is

$$T = \frac{1}{r}\sum_{j=0}^{\infty} p_j \max\{0, j - W\} = \frac{1}{r}\sum_{j=W+1}^{\infty} p_j(j - W)$$

To obtain a closed-form expression for the average delay needed by a packet to get a permit and also to obtain a better model of the practical system, we modify the leaky bucket model slightly so that permits are generated on a per bit basis; this approximates the real situation where messages are broken up into small packets upon arrival at the source node. In particular, we assume that:

1. Credit for admission into the network is generated at a rate r bits/sec for transmission and the size of the bucket (*i.e.*, the maximum credit that can be saved) is W bits.
2. Messages arrive according to a Poisson process with rate λ, and the storage space for messages is infinite. Message lengths are independent and exponentially distributed with mean L bits.

Let $\mu = r/L$, so that $1/\mu$ is the mean time to transmit a message at the credit rate r. Also let $C = W/r$ be the time over which credit can be saved up. The state of the system can be described by the number of bits in the queue and the available amount of credit. At time t, let $X(t)$ be either the number of bits in the queue (if the queue is

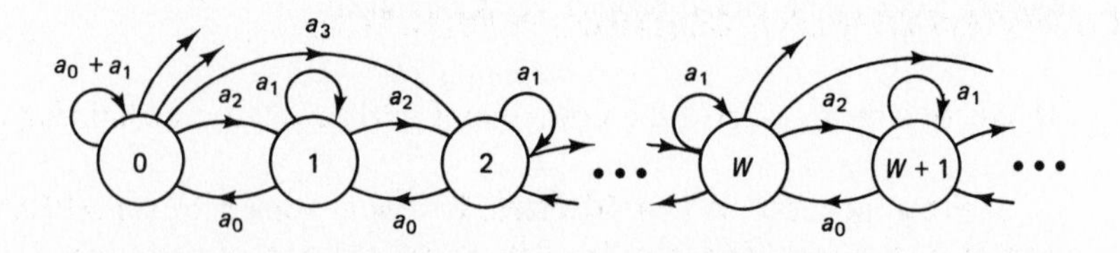

Figure 6.17 Transition probabilities of a discrete Markov chain model for the leaky bucket scheme. Here a_k is the probability of k packet arrivals in $1/r$ seconds. Note that this Markov chain is also the Markov chain for a slotted service $M/D/1$ queue.

nonempty) or minus the available credit. Thus, whenever a message consisting of x bits arrives, $X(t)$ increases by x [one of three things happens: the credit decreases by x; the queue increases by x; the credit decreases to 0 and the queue increases to $X(t)+x$ (this happens if $X(t) < 0$ and $X(t) + x > 0$)].

Letting $Y(t) = X(t) + W$, it can be seen that $Y(t)$ is the unfinished work in a fictitious $M/M/1$ queue with arrival rate λ messages/sec and service rate μ. An incoming bit is transmitted immediately if the size of the fictitious queue ahead of it is less than W and is transmitted C seconds earlier than in the fictitious queue otherwise. Focusing on the last bits of messages, we see that if T_i is the system delay in the fictitious queue for the ith message, $\max\{0, T_i - C\}$ is the delay in the real queue. Using the theory of Section 3.3, it can be verified [a proof is outlined in Exercise 3.11(b)] that the steady-state distribution of the system time T_i at the fictitious queue is

$$P\{T_i \geq \tau\} = e^{-\tau(\mu-\lambda)}$$

Thus letting $T_i' = \max\{0, T_i - C\}$ be the delay of the ith packet in the real queue, we obtain

$$P\{T_i' \geq \tau\} = \begin{cases} 1, & \text{if } \tau \leq 0 \\ e^{-(C+\tau)(\mu-\lambda)}, & \text{if } \tau > 0 \end{cases}$$

From this equation, the average delay of a packet in the real queue can be calculated as

$$T = \int_0^{\infty} P\{T_i' \geq \tau\}\, d\tau = \frac{1}{\mu - \lambda} e^{-C(\mu-\lambda)}$$

The preceding analysis can be generalized for the case where the packet lengths are independent but not exponentially distributed. In this case, the fictitious queue becomes an $M/G/1$ queue, and its system delay distribution can be estimated by using an exponential upper bound due to Kingman; see [Kle76], p. 45.

We finally note that the leaky bucket parameters affect not only the average packet delay to enter the network, which we have just analyzed. They also affect substantially the packet delay *after* the packet has entered the network. The relationship between this delay, the leaky bucket parameters, and the method for implementing the corresponding session rates at the network links is not well understood at present. For some interesting recent analyses and proposals see [Cru91a], [Cru91b], [PaG91a], [PaG91b], [Sas91], and Problem 6.19.

6.4 OVERVIEW OF FLOW CONTROL IN PRACTICE

In this section we discuss the flow control schemes of several existing networks.

Flow control in the ARPANET. Flow control in the ARPANET is based in part on end-to-end windows. The entire packet stream of each pair of machines (known as *hosts*) connected to the subnet is viewed as a "session" flowing on a logical pipe. For each such pipe there is a window of eight messages between the corresponding origin and destination subnet nodes. Each message consists of one or more packets up to a maximum of eight. A transmitted message carries a number indicating its position in the corresponding window. Upon disposing of a message, the destination node sends back a special control packet (permit) to the origin, which in the ARPANET is called RFNM (ready for next message). The RFNM is also used as an end-to-end acknowledgment for error control purposes. Upon reception of an RFNM the origin node frees up a space in the corresponding window and is allowed to transmit an extra message. If an RFNM is not received after a specified time-out, the origin node sends a control packet asking the destination node whether the corresponding message was received. This protects against loss of an RFNM, and provides a mechanism for retransmission of lost messages.

There is an additional mechanism within the subnet for multipacket messages that ensures that there is enough memory space to reassemble these messages at their destination. (Packets in the ARPANET may arrive out of order at their destination.) Each multipacket message must reserve enough buffer space for reassembly at the receiver before it gets transmitted. This is done via a reservation message called REQALL (request for allocation) that is sent by the origin to the destination node. The reservation is granted when the destination node sends an ALL (allocate) message to the origin. When a long file is sent through the network, there is a long sequence of multipacket messages that must be transmitted. It would then be wasteful to obtain a separate reservation for each message. To resolve this problem, ALL messages are piggybacked on the returning RFNMs of multipacket messages, so that there is no reservation delay for messages after the first one in a file. If the reserved buffer space is not used by the origin node within a given timeout, it is returned to the destination via a special message. Single-packet messages do not need a reservation before getting transmitted. If, however, such a message finds the destination's buffers full, it is discarded and a copy is eventually retransmitted by the origin node after obtaining an explicit buffer reservation.

A number of improvements to the ARPANET scheme were implemented in late 1986 [Mal86]. First, the window size can be configured up to a maximum of 127; this allows efficient operation in the case where satellite links are used. Second, there can be multiple independent connections (up to 256) between two hosts, each with an independent window; this provides some flexibility in accommodating classes of traffic with different priorities and/or throughput needs. Third, an effort is made to improve the fairness properties of the current algorithm through a scheme that tries to allocate the available buffer space at each node fairly among all hosts. Finally, the reservation scheme for multipacket messages described above has been eliminated. Instead, the destination node simply reserves space for a multipacket message upon receiving the first packet of the message. If space is not available, the packet is discarded and is retransmitted after a time-out by the origin node.

The ARPANET flow control was supplemented by a rate adjustment scheme in 1989. Each node calculates an upper bound on flow rate for the origin-destination pairs routing traffic through it (the origin and the destination are subnetwork packet switches). This upper bound, also called a *ration*, is modified depending on the utilization of various critical resources of the node (processing power, transmission capacity of incident links, buffer space, etc.). The ration is adjusted up or down as the actual utilization of critical resources falls below or rises above a certain target utilization. The node rations are broadcast to the entire network along with the routing update messages. Each origin then sets its flow rate to each destination to the minimum of the rations of the nodes traversed by the current route to the destination (these nodes become known to the origin through the shortest path routing algorithm). The rates are implemented by using a leaky bucket scheme as discussed in Section 6.3.

Flow control in the TYMNET. Flow control in the TYMNET is exercised separately for each virtual circuit via a sequence of node-by-node windows. There is one such window per virtual circuit and link on the path of the virtual circuit. Each window is measured in bytes, and its size varies with the expected peak data rate (or throughput class) of the virtual circuit. Flow control is activated via the backpressure effect discussed in Section 6.2.2. Fairness is enhanced by serving virtual circuits on a link via a round-robin scheme. This is accomplished by combining groups of bytes from several virtual circuits into data link control frames. The maximum number of bytes for each virtual circuit in a frame depends on the level of congestion on the link and the priority class of the virtual circuit. Flow control permits are piggybacked on data frames, and are highly encoded so that they do not require much bandwidth.

Flow control in SNA. We recall from Section 5.1.2 that the counterpart in SNA of the OSI architecture network layer is called the path control layer, and that it includes a flow control function called virtual route control. The corresponding algorithm, known as the virtual route pacing scheme, is based on an end-to-end window for each virtual circuit (or virtual route in SNA terminology). An interesting aspect of this scheme is that the window size (measured in packets) is dynamically adjusted depending on traffic conditions. The minimum window size is usually equal to the number of links on the path, and the maximum window size is three times as large. Each packet header contains two bits that are set to zero at the source. An intermediate node on the path that is "moderately congested" sets the first bit to 1. If the node is "badly congested," it sets both bits to 1. Otherwise, it does not change the bits. Upon arrival of the packet, the destination looks at the bits and increments the window size for no congestion, decrements the window size for moderate congestion, or sets the window size to the minimum for bad congestion.

Actually, the SNA scheme is a little different from the end-to-end window scheme focused on so far. In the main scheme discussed in Section 6.2.1, the window size is αW, and returning permits result in allocations of α packets each. We concentrated on the case where $\alpha = 1$. In SNA, however, $W = 1$ and α (rather than W) is adjusted between the minimum and maximum window size referred to earlier. Furthermore, the destination node can send a new α-packet allocation message (permit) upon reception of the first packet in an α-packet batch. Thus, full-speed transmission under light load conditions can be maintained even though $W = 1$.

In addition to virtual route control, SNA provides transport layer flow control on a session-by-session basis, which is known as session level pacing. This becomes necessary because a virtual route in SNA may contain several sessions that may have different flow control needs. The main idea here is to prevent the transmitting end of a session from sending data more quickly than the receiving end can process. Session-level pacing is basically a window scheme whereby the transmitting end can introduce a new packet into the subnet upon receiving a permit (called a pacing response in SNA) from the other end. An interesting twist is that pacing responses can be delayed at the node through which the transmitting end of the session accesses the subnet. This provides the subnet with the means to control the rate at which it accepts data from external users.

Flow control in a Codex network. In one of the Codex networks, there is an end-to-end window associated with each virtual circuit. The window size is measured in bytes and is proportional to the number of links on the virtual circuit path and a nominal data rate for the virtual circuit. Returning permits are combined with end-to-end acknowledgments used for error control purposes, so the window scheme does not require additional communication overhead. Data link control (DLC) frames on each link are formed by concatenating groups of bytes from several virtual circuits that have traffic in queue. There is a maximum group size for each virtual circuit. Virtual circuits are served on a round-robin basis and this provides a natural mechanism for maintaining fairness.

There is also a rate control mechanism in the Codex network that is unrelated to the window scheme but plays a complementary role. The idea is to match the transmission rate of a virtual circuit along its incoming and outgoing links at each node on its path under heavy-load conditions. Without going into details (see [HSS86]), this is accomplished by adjusting the maximum number of bytes that a virtual circuit can insert in a DLC frame. As an example, suppose that a virtual circuit is slowed down on a given link due to heavy load. Then the start node of that link sends a special message to the preceding node along the virtual circuit's path, which proceeds to reduce the maximum number of bytes that a DLC frame can carry for this virtual circuit. One effect of this scheme is that when congestion develops downstream, the end-to-end window of a virtual circuit will not pile up entirely at the point of congestion, but rather will be spread more or less evenly along the virtual circuit path starting from the source node and ending at the point of congestion. This, combined with large memory space at the nodes, tends to make buffer overflow unlikely.

Flow control in the PARIS network. PARIS is an experimental high-speed packet switching system for integrated voice, video, and data communications. (PARIS is an acronym for Packetized Automatic Routing Integrated System.) PARIS uses virtual circuits and simple error control to expedite packet processing at the nodes, and achieve high packet throughput. Routes are calculated with an adaptive shortest path routing algorithm similar to the SPF algorithm of the ARPANET. Each link length is based on a measure of the load carried by the link, and is broadcast from time to time through the network using a spanning tree. Source routing is used; each packet carries the sequence of identities of the nodes that it has yet to cross.

Flow control is based on the leaky bucket scheme described in Section 6.3. A session requesting access to the network must provide the corresponding entry node with some information regarding its characteristics, such as average rate, peak rate, and average burst size. The entry node translates this information into an "equivalent capacity," which is a measure of bandwidth required by the session at each link along its path. The entry node then does a routing calculation to obtain a shortest path among the paths that can accommodate the session. If no suitable path is found, the session is rejected. Otherwise the session is accepted and its leaky bucket parameters are determined based on its requirements and the current load of its path. The session may transmit more packets than the ones permitted by its leaky bucket. However, these extra packets are tagged as "red" and the network may discard them much more readily than other packets, which are called "green" and are given preferential access to buffer space. Even though the order of transmission of red and green packets is preserved, the algorithm is operated so that red packets have minimal effect on the loss rate of green packets. The leaky bucket parameters are kept constant during the lifetime of a session. It was found that dynamic adjustment of these parameters was of limited use because congestion information was largely outdated due to the high network speed and the relatively large propagation delays. We refer to [CiG88] and [CGK90] for further details.

Flow control in X.25. As discussed in Section 2.8.3, flow control at the X.25 packet level is implemented by means of a separate window for each virtual circuit. The default window size is 2, but it may be set as high as 7 or 127. Flow control is exercised in both directions [*i.e.*, from the user's machine (DTE) to the entry point of the network (DCE), and also from DCE to DTE]. The implementation of the window strategy is reminiscent of DLC protocols. Each data packet contains a three-bit sequence number and a three-bit request number. (If the window size is 127, these numbers are seven bits long.) The sequence number gives the position of the packet within the sender's window, while the request number is the number of the next packet expected to be received by the sender. Thus, the request number plays the role of a permit allowing the receiver to advance the corresponding window.

There is also a provision in X.25 for flow control between two DTEs communicating through the subnet. The X.25 packet format contains a special bit (called the D bit) that determines whether the piggyback number of a packet received by a DTE relates to the directly attached DCE ($D = 0$) or the remote DTE ($D = 1$). In the latter case, the request number serves as a permit for the advancement of a window maintained for flow control purposes between the two DTEs.

6.5 RATE ADJUSTMENT ALGORITHMS

In this section we look at two systematic formulations of the flow control problem to obtain algorithms for input rate adjustment. In the first approach (Section 6.5.1) we formulate an optimization problem that mathematically expresses the objective of striking a proper balance between maintaining high throughput and keeping average delay per packet at a reasonable level. In the second approach (Section 6.5.2) we emphasize fairness while maintaining average delay per packet at an acceptable level.

6.5.1 Combined Optimal Routing and Flow Control

We consider the possibility of combining routing and end-to-end flow control within the subnet by adjusting optimally *both* the routing variables and the origin–destination (OD) pair input rates. A special case arises when routing is fixed and the only variables to be adjusted are the input rates—a problem of pure flow control.

We adopt a flow model similar to the one discussed in the context of optimal routing in Section 5.4, and we denote by r_w the input rate of an OD pair w. We first formulate a problem of adjusting routing variables together with the inputs r_w so as to minimize some "reasonable" cost function. We subsequently show that *this problem is mathematically equivalent to the optimal routing problem examined in Chapter 5* (in which r_w is fixed), and therefore the optimality conditions and algorithms given there are applicable.

If we minimize the cost function $\sum_{(i,j)} D_{ij}(F_{ij})$ of the routing problem with respect to both the path flows $\{x_p\}$ and the inputs $\{r_w\}$, we unhappily find that the optimal solution is $x_p = 0$ and $r_w = 0$ for all p and w. This indicates that the cost function should include a penalty for inputs r_w becoming too small and leads to the problem

$$\begin{aligned} &\text{minimize} \sum_{(i,j)} D_{ij}(F_{ij}) + \sum_{w \in W} e_w(r_w) \\ &\text{subject to} \sum_{p \in P_w} x_p = r_w, \qquad \text{for all } w \in W \\ &\qquad\qquad x_p \geq 0, \qquad \text{for all } p \in P_w,\ w \in W \\ &\qquad\qquad 0 \leq r_w \leq \overline{r}_w, \qquad \text{for all } w \in W \end{aligned} \tag{6.2}$$

Here the minimization is to be carried out jointly with respect to $\{x_p\}$ and $\{r_w\}$. The given values $\overline{r}_w$ represent the desired input by OD pair w (*i.e.*, the offered load for w, defined as the input for w that would result if no flow control were exercised). As before, F_{ij} is the total flow on link (i,j) (*i.e.*, the sum of all path flows traversing the link). The functions e_w are of the form shown in Fig. 6.18 and provide a penalty for throttling the inputs r_w. They are monotonically decreasing on the set of positive numbers $(0,\infty)$, and tend to ∞ as r_w tends to zero. We assume that their first and second derivatives, e'_w and e''_w, exist on $(0,\infty)$ and are strictly negative and positive, respectively. An interesting class of functions e_w is specified by the following formula for their first derivative:

$$e'_w(r_w) = -\left(\frac{a_w}{r_w}\right)^{b_w}, \qquad \text{for } a_w \text{ and } b_w \text{ given positive constants} \tag{6.3}$$

As explained later in this section, the parameters a_w and b_w influence the optimal mag-

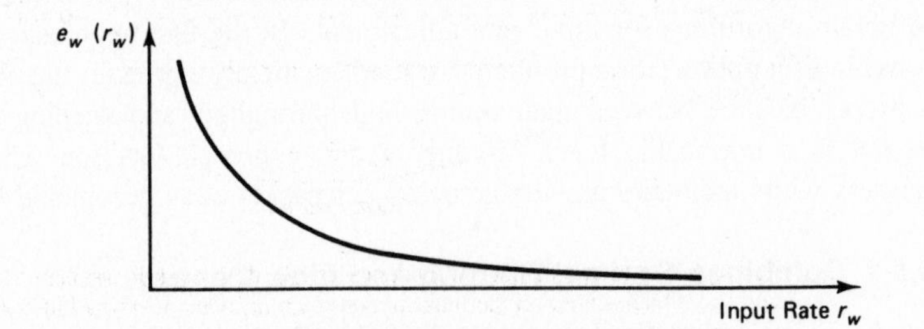

Figure 6.18 Typical form of penalty function for throttling the input rate r_w.

nitude of input r_w and the priority of OD pair w, respectively. The functions D_{ij} are defined and are monotonically increasing with positive second derivative in an interval $[0, C_{ij})$, where C_{ij} is either finite, representing link capacity, or is equal to ∞.

The value of the preceding formulation is enhanced if we adopt a broader view of w and consider it as *a class of sessions sharing the same set of paths* P_w. This allows different priorities (*i.e.*, different functions e_w) for different classes of sessions even if they share the same paths. A problem where P_w consists of a single path for each w can also be considered. This is a problem of pure flow control, namely, choosing the optimal fraction of the desired input flow of each session class that should be allowed into the network.

We now show that the combined routing and flow control problem of Eq. (6.2) is mathematically equivalent to a routing problem of the type considered in Section 5.4. Let us introduce a new variable y_w for each $w \in W$ via the equation

$$y_w = \overline{r}_w - r_w \tag{6.4}$$

We may view y_w as the *overflow* (the portion of $\overline{r}_w$ blocked out of the network), and consider it as a flow on an *overflow link* directly connecting the origin and destination nodes of w as shown in Fig. 6.19. If we define a new function E_w by

$$E_w(y_w) = e_w(\overline{r}_w - y_w) \tag{6.5}$$

the combined routing and flow control problem of Eq. (6.2) becomes, in view of the definition $y_w = \overline{r}_w - r_w$,

$$\begin{aligned} &\text{minimize} \sum_{(i,j)} D_{ij}(F_{ij}) + \sum_{w \in W} E_w(y_w) \\ &\text{subject to} \sum_{p \in P_w} x_p + y_w = \overline{r}_w, \qquad \text{for all } w \in W \\ &\qquad\qquad x_p \geq 0, \qquad \text{for all } p \in P_w,\ w \in W \\ &\qquad\qquad y_w \geq 0, \qquad \text{for all } w \in W \end{aligned} \tag{6.6}$$

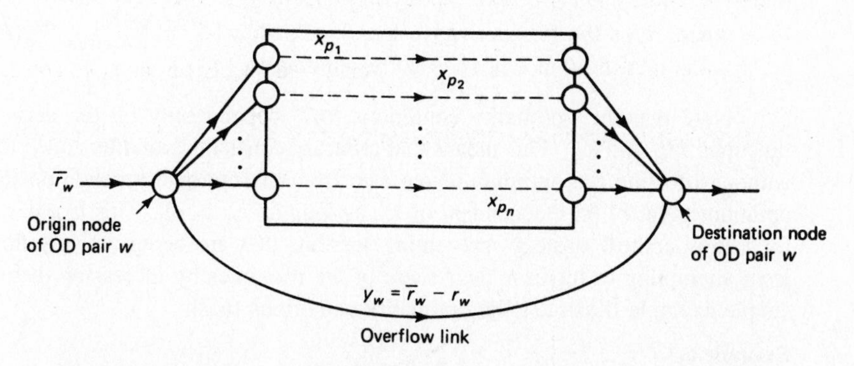

Figure 6.19 Mathematical equivalence of the flow control problem with an optimal routing problem based on the introduction of an artificial overflow link for each OD pair w. The overflow link carries the rejected traffic (*i.e.*, the difference between desired and accepted input flow, $\overline{r}_w - r_w$). The cost of the overflow link is obtained from the cost e_w for throttling the input by a change of variable.

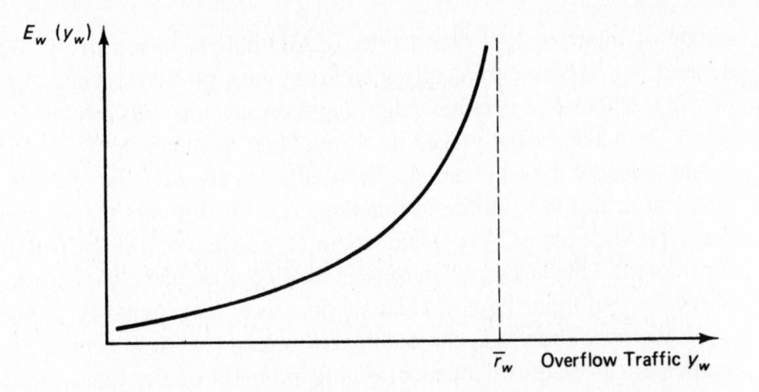

Figure 6.20 Cost function for the overflow link. The cost $E_w(y_w)$ for overflow traffic y_w equals the cost $e_w(r_w)$ for input traffic $r_w = \bar{r}_w - y_w$.

The form of the function E_w of Eq. (6.5) is shown in Fig. 6.20. If $e_w(r_w) \to \infty$ as $r_w \to 0$ (*i.e.*, there is "infinite penalty" for completely shutting off the class of sessions w), then $E_w(y_w) \to \infty$ as the overflow y_w approaches its maximum value—the maximum input $\bar{r}_w$. Thus, E_w may be viewed as a "delay" function for the overflow link, and $\bar{r}_w$ may be viewed as the "capacity" of the link.

It is now clear that the problem of Eq. (6.6) is of the optimal routing type considered in Section 5.4, and that the algorithms and optimality conditions given in Sections 5.5 to 5.7 apply. In particular, the application of the shortest path optimality condition of Section 5.5 (see also Problem 5.24) yields the following result:

A feasible set of path flows $\{x_p^*\}$ and inputs $\{r_w^*\}$ is optimal for the combined routing and flow control problem of Eq. (6.2) if and only if the following conditions hold for each $p \in P_w$ and $w \in W$:

$$x_p^* > 0 \quad \Rightarrow \quad d_p^* \le d_{p'}^*, \quad \text{for all } p' \in P_w, \text{ and } d_p^* \le -e_w'(r_w^*) \tag{6.7a}$$

$$r_w^* < \bar{r}_w \quad \Rightarrow \quad -e_w'(r_w^*) \le d_p^*, \quad \text{for all } p \in P_w \tag{6.7b}$$

where d_p^* is the first derivative length of path p [$d_p^* = \sum_{(i,j)\in p} D_{ij}'(F_{ij}^*)$, and F_{ij}^* is the total flow of link (i,j) corresponding to $\{x_p^*\}$].

Note that the optimality conditions (6.7) depend only on the derivatives of the functions D_{ij} and e_w. This means that arbitrary constants could be added to D_{ij} and e_w without affecting the optimum. Note also from the optimality condition (6.7b) that the optimum point r_w^* is independent of $\bar{r}_w$ as long as $\bar{r}_w > r_w^*$. This is a desirable feature for a flow control strategy, preventing sessions that are being actively flow controlled from attempting to increase their share of the resources by increasing their demands. A simple example illustrates the optimality conditions (6.7).

Example 6.3

Consider the situation in Fig. 6.21 involving a single link connecting origin and destination. The cost function is

$$\frac{r}{C-r} + \frac{a}{r}$$

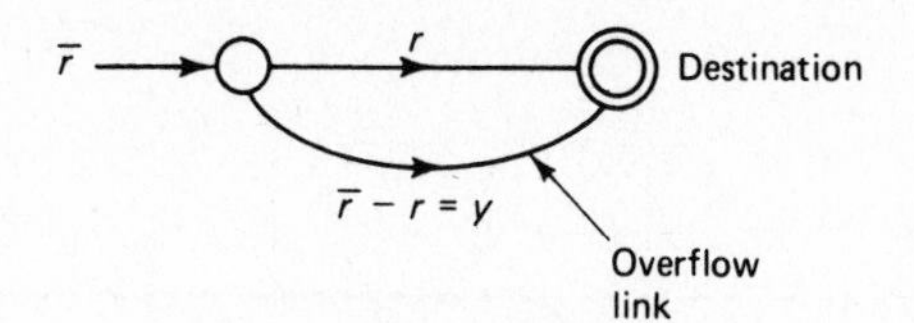

Figure 6.21 Example problem involving an origin and a destination connected by a single link.

where the first term represents a penalty due to large delay [cf. the term $D_{ij}(F_{ij})$ in Eq. (6.2)], and the second term represents a penalty for small throughput [cf. the term $e_w(r_w)$ in Eq. (6.2)]. The constant C is the capacity of the link, while the parameter a is a positive weighting factor. The equivalent routing problem [cf. Eq. (6.6)] is

$$\text{minimize } \frac{r}{C-r} + \frac{a}{\overline{r}-y}$$

$$\text{subject to } r + y = \overline{r}, \qquad r \geq 0,\ y \geq 0$$

where $y = \overline{r} - r$ represents the amount of offered load rejected by flow control (equivalently, the flow on the fictitious overflow link). The optimality conditions (6.7) show that there will be no flow control $(y = 0, r = \overline{r})$ if

$$\frac{C}{(C-\overline{r})^2} < \frac{a}{\overline{r}^2}$$

(*i.e.*, if the first derivative length of the overflow link exceeds the first derivative length of the regular link). Equivalently, there will be no flow control if

$$\overline{r} < C\frac{\sqrt{a}}{\sqrt{a}+\sqrt{C}}$$

According to the optimality condition [Eq. (6.7)], when there is flow control $(y > 0,\ r < \overline{r})$, the two first derivative lengths must be equal, that is,

$$\frac{C}{(C-r)^2} = \frac{a}{(\overline{r}-y)^2}$$

Substituting $y = \overline{r} - r$ and working out the result yields

$$r = C\frac{\sqrt{a}}{\sqrt{a}+\sqrt{C}}$$

The solution as a function of offered load is shown in Fig. 6.22. Note that the throughput is independent of the offered load beyond a certain point, as discussed earlier. The maximum throughput $C\sqrt{a}/(\sqrt{a}+\sqrt{C})$ can be regulated by adjustment of the parameter a and tends to the capacity C as $a \to \infty$.

The meaning of the parameters a_w and b_w in the cost function specified by the formula [cf. Eq. (6.3)]

$$e'_w(r_w) = -\left(\frac{a_w}{r_w}\right)^{b_w}$$

can now be clarified in the light of the optimality condition (6.7b). Consider two distinct classes of sessions, w_1 and w_2, sharing the same paths $(P_{w_1} = P_{w_2})$. Then the condition

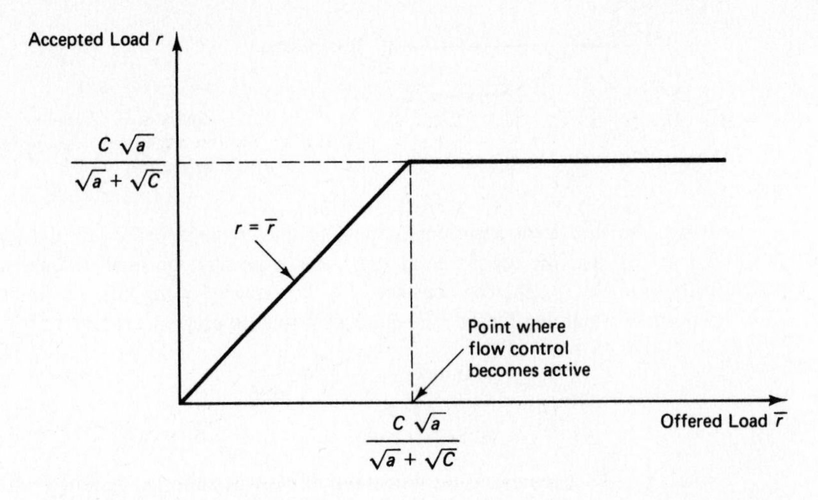

Figure 6.22 Optimal accepted load as a function of offered load in the flow control example. Flow control becomes active when the offered load exceeds a threshold level that depends on the weighting factor a.

(6.7b) implies that at an optimal solution in which both classes of sessions are throttled $(r^*_{w_1} < \bar{r}_{w_1},\ r^*_{w_2} < \bar{r}_{w_2})$

$$-e'_{w_1}(r^*_{w_1}) = -e'_{w_2}(r^*_{w_2}) = \min_{p \in P_{w_1}} \{d^*_p\} = \min_{p \in P_{w_2}} \{d^*_p\} \tag{6.8}$$

If e'_{w_1} and e'_{w_2} are specified by parameters a_{w_1}, b_{w_1} and a_{w_2}, b_{w_2} as in Eq. (6.3), it can be seen that:

1. If $b_{w_1} = b_{w_2}$, then

$$\frac{r^*_{w_1}}{r^*_{w_2}} = \frac{a_{w_1}}{a_{w_2}}$$

   and it follows that the parameter a_w influences the optimal, relative input rate of the session class w.

2. If $a_{w_1} = a_{w_2} = a$ and $b_{w_1} < b_{w_2}$ (see Fig. 6.23), the condition (6.8) specifies that when the input flows must be made small $(r^*_{w_1}, r^*_{w_2} < a)$, the session class w_2 (the one with higher parameter b_w) will be allowed a larger input. It follows that the parameter b_w influences the relative priority of the session class w under heavy load conditions.

6.5.2 Max-Min Flow Control

One of the most difficult aspects of flow control is treating all sessions fairly when it is necessary to turn traffic away from the network. Fairness can be defined in a number of different ways, but one intuitive notion of fairness is that any session is entitled to as much network use as is any other session. Figure 6.24 clarifies some of the ambiguities in this notion. One session flows through the tandem connection of all links, and each

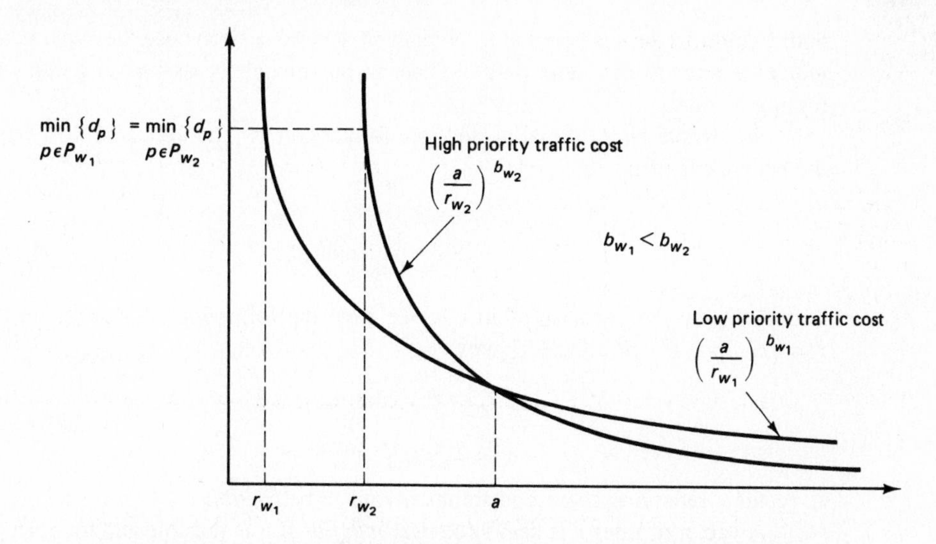

Figure 6.23 Incorporating priorities of session classes in the cost function of the flow control formulation. The session class with larger b_w will be throttled less under heavy load conditions.

other session goes through only one link. It is plausible to limit sessions 0, 1, and 2 to a rate of 1/3 each, since this gives each of these sessions as much rate as the others. It would be rather pointless, however, to restrict session 3 to a rate of 1/3. Session 3 might better be limited to 2/3, since any lower limit would waste some of the capacity of the rightmost link without benefitting sessions 0, 1, or 2, and any higher limit would be unfair because it would further restrict session 0.

This example leads to the idea of maximizing the network use allocated to the sessions with the minimum allocation, thus giving rise to the term *max-min flow control.* After these most poorly treated sessions are given the greatest possible allocation, there might be considerable latitude left for choosing allocations for the other sessions. It is then reasonable to maximize the allocation for the most poorly treated of these other sessions, and so forth, until all allocations are specified. An alternative way to express this intuition, which turns out to be equivalent to the above, is to maximize the allocation of each session i subject to the constraint that an incremental increase in i's allocation does not cause a decrease in some other session's allocation that is already as small as i's or smaller.

We assume a directed graph $G = (\mathcal{N}, \mathcal{A})$ for the network and a set of sessions P using the network. Each session p has an associated fixed path in the network. We use p

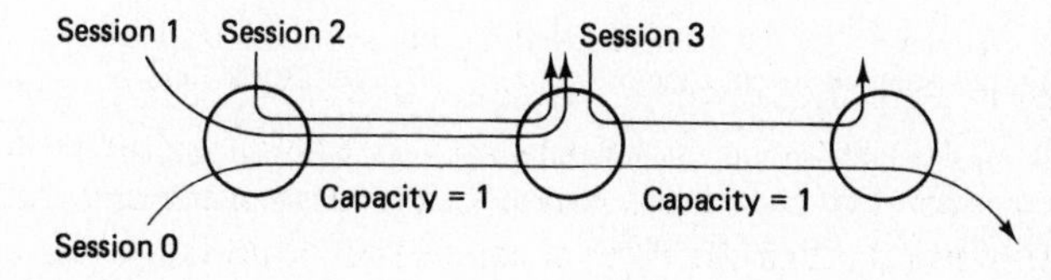

Figure 6.24 The fair solution is to give to sessions 0, 1, and 2 a rate of 1/3 each and to give session 3 a rate of 2/3 to avoid wasting the extra capacity available at the rightmost link.

both to refer to the session and to its path (if several sessions use the same path, several indices p refer to the same path). Thus, in our model we assume a fixed, single-path routing method.

We denote by r_p the allocated rate for session p. The allocated flow on link a of the network is then

$$F_a = \sum_{\substack{\text{all sessions } p \\ \text{crossing link } a}} r_p \tag{6.9}$$

Letting C_a be the capacity of link a, we have the following constraints on the vector $r = \{r_p \mid p \in P\}$ of allocated rates:

$$r_p \geq 0, \qquad \text{for all } p \in P \tag{6.10a}$$

$$F_a \leq C_a, \qquad \text{for all } a \in \mathcal{A} \tag{6.10b}$$

A vector r satisfying these constraints is said to be *feasible*.

A vector of rates r is said to be *max-min fair* if it is feasible and for each $p \in P$, r_p cannot be increased while maintaining feasibility without decreasing $r_{p'}$ for some session p' for which $r_{p'} \leq r_p$. (More formally, r is max-min fair if it is feasible, and for each $p \in P$ and feasible $\overline{r}$ for which $r_p < \overline{r}_p$, there is some p' with $r_p \geq r_{p'}$ and $r_{p'} > \overline{r}_{p'}$.) Our problem is to find a rate vector that is max-min fair.

Given a feasible rate vector r, we say that link a is a *bottleneck link* with respect to r for a session p crossing a if $F_a = C_a$ and $r_p \geq r_{p'}$ for all sessions p' crossing link a. Figure 6.25 provides an example of a max-min fair rate vector and illustrates the concept of a bottleneck link. In this example, every session has a bottleneck link. It turns out that this property holds in general as shown in the following proposition:

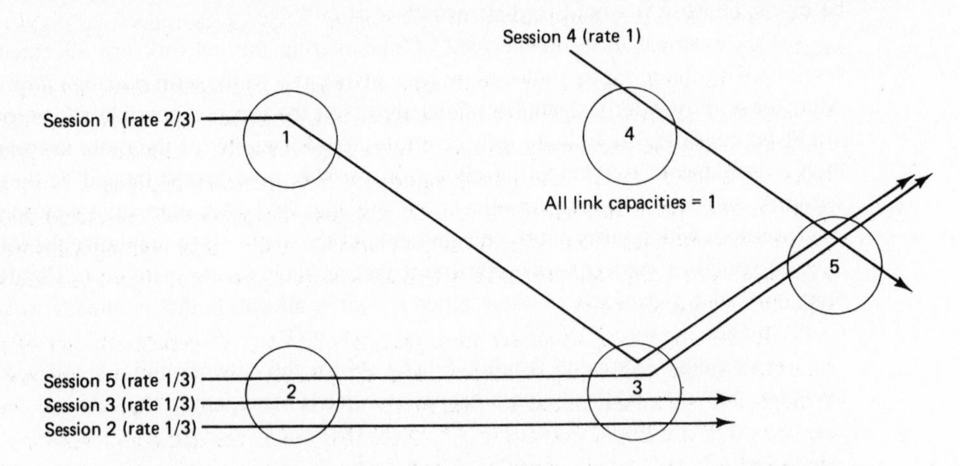

Figure 6.25 Max-min fair solution for an example network. The bottleneck links of sessions 1, 2, 3, 4, and 5 are (3,5), (2,3), (2,3), (4,5), and (2,3), respectively. Link (3,5) is not a bottleneck link for session 5 since sessions 1 and 5 share this link and session 1 has a larger rate than session 5. Link (1,3) is not a bottleneck link of any session since it has an excess capacity of 1/3 in the fair solution.

Proposition. A feasible rate vector r is max-min fair if and only if each session has a bottleneck link with respect to r.

Proof: Suppose that r is max-min fair and, to arrive at a contradiction, assume that there exists a session p with no bottleneck link. Then, for each link a crossed by p for which $F_a = C_a$, there must exist a session $p_a \neq p$ such that $r_{p_a} > r_p$; thus the quantity

$$\delta_a = \begin{cases} C_a - F_a, & \text{if } F_a < C_a \\ r_{p_a} - r_p, & \text{if } F_a = C_a \end{cases}$$

is positive. Therefore, by increasing r_p by the minimum δ_a over all links a crossed by p, while decreasing by the same amount the rates of the sessions r_{p_a} of the links a crossed by p with $F_a = C_a$, we maintain feasibility without decreasing the rate of any session p' with $r_{p'} \leq r_p$; this contradicts the max-min fairness property of r.

Conversely, assume that each session has a bottleneck link with respect to the feasible rate vector r. Then, to increase the rate of any session p while maintaining feasibility, we must decrease the rate of some session p' crossing the bottleneck link a of p (because we have $F_a = C_a$ by the definition of a bottleneck link). Since $r_{p'} \leq r_p$ for all p' crossing a (by the definition of a bottleneck link), the rate vector r satisfies the requirement for max-min fairness. **Q.E.D.**

Next, we give a simple algorithm for computing max-min fair rate vectors. The idea of the algorithm is to start with an all-zero rate vector and to increase the rates on all paths together until $F_a = C_a$ for one or more links a. At this point, each session using a saturated link (*i.e.*, a link with $F_a = C_a$) has the same rate at every other session using that link. Thus, these saturated links serve as bottleneck links for all sessions using them.

At the next step of the algorithm, all sessions not using the saturated links are incremented equally in rate until one or more new links become saturated. Note that the sessions using the previously saturated links might also be using these newly saturated links (at a lower rate). The newly saturated links serve as bottleneck links for those sessions that pass through them but do not use the previously saturated links. The algorithm continues from step to step, always equally incrementing all sessions not passing through any saturated link; when all sessions pass through at least one saturated link, the algorithm stops.

In the algorithm, as stated more precisely below, A^k denotes the set of links not saturated at the beginning of step k, and P^k denotes the set of sessions not passing through any saturated link at the beginning of step k. Also, n_a^k denotes the number of sessions that use link a and are in P^k. Note that this is the number of sessions that will share link a's yet unused capacity. Finally, $\tilde{r}^k$ denotes the increment of rate added to all of the sessions in P^k at the kth step.

Initial conditions: $k = 1$, $F_a^0 = 0$, $r_p^0 = 0$, $P^1 = P$, and $A^1 = A$.

1. n_a^k := number of sessions $p \in P^k$ crossing link a

2. $\tilde{r}^k := \min_{a \in A^k} (C_a - F_a^{k-1}) / n_a^k$

3. $r_p^k := \begin{cases} r_p^{k-1} + \tilde{r}^k & \text{for } p \in P^k \\ r_p^{k-1} & \text{otherwise} \end{cases}$

4. $F_a^k := \sum_{p \text{ crossing } a} r_p^k$

5. $A^{k+1} := \{a \mid C_a - F_a^k > 0\}$

6. $P^{k+1} := \{p \mid p \text{ does not cross any link } a \in A^{k+1}\}$

7. $k := k + 1$

8. If P^k is empty, then stop; else go to 1.

At each step k, an equal increment of rate is added to all sessions not yet passing through a saturated link, and thus at each step k, all sessions in P^k have the same rate. All sessions in P^k passing through a link that saturates in step k have at least as much rate as any other session on that link and hence are bottlenecked by that link. Thus upon termination of the algorithm, each session has a bottleneck link, and by the proposition shown earlier, the final rate vector is max-min fair.

Example 6.4

Consider the problem of max-min fair allocation for the five sessions and the network shown in Fig. 6.25. All links have a capacity of one.

Step 1: All sessions get a rate of $1/3$. Link (2,3) is saturated at this step, and the rate of the three sessions (2, 3, and 5) that go through it is fixed at $1/3$.

Step 2: Sessions 1 and 4 get an additional rate increment of $1/3$ for a total of $2/3$. Link (3,5) is saturated, and the rate of session 1 is fixed at $2/3$.

Step 3: Session 4 gets an additional rate increment of $1/3$ for a total of 1. Link (4,5) is saturated, and the rate of session 4 is fixed at 1. Since now all sessions go through at least one saturated link, the algorithm terminates with the max-min fair solution shown in Fig. 6.25.

Several generalizations can be made to the basic approach described above. First, to keep the flow on each link strictly below capacity, we can replace C_a in the algorithm with some fixed fraction of C_a. Next, we can consider ways to assign different priorities to different kinds of traffic and to make these priorities sensitive to traffic levels. If $b_p(r_p)$ is an increasing function representing the priority of p at rate r_p, the max-min fairness criterion can be modified as follows: For each p, maximize r_p subject to the constraint that any increase in r_p would cause a decrease of $r_{p'}$ for some p' satisfying $b_{p'}(r_{p'}) \leq b_p(r_p)$. It is easy to modify the algorithm above to calculate fair rates with such priorities. Another twist on the same theme is to require that each r_p be upper bounded by $C_a - F_a$ on each link used by path p; the rationale is to maintain enough spare capacity on each link to be able to add an extra session. The problem here, however, is that as the number of sessions on a link grow, the reserve capacity shrinks to zero and the buffer requirement grows with the number of sessions, just like the corresponding growth using windows. This difficulty can be bypassed by replacing the constraint $r_p \leq C_a - F_a$ by a constraint of the form $r_p \leq (C_a - F_a)q_a$, where q_a is a positive scalar factor depending on the number of sessions crossing link a (see Problem 6.18).

There has been a great deal of work on distributed algorithms that dynamically adjust the session rates to maintain max-min fairness as the sessions change. A representative algorithm [Hay81] will help in understanding the situation. In this algorithm, v_a^k represents an estimate of the maximum allowable session rate on link a at the kth iteration, F_a^k is the allocated flow on link a corresponding to the rates r_p^k, and n_a is the number of sessions using link a. The typical iteration of the algorithm is

$$v_a^{k+1} = v_a^k + \frac{C_a - F_a^k}{n_a}$$

$$r_p^{k+1} = \min_{\substack{\text{links } a \text{ on} \\ \text{path } p}} v_a^{k+1}$$

Each iteration can be implemented in a distributed way by first passing the values v_a^{k+1} from the links a to the sessions using those links, and then passing the values r_p^{k+1} from the sessions p to the links used by these sessions. The major problem is that the allocated rates can wander considerably before converging, and link flows can exceed capacity temporarily. There are other algorithms that do not suffer from this difficulty ([GaB84b] and [Mos84]). In the algorithm of [GaB84b], the session rates are iterated at the session sources according to

$$r_p^{k+1} = \min_{\substack{\text{links } a \text{ on} \\ \text{path } p}} \left\{ r_p^k + \frac{C_a - F_a^k - r_p^k}{1 + n_a} \right\}$$

This algorithm aims at solving the max-min fair flow control problem subject to the additional constraints $r_p \leq C_a - F_a$ for each session p and link a crossed by p (see Problem 6.18). By adding the relation

$$r_p^{k+1} \leq r_p^k + \frac{C_a - F_a^k - r_p^k}{1 + n_a}$$

over all sessions p crossing link a, it is straightforward to verify that

$$F_a^{k+1} \leq \frac{n_a C_a}{1 + n_a}$$

so the link flows are strictly less than the link capacities at all times. Extensions of this algorithm to accommodate session priorities are possible (see [GaB84b]).

Max-min fair rates may be implemented using leaky bucket schemes as discussed in Section 6.3. Another approach to implementing max-min fair rates has been explored in [HaG86]. This approach avoids both the problem of communication in a distributed algorithm and also the problem of the ambiguity in the meaning of rate for interactive traffic. The idea is very simple: Serve different sessions on each link in the network on a round-robin basis. This means that *if* a session always has a packet waiting at a node when its turn comes up on the outgoing link, that session gets as much service as any other session using that link. Thus, to achieve max-min fairness, it is only necessary that each session always has a packet waiting at its bottleneck link. In fact, it can be shown that by using node-by-node windowing with a large enough window, a session that always has something to send will always have a packet waiting at its bottleneck link (see [Hah86]).

SUMMARY

In this chapter we identified the major flow control objectives as limiting average delay and buffer overflow within the subnet, and treating sessions fairly. We reviewed the major flow control methods, and we saw that the dominant strategies in practice are based on windows and input rate control. Window strategies combine low overhead with fast reaction to congestion but have some limitations, particularly for networks with relatively large propagation delay. Window strategies are also unsuitable for sessions that require a minimum guaranteed rate. Rate adjustment schemes are usually implemented by means of leaky buckets. However, there remain a number of questions regarding the choice and adjustment of the leaky bucket parameters in response to traffic conditions. We also described two theoretical input rate adjustment schemes. The first scheme extends the optimal routing methodology of Chapter 5 to the flow control context, and combines routing and flow control into a single algorithm. The second scheme assumes fixed, single-path routing for each session, and focuses on maintaining flow control fairness.

NOTES, SOURCES, AND SUGGESTED READING

Section 6.1. Extensive discussions of flow control can be found in the April 1981 special issue of the IEEE Transactions on Communications. An informative survey is [GeK80]. A more recent survey [GeK89] discusses flow control in local area networks.

Section 6.2. The special difficulties of window flow control along satellite links are discussed in [GrB83].

Section 6.3. The leaky bucket scheme is discussed in [Tur86] and [Zha89]. Several rate control schemes are discussed in [Gol90a], [Gol90b], [ELL90a], and [ELL90b].

Section 6.4. Flow control in the ARPANET is described in several sources (*e.g.*, [Kle76] and [KlO77]). The more recent rate control scheme is discussed in [RFS91] and [ELS90]. The TYMNET flow control system is discussed in [Tym81]. For further discussion on SNA, see [Ahu79] and [GeY82], and for more on the Codex network, see [HSS86]. The PARIS flow control scheme and other PARIS protocols are discussed in [CiG88], and [CGK90].

Section 6.5. The combined optimal routing and flow control was formulated in [GaG80] and [Gol80]. For additional material on this subject, see [Ibe81] and [Gaf82]. A simulation together with a discussion of practical implementation schemes is given in [ThC86]. Flow control based on adjusting the rate of encoding of digitized voice has been considered in [BGS80]. The material on fair flow control is based on [Hay81]. For related work, see [Jaf81], [GaB84b], [Mos84], [Hah86], and [HaG86].

PROBLEMS

6.1 Consider a window flow controlled virtual circuit going over a satellite link. All packets have a transmission time of 5 msec. The round-trip processing and propagation delay is 0.5 sec. Find a lower bound on the window size for the virtual circuit to be able to achieve maximum speed transmission when there is no other traffic on the link.

6.2 Suppose that the virtual circuit in Problem 6.1 goes through a terrestrial link in addition to the satellite link. The transmission time on the terrestrial link is 20 msec, and the processing and propagation delay is negligible. What is the maximum transmission rate in packets/sec that can be attained for this virtual circuit assuming no flow control? Find a lower bound to an end-to-end window size that will allow maximum transmission rate assuming no other traffic on the links. Does it make a difference whether the terrestrial link is before or after the satellite link?

6.3 Suppose that node-by-node windows are used in the two-link system of Problem 6.2. Find lower bounds on the window size required along each link in order to achieve maximum speed transmission, assuming no other traffic on the links.

6.4 The three-node network of Fig. 6.26 contains only one virtual circuit from node 1 to 3, and uses node-by-node windows. Each packet transmission on link (1,2) takes 1 sec, and on link (2,3) takes 2 sec; processing and propagation delay is negligible. Permits require 1 sec to travel on each link. There is an inexhaustible supply of packets at node 1. The system starts at time 0 with W permits at node 1, W permits at node 2, and no packets stored at nodes 2 and 3. For $W = 1$, find the times, from 0 to 10 sec at which a packet transmission starts at node 1 and node 2. Repeat for $W = 2$.

6.5 In the discussion of node-by-node window flow control, it was assumed that node i can send a permit back to its predecessor $(i-1)$ once it releases a packet to the DLC of link $(i, i+1)$. The alternative is for node i to send the permit when it receives the DLC acknowledgment that the packet has been received correctly at node $i+1$. Discuss the relative merits of the two schemes. Which scheme requires more memory? What happens when link $(i, i+1)$ is a satellite link?

6.6 Consider a combined optimal routing and flow control problem involving the network of Fig. 6.27 (cf. Section 6.4.1). The cost function is

$$(F_{13})^2 + (F_{23})^2 + (F_{34})^2 + \frac{a}{r_1} + \frac{a}{r_2}$$

where a is a positive scalar parameter. Find the optimal values of the rates r_1 and r_2 for each value of a.

6.7 Consider six nodes arranged in a ring and connected with unit capacity bidirectional links $(i, i+1)$, $i = 1, 2, 3, 4, 5$, and (6,1). There are two sessions from nodes 1, 2, 3, 4, and 5 to node 6, one in the clockwise and one in the counterclockwise direction. Similarly, there are two sessions from nodes 2, 3, and 4 to node 5, and two sessions from node 3 to node 4. Find the max-min fair rates for these sessions.

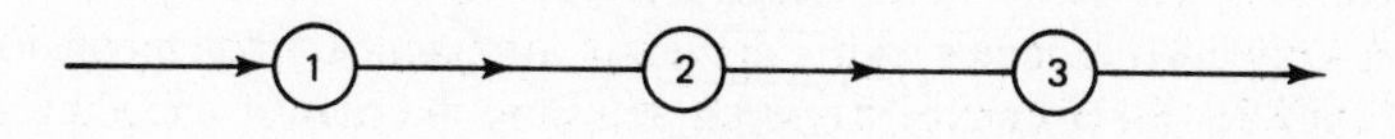

Figure 6.26

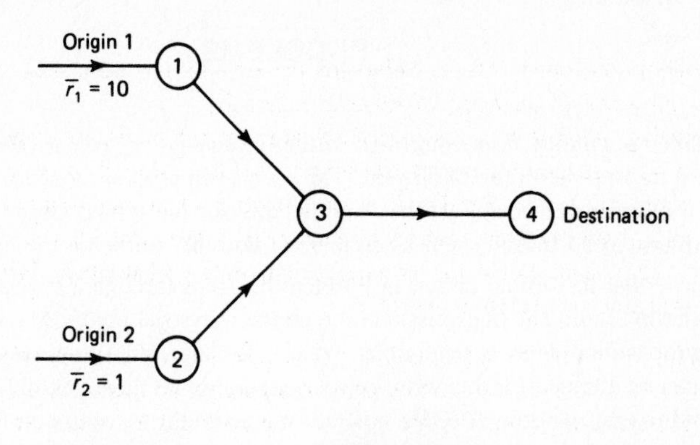

Figure 6.27

6.8 Suppose that the definition of a fair rate vector of Section 6.4.2 is modified so that, in addition to Eq. (6.10), there is a constraint $r_p \leq b_p$ that the rate of a session p must satisfy. Here b_p is the maximum rate at which session p is capable of transmitting. Modify the algorithm given in Section 6.4.2 to solve this problem. Find the fair rates for the example given in this section when $b_p = 1$, for $p = 2, 4, 5$ and $b_p = 1/4$, for $p = 1, 3$. *Hint:* Add a new link for each session.

6.9 The purpose of this problem is to illustrate how the relative throughputs of competing sessions can be affected by the priority rule used to serve them. Consider two sessions A and B sharing the first link L of their paths as shown in Fig. 6.28. Each session has an end-to-end window of two packets. Permits for packets of A and B arrive d_A and d_B seconds, respectively, after the end of transmission on link L. We assume that d_A is exponentially distributed with mean of unity, while (somewhat unrealistically) we assume that $d_B = 0$. Packets require a transmission time on L which is exponentially distributed with mean of unity. Packet transmission times and permit delays are all independent. We assume that a new packet for A (B) enters the transmission queue of L immediately upon receipt of a permit for A (B).

(a) Suppose that packets are transmitted on L on a first-come first-serve basis. Argue that the queue of L can be represented by a Markov chain with the 10 queue states BB, BBA, BAB, ABB, BBAA, BABA, BAAB, ABBA, ABAB, and AABB (each letter stands for a packet of the corresponding session). Show that all states have equal steady-state probability and that the steady-state throughputs of sessions A and B in packets/sec are 0.4 and 0.6, respectively.

(b) Now suppose that transmissions on link L are scheduled on a round-robin basis. Between successive packet transmissions for session B, session A transmits one packet if it has one waiting. Draw the state transition diagram of a five-state Markov chain which

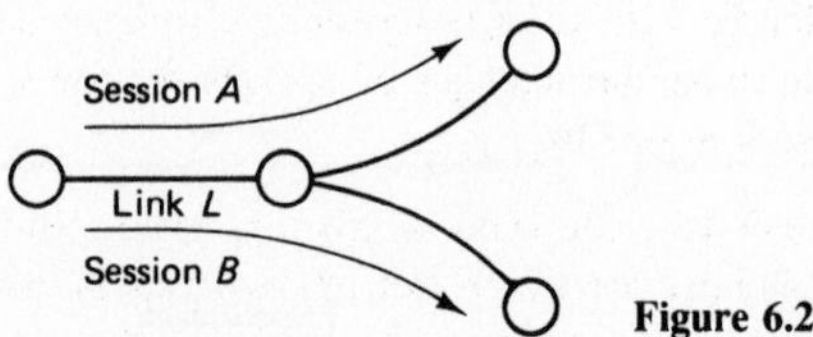

Figure 6.28

models the queue of L. Solve for the equilibrium state probabilities. What are the steady-state throughputs of sessions A and B?

(c) Finally, suppose that session A has nonpreemptive priority over session B at link L. Between successive packet transmissions for session B, session A transmits as many packets as it can. Session B regains control of the link only when A has nothing to send. Draw the state transition diagram of a five-state Markov chain which models L and its queue. Solve for the equilibrium state probabilities. What are the steady-state throughputs of sessions A and B?

6.10 Consider a window between an external source and the node by which that source is connected to the subnet. The source generates packets according to a Poisson process of rate λ. Each generated packet is accepted by the node if a permit is available. If no permit is available, the packet is discarded, never to return. When a packet from the given source enters the DLC unit of an outgoing link at the node, a new permit is instantaneously sent back to the source. The source initially has two permits and the window size is 2. Assume that the other traffic at the node imposes a random delay from the time a packet is accepted at the node to the time it enters the DLC unit. Specifically, assume that in any arbitrarily small interval δ, there is a probability $\mu\delta$ that a waiting packet from the source (or the first of two waiting packets) will enter the DLC; this event is independent of everything else.

(a) Construct a Markov chain for the number of permits at the source.

(b) Find the probability that a packet generated by the source is discarded.

(c) Explain whether the probability in part (b) would increase or decrease if the propagation and transmission delay from source to node and the reverse were taken into account.

(d) Suppose that a buffer of size k is provided at the source to save packets for which no permit is available; when a permit is received, one of the buffered packets is instantly sent to the network node. Find the new Markov chain describing the system, and find the probability that a generated packet finds the buffer full.

6.11 Consider a network with end-to-end window flow control applied to each virtual circuit. Assume that the data link control operates perfectly and that packets are never thrown away inside the network; thus, packets always arrive at the destination in the order sent, and all packets eventually arrive.

(a) Suppose that the destination sends permits in packets returning to the source; if no return packet is available for some time-out period, a special permit packet is sent back to the source. These permits consist of the number modulo m of the next packet awaited by the destination. What is the restriction on the window size W in terms of the modulus m? Why?

(b) Suppose next that the permits contain the number modulo m of *each* of the packets in the order received since the last acknowledgment was sent. Does this change your answer to part (a)? Explain.

(c) Is it permissible for the source to change the window size W without prior agreement from the destination? Explain.

(d) How can the destination reduce the effective window size below the window size used by the source without prior agreement from the source? (By effective window size we mean the maximum number of packets for the source–destination pair that can be in the network at one time.)

6.12 Consider a node-by-node window scheme. In an effort to reduce the required buffering, the designers associated the windows with destinations rather than with virtual circuits. Assume that all virtual circuit paths to a given destination j use a directed spanning tree so that each node $i \neq j$ has only one outgoing link for traffic to that destination. Assume that each node $i \neq j$ originally has permits for two packets for j that can be sent over the outgoing link to j. Each time that node i releases a packet for j into its DLC unit on the outgoing link, it sends a new permit for one packet back over the incoming link over which that packet arrived. If the packet arrived from a source connected to i, the permit is sent back to the source (each source also originally has two permits for the given destination).

(a) How many packet buffers does node i have to reserve for packets going to destination j to guarantee that every arriving packet for j can be placed in a buffer?

(b) What are the pros and cons of this scheme compared with the conventional node-by-node window scheme on a virtual circuit basis?

6.13 Consider a network using node-by-node windows for each virtual circuit. Describe a strategy for sending and reclaiming permits so that buffer overflow never occurs regardless of how much memory is available for packet storage at each node and of how many virtual circuits are using each link. *Hint*: You need to worry about too many permits becoming available to the transmitting node of each link.

6.14 Consider a network using a node-by-node window for each session. Suppose that the transmission capacity of all links of the networks is increased by a factor K and that the number of sessions that can be served also increases by a factor K. Argue that in order for the network to allow for full-speed transmission for each session under light traffic conditions, the total window of the network should increase by a factor K if propagation delay is dominated by packet transmission time and by a factor K^2 if the reverse is true.

6.15 Consider the variation of the leaky bucket scheme where $W > 1$ permits are allocated initially to a session and the count is restored back to W every W/r seconds. Develop a Markov chain model for the number of packets waiting to get a permit. Assume a Poisson arrival process.

6.16 Describe how the gradient projection method for optimal routing can be used to solve in distributed fashion the combined optimal routing and flow control problem of Section 6.5.1.

6.17 Let r be a max-min fair rate vector corresponding to a given network and set of sessions.

(a) Suppose that some of the sessions are eliminated and let $\overline{r}$ be a corresponding max-min fair rate vector. Show by example that we may have $\overline{r}_p < r_p$ for some of the remaining sessions p.

(b) Suppose that some of the link capacities are increased and let $\overline{r}$ be a corresponding max-min fair rate vector. Show by example that we may have $\overline{r}_p < r_p$ for some sessions p.

6.18 *Alternative Formulation of Max-Min Fair Flow Control ([Jaf81] and [GaB84b]).* Consider the max-min flow control problem where the rate vector r is required, in addition, to satisfy $r_p \leq (C_a - F_a)q_a$ for each session p and link a crossed by session p. Here q_a are given positive scalars.

(a) Show that a max-min fair rate vector exists and is unique, and give an algorithm for calculating it.

(b) Show that for a max-min fair rate vector, the utilization factor $\rho_a = F_a/C_a$ of each link a satisfies

$$\rho_a \leq \frac{n_a q_a}{1 + n_a q_a}$$

where n_a is the number of sessions crossing link a.

(c) Show that additional constraints of the form $r_p \leq R_p$, where R_p is a given positive scalar for each session p, can be accommodated in this formulation by adding to the network one extra link per session.

6.19 *Guaranteed Delay Bounds Using Leaky Buckets [PaG91a],[PaG91b].* In this problem we show how guaranteed delay bounds for the sessions sharing a network can be obtained by appropriately choosing the sessions' leaky bucket parameters. We assume that each session i has a fixed route and is constrained by a leaky bucket with parameters W_i and r_i, as in the scheme of Fig. 6.16. We assume a fluid model for the traffic, *i.e.*, the packet sizes are infinitesimal. In particular, if $A_i(\tau, t)$ is the amount of session i traffic entering the network during an interval $[\tau, t]$,

$$A_i(\tau, t) \leq W_i + r_i(t - \tau).$$

Let $T_i^l(\tau, t)$ be the amount of session i traffic transmitted on link l in an interval $[\tau, t]$. Assume that each link l transmits at its maximum capacity C_l whenever it has traffic in queue and operates according to a priority discipline, called *Rate Proportional Processor Sharing*, whereby if two sessions i and j have traffic in queue throughout the interval $[\tau, t]$, then

$$\frac{T_i^l(\tau, t)}{T_j^l(\tau, t)} = \frac{r_i}{r_j} \tag{6.11}$$

We also assume that bits of the same session are transmitted in the order of their arrival. (For practical approximations of such an idealized scheme, see [PaG91a].) Let $S(l)$ be the set of sessions sharing link l and let

$$\rho(l) = \frac{\sum_{i \in S(l)} r_i}{C_l}$$

be the corresponding link utilization. Assume that $\rho(l) < 1$ for all links l.

(a) Let l_i be the first link crossed by session i. Show that the queue length at link l_i for session i is never more than W_i. Furthermore, at link l_i, each bit of session i waits in queue no more than $W_i \rho(l_i)/r_i$ time units, while for each interval $[\tau, t]$ throughout which session i has traffic in queue,

$$T_i^{l_i}(\tau, t) \geq \frac{r_i(t - \tau)}{\rho(l_i)}$$

Hint: The rate of transmission of session i in an interval throughout which session i has traffic in queue is at least $r_i/\rho(l_i)$. If $\overline{\tau}$ and $\overline{t}$ are the start and end of a busy period for session i, respectively, the queue length at times $t \in [\overline{\tau}, \overline{t}]$ is

$$Q_i(t) = A_i(\overline{\tau}, t) - T_i^{l_i}(\overline{\tau}, t) \leq W_i + \left(r_i - \frac{r_i}{\rho(l_i)}\right)(t - \overline{\tau})$$

(b) Let ρ_{max}^i be the maximum utilization over the links crossed by session i and assume that processing and propagation delay are negligible. Show that the amount of traffic of session i within the entire network never exceeds W_i. Furthermore, the time spent inside the network by a bit of session i is at most $W_i \rho_{max}^i / r_i$. *Hint*: Argue that while some link on session i's path has some session i traffic waiting in queue, the rate of departure of session i traffic from the network is at least r_i/ρ_{max}^i.

(c) Consider now a generalization of the preceding scheme called Generalized Processor Sharing. In particular, suppose that at each link l, instead of Eq. (6.11), the following relation holds for all sessions $i, j \in S(l)$

$$\frac{T_i^l(\tau, t)}{T_j^l(\tau, t)} = \frac{\phi_i^l}{\phi_j^l}$$

where ϕ_i^l, ϕ_j^l are given positive numbers. For any session i define

$$g_i = \min_{\substack{\text{all links } l \\ \text{crossed by } i}} \frac{1}{\rho(l)} \frac{\phi_i^l}{\sum_{j \in S(l)} \phi_j^l}$$

Show that if $g_i \geq r_i$, then the amount of traffic of session i within the entire network never exceeds

$$\frac{W_i r_i}{g_i}$$

Furthermore, the time spent by a bit of session i inside the network is at most W_i / g_i.

References

[AaM81] AASHTIANI, H. Z., and MAGNANTI, T. L. 1981. Equilibria on a Congested Transportation Network, *SIAM J. Algebraic Discrete Methods*, 2:213–226.

[Abr70] ABRAMSON, N. 1970. The Aloha System—Another Alternative for Computer Communications, *Proc. Fall Joint Comput. Conf., AFIPS Conf.*, p. 37.

[Ahu79] AHUJA V. 1979. Routing and Flow Control in Systems Network Architecture, *IBM Systems J.*, 18:298–314.

[AkK77] AKINC, U., and KHUMAWALA, B. 1977. An Efficient Branch and Bound Algorithm for the Capacitated Warehouse Location Problem, *Management Sci.*, 23:585–594.

[Ald86] ALDOUS, D. 1986. *Ultimate Instability of Exponential Back-off Protocol for Acknowledgment-Based Transmission Control of Random Access Communication Channels.* Berkeley, CA: University of California, Dept. of Statistics.

[AlM76] ALCOUFFE, A., and MURATET, G. 1976. Optimum Location of Plants, *Management Sci.*, 23:267–274.

[Alt86] ALTES, T. 1986. *Minimum Delay Packet Length* (Report LIDS-TH-1602). Cambridge, MA: MIT Laboratory for Information and Decision Systems.

[AnP86] ANAGNOSTOU, M., and PROTONOTARIOS, E. 1986. Performance Analysis of the Selective Repeat ARQ Protocol, *IEEE Trans. Comm.*, COM-34:127–135.

[Ari84] ARIKAN, E. 1984. Some Complexity Results about Packet Radio Networks, *IEEE Trans. Inform. Theory*, IT-30:681–685.

[Ash90] ASH, G. R. 1990. Design and Control of Networks with Dynamic Nonhierarchical Routing, *IEEE Communications Magazine*, 28:34–40.

[Atk80] ATKINS, J. D. 1980. Path Control: The Transport Network of SNA., *IEEE Trans. Comm.*, COM-28:527–538.

[BaA81] BAKRY, S. H., and ACKROYD, M. H. 1981. Teletraffic Analysis for Single Cell Mobile Radio Telephone Systems, *IEEE Trans. Comm.*, COM-29:298–304.

[BaA82] BAKRY, S. H., and ACKROYD, M. H. 1982. Teletraffic Analysis for Multicell Mobile Radio Telephone Systems, *IEEE Trans. Comm.*, COM-30:1905–1909.

[BaC89] BALLERT, R., and CHING, Y.-C. 1989. SONET: Now It's the Standard Optical Network, *IEEE Comm. Mag.*, 29:8–15.

[Bac88] BACKES, F. 1988. Transparent Bridges for Interconnection of IEEE 802 LANs, *IEEE Network*, 2:5–9.

[Bal79] BALL, M. O. 1979. Computing Network Reliability, *Oper. Res.*, 27:823-838.

[BaP83] BALL, M. O., and PROVAN, J. S. 1983. Calculating Bounds on Reachability and Connectedness in Stochastic Networks, *Networks*, 13:253–278.

[Bar64] BARAN, P. 1964. On Distributed Communication Networks, *IEEE Trans. on Communications Systems*, CS-12:1–9.

[BaS83] BARATZ, L., and SEGALL, A. 1983. *Reliable Link Initialization Procedures* (Report RC10032). Yorktown Heights, NY: IBM Thomas J. Watson Research Center. Also *IEEE Trans. Comm.*, COM-36, 1988, pp. 144–152.

[BeG82] BERTSEKAS, D. P., and GAFNI, E. 1982. Projection Methods for Variational Inequalities with Application to the Traffic Assignment Problem, in D. C. Sorensen and R. J.-B. Wets (Eds.), *Mathematical Programming Studies*, Vol. 17. Amsterdam: North-Holland, pp. 139–159.

[BeG83] BERTSEKAS, D. P., and GAFNI, E. M. 1983. Projected Newton Methods and Optimization of Multicommodity Flows, *IEEE Trans. Automat. Control*, AC-28:1090–1096.

[Ben86] BENJAMIN, R. 1986. Analysis of Connection Survivability in Complex Strategic Communications Networks, *IEEE J. Select. Areas Comm.*, SAC-4:243–253.

[Ber76] BERTSEKAS, D. P. 1976. On the Goldstein–Levitin–Polyak Gradient Projection Method, *IEEE Trans. Automat. Control*, AC-21:174–184.

[Ber79a] BERTSEKAS, D. P. 1979. Algorithms for Nonlinear Multicommodity Network Flow Problems, in A. Bensoussan and J. L. Lions (Eds.), *International Symposium on Systems Optimization and Analysis*. New York: Springer-Verlag, pp. 210–224.

[Ber79b] BERTSEKAS, D. P. 1979. Dynamic Models of Shortest Path Routing Algorithms for Communication Networks with Multiple Destinations, *Proc. 1979 IEEE Conf. Dec. Control*, Ft. Lauderdale, FL, pp. 127–133.

[Ber80] BERTSEKAS, D. P. 1980. A Class of Optimal Routing Algorithms for Communication Networks, *Proc. 5th Internat. Conf. Comput. Comm.*, Atlanta, GA, pp. 71–76.

[Ber82a] BERTSEKAS, D. P. 1982. Distributed Dynamic Programming, *IEEE Trans. Automat. Control*, AC-27:610–616.

[Ber82b] BERTSEKAS, D. P. 1982. Dynamic Behavior of Shortest Path Routing Algorithms for Communication Networks, *IEEE Trans. Automat. Control*, AC-27:60–74.

[Ber82c] BERTSEKAS, D. P. 1982. Projected Newton Methods for Optimization Problems with Simple Constraints, *SIAM J. Control Optim.*, 20:221–246.

[Ber82d] BERTSEKAS, D. P. 1982. *Constrained Optimization and Lagrange Multiplier Methods*. New York: Academic Press.

[Ber83] BERTSEKAS, D. P. 1983. Distributed Asynchronous Computation of Fixed Points, *Math. Programming*, 27:107–120.

[Ber85] BERTSEKAS, D. P. 1985. A Unified Framework for Primal–Dual Methods in Minimum Cost Network Flow Problems, *Math. Programming*, 32:125–145.

[Ber87] BERTSEKAS, D. P. 1987. *Dynamic Programming: Deterministic and Stochastic Models*. Englewood Cliffs, NJ: Prentice Hall.

[Ber91] BERTSEKAS, D. P. 1991. *Linear Network Optimization: Algorithms and Codes*, MIT Press, Cambridge, MA.

[BeT89] BERTSEKAS, D. P., and TSITSIKLIS, J. N. 1989. *Parallel and Distributed Computation: Numerical Methods*, Prentice-Hall, Englewood Cliffs, N.J.

[BGG84] BERTSEKAS, D. P., GAFNI, E. M., and GALLAGER, R. G. 1984. Second Derivative Algorithms for Minimum Delay Distributed Routing in Networks, *IEEE Trans. Comm.*, COM-32:911–919.

[BGS80] BIALLY, T., GOLD, B., and SENEFF, S. 1980. A Technique for Adaptive Voice Flow Control in Integrated Packet Networks, *IEEE Trans. Comm.*, COM-28:325–333.

[BGT84] BERTSEKAS, D. P., GENDRON, R., and TSAI, W. K. 1984. *Implementation of an Optimal Multicommodity Network Flow Algorithm Based on Gradient Projection and a Path Flow Formulation* (Report LIDS-P-1364). Cambridge, MA: MIT Laboratory for Information and Decision Systems.

[BGV79] BERTSEKAS, D. P., GAFNI, E. M., and VASTOLA, K. S. 1979. Validation of Algorithms for Optimal Routing of Flow in Networks, *Proc. 1978 IEEE Conf. Dec. Control*, San Diego, CA.

[Bin75] BINDER, R. 1975. A Dynamic Packet Switching System for Satellite Broadcast Channels, *Proc. ICC*, pp. 41.1–41.5.

[Bla83] BLAHUT, R. E. 1983. *Theory and Practice of Error Control Codes*. Reading, MA: Addison-Wesley.

[BLL84] BURMAN, D. Y., LEHOCZKY, J. P., and LIM, Y. 1984. Insensitivity of Blocking Probabilities in a Circuit-Switching Network, *Advances in Applied Probability*, 21:850–859.

[BoF77] BOORSTYN, R. R., and FRANK, H. 1977. Large-Scale Network Topological Optimization, *IEEE Trans. Comm.*, COM-25:29–47.

[BoM86] BOXMA, O. J., and MEISTER, B. 1986. Waiting-Time Approximations for Cyclic-Service Systems with Switch-Over Times, *Perform. Eval. Rev.*, 14:254–262.

[BoS82] BOCHMANN, G. V., and SUNSHINE, C. A. 1982. A Survey of Formal Methods, in P. Green (Ed.), *Computer Network Architecture*. New York: Plenum.

[BrB80] BRUELL, S. C., and BALBO, G., 1980. *Computational Algorithms for Closed Queueing Networks*. New York: Elsevier/North-Holland.

[BrC91a] BRASSIL, J. T., and CRUZ, R. L. 1991. Nonuniform Traffic in the Manhattan Street Network, *Proceedings of ICC '91*, 3:1647–1651.

[BrC91b] BRASSIL, J. T., and CRUZ, R. L. 1991. Bounds on Maximum Delay in Networks with Deflection Routing, *Proceedings of the 29th Allerton Conference on Communications, Control, and Computing*, Monticello, IL.

[Bux81] BUX, W. 1981. Local Area Networks: A Performance Comparison, *IEEE Trans. Comm.*, COM-29:1465–1473.

[CaC68] CANNON, M. D., and CULLUM, C. D. 1968. A Tight Upper Bound on the Rate of Convergence of the Frank–Wolfe Algorithm, *SIAM J. Control Optim.*, 6:509–516.

[CaH75] CARLEIAL, A. B., and HELLMAN, M. E. 1975. Bistable Behavior of Slotted Aloha-Type Systems, *IEEE Trans. Comm.*, COM-23:401–410.

[Cap77] CAPETANAKIS, J. I. 1977. *The Multiple Access Broadcast Channel: Protocol and Capacity Considerations*, Ph.D. dissertation, MIT, Dept. of Electrical Engineering and Computer Science, Cambridge, MA. Also 1979, *IEEE Trans. Inform. Theory*, IT-25:505–515.

[Car80] CARLSON, D. E. 1980. Bit-Oriented Data Link Control Procedures, *IEEE Trans. Comm.*, COM-28:455–467. (Also in [Gre82].)

[CFL79] CHLAMTAC, I., FRANTA W., and LEVIN, K. D. 1979. BRAM: The Broadcast Recognizing Access Method, *IEEE Trans. Comm.*, COM-27:1183–1190.

[CFN77] CORNUEJOLS, G., FISHER, M. L., and NEMHAUSER, G. L. 1977. Location of Bank Accounts to Optimize Float: An Analytic Study of Exact and Approximate Algorithms, *Management Sci.*, 23:789–810.

[CGK90] CIDON, I., GOPAL, I. S., KAPLAN, M., and KUTTEN, S. 1990. Distributed Control for PARIS, *Proc. of the 9th Annual ACM Symposium on Principles of Computing*, Quebec, Can.

[CiG88] CIDON, I., and GOPAL, I. S. 1988. PARIS: An Approach to Integrated High-Speed Private Networks, *International Journal of Digital and Analog Cabled Systems*, 1:77–85.

[ClC81] CLARK, G. C., and CAIN, J. B. 1981. *Error Correction Coding for Digital Communication*. New York: Plenum.

[CoG89] CONWAY, A. E., and GEORGANAS, N. D. 1989. *Queueing Networks—Exact Computational Algorithms*, MIT Press, Cambridge, MA.

[Com88] COMER, D. 1988. *Internetworking with TCP/IP, Principles, Protocols, and Architectures*, Prentice-Hall, Englewood Cliffs.

[Coo70] COOPER, R. B. 1970. Queues Served in Cyclic Order: Waiting Times, *Bell Systems Tech. J.*, 49:399–413.

[Coo81] COOPER, R. B. 1981, *Introduction to Queueing Theory* (2nd ed.). New York: Elsevier/ North-Holland.

[CPR78] CLARK, D. D., POGRAN, K. T., and REED, D. P. 1978. An Introduction to Local Area Networks. *Proc. IEEE*, pp. 1497–1517.

[Cru91a] CRUZ, R. L. 1991. A Calculus for Network Delay, Part I: Network Elements in Isolation, *IEEE Trans. on Inform. Theory*, IT-37:114–131.

[Cru91b] CRUZ, R. L. 1991. A Calculus for Network Delay, Part II: Network Analysis, *IEEE Trans. on Inform. Theory*, IT-37:132–141.

[CRW73] CROWTHER, W., RETTBURG, R., WALDEN, D., ORNSTEIN, S., and HEART, F. 1973. A System for Broadcast Communication: Reservation Aloha, *Proc. 6th Hawaii Internat. Conf. Syst. Sci.*, pp. 371–374.

[Daf71] DAFERMOS, S. C. 1971. An Extended Traffic Assignment Model with Applications to Two-Way Traffic, *Trans. Sci.*, 5:366–389.

[Daf80] DAFERMOS, S. C. 1980. Traffic Equilibrium and Variational Inequalities, *Trans. Sci.*, 14:42–54.

[Dan63] DANTZIG, G. B. 1963. *Linear Programming and Extensions*. Princeton, NJ: Princeton University Press.

[DeK81] DEMBO, R. S., and KLINCEWICZ, J. G. 1981. A Scaled Reduced Gradient Algorithm for Network Flow Problems with Convex Separable Costs, *Math. Programming Stud.*, 15:125–147.

[DGK79] DIAL, R., GLOVER, F., KARNEY, D., and KLINGMAN, D. 1979. A Computational Analysis of Alternative Algorithms and Labeling Techniques for Finding Shortest Path Trees, *Networks*, 9:215–248.

[DiK85] DISNEY, R. L., and KONIG, D. 1985. Queueing Networks: A Survey of Their Random Processes, *SIAM Rev.*, 27:335–403.

[DiS80] DIJKSTRA, E. W., and SHOLTEN, C. S. 1980. Termination Detection for Diffusing Computations, *Inf. Proc. Lett.*, 11:1–4.

[Dun79] DUNN, J. C. 1979. Rates of Convergence of Conditional Gradient Algorithms near Singular and Nonsingular Extremals, *SIAM J. Control Optim.*, 17:187–211.

[Eis79] EISENBERG, M. 1979. Two Queues with Alternating Service, *SIAM J. Appl. Math.*, 20:287–303.

[ELL90a] ECKBERG, A. E., LUAN, D. T., and LUCANTONI, D. M. 1990. Bandwidth Management: A Congestion Control Strategy for Broadband Packet Networks—Characterizing the Throughput-Burstiness Filter, *Computer Networks and ISDN Systems*, 20:415–423.

[ELL90b] Eckberg, A. E., Luan, D. T., and Lucantoni, D. M. 1990. An Approach to Controlling Congestion in ATM Networks, *International Journal of Digital and Analog Communication Systems*, 3:199–209.

[ELS90] Escobar, J., Lauer, G., and Steenstrup, M. 1990. Performance Analysis of Rate-Based Congestion Control Algorithm for Receiver-Directed Packet-Radio Networks, *Proc. of MILCOM '90.*

[EpB81] Ephremides, A., and Baker, D. J. 1981. The Architectural Organization of a Mobile Radio Network via a Distributed Algorithm, *IEEE Trans. Comm.*, COM-29:1694–1701.

[Eph86] Ephremides, A. 1986. The Routing Problem in Computer Networks, in I. F. Blake and H. V. Poor (Eds.), *Communication and Networks.* New York: Springer-Verlag, pp. 299–324.

[EpV89] Ephremides, A., and Verdu, S. 1989. Control and Optimization Methods in Communication Network Problems, *IEEE Trans. Automat. Control*, 34:930–942.

[EsW66] Essau, L. R., and Williams, K. C. 1966. On Teleprocessing System Design, *IBM Systems J.*, 5:142–147.

[Eve75] Even, S. 1975. An Algorithm for Determining Whether the Connectivity of a Graph Is at Least *k*, *SIAM J. Comput.*, 4:393–396.

[EVW80] Ephremides, A., Varaiya, P., and Walrand, J. 1980. A Simple Dynamic Routing Problem, *IEEE Trans. Automat. Control*, AC-25:690–693.

[FaL72] Farber, D. J., and Larson, K. C. 1972. *The System Architecture of the Distributed Computer System—The Communications System.*, paper presented at the Symposium on Computer Networks, Polytechnic Institute of Brooklyn, New York.

[FaN69] Farmer, W. D., and Newhall, E. E. 1969. An Experimental Distributed Switching System to Handle Bursty Computer Traffic, *Proc. ACM Symp. Problems Optim. Data Comm. Syst.*, pp. 1–33.

[FeA85] Ferguson, M. J., and Aminetzah, Y. J. 1985. Exact Results for Nonsymmetric Token Ring Systems, *IEEE Trans. Comm.*, COM-33:223–231.

[FeA89] Fetterolf, P. C., and Anandalingam, G. 1989. Optimizing Interconnection of Local Area Networks: An Approach Using Simulated Annealing, *Univ. of Pennsylvania Report*; to appear in Annals of Operations Research.

[Fer75] Ferguson, M. J. 1975. A Study of Unslotted Aloha with Arbitrary Message Lengths, *Proc. 4th Data Comm. Symp.*, Quebec, Canada, pp. 5.20–5.25.

[FGK73] Fratta, L., Gerla, M., and Kleinrock, L. 1973. The Flow Deviation Method: An Approach to Store-and-Forward Communication Network Design, *Networks*, 3:97–133.

[Fin79] Finn, S. G. 1979. Resynch Procedures and a Fail-Safe Network Protocol, *IEEE Trans. Comm.*, COM-27:840–845.

[FlN74] Florian, M., and Nguyen, S. 1974. A Method for Computing Network Equilibrium with Elastic Demand, *Transport. Sci.*, 8:321–332.

[FiT84] Fine, M., and Tobagi, F. A. 1984. Demand Assignment Multiple Access Schemes in Broadcast Bus Local Area Networks, *IEEE Trans. Comput.*, C-33:1130–1159.

[FKL86] Fujiwara, T., Kasami, T., and Lin, S. 1986. Error Detecting Capabilities of the Shortened Hamming Codes Adopted for Error Detection in IEEE Standard 802.3, *Int. Symp. I.T.*, Ann Arbor, Mich.

[FoF62] Ford, L. R., Jr., and Fulkerson, D. R. 1962. *Flows in Networks.* Princeton, NJ: Princeton University Press.

[FoS78] Foschini, G. J., and Salz, J. 1978. A Basic Dynamic Routing Problem and Diffusion, *IEEE Trans. Comm.*, COM-26:320–327.

[FrW56] Frank, M., and Wolfe, P. 1956. An Algorithm for Quadratic Programming, *Naval Res. Logist. Quart.*, 3:149–154.

[FuC85] Fuhrmann, S. W., and Cooper, R. B. 1985. Stochastic Decompositions in the $M/G/1$ Queue with Generalized Vacations, *Oper. Res.*, 33:1117–1129.

[GaB81] Gafni, E. M., and Bertsekas, D. P. 1981. Distributed Routing Algorithms for Networks with Frequently Changing Topology, *IEEE Trans. Comm.*, COM-29:11–18.

[GaB83] Gafni, E. M., and Bertsekas, D. P. 1983. *Asymptotic Optimality of Shortest Path Routing* (Report LIDS-P-1307). Cambridge, MA: MIT Laboratory for Information and Decision Systems. Also 1987, *IEEE Trans. Inform. Theory*, IT-33:83–90.

[GaB84a] Gafni, E. M., and Bertsekas, D. P. 1984. Two-Metric Projection Methods for Constrained Optimization, *SIAM J. Control Optim.*, 22:936–964.

[GaB84b] Gafni, E. M., and Bertsekas, D. P. 1984. Dynamic Control of Session Input Rates in Communication Networks, *IEEE Trans. Automat. Control*, AC-29:1009–1016.

[Gaf79] Gafni, E. M. 1979. *Convergence of a Routing Algorithm*, M.S. thesis, University of Illinois, Dept. of Electrical Engineering, Urbana, IL.

[Gaf82] Gafni, E. M. 1982. *The Integration of Routing and Flow Control for Voice and Data in Integrated Packet Networks*, Ph.D. thesis, MIT, Dept. of Electrical Engineering and Computer Science, Cambridge, MA.

[GaG80] Gallager, R. G., and Golestaani, S. J. 1980. Flow Control and Routing Algorithms for Data Networks, *Proc. 5th Internat. Conf. Comput. Comm.*, pp. 779–784.

[GaH83] Gavish, B., and Hantler, S. 1983. An Algorithm for Optimal Route Selection in SNA Networks, *IEEE Trans. Comm.*, COM-31:1154–1161.

[Gal68] Gallager, R. G. 1968. *Information Theory and Reliable Communications.* New York: Wiley.

[Gal77] Gallager, R. G. 1977. A Minimum Delay Routing Algorithm Using Distributed Computation, *IEEE Trans. Comm.*, COM-23:73–85.

[Gal78] Gallager, R. G. 1978. Conflict Resolution in Random Access Broadcast Networks, *Proc. AFOSR Workshop Comm. Theory Appl.*, Provincetown, MA, pp. 74–76.

[Gal81] Gallager, R. G. 1981. Applications of Information Theory for Data Communication Networks, in J. Skwirzynski (Ed.), *New Concepts in Multi-user Communication* (Series E, No. 43). Alphen aan den Rijn, The Netherlands: NATO Advanced Study Institutes (Sijthoff en Noordhoff).

[GaP88] Gallo, G. S., and Pallotino, S. 1988. Shortest Path Algorithms, *Annals of Operations Research*, 7:3–79.

[Gar87] Garcia-Luna-Aceves, J. J. 1987. A New Minimum-Hop Routing Algorithm, *IEEE INFOCOM '87 Proceedings*, pp. 170–180.

[Gav85] Gavish, B. 1985. Augmented Lagrangean-Based Algorithms for Centralized Network Design, *IEEE Trans. Comm.*, COM-33:1247–1257.

[GeK77] Gerla, M., and Kleinrock, L. 1977. On the Topological Design of Distributed Computer Networks, *IEEE Trans. Comm.*, COM-25:48–60.

[GeK80] Gerla, M., and Kleinrock, L. 1980. Flow Control: A Comparative Survey, *IEEE Trans. Comm.*, COM-28:553–574.

[GeK88] Gerla, M., and Kleinrock, L. 1988. Congestion Control in Interconnected LANs, *IEEE Network*, 2:72–76.

[GeP85] Georgiadis, L., and Papantoni-Kazakos, P. 1985. *A 0.487 Throughput Limited Sensing Algorithm.* Storrs, CT: University of Connecticut.

[GeP87] Gelembe, E., and Pujolle, G. 1988. *Introduction to Queueing Networks.* J. Wiley, N.Y.

[GeY82] GEORGE, F. D., and YOUNG, G. E. 1982. SNA Flow Control: Architecture and Implementation, *IBM Syst. J.*, 21:179–210.

[GGM85] GOODMAN, J., GREENBERG, A. G., MADRAS, N., and MARCH, P. 1985. On the Stability of Ethernet, *Proc. 17th Annual ACM Symp. Theory Comput.*, Providence, RI, pp. 379–387.

[GHS83] GALLAGER, R. G., HUMBLET, P. A., and SPIRA, P. M. 1983. A Distributed Algorithm for Minimum-Weight Spanning Trees, *ACM Trans. Programming Language Syst.*, 5:66–77.

[Gol64] GOLDSTEIN, A. A., 1964. Convex Programming in Hilbert Space, *Bull. Am. Math. Soc.*, 70:709–710.

[Gol80] GOLESTAANI, S. J. 1980. *A Unified Theory of Flow Control and Routing on Data Communication Networks*, Ph.D. thesis, MIT, Dept. of Electrical Engineering and Computer Science, Cambridge, MA.

[Gol90a] GOLESTAANI, S. J. 1990. Congestion-Free of Real-Time Traffic in Packet Networks, *Proc. of IEEE INFOCOM '90.* San Francisco, Cal., pp. 527–536.

[Gol90b] GOLESTAANI, S. J. 1990. A Framing Strategy for Congestion Management, *Proc. of SIGCOM '90.*

[Gop85] GOPAL, I. S., 1985. Prevention of Store-and-Forward Deadlock in Computer Networks, *IEEE Trans. Comm.*, COM-33:1258–1264.

[Gra72] GRAY, J. P. 1972. Line Control Procedures, *Proc. IEEE*, pp. 1301–1312.

[GrB83] GROVER, G. A., and BHARATH-KUMAR, K. 1983. Windows in the Sky—Flow Control in SNA Networks with Satellite Links, *IBM Systems J.*, 22:451–463.

[Gre82] GREEN, P. E. 1982. *Computer Network Architectures and Protocols.* New York: Plenum.

[Gre84] GREEN, P. E. 1984. Computer Communications: Milestones and Prophecies, *IEEE Trans. Comm.*, COM-32:49–63.

[Gre86] GREEN, P. E. 1986. Protocol Conversion, *IEEE Trans. Comm.*, COM-34:257–268.

[GrH85] GROSS, D., and HARRIS, C. M. 1985. *Fundamentals of Queueing Theory* (2nd ed.). New York: Wiley.

[GrH89] GREENBERG, A. G., and HAJEK, B. 1989. Deflection Routing in Hypercube Networks, *IEEE Trans. Comm.*, to appear.

[Gun81] GUNTHER, K. D. 1981. Prevention of Deadlocks in Packet-Switched Data Transport Systems, *IEEE Trans. Comm.*, COM-29:512–524.

[HaC90] HAJEK, B., and CRUZ, R. L. 1990. On the Average Delay for Routing Subject to Independent Deflections, *IEEE Trans. of Inform. Theory*, to appear.

[HaG86] HAHNE, E. L., and GALLAGER, R. G. 1986. *Round-Robin Scheduling for Fair Flow Control in Data Communication Networks* (Report LIDS-P-1537). Cambridge, MA: MIT Laboratory for Information and Decisions Systems.

[Hah86] HAHNE, E. 1986. Round-Robin Scheduling for Fair Flow Control in Data Communication Networks Ph.D. thesis, MIT, Dept. of Electrical Engineering and Computer Science, Cambridge, MA.

[Haj82] HAJEK, B. 1982. Birth-and-Death Processes with Phases and General Boundaries, *J. Appl. Problems*, 19:488–499.

[Haj88] HAJEK, B. 1988. Cooling Schedules for Optimal Annealing, *Math of Operations Research*, 13:311–329.

[Haj91] HAJEK, B. 1991. Bounds on Evacuation Time for Deflection Routing, *Distributed Computing*, 5:1–5.

[HaL82] HAJEK, B., and VAN LOON, T. 1982. Decentralized Dynamic Control of a Multiaccess Broadcast Channel, *IEEE Trans. Automat. Control*, AC-27:559–569.

[HaO84] Hajek, B., and Ogier, R. G. 1984. Optimal Dynamic Routing in Communication Networks with Continuous Traffic, *Networks*, 14:457–487.

[Has72] Hashida, O. 1972. Analysis of Multiqueue, *Rev. Electron. Comm. Lab.*, 20:189–199.

[Hay76] Hayes, J. F. 1976. *An Adaptive Technique for Local Distribution* (Bell Telephone Laboratory Technical Memo TM-76-3116-1.) (Also 1978, *IEEE Trans. Comm.*, COM-26:1178–1186.)

[Hay81] Hayden, H. 1981. *Voice Flow Control in Integrated Packet Networks* (Report LIDS-TH-1152). Cambridge, MA: MIT Laboratory for Information and Decision Systems.

[Hay84] Hayes, J. F. 1984. *Modeling and Analysis of Computer Communications Networks*. New York: Plenum.

[HCM90] Hahne, E. L., Choudhury, A., and Maxemchuk, N. 1990. Improving the Fairness of Distributed-Queue-Dual-Bus Networks, *Infocom*, San Francisco, Ca., pp. 175–184.

[HeS82] Heyman, D. P., and Sobel, M. J. 1982. *Stochastic Models in Operations Research*, Vol. 1. New York: McGraw-Hill.

[HlG81] Hluchyj, M. G., and Gallager, R. G. 1981. Multiaccess of a Slotted Channel by Finitely Many Users, *Proc. Nat. Telecomm. Conf.*, New Orleans, LA. (Also LIDS Report P-1131, MIT, Cambridge, MA, August 1981.)

[HSS86] Humblet, P. A., Soloway, S. R., and Steinka, B. 1986. *Algorithms for Data Communication Networks—Part 2*. Codex Corp.

[HuB85] Huang, J.-C., and Berger, T. 1985. Delay Analysis of the Modified 0.487 Contention Resolution Algorithm, *IEEE Trans. Inform. Theory*, IT-31:264–273.

[HuM80] Humblet, P. A., and Mosely, J. 1980. Efficient Accessing of a Multiaccess Channel, *Proc. IEEE 19th Conf. Dec. Control*, Albuquerque, NM. (Also LIDS Report P-1040, MIT, Cambridge, MA, Sept. 1980.)

[Hum78] Humblet, P. A. 1978. *Source Coding for Communication Concentrators* (Report ESL-R-798). Cambridge, MA: MIT.

[Hum83] Humblet, P. A. 1983. A Distributed Algorithm for Minimum Weight Directed Spanning Trees, *IEEE Trans. Comm.*, COM-31:756–762.

[Hum86] Humblet, P. A. 1986. On the Throughput of Channel Access Algorithms with Limited Sensing, *IEEE Trans. Comm.*, COM-34:345–347.

[Hum91] Humblet, P. A. 1991. Another Adaptive Distributed Dijkstra Shortest Path Algorithm, *IEEE Trans. on Comm.*, COM-39:995–1003; also *MIT Laboratory for Information and Decision Systems Report LIDS-P-1775*, May 1988.

[HuS86] Humblet, P. A., and Soloway, S. R. 1986. *Algorithms for Data Communication Networks —Part I*. Codex Corp.

[HuS88] Humblet, P. A. and S. Soloway. 1988/1989. Topology Broadcast Algorithms, *Computer Networks and ISDN Systems*, 16:179–186; also *Codex Corporation Research Report;* revised May 1987 as Report LIDS-P-1692.

[IbC89] Ibe, O. C., and Cheng, X. 1989. Approximate Analysis of Asymmetric Single-Service Token-passing Systems, *IEEE Trans. on Comm.*, COM-36:572–577.

[Ibe81] Ibe, O. C. 1981. *Flow Control and Routing in an Integrated Voice and Data Communication Network*, Ph.D. thesis, MIT, Dept. of Electrical Engineering and Computer Science, Cambridge, MA.

[IEE86] *IEEE Journal on Selected Areas in Communications, Special Issue on Network Performance Evaluation*, SAC-4(6), Sept. 1986.

[IEE88] *IEEE Network*, Vol. 2, No. 1, 1988.

[IEE89] *IEEE Journal on Selected Areas in Communications*, Special Issue on Telecommunications Network Design and Planning, Vol. 7, No. 8, 1989.

[Jac57] JACKSON, J. R. 1957. Networks of Waiting Lines, *Oper. Res.*, 5:518–521.

[Jaf81] JAFFE, J. M. 1981. A Decentralized "Optimal" Multiple-User Flow Control Algorithm, *IEEE Trans. Comm.*, COM-29:954–962.

[Jai68] JAISWAL, N. K. 1968. *Priority Queues.* New York: Academic Press.

[JaM82] JAFFE, J. M., and MOSS, F. M. A. 1982. Responsive Routing Algorithm for Computer Networks, *IEEE Trans. on Comm.*, COM-30:1758–1762.

[JBH78] JACOBS, I. M., BINDER, R., and HOVERSTEN, E. V. 1978. General Purpose Packet Satellite Networks, *Proc. IEEE*, pp. 1448–1468.

[Kap79] KAPLAN, M. 1979. A Sufficient Condition for Nonergodicity of a Markov Chain, *IEEE Trans. Inform. Theory*, IT-25:470–471.

[KaT75] KARLIN, S., and TAYLOR, H. M. 1975. *A First Course in Stochastic Processes.* New York: Academic Press.

[Kau81] KAUFMAN, J. S. 1981. Blocking in a Shared Resource Environment, *IEEE Trans. Comm.*, COM-29:1474–1481.

[KeB83] KERSHENBAUM, A., and BOORSTYN, R. R. 1983. Centralized Teleprocessing Network Design, *Networks*, 13:279–293.

[KeC90] KEY, P. B., and COPE, G. A. 1990. Distributed Dynamic Routing Schemes, *IEEE Communications Magazine*, 28:54–64.

[KeH63] KEUHN, A. A., and HAMBURGER, M. J. 1963. A Heuristic Program for Locating Warehouses, *Management Sci.*, 9:643–666.

[Kei79] KEILSON, J. 1979. *Markov-Chain Models—Rarity and Exponentiality.* New York: Springer-Verlag.

[Kel79] KELLY, F. P. 1979. *Reversibility and Stochastic Networks.* New York: Wiley.

[Kel85] KELLY, F. P. 1985. Stochastic Models of Computer Communication Systems, *J. Roy. Statist. Soc. Ser. B*, 47(1).

[Kel86] KELLY, F. P. 1986. Blocking Probabilities in Large Circuit-Switched Networks, *Adv. Apl. Prob.*, 18:473–505.

[KGB78] KAHN, R. E., GRONEMEYER, S. A., BURCHFIEL, J., and KUNZELMAN, R. C. 1978. Advances in Packet Radio Technology, *Proc. IEEE*, pp. 1468–1496.

[KGV83] KIRKPATRICK, S., GELATT, C. D., Jr., and VECCHI, M. P. 1983. Optimization by Simulated Annealing, *Science*, 220:671–680.

[Khu72] KHUMAWALA, B. M. 1972. An Efficient Branch and Bound Algorithm for the Warehouse Location Problem, *Management Sci.*, 18:B718–B731.

[KhZ89] KHANNA, A., and ZINKY, J. 1989. The Revised ARPANET Routing Metric, *Proc. of SIGCOM '89.*

[Kin62] KINGMAN, J. F. C. 1962. Some Inequalities for the Queue *GI/G/1*, *Biometrika*, 49:315–324.

[Kle64] KLEINROCK, L. 1964. *Communication Nets: Stochastic Message Flow and Delay.* New York: McGraw-Hill.

[Kle69] KLEITMAN, D. 1969. Methods for Investigating the Connectivity of Large Graphs, *IEEE Trans. Circuit Theory*, CT-16:232–233.

[Kle75] KLEINROCK, L. 1975. *Queueing Systems*, Vol. 1. New York: Wiley.

[Kle76] KLEINROCK, L. 1976. *Queueing Systems*, Vol. 2. New York: Wiley.

[KlL75] KLEINROCK, L., and LAM, S. 1975. Packet Switching in Multiaccess Broadcast Channel: Performance Evaluation, *IEEE Trans. Comm.*, COM-23:410–423.

[KlO77] KLEINROCK, L., and OPDERBECK, H. 1977. Throughput in the ARPANET—Protocols and Measurements, *IEEE Trans. Comm.*, COM-25:95–104.

[KlS80] KLEINROCK, L., and SCHOLL. 1980. Packet Switching in Radio Channels: New Conflict Free Multiple Access Schemes, *IEEE Trans. Comm.*, COM-28:1015–1029.

[KlT75] KLEINROCK, L., and TOBAGI, F. A. 1975. Packet Switching in Radio Channels: Part 1: CSMA Modes and Their Throughput-Delay Characteristics, *IEEE Trans. Comm.*, COM-23:1400–1416.

[Kol36] KOLMOGOROV, A. 1936. Zur Theorie der Markoffschen Ketten, *Mathematische Annalen*, 112:155–160.

[Kri90] KRISHNAN, K. R. 1990. Markov Decision Algorithms for Dynamic Routing, *IEEE Communications Magazine*, 28:66–69.

[Kue79] KUEHN, P. J. 1979. Multiqueue Systems with Nonexhaustive Cyclic Service, *Bell System Tech. J.*, 58:671–698.

[KuR82] KUMMERLE, K., and REISER, M. 1982. Local Area Networks—An Overview, *J. Telecomm. Networks*, 1(4).

[KuS89] KUMAR, P. R., and SEIDMAN, T. I. 1989. Distributed Instabilities and Stabilization Methods in Distributed Real-Time Scheduling of Manufacturing Systems, *IEEE Trans. Automat. Control*, AC-35:289–298.

[LaH84] LANPHONGPANICH, S., and HEARN, D. 1984. Simplicial Decomposition of the Asymmetric Traffic Assignment Problems, *Trans. Res.*, 18B:123–133.

[LaK75] LAM, S., and KLEINROCK, L. 1975. Packet Switching in a Multiaccess Broadcast Channel: Dynamic Control Procedures, *IEEE Trans. Comm.*, COM-23:891–904.

[LaL86] LAM, Y. F., and LI, V. O. K. 1986. An Improved Algorithm for Performance Analysis of Networks with Unreliable Components, *IEEE Trans. Comm.*, COM-34:496–497.

[Lam80] LAM, S. 1980. A Carrier Sense Multiple Access Protocol for Local Networks, *Comput. Networks*, 4:21–32.

[Las70] LASDON, L. S. 1970. *Optimization Theory for Large Systems.* New York: Macmillan.

[LeG89] LE GALL, F. 1989. About Loss Probabilities for General Routing Policies in Circuit-Switched Networks, *IEEE Transactions on Comm.*, COM-37:57–59.

[Lei80] LEINER, B. M. 1980. A Simple Model for Computation of Packet Radio Network Communication Performance, *IEEE Trans. Comm.*, COM-28:2020–2023.

[LeP65] LEVITIN, E. S., and POLJAK, B. T. 1965. Constrained Minimization Methods, *USSR Comput. Math. Phys.*, 6:1–50.

[LeY89] LEE, M.-J., and YEE, J. R. 1989. An Efficient Near-Optimal Algorithm for the Joint Traffic and Trunk Routing Problem in Self-Planning Networks, *Proc. IEEE INFOCOM '89*, pp. 127–135.

[LiF82] LIMB, J. O., and FLORES, C. 1982. Description of Fasnet, or Unidirectional Local Area Communications Network, *Bell System Tech. J.*

[Lim84] LIMB, J. O. 1984. Performance of Local Area Networks at High Speed, *IEEE Trans. Comm.*, COM-32:41–45.

[LiS84] LI, V. O. K., and SILVESTER, J. A. 1984. Performance Analysis of Networks with Unreliable Components, *IEEE Trans. Comm.*, COM-32:1105–1110.

[Lit61] LITTLE, J. 1961. A Proof of the Queueing Formula $L = \lambda W$, *Oper. Res. J.*, 18:172–174.

[LMF88] LYNCH, N., MANSOUR, Y., and FEKETE, A. 1988. The Data Link Layer: Two Impossibility Results, *Proceedings of the 7th Annual ACM Symposium on Principles of Distributed Computing.* Toronto, Can., pp. 149–170.

[Luc90] LUCKY, R. 1989. *Information, Man & Machine.* New York: St. Martins Press.

[Lue84] LUENBERGER, D. G. 1984. *Linear and Nonlinear Programming.* Reading MA: Addison-Wesley.

[MaF85] MATHYS, P., and FLAJOLET, P. 1985. *Q*-ary Collision Resolution Algorithms in Random-Access Systems with Free or Blocked Channel Access, *IEEE Trans. Inform. Theory*, IT-31:217–243.

[Mal86] MALIS, A. G. 1986. *PSN End-to-End Functional Specification* (RFC 979). BBN Communications Corp., Network Working Group.

[MaN85] MAXEMCHUK, N. F., and NETRAVALI, A. N. 1985. Voice and Data on a CATV Network, *IEEE J. Select. Areas Comm.*, SAC-3:300–311.

[MaP88] MAGGS, B. M., and PLOTKIN, S. A. 1988. Minimum-Cost Spanning Tree as a Path Finding Problem, *Inf. Proc. Lett.*, 26:291–293.

[Mas80] MASSEY, J. L. 1980. *Collision-Resolution Algorithms and Random Access Communications* (Report UCLA-ENG-8016). Los Angeles: University of California.

[Max87] MAXEMCHUK, N. F. 1987. Routing in the Manhattan Street Network, *IEEE Trans. on Communications*, COM-35:503–512.

[McS77] MCGREGOR, P. V., and SHEN, D. 1977. Network Design: An Algorithm for the Access Facility Location Problem, *IEEE Trans. Comm.*, COM-25:61–73.

[McW77] MCQUILLAN, J. M., and WALDEN, D. C. 1977. The ARPANET Design Decisions, *Networks*, 1.

[MeB76] METCALFE, R. M., and BOGGS, D. R. 1976. Ethernet: Distributed Packet Switching for Local Computer Networks, *Comm. ACM*, 395–404.

[Met73] METCALFE, R. 1973. *Steady State Analysis of a Slotted and Controlled Aloha System with Blocking*, paper presented at 6th Hawaii Conf. System Sci., Honolulu.

[Mik79] MIKHAILOV, V. A. 1979. *Methods of Random Multiple Access*, Candidate Engineering thesis, Moscow Institute of Physics and Technology, Moscow.

[Min89] MINZNER, S. E. 1989. Broadband ISDN and Asynchronous Transfer Mode (ATM), *IEEE Commun. Mag.*, 27:17–24.

[MiT81] MIKHAILOV, V. A., and TSYBAKOV, B. S. 1981. Upper Bound for the Capacity of a Random Multiple Access System, *Problemy Peredachi Inform.* (USSR), 17:90–95.

[MoH85] MOSELY, J., and HUMBLET, P. A. 1985. A Class of Efficient Contention Resolution Algorithms for Multiple Access Channels, *IEEE Trans. Comm.*, COM-33:145–151.

[MoS82] MOSS, F. H., and SEGALL, A. 1982. An Optimal Control Approach to Dynamic Routing in Networks, *IEEE Trans. Automat. Control*, AC-27:329–339.

[MoS86] MONMA, C. L., and SHENG, D. D. 1986. Backbone Network Design and Performance Analysis: A Methodology for Packet Switching Networks, *IEEE J. Select. Areas Comm.*, SAC-4:946–965.

[Mos84] MOSELY, J. 1984. *Asynchronous Distributed Flow Control Algorithms*, Ph.D. thesis, MIT, Dept. of Electrical Engineering and Computer Science, Cambridge, MA.

[MRR78] MCQUILLAN, J. M., RICHER, I., ROSEN, E. C., and BERTSEKAS, D. P. 1978. *ARPANET Routing Algorithm Improvements, Second Semiannual Report* (prepared for ARPA and DCA). : Bolt, Beranek, and Newman, Inc.

[MRR80] MCQUILLAN, J. M., RICHER, I., and ROSEN, E. C. 1980. The New Routing Algorithm for the ARPANET, *IEEE Trans. Comm.*, COM-28:711–719.

[NBH88] NEWMAN, R. M., BUDRIKIS, Z., and HULLETT, J. 1988. The QP5X MAN, *IEEE Commun. Mag.*, 26:20–28.

[Neu81] NEUTS, M. F. 1981. *Matrix-Geometric Solutions in Stochastic Models—An Algorithmic Approach.* Baltimore, MD: The Johns Hopkins University Press.

[NgH87] NG, M. J. T., and HOANG, D. B. 1987. Joint Optimization of Capacity and Flow Assignment in a Packet Switched Communications Network, *IEEE Trans. Comm.*, COM-35:202–209.

[NiS74] NISNEVITCH, L., and STRASBOURGER, E. 1974. Decentralized Priority in Data Communication, *Proc. 2nd Annual Symp. Comput. Architecture.*

[NoT78] NOMURA, M., and TSUKAMOTO, K. 1978. Traffic Analysis of Polling Systems, *Trans. Inst. Electron. Comm. Eng.* (Japan), J61-B:600–607.

[Nyq28] NYQUIST, H. 1928. Certain Topics in Telegraph Transmission Theory, *Trans. AIEE*, 47:617–644.

[PaG91a] PAREKH, A. K., and GALLAGER, R. G. 1991. A Generalized Processor Sharing Approach to Flow Control in Integrated Services Networks: The Single Node Case, Laboratory for Information and Decision Systems Report LIDS-P-2040, M.I.T., Cambridge, MA.

[PaG91b] PAREKH, A. K., and GALLAGER, R. G. 1991. A Generalized Processor Sharing Approach to Flow Control in Integrated Services Networks: The Multiple Node Case, Laboratory for Information and Decision Systems Report LIDS-P-2074, M.I.T., Cambridge, MA.

[Pak69] PAKES, A. G. 1969. Some Conditions for Ergodicity and Recurrence of Markov Chains, *Oper. Res.*, 17:1059–1061.

[Pap74] PAPE, U. 1974. Implementation and Efficiency of Moore-Algorithms for the Shortest Route Problem, *Math. Programming*, 7:212–222.

[PaS82] PAPADIMITRIOU, C. H., and STEIGLITZ, K. 1982. *Combinatorial Optimization: Algorithms and Complexity.* Englewood Cliffs, NJ: Prentice Hall.

[Per83] PERLMAN, R. 1983. Fault-Tolerant Broadcast of Routing Information, *Comput. Networks*, 7:395–405. (Also *Proc IEEE Infocom '83*, San Diego.)

[Per88] PERLMAN, R. 1988. Network Layer Protocols with Byzantine Robustness, Ph.D. Thesis, MIT Dept. of EECS.

[Pip81] PIPPENGER, N. 1981. Bounds on the Performance of Protocols for a Multiple Access Broadcast Channel, *IEEE Trans. Inform. Theory*, IT-27:145–151.

[PiW82] PINEDO, M., and WOLFF, R. W. 1982. A Comparison between Tandem Queues with Dependent and Independent Service Times, *Oper. Res.*, 30:464–479.

[Pol71] POLAK, E. 1971. *Computational Methods in Optimization.* New York: Academic Press.

[PrB84] PROVAN, J. S., and BALL, M. O. 1984. Computing Network Reliability in Time Polynomial in the Number of Cuts, *Oper. Res.*, 32:516–526.

[Pro83] PROAKIS, J. G. 1983. *Digital Communications.* New York: McGraw-Hill.

[Qur85] QURESHI, S. 1985. Adaptive Equalization, *Proc. IEEE*, pp. 1349–1387.

[RaH76] RAUBOLD, E., and HAENLE, J. 1976. A Method of Deadlock-Free Resource Allocation and Flow Control in Packet Networks, *Proc. Internat. Conf. Comput. Comm.*, Toronto.

[RaW83] RAMAMOORTHY, C. V., and WAH, B. W. 1983. The Isomorphism of Simple File Allocation, *IEEE Trans. Comput.*, C-32:221–231.

[ReC90] REGNIER, J., and CAMERON, W. H. 1990. State-Dependent Dynamic Traffic Management for Telephone Networks, *IEEE Communications Magazine*, 28:42–53.

[RFS91] ROBINSON, J., FRIEDMAN, D., and STEENSTRUP, M. 1991. Congestion Control in BBN Packet-Switched Networks, *Computer Communication Reviews.*

[Rin76] RINDE, J. 1976. TYMNET I: An Alternative to Packet Switching, *Proc. 3rd Internat. Conf. Comput. Comm.*

[Riv85] RIVEST, R. L. 1985. *Network Control by Bayessian Broadcast* (Report MIT/LCS/TM-285). Cambridge, MA: MIT, Laboratory for Computer Science.

[Rob72] ROBERTS, L. G. 1972. *Aloha Packet System with and without Slots and Capture* (ASS Note 8). Stanford, CA: Stanford Research Institute, Advanced Research Projects Agency, Network Information Center.

[Roc70] ROCKAFELLAR, R. T. 1970. *Convex Analysis.* Princeton, NJ: Princeton University Press.

[Roc84] ROCKAFELLAR, R. T. 1984. *Network Flows and Monotropic Programming.* New York: Wiley.

[Ros80] ROSS, S. M. 1980. *Introduction to Probability Models.* New York: Academic Press.

[Ros81] ROSEN, E. C. 1981. Vulnerabilities of Network Control Protocols: An Example. *Comput. Comm. Rev.*

[Ros83] ROSS, S. M. 1983. *Stochastic Processes.* New York: Wiley.

[Ros86a] ROSBERG, Z. 1986. Deterministic Routing to Buffered Channels, *IEEE Trans. Comm.* COM-34:504–507.

[Ros86b] ROSS, F. E. 1986. FDDI—A Tutorial, *IEEE Commun. Mag.*, 24:10–17.

[RoT90] ROSS, K., and TSANG, D. 1990. Teletraffic Engineering for Product-Form Circuit-Switched Networks, *Adv. Appl. Prob.*, 22:657–675.

[Ryb80] RYBCZYNSKI, A. 1980. X.25 Interface and End-to-End Virtual Circuit Service Characteristics, *IEEE Trans. Comm.*, COM-28:500–509. (Also in [Gre82].)

[SaH87] SASAKI, G., and HAJEK, B. 1986. Optimal Dynamic Routing in Single Commodity Networks by Iterative Methods, *IEEE Trans. Comm.*, COM-35:1199–1206.

[San80] SANT, D. 1980. Throughput of Unslotted Aloha Channels with Arbitrary Packet Interarrival Time Distribution, *IEEE Trans. Comm.*, COM-28:1422–1425.

[SaO82] SARACHIK, P. E., and OZGUNER, U. 1982. On Decentralized Dynamic Routing for Congested Traffic Networks, *IEEE Trans. Automat. Control*, AC-27:1233–1238.

[Sar82] SARACHIK, P. E. 1982. An Effective Local Dynamic Strategy to Clear Congested Multidestination Networks, *IEEE Trans. Automat. Control*, AC-27:510–513.

[Sas91] SASAKI, G. 1991. Input Buffer Requirements for Round Robin Polling Systems with Nonzero Switchover Times, *Proceedings of the 29th Allerton Conference on Communication, Control, and Computing*, Monticello, IL.

[Sch87] SCHWARTZ, M. 1987. *Telecommunication Networks Protocols, Modeling, and Analysis.* Reading: Addison Wesley.

[ScS80] SCHWARTZ, M., and STERN, T. E. 1980. Routing Techniques Used in Computer Communication Networks, *IEEE Trans. Comm.*, COM-28:539–552.

[Seg81] SEGALL, A. 1981. Advances in Verifiable Fail-Safe Routing Procedures, *IEEE Trans. Comm.*, COM-29:491–497.

[Seg83] SEGALL, A. 1983. Distributed Network Protocols, *IEEE Trans. Inform. Theory*, IT-29:23–34.

[Sha48] SHANNON, C. E. 1948. A Mathematical Theory of Communication, *Bell Syst. Tech J.*, 27:379–423 (Part 1), 623–656 (Part 2). (Reprinted in book form by the University of Illinois Press, Urbana, IL, 1949.)

[Sie86] SIEBERT, W. M. 1986. *Circuits, Signals, and Systems.* Cambridge, MA: MIT Press; and New York: McGraw-Hill.

[SiS81] SIDI, M., and SEGALL, A. 1981. A Busy-Tone-Multiple-Access-Type Scheme for Packet-Radio Networks, in G. Payolk (Ed.), *Performance of Data Communication Systems and Time Applications.* New York: North-Holland, pp. 1–10.

[SoH86] SOLOWAY, S. R., and HUMBLET, P. A. 1986. *On Distributed Network Protocols for Changing Topologies.* Codex Corp.

[SpG89] SPINELLI, J. M., and GALLAGER, R. G. 1989. Event Driven Topology Broadcast without Sequence Numbers, *IEEE Trans. on Comm.*, 37: 468–474.

[Spi85] SPINELLI, J. 1985. *Broadcasting Topology and Routing Information in Computer Networks.* (Report LIDS-TH-1470). Cambridge, MA.: MIT.

[Spi89] SPINELLI, J. M. 1989. Reliable Data Communication in Faulty Computer Networks, *Ph.D. Thesis, MIT Dept. EECS*, June 1989. Also *MIT LIDS Report LIDS-TH-1882.*

[SpM81] SPROULE, D. E., and MELLOR, F. 1981. Routing, Flow and Congestion Control and the Datapac Network, *IEEE Trans. Comm.*, COM-29:386–391.

[Sta85] STALLINGS, W. 1985. *Data Computer Communications.* New York: Macmillan.

[StA85] STUCK, B. W., and ARTHURS, E. 1985. *A Computer Communications Network Performance Analysis Primer.* Englewood Cliffs, NJ: Prentice Hall.

[Sti72] STIDHAM, S., Jr. 1972. $L = \lambda W$: A Discounted Analogue and a New Proof, *Oper. Res.*, 20:1115–1125.

[Sti74] STIDHAM, S., Jr. 1974. A Last Word on $L = \lambda W$, *Oper. Res.*, 22:417–421.

[StK85] STASSINOPOULOS, G. I., and KONSTANTOPOULOS, P. 1985. Optimal Congestion Control in Single Destination Networks, *IEEE Trans. Comm.*, COM-33:792–800.

[Syz90] SYZMANSKI, T. 1990. An Analysis of Hot-Potato Routing in a Fiber Optic Packet Switched Hypercube, *IEEE INFOCOM '90 Proceedings*, 1:918–925.

[Taj77] TAJIBNAPIS, W. D. 1977. A Correctness Proof of a Topology Information Maintenance Protocol for a Distributed Computer Network, *Commun. ACM*, 20:477–485.

[TaK85] TAKAGI, H., and KLEINROCK, L. 1985. Throughput Delay Characteristics of Some Slotted-Aloha Packet Radio Networks, *IEEE Trans. Comm.*, COM-33:1200–1207.

[Tak86] TAKAGI, H., 1986. *Analysis of Polling Systems.* Cambridge, MA: MIT Press.

[Tan89] TANENBAUM, A. S. 1989. *Computer Networks.* (2nd ed.). Englewood Cliffs, NJ: Prentice-Hall.

[TBF83] TOBAGI, F., BORGONOVO, F., and FRATTA, L. 1983. Express-Net: A High Performance Integrated-Services Local Area Network, *IEEE J. Select Areas Comm.*, SAC-1.

[ThC86] THAKER, G. H., and CAIN, J. B. 1986. Interactions between Routing and Flow Control Algorithms, *IEEE Trans. Comm.*, COM-34:269–277.

[Tob74] TOBAGI, F. A. 1974. *Random Access Techniques for Data Transmission over Packet Switched Radio Networks*, Ph.D. thesis, UCLA, Computer Science Dept., Los Angeles.

[ToK75] TOBAGI, F. A., and KLEINROCK, L. 1975. Packet Switching in Radio Channels: Part II: The Hidden Terminal Problem in CSMA and Busy-Tone Solution, *IEEE Trans. Comm.*, COM-23:1417–1433.

[ToW79] TOWSLEY, D., and WOLF, J. K. 1979. On the Statistical Analysis of Queue Lengths and Waiting Times for Statistical Multiplexors with ARQ Retransmission Schemes, *IEEE Trans. Comm.*, COM-27:693–702.

[Tsa89] TSAI, W. K. 1989. Convergence of Gradient Projection Routing Methods in an Asynchronous Stochastic Quasi-Static Virtual Circuit Network, *IEEE Trans. Aut. Control*, 34:20–33.

[TsB86] TSITSIKLIS, J. N., and BERTSEKAS, D. P. 1986. Distributed Asynchronous Optimal Routing in Data Networks, *IEEE Trans. Automat. Control* AC-31:325–331.

[Tsi89] TSITSIKLIS, J. N. 1989. *Markov Chains with Rare Transitions and Simulated Annealing* Math. of Operations Research, Vol. 14, pp. 70–90.

[Tsi87] TSITSIKLIS, J. N. 1987. *Analysis of a Multiaccess Control Scheme, IEEE Trans. Automat. Control*, AC-12:1017–1020.

[TsM78] TSYBAKOV, B. S., and MIKHAILOV, V. A. 1978. Free Synchronous Packet Access in a Broadcast Channel with Feedback, *Problemy Peredachi Inform.* (USSR), 14(4):32–59.

[TsM80] TSYBAKOV, B. S., and MIKHAILOV, V. A. 1980. Random Multiple Access of Packets: Part and Try Algorithm, *Problemy Peredachi Inform.* (USSR), 16:65–79.

[TsR90] TSANG, D., and ROSS, K. 1990. Algorithms to Determine Exact Blocking Probabilities for Multirate Tree Networks, *IEEE Trans. on Comm.*, 38:1266–1271.

[TsS90] TSITSIKLIS, J. N., and STAMOULIS, G. D. 1990. On the Average Communication Complexity of Asynchronous Distributed Algorithms, *MIT LIDS Report LIDS-P-1986.*

[Tsy85] TSYBAKOV, B. S. 1985. Survey of USSR Contributions to Random Multiple-Access Communications, *IEEE Trans. Inform. Theory*, IT-31:143–165.

[Tur86] TURNER, J. 1986. New Directions in Communications (or Which Way to the Information Age), *IEEE Communications Magazine*, 24:8–15.

[Twe82] TWEEDIE, R. L. 1982. Operator-Geometric Stationary Distributions for Markov Chains with Application to Queueing Models, *Adv. Appl. Probab.*, 14:368–391.

[Tym81] TYMES, L. 1981. Routing and Flow Control in TYMNET, *IEEE Trans. Comm.*, COM-29:392–398.

[Var90] VARVARIGOS, E. A. 1990. Optimal Communication Algorithms for Multiprocessor Computers, M.S. Thesis, Dept. of Electrical Engineering and Computer Science, M.I.T., also Center for Intelligent Control Systems Report, CICS-TH-192.

[Vas79] VASTOLA, K. S. 1979. *A Numerical Study of Two Measures of Delay for Network Routing*, M.S. thesis, University of Illinois, Dept. of Electrical Engineering, Urbana, IL.

[VvP83] VVEDENSKAYA, N. D., and PINSKER, M. S. 1983. Non-optimality of the Part-and-Try Algorithm, in *Abstracts of the International Workshop of Convolutional Codes, Multiuser Communication*, Sochi, USSR, pp. 141–148.

[Wal83] WALRAND, J. 1983. Probabilistic Look at Networks of Quasi-reversible Queues, *IEEE Trans. Inform. Theory*, IT-29:825–831.

[Wal88] WALRAND, J. 1988. *An Introduction to Queueing Networks*. Prentice Hall, Englewood Cliffs, N.J.

[WaO90] WATANABE, Y., and ODA, T. 1990. Dynamic Routing Schemes for International Networks, *IEEE Communications Magazine*, 28:70–75.

[Wec80] WECKER, S. 1980. The Digital Network Architecture, *IEEE Trans. Comm.*, COM-28:510–526.

[Wel82] WELDON, E. J. 1982. An Improved Selective Repeat ARQ Strategy, *IEEE Trans. Comm.*, COM-30:480–486.

[Whi83a] WHITT, W. 1983. The Queueing Network Analyzer, *The Bell System Technical Journal*, 62:2779–2815.

[Whi83b] WHITT, W. 1983. Performance of the Queueing Network Analyzer, *The Bell System Technical Journal*, 62:2817–2843.

[WiE80] WIESELTHIER, J. E., and EPHREMIDES, A. 1980. A New Class of Protocols for Multiple Access in Satellite Networks, *IEEE Trans. Automat. Control*, AC-25:865–880.

[Wol82a] WOLFF, R. W. 1982. Poisson Arrivals See Time Averages, *Oper. Res.*, 30:223–231.

[Wol82b] WOLFF, R. W. 1982. Tandem Queues with Dependent Service Times in Light Traffic, *Oper. Res.*, 82:619–635.

[Wol89] WOLFF, R. W. 1989. *Stochàstic Modelling and the Theory of Queues*. Prentice Hall, Englewood Cliffs, N.J.

[Yum81] YUM, T. P. 1981. The Design and Analysis of a Semidynamic Deterministic Routing Rule, *IEEE Trans. Comm.*, COM-29:498–504.

[Zan69] ZANGWILL, W. 1969. *Nonlinear Programming: A Unified Approach*. Englewood Cliffs, N.J.: Prentice Hall.

[Zha89] ZHANG, L. 1989. A New Architecture for Packet Switching Network Protocols, Ph.D. Thesis, *Dept. of Electrical Engineering and Computer Science, MIT*. Cambridge, MA.

[ZVK89] ZINKY, J., VICHNIAC, G., and KHANNA, A. 1989. Performance of the Revised Routing Metric in the ARPANET and MILNET, *Proc. of MILCOM '89*.

Index

A

ABM. *See* Asynchronous balanced mode
ADCPP, 72, 97–103
AD. *See* Amplitude modulation
ANSI. *See* American National Standards Institute
AM. *See* Amplitude modulation
ARM. *See* Asynchronous response mode
ARPANET, 2, 4, 19–20
  ARQ, 84–86
  flow control, 515
  framing, 87
  routing, 374–76, 404–6, 412
ARQ. *See* Automatic repeat request
ASCII code, 21, 31, 58, 86
ATM. *See* Asynchronous transfer mode
Abort capability for frames, 89
Access rights, 31
Acks, 66, 500
  end-to-end, 115–16
Adaptive equalizers, 45, 49, 140
Addressing, 40, 111–14
  in TCP, 124–25
Age field, 422, 425
Aggregation of queues, 254
Airline reservation systems, 7, 9, 10
Allocate message, 500
Aloha, 275–89, 352
  slotted, 277–87
  stabilized, 282–86
  unslotted, 287–89
American National Standards Institute, 97
Amplitude modulation, 48
Application layer, 31–32
Arc, 387, 394
Arrival rate, 12, 152
Arrival theorem, 239, 256
Asynchronous balanced mode, 98–103
Asynchronous character pipes, 38
Asynchronous response mode, 98
Asynchronous transfer mode, 32, 40, 55, 97, 128–39, 141
  adaptation layer, 135–39
  call size, 132
  congestion, 138–39
  CRC on header, 133-34
  flow control, 138–39
  use of SONET, 134–35
  virtual channel and path ID, 133
Attenuation, 56
Automatic repeat request, 39, 64–80, 272
  ARPANET, 84–86
  delay analysis, 190
  go back n, 72–81, 190
  packet radio, 347–48
  ring networks, 323
  selective repeat, 81–84
  stop and wait, 66–71

B

BRAM, 333
BSC. *See* Binary synchronous communication
Backlog, 276, 279
Backlog estimation, 283, 286
Backpressure, 507
Bandwidth, 50–51
Bellman's equation, 399
Bellman-Ford algorithm, 396–400, 404–410, 424, 481, 492
Binary exponential backoff, 286–87, 352
Binary synchronous communication, 87
Birth-death processes, 184, 215, 261
Bisynch, 87
Bit pipes, 20–23, 38, 65, 86
  intermittent synchronous, 38
  synchronous, 38
Bit stuffing, 88–90, 145, 320
Bits, 10
Black box, 17
Blocking probability, 180, 185
Bottleneck link, 526
Branch exchange heuristic, 444, 448
Bridge learning, 384
Bridged local area networks, 382–87
Bridges, 4, 29, 380
Broadband ISDN, 5, 11–12, 32, 55, 128–32
Broadcasting, 369, 375–76, 433, 485
Buffer management, 496–98
Buffer overflow, 496–99
Burke's theorem, 218, 255
Burst detection, 60–61
Bursty sessions, 14–15
Bus systems, 271, 274, 331–33
  undirectional, 334–41
Busy tones, 350–51

C

CCITT, 97
CRC. *See* cyclic redundancy checks
CRC-16 polynomial, 64
CRC-32 IEEE polynomial, 64
CRC-CCITT polynomial, 64, 98
CRP. *See* Collision resolution period
CSMA. *See* Carrier sense multiple access
CSMA/CD. *See* Carrier sense multiple access/collision detection
Cable TV, 56, 341–42
Call blocking, 494
Call-request packet X.25, 119
Capacity assignment, 439–44
Capacity, 51, 150
Capture effects, slotted Aloha, 355
Carrier offset, 53
Carrier sense multiple access, 272, 304–12, 352
  collision avoidance, 333
  collision detection, slotted, 317–18
  collision detection, unslotted, 318–20
  FCFS splitting, 310–12
  nonpersistent, 305
  packet radio, 350–51
  persistent, 305
  P-Persistent, 305, 307
  pseudo-Bayesian stabilization, 307–9
  slotted Aloha, 305–9
  unslotted Aloha, 309–10
  variable delay, 358
Cellular radio system, 247
Character based framing, 86–88
Choke packet, 510
Circuit switching, 14–17, 180, 185, 379
Coaxial cable, 7, 56
Code conversion, 31
Codex networks, 141
  data link control, 518
  flow control, 518
  routing, 378, 417, 476–77
  session identification, 113—15
Coding. *See* Data compression, Encryption, Error detection, or Error correction
Collision resolution, 276–77
  packet radio, 347–49
Collision resolution period, 291, 295–300
Collision-free set, 345
Communication channels, 20, 38, 40–57
  analog, 40
  bandpass, 47–52
  digital, 40, 53–54
  voice-grade, 49–50, 52
Concentrator location problem, 448
Concentrators, 1
Congestion control, 27, 116–17
  in ATM, 138–39 *See also* Flow control
Connected graph, 387, 394
Connectionless service, 26
  in ATM, 137–39
Connectivity, 445
Constrained MST problem, 447
Convergence sublayer in ATM, 136–39
Convex set or function, 453, 486
Convolution, 43
Convolutional codes, 61
Correctness, stop-and-wait, 69–70
Cut-through routing, 373
Cycle, 387
Cyclic redundancy check, 61–64, 140
  on ATM header, 133–34
  end-to-end, 116

D

db. *See* Decibels
DC component, 47, 48
DCE. *See* Data communication equipment
DECNET, 2, 19–20, 91, 93, 404
DLC. *See* Data link control
DLE (data link escape), 87
DNA. *See* Digital network architecture
DQDB. *See* Distributed queue dual bus
DTE. *See* Data terminal equipment
Data bases, remote access, 7–8
Data communication equipment, 21, 118
Data compression, 10, 31
Data link control, 20, 23–24, 37–40, 57–110, 271
  correctness, 76–79, 140
  standards, 97–103, 141
Data terminal equipment, 21, 118
Datagram:
  networks, 115
  routing, 16
  service, 26
Datagram routing, 363
Deadlock, 497–99
Decibels, 51
Decomposition of gueues, 254
Deflection routing, 372
Delay, 13
  CSMA, 308–9
  CSMA/CD, 318, 360
  FCFS splitting algorithm, 300–301

Delay *cont.*
  Fiber distributed data interface, 329–30
  processing, 150
  propagation, 150
  queueing, 150
  satellite systems, 314–15
  slotted Aloha, 284–86
  token buses and polling systems, 331–32
  token ring, 324–25
  transmission, 150
Demand sharing, 274
Descent direction, 456
Designated port, 383
Detailed balance equations, 182, 214, 216, 218, 261, 263
Digital network architecture, 404
Digraph, 394
Dijkstra's algorithm, 401–3, 481
Directed arc, 394
Directed cycle, 394
Directed graph, 394
Directed walk, 394
Directory assistance, 30
Disconnect command, 100
Disconnect protocols, 103–10
Distributed algorithms, 28, 32–34, 39, 139
Distributed queue dual bus, 335–39, 353
Distributed shortest path construction, 404–10
Distributed spanning tree construction, 392
Drift, 264
Drift analysis:
  CSMA, 306
  FCFS splitting algorithm, 300
  slotted Aloha, 280
Dynamic programming, 395
Dynamic routing, 16–17, 436–38, 491
Dynamic window adjustment, 510

E

EXT (end of text), 87
Electronic mail, 8, 9
Embedded chain, 262
Encryption, 31
End-to-end CRC, 116
End-to-end acks, 115–16, 119
End-to-end windows, 506–8, 515–17
Entropy, 91, 140
Ergodic system, 156
Erlang B formula, 179, 167
Erlang C formula, 175, 267
Error correction, 52, 61, 140
Error detection, 39, 53, 57–64, 140
  bursts, 60–61
  random strings, 60–61
Error recovery, 23, 32, 115–16
  in TCP, 125–27
  transport vs. network layer, 117–18
Essau-Williams algorithm, 448
Ethernet, 4, 48, 52, 274, 286, 317–20
  degradation with size and speed, 334
Exponential distribution, 164, 266
Expressnet, 339–41
External site, 19

F

FDDI. *See* Fiber distributed data interface
FDM. *See* Frequency division multiplexing
Facsimile, 8
Feasible direction, 456
Feedback, multiaccess:
  delayed, 303–4
  immediate, 276
Feedback shift register, 62
Fiber distributed data interface, 326–30, 334, 339, 353
  delay, 329–30
  throughput, 330
File transfer, 13
Filtering, 41–53
First derivative length, 452
First-come first-serve splitting, 293–304
  CSMA, 310–12
Flags, 88–90, 320
Flooding, 369, 419–25, 432, 490
Flooding with sequence numbers, 420
Flow control, 25–29, 30, 116–17, 158
  combined with optimal routing, 519–24
  fairness, 498–99, 505, 507, 524–29
  in ARPANET, 515
  in ATM, 138–39
  in Codex network, 518
  in PARIS network, 518
  in SNA, 517
  in TCP, 127–28
  in TYMNET, 517
  in X.25, 519
  input rate adjustment, 519–29
  max-min, 524–29, 534
  means, 494–95
  objectives, 496–99
  optimal, 519–24
  satellite links, 506–8, 517
  transport layer, 509
  window, 500-510
Flow deviation method, 458–64
Flow models, 433–37
Floyd-Warshall algorithm, 403–4
Forwarding database, 383
Fourier transforms, 44, 45
Fragment, 389
Fragmentation, internet protocol, 122
Frames, 23, 52, 65
  internet protocol, 122
  maximum fixed length, 97
  maximum length, 140–41
  maximum variable length, 93–97
Framing, 39, 86–97
  bit-oriented, 88–90
  characted based, 86–88
  length fields, 90–92
  overhead, 87, 89–90, 90–92
  with errors, 92–93
Frank-Wolfe method, 458–64, 471, 488
Free-for-all multiaccess, 272
Frequency division multiplexing, 52, 151, 170, 177, 194
  in satellite channels, 273
Frequency response, 43–46
Front end, 1

G

G/G/1 queue. *See* Queueing System
Gateways, 4, 28–29, 120, 380
Generator polynomials, 62–64, 142
Global balance equations, 168, 260, 263
Go back n ARQ, 72–81, 113, 143–44, 501
  correctness proof, 76–79
  efficiency, 80–81, 144
  ideal, 82
  rules, 74–76
  rules with mod m, 80
Gradient project method, 459, 467–77, 486–87, 489
Graph, 387
Graphics, 8, 13
Guaranteed data rate, 494

H

HDLC, 39, 72, 97–103
  faulty initialization, 106
Header:
  frame, 23, 39, 65
  packet, 25
Hessian matrix, 465
Hierarchy of modules, 17
Homenets, 341–42
Horizontal and vertical parity checks, 58–59
Hot potato routing, 372
Hybrids, 56

I

IEEE 802 standards, 48, 320, 323–24, 332–33, 335
IMP. *See* Interface message processor
IP. *See* Internet protocol
ISDN. *See* Integrated services digital network
ISO. *See* International Standards Organization
Idle fill, 38, 90, 321
Image transmission systems, 9
Implicit tokens, 333
Impulse response, 42
Incident arc, 387
Infinite node assumption, 276
Initialization protocols, 147
  ARQ, 103–10
  balanced, 107–9
  fault in HDLC, 106
  with link failures, 103–9
  master-slave, 104–7
  with node failures, 109–10
  in TCP, 126
  use of stop-and-wait, 104
  Integrated services digital network, 5, 11–12, 54–56
  basic service, 55
  primary service, 55
Interactive sessions, 13, 15
Interface message processor, 2
Internal signaling network, 55
International Consultative Committee on Telegraphy and Telephony. *See* CCITT
International Standards Organization, 19, 97
  terminology, 30
Internet protocol, 40, 120–23, 141
  addressing, 121–22
  datagrams, 120
  fragmentation, 122
  time to live, 123
Internet sublayer, 28–29
Internetworking, 28–29
Intersymbol interference, 42, 45–47, 49
Isarithmic method, 508

J

Jackson's theorem:
  closed networks, 234
  heuristic explanation, 227
  with limit on the number of customers, 256
  multiple classes of customers, 231
  open networks, 223
  state-dependent rates, 230

K

Kleitman's algorithm, 445–47
Kruskal algorithm, 390, 448

L

LAN. *See* Local area network
LAPB, 72, 97–103, 118
Last-come first-serve splitting algorithm, 302–3
Layering, 17–32, 35
Leaky bucket, 245, 511–15, 535
Linear codes, 59
Linear systems, 41–46, 140
Little's theorem, 152–62, 250–51
  slotted Aloha, 280
Liveness:
  go back n, 77–78
  stop-and-wait, 69–70
Load sharing, 8
Local access network, 438
Local access network design, 448–51
Local area networks, 4, 7, 29, 56, 271–72, 317–42, 352–53
Local loop, telephone network, 54, 56

M

M/D/1 queue. *See* Queueing system
M/G/1 queue. *See* Queueing system
M/G/∞ queue. *See* Queueing system
M/M/1 approximation. *See* Networks of queues
M/M/1 queue. *See* Queueing system
M/M/1/m queue. *See* Queueing system
M/M/∞ queue. *See* Queueing system
M/M/m queue. *See* Queueing system
M/M/m/m queue. *See* Queueing system
MAC. *See* Medium access control
MAN. *See* Metropolitan area network
MSAP, 333
Manchester coding, 48
Markov chain:
  aperiodic, 260
  continuous-time, 262–63
  discrete-time, 259–62
  FCFS splitting algorithm, 297
  instability, 265
  irreducible, 260
  M/M/1 system, 166
  M/M/∞ system, 177
  M/M/m system, 174
  M/M/m/m system, 179
  multidimensional, 180-86
  queueing networks, 223
  reversible, 182
  slotted Aloha, 279, 353
  stability, 264
  time reversible, 215
Max-min fairness, 328
Max-min flow control, 524–29, 534
Mean residual service time, 187
Mean value analysis, 238–40
Media access control, 24, 27, 29, 271
Memoryless property, 165
Message switching, 16
Messages, 10
  arrival rate, 12
  delay, 13
  length, 12–13
  ordering, 13
Metering, 214, 436
Metropolitan area networks, 271–72, 326–30, 333–42
Micro-processors, 5
Microwave channels, 56
Min-hop path, 370
Minimum distance, 60
Minimum weight spanning tree, 389, 484
Modems, 21, 38, 40, 48–52
Modularity, 17
Modulation, 140
  amplitude, 48–49
  phase-shift keying, 50
  quadrature amplitude, 49–52
Module, 17
Modulo arithmetic, 58
Multiaccess communication, 24, 195, 271–362
  canonic reservation system, 312
Multidrop telephone lines, 271, 274, 331, 342
Multiplexers, 1
Multiplexing, in TCP, 124–25
Multicap bus systems, 271, 274

N

NRM. *See* Normal response mode
NRZ code, 42, 48
Nak, 66
Network layer, 25–29
  point to point protocols, 110–20
Networks of queues:
  acyclic, 221
  closed, 233
  Jackson's theorem, 212, 221, 434
  Kleinrock independence approximation, 212, 221, 434, 440
  M/M/1 approximation, 213, 221, 434, 440, 453, 476
  multiple classes of customers, 230
  state dependent service rates, 229
  tandem queues, 209, 220
  with multiple classes of customers, 222
Newton's method, 467
Node of a graph, 387
Node-by-node windows, 501–6, 517
Noise, 45, 51
Nonlinear distortion, 53
Normal response mode, 98–101
Nyquist criterion, 47

O

OD pair. *See* Origin-destination pair
OSI. *See* Open systems interconnection
On-off flow, 11
Open systems interconnection, 19–20, 35
Optical fibers, 6–7, 15, 56, 326, 333
Optimal routing:
  algorithms, 455–75
  dynamic, 436–38, 491
  formulation, 433–36
Origin-destination pair, 434

P

PARIS network, 392, 518
PERT networks, 395
PSK. *See* Phase-shift keying
Packet blocking, 495
Packet discarding, 495
Packet radio nets, 57, 273, 275, 286, 344–51
  collision resolution, 347–49
  TDM, 346–47
  transmission radii, 349–50, 353
Packet scheduling, 495
Packet switching, 131
  why used in ATM, 16
Packets, 10, 39
  ordering, 13, 86
Parity check codes, 59–64, 140
Parity checks, 58
Partial balance equations, 262
Path in a graph, 387
Peer process (modules), 18, 23, 25, 28, 39
Periodic routing updates, 422
Permit, 117, 500
Permit (window) in TCP, 127
Phase jitter, 53
Phase-shift keying, 50
Physical layer, 20–23, 37–38, 40–57
Piggybacked requests, 67
Pipelining, 94–96
Point to point protocols, 37
Poisson distributing, 164, 266
Poisson process, 11, 164–65, 172–73, 275
  combined, 244
  merged, 165
  properties, 243
  split, 165, 245
Poll final bit, 99
Pollaczek-Khinchin formula, 186
Polling, 274, 331–32, 342–44
  generalized, 342–44
  hub, 331
  queueing analysis, 195–203
Polynomial division, 62
Port, 382
  in TCP, 124
Positive definite matrix, 465
Positive semidefinite matrix, 465
Presentation layer, 31
Prim-Dijkstra algorithm, 390, 448, 484
Primitive polynomials, 64
Priorities, FDDI, 326–27
Priority queueing system. *See* Queueing system
Product form, 182
Product form solution, 220, 223–25
Promiscuous mode, 382
Protocols. *See* Automatic reqeat request, Distributed algorithms, HDLC, X.25, X.21
Pseudo-Bayesian algorithm, 283–86, 352
  CSMA slotted Aloha, 307–9
  CSMA unslotted Aloha, 359
Pure Aloha, 287–89

Q

QAM. *See* Quadrature amplitude modulation
Quasistatic assumption, 177
Queueing system:
  closed, 158, 233, 256–58
  discrete-time M/M/1, 246
  G/G/1, 206–9, 253, 269
  last-come first-serve, 232
  last-come first-serve M/G/1 system, 253
  M/G/1 system, 267
    with arbitrary order of service, 248
    with batch arrivals, 251
    with busy period overhead, 251
    with delay for new busy period, 252
    with vacations, 192–95, 252, 268
  M/G/m/m, 179
  M/G/∞, 252
  M/M/1, 162–73, 175, 216, 266
    shared service, 247
    with multiple classes of customers. 258
    with state-dependent rates, 246
  M/M/1/m, 247, 255

M/M/2 with heterogeneous servers, 247
M/M/∞, 177–78, 216, 247
M/M/m, 174–77, 216, 266
M/M/m/m, 175, 179, 216, 267
networks. *See* Networks of queues
nonpreemptive priority, 203–5, 249–50, 269
with phase-type distributions, 232
polling, 160, 195, 268
preemptive resume priority, 205–6, 249, 269
priority with multiple servers, 249
processor sharing, 232
reservation, 195–204, 268
time reversible, 214

R

RN. *See* Request number
RS-232-c interface, 21–22
Radio channels, 56, 271 *See also* Packet radio nets)
Randomization, 214, 222, 436
Rate control schemes, 510–15
Real time computation, 8
Real time control, 13
Reassembly, 29
Receive-not-ready:
supervisory frame, 99–100
X.25 packet, 119
Receive-ready:
supervisory frame, 99–100
X.25 packet, 119
Register insertion rings, 330–31
Regular chain, 262
Reject supervisory frame, 99–100
Reject X.25 packet, 119
Reliability, 9, 13, 29, 443–48
Repeaters, 53–54
Request number, 67
Reservation system:
canonic multiaccess system, 312
exhaustive, 195
gated, 195
limited service, 201, 249
multiaccess, 272, 312–44
multiuser, 198
partially gated, 195
satellites 313–17
single-user 196
Reset command, 103
Reversibility, 214–21
Ring networks, 320–31
Ringing, 48
Round-robin scheduling, 495
Router, 380
Routing, 25–29, 363–491
adaptive, 368, 410–17, 455
asynchronous, 375, 404–10, 425
broadcast, 487
centralized, 367, 418
distributed, 367
hierarchical, 379–82
in the ARPANET, 374
in Codex network, 476–77
in circuit switching networks, 379
in SNA, 378
in the TYMNET, 376
interaction with flow control, 365
interconnected network, 379–87
main issues, 365–68
nonhierarchical, 380–82
optimal, 372, 433, 451–75
oscillations, 371, 374, 410–17
shortest path, 374–76, 387–417
shortest path first, 375, 382
static, 368
Routing *cont.*
tables, 365, 375–76, 377, 382
with frequently changing topology, 491

S

SDLC, 72, 97–103
SN. *See* Sequence number
SNA. *See* System network architecture
SONET, 54, 141
use in ATM, 134–35
SPTA. *See* Shortest path topology algorithm
STX (start of text), 87
SYN (synchronous idle), 86
Sampling theorem, 46–47, 51
Satellite channels, 56–57, 271, 273–74, 313–17, 352–53
Saturated cut method, 81–84, 140–41, 444
Scheduling for multiaccess, 272 (*See also* Reservations)
Segmentation and reassembly, in ATM, 136–39
Selective repeat ARQ, 144
ideal, 82
Selective-reject supervisory frame, 99–100
Sequence number, 66
Service rate, 152
Session holding time, 12
Session identification, 111–14
Session initiation, 31
Session layer, 30–31
Sessions, 11–14
Shannon capacity theorem, 50–51, 53, 140
Shortest path problem, 394
Shortest path routing, 370
Shortest path spanning tree, 400
Shortest path topology algorithm, 425–33, 483
Signal constellations, 50–51
Signal power, 51
Simulated annealing, 443
Sliding window, 72 (*See also* Go back n ARQ)
Slotted Aloha:
capture, 355
delay, 354
stability, 353
Slotted frequency-division multiplexing, 194
Slotted multiaccess sytems, 275–88
Slotted ring, 330
Socket in TCP, 124
Solid state technology, 5
Source coding theorem, 91–92, 140
Source routing, 385–87
Spanning tree, 369, 388
Spanning tree routing, 383–85
Spanning tree topoligy design, 447–48
Splitting algorithms, 289–304
first-come first-serve, 293–304, 352
LCFS, 302–3
round robin, 304
with delayed feedback, 303–4
zero or positive feedback, 342–44
Stability:
CSMA, 307–12
datagram networks, 410–14
FCFS splitting algorithm, 300
shortest path routing algorithm, 410–417
slotted Aloha, 281–86, 352
unslotted Aloha, 288, 359
virtual circuit networks, 414–17
Stack algorithms, 291 (*See also* Tree algorithms)
unblocked stack, 293
Standardization, 19–20, 32
Star configuration for rings, 324
Stationary distribution, 260, 263
Statistical multiplexing, 113, 150, 170, 176, 214
Steepest descent method, 465
Stop-and-wait, 142–43
ARQ, 66–69
correctness proof, 69–70
in initialization protocols, 104
Store-and-forward switching, 14–17
Strongly connected graph, 394
Subgraph, 388
Sublinear convergence rate, 461
Subnet, 2–3, 10
Subnet design, 439–48
Supercomputers, 8
Supervisory frames, 99–100
System network architecture (SNA), 19–20
explicit route control, 378
flow control, 501, 517
path control layer, 378
routing, 378
SDLC, 72, 97–103
session level pacing, 517
transmission control layer, 378
transmission group control, 378
virtual route control, 378, 517
virtual route pacing scheme, 501, 517

T

TCP. *See* Transport control protocol
TCP/IP. *See* Internet protocol, Transport control protocol
TDM. *See* Time-division multiplexing
TP4 (ISO transport protocol), 128
T1 carrier, 53–54
T3 carrier, 54
TYMNET, 2, 4, 19–20, 141
flow control, 517
routing, 376–78
session identification, 112–13
Technology:
communication, 6–7
computer, 5–6
Three army problem, 33–34, 71
Throughput:
CSMA FCFS splitting, 311
CSMA slotted Aloha, 306
CSMA unslotted Aloha, 310
CSMA/CD:
slotted, 318
unslotted, 319
FCFS splitting algorithm, 300–301
slotted Aloha, 282–83
token buses and polling systems, 331–32
unslotted Aloha, 288
Time invariance, 41
Time sharing system, 160, 236
Time to live, internet protocol, 123
Time window flow control, 511
Time-division multiplexing, 52, 151, 170, 177, 194, 214, 273, 277
for packet radio, 346–47
Time-outs, 67–70, 73
Token buses, 331–33
degradation with size and speed, 334
implicit tokens, 333
Token rings, 4, 320–30
Topological design, 437–51, 464
Topology, 418

Trailer frame, 23, 39, 65
Transition rate, 262
Transmission delay. *See* Delay
Transparent mode, 87
Transport control protocol, 40, 124–28, 141
  addressing, 124–25
  error recovery, 117–18, 125–27
  flow control, 127–28
  initialization, 126
  multilexing, 124–25
Transport layer, 29–30, 123–28
  standards, 123–24
Transport layer flow control, 509
Tree, 388
Tree algorithms, 290–93, 352
Truncation of Markov chains, 182, 254
Twisted pair, 7, 56

U

Unary-binary encoding, 91, 114
Undetectable errors, 60, 64, 114–15
Unnumbered acks, 100
Unnumbered frames, 100
Unslotted Aloha, 287–89
  precise feedback, 356
  throughput, 357
Unstable equilibrium for Aloha, 281
Utilization factor, 157, 169

V

VLSI. *See* Very large scale integration
Vacation Queueing system. *See* Queueing system
Very large scale integration, 5
Video conferencing, 12
Virtual channel number, X.25, 118
Virtual channels, 111
Virtual circuit routing, 16, 26, 111, 116, 363, 476–77, 490
Voice:
  digitized, 11
  packetized, 13–14
Voice mail, 8
Voice network, 5, 8
Voice-grade. *See* Communication channels

W

WAN. *See* Wide area network
Walk in a graph, 387
Wide area network, 4, 6
Window (permit) in TCP, 127
Window flow control, 117, 158, 500–10

X

X.21 interface, 21–22, 118
X.25 standard, 40, 118–20, 500-10
  LAPB, 97–103